THERMOELECTRICS AND ITS ENERGY HARVESTING

MODULES, SYSTEMS, AND APPLICATIONS IN THERMOELECTRICS

THERMOELECTRICS AND ITS ENERGY HARVESTING

MODULES, SYSTEMS, AND APPLICATIONS IN THERMOELECTRICS

Edited by **D. M. Rowe**
OBE, DSc, PhD

CRC Press
Taylor & Francis Group
Boca Raton London New York

CRC Press is an imprint of the
Taylor & Francis Group, an **informa** business

CRC Press
Taylor & Francis Group
6000 Broken Sound Parkway NW, Suite 300
Boca Raton, FL 33487-2742

© 2012 by Taylor & Francis Group, LLC
CRC Press is an imprint of Taylor & Francis Group, an Informa business

No claim to original U.S. Government works

Version Date: 20120316

International Standard Book Number: 978-1-4398-7472-1 (Hardback)

Library of Congress Cataloging-in-Publication Data

Thermoelectrics and its energy harvesting / edited by David Michael Rowe.
  v. cm.
  Summary: "This book includes updated theoretical considerations which provide an insight into avenues of research most likely to result in further improvements in material performance. It details the latest techniques for the preparation of thermoelectric materials employed in energy harvesting, together with advances in the thermoelectric characterisation of nanoscale material. The book reviews the use of neutron beams to investigate phonons, whose behaviour govern the lattice thermal conductivity and includes a chapter on patents"-- Provided by publisher.
  Includes bibliographical references and index.
  Contents: v. 1. Materials, preparation, and characterization in thermoelectrics -- v. 2. Modules, systems, and applications in thermoelectrics.
  ISBN 978-1-4398-7470-7 (hardback : v. 1) -- ISBN 978-1-4398-7472-1 (hardback : v. 2)
  1. Thermoelectric apparatus and appliances. I. Rowe, David Michael.

TK2950.T533 2012
621.31'243--dc23
                                                                        2011052390

Visit the Taylor & Francis Web site at
http://www.taylorandfrancis.com

and the CRC Press Web site at
http://www.crcpress.com

Contents

v

SECTION II   Thermoelectric Modules, Devices, and Applications

SECTION III   Thermoelectric Systems and Applications

Preface

Nanotechnology has had a profound positive impact on thermoelectrics. It has contributed substantially to the development of advanced materials with improved thermoelectric and physical properties. This in turn has resulted in significantly increasing the economic competitiveness of thermoelectric energy harvesting.

The Handbook, *Modules, Systems, and Applications in Thermoelectrics* is comprised of 26 chapters by 75 contributors who are leading experts in their respective fields. As in previous CRC Thermoelectric Handbooks, the majority of chapters are reviews. The book is divided into three sections. Section I deals with Thermoelectric Materials and Properties. Section II discusses Thermoelectric Modules, Devices, and Applications, and Section III explains Thermoelectric Systems and Applications.

Section I is the largest, comprising 16 chapters, with the first chapter "Nanostructured Thermoelectric Materials," by Ren, Chen, and Dresselhaus, setting the scene for this all embracing materials section. The five chapters of Section II provide a comprehensive overview of progress made in the development of modules and their application in devices. Highlighted is the dramatic progress made in device miniaturization and its successful commercialization in energy harvesting. Section III, also comprised of five chapters, is devoted to thermoelectric energy harvesting and contains comprehensive reviews of the solar, automotive, and medical harvesting of waste heat.

Again, I have tried to preserve the international flavor of previous Handbooks and made minor changes only where necessary to clarify the meaning. The compilation of the two volume set, *Thermoelectrics and Its Energy Harvesting*, has been made possible by authoritative contributions from fellow thermoelectricians. The thermoelectric fraternity is deeply indebted to these authors who have unselfishly devoted their time to this project.

Finally, I am thankful to my wife Barbara, whose help has enabled me to bring this labor of love to a successful conclusion.

MATLAB® is a registered trademark of The MathWorks, Inc. For product information, please contact:

The MathWorks, Inc.
3 Apple Hill Drive
Natick, MA 01760-2098, USA
Tel: 508-647-7000
Fax: 508-647-7001
E-mail: info@mathworks.com
Web: www.mathworks.com

Editor

David Michael Rowe is an emeritus professor in the School of Engineering, Cardiff University and research director at Babrow Thermoelectric Consultants Ltd.

He obtained a BSc in both pure mathematics and in applied physics from Swansea University, an MSc from Bristol University, and a PhD from University of Wales Institute of Science and Technology. In 1987, he was awarded a DSc by the University of Wales for "The Development of Semiconductor Alloys with Improved Thermoelectric Figures of Merit."

Professor Rowe started his career as a Harwell United Kingdom Atomic Energy Fellow researching thermoelectric materials for application in nuclear-powered cardiac pacemaker batteries. In 1966, he joined the academic staff at Cardiff University and was awarded a Personal Chair in 1995. He became a distinguished research professor in 2005, and in 2010 received the title emeritus professor.

Professor Rowe's research interests are in thermoelectric materials and their applications. In 1981, he was the first to demonstrate that the thermal conductivity of thermoelectric materials was reduced by phonon-boundary scattering in very small grain size compacted materials, and in 1988 patented the first miniature thermoelectric generator fabricated using ITC technology. Professor Rowe pioneered the thermoelectric recovery of low-temperature heat and a generator that resulted from his research was demonstrated at the Kyoto Energy Summit as an example of green technology.

Forty years of research effort is embodied in more than 350 publications, author-/coauthorship of three books, editorship of three International Conference Proceedings, and of the best-selling definitive texts—the *CRC Handbook of Thermoelectrics* (1995) and (2005). *Thermoelectrics Handbook: Macro to Nano* (2012).

He is a member of the editorial board of several international journals and serves as an expert assessor for European and U.S. funding agencies. As research director of Babrow Consultants, he is actively involved in consultancy roles for a number of national and international thermoelectric technology projects.

In 1994, Professor Rowe was elected as a Fellow of the Institute of Physics and Fellow of Institution of Electrical Engineers, and in 1995 he founded Cardiff's NEDO Laboratory of Thermoelectric Engineering and served as its director. During 1997–2001, he was president of the International Thermoelectric Society (ITS) and since 2004 has been its Secretary. On May 23, 2007, Professor Rowe received the Order of Knighthood called "The most excellent order of the British Empire" (OBE) from Her Royal Highness Queen Elizabeth II at Buckingham Palace. In 2008, he was a recipient of the 2008 Gold Prize from the International Thermoelectric Academy for contributions to thermoelectricity and in the same year was elected an Honorary Academician of the Institution of Refrigeration. In 2009, Professor Rowe was elected as president of the European Thermoelectric Society.

Contributors

M. Anderson
Department of Chemistry
University of Oregon
Eugene, Oregon

R. Atkins
Department of Chemistry
University of Oregon
Eugene, Oregon

Phil Barletta
Center for Solid State Energetics
RTI International
Research Triangle Park
Durham, North Carolina

M. Beekman
Department of Chemistry
University of Oregon
Eugene, Oregon

Jesus Mª Blanco
Applied Physics Department I
University of the Basque
  Country
San Sebastian, Spain

Harald Böttner
Fraunhofer Institute for
  Physical Measurement
  Techniques
Freiburg, Germany

Gary Bulman
Center for Solid State
  Energetics
RTI International
Research Triangle Park
Durham, North Carolina

E.D. Case
Department of Chemical
  Engineering and Materials
  Science
Michigan State University
East Lansing, Michigan

Alic Chen
Department of Mechanical
  Engineering
University of California,
  Berkeley
Berkeley, California

Gang Chen
Department of Mechanical
  Engineering
Massachusetts Institute of
  Technology
Cambridge, Massachusetts

Lidong Chen
Shanghai Institute of
  Ceramics
Chinese Academy of
  Sciences
Shanghai, People's Republic of
  China

Thomas Colpitts
Center for Solid State Energetics
RTI International
Research Triangle Park
Durham, North Carolina

Douglas T. Crane
BSST LLC
Irwindale, California

Mildred Dresselhaus
Department of Physics
Massachusetts Institute of
  Technology
Cambridge, Massachusetts

Jesus Esarte
CEMITEC-R&D Center
Noain, Spain

Matthias Falmbigl
Institute of Physical Chemistry
University of Vienna
Vienna, Austria

M.I. Fedorov
Ioffe Physical-Technical
  Institute of the Russian
  Academy of Sciences
St. Petersburg, Russia

C. Heideman
Department of Chemistry
University of Oregon
Eugene, Oregon

Terry J. Hendricks
Pacific Northwest National
  Laboratory
Battelle Memorial Institute
Corvallis, Oregon

Y.V. Ivanov
Ioffe Physical-Technical
  Institute of the Russian
  Academy of Sciences
St. Petersburg, Russia

D.C. Johnson
Department of Chemistry
University of Oregon
Eugene, Oregon

David Koester
Nextreme Thermal
   Solutions
Durham, North Carolina

P.P. Konstantinov
Ioffe Physical-Technical
   Institute of the Russian
   Academy of Sciences
St. Petersburg, Russia

K. Koumoto
Graduate School of Engineering
Nagoya University
Nagoya, Japan
and
CREST
Japan Science and
   Technology Agency
Kawaguchi, Japan

Daniel Kraemer
Department of Mechanical
   Engineering
Massachusetts Institute of
   Technology
Cambridge, Massachusetts

Y.A. Kumzerov
Ioffe Physical-Technical
   Institute of the
   Russian Academy
   of Sciences
St. Petersburg, Russia

Ken Kurosaki
Graduate School of
   Engineering
Osaka University
Osaka, Japan

M. Kusunoki
EcoTopia Science Institute
Nagoya University
Nagoya, Japan

V.A. Kutasov
Ioffe Physical-Technical
   Institute of the Russian
   Academy of Sciences
St. Petersburg, Russia

D.R. Leadley
Department of Physics
University of Warwick
Coventry, United Kingdom

Q. Lin
Department of Chemistry
University of Oregon
Eugene, Oregon

Ruiheng Liu
Shanghai Institute of
   Ceramics
Chinese Academy of Sciences
Shanghai, People's Republic of
   China

L.N. Lukyanova
Ioffe Physical-Technical
   Institute of the Russian
   Academy of Sciences
St. Petersburg, Russia

Kenneth McEnaney
Department of Mechanical
   Engineering
Massachusetts Institute of
   Technology
Cambridge, Massachusetts

Donald T. Morelli
Department of Chemical
   Engineering and Materials
   Science
Michigan State University
East Lansing, Michigan

Takao Mori
National Institute for
   Materials Science
Tsukuba, Japan

N. Nguyen
Department of Chemistry
University of Oregon
Eugene, Oregon

W. Norimatsu
EcoTopia Science Institute
Nagoya University
Nagoya, Japan

Joachim Nurnus
Micropelt GmbH
Freiburg, Germany

Yoichi Okamoto
Department of Materials
   Science and Engineering
National Defense Academy
Kanagawa, Japan

Brooks O'Quinn
Nextreme Thermal Solutions
Durham, North Carolina

E.H.C. Parker
Department of Physics
University of Warwick
Coventry, United Kingdom

Jonathan Pierce
Center for Solid State
   Energetics
RTI International
Research Triangle Park
Durham, North Carolina

V.V. Popov
Ioffe Physical-Technical
   Institute of the Russian
   Academy of Sciences
St. Petersburg, Russia

John Posthill
Center for Solid State
   Energetics
RTI International
Research Triangle Park
Durham, North Carolina

M.J. Prest
Department of Physics
University of Warwick
Coventry, United Kingdom

David Prieto
University of the Basque
    Country
San Sebastian, Spain

Y.E. Putri
Graduate School of
    Engineering
Nagoya University
Nagoya, Japan

Zhifeng Ren
Department of Physics
Boston College
Chestnut Hill,
    Massachusetts

Peter F. Rogl
Institute of Physical
    Chemistry
University of Vienna
Vienna, Austria

David Michael Rowe
Cardiff School of
    Engineering
Cardiff University
Newport, United Kingdom

H. Scherrer
Physics of Materials
    Laboratory
Ecole des Mines
Nancy, France

S. Scherrer
Physics of Materials
    Laboratory
Ecole des Mines
Nancy, France

Vladimir A. Semenyuk
Thermion Company
Silverdale, Washington

Yoshikazu Shinohara
National Institute for
    Materials Science
Tsukuba, Japan

Edward Siivola
Nextreme Thermal Solutions
Durham, North Carolina

M. Smeller
Department of Chemistry
University of Oregon
Eugene, Oregon

Francis R. Stabler
Future Tech LLC
Troy, Michigan

David Stokes
Center for Solid State Energetics
RTI International
Research Triangle Park
Durham, North Carolina

Hiroaki Takiguchi
Department of Materials
    Science and Engineering
National Defense Academy
Kanagawa, Japan

Ctirad Uher
Department of Physics
University of Michigan
Ann Arbor, Michigan

O.N. Uryupin
Ioffe Physical-Technical
    Institute of the Russian
    Academy of Sciences
St. Petersburg, Russia

M.V. Vedernikov
Ioffe Physical-Technical
    Institute of the Russian
    Academy of Sciences
St. Petersburg, Russia

Rama Venkatasubramanian
Center for Solid State
    Energetics
RTI International
Research Triangle Park
Durham, North Carolina

C.L. Wan
Graduate School of Engineering
Nagoya University
Nagoya, Japan
and
CREST
Japan Science and Technology
    Agency
Kawaguchi, Japan

N. Wang
Graduate School of
    Engineering
Nagoya University
Nagoya, Japan

Y.F. Wang
Graduate School of Engineering
Nagoya University
Chikusaku, Nagoya, Japan

T.E. Whall
Department of Physics
University of Warwick
Coventry, United Kingdom

Cecilia Wolluschek
CEMITEC-R&D Center
Noain, Spain

Paul K. Wright
Department of Mechanical
    Engineering
University of California,
    Berkeley
Berkeley, California

Zhen Xiong
Shanghai Institute of Ceramics
Chinese Academy of Sciences
Shanghai, People's Republic
    of China

Shinsuke Yamanaka
Graduate School of Engineering
Osaka University
Osaka, Japan

Jihui Yang
Electrochemical Energy
    Research Lab
General Motors R&D Center
Warren, Michigan

V.K. Zaitsev
Ioffe Physical-Technical
    Institute of the Russian
    Academy of Sciences
St. Petersburg, Russia

Wenqing Zhang
Shanghai Institute of Ceramics
Chinese Academy of Sciences
Shanghai, People's Republic of
    China

I

Thermoelectric Materials and Properties

1

Nanostructured Thermoelectric Materials

Zhifeng Ren
Boston College

Gang Chen
Massachusetts Institute of Technology

Mildred Dresselhaus
Massachusetts Institute of Technology

1.1 Introduction

Thermoelectricity has become important in recent years because of the societal needs for energy sustainability on our planet. A recent Physics 2010 National Academy decadal survey on the challenges for materials research for this decade titled "Condensed Matter and Materials Physics: The Science of the World Around Us" cites six grand challenges to this important field of science:

1. How Do Complex Phenomena Emerge from Simple Ingredients?
2. How Will the Energy Demands of Future Generations Be Met?
3. What Is the Physics of Life?
4. What Happens Far from Equilibrium and Why?
5. What New Discoveries Await Us in the Nanoworld?
6. How Will the Information Technology Revolution Be Extended?

It is interesting that challenge #2 speaks explicitly about the human need for sustainable energy, while most of the other challenges provide a means by which this challenge can be met. For example, nanostructures (challenge #5) provide a means for independent variation of the parameters that control the thermoelectric (TE) figure of merit, and large-scale modeling requires advances in information technology (challenge #6) which are needed to make predictions for the convergent optimization of the parameters that control TE performance. Most of the composite materials used for high TE performance are conceived via processes that are far from equilibrium (challenge #4), while living systems (challenge #3) give us examples of how nature provides us with systems that achieve energy sustainability during their lifetimes. The inorganic world tells us how the interactions between the constituents of matter (electrons, atoms, and ions) interact with each other so that their collective behavior can be different from that of their individual behavior (challenge #1), thereby giving us a means that can be used to design and constrain systems that mimic what nature can achieve. Thus we can expect the science of the world around us to furnish us with the necessary tools to provide the energy needed annually for sustainability (15 TW) from the much larger annual amount supplied by the sun (36,000 TW) to the land surface of the world.

TE materials are capable of directly converting heat into electricity and vice versa. If a temperature gradient is imposed on a TE junction, a voltage gradient will form in response to the Seebeck effect, discovered by Thomas Johann Seebeck in 1821. Likewise, a current flowing across a TE junction will produce cooling or heating at the junction via the Peltier effect, discovered by Jean Charles Athanase Peltier in 1834. Solid-state TE devices based on these fundamental principles can be used in a wide range of applications such as temperature measurement, waste heat recovery, air conditioning, and refrigeration.[1-11] Especially, solar TE energy conversion has emerged as a promising application as recently reported [12] and also as summarized in the review chapter in this handbook by Kraemer et al. TE devices have attracted extensive interest for several decades because of their unique features: no moving parts, quiet operation, low environmental impact, and high reliability.[1-4,7,8,10] The efficiency of the TE materials is determined by a dimensionless figure-of-merit (ZT), defined as[1,3,4,13,14]

$$ZT = (S^2\sigma/\kappa)T \tag{1.1}$$

where S, σ, κ, and T are the Seebeck coefficient, the electrical conductivity, the thermal conductivity, and the absolute temperature at which the properties are measured, respectively. In these measurements it is important that all the temperature gradient and S, σ, and κ measurements are made in the same directions and on the same material. The efficiency of a TE device is directly related to ZT. For power generation, the efficiency is given by

$$\varepsilon = \frac{T_h - T_c}{T_h} \frac{\sqrt{1 + ZT^*} - 1}{\sqrt{1 + ZT^*} + T_c/T_h} \tag{1.2}$$

and for air conditioning and refrigeration, we evaluate the effectiveness of devices in terms of the coefficient of performance given by

$$\eta = \frac{T_c}{T_h - T_c} \frac{\sqrt{1 + ZT^*} - T_h/T_c}{\sqrt{1 + ZT^*} + 1} \tag{1.3}$$

where T_h and T_c are the hot-end and cold-end temperatures of the TE materials, respectively, and T^* is the average temperature of the materials. Thus, it is important to use materials with a high ZT value for practical applications.

The low ZT values of today's commercially available TE materials limit the applications of TE devices. Metals and metal alloys whose ZT values are low ($ZT \ll 1$) can only be applied in thermocouples to measure temperature and radiant energy.[2] Semiconducting TE materials, such as Bi_2Te_3 and SiGe alloys with $ZT \sim 1$,[6,10] are presently used commercially in low-power cooling and low-power TE power generators, such as beverage coolers and laser diode coolers, and power generators in space missions.

To make TE devices competitive in large scale and high-power commercial applications, materials with significantly higher ZT values in the application temperature range of 250–1000°C are required.[1,3,4,10,15] Since the 1960s, much research has been devoted to identifying TE materials which could satisfy this requirement.

The traditional method to improve ZT has been to discover new TE materials. Since the TE effect was discovered, many TE materials have been identified, such as Bi_2Te_3, skutterudites Co_4Sb_{12}, SiGe alloys, PbTe, $CsBi_4Te_6$,[16] Tl_9BiTe_6,[17] clathrate $(Ba,Sr)_8(Al,Ga)_{16}(Si,Ge,Sn)_{30}$,[18-21] PbTe-PbS,[22] lead anti-mony silver tellurium based materials[23-27] (such as $AgPb_mSbTe_{2+m}$ (LAST),[23] $Ag(Pb_{1-x}Sn_x)_mSbTe_{2+m}$ (LASTT),[24] $Na_{1-x}Pb_mSb_yTe_{m+2}$ (SALT),[25] and $NaPb_{18-x}Sn_xSbTe_{20}$ (SALTT)[26]), and $In_4Se_{3-\delta}$.[28] Many of these materials are alloys which help to reduce the phonon thermal conductivity, following early work of Ioffe.[13] Additional TE materials can be found in other review papers and books.[6,27,29-31] Some of the above materials have been incorporated in commercially available devices, as for example $Bi_2Te_{3(1-x)}Se_{3x}$.

A second method to improve ZT is through nanostructuring. Experiments show that the thermal conductivity decreases with grain size in TE bulk materials.[14,32-40] According to Equation 1.1, ZT will be increased when the thermal conductivity decreases so long as the power factor, $S^2\sigma$, is not strongly reduced at the same time. More details are reviewed in the literature.[4,41-43] In the nanostructuring approach, numerous boundaries or interfaces are introduced throughout the TE materials, and a higher density of defects are also introduced in the interface region, such that phonons are highly scattered, reducing the thermal conductivity to very low values. Of course, for the strategy to be successful, it must be again emphasized that the electrical conductivity and Seebeck coefficient (and resulting power factor) should not at the same time be significantly reduced, as stated above.

The concept of low-dimensional nanostructured TE materials was introduced in the 1990s.[44] Both an increase in the electron power factor ($S^2\sigma$) and a reduction in the lattice thermal conductivity are possible simultaneously in nanostructures. Thus far, theories and experiments indicated that a larger reduction in the thermal conductivity can be achieved in nanometer-sized low-dimensional structures as well as in bulk nanograined materials, arising from similar boundary and interface phonon scattering mechanisms.[34,45-47] In recent years, many experimental studies have shown that the nanostructuring approach is indeed effective in improving ZT.[47-57] The lattice thermal conduc-tivity κ_l has been reduced via the increased phonon scattering at the interfaces in one-dimensional nanotubes and nanowires,[47,58-61] in two-dimensional superlattices,[44-46,51-56,62-71] such as GaAs/AlAs superlattices,[62-64] Bi_2Te_3/Sb_2Te_3 superlattices,[51-54] $PbSe_{0.98}Te_{0.02}$/PbTe quantum dot superlattices,[55,56] and SiGe/Si superlattices,[66,69-71] and in nanostructured materials which consist of three-dimensional nanograins. (Hereafter, when we refer to "nanocomposites" we assume that they are nanostructured materials, defined as an agglomeration of nanograins with either different compositions or crystal-line structures, or the same material with precipitates or inclusions.[47,72-77]) The significant enhance-ment of ZT in all these nanostructured materials systems is believed to result primarily from the reduction of the thermal conductivity by scattering phonons more effectively than electrons at the interfaces of the superlattices and within the layers of the superlattice by introducing disorder within the layers, or at grain boundaries within the nanostructures and between the nanostructures.[45,46,48-57,72]

According to Equation 1.1, ZT can be increased by increasing the power factor at the same time as decreasing the thermal conductivity. In the literature, there are reports that nanostructuring has increased ZT by increasing the Seebeck coefficient.[68,78-84] At first sight, increasing of power factor by introducing nanostructures may be surprising, since the interfacial scattering would be expected to reduce the electrical conductivity. However, TE materials are usually heavily doped and the electron mean free path is already very short. As long as the barrier height between the interfaces is not too large, the electrical conductivity will not suffer much reduction, and this effect was compensated in the past by increasing the doping. In fact, the Seebeck coefficient can potentially be enhanced by the preferential scattering of low-energy electrons at grain boundaries. In addition to the nanostructuring approach, an increase in the electrical power factor can also be achieved in bulk materials via creating sharp features in the density of states near the Fermi level, a general principle as predicted theoretically.[84]

In this chapter, we mainly focus on the *ZT* enhancements achieved in bulk nanocomposites consisting of three-dimensional nanograins. The very small size of the nanograins introduces a high density of grain boundary interfaces in the nanocomposites. These interfaces as designed in some way to scatter phonons more strongly than charge carriers and to thereby decrease the thermal conductivity. Reviews of other work on enhancing *ZT* can be found in the literature for one-dimensional nanowires[4,47,85,86] and for two-dimensional superlattices.[4,11,61,67]

The higher *ZT* of the nanocomposites makes them attractive for cooling and low-grade waste heat recovery for energy-harvesting applications. Different nanostructured materials can also be integrated into segmented TE devices for TE power generation which operate at high temperatures and over a wide temperature range. In addition to the high *ZT* values, the nanocomposites are isotropic and show better mechanical properties than single crystals, which are often brittle. The nanocomposites do not suffer from the cleavage problem that is common in ingots made from traditional zone melting techniques, thereby leading to easier device fabrication and system integration and a potentially longer device lifetime.

1.2 Nanocomposite Preparation

In order to produce nanocomposites, TE nanoparticles are usually prepared first and then assembled into dense bulk solids. This topic is now discussed.

1.2.1 Nanoparticle Preparation

TE nanoparticles can be produced by many techniques, such as hydrothermal methods,[87-95] wet chemical reactions,[96-99] and ball-milling,[34,37,72-77,86,100-110] among others.

Among these methods, ball-milling is an effective top-down industrial approach to obtain fine particles. Conventional ball-milling has been employed to produce large quantities of fine particles with a size of one to several microns, and this process can be readily scaled up for commercial use at reasonable cost. High-energy ball-milling, developed in the 1970s as an industrial process, can create nanoparticles with a size as small as several nanometers.[111] It has been proven that high-energy ball-milling is an effective and powerful processing technique to produce large quantities of TE nanoparticles in a short time.[77] Figure 1.1a,b shows crystalline ingots and the crystallinity of a *p*-type $(Bi,Sb)_2Te_3$ ingot before ball-milling. The ingot is a bulk assembly of single crystal material. After ball-milling, nanoparticles can be obtained (Figure 1.1c,d). Besides grinding down crystalline Bi_2Te_3 ingots, TE nanoparticles can also be prepared directly from the individual elements Bi and Te using a ball-milling method.[73-77,102,112-114]

1.2.2 Bottom-Up Methods to Produce Nanocomposites

In order to obtain TE nanocomposite materials, the TE nanoparticles are assembled into a dense solid using various bottom-up methods, such as spark plasma sintering,[98,99,103,107-109,115] cold-pressing,[100,101,104,116] sintering,[37,101,117,118] hot-pressing,[72,73,96,105,119-121] and extrusion methods.[101,122]

Among these methods, cold-pressing only mechanically compacts the nanoparticles and thus the density of cold-pressed composites tends to be low, resulting in a material with poor mechanical properties. To improve the mechanical and electrical properties of the composites, the cold-pressing can be followed by sintering at appropriate temperatures. However, to reliably create composites with a density

FIGURE 1.1 (a) Optical image of crystalline ingots before ball milling. (b) TEM image of a p-type $(Bi,Sb)_2Te_3$ ingot. Insets are an HRTEM image and an SAED pattern of the ingot. (c) Optical image of ball-milled nanopowder. (d) HRTEM image of a typical $(Bi,Sb)_2Te_3$ nanoflake produced by high-energy ball milling. (e) Optical image of hot-pressed nanocomposites made from the ball-milled nanopowders shown in (c). (f) HRTEM image of several nanograins in a p-type $(Bi,Sb)_2Te_3$ nanocomposite produced by the ball-milling and hot-pressing method. The ingots, ball-milled nanopowder, and hot-pressed nanostructured samples shown in (a), (c), and (e), respectively, are prepared as described in Ref. [72]. (Reprinted with permission from Y. C. Lan et al., *Adv. Func. Mater.* 20, 357–376. Copyright 2010, Wiley-VCH Verlag GmbH & Co. KGaA.)

of 95–100% of the theoretical value, the hot-pressing method and the spark plasma sintering method are more commonly employed.[34,72,98,99,103,123,124]

Hot-pressing was used to prepare PbSe composites as early as 1960[125] and this process was later employed to produce SiGe polycrystalline materials for NASA space missions. Currently, hot-pressing is a mature technique that can also be used commercially to produce TE nanocomposites.[47] Most of the recently reported nanocomposites with high ZT have been prepared using the hot-pressing technique described in Section 1.2.3.

The technique used to produce nanocomposites can be any combination of a nanoparticle preparation method and a bottom-up nanocomposite assembly method. Based on the various combinations, many nanocomposite preparation methods are possible. Table 1.1 lists some typical nanocomposite preparation techniques. Under appropriate processing conditions, the nanocomposite TE materials thus produced have a reduced thermal conductivity and a higher ZT. For example, a peak ZT of 1.4 has been achieved in p-type $Bi_xSb_{2-x}Te_3$ nanocomposites produced by the ball-milling and hot-pressing method,[72] compared to a peak $ZT = 1.0$ for the conventional bulk materials.

1.2.3 Ball-Milling and Hot-Pressing Method

The ball-milling and hot-pressing method has been employed to produce TE composites since the 1960s.[126] Interested readers are referred to earlier articles on ball-milling and hot-pressing methods in the literature.[34,113,124,126] In earlier works, only micron-sized particles were obtained from ball-milling. The grain size of these composites varies from several microns to 100 μm. As a result, the thermal conductivity of these hot-pressed composites was decreased compared with that of the single crystal,[34,36] and the peak ZT of the resulting SiGe composites, consisting of micron-sized grains, was increased, but only by less than 20%.

Since the invention of high-energy ball-milling, large quantities of nanoparticles can be produced. After hot-pressing, the grains in the nanocomposites still have a size in the nanometer range, although grain growth can take place during hot-pressing (Figure 1.1f).

The combination of high-energy ball-milling and hot-pressing is very attractive from a commercial point of view because large quantities of nanopowders can be ball-milled and hot-pressed in a short time. From the research point of view, this method can prepare many TE materials with different chemical compositions and doping levels in a short time, allowing systematic studies to be carried out efficiently.

Figure 1.1e shows several hot-pressed nanocomposite samples made from the ball-milled nanoparticles. Figure 1.1f shows the microstructure of a p-type $(Bi,Sb)_2Te_3$ nanocomposite made from ball-milling and hot-pressing. The grain size is below 1 μm. Compared with other bottom-up methods, the ball-milling and hot-pressing method can produce dense bulk samples with a high density. For example, the density of the hot-pressed nanocomposites is higher than 98%,[34] and even a density of 100%[72] compared to that of single crystalline material has been achieved. The hot-pressed nanocomposites are also thermally stable at high temperature and thermoelectrically isotropic.[72-74]

The high-energy ball-milling and hot-pressing method has been employed to prepare many TE nanocomposites, for example, Bi_2Te_3,[39,72,74,105] SiGe alloys,[73,75,76] and skutterudite $CoSb_3$[127-129] nanocomposite materials. Table 1.1 summarizes some reported n- and p-type TE nanocomposites prepared by the ball-milling and hot-pressing method and some pertinent information about each is given in this table.

1.2.4 Chemical Synthesis and the Spark Plasma Sintering Method

Besides the ball-milling and hot-pressing method, some nanocomposites have been successfully produced by a chemical synthesis and spark plasma sintering method. In this method, the TE nanoparticles are firstly synthesized by a chemical method. The size distribution, nanoparticle shape, and quality of the nanoparticles can be finely tuned by the parameters of the preparation process.[130] After spark plasma sintering, fully dense nanocomposites are produced.

Hot-pressing is a pressure-assisted sintering process, in which a direct-current or alternating current produces the required high temperature in the sample heating process. Spark plasma sintering is another kind of pressure-assisted sintering process, in which a pulsed direct-current source produces spark discharges to heat samples under high pressure. Both techniques have successfully produced fine-grained nanostructured TE bulk materials with high density.

TABLE 1.1   TE Properties of Some Typical Nanocomposites

Nanocomposites	Carrier Type	ZT^a	Thermal Conductivity $(W\ m^{-1}\ K^{-1})^{a,b}$	Methods[c,d]	References
$Si_{80}Ge_{20}B_x$	p	0.95 (0.5) at 800–900°C	2.5 (5.0)	BM (10–60 h) + HP (950–1200°C)	[54]
$Si_{80}Ge_{20}P_2$	n	1.3 (0.93) at 900°C	2.5 (4.6)	BM + HP (1000–1200°C)	[56]
$Si_{63.5}Ge_{36.5}P_x$	n	0.128 (0.0937) at RT	3.82 (5.07)	BM + HP	[28]
$Si_{95}Ge_5$	p	0.95 at 900°C	5 (10)	BM + HP	[57]
SiP_x	n	0.7 at 1000°C	12	BM (1 h) + HP (>1000°C)	[100]
$Bi_{0.5}Sb_{1.5}Te_3$	p	0.8 at RT	0.9	BM (300 rpm, 20 h) + HP (300–550°C, 30 min)	[33]
$(Bi,Sb)_2Te_3$	p	1.4 (1.0) at 100°C	1.1 (1.4)	BM + HP	[53]
$(Bi,Sb)_2Te_3$	p	1.3 (1.0) at 75–100°C	1.25 (1.4)	BM + HP	[55]
$(Bi,Sb)_2Te_3$	p	0.9 at RT		BM (400 cycles) + HP (420°C, 1.2 GPa, 1 h)	[101]
$(Bi,Sb)_2(Te,Se)_3$	p	0.7–0.9 at RT	1.0–1.5	BM (1200 rpm, 6.5 h) + HP (550°C, 30 min)	[102]
$Bi_2(Te,Se)_3$	n	1.0 at 100°C		BM + HP	Unpublished
$Bi_2(Te,Se)_3$	n	0.5 at RT	1.0	BM (400 rpm,10 h) + HP (500°C, 4 h)	[79]
$Bi_2Te_{2.7}Se_{0.3}$	n	0.9 at RT 1.04 at 125°C	1.16	BM + HP + re-pressing	[147]
$(Bi,Sb)_2Te_3$	p	1.35 (0.8) at RT	0.6 (1.4)	MS + SPS (510°C, 10 min)	[103]
$(Bi,Sb)_2Te_3$	p			BM (800 cycle) + extrusion	[78]
Co_4Sb_{12}	p	0.05 at RT	2.9(4.3)	BM (450 rpm, 15 h) + SPS (50 MPa, 300°C, 5 min)	[82]
Co_4Sb_{12}	p	<0.01 at RT	1.7	CM + HP (100 MPa, 450°C)	[104]
Co_4Sb_{12}		0.11 at 723 K	4.7	CM + HP (100 MPa, 600°C, 30 min)	[95]
$Yb_{0.35}Co_4Sb_{12}$	n	1.2 at 550°C	1.52	BM + HP	[105]
$Yb_{0.19}Co_4Sb_{12}$	n	0.26 at RT	2.6	BM + HP (26PSI, 650°C, 2 h)	[106]
$Yb_{0.29}Co_4Sb_{12}$	n	1.3 at 800 K	2.5	MS + SPS (550°C, 5 min)	[89]
$(Ni_{0.09}Co_{0.91})_4Sb_{12}$	n	0.75 at RT	3	BM (20–50 h) + HP (60–160 MPa, 550–780°C, 1–6 min)	[107]
$(Ni_{0.08}Co_{0.92})_4Sb_{12}$		0.065 at 450 K	5	CM + HP (100 MPa, 450°C)	[72]
$(Ni_{0.8}Co_{0.2})_4Sb_{12}$		0.1 at RT	3.0	BM (39 h) + HP (80–160 MPa, 500°C, 1–5 min)	[108]
$Co_{4-x}Fe_xSb_{12}$				BM (230 rpm) + sintering	[91]
$Fe_{1.5}Co_{2.5}Sb_{12}$	p	0.32 at 600 K	2	BM (100 h) + HP (60 MPa, 550°C, 2 h)	[109]
$SnFe_3Co_5Sb_{24}$	n	0.15 at RT	1.5	BM (100 h) + HP (60 MPa, 550°C, 2 h)	[110]
$La_x(Ni,Co)_3Sb_{12}$		0.12 at RT	3.2	BM (400 rpm, 10 h) + HP (50 MPa, 600°C, 2 h)	[111]
$La_{1.5}Fe_4Sb_{12}$		0.41 at 750 K	2	BM (20 h) + HP (50 MPa, 650°C, 2 h)	[112]
$La_{0.4}FeCo_3Sb_{12}$		0.04 at RT	3	BM (400 rpm, 14 h) + HP (70 MPa, 700°C, 3 h	[113]

continued

TABLE 1.1 (continued) TE Properties of Some Typical Nanocomposites

Nanocomposites	Carrier Type	ZT^a	Thermal Conductivity (W m^{-1} K^{-1})^{a,b}	Methods^{c,d}	References
β-FeSi$_2$		0.018 at RT	4.5	BM (4 h) + HP (80–100 MPa, 950°C, 30 min)	[114]
Half-Heusler	p	0.8 at 700°C	4.7	BM + HP	[216]
Fe$_{0.95}$Co$_{0.05}$Si$_2$	n	0.18 at 923 K	4.5	BM (300 rpm, 30 min) + HP (50 MPa, 950°C, 30 min)	[115]
Mg$_2$Si$_{0.4}$Sn$_{0.6}$		0.13 at 653 K	2.1	BM (600 cycle) + HP	[116]
Mg$_2$Si$_{0.6}$Ge$_{0.4}$	p	0.21 at 610 K	4.3	BM (600 cycle) + HP (1 GPa, 500°C)	[117]
TiNiSn$_{0.95}$Sb$_{0.05}$			4 (10)	BM (18 h) + shock compaction (>5 GPa)	[118]
TiO$_x$	n	<0.025 at RT	2.7–5.7	HP (114 MPa, 1200°C, 2 min)	[119]
Bi$_{85}$Sb$_{15}$		0.3 at 150 K	2.1	BM (236 rpm) + extrusion	[75]
PbTe	n	0.07 at RT	4	BM + SPS (40 MPa, 310–510°C, 1–10 min)	[77]
GaSb$_{10}$Te$_{16.5}$	p	0.6 at RT	1.2	BM (350 rpm, 5 h) + SPS (40 MPa)	[120]
Ag$_{0.8}$Pb$_{22}$SbTe$_{20}$		1.37 at 673 K		BM + SPS (50 MPa, 10 min)	[81]
2 wt%ZrO$_2$ + β-FeSi$_2$		0.14 (0.35) at 700 K	4.5 (5.0)	BM (20 h) + HP (25 MPa, 900°C, 1 h)	[121]
2 wt%Y$_2$O$_3$ + β-FeSi$_2$		0.064 (0.035) at 700 K	4.5 (5.5)	BM (20 h) + HP (25 MPa, 900°C, 1 h)	[122]
5 wt%TiB$_2$ + β-FeSi$_2$		0.015 (0.018) at RT	5.2 (4.5)	BM (4 h) + HP (80–100 MPa, 950°C, 30 min)	[114]
50% Bi$_2$Te$_3$ + 50% Sb$_2$Te$_3$		1.47 at 450 K	1.1	HS + HP (75 MPa, 350°C, 15 min)	[71]
10 wt%Bi$_2$Te$_3$ nanoparticle + Bi$_{0.5}$Sb$_{1.5}$Te$_3$ microparticle		0.65 at RT	0.55	BM + HP (50 MPa, 310°C, 30 min)	[123]
10 wt%Bi$_2$Te$_3$ nanoparticle + Bi$_2$Te$_{2.85}$Se$_{0.15}$ microparticle		0.83 at 350 K	0.55	BM + HP (50 MPa, 310°C, 30 min)	[123]
7 wt%BN + (Bi$_{0.2}$Sb$_{0.8}$)$_2$Te$_3$		0.54 (1.0) at RT	1.2 (1.5)	BM (1200 rpm, 5 h) + HP (425 MPa, 550°C, 30 min)	[124]
7 wt%WO$_3$ + (Bi$_{0.2}$Sb$_{0.8}$)$_2$Te$_3$		0.75 (1.0) at RT	1.3 (1.5)	BM (1200 rpm, 5 h) + HP (425 MPa, 550°C, 30 min)	[124]
15 wt% nanoplate Bi$_2$Te$_3$ + Bi$_2$Te$_3$		0.39 at RT	1.1	Plasma sintering	[125]
25.4 vol%TiB$_2$ + B$_4$C		0.002 (0.01) at 1050 K	16 (9)	BM (10 h) + HP (35 MPa, 1900°C, 30 min)	[165]

^{a} Data within parentheses are the values of single crystals or of nanocomposites without addition of second phase material.

^{b} At room temperature (RT).

^{c} BM, ball-milling; HS, hydrothermal synthesis; CM, chemical method; MS, melt spinning; HP, hot-pressing; CP, cold-pressing; SPS, spark plasma sintering.

^{d} Experimental details are listed within parentheses.

1.3 Microstructures and TE Properties of Nanocomposites

The microstructures created during the fabrication process affect the phonon transport and electron transport in the nanocomposites, altering their thermal conductivity, electrical conductivity, and Seebeck coefficient, and thereby changing ZT. Below, we introduce the microstructures of some TE nanocomposites in detail, showing how the microstructures affect the TE properties of the nanocomposites.

1.3.1 $Bi_xSb_{2-x}Te_3$ Nanocomposites

Bi_2Te_3 is the most widely commercially used TE material near room temperature, and has great commercial applications in refrigeration and waste heat recovery up to 200°C. The maximum ZT of the commercial Bi_2Te_3 ingots remained at about $ZT = 1.0$ for many years (1960–1995). The peak ZT of Bi_2Te_3 nanocomposites has more recently been enhanced to 1.3–1.4 at about 100°C,[72,74] and advances for both p- and n-type nanocomposites have been made, and furthermore alloys of $Bi_xSb_{2-x}Te_3$ are most commonly used to lower the thermal conductivity while maintaining a high power factor, as described below.

1.3.1.1 p-Type $Bi_xSb_{2-x}Te_3$ Nanocomposites

Nanostructured p-type $Bi_xSb_{2-y}Te_3$ was the first materials system in which an enhanced ZT was reported.[72] We use the possibility that $x \neq y$ so that some local regions of non-stoichiometry are found in the actual materials. Using the ball-milling and hot-pressing approach, ZT is typically increased by 30–40% compared with the starting crystalline ingots.[72,74]

The nanoparticles are produced by ball-milling crystalline ingots. High-energy ball-milling grinds the crystalline ingots into nanocrystals with sizes ranging from 5 to 50 nm, and with an average size of about 20 nm. The milled nanoparticles are very good crystals with clean surfaces (Figure 1.1d), and no nanoprecipitates or amorphous nanoregions are observed in HRTEM images of the nanoparticles.

During hot-pressing at high temperature, the nanoparticles are compressed into a solid nanocomposite bulk material. The final nanocomposites consist of crystalline nanograins. Statistical analysis of thousands of nanograins shows that the majority of the nanograins have a diameter below 1.0 µm, with 12% of the nanograins having a diameter of below 20 nm and about 5% of the nanograins having a diameter of 20–40 nm. Some nanograins are larger, with a diameter of several hundred nanometers (Figure 1.2a) while some are very small, with a diameter of several nanometers (Figure 1.1d). The grains with a wide size distribution, as shown in Figure 1.2b, scatter phonons with a variety of mean free paths (which are a function of the phonon velocity and phonon wavelength) comparable to or greater than the grain size, thus resulting in interface scattering, thereby reducing the thermal conductivity.

High-resolution transmission electron microscope (HRTEM) images such as in Figure 1.2c indicate a preponderance of high-angle grain boundaries with random orientation. Selected area electron diffraction (SAED) measurements show that the grains are single crystals. The adjacent grains, with random orientations and sizes, scatter phonons with a wide range of mean free paths and phonon wave vectors effectively, which is highly desirable for good TE performance.

In addition to the clean boundaries between grains shown in Figure 1.2c, some grains in the nanocomposite are surrounded by nanometer thick interface regions (Figure 1.2d). The interface region in Figure 1.2d is slightly bismuth-rich, with 1.0 ± 0.5 atomic% Bi higher than that of the grains.[131] This bismuth-rich region builds up charges and thus increases the hole concentration in the grains. At the same time, the interface region scatters phonons in addition to scattering hole carriers. Thus the phonon scattering in the bismuth-rich interface region is strong and blocks phonon transport while favoring charge carrier transport.

Besides the unique grain and grain boundary structures in the nanocomposites, most of the individual nanograins have abundant microstructural defects, which also scatter phonons. Usually there are many three-dimensional nanoprecipitates embedded in the nanograins, as shown in Figure 1.3a–d.

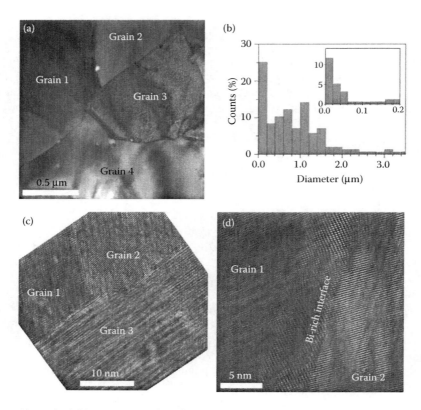

FIGURE 1.2 (a) Bright-field TEM image of multigrains and (b) grains size distribution. HRTEM images of (c) clean grain boundaries and (d) an interface region between two grains in a p-type $Bi_xSb_{2-x}Te_3$ nanocomposite. The inset in (b) is a zoom of the small diameter region of the larger figure. (Reprinted with permission from Y. C. Lan et al., *Nano Lett.* 9, 1419–1422. Copyright 2009, American Chemical Society.)

In $Bi_xSb_{2-x}Te_3$ nanocomposites, at least four kinds of nanoprecipitates, as enumerated below, are observed in the crystalline nanograins. A common type of nanoprecipitates is an antimony-rich nanoprecipitate (Figure 1.3a). There are no obvious boundaries between the nanoprecipitates and nanograin matrix, with even no obvious lattice distortion between them. The lattice is continuous and the fringe spacing is almost the same. In the HRTEM images in Figure 1.3, the nanoprecipitates are brighter than the surrounding crystalline matrix. Another type of nanoprecipitate is a nanodot with an orientation that is twisted with respect to that of the matrix. The chemical composition of the nanodot is the same as that of the surrounding matrix. The third type of nanoprecipitate (Figure 1.3c) is also twisted from the surrounding matrix and is also antimony rich. This type of nanoprecipitate is the combination of type one and type two nanoprecipitates. The fourth type is a pure tellurium nanodot (Figure 1.3d). These polygonal tellurium precipitates are easily identified structurally from the other three types of irregular nanoprecipitates.

The types of nanoprecipitates described above are very similar to those observed in $Ag_{1-x}Pb_mSbTe_{m+2}$ alloys.[23-26] It is believed that the presence of these nanoprecipitates as scattering centers are responsible for the ZT enhancement in these alloys. Experiments have shown that the nanometer-sized precipitates embedded in a crystalline host, like 30–40 nm Pb or Au metal particles in crystalline PbTe samples, can at the same time increase the TE power factor and reduce the thermal conductivity.[57,79] In the $Bi_xSb_{2-x}Te_3$ nanocomposites, it is expected that these nanoprecipitates predominately scatter phonons and reduce the thermal conductivity. We therefore need to look for more opportunities to further increase the power factor and thereby increase ZT.

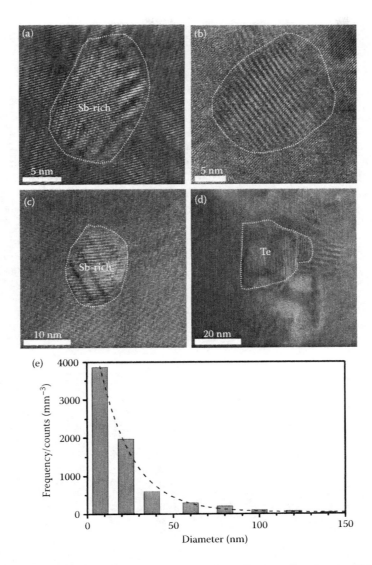

FIGURE 1.3 Nanoprecipitates embedded in the nanograins of a $Bi_xSb_{2-x}Te_3$ nanocomposite. HRTEM images of (a) an antimony-rich precipitate without a boundary, (b) a nanoprecipitate with a twisted boundary, with the same chemical composition as the surrounding matrix, (c) an antimony-rich nanoprecipitate with a twisted boundary, and (d) a tellurium nanoprecipitate with a high-angle boundary. (e) The nanoprecipitate size (diameter) distribution. (Reprinted with permission from Y. C. Lan et al., *Nano Lett.* **9**, 1419–1422. Copyright 2009, American Chemical Society.)

Nanoprecipitates with different chemical compositions are expected to scatter both electrons and phonons effectively. These nanoprecipitates increase the carrier concentration in the nanograins. Hall effect measurements at room temperature indicate that the hole concentration ($p = 2.5 \times 10^{19}$ cm^{-3}) of the nanocomposites is about 39% higher than that ($p = 1.8 \times 10^{19}$ cm^{-3}) of the ingots. The increased carrier concentration is partly due to the nanoprecipitates and partly due to the nanointerfaces between nanograins.

All of the grain boundaries, nanointerfaces between nanograins, and nanoprecipitates affect the transport of electrons and phonons. More discussions of the interfacial structure effects are available in the literature.[132]

Besides the nanoprecipitates, there are two-, one-, and zero-dimensional defects in the nanograins of the nanocomposites.[131] TEM investigation indicates that stacking faults exist in the nanograins. The threading dislocation concentration is ~10^{11} cm^{-2} in nanograins, at least 10 times higher than that in the crystalline ingots (~5×10^9 cm^{-2}). The point defect concentration in nanograins is 2–3 orders of magnitude higher than that in the ingot. Structural modulations are also observed in the nanograins even though the concentration of structural defects is almost the same as that in the ingots. All these defects would scatter phonons more effectively and decrease the thermal conductivity of the nanocomposites.

p-Type Bi$_x$Sb$_{2-x}$Te$_3$ nanoparticles can also be produced from the individual elements directly through mechanical alloying during high-energy ball-milling. Mechanical alloying can synthesize Bi$_x$Sb$_{2-x}$Te$_3$ nanoparticles with a single structural phase (see the x-ray diffraction pattern in Figure 1.4a) starting directly from the element Bi, Sb, and Te.[74] The size of the synthesized nanoparticles is about 5–20 nm with an average size of about 10 nm. The crystalline quality of the alloyed nanoparticles obtained from the mechanical alloying of nanopowders of the elements (Figure 1.4b) is almost the same as that obtained from ball milling the Bi$_x$Sb$_{2-x}$Te$_3$ ingots.

It is reported that trace amounts of Sb and Te phases are observed in the ball-milled Bi$_x$Sb$_{2-x}$Te$_3$ nanopowders although a Bi$_x$Sb$_{2-x}$Te$_3$ single phase is obtained after hot-pressing.[133] The un-alloyed Sb and Te impurity could be due to insufficient milling time or insufficient transfer energy during ball-milling. If the transfer energy is not high enough, the ball-milling will only grind and mix the elements, and not create alloyed nanoparticles.

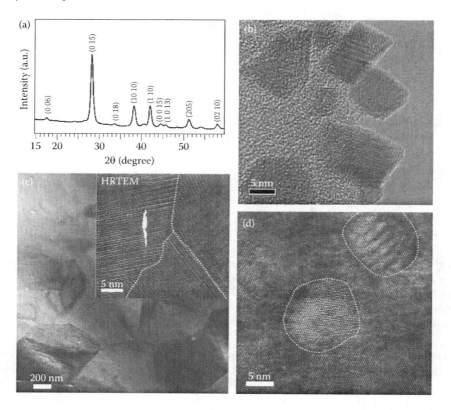

FIGURE 1.4   (a) An x-ray diffraction pattern and (b) an HRTEM image of a mechanically alloyed Bi$_x$Sb$_{2-x}$Te$_3$ nanopowder sample prepared from the chemical elements. (c) A BF-TEM image and (d) an HRTEM image of nanocomposites made from nanoparticles shown in (b). The white dots outline the grain boundaries or nanoprecipitates embedding the nanograins. (Reprinted with permission from Y. Ma et al., *Nano Lett. 8*, 2580–2584. Copyright 2008, American Chemical Society.)

The microstructure of the nanocomposite made from the elements is very similar to that of the nanocomposite made from ingots of the same elemental composition for the starting materials. The nanograin size is in the sub-micrometer range and the nanocomposites are densely packed (Figure 1.4c). Compared with the starting mechanically alloyed nanopowders (Figure 1.4b), the nanograins in nanocomposites grow to larger sizes after hot-pressing (Figure 1.4c). Detailed TEM studies indicate that nanoprecipitates are also formed and embedded inside the $Bi_xSb_{2-x}Te_3$ nanograins (Figure 1.4d), very similar to the nanoscale morphology of the hot-pressed nanocomposites made from the ingots (Figure 1.3a–d).

These unique microstructures (shown in Figures 1.2 through 1.4) are important for obtaining superior TE properties of the nanocomposites. In Figure 1.5, we compare the transport properties of three samples: the hot-pressed nanocomposite made from the Bi, Sb, and Te elements, the hot-pressed nanocomposite made from the crystalline $Bi_xSb_{2-x}Te_3$ ingots, and the crystalline ingot itself.[74]

The electrical conductivity of the $Bi_xSb_{2-x}Te_3$ nanocomposites is always higher than that of the crystalline ingots (Figure 1.5a) because of the higher carrier concentration (2.9×10^{19} cm^{-3} for the nanocomposites and 1.8×10^{19} cm^{-3} for the ingots at room temperature).[74] Here the nanoprecipitates embedded in the nanograins and nanointerfaces should contribute to the increase of the electrical conductivity. The enhancement of electrical conductivity is very important to improve the ZT of nanocomposites. Since the 1960s, much work has been carried out in low-dimensional composites, and ZT could not be increased because of the decrease of the electrical conductivity. Although phonons are scattered greatly by the grain boundaries, resulting in a decrease in the thermal conductivity, electrons are also scattered by the grain boundaries, resulting in a decrease in the electrical conductivity. According to Equation 1.1, ZT increases or decreases depending on the relative values of the thermal conductivity and the electrical conductivity when the Seebeck coefficient remains constant.

The Seebeck coefficient (Figure 1.5b) of both $Bi_xSb_{2-x}Te_3$ nanocomposites is lower than that of the crystalline ingot sample below 150°C but is higher above 150°C.[74] The smaller magnitude of the Seebeck coefficient near room temperature is due to a higher carrier concentration, while the larger Seebeck coefficient at higher temperatures is due to the suppression of minority carrier (electron) excitation in heavily doped samples.

The thermal conductivities of both types of nanocomposites in Figure 1.5 are significantly lower than that of the ingot because of the presence of abundant grain boundaries in the nanocomposites, and the thermal conductivities of both types of nanocomposites increase more slowly with increasing temperature than does the thermal conductivity of the ingot (Figure 1.5d).[74] Comparing the properties of the nanocomposite made from the elements with that made from the ingot, the thermal conductivity of the nanocomposite made from the elements is systematically higher than that of the nanocomposite made from ingot. The difference is probably due to compositional differences arising from the lack of some trace amount of elements that were used in the process of making the ingot and to some structural differences in the two types of samples.

The nanocomposite made from $Bi_xSb_{2-x}Te_3$ ingots has a power factor comparable to that of the crystalline ingot below 100°C but is higher above 100°C. The nanocomposites made from the Bi, Sb, and Te elements in Figure 1.5 have a slightly lower power factor than those made from the ingots. The power factor increase is likely due to some energy filtering effect at grain boundaries.

According to Equation 1.1, ZT for the nanocomposite $Bi_xSb_{2-x}Te_3$ should be enhanced because the thermal conductivity decreases significantly while the electrical conductivity increases and the Seebeck coefficient changes only slightly. Figure 1.5e shows the resulting temperature dependence of ZT for the hot-pressed nanocomposites made from the Bi, Sb, and Te elements and from the ingots, in comparison with that of the commercial $Bi_xSb_{2-x}Te_3$ ingots. The peak ZT values for both types of nanocomposites shift to a higher temperature and remain significantly higher than that of the ingots at all temperatures.[74] The peak ZT values of the hot-pressed nanocomposites made from the elements and ingots are

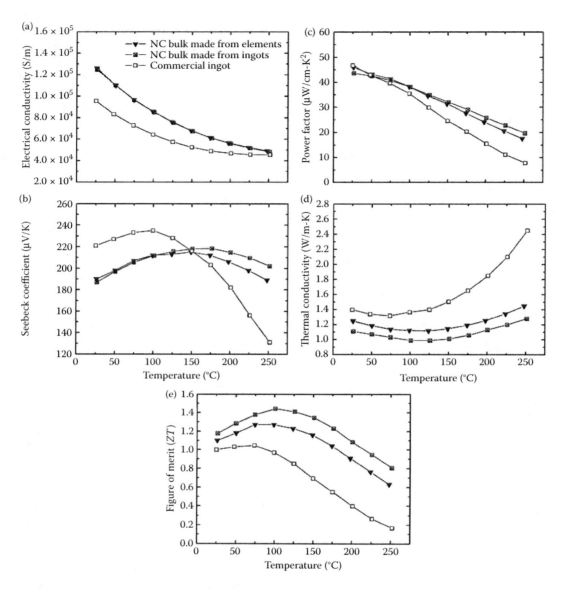

FIGURE 1.5 The temperature dependences of the (a) electrical conductivity, (b) Seebeck coefficient, (c) power factor, (d) thermal conductivity, and (e) the temperature dependence of the ZT of hot-pressed $Bi_xSb_{2-x}Te_3$ nanocomposites made from the constituent elements (filled triangles), made from ingots (crossed squares), in comparison with a commercial ingot (open squares). (Reprinted with permission from Y. Ma et al., *Nano Lett.* **8**, 2580–2584. Copyright 2008, American Chemical Society.)

about 1.3 and 1.4 at 100°C, respectively, both of which are significantly higher than that of the best $Bi_xSb_{2-x}Te_3$ ingots ($ZT \sim 1.0$).

As shown in Figure 1.5e, the ZT value of the $Bi_xSb_{2-x}Te_3$ ingots starts to drop above 75°C and is below 0.25 at 250°C, whereas the nanocomposites are still higher than 0.6 at 250°C.[74] The good ZT at high temperatures is very important for power generation applications, since there are no other materials with a similar high ZT in this temperature range of 25–250°C.

Using the ball-milling and hot-pressing method, other groups have also prepared *p*-type $Bi_xSb_{2-x}Te_3$ nanocomposites[34,39,133–138] and *p*-type $(Bi,Sb)_2(Te,Se)_3$ nanocomposites.[139] In these works, *ZT* is less than 1.0 and lower than that of the crystalline ingots. There are many reasons why *ZT* could decrease, with the most likely cause being a decrease in the electrical conductivity. The electrical conductivity of the nanocomposites is very sensitive to oxygen, moisture, and other environmental factors. A decrease in the electrical conductivity results in a cancellation of the benefits derived from the decreased thermal conductivity of these nanostructures to *ZT*. In order to achieve good electrical conductivity, the TE materials should be kept away from oxygen and moisture.

Using a ball-milling and hot-pressing method, *p*-type $(Bi,Sb)_2Te_3$ composites composed of 100 μm particles and 15–30 nm Bi_2Te_3 nanoparticles were also obtained.[140] For each samples, a peak *ZT* of 0.83 was achieved. This result shows the importance of reducing the amount of large size particles in the composite.

p-Type $(Bi,Sb)_2Te_3$ nanocomposites have also been prepared by other methods, such as by a shear extrusion method at high temperatures from ball-milled powder,[104] by a hot-pressing method from hydrothermally synthesized nanoparticles,[95] and by melting spinning and spark plasma sintering.[141] Results obtained with other nanocomposites and detailed information about the results are listed in Table 1.1. Among them, one interesting nanocomposite is a *p*-type Bi_2Te_3 nanocomposite with a layered nanostructure.[141] The nanocomposite bulk material that is obtained in this case consists of Bi_2Te_3 nano-layers with a thickness of 10–40 nm, similar to the structure of the Bi_2Te_3/Sb_2Te_3 superlattices prepared by molecular-beam epitaxy.[51–54] The large number of interfaces in the superlattice samples reduce the lattice thermal conductivity extraordinarily to the low value of 0.58 W m^{-1} K^{-1}. A peak *ZT* of 1.35 is obtained at 300 K when using this nanocomposite, and an increase of 73% is obtained over that for the *ZT* of the ingot, showing the effectiveness of this method of sample preparation. However, the scale-up of this method for commercial scale production would be challenging.

1.3.1.2 *n*-Type $Bi_2Te_{3-y}Se_y$ Nanocomposites

n-Type $Bi_2Te_{3-y}Se_y$ nanocomposites can also be prepared using the various methods, as mentioned below. One notable attribute about the Bi_2Te_3-based single-crystal nanostructural bulk samples is the lamellar structure and the weak van der Waals bonding between $Te^{(1)}$ atoms [locate at (1/3, 2/3, 0.457) atomic position in the unit cell] and $Te^{(1)}$ atoms [locate at (2/3,1/3,0.523) atomic position in the unit cell], which is responsible for the easy cleavage along the planes perpendicular to the *c*-axis.[3,142] Originating from this unique structural anisotropy of Bi_2Te_3, the TE properties of *n*-type $Bi_2Te_{3-y}Se_y$ single-crystal solid solutions prepared by the traveling heater method shows a strong anisotropy.[143] The electrical and thermal conductivities along the cleavage planes (perpendicular to the *c*-axis) are, respectively, about four and two times larger than those along the *c*-axis. Even though the Seebeck coefficient in Bi_2Te_3 is nearly isotropic, the TE figure-of-merit *Z* along the cleavage planes is approximately two times as large as that along the *c*-axis. At room temperature a maximum dimensionless TE figure-of-merit was achieved at $ZT = 0.85$ ($Z = 2.9 \times 10^{-3}$ K^{-1}) for solid solutions with a 2.5 at% Se replacing Te in the $Bi_2Te_{2.925}Se_{0.075}$ alloy that has a power factor $S^2\sigma$ of 47×10^{-4} W m^{-1} K^{-2} and a thermal conductivity of 1.65 W m^{-1} K^{-1} in which the lattice contribution is 1.27 W m^{-1} K^{-1} (see Ref. 142).

In principle, *ZT* could be greatly improved if we can decrease the thermal conductivity by breaking a single crystal into individual nano size grains to thus greatly increase its phonon scattering due to the significantly increased density of grain boundaries of the resulting nano grains[34,144] while at the same time maintaining the high power factor by retaining the preferential orientation of the grains.[145,146] We have successfully synthesized *n*-type $Bi_2Te_{2.7}Se_{0.3}$ bulk samples by ball milling and dc hot pressing, and we have achieved a significantly lower thermal conductivity of 1.06 W m^{-1} K^{-1} (with a lattice contribution of 0.7 W m^{-1} K^{-1}, much lower than the 1.27 W m^{-1} K^{-1} in single crystals) due to the increased phonon scattering by grain boundaries, in comparison to the 1.65 W m^{-1} K^{-1} value in the case of single-crystal bulk samples.[147] However, *ZT* was not enhanced at all because of a much lower power factor of 25×10^{-4} W m^{-1} K^{-2},[147] in comparison to the 47×10^{-4} W m^{-1} K^{-2} in single-crystal bulk samples. We suspect that the reason for the lower power factor is due to increased carrier scattering arising from the randomness of the small grains. Therefore, the challenge is to improve the power factor to a level close to that of the single-crystal-like nanostructural nanocomposite bulk samples while retaining the low thermal conductivity due to the fine grains which scatter both phonons and electrons strongly. In the literature, there have been a few methods reported on how to prepare polycrystalline Bi_2Te_3-based bulk alloys with preferred grain orientations, such as by hot pressing[148,149] and by hot extrusion.[104,105,150] However, those reports were not about preparing samples with small grains of less than a couple of micrometers, but rather with large grains of many micrometers. There were also reports about melt spinning to improve the *ZT*.[151,152] However, the important detailed aspects of their structural and property anisotropies were not discussed.

1.3.1.3 Partially Aligned

In this section we describe the experimental studies on the physical properties of samples with partial alignment of the small grains.[147] To achieve grain alignment, we re-pressed the as-pressed *n*-type $Bi_2Te_{2.7}Se_{0.3}$ bulk nanocomposite sample in a bigger diameter die at a higher temperature so that lateral flow of the small grains takes place in the disk plane to achieve a certain orientation of the *ab* plane of the sample in the disk plane of the die. As a result, we increased the power factor by 40% in the disk plane direction from 25×10^{-4} to 35×10^{-4} W m^{-1} K^{-2} without too much penalty on the thermal conductivity, leading to a peak *ZT* enhancement of 22% from 0.85 to 1.04.[147] In contrast, the *ZT* along the press direction is decreased, which does not affect the application of TE materials since only the direction with the highest *ZT* is used in devices.

Figure 1.6a,b shows the XRD patterns of both the planes perpendicular ($\perp$) and parallel (//) to the press direction, respectively, and Figure 1.6c,d shows the SEM images of the freshly fractured surfaces in the $\perp$ and // directions, respectively. These two XRD spectra look the same for both the 2θ position and the intensity of each peak, indicating that there is no significant grain orientation anisotropy. The ratios of the integrated intensities of planes (006) and (00$\overline{15}$) to that of the strongest peak (015) are shown in the insets. The minor larger ratios of $I_{(006)}/I_{(015)}$ (15% vs. 12%) and $I_{(00\overline{15})}/I_{(015)}$ (12% vs. 8%) for the plane that is $\perp$ to the

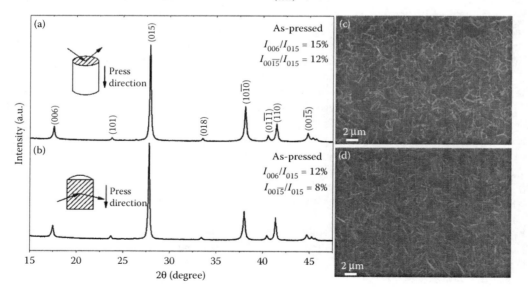

FIGURE 1.6   XRD patterns of the as-pressed $Bi_2Te_{2.7}Se_{0.3}$ bulk nanocomposite samples for the planes (a) perpendicular ($\perp$) and (b) parallel (//) to the press direction, and SEM images of the freshly fractured surfaces of as-pressed samples in the direction (c) perpendicular ($\perp$) and (d) parallel (//) to the press direction. (Reprinted with permission from X. Yan et al., *Nano Lett. 10*, 3373–3378. Copyright 2010, American Chemical Society.)

press direction (Figure 1.6a) mean that there is a little bit more *ab* orientation in the ⊥ direction, which is reflected by the physical properties (discussed below). From the SEM images (Figure 1.6c,d), we can see that the physical properties of these samples are consistent with the XRD result (Figure 1.6a,b): the grains are random without a significant preferred crystal orientation. The average grain size is about 1–2 μm even though there are many grains smaller than 1 μm.

In Figure 1.7, we show the transport properties of the as-pressed samples in both the directions perpendicular (⊥) and parallel (//) to the press direction. The electrical conductivity $\sigma_\perp$ is about 12–15% higher than $\sigma_{//}$ for

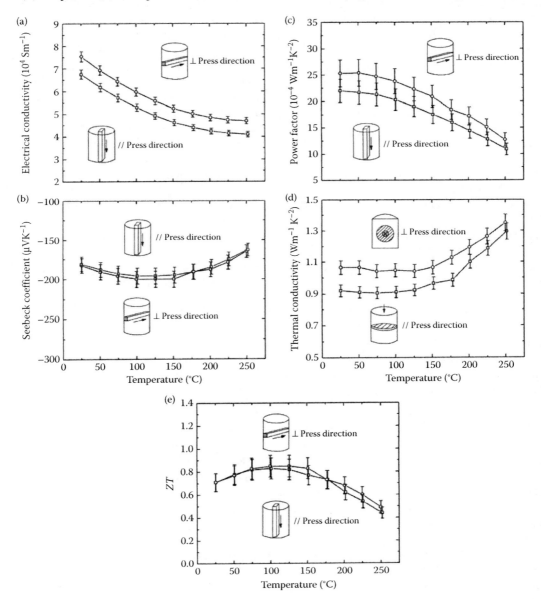

FIGURE 1.7 The temperature dependence from 50 < *T* < 250°C of the (a) electrical conductivity, (b) Seebeck coefficient, (c) power factor, (d) thermal conductivity, and (e) *ZT* of the as-pressed samples in both the directions perpendicular (⊥) and parallel (//) to the press direction. (Reprinted with permission from X. Yan et al., *Nano Lett.* *10*, 3373–3378. Copyright 2010, American Chemical Society.)

the whole temperature range (Figure 1.7a), which clearly shows that there is a little bit more ab plane orientation in the $\perp$ direction, consistent with the results shown by the XRD (Figure 1.6a). The Seebeck coefficients of the as-pressed sample are very similar to each other in both directions (Figure 1.7b), consistent with the fact that the Seebeck coefficient of single crystals is nearly isotropic in different crystal orientations.[143] The corresponding power factors of the as-pressed sample in both directions are shown in Figure 1.7c. A power factor ($S^2\sigma$) of 25×10^{-4} W m^{-1} K^{-2} at room temperature was obtained for the $\perp$ direction and the power factor drops down to 13×10^{-4} W m^{-1} K^{-2} at 250°C, while that in the $//$ direction is about 15–19% lower in the temperature range. It is very important to note that the power factor of the as-pressed sample in either direction is much lower than the 47×10^{-4} W m^{-1} K^{-2} in the ab plane of the single crystals,[143] which led us to think of a possibility for further improving ZT by enhancing the power factor through aligning the ab planes into the disk plane while keeping the thermal conductivity at the same low value. The thermal conductivity dependences on temperature of the as-pressed samples in both directions are shown in Figure 1.7d. Over the whole temperature range, the thermal conductivity is much lower (1.06 W m^{-1} K^{-1} in the $\perp$ direction and 0.92 W m^{-1} K^{-1} in the $//$ direction with a lattice contribution of 0.7 and 0.6 W m^{-1} K^{-1}, respectively) (a Lorenz number of about 1.6×10^{-8} J^2 K^{-2} C^{-2} is used for the nano samples) than 1.65 W m^{-1} K^{-1} [with a lattice contribution of 1.27 W m^{-1} K^{-1} (see Ref. 142)] in the ab plane of the single crystals, a clear indication of the benefits of the stronger phonon scattering resulting from the smaller grains in the samples with partially aligned grains. Below 150°C, the thermal conductivity does not increase too much, but κ increases above 150°C due to the bipolar effect, and the differences in the thermal conductivity between the samples in the two directions become smaller and smaller with increasing temperature. Similar to the electrical conductivity, the thermal conductivity is also higher in the direction perpendicular to the press direction, which resulted in almost the same ZT dependence on temperature all the way up to 250°C in both directions, indicating an isotropic ZT (Figure 1.7e). The peak ZT is about 0.85 and is about the same as that of the single crystals at room temperature.

Considering the potentially high power factor of 47×10^{-4} W m^{-1} K^{-2} in the ab plane of the single crystals, it is possible to enhance ZT by orienting the ab planes of the small grains into the disk plane. One way to orient the grains is to press the as-pressed samples in a bigger diameter die so that lateral flow takes place.[153]

Figure 1.8 shows the XRD patterns of the re-pressed samples in the planes perpendicular ($\perp$) (Figure 1.8a) and parallel ($//$) (Figure 1.8b) to the press direction and the SEM images of the freshly fractured surfaces of the re-pressed samples in the plane perpendicular ($\perp$) (Figure 1.8c) and parallel ($//$) (Figure 1.8d) to the press direction. From the much stronger diffraction intensities of the (006) and $(00\overline{15})$ peaks (Figure 1.8a) and the much weaker intensities of the same peaks in Figure 1.8b, it is very clear that reorientation of the ab planes of the grains into the disk plane took place during the re-pressing process. The ratios of the integrated intensities of the (006) and $(00\overline{15})$ peaks to that of the strongest peak (015) in the $\perp$ direction are 55% ($I_{(006)}/I_{(015)}$) and 44% ($I_{(00\overline{15})}/I_{(015)}$) (inset in Figure 1.8a), which are much higher than those from the as-pressed samples (inset in Figure 1.6a), and much stronger than the 6% ($I_{(006)}/I_{(015)}$) and 3% ($I_{(00\overline{15})}/I_{(015)}$) for the plane parallel to the press direction (inset in Figure 1.8b), respectively. On the one hand, the (006) and $(00\overline{15})$ peak intensities increased a lot in the disk plane after re-pressing (Figure 1.8a vs. 1.6a), and on the other hand, the intensity of the same peaks decreased a lot in the plane parallel to the press direction (Figure 1.8b vs. 1.6b). The significant diffraction peak intensity difference shows that we have successfully developed a significant anisotropy in the re-pressed samples, which should clearly show up in the microstructures and physical properties (discussed below). From the SEM images (Figure 1.8c,d), we can clearly see that the grains are plate-like and with the ab plane preferentially oriented in the disk plane, which confirms the much increased (00l) peak XRD intensities in the disk plane and the decreased (00l) peak intensities in the plane parallel to the press direction. Such a re-orientation of the ab planes is clearly the result of the lateral flow during the re-pressing process.[153] The lateral flow also makes the plates thinner due to the shear force. However, these plates are still quite large (up to 5 μm) and thick (up to 0.5 μm). Compared to the size before re-pressing, we can also see that a significant grain growth had occurred (in plane). The detailed

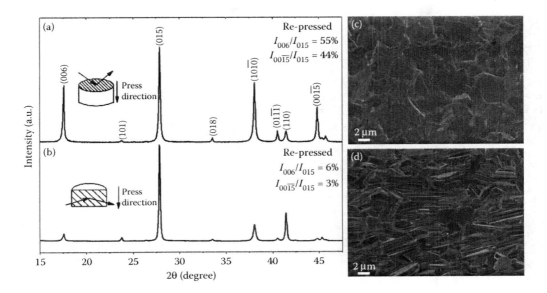

FIGURE 1.8 XRD patterns of the re-pressed $Bi_2Te_{2.7}Se_{0.3}$ bulk nanocomposite samples of the planes (a) perpendicular ($\perp$) and (b) parallel (//) to the press direction, and SEM images of the freshly fractured surfaces of re-pressed samples in the direction of (c) perpendicular ($\perp$) and (d) parallel (//) to the press direction. (Reprinted with permission from X. Yan et al., *Nano Lett. 10*, 3373–3378. Copyright 2010, American Chemical Society.)

microstructure investigations of these samples by TEM are in progress and will be reported when the data are available.

In Figure 1.9, we show the results of the temperature dependence of the electrical conductivity (σ), the Seebeck coefficient (S), the power factor ($S^2\sigma$), the thermal conductivity (κ), and the ZT of the re-pressed samples in both the direction perpendicular ($\perp$) and parallel (//) to the press direction. It is clearly shown that the electrical conductivity ($\sigma_\perp$) of the disk plane is increased while $\sigma_{//}$ is decreased in comparison with those of the as-pressed samples (Figure 1.7a). A ratio ($\sigma_\perp/\sigma_{//}$) of about 2.3 of the electrical conductivity along these two directions is maintained from room temperature to 250°C, which is however much smaller than the ratio of 4.3 found in single crystals.[143] As expected, the Seebeck coefficients of the re-pressed samples in both directions are very similar (Figure 1.9b), confirming the fact that the Seebeck coefficient is basically isotropic like that in single crystals.[6] Due to the electrical conductivity increase, the power factor $S^2\sigma$ of the re-pressed samples in the $\perp$ direction is increased to 35×10^{-4} W m^{-1} K^{-2} at room temperature and drops with temperature to 16×10^{-4} W m^{-1} K^{-2} at 250°C while that in the // direction is decreased to 15×10^{-4} W m^{-1} K^{-2} at room temperature and 6×10^{-4} W m^{-1} K^{-2} at 250°C. With the changes in electrical conductivity in both directions, the thermal conductivity of the re-pressed samples also is changed (Figure 1.9d): $\kappa_\perp$ increased from 1.06 to 1.16 W m^{-1} K^{-1} in that the lattice contribution remained at 0.7 W m^{-1} K^{-1} and $\kappa_{//}$ decreased from 0.92 to 0.78 W m^{-1} K^{-1} in that the lattice contribution is about 0.58 W m^{-1} K^{-1}. A ratio ($\kappa_\perp/\kappa_{//}$) of 1.5 is observed, which is smaller than the ratio of 2 in single crystals.[143] ZT of the re-pressed samples in the disk plane is about 0.9 at room temperature and reaches the peak value of 1.04 at 125°C (Figure 1.9e), which is about a 22% improvement over the peak ZT (0.85) of the as-pressed samples. In contrast, ZT of the re-pressed samples in the // direction remains very low for all temperatures (Figure 1.9e): 0.6 at room temperature and 0.3 at 250°C.

It is very clear that re-pressing improves the ZT in the direction perpendicular to the press direction by enhancing the power factor due to the reorientation of the *ab* planes of some of the crystals. The as-pressed samples consist of small crystals, and each of these crystals can be regarded as a single crystal and thus can show anisotropic TE properties. The shape of each grain is flake-like because of its cleavage

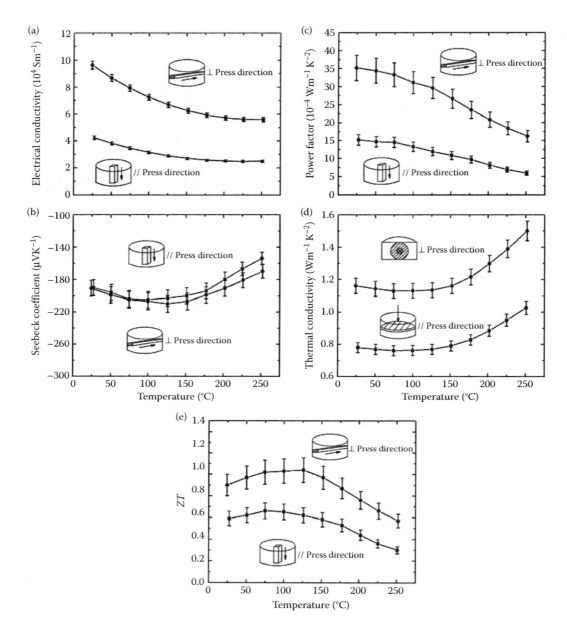

FIGURE 1.9 The temperature dependence from $50 < T < 250°C$ of the (a) electrical conductivity, (b) Seebeck coefficient, (c) power factor, (d) thermal conductivity, and (e) ZT dependence on temperature of the re-pressed samples in both the directions perpendicular ($\perp$) and parallel ($//$) to the press direction. (Reprinted with permission from X. Yan et al., *Nano Lett. 10*, 3373–3378. Copyright 2010, American Chemical Society.)

characteristics between the $Te^{(3)}$ and $Te^{(1)}$ layers. During re-pressing, grains easily slip along the cleavage planes and their *ab* planes tend to be oriented preferentially perpendicular to the press direction. The more the reorientation occurs, the higher the power factor and ZT since the thermal conductivity does not increase so much due to the grain boundary scattering. The overall degree of reorientation is determined mainly by the amount of lateral flow of the grains during re-pressing. However, we have not reached the limit yet since (1) the grains are still not completely reoriented to the disk planes, resulting in a nonoptimal power factor, and (2) the grains are still large (average size of 1–2 µm) and hence the lattice thermal conductivity of 0.7 W m^{-1} K^{-1} is still relatively high.

n-Type Bi$_2$Te$_3$ nanocomposites are also prepared by other groups using the ball-milling and hot-pressing method.[105,106,154,155] The grain size of these nanocomposites varies from nanometers to micrometers. The reported peak *ZT* values for these samples are less than 1.0.

1.3.2 SiGe Alloy Nanocomposites

Silicon–germanium alloys have been the primary TE materials in power generation devices operating in the temperature range of 600–1000°C,[3,6] having long been used in radio-isotope TE generators (RTGs) for deep-space missions to convert radio-isotope heat into electricity.[3,6] SiGe alloys also hold promise in terrestrial applications such as for waste heat recovery.[13,14,156]

Since the 1960s, efforts have been made to improve the *ZT* of SiGe alloys,[14,34,156–160] with the peak *ZT* value obtained for *n*-type SiGe reaching 1 at 900–950°C[14,34,156–160] and 0.65 in *p*-type SiGe alloys.[35,36,159] Through a ball-milling and hot-pressing method, a significant increase in *ZT* was achieved in *p*-type SiGe nanocomposites, reaching a peak value of about 0.95 at 900–950°C.[73] The *ZT* improvement was about 90% over that of the RTG samples (peak *ZT* = 0.5), and 50% over the previous highest record (0.65).[35] A peak *ZT* of about 1.3 at 900°C was also achieved in *n*-type SiGe nanocomposites using the ball-milling and hot-pressing method.[75]

1.3.2.1 *p*-Type Si$_{80}$Ge$_{20}$ Nanocomposites

SiGe alloys consisting of micron-sized particles have been prepared using the ball-milling and hot-pressing method since the 1960s.[36] A 20% increase in *ZT* of *p*-type SiGe with an optimal grain size in the 2–5 µm range was reported.[36] Presently the typical grain size is 1–10 µm in the RTG samples used for NASA space missions. In heavily doped SiGe alloys, the mean free paths of electrons and phonons are different, with electrons having an average mean free path of about 5 nm and phonons having an average mean free path of 200–300 nm at room temperature. Thus, nanocomposites can significantly reduce the phonon thermal conductivity (2.5 W m^{-1} K^{-1} in *p*-type SiGe nanocomposites vs. 4.6 Wm^{-1} K^{-1} in RTG bulk samples) without significantly reducing the electrical conductivity, resulting in a higher *ZT*.[23–26] In some early reports, *ZT* was found to decrease when the grain size is reduced below micron size,[14] because of a reduction of the electrical conductivity besides a decrease in the thermal conductivity. The reason for the electrical conductivity decrease was not clearly stated, but we guess that it is because of massive defect creation and some contamination that occurred during the ball milling process.

p-Type SiGe nanocomposites with high *ZT* are prepared using the same technique as that used for Bi$_2$Te$_3$ nanocomposites. Here *p*-type (boron doped) Si$_{80}$Ge$_{20}$ single phase nanoparticles are first prepared by a high-energy ball-mill technique from the individual elements.[73] The SiGe nanoparticles thus prepared are made from a single-phase boron-doped Si$_{80}$Ge$_{20}$ alloy (Figure 1.10a). The crystallinity of the mechanically alloyed SiGe nanoparticles is different from that of the ball-milled Bi$_2$Te$_3$ nanoparticles. More specifically, the mechanically alloyed Si$_{80}$Ge$_{20}$ nanoparticles are polycrystalline (Figure 1.10c,d), consisting of several sub-nanograins (Figure 1.10d), while the mechanically alloyed Bi$_2$Te$_3$ nanoparticles are single crystals (Figure 1.4). Similar polycrystalline structures are also observed in ball-milled pure Si nanopowders. X-ray diffraction (XRD) and TEM images (Figure 1.10) indicate that the agglomerated particles vary in size from 20 to 200 nm. The microstructure can be explained by high-energy ball-milling theory.[161,162]

There are many defects inside individual sub-nanograins (Figure 1.10d) because the nanograins were formed by a low-temperature mechanical alloying process and not by high-temperature melting and solidification. High stresses are expected in such ball-milled SiGe nanoparticles. Dislocations and atomic level strains are usually produced in ball-milled nanoparticles[163] and detailed information can be found in articles on high-energy ball-milling.[111,112,163]

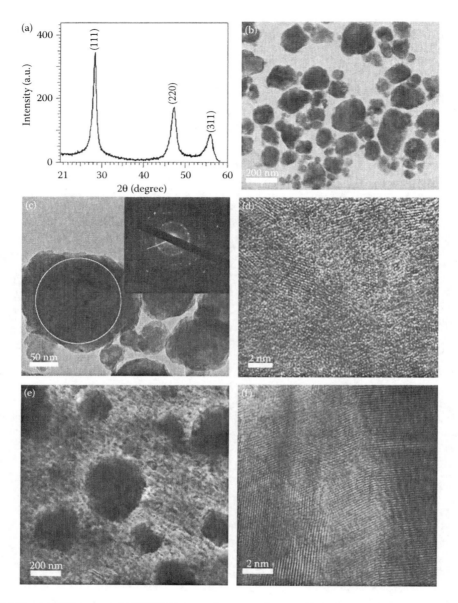

FIGURE 1.10 (a) XRD pattern, (b) BF-TEM image at low magnification, (c) BF-TEM image at medium magnification, and (d) HRTEM image of *p*-type boron-doped $Si_{80}Ge_{20}$ nanoparticles prepared by ball milling. (e) BF-TEM image and (f) HRTEM image of the hot-pressed *p*-type boron-doped $Si_{80}Ge_{20}$ nanocomposites. The inset in (c) is a SAED image of an individual nanograin indicating the polycrystalline nature of the nanograin. (Reprinted with permission from G. Joshi et al., *Nano Lett.* 8, 4670–4674. Copyright 2008, American Chemical Society.)

After hot-pressing, the grain size in the *p*-type $Si_{80}Ge_{20}$ nanocomposites is about 20–100 nm, which is about 2–5 times the size of the initial nanopowders, indicating that some grain growth occurred during the hot-pressing process. These nanograins are highly crystalline, completely random, closely packed (Figure 1.10f), and have clean boundaries, consistent with the measured density (2.88 g cm^{-3}, being equal to the theoretical density value).

The Seebeck coefficient of the *p*-type $Si_{80}Ge_{20}$ nanocomposites (Figure 1.11b) is comparable to that of the RTG samples while the electrical conductivity (Figure 1.11c) is higher than that of the RTG samples

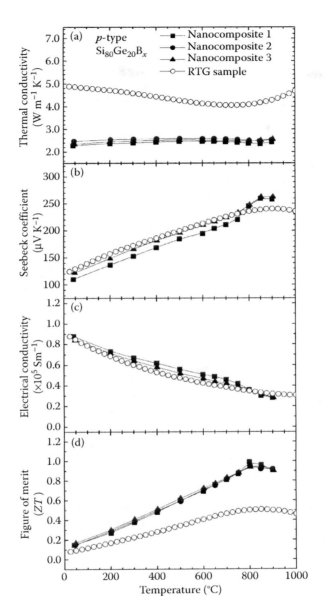

FIGURE 1.11 Temperature dependence of the (a) thermal conductivity, (b) Seebeck coefficient, (c) electrical conductivity, and (d) ZT of three hot-pressed p-type $Si_{80}Ge_{20}$ nanocomposite samples (squares, circles, and triangles), in comparison with p-type SiGe bulk samples used in RTGs for space power missions (open circles). (Reprinted with permission from G. Joshi et al., *Nano Lett. 8*, 4670–4674. Copyright 2008, American Chemical Society.)

over the entire temperature range. More importantly, the thermal conductivity of the nanostructured bulk samples is much lower than that of the RTG samples over the whole temperature range up to 1000°C, which leads to a peak ZT of about 0.95 in the ball-milled and hot-pressed p-type $Si_{80}Ge_{20}$ nanocomposites (Figure 1.11). Such a peak ZT value is about a 90% increase over that of the p-type RTG SiGe samples (peak $ZT = 0.5$) presently used in space missions and 50% above that of the reported RTG record value (peak $ZT = 0.65$).[35]

The significant reduction of the thermal conductivity in the nanostructured samples is mainly due to the increased phonon scattering at the numerous interfaces of the nanocomposites. Since the electrical conductivity of the nanocomposites is similar to that of the RTG samples, the actual lattice thermal conductivity is reduced by a factor of 2 based on the experimental data shown in Figure 1.11a.

Results for p-type SiGe composites are also reported by other groups.[120,124,126,164] The grain size of these reported composites is in the micron range. The thermal conductivity is decreased, and ZT is enhanced effectively despite the decrease of the electrical conductivity and Seebeck coefficient.

1.3.2.2 n-Type $Si_{80}Ge_{20}$ Nanocomposites

The microstructure of the mechanically alloyed n-type $Si_{80}Ge_{20}$ nanoparticles is very similar to that of the mechanically alloyed p-type nanoparticles.[75] Phosphorus-doped $Si_{80}Ge_{20}$ single phase nanoparticles are synthesized using ball-milling (Figure 1.12a). The particle size of the as-prepared ball-milled nanopowders is in the range of 30–200 nm (Figure 1.12b,c). These nanoparticles are composed of many small crystalline sub-nanograins, similar to the p-type mechanically alloyed SiGe nanoparticles. The size of the sub-nanograins is in the range of 5–15 nm (Figure 1.12d) with an average size of 12 nm.

After hot pressing, the average grain size in the hot-pressed $Si_{80}Ge_{20}$ nanocomposites is 22 nm, indicating that the grain size is almost doubled after hot pressing, but is still very small.[73] These nanometer-sized grains with random crystalline orientations promote phonon scattering much more effectively than the micron-sized grains in the bulk RTG SiGe materials. The stress of the hot-pressed nanocomposites is much smaller than that of the as-prepared ball-milled nanoparticles, with the strain value being 10 times lower in the hot-pressed nanocomposites. This smaller strain in the nanocomposites is understandable since the hot-pressing temperature is above 1000°C, allowing the stresses built up in the nanoparticles during the ball-milling process to be relaxed.

Examining the temperature-dependent thermal conductivity, we can clearly see that the n-type $Si_{80}Ge_{20}$ nanocomposites in Figure 1.13a have a much lower thermal conductivity than that of the RTG samples with grain size of 1–10 μm. Based on the measured electrical conductivity of the n-type $Si_{80}Ge_{20}$ nanocomposites and RTG samples, $\kappa_e = 0.77$ W m^{-1} K^{-1} at room temperature for RTG samples, whereas $\kappa_e = 0.55$ W m^{-1} K^{-1} at room temperature for a typical nanocomposite.[75] By subtracting the electronic contribution κ_e from the total thermal conductivity κ, a lattice thermal conductivity of the nanocomposite samples of ~1.8 W m^{-1} K^{-1} at room temperature is obtained, which is about 47% of the RTG samples (~3.8 W m^{-1} K^{-1}). The decrease of the lattice thermal conductivity is mainly due to a stronger boundary phonon scattering in the nanostructured samples.[75]

The electrical conductivity of the n-type nanocomposites shown in Figure 1.13c is normally lower than that of the RTG samples in the low-temperature region, but is similar above 750°C, although the carrier concentrations for both types of samples determined from Hall effect measurements are almost the same at room temperature (~2.2 × 10^{20} cm^{-3}), indicating a lower electron mobility in the nanocomposite samples, which is reasonable considering the large number of grain boundaries in the nanocomposite samples. The Seebeck coefficient of the nanocomposite samples shown in Figure 1.13b is similar to that of the RTG samples below 400°C and above 700°C, and is slightly higher than that of the RTG samples between 400°C and 700°C.[75]

As a result, for the n-type $Si_{80}Ge_{20}$ nanocomposites, the ZT value in Figure 1.13d shows a maximum of about 1.3 at 900°C which is about 40% higher than that of the RTG reference (0.93). The significant enhancement of ZT is mainly attributed to the thermal conductivity reduction, which is strongly correlated with the nanostructural features in the n-type $Si_{80}Ge_{20}$ nanocomposites.[75]

1.3.2.3 n-Type Si and $Si_{95}Ge_5$

$Si_{95}Ge_5$ nanoparticles were also synthesized by high-energy ball-milling from Si, Ge, P, and GaP powders. These ball-milled nanoparticles are single phase (Figure 1.14a) with a size in the 20–150 nm range

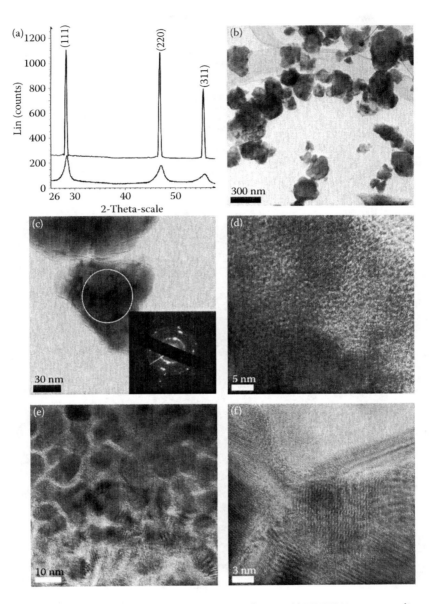

FIGURE 1.12 (a) XRD pattern, (b) BF-TEM image at low magnification, (c) BF-TEM image at medium magnification, and (d) HRTEM image of *n*-type phosphorus-doped $Si_{80}Ge_{20}$ nanoparticles. (e) BF-TEM and (f) HRTEM images of hot-pressed *n*-type $Si_{80}Ge_{20}$ nanocomposites made from the nanoparticles shown in (b). The inset of (c) shows a SAED pattern of an individual nanoparticle. (Reprinted with permission from Y. Ma et al., *Nano Lett. 8*, 2580–2584. Copyright 2008, American Chemical Society.)

(Figure 1.14b). Selected area electron diffraction (SAED) patterns and HRTEM images (insets in Figure 1.14b) indicate that the ball-milled nanoparticles also consist of sub-nanograins, similar to what was observed in the heavily doped *p*-type $Si_{80}Ge_{20}$ nanoparticles (Figure 1.10) and *n*-type $Si_{80}Ge_{20}$ nanoparticles (Figure 1.12).

After hot pressing, most of the nanograins of the ball-milled $Si_{95}Ge_5$ nanocomposite are in the 10–30 nm range, larger than the 5–20 nm of the initial nanoparticles due to grain growth during the

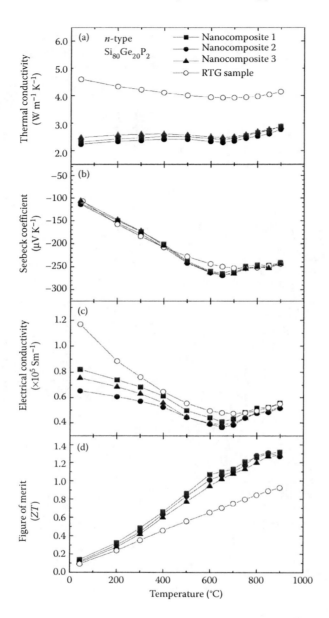

FIGURE 1.13   Temperature dependence in the range 50 < *T* < 900°C of the (a) thermal conductivity, (b) Seebeck coefficient, (c) electrical conductivity, and (d) dimensionless *ZT* of three hot-pressed *n*-type $Si_{80}Ge_{20}$ nanocomposite samples (squares, triangles, and circles), in comparison with the RTG reference sample (open circles). (Reprinted with permission from X. W. Wang et al., *Appl. Phys. Lett. 93*, 193121. Copyright 2008, American Institute of Physics.)

hot-pressing process. The nanograins are highly crystalline and randomly oriented after hot pressing (Figure 1.14d).

Figure 1.15 shows the comparative TE property measurement results for the nanostructured Si, $Si_{95}Ge_5$ nanocomposite, bulk Si, and bulk $Si_{80}Ge_{20}$ RTG alloy samples. Both the nanostructured Si and the nanostructured $Si_{95}Ge_5$ samples show a higher electrical conductivity (Figure 1.15a) but a lower absolute Seebeck coefficient (Figure 1.15b) than that of the bulk $Si_{80}Ge_{20}$ RTG sample. This is mainly

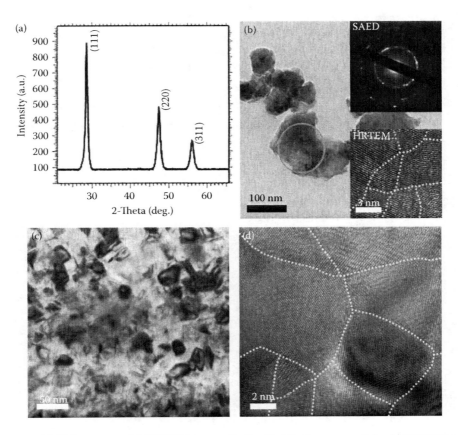

FIGURE 1.14 (a) XRD pattern and (b) BF-TEM image of ball-milled $Si_{95}Ge_5$ nanoparticles. (c) BF-TEM image and (d) HRTEM image of a hot-pressed $Si_{95}Ge_5$ nanocomposite sample. The insets in (b) are an SAED pattern and an HRTEM image of the circled region in (b). (Reprinted with permission from G. H. Zhu et al., *Phys. Rev. Lett. 102*, 196803. Copyright 2009, American Physical Society.)

attributed to the higher solubility limit of P and the lower alloy scattering of charge carriers in nanostructured Si and nanostructured $Si_{95}Ge_5$ samples in comparison with the bulk $Si_{80}Ge_{20}$ RTG sample. Up to 600°C the power factors for both the nanostructured samples (Figure 1.15c) are slightly lower than the values calculated for bulk materials with the same carrier concentration values as measured for the nanostructured samples. The power factor of the nanostructured $Si_{95}Ge_5$ sample is much higher than that of the bulk $Si_{80}Ge_{20}$ RTG sample (Figure 1.15c), especially at temperatures above 300°C. Figure 1.15d shows the temperature-dependent thermal conductivity of the nanostructured Si and nanostructured $Si_{95}Ge_5$ samples in comparison with bulk Si and bulk $Si_{80}Ge_{20}$ RTG samples. The thermal conductivity of the nanostructured Si shows a significant reduction (by about a factor of 10) compared with that of the heavily doped bulk Si, which is around 100 W $m^{-1}K^{-1}$, at room temperature, a clear demonstration of the nanosize effect on phonon scattering. Moreover, with a 5 at% replacement of Si by Ge, the thermal conductivity value of the nanostructured $Si_{95}Ge_5$ is even lower, close to that of the bulk $Si_{80}Ge_{20}$ RTG sample, due to both the nanosize and point defect scattering effects in nanostructured $Si_{95}Ge_5$. Since the bulk $Si_{80}Ge_{20}$ RTG sample has 20 at% Ge and the $Si_{95}Ge_5$ nanostructure has only 5 at% Ge, a weaker alloy phonon scattering effect is expected in $Si_{95}Ge_5$. When the Ge concentration is increased from 5 to 20 at%, the thermal conductivity is decreased by another factor of 2 to about 2–3 W m^{-1} K^{-1}, but the power factor is also decreased[75] accordingly because of the reduced charge mobility due to the alloy scattering of charge carriers.

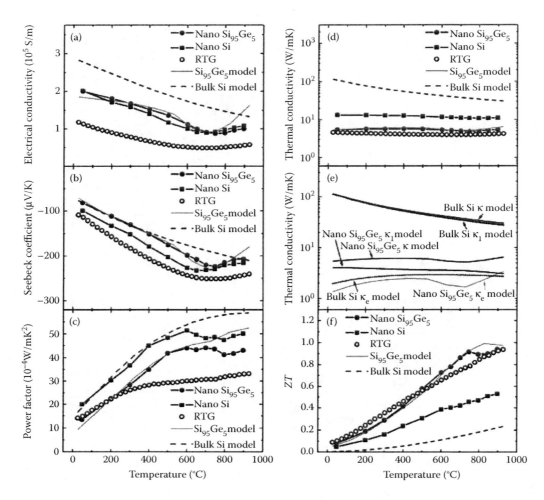

FIGURE 1.15 Temperature-dependence in the range $50 < T < 900°C$ of the (a) electrical conductivity, (b) Seebeck coefficient, (c) power factor, (d) thermal conductivity, (e) electron (κ_e), phonon (κ_l), and total (κ) thermal conductivity by modeling, and (f) ZT of nanostructured Si (filled squares), $Si_{95}Ge_5$ nanocomposite (filled circles for experiment and solid line for model), bulk Si model (dashed line), and $Si_{80}Ge_{20}$ RTG samples (open circles). (Reprinted with permission from G. H. Zhu et al., *Phys. Rev. Lett.* **102**, 196803. Copyright 2009, American Physical Society.)

Figure 1.15e shows that the calculated thermal conductivity of the $Si_{95}Ge_5$ nanocomposite matches quite well with the experimental results. The electron contribution to the thermal conductivity is calculated from the electrical conductivity measurement results using the Wiedemann–Franz Law. The Lorenz number is calculated using the bulk model. Our modeling results show that the Lorenz number in the bulk SiGe alloy varies from 1.3 to 2.2×10^{-8} W Ω K^{-2} from 25°C to 1000°C, and that variation within any specific temperature is 0.2 for the range of the doping concentration in our samples. The calculated phonon thermal conductivity for $Si_{95}Ge_5$ is below 4 W m^{-1} K^{-1} at room temperature and reaches 3 W m^{-1} K^{-1} at 900°C (Figure 1.15e). The low thermal conductivity for the $Si_{95}Ge_5$ system is mainly attributed to both the enhanced boundary phonon scattering and the alloy scattering effect. Thus, due to the significant thermal conductivity reduction without reduction of the power factor, ZT of the $Si_{95}Ge_5$ nanocomposites shows a maximum value of 0.95 at 900°C, which is about the same as that of the bulk n-type $Si_{80}Ge_{20}$ RTG sample (Figure 1.15f), which is an interesting result.

In nanostructured bulk Si, it is reported that the grain boundaries with very high density do not strongly affect the electron mobility but dramatically reduce the lattice thermal conductivity.[165] The combined transport effects produce an increase in the ZT of nanostructured bulk Si by a factor of nearly 3.5, reaching a ZT value of 0.5 at 900°C.

1.3.3 Skutterudite Nanocomposites

Skutterudites have been widely studied for their promising TE properties,[166–183] and are regarded as potential candidates for the next-generation TE materials for electrical power generation using either solar energy or waste heat as energy sources. Undoped $CoSb_3$ is p-type with a ZT of about 0.2.[128] After doping with rare earth elements, ZT increases to 0.52 at 600 K for n-type bulk crystals.[184] A peak ZT value of 0.7 was achieved in n-type $(Ni,Co)_4Sb_{12}$ nanocomposites produced by the ball-milling and hot-pressing method.[128]

Another remarkable feature of skutterudites is that the cage-like open structure of the compounds can be filled with foreign atoms acting as phonon rattlers. The filled foreign atoms scatter phonons strongly and drastically reduce the thermal conductivity of the skutterudite compounds.[166–183] With various kinds of atoms filling the cages, an increased ZT can be achieved in $R_xCo_4Sb_{12}$ (R = Ce,[185] La,[186] Ca,[187] Ba,[188] and Yb[115,189,190]). For example, Yb filled n-type $Yb_{0.19}Co_4Sb_{12}$ with a peak ZT close to 1 at 373°C[189] and $Yb_{0.15}Co_4Sb_{12}$ with a peak ZT of about 0.7 at 400°C [190] have been reported. The peak ZT of $Yb_xCo_4Sb_{12}$ nanocomposites made from the ball-milling and hot-pressing method was increased to 1.2.[127]

1.3.3.1 n-Type $Yb_xCo_4Sb_{12}$ Nanocomposites

We have tried to obtain $Yb_xCo_4Sb_{12}$ nanopowders by ball milling from elements. However, unlike the Bi_2Te_3 and SiGe systems, only a small portion of the ball-milled nanopowders are mechanically alloyed after high-energy ball milling, regardless of the ball-milling time. The same phenomenon is also observed in other skutterudites, such as $CoSb_3$,[191] $(La,Ni,Co)_4Sb_{12}$,[192] $Fe_xCo_4Co_{12}$,[117,191] $Co(Sb,Te)_3$,[109] and $FeCo_3Sb_{12}$,[193] and in other material systems such as the $Ag_{0.8}Pb_{18+x}SbTe_{20}$ system.[107] This phenomenon is common in mechanical alloying. For some TE compounds, like $FeSi_2$, the ball-milling process does not mechanically alloy Fe and Si elements.[194] However, the ball-milling process does mix the elements uniformly at the nanometer scale, facilitating the alloying via a chemical reaction during the hot-pressing process.

After the hot pressing, the nanopowders are completely transformed into a single-phase skutterudite. The average grain size is about 200–500 nm (Figure 1.16). There are two types of nanograins in the skutterudite nanocomposites: larger ones of about 1 μm in a bar shape (in the middle of Figure 1.16a), and smaller ones of about 200–500 nm in a spherical shape. The larger ones could grow from the mechanically alloyed small particles while the smaller ones could form from un-alloyed powders.[128] Both types of nanograins are well crystallized with clear facets. The crystallized grains are closely packed, implying a high density, consistent with the theoretical density of $Yb_{0.35}Co_4Sb_{12}$ (7.6 g cm^{-3}).

All the nanograins in the nanocomposites are well crystallized, and the large-angle grain boundaries in Figure 1.16b are clean. The excellent crystallinity and clean grain boundaries are needed to achieve good electrical transport properties, whereas the large-angle grain boundaries also benefit phonon scattering.

A series of $Yb_xCo_4Sb_{12}$ (x = 0.3, 0.35, 0.4, 0.5, and 1.0) nanostructured samples were prepared by ball milling and hot pressing. Of these, $Yb_{0.35}Co_4Sb_{12}$ has an optimized lowest thermal conductivity with a minimum value of 2.7 W m^{-1} K^{-1}, which leads to the highest observed ZT value among all the samples.[127] Below we focus on the $Yb_{0.35}Co_4Sb_{12}$ nanostructured sample.

The TE properties of $Yb_{0.35}Co_4Sb_{12}$ are next compared with other reported properties of nanocomposites prepared by the ball-milling and hot-pressing method. The nanocomposites produced here show a much higher electrical conductivity than those reported previously,[115,189] presumably due to the large electron doping effect from the high Yb concentration. The thermal conductivity of the

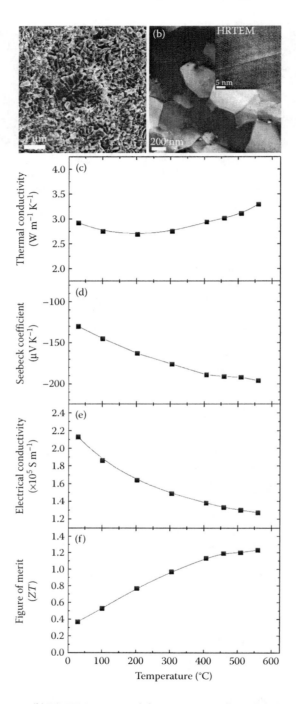

FIGURE 1.16 (a) SEM image, (b) BF-TEM image, and the temperature dependence in the range $50 < T < 560°C$ of the: (c) thermal conductivity, (d) Seebeck coefficient, (e) electrical conductivity, and (f) ZT of hot-pressed n-type skutterudite $Yb_{0.35}Co_4Sb_{12}$ nanocomposite samples. The inset in (b) is an HRTEM image of a grain boundary. (Reprinted with permission from J. Yang et al., *Phys. Rev. B 80*, 115329. Copyright 2009, American Physical Society.)

nanocomposites (2.7–3.3 W m^{-1} K^{-1} from room temperature to 600°C) is almost the same as the thermal conductivity of those prepared by spark plasma sintering (2.4–3.9 W m^{-1} K^{-1}). The low thermal conductivity of these *n*-type Yb$_x$Co$_4$Sb$_{12}$ nanocomposites should come from the stronger phonon scattering resulting from the presence of a large concentration of rattlers and an increased concentration of grain boundaries in the nanocomposites. Therefore, the nanocomposites have a higher *ZT* value than the doped Co$_4$Sb$_{12}$ reference samples without rattlers (*ZT* = 0.52 at 600 K[184]) as a result of both an enhanced power factor and a reduced thermal conductivity of the nanocomposites.

ZT increases with temperature and reaches a maximum value (~1.2) at around 550°C as shown in Figure 1.16f. Similar enhancements have recently been reported by various groups in *n*-type skutterudite nanocomposites.[115,128,189,190,195]

The thermal conductivity decreases with increased Yb doping content in the nanocomposites.[127] Yb$_{0.35}$Co$_4$Sb$_{12}$ has an optimized lowest thermal conductivity with a minimum of 2.7 W m^{-1} K^{-1} at room temperature, which leads to the highest observed *ZT* value.

n-Type skutterudite CoSb$_3$ nanocomposites are also reported using the ball-milling and hot-pressing method.[117,138,192,196] *ZT* = 1.1 can be achieved in *n*-type Ba$_{0.24}$Co$_4$Sb$_{12}$ nanocomposites prepared by ball milling, followed by cold pressing and sintering,[188] with *ZT* = 1.3 in Yb$_x$Co$_4$Sb$_{12+y}$ nanocomposites obtained by the spark plasma sintering method,[115,197] and *ZT* = 1.1 in CoSb$_{3-x}$Te$_x$ nanocomposites obtained by the ball milling and spark plasma sintering method.[109] The thermal properties of these nanocomposites are listed in Table 1.1.

1.3.3.2 *p*-Type (La,Ce,Nd)Fe$_{4-x}$Co$_x$Sb$_{12}$ Nanocomposites

Rare-earth-filled and Fe-substituted skutterudites are *p*-type. Many *p*-type skutterudite nanocomposites can also be prepared using the ball-milling and hot-pressing method.[166–171,193,198–202] The ball-milled nanoparticles are a mixture of Sb, skutterudites, and (Co,Fe)Sb$_2$ compounds. After hot pressing, nanocomposites with a single phase are obtained. The thermal conductivity decreases significantly because of the rattling effect from the rare-earth atoms in the voids and the substitution of Fe for Co. The lattice thermal conductivity decreases with the rare-earth filling content. The thermal conductivity can be reduced down to 4 W m^{-1} K^{-1} at room temperature (one-third of that of CoSb$_3$ samples).[184]

La$_x$(Ni,Co)$_4$Sb$_{12}$ nanocomposites can also be produced by ball milling and spark plasma sintering, and this material also exhibits a lower thermal conductivity than the bulk material.[192] Other skutterudite nanocomposites are also reported, such as Sn$_y$Fe$_3$Co$_5$Sb$_{24}$ nanocomposites,[203] rare-earth-filled Fe$_4$Sb$_{12}$ skutterudite nanocomposites,[198] and FeCo$_3$Sb$_{12}$ nanocomposites.[193,199]

1.3.4 PbTe Nanocomposites

Lead telluride (PbTe) is one of the best TE materials at intermediate temperatures (450–800 K). The peak *ZT* of PbTe is 0.7 at 700 K. Recently, *ZT* was doubled to about 1.5 at 773 K in thallium-doped PbTe through an enhancement of the Seebeck coefficient by the introduction of a resonance in the density of the electronic states near the Fermi level.[63]

For the ball-milled Tl-doped PbTe samples, the PbTe nanoparticles are prepared by ball-milling thallium, lead, and tellurium.[83] The microstructure of the ball-milled nanoparticles (Figure 1.17a) is similar to that of SiGe alloys. During hot pressing, the nanoparticles grow quickly, and the grain size in the hot-pressed PbTe is bigger than 1 μm (Figure 1.17b). Compared with the properties of the ingots, the thermal conductivity of the hot-pressed nanostructured samples is slightly decreased over the entire temperature range (Figure 1.17c) while the electrical conductivity is at the same time increased (Figure 1.17e). The Seebeck coefficient is slightly decreased at higher temperatures (Figure 1.17d). The temperature dependence of *ZT* is the same for the nanocomposite PbTe sample as for the values reported by the ingot-derived ball-milled sample as shown in Figure 1.17f.[82–84]

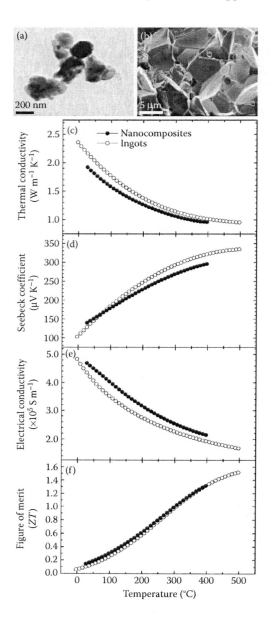

FIGURE 1.17 (a) TEM image of ball-milled PbTe nanopowder. (b) SEM image of hot-pressed PbTe nanocomposites. Temperature dependence of the (c) thermal conductivity, (d) Seebeck coefficient, (e) electrical conductivity, and (f) ZT of hot-pressed PbTe nanocomposites. The corresponding data for PbTe bulk ingots[83] are also plotted for comparison. (Reprinted with permission from B. Yu et al., J. Appl. Phys. *108*, 016104. Copyright 2010, American Institute of Physics.)

A smaller grain size can lead to further thermal conductivity reduction and a higher *ZT*. More work is being carried out to decrease the grain size while preserving the electrical conductivity and the Seebeck coefficient.

PbTe nanocomposites with 95% theoretical density are also prepared by the spark plasma sintering method from 100 to 150 nm nanoparticles.[98,99] The reported *ZT* of the PbTe nanocomposites thus produced is 0.1 at room temperature.

1.3.5 Half-Heusler Nanocomposites

The potential of half-Heusler alloys for high-temperature TE power generation has been discussed.[204,205] Half-Heuslers have the cubic MgAgAs type of structure, forming three interpenetrating face-centered-cubic (fcc) sublattices and one vacant sublattice.[206] Until recently the reported p-type half-Heusler alloys are mostly based on the formula $MCoSb$, where M is Ti, Zr, or Hf.[207–210] The high substitutability of the three lattice sites (M, Co, and Sb) provides many opportunities for tuning the electronic and lattice properties of the half-Heuslers. Through partial elemental substitution, the large lattice thermal conductivity of half-Heusler compounds is expected to be greatly reduced due to mass fluctuation and strain field effects.[207–210] However, the thermal conductivity of p-type half-Heusler alloys reported so far still remains very high relatively (higher than 4 W m^{-1} K^{-1}),[207–210] which prevents ZT from reaching a meaningful value for a TE material. The highest ZT of about 0.5 in p-type $Zr_{0.5}Hf_{0.5}CoSb_{0.8}Sn_{0.2}$ was achieved at 1000 K with a thermal conductivity of 4.1 W m^{-1} K^{-1} at 300 K and 3.6 W m^{-1} K^{-1} at 1000 K.[211] Such a thermal conductivity is still very high due mainly to the relatively large contribution of the lattice thermal conductivity. Besides alloy scattering via partial elemental substitution, boundary scattering can be introduced to further reduce the thermal conductivity.[212] Even though ball milling was attempted to reduce the thermal conductivity by decreasing the average grain size, the ZT improvement was minimal due to the still too large average grain size of at least 1 micrometer.[213–215] Based on our success in achieving a grain size much smaller than 1 μm in a number of materials by ball milling and hot pressing,[72–76,127] we have succeeded in achieving grain sizes smaller than 200 nm in p-type half-Heusler samples with the composition of $Zr_{0.5}Hf_{0.5}CoSb_{0.8}Sn_{0.2}$ by ball milling the alloyed ingot into nanopowders and then hot pressing them into dense bulk samples, resulting in a simultaneous increase in the Seebeck coefficient and a significant decrease in the thermal conductivity, which led to a 60% increase in the peak ZT from 0.5 to 0.8 at 700°C.[216]

In a typical experiment, the alloyed ingot with the composition of $Zr_{0.5}Hf_{0.5}CoSb_{0.8}Sn_{0.2}$ was loaded into a jar with grinding balls and then subjected to a mechanical ball milling process.[216]

Figure 1.18 shows the TEM images of the ball-milled nanopowders. The low (Figure 1.18a) and medium (Figure 1.18b) magnification TEM images show that the average cluster size of the nanopowders ranges from 20 to 500 nm. However, those big clusters are actually agglomerates of many much smaller crystalline nanograins, which is confirmed by the corresponding selected area electron diffraction (SAED) patterns (Figure 1.18c) obtained inside a single cluster (Figure 1.18b). The high-resolution TEM image (Figure 1.18d) shows that the sizes of the small grains are in the range of 5–10 nm.

Figure 1.19 displays the TEM images of the as-pressed bulk samples pressed from the ball-milled powder. The low-magnification TEM image is presented in Figure 1.19a, from which we can see that the grain sizes are in the range of 50–300 nm with an estimated average size being about 100–200 nm. Therefore, there is a significant grain growth during the hot-pressing process. The SAED pattern (inset of Figure 1.19a) of each individual grain indicates that the individual grains are single crystalline. The high-resolution TEM image (Figure 1.19b) demonstrates the good crystallinity inside each individual grain. Figure 1.19c shows one nanodot embedded inside the matrix. Such dots are commonly observed in most of the grains. The compositions of both the nanodot and its surrounding areas are checked by energy-dispersive x-ray spectroscopy (EDS), showing Hf-rich and Co-deficient regions for the nanodot. Another feature pertaining to our sample is that small grains (~30 nm) are also common (Figure 1.19d), which have similar composition as the surrounding bigger grains determined by EDS. We suspect that the nonuniformity in both the grain sizes and the compositions all contribute to the reduction of the thermal conductivity.

The temperature-dependent TE properties of the hot-pressed $Zr_{0.5}Hf_{0.5}CoSb_{0.8}Sb_{0.2}$ bulk samples in comparison with that of the ingot are plotted in Figure 1.20. For all of the samples examined, the temperature dependence of the electrical conductivity was found to exhibit semimetallic or degenerate semiconductor behavior (Figure 1.20a). Specifically, the electrical conductivities of all the ball-milled

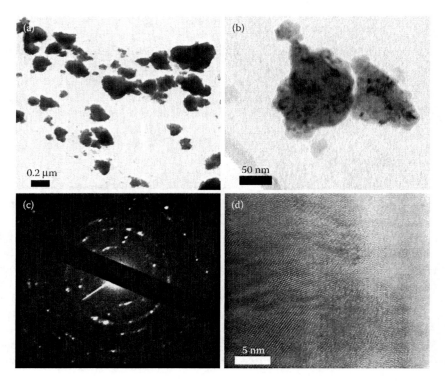

FIGURE 1.18 (a) Low and (b) medium magnification TEM images of (c) a selected area electron diffraction pattern and of (d) a high-magnification TEM image of the ball-milled nano powders of *p*-type half-Heusler alloys. The selected area electron diffraction pattern in (c) shows the multicrystalline nature of an agglomerated cluster in (b). (Reprinted with permission from X. Yan et al., *Nano Lett. 11,* 556–560. Copyright 2011, American Chemical Society.)

and hot-pressed samples are lower than that of the ingot. We have measured the mobility and carrier concentration at room temperature to be 3.86 cm^2 V^{-1} s^{-1} and 1.6 × 10^{21} cm^{-3}, respectively. The mobility is lower than the previously reported value while the carrier concentration is higher.[211] The electrical conductivities of our ball-milled samples decrease slowly in the higher temperature range. The Seebeck coefficients (Figure 1.20b) of our ball-milled samples are significantly higher than that of the ingot for the whole temperature range. These facts strongly indicate that grain boundaries may be trapping electrons, leading to increased hole concentrations in the sample and an energy-filtering effect[68] where low-energy holes are preferentially scattered at the grain boundaries. As a result of the enhancement in the Seebeck coefficient and a slight decrease in the electrical conductivity, the power factors (Figure 1.20c) of our ball-milled and hot-pressed samples are higher than that of the ingot. The total thermal conductivity of our ball-milled and hot-pressed samples (Figure 1.20d) decreases gradually with temperature up to 500°C and does not change too much after that, which shows a much weaker bi-polar effect, consistent with our earlier report in other TE materials with nanostructures.[72] The reduction of the thermal conductivity in our ball-milled and hot-pressed nanostructured samples compared with the ingot is mainly due to the increased phonon scattering at the numerous interfaces of the random nanostructures. To get a quantitative view of the effect of ball milling and hot pressing on phonon transport, the lattice thermal conductivity (κ_l) was estimated by subtracting the electronic contribution (κ_e) from the total thermal conductivity (κ). The electronic contribution to the thermal conductivity (κ_e) can be estimated using the Wiedemann–Franz law. The Lorenz number can be obtained from the reduced Fermi energy, which can be calculated from the Seebeck coefficient at room temperature and the two band

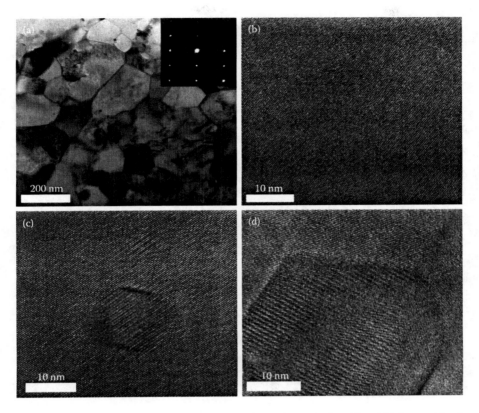

FIGURE 1.19 TEM images of hot-pressed nanostructured samples under low (a) and high magnifications (b, c, d). The inset in (a) is the selected area electron diffraction pattern of *p*-type half-Heusler alloys showing the single crystalline nature of the individual grains. (Reprinted with permission from X. Yan et al., *Nano Lett. 11*, 556–560. Copyright 2011, American Chemical Society.)

theory.[217] Within expectation, the lattice part of the thermal conductivity (Figure 1.20e) decreases with temperature. For the ingot sample, we obtained $\kappa_e = 0.7$ W m^{-1} K^{-1} and $\kappa_l = 4.01$ W m^{-1} K^{-1} at room temperature, whereas for the ball-milled and hot-pressed samples $\kappa_e = 0.54$ W m^{-1} K^{-1} due to a lower electrical conductivity and $\kappa_l = 2.86$ W m^{-1} K^{-1} at room temperature. The lattice thermal conductivity of the ball-milled and hot-pressed samples at room temperature is about 29% lower than that of the ingot, which is mainly due to a stronger boundary scattering in the nanostructured sample. It appears that the lattice part κ_l is still a large portion of the total thermal conductivity. If an average grain size below 100 nm can be achieved during hot pressing, the thermal conductivity can be expected to be further reduced. The slightly improved power factor, coupled with the significantly reduced thermal conductivity, makes the *ZT* (Figure 1.20f) of our ball-milled and hot-pressed samples greatly improved in comparison with that of the ingot. The peak *ZT* value of all our ball-milled and hot-pressed samples reached 0.8 at 700°C, a 60% improvement over the highest reported *ZT* value of 0.5 obtained in the ingot,[211] showing great promise for these ball-milled/hot-pressed nanocomposites as *p*-type TE materials for high-temperature applications.

We also show the specific heat (Figure 1.21a) and thermal diffusivity (Figure 1.21b) of our ball-milled and hot-pressed half-Heusler samples in comparison with those of the corresponding ingot sample. The specific heat (Figure 1.21a) of both the ingot and the ball-milled and hot-pressed samples increases steadily with temperature up to 600°C (the limit of our differential scanning calorimetry

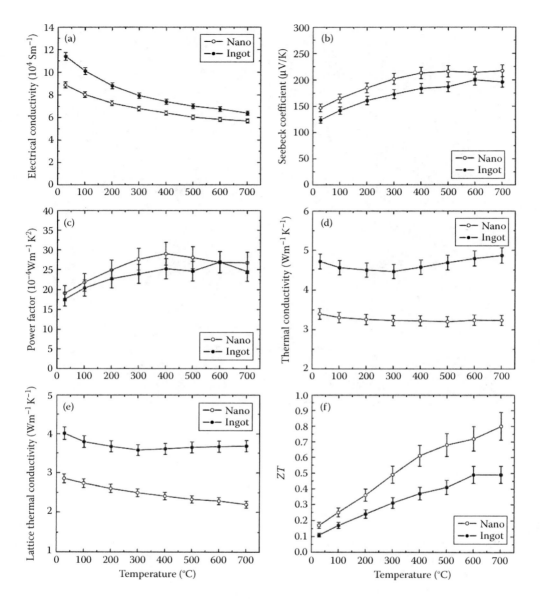

FIGURE 1.20   Temperature-dependent (a) electrical conductivity, (b) Seebeck coefficient, (c) power factor, (d) total thermal conductivity, (e) lattice part of thermal conductivity, and (f) ZT of ball-milled and hot-pressed sample of a *p*-type half-Heusler alloy in comparison with that of the ingot of the same chemical composition. (Reprinted with permission from X. Yan et al., *Nano Lett. 11,* 556–560. Copyright 2011, American Chemical Society.)

measurement instrument). The specific heat value at 700°C was obtained by a reasonable extrapolation. The specific heat difference of about 3% is within the experimental error of the measurement. It is very clear that the major decrease is in the thermal diffusivity (Figure 1.21b) with our ball-milled and hot-pressed sample being consistently lower than that of the ingot sample for the whole temperature range, which provides solid evidence showing the strong effect of grain boundaries on phonon scattering.

By comparison with the data on ingot samples previously reported,[211] we found that the resistivity of our ingot half-Heusler samples is almost the same as the reported value, and the Seebeck coefficient is

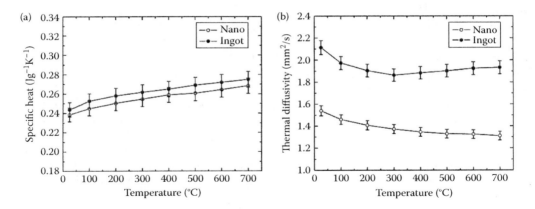

FIGURE 1.21 Temperature-dependent specific heat (a) and thermal diffusivity (b) of ball-milled and hot-pressed sample in comparison with that of the ingot of a *p*-type half-Heusler alloy material. (Reprinted with permission from X. Yan et al., *Nano Lett. 11*, 556–560. Copyright 2011, American Chemical Society.)

higher, which leads to a higher power factor. However, the thermal conductivity of our ingot sample is proportionally higher than the reported value,[211] which leads to the same *ZT* of our ingot samples as the reported value.[211] These small differences in individual properties may be due to some minor differences in sample preparation procedures, which is very reasonable and understandable.

Since the ingot samples are made and measured by the same person on the same machine with the nanostructured samples, we are confident that the enhancement of *ZT* in the nanostructured samples made by ball milling and hot pressing in this study is real and significant.

Although we have achieved a significant enhancement in *ZT* of *p*-type half-Heusler alloys, there remains much room for further improvement. The average grain size of 100–200 nm of our hot-pressed bulk samples is much larger than the 5–10 nm of the ball-milled precursor nanopowders, and this large grain size is why the lattice thermal conductivity of our *p*-type half-Heusler nanocomposites is still very high. If we can preserve the grain size of the original nanopowders, a much lower thermal conductivity and thus a much higher *ZT* can be expected. Besides boundary scattering, minor dopants may also be introduced to enhance the alloy scattering, provided that this enhanced alloy scattering does not cause deterioration to the electronic properties. The *ZT* values we report here are very reproducible within 5% from run to run on more than 10 samples made under similar conditions.

1.3.6 Other Nanocomposites

The ball-milling and hot-pressing method has been widely applied to produce TE nanocomposites. Besides the above reviewed materials (*n*-type and *p*-type Bi_2Te_3 nanocomposites, *n*-type and *p*-type SiGe nanocomposites, *n*-type and *p*-type skutterudite Co_4Sb_{12} nanocomposites, *p*-type PbTe nanocomposites, half-Heusler nanocomposites), the technique has also been applied to other materials systems. Table 1.1 lists some typical nanocomposites produced by the ball-milling and hot-pressing method. The thermal conductivity of these nanocomposites is lower than that of the present single crystalline bulk materials. In some nanocomposites prepared by the ball-milling and hot-pressing method, the power factor is also enhanced. It is reported that the power factor of *n*-type $Bi_{85}Sb_{14}Ag$ nanocomposites is 2.98×10^{-3} W m^{-1} K^{-2} at 255 K,[218] which is about three times higher than that of $Bi_{0.85}Sb_{0.15}$ single crystals. The power factor of a $Bi_{0.85}Sb_{0.13}Pr_{0.02}$ nanocomposite is 3.8×10^{-3} W m^{-1} K^{-2} at 235 K,[219] which is about four times higher than that of $Bi_{0.85}Sb_{0.15}$ single crystals.

Besides the ball-milling and hot-pressing method, TE nanocomposites are also prepared by other methods. For example, $Ga_mSb_nTe_{1.5+m+n}$ nanocomposites with peak *ZT* = 0.98 values are prepared by the spark plasma sintering method from ball-milled nanoparticles.[220] The nanograin size is small (<30 nm). Zn_4Sb_3 nanocomposites are produced by sintering cold-pressed samples.[118] The thermal conductivity of $Ag_{0.008}Pb_{0.18+x}SbTe_{0.20}$ nanocomposites decreases by 40% after ball-milling and spark plasma sintering and a peak ZT value of 1.3 was reported for this material at 300°C.[107] $Bi_{1-x}Sb_x$ nanocomposites obtained by a ball-milling and extrusion method showed a thermal conductivity decrease of 75%.[101] These nanocomposites are also listed in Table 1.1.

In all of these TE nanocomposites, the thermal conductivity is decreased. Despite this, ZT is not necessarily enhanced, and could even be decreased in these nanocomposites, depending on the change of the electrical conductivity and the Seebeck coefficient.

To further decrease the thermal conductivity, additional phases have been dispersed into the nanocomposites. It is postulated that the addition of second-phase nanoparticles will work as phonon scattering centers and further decrease the thermal conductivity. ZnO_2 and rare-earth oxides have been added into β-$FeSi_2$ nanocomposites.[221] It is reported that the dispersion of Y_2O_3 nanoparticles doubles the ZT of n-type β-$FeSi_2$ nanocomposites from 0.35 to 0.63 when 2 wt% Y_2O_3 is added to the material being nanostructured.[222] The thermal conductivity then decreases to 0.7 from 4.5 W m^{-1} K^{-1} after the addition of the nanoparticles.

TiB_2 has also been dispersed into a β-$FeSi_2$ nanocomposite material using the ball-milling and hot-pressing method.[223] However, increasing the TiB_2 content (0–30 vol%) results in a decreased power factor and ZT. TiB_2 has also been dispersed into B_4C nanocomposites using the ball-milling and hot-pressing method.[224] B_4C works as a TE host. Both the thermal conductivity and the electrical conductivity decrease with increasing TiB_2 content from 0 to 25 vol% and ZT also decreases.

BN and WO_3 have been dispersed into p-type $(Bi,Sb)_2Te_3$ nanocomposites using the ball-milling and hot-pressing method.[225] The thermal conductivity decreases slightly from 1.5 to 1.2 W m^{-1} K^{-1} when the volume fraction of BN and WO_3 increases from 0 to 7 vol%. ZT decreases with increasing BN and WO_3 content because of a rapid decrease of the electrical conductivity.

In all of these nanocomposites with additions of second-phase nanoparticles, the thermal conductivity indeed decreases with increasing volume fraction of the second phase because of enhanced phonon scattering. However, the addition of the second phase also decreases the electrical conductivity and Seebeck coefficient. ZT usually decreases because of a greater reduction in electrical conductivity and Seebeck coefficient. In order to increase the ZT of the nanocomposites with the addition of a second-phase material, it is necessary to avoid a decrease of the electrical conductivity and Seebeck coefficient of the mixture after the addition of a second phase. One possible cause for the decreased electrical conductivity and/or Seebeck coefficient is due to a large band structure mismatch between the added second phase and the host materials, leading to either a too high or too low carrier concentration. Another cause is that the second phase reacts with the host materials to change fundamentally the electronic band structures. It is also possible that these second phases form a percolating conducting phase that significantly changes the transport properties. More work is needed to clarify these mechanisms before this approach can be used successfully.

It is also suggested that an amorphous structure and lattice distortion in the TE materials can contribute to a decrease in the thermal conductivity,[220] as well as atomic disorder decreasing the thermal conductivity.[226] This is another approach to produce an increase in ZT through introducing amorphous or defect phonon scattering centers. Further investigation into this method for possible enhancement of ZT is expected in the future.

1.4 Phonon Transport in Nanocomposites

Since most nanocomposites to date have achieved their high ZT values by a reduction in lattice thermal conductivity, one might expect that phonon transport in nanocomposites is fairly well understood. However, the situation turns out to be just the opposite. Even in bulk materials, there is much uncertainty regarding the values of key physical parameters and the physics behind these values, for parameters such as the phonon mean free path. Furthermore, the precise way in which interfaces affect phonon transport to reduce the thermal conductivity in nanocomposites remains poorly understood. Theoretical predictions for the lower limit of the thermal conductivity turn out not to be applicable to nanostructures, as was experimentally demonstrated by Cahill.[227] Understanding phonon transport is crucial to further reducing the thermal conductivity, however. We highlight here what is known about phonon transport and what we now seek to understand better.

It has been shown that nanocomposites are able to achieve a thermal conductivity lower than that of their bulk counterparts. Before the actual preparation of nanostructured materials in the laboratory, it was believed that the lower limit of the thermal conductivity was set by the value obtained by an alloy, as in SiGe, and this lower value of the thermal conductivity was termed the "alloy limit." While this "limit" of the thermal conductivity remained in the literature for 50 years, nanocomposites have recently exhibited values for the thermal conductivity that are lower than that of their corresponding alloy.[73,75]

The reason why this is possible is now fairly well understood. Phonons in a material have a spectrum of wavelengths and mean free paths, each of which contributes to the total thermal conductivity. Introducing impurity atoms to make an alloy causes phonon scattering by these impurity atoms, which are only effective in scattering short wavelength, high-frequency phonons and not very effective in scattering long wavelength phonons. The phonon scattering process is very similar to Rayleigh scattering of light, where high-frequency light (short wavelength blue light) is scattered preferentially by atmospheric molecules relative to lower frequency red light. Thus while high-frequency phonons are strongly scattered in alloys, mid- to long-wavelength phonons are still able to transport heat. Furthermore, it has been shown that in bulk Si these mid- to long-wavelength modes can actually carry a substantial fraction of the heat.[228] By incorporating structures with a larger characteristic length than that of an impurity atom, an interface scattering mechanism is introduced which scatters mid- to long-wavelength phonons, resulting in a further reduction in the thermal conductivity and thereby yielding κ values that are lower than the "alloy limit." This scattering mechanism is most effective when the structures are smaller but comparable to the phonon mean free path, which is approximately 100 nm in the $Si_{1-x}Ge_x$ system.

This physical picture gives a good qualitative understanding of why nanocomposites are able to beat the alloy limit. Simple phonon models based on the Callaway model for the thermal conductivity,[229,230] which uses the Debye model, along with an additional interface scattering term has been able to explain the thermal conductivity data for $In_{0.53}Ga_{0.47}As/ErAs$ nanostructures[57] and Si nanowires.[58] However, while these models can fit the data by adjusting various fitting parameters, a more careful examination shows that many of the model's fundamental predictions are not correct. Figure 1.22a,b shows the cumulative distribution function of the thermal conductivity with respect to the phonon wavelength and mean free path for the Callaway model in comparison to a more exact molecular dynamics simulation[228] in bulk, undoped Si at 300 K. From Figure 1.22 we see that while the simple Callaway model can be adjusted to give the correct total thermal conductivity, the more quantitative results, such as the specific phonon mean free path and wavelength contributions to the thermal conductivity for nanomaterials are not well handled by the Callaway model. These models in fact are not consistent even for the bulk case where the spectral dependence of the thermal conductivity shows only fair agreement with the molecular dynamics result, while the mean free path accumulation for specific phonons in this case does not agree well with the molecular simulations as shown in Figure 1.22b. Crucially, the key result is that the long mean free path phonons carry a large fraction of the heat in a material like undoped Si. In fact, Dames has indicated that the effective phonon mean free path is actually smaller than the particle size for nanostructured materials and that the mean free path for boundary scattering is increasing proportional to the phonon frequency.[231] These are important facts to know when designing TE materials, and this fact is not predicted by the Callaway model. Thus there remain important aspects of phonon transport in bulk materials that are not yet well understood, and new discoveries are still taking place in this field for a variety of materials, not only for silicon. For example, Morelli et al. recently reported an

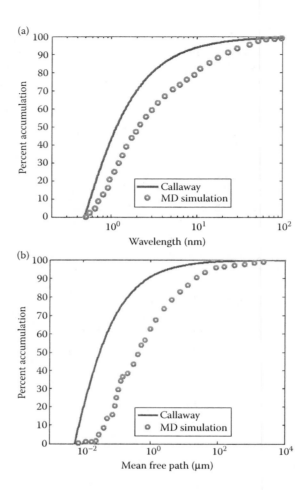

FIGURE 1.22 Cumulative distribution function of the lattice thermal conductivity with respect to (a) phonon wavelength and (b) phonon mean free path predicted by the Callaway model and by molecular dynamics simulations in bulk, undoped Si at 300 K. (After A. S. Henry, G. Chen, *J. Comput. Theor. Nanosci. 5*, 141–152. Copyright 2008, American Scientific Publishers.)

unusually low thermal conductivity in bulk cubic $AgSbTe_2$ due to the extremely anharmonic bonds in this material.[232]

Nanostructures add another layer of complexity to this problem because they contain many interfaces with a spacing smaller than the phonon mean free path, and such interfaces introduce a thermal boundary resistance between different regions of the nanocomposite. While researchers have been able to use a thermal boundary resistance to obtain a low thermal conductivity, there is currently a limited ability to predict the thermal boundary resistance despite many years of research.[233,234] A thermal boundary resistance has also been identified as a key mechanism for the low thermal conductivity in superlattices, rather than the periodicity of the superlattices.[45,235,238] Such understanding led to the concept of using a nanocomposite to enhance the *ZT*. Efforts have been made to calculate the thermal boundary resistance using the Boltzmann equation[237] and Monte Carlo simulations,[238] and Minnich and Chen introduced a modified theory to analytically compute the thermal conductivity of nanocomposites.[239] Similarly, the Peierls instability can also lead to a low thermal conductivity and to *ZT* enhancement.[28]

However, existing models cannot yet give an accurate prediction for the phonon transmission at a single interface, which is a fundamental parameter in predicting the thermal boundary resistance of an interface. Nanoparticles inside the grains of composites have multiple interfaces, raising questions about phonon scattering by these particles. In the past, scattering models based on Rayleigh scattering[57] and acoustic Mie scattering theory[240] have been used to treat nanoparticle scattering. These models, however, do not include interfacial roughness effects and the effects of scattering inside the particle. Although molecular dynamics[241] may provide a potential tool, available computation power at this stage limits the size of the nanoparticles that can be treated in such model calculations.

As discussed in the introduction to this chapter, this lack of understanding affects our ability to design and optimize nanostructured materials to have low thermal conductivity. For example, we are currently unable to answer questions such as which phonon modes are the dominant heat carriers in nanocomposites, what is the optimal size distribution of nanostructures, what type of interface leads to the strongest phonon scattering, or what is the interfacial resistance and reflectivity of a given nanocomposite interface. In fact, it is not even clear what is the lower limit to the thermal conductivity in a nanostructured material. For bulk materials, the criterion that the phonon mean free path must be at least half the phonon wavelength[242,243] sets this lower limit. This might not be the case in nanostructured materials, however. Chen has suggested that the minimum thermal conductivity in nanostructures is likely to be lower than in bulk materials because the minimum thermal conductivity theory for bulk materials is based on isotropic scattering, while in nanostructures, interfacial scattering is highly anisotropic.[244] Cahill has experimentally demonstrated that the thermal conductivity of layered WSe_2 is six times lower than the value predicted by theory and only twice that of air, an experimental verification that nanostructures can affect phonon transport in ways that are not accounted for by current theories.[227,245]

Studies on specific materials for which experimental data are available has been one promising strategy for advancing our understanding of the key factors for enhancing ZT, such as the control of grain boundary scattering to lower the thermal conductivity, taking both electron and phonon grain boundary scattering effects into account. In this context, Minnich, Chen, and coworkers[246] have used the Boltzmann equation under the relaxation time approximation to model the TE properties of nanocomposite $Si_{1-x}Ge_x$ alloys, with particular emphasis given to the $Si_{80}Ge_{20}$ composition. For typical experimental samples, both n- and p-type nanocomposites show dopant precipitation at grain boundaries with a boundary scattering length that is significantly smaller than the average grain size.[231] The Boltzmann-equation-based calculations[246] generally account for the observed temperature-dependent electrical conductivity, Seebeck coefficient, thermal conductivity and ZT of $Si_{1-x}Ge_x$ quite well. Two interesting discrepancies between the model and the experimental data obtained for a $Si_{80}Ge_{20}$ nanocomposite material are especially informative. First, the calculated hole effective mass has to be increased for the nanocomposites compared to bulk values with the same chemical composition and the calculated grain boundary barrier height of the Poisson equation has to be increased by a factor of several to fit the experimental data. Model calculations of this type have been found to be effective in clarifying the most important effects introduced by the nanostructuring of the constituents of nanocomposite TE materials to guide experiments toward optimizing the performance of the actual materials that we can produce.

It is clear that there are many unanswered questions in phonon transport theory that critically affect how we design nanostructured materials to exhibit a reduced thermal conductivity. Gaining a better understanding of the fundamentals of phonon transport in nanostructured materials is thus a key challenge. There is much work to be done, but it is expected that with a better understanding of phonon transport, further lattice thermal conductivity reductions in nanocomposite systems should be possible.

In addition to further reducing phonon thermal conductivity, improving electron transport is also very important for further ZT enhancements. The idea of sharp features in density of states[44,84] for improved power factor has recently been realized in bulk materials via resonant states caused by impurities.[82] Another example of such a strategy is the work of Zebarjadi and coworkers,[247] who have taken the modulation doping approach from semiconductor physics[248] and adapted this approach to bulk nanostructured TE materials.[247] According to this approach, the modulation-doped sample is prepared by

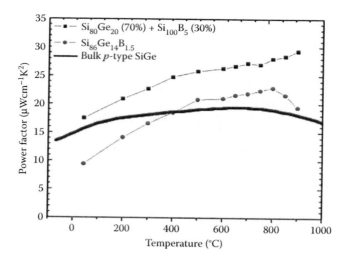

FIGURE 1.23 Temperature dependence of the power factor of the two-phase nanocomposite sample $Si_{80}Ge_{20}$ (70%) + $Si_{100}B_5$ (30%) (squares) in comparison to the single-phase uniformly alloyed nanocomposite sample $Si_{86}Ge_{14}B_{1.5}$ (circles) and to the p-type SiGe bulk alloy used in RTGs for space power missions (solid line). (Reprinted with permission from M. Zebarjadi et al., *Nano Lett. 11*, 2225–2230. Copyright 2011, American Chemical Society.)

using two types of nanograins. In one type of nanograin, dopants are incorporated into the grain, while the second type of nanograin has no added dopants. In this way, the charge carriers are separated from their parent atoms, allowing the ionized impurity scatterers to be separated from their associated carriers. This gives rise to dissimilar grains with large interface resistances and results in an increase in the ZT of the nanocomposite over that of its nano-structured constituents. Second, the modulation doping approach allows the carrier mobility of the modulation-doped sample to be larger than that of either of the constituent nanograins, thereby resulting in a power factor for the modulation-doped sample that is larger than that of either constituent,[247] as shown in Figure 1.23. The work of Zebarjadi et al.[243] shows that the concept of modulation doping can be applied to TE materials and can enhance the resulting ZT value over that of its individual constituents. This new approach now needs to be applied to two nanograined materials that are themselves good TE materials and the approach needs to be optimized to show the full potential of this approach.

1.5 Conclusions and Outlook

TE nanocomposites with high ZT values have been successfully produced using a nanostructuring approach. The nanocomposites consist of a high density of nanograins which themselves contain structural defects which scatter phonons. The unique nanostructures of the nanocomposites reduce the thermal conductivity by increasing the phonon scattering from the high density of imperfections within the nanocomposites. Simultaneously, the high electrical conductivity and power factor of the nanocomposites are largely preserved, while the phonon thermal conductivity is significantly reduced, resulting in a high ZT. In some cases, both the electrical conductivity and the Seebeck coefficient are increased slightly in the nanocomposite materials, while at the same time the thermal conductivity is decreased, thereby greatly enhancing ZT. High ZT has been achieved in many TE nanocomposites. For example, the peak ZT is increased to 1.4 in p-type $Bi_xSb_{2-x}Te_3$ nanocomposites from a value $ZT = 1.0$ in the corresponding bulk material, and to $ZT = 1.1$ in n-type $Bi_2Te_{2.7}Se_{0.3}$ from 0.85 in the corresponding bulk material. In p-type $Si_{80}Ge_{20}$ an increase in ZT to 0.95 has been achieved in nanocomposites from $ZT = 0.65$ in the corresponding bulk material, and to $ZT = 1.3$ in n-type $Si_{80}Ge_{20}$ nanocomposites from $ZT = 0.9$ in the corresponding bulk material. Similarly, an increase in ZT values to $ZT = 0.8$ has been achieved in p-type nanocomposite half-Heusler alloys from $ZT = 0.5$ in the corresponding bulk material. All of these nano-composites with enhanced ZT values have significant commercial potential for cooling, industrial waste heat recovery, space power generation, and solar power conversion applications.

Further investigation is needed to achieve even greater enhancement in ZT which is believed to be possible by researchers working in this field. It is believed that this will come both from improvement in the processing of currently used materials and from development of better materials and the introduction of new concepts. In the present nanocomposites, the grain size is usually between 10 and 1000 nm, which is relatively large in comparison with the phonon mean free path. This is why it is expected that it is still possible to reduce the thermal conductivity of common nanocomposite materials. It should be possible to prepare nanocomposites with smaller grain sizes by limiting the grain growth during the densification process of the nanostructured material. The smaller grain size will hopefully reduce the thermal conductivity to 1 W m^{-1} K^{-1} or less, thereby further enhancing ZT. In order to decrease the thermal conductivity to 1 W m^{-1} K^{-1}, phonon transport in the nanocomposites needs to be understood in greater detail.

It is reported above that the power factor can be increased in some nanocomposites. However, the thermal conductivity in these nanocomposites is still high, which gives us hope to further improve ZT if the thermal conductivity can be reduced significantly in such cases. In the coming years, it is important to determine effective strategies to simultaneously improve the power factor and reduce the thermal conductivity.

There is hope that a ZT value of 2 can be achieved in these nanocomposites if the thermal conductivity is reduced to below 1 W m^{-1} K^{-1} and the power factor is further enhanced. The TE devices fabricated from these high ZT nanocomposites are expected to find widespread use in applications such as refrigeration, air conditioning, and high-power commercial generators. It is expected that in the near future, TE devices could contribute significantly to a renewable energy supply in the global marketplace, ensuring a future source of clean, reliable energy while protecting our planet's environment.

Acknowledgments

This work has been funded by DOE DE-FG02-00ER45805 (ZFR), DOE DE-FG02-08ER46516 (MD, GC, and ZFR), and DOE DE-SC0001299/DE-FG02-09ER 46577 (GC, MD, and ZFR).

References

1. H. J. Goldsmid, *Thermoelectric Refrigeration*, Plenum Press, New York, **1964**.
2. D. D. Pollock, *Thermocouples: Theory and Properties*, CRC Press, Boca Raton, FL, **1991**.
3. D. M. Rowe, ed. *CRC Handbook of Thermoelectrics*, CRC Press, Boca Raton, FL, **1995**.
4. T. M. Tritt, ed. *Recent Trends in Thermoelectric Materials Research III*, Vol. 71, Academic Press, San Diego, CA, **2001**.
5. D. M. Rowe, ed. *Thermoelectrics Handbook: Macro to Nano*, CRC/Taylor & Francis, Boca Raton, FL, **2006**.
6. C. Wood, *Rep. Prog. Phys.* **1988**, *51*, 459–539; M. Zebarjadi, K. Esfarjani, M. S. Dresselhaus, Z. F. Ren, G. Chen, *Energy Environ. Sci.* **2012**, *5*, 5147–5162.
7. F. J. DiSalvo, *Science* **1999**, *285*, 703–706.
8. B. C. Sales, *Science* **2002**, *295*, 1248–1249.
9. S. B. Riffat, X. Ma, *Appl. Therm. Eng.* **2003**, *23*, 913–935.
10. L. E. Bell, *Science* **2008**, *321*, 1457–1461.
11. G. J. Snyder, E. S. Toberer, *Nat. Mater.* **2008**, *7*, 105–114.
12. D. Kraemer, B. Poudel, H.-P. Feng, J. C. Caylor, B. Yu, X. Yan, Y. Ma et al., *Nat. Mater.* **2011**, *10*, 532–538.
13. A. F. Ioffe, *Physics of Semiconductors*, Academic Press, New York, **1960**.

14. C. B. Vining, *J. Appl. Phys.* **1991**, *69*, 331–341.

15. J. Yang, T. Caillat, *MRS Bulletin* **2006**, *31*, 224–229.

16. D. Y. Chung, T. Hogan, P. Brazis, M. Rocci-Lane, C. Kannewurf, M. Bastea, C. Uher, M. G. Kanatzidis, *Science* **2000**, *287*, 1024–1027.

17. B. Wölfing, C. Kloc, J. Teubner, E. Bucher, *Phys. Rev. Lett.* **2001**, *86*, 4350–4353.

18. V. L. Kuznetsov, L. A. Kuznetsova, A. E. Kaliazin, D. M. Rowe, *J. Appl. Phys.* **2000**, *87*, 7871–7875.

19. J. L. Cohn, G. S. Nolas, V. Fessatidis, T. H. Metcalf, G. A. Slack, *Phys. Rev. Lett.* **1999**, *82*, 779–782.

20. X. F. Tang, P. Li, S. K. Deng, Q. J. Zhang, *J. Appl. Phys.* **2008**, *104*, 013706.

21. S. K. Deng, X. F. Tang, P. Li, Q. J. Zhang, *J. Appl. Phys.* **2008**, *103*, 073503.

22. J. Androulakis, C. H. Lin, H. J. Kong, C. Uher, C. Wu, T. Hogan, B. A. Cook, T. Caillat, K. M. Paraskevopoulos, M. Kanatzidis, *J. Am. Chem. Soc.* **2007**, *129*, 9780–9788.

23. K. F. Hsu, S. Loo, F. Guo, W. Chen, J. S. Dyck, C. Uher, T. Hogan, E. K. Polychroniadis, M. G. Kanatzidis, *Science* **2004**, *303*, 818–821.

24. J. Androulakis, K. F. Hsu, R. Pcionek, H. Kong, C. Uher, J. J. D'Angelo, A. Downey, T. Hogan, M. G. Kanatzidis, *Adv. Mater.* **2006**, *18*, 1170–1173.

25. P. F. P. Poudeu, J. D'Angelo, A. D. Downey, J. L. Short, T. P. Hogan, M. G. Kanatzidis, *Angew. Chem. Int. Ed.* **2006**, *45*, 3835–3839.

26. A. Guéguen, P. F. P. Poudeu, C. P. Li, S. Moses, C. Uher, J. He, V. Dravid, K. M. Paraskevopoulos, M. G. Kanatzidis, *Chem. Mater.* **2009**, *21*, 1683–1694.

27. G. S. Nolas, J. Poon, M. Kanatzidis, *MRS Bull.* **2006**, *31*, 199–205.

28. J. S. Rhyee, K. H. Lee, S. M. Lee, E. Cho, S. I. Kim, E. Lee, Y. S. Kwon, J. H. Shim, G. Kotliar, *Nature* **2009**, *459*, 965–968.

29. G. S. Nolas, J. Sharp, H. J. Goldsmid, *Thermoelectrics: Basic Principles and New Materials Developments*, Springer, New York, **2001**.

30. M. G. Kanatzidis, S. D. Mahanti, T. P. Hogan, eds. *Chemistry, Physics, and Materials Science of Thermoelectric Materials: Beyond Bismuth Telluride*, Kluwer Academic/Plenum Publishers, New York, **2003**.

31. K. Koumoto, I. Terasaki, R. Funahashi, *MRS Bull.* **2006**, *31*, 206–210.

32. N. Savvides, H. J. Goldsmid, *J. Phys. C: Solid State Phys.* **1980**, *13*, 4657–4670.

33. N. Savvides, H. J. Goldsmid, *J. Phys. C: Solid State Phys.* **1980**, *13*, 4671–4678.

34. D. M. Rowe, V. S. Shukla, N. Savvides, *Nature* **1981**, *290*, 765–766.

35. C. B. Vining, W. Laskow, J. O. Hanson, R. R. V. der Beck, P. D. Gorsuch, *J. Appl. Phys.* **1991**, *69*, 4333–4340.

36. D. M. Rowe, L. W. Fu, S. G. K. Williams, *J. Appl. Phys.* **1993**, *73*, 4683–4685.

37. A. A. Joraide, *J. Mater. Sci.* **1995**, *30*, 744–748.

38. C. W. Nan, R. Birringer, *Phys. Rev. B* **1998**, *57*, 8264–8268.

39. D. B. Hyun, J. S. Hwang, J. D. Shim, *J. Mater. Sci.* **2001**, *36*, 1285–1291.

40. D. H. Kim, T. Mitani, *J. Alloy Compd.* **2005**, *399*, 14–19.

41. G. Chen, 2001, in *Semiconductors and Semimetals, Recent Trends in Thermoelectric Materials Research III*, Vol. 71 (Ed. T. Tritt), Academic Press, San Diego, CA, **2001**, pp. 203–259.

42. G. Chen, M. S. Dresselhaus, J.-P. Fleurial, T. Caillat, *Int. Mat. Rev.* **2003**, *48*, 45–66.

43. G. Chen, A. Shakouri, *ASME J. Heat Transfer* **2002**, *124*, 242–252.

44. L. D. Hicks, M. S. Dresselhaus, *Phys. Rev. B* **1993**, *47*, 16631–16634.

45. G. Chen, *Phys. Rev. B* **1998**, *57*, 14958–14973.

46. H. Q. Liu, Y. Song, S. N. Zhang, X. B. Zhao, F. P. Wang, *J. Phys. Chem. Solids* **2009**, *70*, 600–603.

47. M. S. Dresselhaus, G. Chen, M. Y. Tang, R. G. Yang, H. Lee, D. Z. Wang, Z. F. Ren, J. P. Fleurial, P. Gogna, *Adv. Mater.* **2007**, *19*, 1043–1053.

48. T. Koga, S. B. Cronin, M. S. Dresselhaus, J. L. Liu, K. L. Wang, *Appl. Phys. Lett.* **2000**, *77*, 1490–1492.

49. X. B. Zhao, S. H. Yang, Y. Q. Cao, J. L. Mi, Q. Zhang, T. J. Zhu, *J. Electron. Mater.* **2009**, *38*, 1017–1024.
50. Q. Zhang, J. He, T. J. Zhu, S. N. Zhang, X. B. Zhao, T. M. Tritt, *Appl. Phys. Lett.* **2008**, *93*, 102109.
51. R. Venkatasubramanian, E. Siivola, T. Colpitts, B. O'Quinn, *Nature* **2001**, *413*, 597–602.
52. Q. Zhang, J. He, X. B. Zhao, S. N. Zhang,T. J. Zhu, H. Yin, T. M. Tritt, *J. Phys. D: Appl. Phys.* **2008**, *41*, 185103.
53. Q. Zhang, X. B. Zhao, H. Yin, T. J. Zhu, *J. Alloys. Compd.* **2008**, *464*, 9–12.
54. Y. Q. Cao, T. J. Zhu, X. B. Zhao, X. B. Zhang, J. P. Tu, *Appl. Phys. A* **2008**, *92*, 321–324.
55. T. C. Harman, P. J. Taylor, M. P. Walsh, B. E. LaForge, *Science* **2002**, *297*, 2229–2232.
56. W. J. Xie, X. F. Tang, Y. G. Yan, Q. J. Zhang, T. M. Tritt, *Appl. Phys. Lett.* **2009**, *94*, 102111.
57. W. Kim, J. Zide, A. Gossard, D. Klenov, S. Stemmer, A. Shakouri, A. Majumdar, *Phys. Rev. Lett.* **2006**, *96*, 045901.
58. A. I. Hochbaum, R. Chen, R. D. Delgado, W. Liang, E. C. Garnett, M. Najarian, A. Majumdar, P. Yang, *Nature* **2008**, *451*, 163–167.
59. A. I. Boukai, Y. Bunimovich, J. Tahir-Kheli, J. K. Yu, W. A. Goddard III, J. R. Heath, *Nature* **2008**, *451*, 168–171.
60. D. Li, Y. Wu, P. Kim, L. Shi, P. Yang, A. Majumdar, *Appl. Phys. Lett.* **2003**, *83*, 2934.
61. J. Zhou, Q. Jin, J. H. Seol, X. Li, L. Shi, *Appl. Phys. Lett.* **2005**, *87*, 133109.
62. T. Yao, *Appl. Phys. Lett.* **1987**, *51*, 1798–1800.
63. G. Chen, C. L. Tien, X. Wu, J. S. Smith, *J. Heat Transfer*, **1994**, *116*, 325–331.
64. X. Y. Yu, G. Chen, A. Verma, J. S. Smith, *Appl. Phys. Lett.*, **1995**, *67*, 3554–3556.
65. G. Chen, C. L. Tien, *J. Thermophys. Heat Transfer*, **1993**, *7*, 311–318.
66. S. Lee, D. Cahill, R. Venkatasubramanian, *Appl. Phys. Lett.* **1997**, *70*, 2957–2959.
67. H. Böttner, G. Chen, R. Venkatasubramanian, *MRS Bull.* **2006**, *31*, 211–217.
68. J. P. Heremans, C. M. Thrush, D. T. Morelli, *Phys. Rev. B* **2004**, *70*, 115334.
69. S. M. Lee, D. G. Cahill, R. Venkatasubramanian, *Appl. Phys. Lett.* **1997**, *70*, 2957–2959.
70. B. Yang, W. L. Liu, J. L. Liu, K. L. Wang, G. Chen, *Appl. Phys. Lett.* **2002**, *81*, 3588–3590.
71. Y. Zhang, J. Christofferson, A. Shakouri, D. Li, A. Majumdar, Y. Wu, R. Fan, P. Yang, *IEEE Trans. Nanotechnol.* **2006**, *5*, 67–74.
72. B. Poudel, Q. Hao, Y. Ma, Y. C. Lan, A. Minnich, B. Yu, X. Yan et al., *Science* **2008**, *320*, 634–638.
73. G. Joshi, H. Lee, Y. C. Lan, X. W. Wang, G. H. Zhu, D. Z. Wang, R. W. Gould et al., *Nano Lett.* **2008**, *8*, 4670–4674.
74. Y. Ma, Q. Hao, B. Poudel, Y. C. Lan, B. Yu, D. Z. Wang, G. Chen, Z. F. Ren, *Nano Lett.* **2008**, *8*, 2580–2584.
75. X. W. Wang, H. Lee, Y. C. Lan, G. H. Zhu, G. Joshi, D. Z. Wang, J. Yang et al., *Appl. Phys. Lett.* **2008**, *93*, 193121.
76. G. H. Zhu, H. Lee, Y. C. Lan, X. W. Wang, G. Joshi, D. Z. Wang, J. Yang et al., *Phys. Rev. Lett.* **2009**, *102*, 196803.
77. Z. F. Ren, B. Poudel, Y. Ma, Q. Hao, Y. C. Lan, A. Minnich, A. Muto et al., *Mater. Res. Soc. Symp. Proc.*, **2009**, *1166*, 1166.
78. S. Bhattacharya, A. L. Pope, R. T. L. IV, T. M. Tritt, V. Ponnambalam, Y. Xia, S. J. Poon, *Appl. Phys. Lett.* **2000**, *77*, 2476–2478.
79. J. P. Heremans, C. M. Thrush, D. T. Morelli, *J. Appl. Phys.* **2005**, *98*, 063703.
80. J. M. O. Zide, D. Vashaee, Z. X. Bian, G. Zeng, J. E. Bowers, A. Shakouri, A. C. Gossard, *Phys. Rev. B* **2006**, *74*, 205335.
81. H. Ohta, S. Kim, Y. Mune, T. Mizoguchi, K. Nomura, S. Ohta, T. Nomura et al., *Nat. Mater.* **2007**, *6*, 129–134.
82. J. P. Heremans, V. Jovovic, E. S. Toberer, A. Saramat, K. Kurosaki, A. Charoenphakdee, S. Yamanaka, G. J. Snyder, *Science* **2008**, *321*, 554–557.

83. B. Yu, Q. Y. Zhang, H. Wang. X. W. Wang, H. Z. Wang, D. Z. Wang, H. Wang, G. J. Snyder, G. Chen, Z. F. Ren, *J. Appl. Phys.* **2010**, *108*, 016104.

84. G. D. Mahan, J. O. Sofo, *Proc. Natl. Acad. Sci. USA* **1996**, *93*, 7436–7439.

85. M. Dresselhaus, G. Dresselhaus, X. Sun, Z. Zhang, S. Cronin, T. Koga, *Phys. Solid State* **1999**, *41*, 679–682.

86. M. S. Dresselhaus, G. Chen, M. Y. Tang, R. G. Yang, H. Lee, D. Wang, Z. F. Ren, J. P. Fleurial, P. Gogna, *Mater. Res. Soc. Symp. Proc.*, **2006**, *886*, 0886.

87. H. Ni, T. Zhu, X. Zhao, *Physica B* **2005**, *364*, 50–54.

88. T. J. Zhu, Y. Q. Liu, X. B. Zhao, *Mater. Res. Bull.* **2008**, *43*, 2850–2854.

89. W. Z. Wang, B. Poudel, J. Yang, D. Z. Wang, Z. F. Ren, *J. Am. Chem. Soc.* **2005**, *127*, 13792–13793.

90. J. L. Mi, X. B. Zhao, T. J. Zhu, J. P. Tu, *Mater. Lett.* **2008**, *62*, 2363–2365.

91. Y. Q. Cao, T. J. Zhu, X. B. Zhao, *J. Alloys. Compd.* **2008**, *449*, 109–112.

92. W. Z. Wang, B. Poudel, D. Z. Wang, Z. F. Ren, *Adv. Mater.* **2005**, *17*, 2110–2114.

93. B. Poudel, W. Z. Wang, D. Z. Wang, J. Y. Huang, Z. F. Ren, *J. Nanosci. Nanotechnol.* **2006**, *6*, 1050–1053.

94. W. Z. Wang, X. Yan, B. Poudel, Y. Ma, Q. Hao, J. Yang, G. Chen, Z. F. Ren, *J. Nanosci. Nanotechnol.* **2008**, *8*, 452–456.

95. Y. Q. Cao, X. B. Zhao, T. J. Zhu, X. B. Zhang, J. P. Tu, *Appl. Phys. Lett.* **2008**, *92*, 143106.

96. L. Bertini, C. Stiewe, M. Toprak, S. Williams, D. Platzek, A. Mrotzek, Y. Zhang et al., *J. Appl. Phys.* **2003**, *93*, 438–447.

97. T. Sun, X. B. Zhao, T. J. Zhu, J. P. Tu, *Mater. Lett.* **2006**, *60*, 2534–2537.

98. J. Martin, G. S. Nolas, W. Zhang, L. Chen, *Appl. Phys. Lett.* **2007**, *90*, 222112.

99. Y. Y. Zheng, T. J. Zhu, X. B. Zhao, J. P. Tu, G. S. Cao, *Mater. Lett.* **2005**, *59*, 2886–2888.

100. K. Sridhar, K. Chattopadhyay, *J. Alloy Compd.* **1998**, *264*, 293–298.

101. R. Martin-Lopez, A. Dauscher, H. Scherrer, J. Hejtmanek, H. Kenzari, B. Lenoir, *Appl. Phys. A: Mater. Sci. Process.* **1999**, *68*, 597–602.

102. J. Schilz, M. Riffel, K. Pixius, H. J. Meyer, *Powder Technol.* **1999**, *105*, 149–154.

103. K. Kishimoto, T. Koyanagi, *J. Appl. Phys.* **2002**, *92*, 2544–2549.

104. S. S. Kim, S. Yamamoto, T. Aizawa, *J. Alloy Compd.* **2004**, *375*, 107–113.

105. J. Yang, X. Fan, R. Chen, W. Zhu, S. Bao, X. Duan, *J. Alloy Compd.* **2006**, *416*, 270–273.

106. J. Yang, R. Chen, X. Fan, S. Bao, W. Zhu, *J. Alloy Compd.* **2006**, *407*, 330–333.

107. H. Wang, J. F. Li, C. W. Nan, M. Zhou, W. Liu, B. P. Zhang, T. Kita, *Appl. Phys. Lett.* **2006**, *88*, 092104.

108. W. S. Liu, B. P. Zhang, J. F. Li, L. D. Zhao, *J. Phys. D: Appl. Phys.* **2007**, *40*, 566–572.

109. W. S. Liu, B. P. Zhang, L. D. Zhao, J. F. Li, *Chem. Mater.* **2008**, *20*, 7526–7531.

110. M. Zakeri, M. Allahkarami, G. Kavei, A. Khanmohammadian, M. Rahimipour, *J. Mater. Process. Technol.* **2009**, *209*, 96–101.

111. C. Suryanarayana, *Prog. Mater. Sci.* **2001**, *46*, 1–184.

112. P. S. Gilman, J. S. Benjamin, *Annu. Rev. Mater. Res.* **1983**, *13*, 279–300.

113. R. Davis, C. Koch, *Scripta Metall.* **1987**, *21*, 305–310.

114. C. C. Koch, *Annu. Rev. Mater. Res.* **1989**, *19*, 121–143.

115. H. Li, X. F. Tang, X. L. Su, Q. J. Zhang, *Appl. Phys. Lett.* **2008**, *92*, 202114.

116. T. Kumpeerapun, H. Scherrer, J. Khedari, J. Hirunlabh, S. Weber, A. Dauscher, B. Lenoir, B. Zighmati, H. M. Jahed, V. Kosalathip, in *Proc. of 25th International Conference on Thermoelectrics*, Vienna, Austria, **2006**, pp. 136–140.

117. J. Peng, J. Yang, T. Zhang, X. Song, Y. Chen, *J. Alloy Compd.* **2004**, *381*, 313–316.

118. S. C. Ur, I. H. Kim, P. Nash, *J. Mater. Sci.* **2007**, *42*, 2143–2149.

119. C. J. Liu, H. Yamauchi, *Phys. Rev. B* **1995**, *51*, 11826–11829.

120. J. L. Harringa, B. A. Cook, *Mater. Sci. Eng. B* **1999**, *60*, 137–142.

121. Z. He, C. Stiewe, D. Platzek, G. Karpinski, E. Muller, S. Li, M. Toprak, M. Muhammed, *J. Appl. Phys.* **2007**, *101*, 053713.

122. J. M. Schultz, J. P. McHugh, W. A. Tiller, *J. Appl. Phys.* **1962**, *33*, 2443–2450.

123. R. Lefever, G. McVay, R. Baughman, *Mater. Res. Bull.* **1974**, *9*, 863–872.

124. N. Savvides, H. J. Goldsmid, *J. Mater. Sci.* **1980**, *15*, 594–600.

125. J. F. Miller, R. C. Himes, *J. Electrochem. Soc.* **1960**, *107*, 915–919.

126. D. M. Rowe, R. W. Bunce, *J. Phys. D: Appl. Phys.* **1969**, *2*, 1497–1502.

127. J. Yang, Q. Hao, H. Wang, Y. C. Lan, Q. Y. He, D. Z. Wang, J. A. Harriman et al., *Phys. Rev. B* **2009**, *80*, 115329.

128. Q. Y. He, Q. Hao, X. W. Wang, J. Yang, Y. C. Lan, X. Yan, B. Yu et al., *J. Nanosci. Nanotechnol.* **2008**, *8*, 4003–4006.

129. Q. Y. He, S. J. Hu, X. G. Tang, Y. C. Lan, J. Yang, X. W. Wang, Z. F. Ren, Q. Hao, G. Chen, *Appl. Phys. Lett.* **2008**, *93*, 042108.

130. C. B. Murray, C. R. Kagan, M. G. Bawendi, *Annu. Rev. Mater. Sci.* **2000**, *30*, 545–610.

131. Y. C. Lan, B. Poudel, Y. Ma, D. Z. Wang, M. S. Dresselhaus, G. Chen, Z. F. Ren, *Nano Lett.* **2009**, *9*, 1419–1422.

132. D. Medlin, G. Snyder, *Curr. Opin. Colloid Interface Sci.* **2009**, *14*, 226–235.

133. J. Yang, T. Aizawa, A. Yamamoto, T. Ohta, *J. Alloy Compd.* **2000**, *309*, 225–228.

134. J. Yang, T. Aizawa, A. Yamamoto, T. Ohta, *Mater. Chem. Phys.* **2001**, *70*, 90–94.

135. N. Miyashita, T. Yano, R. Tsukuda, I. Yashima, *J. Ceram. Soc. Jpn.* **2003**, *111*, 386–390.

136. H. J. Im, D. H. Kim, T.Mitani, K. C. Je, *Jpn. J. Appl. Phys.* **2004**, *43*, 1094–1099.

137. X. Fan, J. Yang, R. Chen, W. Zhu, S. Bao, *Mater. Sci. Eng. A* **2006**, *438–440*, 190–193.

138. X. Zhao, S. Yang, Y. Cao, J. Mi, Q. Zhang, T. Zhu, *J. Electron. Mater.* **2009**, *38*, 1017–1024.

139. H. C. Kim, T. S. Oh, D. B. Hyun, *J. Phys. Chem. Solids* **2000**, *61*, 743–749.

140. H. Ni, X. Zhao, T. Zhu, X. Ji, J. Tu, *J. Alloy Compd.* **2005**, *397*, 317–321.

141. X. Tang, W. Xie, H. Li, W. Zhao, Q. Zhang, M. Niino, *Appl. Phys. Lett.* **2007**, *90*, 012102.

142. O. B. Sokolov, S. Y. Skipidarov, N. I. Duvankov, *J. Cryst. Growth* **2002**, *236*, 181.

143. M. Carle, P. Pierrat, C. Lahalle-Gravier, S. Scherrer, H. Scherrer, *J. Phys. Chem. Solids* **1995**, *56*, 201.

144. R. Yang, G. Chen, *Phys. Rev. B* **2004**, *69*, 195316.

145. O. Ben-Yehuda, Y. Gelbstein, Z. Dashevsky, R. Shuker, M. P. Dariel, *Proc. of the 25th ICT* **2006**, 492.

146. L. D. Zhao, B.-P. Zhang, J.-F. Li, H. L. Zhang, W. S. Liu, *Solid State Sci.* **2008**, *10*, 651.

147. X. Yan, B. Poudel, Y. Ma, W. S. Liu, G. Joshi, H. Wang, Y. C. Lan, D. Z. Wang, G. Chen, Z. F. Ren, *Nano Lett.* **2010**, *10*, 3373–3378.

148. J. Jiang, L. D. Chen, S. Q. Bai, Q. Yao, Q. Wang, *Mater. Sci. Eng. B* **2005**, *117*, 334.

149. S. J. Hong, Y. S. Lee, J. W. Byeon, B. S. Chun, *J. Alloys Compd.* **2006**, *414*, 146.

150. J. T. Im, K. T. Hartwig, J. Sharp, *Acta Mater.* **2004**, *52*, 49.

151. T. S. Kim, B. S. Chun, *Mater. Sci. Forum* **2007**, *161*, 534–536.

152. W. J. Xie, X. F. Tang, G. Chen, Q. Jin, Q. J. Zhang, *Proc. of the 26th ICT* **2007**, 23.

153. Z. F. Ren, J. H. Wang, D. J. Miller, K. C. Goretta, *Physica C* **1994**, *229*, 137.

154. I. Yashima, H. Watanave, T. Ogisu, R. Tsukuda, S. Sato, *Jpn. J. Appl. Phys.* **1998**, *37*, 2472–2473.

155. J. Seo, K. Park, D. Lee, C. Lee, *Mater. Sci. Eng. B* **1997**, *49*, 247–250.

156. G. A. Slack, M. A. Hussain, *J. Appl. Phys.* **1991**, *70*, 2694–2718.

157. B. Abeles, *Phys. Rev.* **1963**, *131*, 1906–1911.

158. N. K. Abrikosov, V. S. Zemskov, E. K. Iordanishvili, A. V. Petrov, V. V. Rozhdestvenskaya, *Sov. Phys. Semicond.* **1968**, *2*, 1762.

159. J. P. Dismukes, L. Ekstrom, E. F. Steigmeier, I. Kudman, D. S. Beers, *J. Appl. Phys.* **1964**, *35*, 2899–2907.

160. C. M. Bhandari, D. M. Rowe, *Contemp. Phys.* **1980**, *21*, 219–242.

161. M. Abdellaoui, E. Gaffet, *J. Phys. IV France* **1994**, *4*, C3-291–C3-296.

162. E. Gaffet, L. Yousfi, *Mater. Sci. Forum* **1992**, *88–90*, 51–58.

163. H. Bakker, L. M. Di, *Mater. Sci. Forum* **1994**, *88–90*, 27–34.

164. D. M. Rowe, *J. Phys. D: Appl. Phys.* **1974**, *7*, 1843–1846.

165. S. K. Bux, R. G. Blair, P. K. Gogna, H. Lee, G. Chen, M. S. Dresselhaus, R. B. Kaner, J. P. Fleurial, *Adv. Funct. Mater.* **2009**, *19*, 2445–2452.

166. B. C. Sales, D. Mandrus, R. K. Williams, *Science* **1996**, *272*, 1325–1328.

167. J. L. Mi, X. B. Zhao, T. J. Zhu, J. P. Tu, *J. Phys. D: Appl. Phys.* **2008**, *41*, 205403.

168. J. L. Mi, X. B. Zhao, T. J. Zhu, J. Ma, *J. Alloys Compd.* **2008**, *452*, 225–229.

169. J. L. Mi, X. B. Zhao, T. J. Zhu, J. P. Tu, *Appl. Phys. Lett.* **2007**, *91*, 172116.

170. H. Li, X. Tang, Q. Zhang, *J. Electron. Mater.* **2009**, *38*, 1224–1228.

171. X. F. Tang, H. Li, Q. J. Zhang, M. Niino, T. Goto, *J. Appl. Phys.* **2006**, *100*, 123702.

172. G. S. Nolas, J. L. Cohn, G. A. Slack, *Phys. Rev. B* **1998**, *58*, 164–170.

173. Y. Z. Pei, J. Yang, L. D. Chen, W. Zhang, J. R. Salvador, J. Yang, *Appl. Phys. Lett.* **2009**, *95*, 042101.

174. S. Q. Bai, Y. Z. Pei, L. D. Chen, W. Q. Zhang, X. Y. Zhao, J. Yang, *Acta Mater.* **2009**, *57*, 3135–3139.

175. Y. Z. Pei, S. Q. Bai, X. Y. Zhao, W. Zhang, L. D. Chen, *Solid State Sci.* **2008**, *10*, 1422–1428.

176. X. Y. Zhao, X. Shi, L. D. Chen, W. Q. Zhang, S. Q. Bai, Y. Z. Pei, X. Y. Li, T. Goto, *Apply. Phys. Lett.* **2006**, *89*, 092121.

177. X. F. Tang, Q. J. Zhang, L. D. Chen, T. Goto, T. Hirai, *J. Appl. Phys.* **2005**, *97*, 093712.

178. C. Uher in *Advances in Thermoelectric Materials I*, Vol. 69 (Ed.: T. Tritt), Academic Press, New York **2001**, pp. 139–253.

179. Z. G. Mei, J. Yang, Y. Z. Pei, W. Zhang, L. D. Chen, J. Yang, *Phys. Rev. B* **2008**, *77*, 045202.

180. X. Shi, L. D. Chen, S. Q. Bai, X. Y. Huang, X. Y. Zhao, Q. Yao, U. Uher, *J. Appl. Phys.* **2007**, *102*, 103709.

181. X. Y. Zhao, X. Shi, L. D. Chen, W. Q. Zhang, W. B. Zhang, Y. Z. Pei, *J. Appl. Phys.* **2006**, *99*, 053711.

182. X. Y. Li, L. D. Chen, J. F. Fan, W. B. Zhang, T. Kawahara, T. Hirai, *J. Appl. Phys.* **2005**, *98*, 083702.

183. X. F. Tang, L. D. Chen, J. Wang, Q. J. Zhang, T. Goto, T. Hirai, *J. Alloys Compd.* **2005**, *394*, 259–264.

184. T. Caillat, A. Borshchevsky, J. P. Fleurial, *J. Appl. Phys.* **1996**, *80*, 4442–4449.

185. B. Chen, J. H. Xu, C. Uher, D. T. Morelli, G. P. Meisner, J. P. Fleurial, T. Caillat, A. Borshchevsky, *Phys. Rev. B* **1997**, *55*, 1476–1480.

186. V. Keppens, D. Mandrus, B. C. Sales, B. C. Chakoumakos, P. Dai, R. Coldea, M. B. Maple, D. A. Gajewski, E. J. Freeman, S. Bennington, *Nature* **1998**, *395*, 876–878.

187. M. Puyet, B. Lenoir, A. Dauscher, M. Dehmas, C. Stiewe, E. Muller, *J. Appl. Phys.* **2004**, *95*, 4852–4855.

188. L. D. Chen, T. Kawahara, X. F. Tang, T. Goto, T. Hirai, J. S. Dyck, W. Chen, C. Uher, *J. Appl. Phys.* **2001**, *90*, 1864–1868.

189. G. S. Nolas, M. Kaeser, R. T. Littleton IV, T. M. Tritt, *Appl. Phys. Lett.* **2000**, *77*, 1855–1857.

190. H. Y. Geng, S. Ochi, J. Q. Guo, *Appl. Phys. Lett.* **2007**, *91*, 022106.

191. J. Yang, Y. Chen, J. Peng, X. Song, W. Zhu, J. Su, R. Chen, *J. Alloy Compd.* **2004**, *375*, 229–232.

192. J. Yang, Y. Chen, W. Zhu, S. Bao, J. Peng, X. Fan, *J. Phys. D: Appl. Phys.* **2005**, *38*, 3966–3969.

193. S. Bao, J. Yang, X. Song, J. Peng, W. Zhu, X. Fan, X. Duan, *Mater. Sci. Eng. A* **2006**, *438–440*, 186–189.

194. S. C. Ur, I. H. Kim, *Mater. Lett.* **2002**, *57*, 543–551.

195. X. Shi, H. Kong, C. P. Li, C. Uher, J. Yang, J. R. Salvador, H. Wang, L. Chen, W. Zhang, *Appl. Phys. Lett.* **2008**, *92*, 182101.

196. S. C. Ur, J. C. Kwon, I. H. Kim, *J. Alloy Compd.* **2007**, *442*, 358–361.

197. H. Li, X. F. Tang, Q. J. Zhang, C. Uher, *Appl. Phys. Lett.* **2008**, *93*, 252109.

198. S. Bao, J. Yang, W. Zhu, X. Fan, X. Duan, J. Peng, *Mater. Lett.* **2006**, *60*, 2029–2032.

199. S. Bao, J. Yang, J. Peng, W. Zhu, X. Fan, X. Song, *J. Alloy Compd.* **2006**, *421*, 105–108.

200. B. C. Sales, D. Mandrus, B. C. Chakoumakos, V. Keppens, J. R. Thompson, *Phys. Rev. B* **1997**, *56*, 15081–15089.

201. S. Bao, J. Yang, W. Zhu, X. Fan, X. Duan, *J. Alloy Compd.* **2009**, *476*, 802–806.

202. X. Song, J. Yang, J. Peng, Y. Chen, W. Zhu, T. Zhang, *J. Alloys Compd.* **2005**, *399*, 276–279.

203. S. C. Ur, J. C. Kwon, I. H. Kim, *Met. Mater. Int.* **2008**, *14*, 625–630.

204. C. Uher, J. Yang, S. Hu, D. T. Morelli, G. P. Meisner, *Phys. Rev. B* **1999**, *59*, 8615.

205. S. J. Poon, *Recent Trends in Thermoelectric Materials Research II*, Semiconductors and Semimetals, edited by T. M. Tritt (Academic, New York, **2001**), 70, 37.

206. W. Jeischko, *Metall. Trans.* **1970**, *1*, 3159.

207. Y. Xia, S. Bhattacharya, V. Ponnambalam, A. L. Pope, S. J. Poon, T. M. Tritt, *J. Appl. Phys.* **2000**, *88*, 1952.

208. T. Sekimoto, K. Kurosaki, H. Muta, S. Yamanaka, *J. Alloys Compd.* **2006**, *407*, 326.

209. T. Wu, W. Jiang, X. Y. Li, Y. F. Zhou, L. D. Chen, *J. Appl. Phys.* **2007**, *102*, 103705.

210. V. Ponnambalam, P. N. Alboni, J. Edwards, T. M. Tritt, S. R. Culp, S. J. Poon, *J. Appl. Phys.* **2008**, *103*, 063716.

211. S. R. Culp, J. W. Simonson, S. J. Poon, V. Ponnambalam, J. Edwards, T. M. Tritt, *Appl. Phys. Lett.* **2008**, *93*, 022105.

212. J. W. Sharp, S. J. Poon, H. J. Goldsmid, *Phys. Status Solidi* **2001**, *187*, 507.

213. T. M. Tritt, S. Bhattacharya, Y. Xia, V. Ponnambalam, S. J. Poon, N. Thadhani, *ICT'01*, **2001**, 7.

214. S. Katsuyama, T. Kobayashi, *Mater. Sci. Eng., B* **2010**, *166*, 99.

215. W. J. Xie, X. F. Tang, Q. J. Zhang, *Chin. Phys.* **2007**, *16*, 3549.

216. X. Yan, G. Joshi, W. S. Liu, Y. C. Lan, H. Wang, S. Lee, J. W. Simonson et al., *Nano Lett.* **2011**, *11*, 556–560.

217. W. S. Liu, B.-P. Zhang, J.-P. Li, H.-L. Zhang, L.-D. Zhao, *J. Appl. Phys.* **2007**, *102*, 103717.

218. W. Xu, L. Li, R. Huang, M. Zhou, L. Zheng, L. Gong, C. Song, *Front. Energy Power Eng. China* **2009**, *3*, 90–93.

219. R. J. Huang, L. F. Li, W. Xu, L. H. Gong, *Solid State Commun.* **2009**, *149*, 1633–1636.

220. J. Cui, X. Liu, W. Yang, D. Chen, H. Fu, P. Ying, *J. Appl. Phys.* **2009**, *105*, 063703.

221. M. Ito, T. Tada, S. Katsuyama, *J. Alloy Compd.* **2003**, *350*, 296–302.

222. M. Ito, T. Tada, S. Hara, *J. Alloy Compd.* **2006**, *408-412*, 363–367.

223. K. Cai, E. Mueller, C. Drasar, C. Stiewe, *Solid State Commun.* **2004**, *131*, 325–329.

224. K. F. Cai, C. W. Nan, Y. Paderno, D. S. McLachlan, *Solid State Commun.* **2000**, *115*, 523–526.

225. J. S. Lee, T. S. Oh, D. B. Hyun, *J. Mater. Sci.* **2000**, *35*, 881–887.

226. G. J. Snyder, M. Christensen, E. Nishibori, T. Caillat, B. B. Iversen, *Nat. Mater.* **2004**, *3*, 458–463.

227. C. Chiritescu, D. G. Cahill, N. Nguyen, D. Johnson, A. Bodapati, P. Keblinski, P. Zschack, *Science* **2007**, *315*, 351–353.

228. A. S. Henry, G. Chen, *J. Comput. Theor. Nanosci.* **2008**, *5*, 141–152.

229. J. Callaway, *Phys. Rev.* **1959**, *113*, 1046–1051.

230. E. F. Steigmeier, B. Abeles, *Phys. Rev.* **1964**, *136*, A1149–A1155.

231. Z. Wang, J. E. Alaniz, W. Jang, J. E. Garay, C. Dames, *Nano Lett.* **2011**, *11*, 2206–2213.

232. D. T. Morelli, V. Jovovic, J. P. Heremans, *Phys. Rev. Lett.* **2008**, *101*, 035901.

233. P. L. Kapitza, *J. Phys. (USSR)* **1941**, *4*, 181.

234. E. Schwartz, R. Pohl, *Rev. Mod. Phys.* **1989**, *61*, 605–668.

235. G. Chen, *J. Heat Transfer* **1997**, *119*, 220–229.

236. G. Chen, T. Zeng, *Microscale Thermophys. Eng.* **2001**, *5*, 71–88.

237. R. Prasher, *Int. J. Heat Mass Trans.* **2005**, *48*, 4942–4952.

238. H. Zhong, J. R. Lukes, *Phys. Rev. B* **2006**, *74*, 125403-1–125403-10.

239. A. Minnich, G. Chen, *Appl. Phys. Lett.* **2007**, *91*, 073105-1–073105-3.

240. A. Khitun, K. L. Wang, G. Chen, *Nanotechnology* **2000**, *11*, 327–331.

241. Z. Neil, R. L. Jennifer, *Phys. Rev. B* **2008**, *77*, 094302.

242. G. A. Slack in *Solid State Physics*, Vol. 34 (Eds.: F. Seitz, D. Turnbull, H. Ehrenreich), Academic Press, New York, **1979**, pp. 1–71.

243. D. G. Cahill, S. K. Watson, R. O. Pohl, *Phys. Rev. B* **1992**, *46*, 6131–6140.

244. G. Chen in *Phonon Transport in Low-dimensional Structures,* Vol. 71 (Ed.: T. M. Tritt), Academic Press, San Diego, CA, **2001**, pp. 203–259.

245. A. J. Minnich, M. S. Dresselhaus, Z. F. Ren, G. Chen, *Energy Environ. Sci.* **2009**, *2*, 466–479.

246. A. J. Minnich, H. Lee, X. W. Wang, G. Joshi, M. S. Dresselhaus, Z. F. Ren, G. Chen, D. Vashaee, *Phys. Rev. B* **2009**, *80*, 155327.

247. M. Zebarjadi, G. Joshi, G. H. Zhu, B. Yu, A. Minnich, Y. C. Lan, X. W. Wang, M. Dresselhaus, Z. F. Ren, G. Chen, *Nano Lett.* **2011**, *11*, 2225–2230.

248. R. Dingle, H. L. Störmer, A. C. Gossard, W. Wiegmann *Appl. Phys. Lett.* **1978**, *33*, 665–667.

249. Y. Lan, A. J. Minnich, G. Chen, and Z. *Ren Adv. Func. Mater.* **2010**, *20*, 357–376. Wiley-VCH Verlag GmbH & Co. KGaA.

2

Design and Realization of Nanostructured Inorganic Intergrowths

M. Beekman
University of Oregon

C. Heideman
University of Oregon

M. Anderson
University of Oregon

M. Smeller
University of Oregon

R. Atkins
University of Oregon

Q. Lin
University of Oregon

N. Nguyen
University of Oregon

D.C. Johnson
University of Oregon

2.1 Introduction

Recent experimental and theoretical advances in the field of thermoelectric (TE) materials research have resulted in the proven ability to achieve unprecedented TE properties via nanostructured approaches [1–6]. The potential to decouple interrelated materials properties is a key opportunity offered by nanostructured approaches to TE materials development. While significant progress has been made toward addressing the long-standing challenge of identifying higher performance TE materials, outlined in a number of recent reviews [7–13] including several chapters within this Handbook, the scientific understanding of and control over the synthesis, structure, and physical properties of nanostructured TE materials continues to be developed. New synthetic methods are required that allow access to material systems in which composition and nanostructure can be precisely controlled, such that the mechanisms for TE enhancement can be studied, understood, and applied. Toward this aim, we have developed an approach for preparing new classes of nanostructured intergrowth compounds and nanolaminates via self-assembly from elemental nanolaminate precursors. In this chapter, we describe the principles and implementation of the synthetic approach, review the unusual structural features of the intergrowths that are formed, and discuss the current understanding of the transport properties in some of these materials.

2.2 Synthetic Approach

Metastable superlattice structures formed from components with epitaxial relationships have been grown by various techniques, including molecular beam epitaxy, chemical vapor deposition, and solid- and liquid-phase epitaxy. These layer-by-layer deposition techniques create heterostructures by controlling surface reaction kinetics and species mobilities through control of reactant fluxes and substrate temperature *in situ* during the growth process. Using these methods, high-quality heterostructures can be produced. Material systems that are accessible to these techniques are typically limited to components with epitaxial relationships.

There are also thermodynamically stable nanostructured solids that consist of two structures or components interleaved on a nanometer or subnanometer length scale, often with a structural misfit between the constituent layers. One of the first such examples was discovered by Cowley and Ibers in 1956 when they examined the structure of $FeCl_3$ intercalated graphite [14]. While several different classes of misfit compounds have been discovered, one of the most extensively studied ones consists of intergrowths of monochalcogenides with distorted rock salt structures and layered transition metal dichalcogenides, formulated as $[(MX)_{1+\delta}]_m[TX_2]_n$ [15,16]. Approximately 100 of these misfit-layered compounds have been prepared by conventional synthesis techniques, where M is either a divalent or trivalent metal (M = rare earth, Sn, Pb, Bi, or Sb), T is an early transition metal (T = Ti, V, Cr, Nb, or Ta), X is either sulfur or selenium, and m and n denote the number of contiguous units (i.e., layers) of each substructure constituent [15,16]. For the vast majority of misfit-layered compounds prepared by traditional synthesis techniques, m and n have an empirical upper bound of 1 or 2 [16]. The quantity $1 + \delta$ represents the extent of the misfit between the in-plane lattice parameters of the components, where δ is typically between 0.08 and 0.23 [16]. The surprising stability of the known misfit compounds is thought to be a consequence of electron transfer from the donor MX layer to the TX_2 acceptor layer [15,16] (with analogy to modulation doping or intercalation compounds) and/or the mixing of cations between the constituents [17], but this intriguing question still stands without a well-established and universally accepted explanation.

The synthesis of misfit-layered compounds with larger values of m and n is historically not accomplished using traditional synthetic procedures, presumably due to small but significant free energy differences between family members with different m and n values. The traditional synthesis approach [16], in which stoichiometric amounts of the constituent elements are combined in a sealed ampoule and typically annealed at temperatures between 700°C and 1000°C for several days, provides little control of the reaction pathway and little ability to select for potential products. To obtain single crystals, trace amounts of iodine or another vapor transport agent can be added to the charge, which is annealed to promote crystal growth either using chemical vapor transport or flux methods [16]. Although a diverse collection of compositions have been prepared by traditional synthetic routes, they have several limitations. Typically only one or two members, where $m, n \leq 2$, of a potentially infinite family of compounds can be prepared, for example, $[(EuS)_{1.15}]_1[NbS_2]_1$ and $[(EuS)_{1.15}]_{1.5}[NbS_2]_1$. When the synthesis of other members is attempted, the result is typically a mixture of different family members and/or the parent binary compounds are obtained. Often, the stoichiometry of the resulting compound does not correspond to the composition of the mixture used to prepare it [18].

To access higher-order members of intergrowth compounds such as these, our laboratory has developed a different synthetic approach, whereby annealing layered, elemental nano-laminate precursors at low temperatures kinetically traps metastable intergrowth compounds. Our approach is guided by a relatively simple design principle: Presuming the interface between them is stable, two different components interleaved on a nanometer length scale should reside at local free energy minima and hence may be at least kinetically stable. This design principle is based on the hypothesis that fulfilling the local bonding requirements of the atoms in each constituent may be sufficient to stabilize the intergrowth in spite of the interfaces between the constituents. A reactant is designed that closely approximates the desired final structure, or, more precisely, the spatially modulated compositional waveform of the desired final product, such that upon low-temperature annealing it self-assembles into the targeted compound. Figuratively, the reactant sits in the free energy cone above the desired compound and is diffusionally constrained to lower its free energy by self-assembling into the targeted structure. We have initially tested this design principle by capitalizing on known phase relationships and systems possessing stable interfaces such as the misfit compounds introduced above and discussed in more detail subsequently. Remarkably, entire families of kinetically stable compounds have been isolated.

It is useful at this point to define the terms we use to describe our precursors and the final products. The precursor is a compositionally modulated multilayer thin film prepared by sequentially evaporating the constituent elements [19]. A fragment of such a precursor is shown schematically in Figure 2.1. The precursor is built from a periodic repeating unit with a compositional wavelength, λ_c. We refer to the compositional profile within the wavelength of the repeating unit, that is, the average position dependence of atomic concentrations, as the compositional waveform, ϖ. These precursors are intended to have precisely the needed elemental compositions in a spatial arrangement that is very close to the desired product, which forms upon gentle annealing via a self-assembly process. In this respect, very little solid-state diffusion is required. This key aspect of this approach inhibits the formation of other more kinetically or thermodynamically stable products.

Given that the intergrowth component thicknesses are of the order of several angstroms, the practical challenge in this synthesis approach is preparing precursors where the compositional waveform in the deposited precursor contains precisely the correct number of atoms in each spatial region to self-assemble into the repeating structural units of the constituents. In our laboratory, films are deposited with custom-built high-vacuum (10^{-7}–10^{-8} Torr) deposition chambers [20], using elemental sources evaporated from electron beam guns or effusion cells, depending on the melting point and vapor pressure of the particular elemental source. Constant deposition rates are maintained by feedback loops employing quartz crystal rate monitors, while the source exposure and sequence of layers are controlled by computer-activated shutters. As an example, the steps necessary to synthesize a family of intergrowths with general chemical formula $[(MSe)_{1+\delta}]_m[TSe_2]_n$ from a precursor with the composition $[(MSe_x)_y]_p[TSe_z]_q$ are outlined below. Here, p and q are the number of sequential M–Se and T–Se layers in ϖ, respectively, and m and n are the number of MSe and TSe_2 layers in the intergrowth compound repeating unit. Depending on the nucleation and diffusion characteristics of the material system, p and q may or may not be chosen to be equal to m and n, respectively.

The first step in the procedure is to calibrate the deposition process to obtain the correct ratio of the elements for each individual component of the desired intergrowth. This can be accomplished by making six multilayer specimens: three containing a repeating unit with a fixed thickness of element "M" and varying thickness of Se, and three with a fixed thickness of Se and varying thickness of "M." Low angle (typically less than $2\theta = 10°$ with Cu Kα radiation) x-ray diffraction of these specimens yields the repeat layer thicknesses via the position of the Bragg reflections resulting from the periodic electron density in the otherwise amorphous multilayer [19]. Plotting the measured thickness of each set of three specimens against the intended thickness of the layer that is varied yields the thickness of the

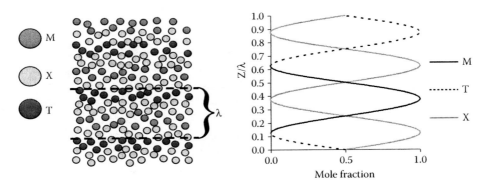

FIGURE 2.1 A schematic (left) showing three repeating sequences of a precursor of amorphous elemental layers designed to form $[(MSe)_{1+\delta}]_1[TSe_2]_1$. The precursor compositional wavelength is denoted λ and the components of the compositional waveform, ϖ, are shown on the right.

component layers. This is used to calculate an effective tooling factor which quantifies the relationship between the nominal and actual layer thicknesses for the particular material system and deposition chamber. Electron probe microanalysis (EPMA) data collected as a function of accelerating voltage are used to determine the compositions of the films [21]. The linear dependence of the atomic ratio on the ratio of deposited layer thicknesses yields an estimate for the latter that is required to obtain the desired composition. This information is used to create three specimens with compositions close to the stoichiometry of the desired component compound with a range of binary layer thicknesses, that is, different values of λ_c for the binary. These specimens are then annealed at low temperatures to yield the desired crystalline compound, obtaining both processing information and the amount the film thickness decreases upon crystallization of the constituent. The value of λ_c required to form an amorphous intermediate and avoid interfacial nucleation of competing compounds can also be determined in this step [19]. Preparing specimens with λ_c below this value permits the desired constituent to be formed directly from the precursor. By determining the annealed film thickness, it is possible to estimate the value of λ_c required to form a single unit cell of the desired compound. This process is then repeated for the second constituent of the targeted intergrowth, in this example a transition metal dichalcogenide TSe_2.

The second step in the process is to prepare two sets of three specimens each that contain interleaved elemental layers designed to evolve into the desired constituent compounds. In this step, the deposition parameters needed to synthesize the intergrowth are further refined. An effective procedure is to prepare three $[(MSe_x)_y]_p[TSe_z]_q$ films with each pair of elements (M–Se and T–Se) deposited in the ratio determined in the first step, where either p or q is held fixed at 1 while the other is varied. The precursors can be prepared by repeating the deposition sequence illustrated in Figure 2.2b for the case $[(PbSe_x)_y]_p[MoSe_z]_1$. The slope of a linear fit in a plot of the repeat unit thickness (determined from x-ray

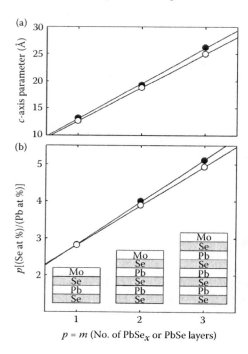

FIGURE 2.2 (a) C-axis parameter for the repeating unit as determined from x-ray diffraction for as-deposited $[(PbSe_x)_y]_m[MoSe_z]_n$ precursors (•) and the resulting $[(PbSe)_{1.00}]_m[MoSe_2]_n$ misfit-layered compounds self-assembled by annealing at 400°C (○), with $(m,n) = (1,1), (2,1),$ and $(3,1)$. (b) Chemical composition as determined by electron microprobe analysis for the same series of specimens. The insets illustrate the deposition sequences used to form the respective compositional waveforms, ϖ.

diffraction), that is, the corresponding λ_c against p for of the first set of films yields the thickness of the Pb–Se contribution to the total as deposited repeat thickness, λ_c, and for the second set (plotting against q) the thickness of the T–Se contribution. This is illustrated in Figure 2.2a for the case of changing $p = m$ in $[(PbSe_x)_y]_p[MoSe_z]_1$, indicating an amorphous $PbSe_x$ layer thickness of 0.651 nm. This is slightly larger than the resulting crystallized PbSe layer thickness of 0.617 nm, obtained from the analogous data for $[(PbSe)_{1.00}]_m[MoSe_2]_1$. EPMA data are collected from these series of specimens, from which the composition of the respective components and their ratios, that is, x, y, and z, can be determined by appropriately plotting mole ratios of the elements against p, q, n, or m. As illustrated in Figure 2.2b, p times the Se to Pb atomic ratio is linearly related to x. The fits yield $x = 1.14$ for these precursors (i.e., slightly Se rich), and $x = 1.06$ in the annealed product, in agreement with the stoichiometric composition PbSe within experimental error. The thickness and composition information gained from these steps are used to make any needed adjustments to the amount of material deposited in the respective layers. While not always the case, the precursor is often designed so that $p = m$ and $q = n$, that is, thicknesses and compositions of the T–Se and M–Se precursor layers closely correspond to those required to form a crystallographic unit of the constituent structures upon annealing, yielding the desired $[(MSe)_{1+\delta}]_m[TSe_2]_n$ intergrowths. Small adjustments to the thickness and composition of the M–Se and T–Se layers are typically needed to improve the quality of the self-assembled compounds, which is judged empirically by x-ray diffraction. Often a precursor that is chalcogen rich relative to the final product is required for optimized synthesis, attributed to the relatively high chalcogen vapor pressure resulting in a slight loss of this constituent during annealing. Cross-section TEM images can also be very useful at this stage of the process to confirm that the desired intergrowth is being prepared, with the appropriate m and n values. Once this optimization is completed, any desired family member with specific m and n, including compositional isomers, can in principle be prepared by depositing the elements M and Se m times and depositing T and Se n times in the appropriate sequence, and repeating the required deposition sequence until the desired film thickness is reached. The successful synthesis of a family of compounds, say $[(MSe)_{1+\delta}]_m[TSe_2]_n$, makes the optimization of a new constituent compound, say $T'Se_2$, significantly easier because the optimal deposition thicknesses for the elements M and Se to yield M–Se are already known. This reduces the number of calibration process steps, as only T' and Se need to be optimized to form the new $[(MX)_{1+\delta}]_m[T'X_2]_n$ compound.

The optimal annealing temperature to effectively self-assemble the precursor into the desired intergrowth while avoiding decomposition is determined empirically. Diffraction data collected in the specular geometry as a function of temperature and time (Figure 2.3) allow the increase in order along

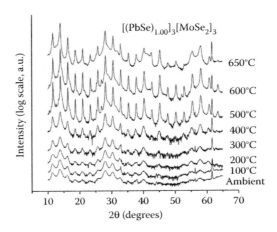

FIGURE 2.3   X-ray diffraction patterns for $[(PbSe)_{1.00}]_3[MoSe_2]_3$ collected after annealing at progressively increasing annealing temperatures.

the axis perpendicular to the substrate to be tracked. This increase in order reflects the formation of the individual constituent structures with abrupt, planar interfaces between them. An increase in the angular range where Kiessig fringes are observed in x-ray reflectivity scans [19] implies an increase in film thickness uniformity, and is directly related to the planar nature of the interfaces, which constrain the film to possess extraordinary uniformity in thickness after self-assembly. Rocking curve scans collected as a function of annealing temperature and time also provide information about the self-alignment of the intergrowth to the substrate with annealing.

2.3 Structural Characteristics

We now discuss in more detail the structural characteristics of new intergrowth compounds that are formed by annealing the precursors described in Section 2.2. We will emphasize that, in spite of the fact that in-plane crystallinity, crystallographic orientation, and atomically abrupt interfaces produce well-defined out-of-plane diffraction patterns with reflections observed to very high order, these materials are not crystals in the conventional sense. Instead, they can be described as uniquely poised between the amorphous and crystalline states. The weak interlayer interaction and difference between the sizes of the crystal lattices of the constituents results in the unusual stoichiometries of these compounds and a layer-to-layer misregistration that limits long-range order. This misregistration results in a random in-plane rotational/translational disorder between adjacent layers. Similar features are referred to as turbostratic disorder in the clay literature. We propose a new term, *ferecrystals*, for what is essentially a new state of solid matter: precisely layered structures, with in-plane "crystallinity" and order, abrupt interfaces, layer-to-layer misregistration, and turbostratic disorder. The term is derived from the Latin *fere*, meaning *almost*. The following paragraphs present conventional and synchrotron x-ray diffraction data and refinement results, transmission electron microscopy images, and describe the structure of several representative ferecrystals containing constituents with different crystal structures.

2.3.1 Transition Metal Dichalcogenide Superlattices

The ability to self-assemble intergrowth compounds from a rationally designed layered elemental precursor was first demonstrated by Noh et al. [22,23], who prepared superlattices of distinct transition metal dichalcogenide compounds. MX_2 compounds (M = Ti, Zr, Hf, V, Nb, Ta, Mo, W; X = S, Se, and Te) form with layered crystal structures, in which the transition metal is covalently bonded to six chalcogen atoms in an octahedral or trigonal prismatic coordination within a single X–M–X trilayer. Due to relatively small free energy differences, various polytypes are observed [24] depending upon the stacking sequence of the X–M–X layers, which interact through weaker van der Waals forces. Small adjustments to reaction conditions are traditionally used to control the formation of particular polytypes. When reaction mixtures containing two or more different transition metals and elemental chalcogen are heated for short times, typically mixtures of the different compounds are formed. After annealing for extended periods, solid solutions are obtained with random distributions of the transition metals in the crystal lattice.

While investigating the reaction mechanism of reactants containing alternating layers of elemental niobium and selenium, Fukuto and coworkers [25] discovered that niobium diselenide preferentially nucleated at the interface between niobium and selenium layers at low temperature even for reactants with sub-nanometer elemental layers in selenium-rich reactants. This finding inspired Noh and coworkers to explore designed reactants with a repeating unit containing alternating titanium, selenium, and niobium layers. The hypothesis was that interfacial nucleation would result in the formation of interwoven layers of titanium diselenide and niobium diselenide at temperatures insufficient to interdiffuse the transition metal cations, which would result in a solid solution as opposed to an intergrowth. After establishing a procedure to control the appropriate amount of the elements required to form a crystallographic unit of each of the constituents, Noh et al. [22,23]successfully prepared a series of new

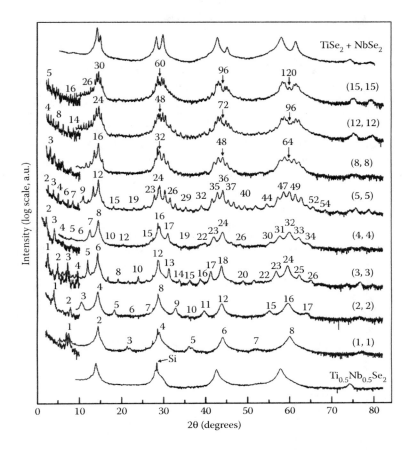

FIGURE 2.4 Diffraction patterns for several niobium diselenide-titanium diselenide ferecrystals, a niobium/titanium diselenide alloy, and a physical mixture of the two constituents.

compounds within the family of compounds $[(TiSe_2)_{1+\delta}]_m(NbSe_2)_n$. Figure 2.4 contains representative x-ray diffraction patterns for several such members, showing $00l$ reflections which can be indexed precisely to the indicated structures. Nguyen et al. [26] demonstrated that this concept could be generally applied to other transition metal dichalcogenides and refined the synthesis procedure to prepare the first 16 members of the $[(VSe_2)_{1.06}]_m(TaSe_2)_n$ family of compounds, where n and m are less than or equal to four. The enhanced scattering contrast between V and Ta enabled the extent of interdiffusion between the constituents to be quantified. Minimal cation mixing was observed for short annealing times and low annealing temperatures, whereas at higher temperatures it is anticipated that the entropy of mixing would drive the interdiffusion of the cations.

2.3.2 Bi_2Te_3-Based Intergrowths

Harris et al. [27,28] and subsequently Mortensen et al. [29] targeted the synthesis and characterization of $(Bi_{2-x}Sb_xTe_3)_m(TiTe_2)_n$ ferecrystals, and showed that intergrowths could be kinetically trapped containing approximately 1 nm-thick Te–M–Te–M–Te (M = Sb, Bi) structural units interleaved with titanium telluride trilayers Te–Ti–Te. To our knowledge, there was no previous precedent for the intergrowth of these two structure types. Regular increases in period consistent with the formation of the desired ferecrystals were observed along the c-axis, while a–b plane diffraction patterns could be indexed as the sum of the diffraction patterns of each of the binary constituents, with unit cell parameters that did not vary as m or n was changed in either the $(Bi_2Te_3)_m(TiTe_2)_n$ or $(Sb_2Te_3)_m(TiTe_2)_n$ family of compounds. Ferecrystals containing a $Bi_{2-x}Sb_xTe_3$ alloy intergrown with $TiTe_2$ can further be prepared by depositing separate elemental Bi and Sb layers, as these elements readily interdiffuse at low temperatures. All diffraction maxima observed from in-plane x-ray diffraction scans can be indexed as $hk0$ reflections coming from $Bi_{2-x}Sb_xTe_3$ alloy and $TiTe_2$ components; other mixed hkl reflections are not observed. The calculated a-lattice parameters of $TiTe_2$ (3.77 Å) and $(Bi_{0.5}Sb_{0.5})_2Te_3$ (3.28 Å) agree very well with previously published lattice parameters for these compounds, reflecting the independent nature of the component structures. Cross-sectional TEM images [27,28] clearly showed the formation of the different crystallographic subunits with the period determined by the design of the initial precursor, in agreement with the c-lattice parameters determined from the $00l$ diffraction data. These ferecrystals were not very kinetically stable, decomposing to a mixture of the binary constituents at temperatures above 300°C.

2.3.3 $[(MX)_{1+\delta}]_m[TX_2]_n$ Misfit-Layered Compounds

The discovery that multicomponent-layered intergrowths can be readily self-assembled from appropriately designed elemental nanolaminate precursors prompted the exploration of more complex materials systems that may be "engineered" by controlling intergrowth nanostructure. The misfit-layered compounds introduced above, with general chemical formula $[(MX)_{1+\delta}]_m[TX_2]_n$, comprise layered intergrowths of transition metal dichalcogenides (TX_2, e.g., TiS_2, $NbSe_2$, $TaSe_2$) with group 14, pnicogen, or rare earth monochalcogenides (MX, e.g., PbSe, BiS, LaS). The subscripts m and n denote the number of consecutive layers of the respective component along the intergrowth direction. The TX_2 and MX components do not possess an epitaxial relationship. This fact is reflected in the "misfit parameter" δ in the chemical composition, which is calculated from the number density ratio of atoms for MX and TX_2 components projected onto the a–b plane.

The existence of these intergrowths, with respect to macroscopic two-phase mixtures of the component compounds or other ternary phases, is directly linked to the remarkably stable interface between the TX_2 and MX components. Capitalizing on this interface stability, Heideman, Lin, and coworkers [30–32] have shown that entire families of new intergrowths with essentially arbitrary m or n and high kinetic stability can be synthesized from carefully designed elemental nanolaminates. By using the TX_2 and MX components as structural "building blocks," intergrowths with predictable structure and composition can be prepared. Figure 2.5a shows thin film x-ray diffraction collected in specular geometry on a series of $[(PbSe)_{0.99}]_1[WSe_2]_n$ compounds, where n is varied between 1 and 8. Excluding reflections attributed to the substrate and specimen stage, all reflections can be indexed in the $00l$ family of reflections for each specimen. The variation in c-axis lattice parameters for the first 64 members of the $[(PbSe)_{0.99}]_m[WSe_2]_n$ intergrowths, shown in Figure 2.5b, are linearly dependent upon m and n, with slopes and intercepts that agree well with the expected layer thicknesses expected from the corresponding bulk compounds [30–32].

Figure 2.6 contains a low-angle x-ray diffraction pattern from a $[(PbSe)_{0.99}]_1[WSe_2]_1$ specimen. The subsidiary maxima between the Bragg diffraction peaks (Kiessig fringes) result from the interference of x-rays scattered from the top and the bottom of the ferecrystal film. The observation of these Kiessig fringes beyond $2\theta = 25°$ using Cu $K\alpha$ radiation requires that the thickness of the film is uniform to within 0.1 nm [33]. Rocking curve data (not shown) collected at the positions of the Bragg peaks confirmed the near-perfect texture of the specimen, with the c-axis perpendicular to the substrate.

The diffraction patterns shown in Figures 2.5 and 2.6 are indicative of intergrowths with highly precise layering. Since only $00l$ reflections are observed, these diffraction patterns collected in specular geometry only contain information about the periodic electron density along the c-direction (direction perpendicular to the substrate). The brilliance and tunable energy of a synchrotron x-ray source have been important tools to structurally characterize these materials in order to further elucidate the nature of structure and nanostructure [34]. Area diffraction collected from films with a grazing incidence geome-

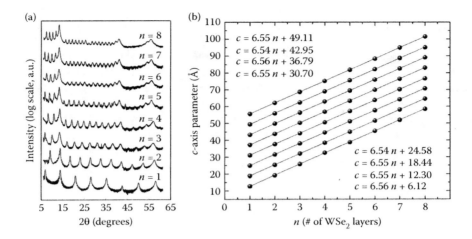

FIGURE 2.5 (a) X-ray diffraction patterns for $[(PbSe)_{0.99}]_1[WSe_2]_n$ thin films, revealing only $00l$ reflections. (b) C-axis parameter for the first 64 $[(PbSe)_{0.99}]_m[WSe_2]_n$ intergrowths, illustrating the ability to prepare all possibilities of $1 \leq m, n \leq 8$. (Reprinted with permission from Q. Lin et al., Rational synthesis and characterization of a new family of low thermal conductivity misfit layer compounds $[(PbSe)_{0.99}]_m(WSe_2)_n$, *Chem. Mater.* **22**, 1002–1009, 2010. Copyright 2010, American Chemical Society.)

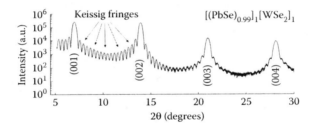

FIGURE 2.6 Low-angle x-ray diffraction from a $[(PbSe)_{0.99}]_1[WSe_2]_1$ thin film.

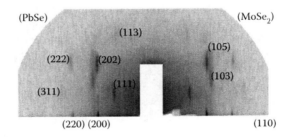

FIGURE 2.7 Synchrotron x-ray diffraction, in grazing incidence geometry, collected from $[(PbSe)_{1.00}]_2[MoSe_2]_2$. The intensity maxima can be independently indexed to the PbSe and $MoSe_2$ components, as indicated on the left and right side of the diffraction pattern, respectively.

try ($\theta \sim 0.6°$), an example of which is shown for a $[(PbSe)_{1.00}]_2[MoSe_2]_1$ film in Figure 2.7, illustrates key features of the nanostructure in these materials. In this experimental geometry, a powder-like specimen would produce rings of intensity, whereas a single-crystal specimen oriented at a diffracting condition would produce localized spots. Neither of these features is observed in the pattern of Figure 2.7, which instead reveals diffuse streaks which can be separately and precisely indexed to the allowed hkl reflections of the individual constituents of the intergrowth, in this case PbSe and $MoSe_2$ as labeled in the figure. No mixed reflections due to the superlattice structure are observed, only reflections expected from PbSe and $MoSe_2$ phases with their "crystallographic" c-axes-oriented perpendicular to the substrate are observed. The diffracted intensity is independent of the rotation of the specimen about the azimuthal angle, φ, indicating a random rotational "powder-like" character of the components within the plane of the layers. These data can be explained by a model in which the individual PbSe and $MoSe_2$ components are oriented with their c-axes along the period of the intergrowth, but possess little to no registry between the individual layers. The mixed hkl reflections originating from diffraction on the components are very broad, appearing as streaks along the 00L direction in reciprocal space, indicative of short coherence lengths and constitutes further evidence of the lack of interlayer registry. The diffracting domain size can be estimated by scanning through these reflections in reciprocal space along appropriate directions and analyzing the resulting peak widths [34]. Typical component in-plane domain sizes are of the order of 5–10 nm, and characteristically a few nanometers larger for the MSe component than for the TSe_2 component [30–32,34]. In-plane diffraction data collected from these materials similarly contain the $hk0$ diffraction peaks expected for both the transition metal dichalcogenide and the rock salt constituent of the structure [32]. The lattice parameters for each constituent do not vary systematically as m and n are varied, consistent with the nonepitaxial relationship between the constituents and weak interlayer interactions.

The need for super-space crystallographic approaches [16] (in general, $1 + \delta$ is an irrational number, making these misfit-layered structures incommensurate) and challenges in preparing single crystals has historically made precise structural determinations of bulk misfit-layered compounds challenging. Such detailed crystal structure determinations or investigations have been reported for less than 20% of the known misfit-layered compositions. The rotational disorder in the new materials discussed above presents additional challenges to extracting detailed structural information from diffraction data. As noted above, out-of-plane diffraction patterns only contain $00l$ reflections, and therefore only contain information about the electron density profile projected onto the c-axis, while no mixed hkl reflections originating from the superlattice structure exist. In-plane diffraction data provide additional information about the individual components, similar to that obtained from area diffraction data presented above. We have used Rietveld analysis in order to extract more detailed information from thin-film diffraction data. A complete exposition of these methods is forthcoming [35], and we focus here only briefly on an approach to Rietveld refinement against $00l$ diffraction data. Utilizing the tunable and highly monochromatic radiation available at beamlines 33-BM and 33-ID of the Advanced Photon Source, diffraction patterns in various geometries are collected using x-ray energies above and below constituent absorption edges, allowing scattering contrast for the element of interest which enhances the ability to differentiate between potential structural models. Since the data only contain information about the z position of the atoms, the model used to describe the $00l$ intensity places the atoms at crystallographic positions $(0, 0, z)$ in space group $P\bar{3}m1$, representing their projection onto the c-axis. The unit cell is constructed to take advantage of symmetry to reduce the number of refined structural parameters. The in-plane misfit of the two components, which is at first glance lost by projection onto the c-axis, is accounted for by the fractional site occupancies. For example, the ratio of the cation site occupancies for the case of $n = m = 1$, assuming full occupancy of all sites and no mixing, would be numerically equal to $1 + \delta$. Figure 2.8 shows representative diffraction patterns (observed, calculated, and difference) for the $[(PbSe)_{0.99}]_1[WSe_2]_1$ intergrowth, obtained from Rietveld refinement [32,34]. The oscillatory features in the difference pattern near the Bragg peak positions originate from the Kiessig fringes, which are not accounted for by traditional Rietveld refinement software (in this case, GSAS [36]) designed for powder x-ray diffraction. By refining against data sets collected above and below the Pb absorption edge, stable refinement of the atomic positions z was achieved, allowing the determination of the extent of structural distortion in the PbSe layer from the ideal rock salt structure [32,34]. The interplane distances in the component structures implied by this model are shown in the inset to Figure 2.8. Information about interlayer bonding can be inferred, suggesting the interlayer interactions in this turbostratic misfit-layered compound are similar in nature to the bulk counterparts.

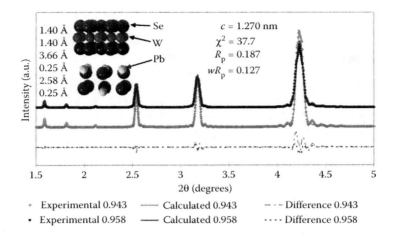

FIGURE 2.8 Observed, calculated, and difference x-ray diffraction patterns obtained from Rietveld refinement for $[(PbSe)_{0.99}]_1[WSe_2]_1$. The inset shows the corresponding model inferred from the structure refinement.

The above interpretation of diffraction data is corroborated by scanning transmission electron microscope images that have been collected from these materials. Figure 2.9 shows a cross-sectional scanning transmission electron microscope high-angle annular dark field (STEM-HAADF, Z contrast) image of a $[(PbSe)_{0.99}]_2[WSe_2]_1$ intergrowth. These data corroborate the above-described model of the nanostructure of the MER-prepared misfit-layered intergrowths in a number of respects. First, the precise layering of the PbSe and WSe_2 components parallel to the substrate is clearly discernable. Starting from the left-hand side of the image, the PbSe layer is in a <100> zone axis orientation, however, progressing to the next adjacent PbSe layer, the same orientation is not observed. This behavior, which continues throughout the imaged region in both the "*c*-axis" direction as well as the in-plane direction, is the turbostratic

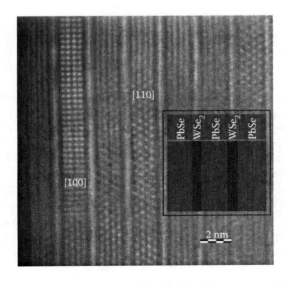

FIGURE 2.9 STEM-HAADF image collected from a $[(PbSe)_{0.99}]_2[WSe_2]_1$ intergrowth. (Reprinted with permission from Q. Lin et al., Rational synthesis and characterization of a new family of low thermal conductivity misfit layer compounds $[(PbSe)_{0.99}]_m(WSe_2)_n$, *Chem. Mater.* **22**, 1002–1009, 2010. Copyright 2010, American Chemical Society.)

disorder revealed by diffraction above. In extensive STEM investigations, we observe TX_2 zone axes in these materials far less frequently than for the MX component, likely due to the smaller in-plane domain size of the TX_2 component which results in more significant orientation averaging through the thickness of the imaged region.

The principles underlying the formation of the unusual disorder present in these materials are not yet well understood. They are likely governed by the nucleation and growth mechanisms that occur during self-assembly, highlighting the need for a better understanding of how the compositionally modulated multilayer precursor evolves into the final product. A reasonable hypothesis is that each constituent layer nucleates and grows independently with a high density of independent nucleation sites within each layer. This may explain both the interlayer rotational disorder and the nanometer in-plane domain sizes within each constituent layer. The remarkable smoothness of the interfaces between layers is also a consequence of the self-assembly process, and suggests that there is a relatively high-energy penalty to pay for a partial or unaligned layer that would result in local roughness.

2.3.4 Skutterudite Superlattices

The facile formation of interleaved layers of dichalcogenides prompted Williams et al. to explore the formation of interleaved layers of compounds with the cubic skutterudite structure [37]. Prior work by Hornbostel et al. [38] showed that M = Fe or Co layers, when interleaved with antimony layers thick enough to make the composition antimony rich relative to the stoichiometry of the desired MSb_3 compound, formed the MSb_3 compounds via an amorphous intermediate when the layering was thinner than 4 nm. Thicker layers initially nucleated MSb_2. Williams discovered that he could form $(CoSb_3)_m(IrSb_3)_n$ superlattices by layering the constituents with layers thinner than 4 nm, but the structure appeared to be coherent across the interfaces. The structure consisted of randomly orientated crystals, with compositional modulation only perpendicular to the substrate, where each crystallographic orientation comprised a different superlattice structure. Analysis of the diffraction data from these films to obtain detailed structural information is exceedingly difficult. These experimental results serve to illustrate the differences in the formation of intergrowths characterized by different chemical bonding: the structure formation of cubic skutterudite supperlattices in random crystallographic orientations contrasts the formation of intergrowths with highly anisotropic bonding such as the transition metal dichalcogenides and misfit-layered compounds, which form with precise crystallographic orientation of the individual constituents relative to the initial precursor. The coherent structure obtained for each grain also suggests that the formation of ferecrystals, characterized by strong rotational misorienation of crystallographical alligned domains, may require a substantial structural mismatch between the constituents, highly anisotropic bonding with relatively weak interaction between the layers, and/or multiple, independent nucleation events for each constituent.

2.4 Thermal Transport Properties

Now we discuss the thermal transport properties of some of the intergrowth compounds that have been synthesized from modulated elemental precursors. The thermal conductivity is one of the most striking physical properties in these materials investigated so far. It was recently demonstrated [39] that the cross-plane thermal conductivity of turbostratically disordered, layered WSe_2 obtains values as low as 0.03 W m^{-1} K^{-1}, two orders of magnitude lower than that for single-crystal WSe_2 and several times lower than the predicted minimum thermal conductivity for this composition based on the concepts proposed by Slack [40] and Cahill [41]. These thermal conductivity values are lower than any previously reported for a fully dense solid and cannot be accounted for by conventional theories of thermal transport in solids [42]. The unusual thermal transport is attributed to the random rotational orientation of precisely layered basal plane units of WSe_2, which effectively reduces long-range order in the cross-plane direction to a length scale comparable to the thickness of a single WSe_2 c-axis unit cell [39]. As a result of the weak interlayer bonding and lack of long-range order between precisely layered sheets [43], extended vibrational modes in the cross-plane direction are not expected [39]. It is to this phonon localization that the ultra-low thermal conductivity is attributed. This interpretation is supported by measurements of the in-plane thermal conductivity of suspended WSe_2 films, which revealed anisotropy by a factor of 30 [44].

Synchrotron x-ray diffraction and transmission electron microscopy investigations on misfit-layered intergrowths, described in Section 2.3.3, suggest that turbostratic disorder may be a general feature of layered compounds synthesized by the method of modulated elemental reactants. The cross-plane thermal conductivities of a number of $[(PbSe)_{0.99}]_m[WSe_2]_n$ and $[(PbSe)_{1.00}]_m[MoSe_2]_m$ compounds have been determined by Chiritescu et al. [45] by time-domain thermoreflectance spectroscopy, and are shown in Figure 2.10. These data reveal extremely low thermal conductivities for all compositions studied.

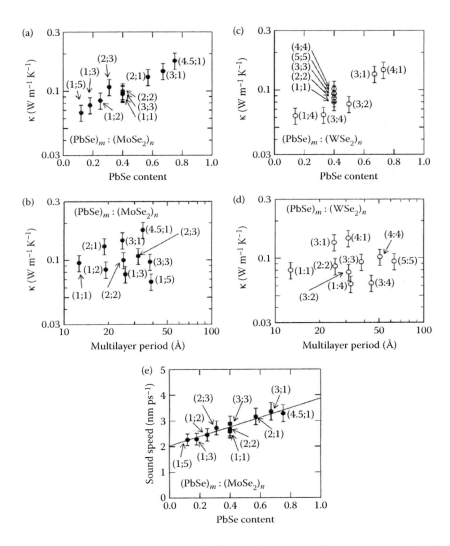

FIGURE 2.10 Cross-plane thermal conductivity of $[(PbSe)_{1.00}]_m[MoSe_2]_n$ thin films as a function of (a) PbSe content and (b) multilayer period. Cross-plane thermal conductivity of $[(PbSe)_{0.99}]_m[WSe_2]_n$ thin films as a function of (c) PbSe content and (d) multilayer period. (e) Measured speed of sound for $[(PbSe)_{1.00}]_m[MoSe_2]_n$. (Reprinted with permission from C. Chiritescu et al., Low thermal conductivity in nanoscale layered materials synthesized by the method of modulated elemental reactants, *J. Appl. Phys.* **104**, 033533, 2008. Copyright 2008, American Institute of Physics.)

The lowest value obtained—0.06 W m^{-1} K^{-1} for [(PbSe)$_{0.99}$]$_1$[WSe$_2$]$_5$—is several times lower than the predicted minimum thermal conductivities of either PbSe and WSe$_2$ based on the bulk properties of those compositions. Interestingly, the thermal conductivity is relatively insensitive to the period of the multilayer. The thermal conductivity does, however, decrease monotonically with increasing TSe$_2$ content, which STEM and synchrotron XRD investigations indicate is the greatest source of disorder in these materials, possessing shorter coherence lengths in the cross-plane direction (cf. Section 2.3.3). Measurements of the cross-plane longitudinal speed of sound in [(PbSe)$_{1.00}$]$_m$[MoSe$_2$]$_n$ films (Figure 2.10e) indicate a softening of the effective lattice occurs with increasing MoSe$_2$, with extrapolated values in agreement with those expected for the pure binary components [45].

Mavrokefalos et al. studied the in-plane thermal conductivity of [(PbSe)$_{0.99}$]$_n$[WSe$_2$]$_n$ (n = 2, 3, and 4) films [46], that is, intergrowths with equal number of PbSe and WSe$_2$ layers. Films were mechanically separated from the substrate, and suspended across a custom measurement microdevice. Remarkably, the in-plane thermal conductivity in this case was found to be nearly three times lower than the in-plane thermal conductivity for WSe$_2$ prepared by the MER approach [44], in spite of the observation that the cross-plane thermal conductivity of the latter is lower than the former [39,45]. The low measured thermal conductivity values for these films, which were found to be relatively independent of both temperature as well as the density of interfaces or c-axis period, and was attributed to interface scattering of phonons. It is expected that a transition at larger layer thicknesses than n = 4 should occur in these materials, whereby interface effects might be reduced.

The ability to synthesize compounds with well-defined and controllable nanostructure presents an opportunity to probe mechanisms for thermal conductivity reduction in TE materials. As discussed in Section 2.3.2, using compositionally modulated precursors, nanostructured Bi$_2$Te$_3$, Sb$_2$Te$_3$, Bi$_{2-x}$Sb$_x$Te$_3$, (Bi$_2$Te$_3$)$_m$[(TiTe$_2$)$_{1.36}$]$_n$, and (Bi$_{2-x}$Sb$_x$Te$_3$)$_m$[(TiTe$_2$)$_{1.36}$]$_n$ specimens have been prepared [29]. In these materials, the effects of grain size, alloying, and turbostratic disorder on the cross-plane thermal conductivity has been systematically studied by independently varying the composition and nanostructure in the specimens. By varying the annealing time and temperature and tracking grain size by x-ray diffraction, single-phase Bi$_2$Te$_3$, Sb$_2$Te$_3$, and Bi$_{2-x}$Sb$_x$Te$_3$ thin films with different average grain sizes are obtained, allowing the influence of grain size on the thermal conductivity to be assessed. Moreover, otherwise unattainable effective grain sizes can be realized in materials such as (Bi$_2$Te$_3$)$_m$[(TiTe$_2$)$_{1.36}$]$_n$ by kinetically "trapping" ultra-thin layers of crystallographically aligned Bi$_2$Te$_3$ between TiTe$_2$ barriers. By using an effective medium approach [29] the contribution of the Bi$_2$Te$_3$ component can be extracted, enabling the study of the effects on thermal conductivity in the c-direction at grain sizes less than 10 nm. These data for Bi$_2$Te$_3$ are shown in Figure 2.11 as a function of effective grain size, along with the predicted dependence based on a Debye–Calloway model [29]. Access to very small effective grain sizes allows such

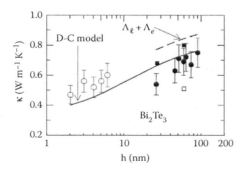

FIGURE 2.11 Cross-plane thermal conductivity of homogeneous nanograin Bi$_2$Te$_3$ (●) and effective thermal conductivity of Bi$_2$Te$_3$ in (Bi$_2$Te$_3$)$_m$[(TiTe$_2$)$_{1.36}$]$_n$ intergrowths (○), as a function of the effective grain size. The solid line is the grain size dependence predicted by a Debye–Calloway model of lattice thermal conductivity.

models to be tested in regimes that are difficult to access due to the challenges in preparing materials with controlled and well-defined nanostructure.

2.5 Electrical Transport Properties

Diverse behavior can be obtained in the electrical properties of intergrowth compounds, making this aspect of these materials of particular interest. For the thermodynamically stable misfit-layered compounds that have been investigated, the interaction between the layered constituents depends on the composition and chemical characteristics of the components, and this determines the resulting electrical transport properties [15,16]. The electrical conduction in $[(MX)_{1+\delta}]_m[TX_2]_n$ compounds has been explained in terms of charge transfer between the components, by regarding these materials as intercalation compounds of the transition metal dichalcogenide, with the MX playing the role of intercalant [15,16]. The available experimental data indicate that the majority of these materials behave as metallic-like conductors. As illustrated in Figure 2.12, the electrical resistivity of a $[(PbSe)_{1.10}]_1[NbSe_2]_1$ thin film prepared by MER exhibits metallic behavior as does the bulk analog [47], and can be understood within a rigid-band model: charge transferred from the PbSe component results in incomplete filling of Nb-derived d-bands, resulting in metallic behavior. Although the majority of the known misfit-layered compounds behave as metallic-like conductors, the MER synthetic route allows access to new compositions possessing characteristic bonding and valence electron counts that are favorable for formation of semiconductors. For example, the bulk parent compounds PbSe and WSe$_2$ are both characterized as semiconductors with filled valence bands, thus a significant amount of inter-layer charge transfer is not expected. In contrast to $[(PbSe)_{1.10}]_1[NbSe_2]_1$, the electrical resistivity of a $[(PbSe)_{0.99}]_1[WSe_2]_1$ thin film prepared by MER (Figure 2.12) shows activated behavior typical of a semiconducting composition. Density functional theory (DFT) calculations within the local density approximation on commensurate approximate structures of $[PbSe]_1[WSe_2]_1$ and $[PbSe]_1[WSe_2]_2$ showed [48] that these materials may be characterized as semiconductors, in which the band gap is increased slightly by increasing n from 1 to 2. By appropriate choice of MX and TX$_2$ components, additional semiconducting compositions can also be prepared [49].

Considering the ability to precisely control the individual thicknesses and therefore also the ratio of both components through the above-described precursor approach, a unique pathway exists to prepare materials with well-defined composition and nanostructure for systematic investigation of the influence on electrical transport. However, a significant challenge encountered in assessing the "intrinsic" effects of controlled nanostructure in metastable materials such as these are separating the "extrinsic" influence of defects, which can contribute significantly to and in some cases dominate the electrical transport properties. The nonequilibrium synthetic route discussed above is highly effective in kinetically trapping an infinite number of intergrowths, but a distribution of structural and composition defects is

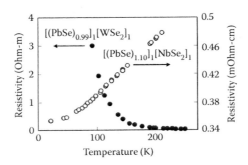

FIGURE 2.12 Temperature-dependent electrical resistivity for $[(PbSe)_{0.99}]_1[WSe_2]_1$ (●, left axis) and $[(PbSe)_{1.10}]_1[NbSe_2]_1$ (○, right axis).

concurrently trapped in the thin film specimens, typically resulting in high carrier concentration in the films upon self-assembly of the intergrowth.

One method that has been used for reducing defects as well as fixing carrier concentrations in bulk single-crystalline specimens can also be applied to thin film specimens such as those discussed here. The method entails prolonged annealing of the specimen under a controlled atmosphere of the most volatile constituent in the material, typically using secondary bulk material of known composition as the vapor source. Given a fixed temperature and sufficient time, the chemical potential of the source material, vapor, and specimen should equilibrate. Thus, the composition of the source can be fixed for bulk or single-crystal specimens, simultaneously reducing all but equilibrium defects. An effective application of this technique was demonstrated by Fleurial et al. [50], who used "saturation annealing" to fix compositions in Bi_2Te_3 single crystals of arbitrary composition by annealing with a source composition near the Bi- or Te-rich solidus lines, such that the liquid, solid, and vapor phases were in equilibrium allowing efficient mass exchange between Bi_2Te_3 source material and single crystals via the vapor phase. A modified low-temperature version of this approach was also shown to be effective in fixing carrier concentrations in nanocrystalline Bi_2Te_3 thin films prepared the MER method [51].

Motivated by the relatively unexplored potential to reduce kinetically trapped defects in kinetically stable films through similar vapor annealing approaches, Lin et al. have investigated the influence of annealing $[(PbSe)_{0.99}]_m[WSe_2]_n$ films under a controlled vapor of selenium using WSe_2 as a vapor source [52]. The effects of annealing time on the Seebeck coefficient, hole concentration, and calculated Hall mobility (calculated using the measured electrical resistivity, not shown) for a semiconducting $[(PbSe)_{0.99}]_1[WSe_2]_1$ film annealed isothermally at 550°C (specimen and WSe_2 source held at the same temperature) are shown in Figure 2.13 as a function of annealing time. The hole concentration decreases and Seebeck coefficient increases with annealing time, until a plateau is reached and the values remain unchanged upon further annealing treatments. Conventional and synchrotron x-ray diffraction collected from $[(PbSe)_{0.99}]_1[WSe_2]_1$ postannealing revealed no change in texture or in-plane domain size. The nearly two orders of magnitude decrease in carrier concentration by the relatively simple annealing process is attributed to the reduction of electrically active compositional and/or structural defects in the film. The vapor annealing is also effectively increased the Hall mobility, which stabilizes above 60 cm² V⁻¹ s⁻¹ in spite of the nanometer-sized domains of the components. These postprocessing annealing methods will likely play an important role in the determination of "intrinsic" properties of these interesting materials. Remarkably, diffraction patterns collected after prolonged annealing on a variety of metastable $[(MX)_{1+\delta}]_m[TX_2]_n$ compositions prepared by MER show no detectable specimen degradation or decomposition, illustrating these nanostructured materials possess remarkable thermal stability.

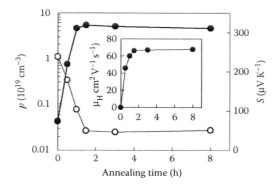

FIGURE 2.13 Hole concentration (○, left axis) and Seebeck coefficient (●, right axis) as a function of annealing time for a $[(PbSe)_{0.99}]_1[WSe_2]_1$ specimen, annealed in a closed volume with bulk WSe_2 powder. Inset: Hall mobility for the same specimen as a function of annealing time (same temporal units as main graph).

2.6 Concluding Remarks

We have presented an overview of the current understanding of the synthesis, structure, and properties of novel classes of solid state intergrowths that are self-assembled from nanolaminate precursors. Structurally, these materials cannot be accurately described as crystalline or amorphous. We propose a new term—"ferecrystals"—which reflects their quasi-ordered and quasi-disordered nature that is nearly crystalline without possessing true long-range order. Remarkably, this unusual balance between order and disorder results in ultralow thermal conductivity values that are unprecedented for fully dense solids, and appears to be a general phenomenon. The microscopic nature of the "lattice" dynamics in these materials is yet to be experimentally examined.

The examples above illustrate that materials with well-defined composition and nanostructure can be prepared by following a relatively simple design principle, whereby stable intergrowths are synthesized by capitalizing on the stable interface between the components. We believe that this approach is not limited to the material systems discussed above, and should in principle be applicable to a wide variety of systems presuming prerequisite phase relationships exist and a stable interface can be realized. An important implication is the possibility of designing materials in which charge transfer or quantum confinement might be used to tune electrical transport, of particular interest in the context of TE materials research.

While we have substantial experimental data demonstrating that the designed precursors with appropriate composition wavelengths and composition profiles self-assemble into the desired ferecrystals with precise values of m and n, we are yet to develop a fundamental understanding of the actual self-assembly process. This includes an understanding of the driving forces that control the orientation between the constituents, as well as the order in which the individual constituents nucleate (e.g., does the second constituent nucleate from the first, or do they nucleate simultaneously together). The answers to basic questions such as these will likely help to guide the design of other novel "ferecrystalline" compounds.

Acknowledgments

The research was funded by National Science Foundation under Grant DMR 0907049 and supported in part by ONR Grant No. N00014-07-1-0358, by the Air Force Research Laboratory under agreement number FA8650-05-1-5041, by the Army Research Laboratory under agreement number W911NF-07-2-0083, and by the National Science Foundation through CCI Grant No. CHE-0847970. CLH and MDA acknowledge support, in part, by the University of Oregon's National Science Foundation IGERT Fellowship Program under Grant No. DGE-0549503. The authors would like to thank Ms Jenia Karapetrova at the Advanced Photon Source for assistance working on the beam line. Research at the Advanced Photon Source was supported by the U.S. Department of Energy, Office of Science, Office of Basic Energy Sciences, under the contract No. W-31-109-ENG-38. The authors also gratefully acknowledge Professor David G. Cahill, Professor Li Shi, Dr. Paul Zschack, Dr. Ian Anderson, and Professor Douglas A. Kezsler for their collaboration and contributions to the work described above.

References

1. R. Venkatasubramanian, E. Siivola, T. Colpitts, and B. O'Quinn, Thin-film thermoelectric devices with high room-temperature figures of merit, *Nature* **413**, 597–602, 2000.

2. T.C. Harman, P.J. Taylor, M.P. Walsh, and B.E. LaForge, Quantum dot superlattice thermoelectric materials and devices, *Science* **297**, 2229–2232, 2002.

3. K.F. Hsu, S. Loo, F. Guo, W. Chen, J.S. Dyck, C. Uher, T. Hogan, E.K. Polychroniadis, and M.G. Kanatzidis, Cubic $AgPb_mSbTe_{2+m}$: Bulk thermoelectric materials with high figure of merit, *Science* **303**, 818–821, 2004.

4. A.I. Boukai, Yu. Bunimovich, J. Tahir-Kheli, J.-K. Yu, W.A. Goddard, III, and J.R. Heath, Silicon nanowires as efficient thermoelectric materials, *Nature* **451**, 168, 2008.

5. A.I. Hochbaum, R. Chen, R.D. Delgado, W. Liang, E.C. Garnett, M. Najarian, A. Majumdar, and P. Yang, Enhanced thermoelectric performance of rough silicon nanowires, *Nature* **451**, 163, 2008.

6. B. Poudel, Q. Hao, Y. Ma, Y. Lan, A. Minnich, B. Yu, X. Yan et al., High-thermoelectric performance of nanostructured bismuth antimony telluride bulk alloys, *Science* **320**, 634–638, 2008.

7. M.S. Dresselhaus and J.P. Heremans, Recent developments in low-dimensional thermoelectric materials, in *Thermoelectrics Handbook: Macro to Nano*, Ed. D.M. Rowe, CRC Press, Boca Raton, FL, 2006.

8. M.S. Dresselhaus, G. Chen, M.Y. Tang, R. Yang, H. Lee, D. Wang, Z. Ren, J.-P. Fleurial, and P. Gogna, New directions for low-dimensional thermoelectric materials, *Adv. Mater.* **19**, 1043, 2007.

9. Y. Lan, A.J. Minnich, G. Chen, and Z. Ren, Enhancement of thermoelectric figure-of-merit by a bulk nanostructuring approach, *Adv. Funct. Mater.* **19**, 1, 2009.

10. A.J. Minnich, M.S. Dresselhaus, Z.F. Ren, and G. Chen, Bulk nanostructured thermoelectric materials: Current research and future prospects, *Energy Environ. Sci.* **2**, 466, 2009.

11. M.G. Kanatzidis, Nanostructured thermoelectrics: The new paradigm? *Chem. Mater.* **22**, 648, 2010.

12. C.J. Vineis, A. Shakouri, A. Majumdar, and M.G. Kanatzidis, Nanostructured thermoelectrics: Big efficiency gains from small features, *Adv. Mater.* **22**, 3970, 2010.

13. H.J. Goldsmid, A new upper limit to the thermoelectric figure-of-merit, in *Thermoelectrics Handbook: Macro to Nano*, Ed. D.M. Rowe, CRC Press, Boca Raton, FL, 2006.

14. J.M. Cowley and J.A. Ibers, The structures of some ferric chloride-graphite compounds, *Acta Cryst.* **9**, 421–431, 1956.

15. J. Rouxel, A. Meerschaut, and G.A. Wiegers, Chalcogenide misfit layer compounds, *J. Alloys Comp.* **229**, 144–157, 1995.

16. G.A. Wiegers, Misfit layer compounds: Structures and physical properties, *Prog. Solid State Chem.* **24**, 1–139, 1996.

17. M. Kalläne, K. Rossnagel, M. Marczynski-Bühlow, L. Kipp, H. Starnberg, and S.E. Stoltz, Stabilization of the misfit layer compound $(PbS)_{1.13}TaS_2$ by metal cross substitution, *Phys. Rev. Lett.* **100**, 065502, 2008.

18. N. Giang, Q. Xu, Y.S. Hor, A.J. Williams, S.E. Dutton, H.W. Zandbergen, and R.J. Cava, Superconductivity at 2.3 K in the misfit compound $(PbSe)_{1.16}(TiSe_2)_2$, *Phys. Rev. B* **82**, 024503, 2010.

19. L. Fister, T. Novet, C.A. Grant, and D.C. Johnson, Controlling solid state reactions via design of superlattice reactants, in *Advances in the Synthesis and Reactivity of Solids*, Ed. T.E. Mallouk, Vol. 2, JAI Press, Greenwich, CT, 1994.

20. L. Fister, X.-M. Li, J. McConnell, T. Novet, and D.C. Johnson, Deposition system for the synthesis of modulated, ultrathin-film composites, *J. Vac. Sci. Technol. A* **11**, 3014, 1993.

21. T.M. Phung, J.M. Jensen, D.C. Johnson, J.J. Donovan, and, B.G. McBurnett, Determination of the composition of ultra-thin Ni–Si films on Si: Constrained modeling of electron probe microanalysis and x-ray reflectivity data, *X-ray Spectrom.* **37**, 608, 2008.

22. M. Noh, J. Thiel, and D.C. Johnson, Synthesis of crystalline superlattices by controlled crystallization of modulated reactants, *Science* **270**, 1181, 1995.

23. M. Noh and D.C. Johnson, Designed synthesis of solid state structural isomers from modulated reactants, *J. Am. Chem. Soc.* **118**, 9117–9122, 1996.

24. J.A. Wilson, F.J. DiSalvo, and S. Mahajan, Charge-density waves and superlattices in the metallic layered transition metal dichalcogenides, *Adv. Phys.* **24,** 117, 1975.

25. M. Fukuto, M.D. Hornbostel, and D.C. Johnson, Use of superlattice structure to control reaction mechanism: Kinetics and energetics of $NbsSe_4$ formation, *J. Am. Chem. Soc.* **116**, 9136–9140, 1994.

26. N.T. Nguyen, B. Howe, J.R. Hash, N. Liebrecht, P. Zschack, and D.C. Johnson, Synthesis of a family of $\{[(VSe_2)_n]1.06(TaSe_2)_m\}_z$ compounds, *Chem. Mater.* **19**, 1923–1930, 2007.

27. F.R. Harris, S. Standridge, C. Feik, and D.C. Johnson, Design and synthesis of $[(Bi_2Te_3)_x(TiTe_2)_y]$ superlattices, *Angew. Chem.* **42**, 5296–5299, 2003.

28. F.R. Harris, S. Standridge, and D.C. Johnson, The synthesis of $[(Bi_2Te_3)_x\{(TiTe_2)_y\}_{1.36}]$ superlattices from modulated elemental reactants, *J. Am. Chem. Soc.* **127**, 7843–7848, 2005.

29. C. Chiritescu, C. Mortensen, D.G. Cahill, D. Johnson, and P. Zschack, Lower limit to the lattice thermal conductivity of nanostructured Bi_2Te_3-based materials, *J. Appl. Phys.* **106**, 073503, 2009.

30. C. Heideman, N. Nguyen, J. Hanni, Q. Lin, S. Duncombe, D.C. Johnson, and P. Zschack, The synthesis and characterization of new $[(BiSe)_{1.10}]_m[NbSe_2]_n$, $[(PbSe)_{1.10}]_m[NbSe_2]_n$, $[(CeSe)_{1.14}]_m[NbSe_2]_n$, and $[(PbSe)_{1.12}]_m[TaSe_2]_n$ misfit layered compounds, *J. Solid State Chem.* **181**, 1701–1706, 2008.

31. Q. Lin, C.L. Heideman, N. Nguyen, P. Zschack, C. Chiritescu, D.G. Cahill, and D.C. Johnson, Designed synthesis of families of misfit-layered compounds, *Eur. J. Inorg. Chem.* **2008**, 2382–2385, 2008.

32. Q. Lin, M. Smeller, C.L. Heideman, P. Zschack, M. Koyano, M.D. Anderson, R. Kykyneshi, D.A. Keszler, I.M. Anderson, and D.C. Johnson, Rational synthesis and characterization of a new family of low thermal conductivity misfit layer compounds $[(PbSe)_{0.99}]_m(WSe_2)_n$, *Chem. Mater.* **22**, 1002–1009, 2010.

33. N. Wainfan and L.G. Parrat, X-ray reflection studies of the anneal and oxidation of some thin solid films, *J. Appl. Phys.* **31**, 1331, 1960.

34. P. Zschack, C. Heideman, C. Mortensen, N. Nguyen, M. Smeller, Q. Lin, and D.C. Johnson, X-ray characterization of low-thermal conductivity thin-film materials, *J. Electron. Mater.* **38**, 1401, 2009.

35. M. Smeller, P. Zschack, D.C. Johnson et al., unpublished.

36. A.C. Larson and R.B. Von Dreele, General Structure Analysis System (GSAS), Los Alamos National Laboratory Report LAUR 86–748, 2004.

37. J.R. Williams, A.L.E. Smalley, H. Sellinschegg, C. Daniels-Hafer, J. Harris, M.B. Johnson, and D.C. Johnson, Synthesis of crystalline skutterudite superlattices using the modulated elemental reactant method, *J. Am. Chem. Soc.* **125**, 10335–10341, 2003.

38. M.D. Hornbostel, E.J. Hyer, J. Thiel, and D.C. Johnson, Rational synthesis of metastable skutterudite compounds using multilayer precursors, *J. Am. Chem. Soc.* **119**, 2665–2668, 1997.

39. C. Chiritescu, D.G. Cahill, N. Nguyen, D. Johnson, A. Bodapati, P. Keblinski, and P. Zschack, Ultralow thermal conductivity in disordered, layered WSe_2 crystals, *Science* **315**, 351, 2007.

40. G.A. Slack, The thermal conductivity of nonmetallic solids, *Solid State Phys.* **34**, 1, 1979.

41. D.G. Cahill, S.K. Watson, and R.O. Pohl, Lower limit to the thermal conductivity of disordered crystals, *Phys. Rev. B* **46**, 6131, 1992.

42. K.E. Goodson, Ordering up the minimum thermal conductivity of solids, *Science* **315**, 342, 2007.

43. N.T. Nguyen, P.A. Berseth, Q. Lin, C. Chiritescu, D.G. Cahill, A. Mavrokefalos, L. Shi et al., Synthesis and properties of turbostratically disordered, ultrathin WSe_2 films, *Chem. Mater.* **22**, 2750, 2010.

44. A. Mavrokefalos, N.T. Nguyen, M.T. Pettes, D.C. Johnson, and L. Shi, In-plane thermal conductivity of disordered layered WSe_2 and $(W)_x(WSe_2)_y$ superlattice films, *Appl. Phys. Lett.* **91**, 171912, 2007.

45. C. Chiritescu, D.G. Cahill, C. Heideman, Q. Lin, C. Mortensen, N.T. Nguyen, D. Johnson, R. Rostek, and H. Böttner, Low thermal conductivity in nanoscale layered materials synthesized by the method of modulated elemental reactants, *J. Appl. Phys.* **104**, 033533, 2008.

46. A. Mavrokefalos, Q. Lin, M. Beekman, J.-H. Seol, Y.J. Lee, H. Kong, M.T. Pettes, D.C. Johnson, and L. Shi, In-plane thermal and thermoelectric properties of misfit-layered $[(PbSe)_{0.99}]_x(WSe_2)_x$ superlattice thin films, *Appl. Phys. Lett.* **96**, 181908, 2010.

47. C. Auriel, R. Roesky, A. Meerschaut, and J. Rouxel, Structure determination and electrical properties of a new misfit layered selenide $[(PbSe)_{1.10}NbSe_2]$, *Mater. Res. Bull.* **28**, 247, 1993.

48. L. Zhang and D.J. Singh, Electronic structure and thermoelectric properties of layered PbSe-WSe$_2$ materials, *Phys. Rev. B* **80**, 075117, 2009.

49. M. Beekman and D.C. Johnson, unpublished.

50. J.P. Fleurial, L. Galliard, R. Triboulet, H. Scherrer, and S. Scherrer, Thermoelectric properties of $(Bi_xSb_{1-x})_2Te_3$ single crystal solid solutions grown by the T.H.M. method, *J. Phys. Chem. Sol.* **49**, 1237, 1988.
51. A. Taylor, C. Mortensen, R. Rostek, N. Nguyen, and D.C. Johnson, Vapor annealing as a post-processing technique to control carrier concentrations of Bi_2Te_3 thin films, *J. Electron. Mater.* **39**, 1931, 2009.
52. Q. Lin, S. Tepfer, C. Heideman, C. Mortensen, N. Nguyen, P. Zschack, M. Beekman, and D.C. Johnson, Influence of selenium vapor post-annealing on the electrical transport properties of $PbSe$-WSe_2 nanolaminates, *J. Mater. Res.* **26**, 1866, 2011.

3

Bulk Nanocomposites of Thermoelectric Materials

Lidong Chen
Chinese Academy of Sciences

Zhen Xiong
Chinese Academy of Sciences

Ruiheng Liu
Chinese Academy of Sciences

Wenqing Zhang
Chinese Academy of Sciences

3.1 Introduction

The phenomenon of thermoelectricity (TE), which was based on the Seebeck effect and the Peltier effect, was initially discovered in metals in the early 1800s. During the mid-twentieth century, with the discovery of semiconductors, the first-generation thermoelectric materials, such as Bi_2Te_3, PbTe, and SiGe, gave rise to commercial TE devices. However, wide application was obstructed by the low performance of the TE materials, since it was a great challenge to improve the ZT value ($ZT = S^2 \sigma T / \kappa$, where S, σ, T, κ are Seebeck coefficient, conductivity, absolute temperature, and thermal conductivity, respectively) above 1 in the following 50 years. Until the 1990s, the emerging of some new compounds, such as skutterudite and clathrate, brought great hope to overcome this problem. Besides that, nanotechnology has brought about a great change to the traditional TE materials. Various nanostructured materials, such as nanocomposites,[1] superlattice,[2–3] quantum well, and dots[4–6] have improved the ZT values to 2, which arouses great enthusiasm among TE researchers. Figure 3.1 shows the history of development over the past 60 years.

Up to now, the nanostructured TE materials mainly include three kinds of materials, that is, low-dimensional materials, nanograin-sized materials, and nanocomposites. From the traditional view, thermoelectric materials are expected to be pure semiconductors without any secondary phase to avoid the impurity scattering to electrons. It has been proved that, by both theoretical predication and experimental results, the TE performance of large-scale composites is just a compromise of the two (or more) phases. However, with the progress of fabrication technique and microstructure characterization, when the second phase goes into nanosize, TE performance can be improved if the nanoinclusions scattering is applied to appropriately tuning the transport of electrons and phonons. And bulk nanocomposites, which contain nanoinclusions dispersing in the matrix, are expected to be one of the most potential thermoelectric materials for wide use in the future.

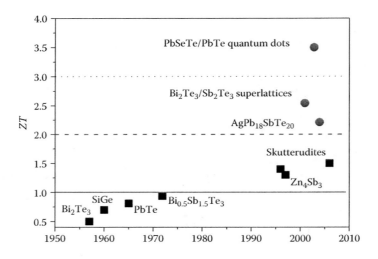

FIGURE 3.1 Nanostructured materials in the thermoelectric history.

ZT improvements brought by nanocomposites have been confirmed in many TE systems. One typical example is PbTe, which was a traditional thermoelectric material in moderate temperature used by NASA in the 1960s. Doping was the main approach to optimize the thermoelectric performance for a long time, and $ZT \sim 0.8$ was obtained. In 2004, Kanatzidis et al.[1] reported AgPb$_m$SbTe$_{2+m}$ nanocomposite containing Ag$^+$–Sb^{3+}-rich nanodots, and the maximal $ZT \sim 2.2$ at 800 K was achieved. From then on, PbTe-based nanocomposites dispersing with PbS[7] and Na–Sb[8]-rich nanodots were fabricated, and found that the microstructure fluctuation contributes to the enhanced thermoelectric performance. Another example is the Bi$_2$Te$_3$-based alloy, which functions around room temperature. The thermoelectric performance of zone-melted Bi$_2$Te$_3$ was around $ZT = 1$ in the earlier decades. Recently, Chen and Ren et al.[9] reported nanostructure p-Bi$_2$Te$_3$ with high ZT value made by ball milling. It is concluded that the enhancement of ZT originated from the greatly depressed lattice thermal conductivity. Tang and Tritt et al.[10–12] fabricated nanostructure Bi$_2$Te$_3$ by melt spinning. The contact face is nanograined or amorphous, and the free face grain is in microsize. Due to the micro–nano composite microstructure, the lattice thermal conductivity is obviously depressed, and the ZT values are greatly improved to 1.56. In the following reports, skutterudites with nanostructures Yb$_2$O$_3$,[13] InSb,[14] and GaSb[15] are fabricated by different routes, and the ZT value approaches 1.5, that is, about 25% enhancement comparing with $ZT \sim 1.2$ of the matrix.

In this chapter, we will introduce the recent developments of thermoelectric nanocomposites in various systems, including fabrication and microstructure analysis of nanoinclusions, the effect of nanoinclusions on the transport character and the mechanisms underlying. The investigations of microstructure stability in skutterudites are also discussed.

3.2 Fabricating Methods and the Microstructure of Nanocomposites

Up to now, semiconductors and insulators have been found to be suitable when choosing nano dispersions, because these compounds are inert to the matrix when the TE materials are subjected to high temperature. Some active metal elements and compounds with low melting point or high vapor pressure are usually excluded. The following factors would be considered when fabricating nanocomposites. (1) The grain size of nano dispersions is expected to be below 100 nm, for the mean free path of phonons is just about several nanometers. (2) Homogeneous distribution in the matrix of the nano dispersions is desired to scatter phonons on a large scale. (3) The integrality of the matrix, especially the electron transport channels, would not be destroyed by the introduction of nanodispersions. (4) The

(a) (b)

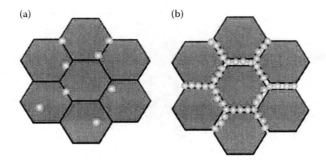

FIGURE 3.2 Microstructure schematic of the nanocomposite. (a) Nanodispersions are homogeneously dispersed; (b) nanodispersions are gathered at the grain boundary.

thermodynamically stability of nanodispersions is also an important issue for the application of devices. The microstructure schematic of the nanocomposite is shown in Figure 3.2a. Nanoinclusions are dispersed homogeneously, not only locating at the grain boundaries, but also inside the grains.

For the above reasons, the fabrication of nanocomposites fulfilling the above-mentioned features integrates mechanical engineering, kinetics, phase diagrams, and some other materials sciences. According to the formation of the nanoinclusions, two major routes are applied to fabricate the thermoelectric nanocomposite: the *ex situ* method, involving mechanical mixing[16,17] and liquid–solid compositing,[18,19] and the *in situ* method, involving phase decompositions[20–24] and nanosize precipitations.[25–27]

3.2.1 *Ex Situ* Route

The first step of fabricating nanocomposites via the *ex situ* route is to prepare the nanosized second particles, and then disperse them into bulk matrix materials by following the mixing and sintering process. The sketch map is shown in Figure 3.3. Initially, skutterudite-based nanocomposites were produced via high-energy ball milling, which is a typical *ex situ* route. High-energy ball milling could produce large amounts of particles with a grain size as small as several nanometers. The nanosized second phase $CoSb_3/FeSb_2$,[28] $CoSb_3/NiSb$,[29] $CoSb_3/C_{60}$,[30] $Ba_{0.44-x}Co_4Sb_{12}/Ba_xC_{60}$,[31] and $CoSb_3/ZrO_2$[32–33] were fabricated via this approach. Most nanodispersions are distributed in the grain boundaries while some are enclosed inside the grains during the sintering process.

Research conducted also included half-Heusler compounds, which have high thermal conductivity. Chen et al.[34] introduced nanoparticles ZrO_2 to the alloys $Zr_{0.5}Hf_{0.5}Ni_{0.8}Pd_{0.2}Sn_{0.99}Sb_{0.01}$. The microstructure analysis is shown in Figure 3.4.

Another *ex situ* method, liquid–solid compositing, was applied as an effective *ex situ* method to introduce nanodispersions in the matrix. During the period of hydroxylation, nanodispersions are precipitated on the surface of matrix particles, which will make a homogenous distribution of the nanodispersion in the matrix. And the size of nanoparticles could be more uniform compared to

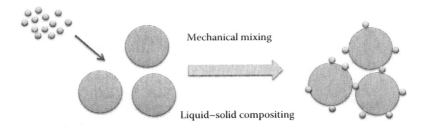

Mechanical mixing

Liquid–solid compositing

FIGURE 3.3 Sketch map of the *ex situ* compositing method.

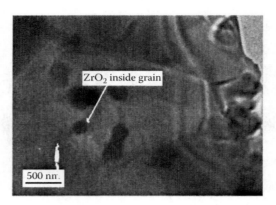

FIGURE 3.4 TEM image of the nanocomposite $Zr_{0.5}Hf_{0.5}Ni_{0.8}Pd_{0.2}Sn_{0.99}Sb_{0.01}/ZrO_2$. (Adapted from L. D. Chen et al. 2006. *J. Appl. Phys.* **99**, 064305.)

the ball-milling process. In the series $La_{0.9}CoFe_3Sb_{12}$-$CoSb_3$,[19] $CoSb_3$(micro)/$CoSb_3$(nano),[35] and $Ba_{0.22}Co_4Sb_{12}/TiO_2$,[18] nanodispersions were both introduced in the liquid circumstance. Microstructure of the $Ba_{0.22}Co_4Sb_{12}/TiO_2$ nanocomposite was shown in Figure 3.5.

3.2.2 *In Situ* Route

The *in situ* route is a way in which the nanosized particles form in the homogeneous matrix via nucleation and growth. The *in situ* route, which means at least one component of the nanodispersions come from the matrix, usually resulting in nanocomposites with a homogeneous microstructure. One typical *in situ* route is that nanophase is precipitated in the supersaturated matrix during the cooling process.[20,23,36,37] The size of the second phase can be controlled by a certain heat-treatment process, and the distribution is greatly enhanced compared with the *ex situ* route. *In situ* oxidizing, nonuniform growth via melting, spinning, metastable voiding, and filling have been explored for skutterudites-based nanocomposites in recent years.

3.2.2.1 *In Situ* Oxidization

In the process of fabricating thermoelectric compounds, some components are added in excess for the purpose of being oxidized in the following period, called *in situ* oxidization. *In situ* oxidization was used in the preparation of $Yb_xCo_4Sb_{12}/Yb_2O_3$ nanocomposites. In the annealing period (1000–1100 K), the rare earth metal Yb has a higher filling fraction in the voids of $CoSb_3$ than that at room temperature. As the temperature is lowered, the *in situ* precipitated Yb is oxidized to Yb_2O_3. Some nanosized Yb_2O_3 are observed inside the matrix grain and some dispersion with micro size are located in the grain boundary as well. The sketch map and the microstructure are shown in Figures 3.6 and 3.7, respectively. The κ_L was drastically depressed due to the existence of Yb_2O_3 nano-oxides and a ZT of 1.3 was achieved.

In the preparation of half-Heusler compound $TiFe_{0.2}Co_{0.8}Sb$, L. D. Chen et al.[38] have added trace oxidant Fe_2O_3 into the matrix. In the arc-melting process, excess Ti is oxidized into TiO_2 by oxidant Fe_2O_3 as the following reaction:

$$3Ti + 2Fe_2O_3 - 3TiO_2 + 4Fe$$

The *in situ* oxidized TiO_2 nanoinclusions are dispersed in the $TiFe_{0.2}Co_{0.8}Sb$ matrix, as shown in Figure 3.8.

3.2.2.2 Nonuniform Growth

Melt-spinning is an ultrafast cooling technique, which is usually used to produce metal glass because the nucleation and growth of grains is prevented by the super-cooling.[39–42] The grain in the contact face

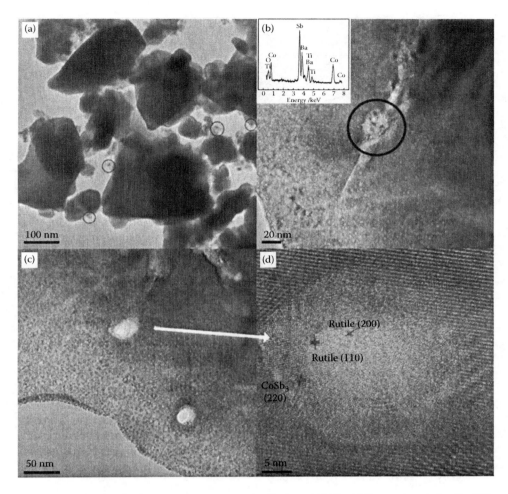

FIGURE 3.5 TEM images for the nanocomposite $Ba_{0.22}Co_4Sb_{12}/0.4vol.\%TiO_2$. (a) Composite powder. (b–d) Sintered bulk. The inset of (b) is the EDS results of the circled area.

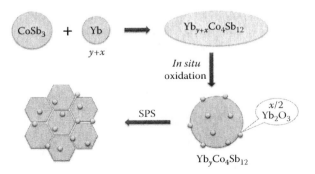

FIGURE 3.6 Sketch map of *in situ* oxidization for Yb in $Yb_yCo_4Sb_{12}$.

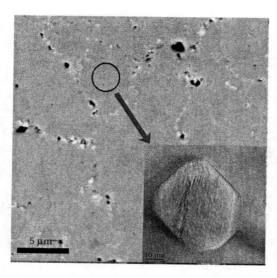

FIGURE 3.7 Backscattering electron image of the $Yb_xCo_4Sb_{12}/Yb_2O_3$ composite and the inset in the TEM image for Yb_2O_3 nanodispersion.

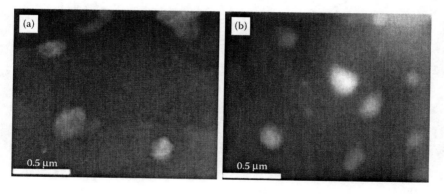

FIGURE 3.8 TEM image of the nanocomposite (a) $TiFe_{0.1}Co_{0.9}Sb$ and (b) $TiFe_{0.2}Co_{0.8}Sb$. (Adapted from T. Wu, 2007. *J. Appl. Phys.* **102**, 103705.)

is nanosized or amorphous, while the grain in the free face is microsized. This allows the nonuniform growth of the grains to be possible in the following SPS or hot press sintering process. The sketch map of melt spinning is shown in Figure 3.9. Tang et al.[43] added excess Sb when fabricating $Yb_yCo_4Sb_{12}$, and after melt-spinning, found that the excess Sb were locating on the grain boundaries. The κ_L was observed to be depressed as other nanodispersion containing series. Tang and Tritt et al.[11,12] prepared Bi_2Te_3-based nanocomposite by melt spinning, and found that nanograins were dispersed in the micro-sized matrix with coherent boundaries, which is helpful for reducing the lattice thermal conductivity while maintaining the electrical conductivity.

3.2.2.3 Precipitation from Metastable Phase

So far, a variety of atoms have been successfully filled into the voids of $CoSb_3$-based skutterudites, such as rare earth elements, alkaline earth elements, and alkaline elements.[44–46] When Chen and Zhang et al. studied the filling fraction of the available fillers, they found that some elements may fill into the void in a metastable state.[15] They focused their attention on the IIIA elements, that is, Ga, In, and Tl, which

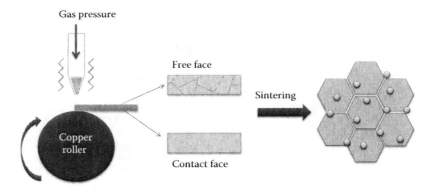

FIGURE 3.9 Sketch map of the melt-spinning method.

remain ambiguous in skutterudite systems.[47,48] As shown in Figure 3.10, *ab initio* calculations reveal that the ΔG of Ga-filled $CoSb_3$ turns to positive when the temperature is below 1100 K, which implies a possible metastable void filling in $CoSb_3$. This suggests a precipitation route to produce nanocomposite, that is, filling Ga into the void at high temperature and driving it out when the temperature is lowered, forming related nanoinclusions in an *in situ* way.[22,26] The sketch map is shown in Figure 3.11. A TEM image of the quenched sample (Figure 3.12) $Yb_{0.26}Ga_{0.2}Co_4Sb_{12.2}$-q combining with the EDS results shows a homogeneous microstructure, which proves that Ga filling to the $CoSb_3$ voids at high temperature. For a furnace cooling procedure, the metastable filling Ga was driven out of the voids and formed GaSb nanoinclusions in an *in situ* way. As shown in Figure 3.13, GaSb nanoinclusions with grain size ~11 nm are dispersed homogeneously in the nanocomposite $Yb_{0.26}Co_4Sb_{12}$/0.2GaSb. To control the grain size of nanoinclusions, the cooling procedure of the Ga–Yb dual filling samples at high temperature is adjusted, then the GaSb grain size is coarsened to ~88 nm in the nanocomposite $Yb_{0.26}Co_4Sb_{12}$/0.2GaSb-c, as shown in Figure 3.14.

Samples of $Yb_{0.26}Co_4Sb_{12}$/yGaSb were fabricated and proved that the GaSb nanodispersion in the grain size of 5–20 nm are locating homogeneously in the matrix of Yb-filled $CoSb_3$. The Seebeck coefficients of GaSb containing samples were enhanced at the same carrier concentration level. As a result, the *ZT* values are increased in the whole temperature range.

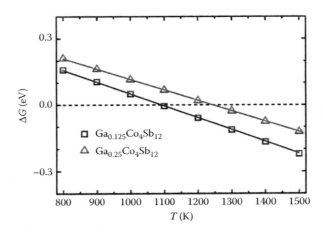

FIGURE 3.10 Predicted temperature-dependent Gibbs free energy ΔG in filled skutterudite $Ga_xCo_4Sb_{12}$.

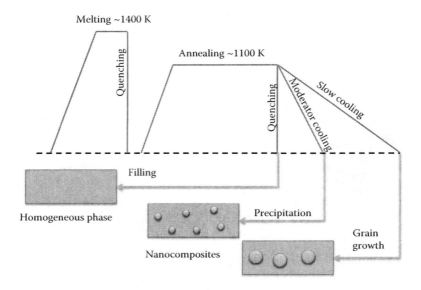

FIGURE 3.11   Fabrication process of metastable void filling method.

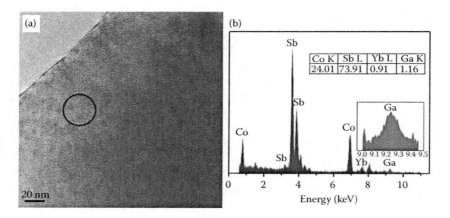

FIGURE 3.12   (a) TEM image for the quenched sample $Yb_{0.26}Ga_{0.2}Co_4Sb_{12.2}$. (b) The corresponding EDS results for the marked circle in (a). The inset pattern is the local enlargement of Ga peak, and the inset table is the mole composition by EDS.

3.3 Transport Characteristics for the Skutterudite-Based Nanocomposites

For the state-of-the-art TE materials, wavelength (λ) and mean free path (l) are two critical factors for both electron and phonon transport. In a Boltzmann approximation, the lattice thermal conductivity and carrier mobility are in proportion to the mean free path of phonons l_{ph}, and that of carrier l_e, accordingly. Based on the scattering theory, the waves will be scattered strongly by the objective possessing the equivalent size as the wavelength. For an electron, λ_e and l_e occur in the scale of nanometers, while those of a phonon (λ_{ph} and l_{ph}) distribute in a broader range. The difference of λ_e and λ_{ph} make it possible to selective scattering brought by the suitable-sized nanoparticles, which would effectively reduce l_{ph} but not l_e.[49–53] Usually, comparing with the pure thermoelectric matrix, three changes in

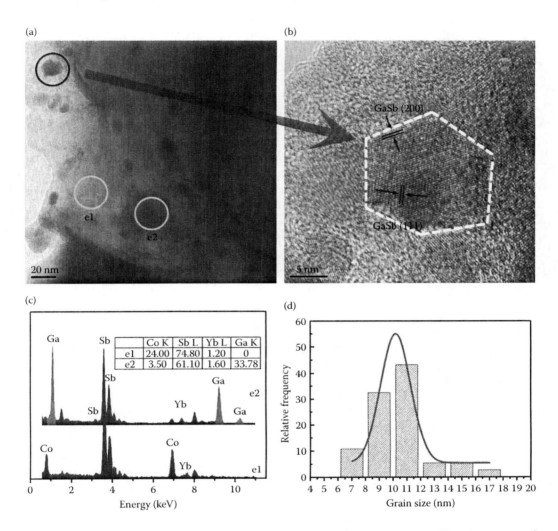

FIGURE 3.13 (a) TEM image for the sample $Yb_{0.26}Co_4Sb_{12}/0.2GaSb$ nanocomposite. (b) HRTEM image for selected nanodot in (a). (c) The corresponding EDS results for selected areas in (a). (d) Grain size distribution of the nanodispersions in (a).

thermoelectric properties are expected for the nanocomposites. (1) The κ_L is decreased due to the scattering effects to phonons from the scattering centers both of nanoinclusion and the interface.[49,54–56] (2) The Seebeck coefficient is expected to increase by a degree, which may be attributed from the filtering effects to electrons of low energy, resulting in the increasing change of carrier DOS dependent on energy near the Fermi level.[51,57,58] (3) The electrical conductivity may be influenced, depending on the characteristic of the nanodispersions. The Hall mobility may be affected for the nanoscattering to electrons. Nevertheless, the *ZT* value could be enhanced if the microstructure is carefully manipulated to minimize the effect to mobility. Tremendous experiments have revealed the promising potential.

3.3.1 The Impact on Phonon Scattering and Thermal Conductivity

For unfilled skutterudites, the improvement of TE performance after compositing with nanoinclusions is considerable. The phonons have very large mean free path values without the "rattling" effects of

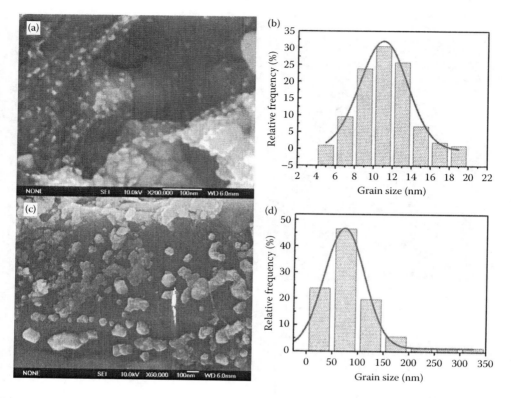

FIGURE 3.14 FESEM images for fractured surface of the sample $Yb_{0.26}Co_4Sb_{12}/0.2GaSb$ nanocomposite (a) and $Yb_{0.26}Co_4Sb_{12}/0.2GaSb$-c nanocomposite (c) with large GaSb dispersions. (b) and (d) are the corresponding grain size distribution for (a) and (c), respectively.

fillers to scatter phonons; therefore, once the scattering center such as nanoscale or larger-scale defects are introduced, the depression of the lattice thermal conductivity is prominent. For instance, κ_L among the nanocomposite $CoSb_3/6.54$ wt.%C_{60} has a factor of 2 smaller than that of the pure $CoSb_3$. And composites of $CoSb_3/FeSb_2$, $CoSb_3/NiSb$, $CoSb_3/C_{60}$, and $CoSb_3/ZrO_2$ were also fabricated to explore the large defect scattering effects to phonons. The improvement of ZT is mainly attributed from the depression of κ_L.

For filled skutterudites, as the fillers could dramatically impede the transportation of phonons with short–middle wavelength,[59,60] the nanoscattering mainly takes effect in long wavelength spectrum range. For instance, κ_L of the nanocomposite $Ba_{0.22}Co_4Sb_{12}/0.4vol.\%TiO_2$ decrease by 12% comparing with that of the matrix; however, κ_L of the nanocomposite $Ba_{0.22}Co_4Sb_{12}/1.8vol.\%TiO_2$ increases by 28%. As the interface effects turn to volume effects, the total κ_L is a compromise of the matrix and the second phase. Since the nanodispersions, including oxides and semiconductors, have higher thermal conductivities than the filled skutterudite, the total κ_L will increase once the nanodispersions exceed the optimal content.

Yb is effective in reducing the κ_L due to its heavy element mass and small radius, but the presence of Yb_2O_3 can scatter the phonons more drastically. The κ_L for the nanocomposite $Yb_{0.25}Co_4Sb_{12}/Yb_2O_3$ at room temperature is depressed to 1.72 W m^{-1} K^{-1} with the Lorentz number 2.0 E^{-8}V^2 K^{-2}. The κ_L of some typical systems nanocomposites are displayed in Figure 3.15.

The scattering mechanism of nanoinclusion to phonon is complicated. Two aspects should be considered. The scattering comes from the nanoinclusion and from the interface. Interface is a discontinuous part in the nanocomposites, at which the elastic modulus, crystal plane matching, and Fermi level, and so on are all changed abruptly. During the transportation, phonons will be scattered at the

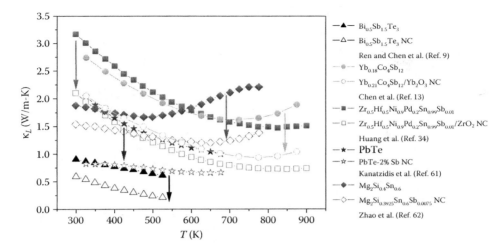

FIGURE 3.15 Lattice thermal conductivities of several nanocomposites and the matrix thereof at room temperature. (Adapted from B. Poudel et al. 2008. *Science* **320**, 634; X. Y. Zhao et al. 2006. *Appl. Phys. Lett.* **89**, 092101; L. D. Chen et al. 2006. *J. Appl. Phys.* **99**, 064305; J. R. Sootsman et al. 2006. *Chem. Mater.* **18**, 4993; Q. Zhang et al. 2008. *Appl. Phys. Lett.* **93**, 102109.)

interface due to the above reasons. The interface of inclusion and the matrix can be regarded as several phonon scattering centers. Regarding the phonons as particles during transport in the solid materials, it would be scattered strongly by the inclusions possessing the equivalent grain size, then the inclusion serves as phonon scattering centers. Comparing with point defects, which are in the atomic level, nanoinclusion mainly scatters the mid-to-long wavelength phonons, and then the lattice thermal conductivity.

3.3.2 The Effect on Electrical Transportation

The change of electrical conductivities is strongly dependent on the nature of nanodispersions. For n-type skutterudite nanocomposites, nanoinclusions of metal will increase the carrier concentration, and usually, the enhancement of electrical conductivity is difficult to offset the decrease of Seebeck and the increase of electron thermal conductivity. The insulator and semiconductor are usually chosen as the candidates for nano-inclusions due to their stable characteristics. Oxides, being inert to the matrix and only worked as phonon scattering centers, usually exert negative influence on the electrical conductivity. As the concentration exceeds the optimal content, in other words, the nano-dispersions are connected with each other and form a barrier layer in the grain boundary, the electrical conductivity will be depressed on a large scale. For filled skutterudites, the heavily doped semiconducting behavior will be changed to an intrinsic semiconducting pattern.[18,30,32] The electrons have to cross the insulator layer for electrical conductivity to take place, which is strongly dependent on the temperature. Thus, for material design, the most important issue is to avoid aggregation of the second phase.

The dispersed inclusions can also scatter electrons in a degree, behaving as the mobility is lowered. For instance, the electron mobility of $Ba_{0.22}Co_4Sb_{12}/TiO_2$ nanocomposite is declined when compared with the matrix, which can also be found in the nanocomposites $CoSb_3/C_{60}$ and $Yb_{0.26}Co_4Sb_{12}/GaSb$. The electron–phonon scattering mode is dominant for skutterudites, nanocomposites also obey the $\mu_H \sim T^{-3/2}$ relationship largely once the nanoinclusion content is not high enough. The departure implies the change of scattering parameters as shown from the results of the nanocomposites $Ba_{0.22}Co_4Sb_{12}/TiO_2$ and $CoSb_3/C_{60}$ in Figure 3.16. Thus, what we could do is to control the content and distribution of the nanodispersions not to reduce the mobility much, at least, smaller than the beneficial effect for Seebeck coefficient.

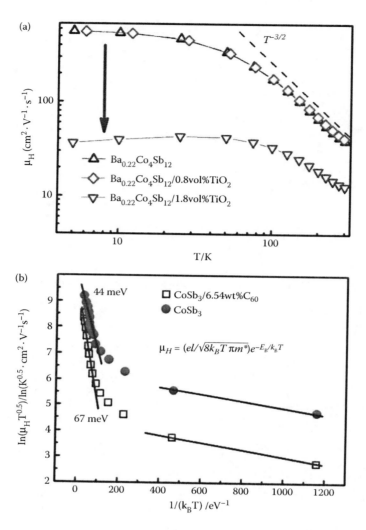

FIGURE 3.16 (a) Temperature dependence of μ_H (hall mobility) for the nanocomposite $Ba_{0.22}Co_4Sb_{12}/TiO_2$; (b) Arrhenius plot of $\ln(\mu_H T^{0.5})$ vs $1/(k_B T)$ for the nanocomposite $CoSb_3/C_{60}$.

At the interface of nanocomposites, a barrier is formed due to the different Fermi level between the nanoinclusion and the matrix. It is generally believed that such a barrier can filter electrons with low energy. As a result, the change of carrier density depending on energy near the Fermi level could be increased, which is helpful for enhancing the Seebeck coefficients.[51,63] It is indeed that the absolute Seebeck coefficients are improved in the reported nanocomposites $CoSb_3/C_{60}$, $In_xCe_yCo_4Sb_{12}/InSb$, and $Ba_{0.22}Co_4Sb_{12}/TiO_2$. For the nanocomposite $Yb_{0.26}Co_4Sb_{12}/GaSb$, the Seebeck coefficient is enhanced at the same carrier concentration, as shown in Figure 3.17.

For semiconductors with a single parabola band mode, the power factor ($S^2\sigma$) usually has a parabola-like trend depending on the carrier concentration.[66-68] In nanocomposites, the carefully manipulated nanoinclusions will decrease the electrical conductivity and increase the Seebeck coefficient at the same time. The resulted power factor is elevated up to the normal trend line. Combined with the reduction of lattice thermal conductivity, the ZT values of several nanocomposites are enhanced markedly, as shown in Figure 3.18.

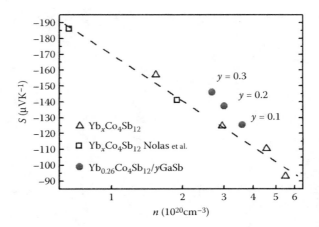

FIGURE 3.17 Seebeck coefficients of $Yb_xCo_4Sb_{12}$ and $Yb_{0.26}Co_4Sb_{12}/yGaSb$ virus carrier concentration. (Adapted from G. S. Nolas et al. 2000. *Appl. Phys. Lett.* **77**, 1855; G. A. Lamberton et al. 2005. *J. Appl. Phys.* **97**, 113715; Xiong Z, J. D. et al. unpublished.)

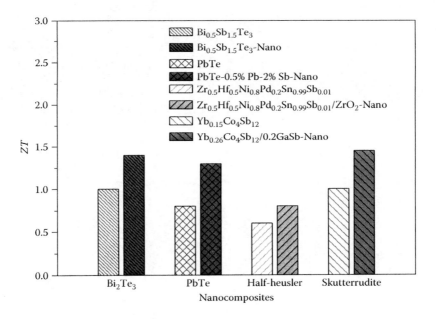

FIGURE 3.18 *ZT* value of several nanocomposites and the matrix thereof. (Adapted from B. Poudel et al. 2008. *Science* **320**, 634; Z. Xiong et al. 2010. *Acta Mater.* **58**, 3995; L. D. Chen et al. 2006. *J. Appl. Phys.* **99**, 064305; J. R. Sootsman et al. 2008. *Angew Chem Int Edit* **47**, 8618.)

3.4 Microstructure Stability of Thermoelectric Nanocomposites

The application situation demands that TE materials should be able to function at a relatively high temperature for a long term. And the elements may also diffuse from a high concentration to a low concentration due to the temperature gradient. Therefore, the stability of the thermodynamics and microstructure is vital to the thermoelectric performance. Chen and Gu et al.[70] focused on the microstructure stability of skutterudite nanocomposites $Yb_yCo_4Sb_{12}/Yb_2O_3$, in which nano-Yb_2O_3 inclusions were proved to be effective phonon scattering centers. The composite $Yb_{0.6}Co_4Sb_{12}/Yb_2O_3$ was heat treated at 600° for a

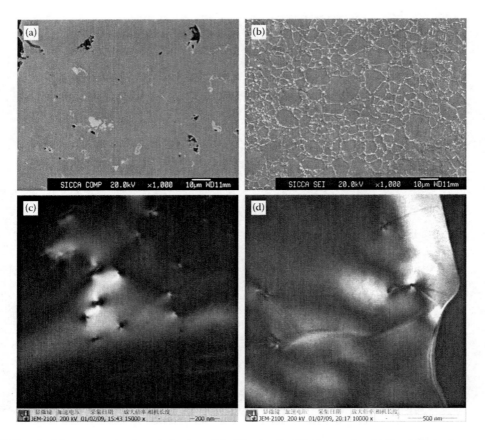

FIGURE 3.19   Microstructure characterization of the $Yb_{0.6}Co_4Sb_{12}/Yb_2O_3$ samples. (a) Shows the back scattering image of the polished surface for the sample heat treated for 0 day. (b) The SEM image of the polished surface for the sample heat treated for 30 days. (c–d) Shows the TEM image for the sample heat treated for 30 days.

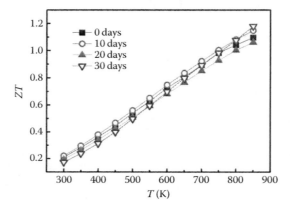

FIGURE 3.20 ZT values for the $Yb_{0.6}Co_4Sb_{12}/Yb_2O_3$ samples heat treated for 0, 10, 20, 30 days, respectively.

duration of 10, 20, and 30 days, respectively. The TEM images reveal that the microstructure of $Yb_{0.6}Co_4Sb_{12}/Yb_2O_3$ nanocomposite is stable even when the sample was heat treated at 600°C for 30 days. As shown in Figure 3.19c,d, dislocations are estimated to be the reason for microstructure stabilization. The stress induced by dislocation can prevent the filling Yb from diffusion. Figure 3.20 shows the *ZT* values for the $Yb_{0.6}Co_4Sb_{12}/Yb_2O_3$ nanocomposite heat treated for different duration. The results indicate that the *ZT* deviation is limited to be 10%.

3.5 Summary and Future

In the past decade, various nanocomposites with enhanced *ZT* values have been successfully produced, and nanostructuring has become a common strategy to improve the performance of TE materials. At present, most enhancements of TE performance are mainly attributed to the reductions of lattice thermal conductivity, and the comprehension that nanoinclusion effectively scattering the mid-to-long wavelength phonons has been confirmed. However, further improvement of *ZT* requires not only the decrease in thermal conductivity, but also the enhancement of power factor. Although theoretical studies by Dresselhaus et al.[71,72] were impressive, however, researchers have realized that nanoinclusion could change the transport behavior of the carriers,[15,69] the scattering mechanism is still very ambiguous, and it remains a great challenge to control the scattering process to improve the power factor.

Besides comprehensive problems of the mechanism, other challenges mainly are concentrated in how to fabricate nanocomposites with a desired microstructure, that is, homogeneous distribution of nanoinclusions with suitable size. Precipitation caused by thermokinetics is expected to be the ideal approach to realize that. However, the studies on the thermodynamics phase diagrams of TE systems seem to be insufficient. The related works would be very necessary and helpful for TE nanocomposites engineering in the future. In the development history of TE materials, the TE performance is approaching limitation before the nanostructuring concept. With the progress of theory and fabricating technology, we hope a *ZT* of 2 or even higher values could be realized in TE nanocomposites.

References

1. K. F. Hsu, S. Loo, F. Guo, W. Chen, J. S. Dyck, C. Uher, T. Hogan, E. K. Polychroniadis, and M. G. Kanatzidis, 2004. Cubic $AgPb_mSbTe_{2+m}$: Bulk thermoelectric materials with high figure of merit. *Science.* 303, 5659.
2. H. Beyer, J. Nurnus, H. Bottner, A. Lambrecht, E. Wagner, and G. Bauer, 2002. High thermoelectric figure of merit *ZT* in PbTe and Bi_2Te_3-based superlattices by a reduction of the thermal conductivity. *Physica E.* **13**, 965.
3. T. C. Harman, D. L. Spears, and M. P. Walsh, 1999. PbTe/Te superlattice structures with enhanced thermoelectric figures of merit. *J. Elec. Mater.* **28**, L1.
4. T. C. Harman, D. L. Spears, and M. J. Manfra, 1996. High thermoelectric figures of merit in PbTe quantum wells. *J. Elec. Mater.* **25**, 1121.
5. T. C. Harman, P. J. Taylor, D. L. Spears, and M. P. Walsh, 2000. Thermoelectric quantum-dot superlattices with high *ZT*. *J. Elec. Mater.* **29**, L1.
6. R. Venkatasubramanian, E. Siivola, T. Colpitts, and B. O'Quinn, 2001. Thin-film thermoelectric devices with high room-temperature figures of merit. *Nature* **413**, 597.
7. S. N. Girard, J. Q. He, C. P. Li, S. Moses, G. Y. Wang, C. Uher, V. P. Dravid, and M. G. Kanatzidis, 2010. *In situ* nanostructure generation and evolution within a bulk thermoelectric material to reduce lattice thermal conductivity. *Nano Lett.* **10**, 2825.
8. P. F. R. Poudeu, J. D'Angelo, A. D. Downey, J. L. Short, T. P. Hogan, and M. G. Kanatzidis, 2006. High thermoelectric figure of merit and nanostructuring in bulk p-type $Na_{1-x}Pb_mSb_yTe_{m+2}$. *Angew Chem. Int. Edit.* 45, 3835.

9. B. Poudel, Q. Hao, Y. Ma, Y. C. Lan, A. Minnich, B. Yu, X. A. Yan et al. 2008. High-thermoelectric performance of nanostructured bismuth antimony telluride bulk alloys. *Science* **320**, 634.

10. X. F. Tang, W. J. Xie, H. Li, W. Y. Zhao, Q. J. Zhang, and M. Niino, 2007. Preparation and thermoelectric transport properties of high-performance p-type Bi_2Te_3 with layered nanostructure. *Appl. Phys. Lett.* **90**, 012102.

11. W. J. Xie, J. A. He, H. J. Kang, X. F. Tang, S. Zhu, M. Laver, S. Y. Wang et al. 2010. Identifying the specific nanostructures responsible for the high thermoelectric performance of $(Bi,Sb)_2Te_3$ nano-composites. *Nano Lett.* **10**, 3283.

12. W. J. Xie, X. F. Tang, Y. G. Yan, Q. J. Zhang, and T. M. Tritt, 2009. Unique nanostructures and enhanced thermoelectric performance of melt-spun BiSbTe alloys. *Appl. Phys. Lett.* **94**, 102111.

13. X. Y. Zhao, X. Shi, L. D. Chen, W. Q. Zhang, S. Q. Bai, Y. Z. Pei, X. Y. Li, and T. Goto, 2006. Synthesis of $Yb_yCo_4Sb_{12}/Yb_2O_3$ composites and their thermoelectric properties. *Appl. Phys. Lett.* **89**, 092101.

14. H. Li, X. F. Tang, Q. J. Zhang, and C. Uher, 2009. High performance $In_xCe_yCo_4Sb_{12}$ thermoelectric materials with *in situ* forming nanostructured InSb phase. *Appl. Phys. Lett.* **94**, 102114.

15. Z. Xiong, X. Chen, X. Huang, S. Bai, and L. Chen, 2010. High thermoelectric performance of $Yb_{0.26}Co_4Sb_{12}/yGaSb$ nanocomposites originating from scattering electrons of low energy. *Acta Mater.* **58**, 3995.

16. G. Rogl, A. Grytsiv, E. Bauer, P. Rogl, and M. Zehetbauer, 2010. Thermoelectric properties of novel skutterudites with didymium: $DD_y(Fe_{1-x}Co_x)_4Sb_{12}$ and $DD_y(Fe_{1-x}Ni_x)_4Sb_{12}$. *Intermetallics* 18, 57.

17. G. Joshi, H. Lee, Y. C. Lan, X. W. Wang, G. H. Zhu, D. Z. Wang, R. W. Gould et al. 2008. Enhanced thermoelectric figure-of-Merit in nanostructured p-type silicon germanium bulk alloys. *Nano Lett.* **8**, 4670.

18. Z. Xiong, X. H. Chen, X. Y. Zhao, S. Q. Bai, X. Y. Huang, and L. D. Chen, 2009. Effects of nano-TiO_2 dispersion on the thermoelectric properties of filled-skutterudite $Ba_{0.22}Co_4Sb_{12}$. *Solid State Sci.* **11**, 1612.

19. P. N. Alboni, X. Ji, J. He, N. Gothard, and T. M. Tritt, 2008. Thermoelectric properties of $La_{0.9}CoFe_3Sb_{12}$-$CoSb_3$ skutterudite nanocomposites. *J. Appl. Phys.* **103**, 113707.

20. T. Ikeda, E. S. Toberer, V. A. Rıvi, G. J. Snyder, S. Aoyagi, E. Nishibori, and M. Sakata, 2009. *In situ* observation of eutectoid reaction forming a $PbTe$-Sb_2Te_3 thermoelectric nanocomposite by synchrotron x-ray diffraction. *Scripta Mater.* **60**, 321.

21. T. Ikeda, S. M. Haile, V. A. Ravi, H. Azizgolshani, F. Gascoin, and G. J. Snyder, 2007. Solidification processing of alloys in the pseudo-binary $PbTe$-Sb_2Te_3 system. *Acta Mater.* **55**, 1227.

22. T. Ikeda, L. A. Collins, V. A. Ravi, F. S. Gascoin, S. M. Haile, and G. J. Snyder, 2007. Self-assembled nanometer lamellae of thermoelectric $PbTe$ and Sb_2Te_3 with epitaxy-like interfaces. *Chem. Mater.* **19**, 763.

23. D. L. Medlin and G. J. Snyder, 2009. Interfaces in bulk thermoelectric materials. *Curr Opin Coll. Inter.& Sci.* **14**, 226.

24. Z. Xiong, X. Y. Huang, X. H. Chen, J. Ding, and L. D. Chen, 2010. Realizing phase segregation in the $Ba_{0.2}(Co_{1-x}Ir_x)_4Sb_{12}$ $(x = 0,0.1,0.2)$ filled skutterudite system. *Scripta Mater.* 62, 93.

25. Y. Pei, A. F. May, and G. J. Snyder, 2011. Self-tuning the carrier concentration of $PbTe/Ag_2Te$ composites with excess Ag for high thermoelectric performance. *Adv. Energy Mater.* **1**, 291.

26. Y. Z. Pei, J. Lensch-Falk, E. S. Toberer, D. L. Medlin, and G. J. Snyder, 2011. High thermoelectric performance in $PbTe$ due to large nanoscale Ag_2Te precipitates and La doping. *Adv. Funct. Mater.* **21**, 241.

27. X. Shi, Z. Zhou, W. Zhang, L. D. Chen, J. Yang, and C. Uher, 2007. Solid solubility of Ir and Rh at the Co sites of skutterudites. *J. Appl. Phys.* **101**, 123525.

28. S. Katsuyama, Y. Kanayama, M. Ito, K. Majima, and H. Nagai, 2000. Thermoelectric properties of $CoSb_3$ with dispersed $FeSb_2$ particles. *J. Appl. Phys.* **88**, 3484.

29. S. Katsuyama, M. Watanabe, M. Kuroki, T. Maehata, and M. Ito, 2003. Effect of NiSb on the thermoelectric properties of skutterudite $CoSb_3$. *J. Appl. Phys.* **93**, 2758.

30. X. Shi, L. Chen, J. Yang, and G. P. Meisner, 2004. Enhanced thermoelectric figure of merit of $CoSb_3$ via large-defect scattering. *Appl. Phys. Lett.* **84**, 2301.

31. X. Shi, L. D. Chen, S. Q. Bai, X. Y. Huang, X. Y. Zhao, Q. Yao, and C. Uher, 2007. Influence of fullerene dispersion on high temperature thermoelectric properties of $Ba_yCo_4Sb_{12}$-based composites. *J. Appl. Phys.* **102**, 103709.

32. Z. M. He, C. Stiewe, D. Platzek, G. Karpinski, E. Muller, S. H. Li, M. Toprak, and M. Muhammed, 2007. Effect of ceramic dispersion on thermoelectric properties of nano-ZrO_2/$CoSb_3$ composites. *J. Appl. Phys.* **101**, 043707.

33. Z. M. He, C. Stiewe, D. Platzek, G. Karpinski, E. Mueller, S. H. Li, M. Toprak, and M. Muhammed, 2007. Nano ZrO_2/$CoSb_3$ composites with improved thermoelectric figure of merit. *Nanotechnology* **18**, 235602.

34. L. D. Chen, X. Y. Huang, M. Zhou, X. Shi, and W. B. Zhang, 2006. The high temperature thermoelectric performances of $Zr_{0.5}Hf_{0.5}Ni_{0.8}Pd_{0.2}Sn_{0.99}Sb_{0.01}$ alloy with nanophase inclusions. *J. Appl. Phys.* **99**, 064305.

35. J. L. Mi, X. B. Zhao, T. J. Zhu, and J. P. Tu, 2007. Improved thermoelectric figure of merit in n-type $CoSb_3$ based nanocomposites. *Appl. Phys. Lett.* **91**, 172116.

36. T. Ikeda, V. Ravi, L. A. Collins, S. M. Haile, and G. J. Snyder, 2006. Development of nanostructures in thermoelectric Pb–Te–Sb alloys. *ICT'06: XXV International Conference on Thermoelectrics, Proceedings*, Vienna, Austria, 172.

37. F. Yang, T. Ikeda, G. J. Snyder, and C. Dames, 2010. Effective thermal conductivity of polycrystalline materials with randomly oriented superlattice grains. *J. Appl. Phys.* **108**, 034310.

38. T. Wu, W. Jiang, X. O. Li, Y. F. Zhou, and L. D. Chen, 2007. Thermoelectric properties of p-type Fe-doped TiCoSb half-Heusler compounds. *J. Appl. Phys.* **102**, 103705.

39. B. Cantor, K. B. Kim, and P. J. Warren, 2002. Novel multicomponent amorphous alloys. *Metastable, Mech. Alloyed Nanocry. Mater.* **386–3**, 27.

40. R. Li, Q. Yang, S. J. Pang, C. L. Ma, and T. Zhang, 2008. Misch metal based metallic glasses. *J. Alloy Comp.* **450**, 181.

41. A. Michalcova, D. Vojtech, G. Schumacher, P. Novak, M. Klementova, J. Serak, M. Mudrova and J. Valdaufova, 2010. Influence of cooling rate and cerium addition on rapidly solidified Al-TM alloys. *Kovove Mater.* **48**, 1.

42. R. E. Napolitano and H. Meco, 2004. The role of melt pool behavior in free-jet melt spinning. *Metall Mater Trans.* A **35A**, 1539.

43. H. Li, X. F. Tang, X. L. Su, Q. J. Zhang, and C. Uher, 2009. Nanostructured bulk $Yb_xCo_4Sb_{12}$ with high thermoelectric performance prepared by the rapid solidification method. *J. Phys. D: Appl. Phys.* **42**, 145409.

44. Y. Z. Pei, L. D. Chen, W. Zhang, X. Shi, S. Q. Bai, X. Y. Zhao, Z. G. Mei, and X. Y. Li, 2006. Synthesis and thermoelectric properties of $K_yCo_4Sb_{12}$. *Appl. Phys. Lett.* **89**, 221107.

45. G. S. Nolas, M. Kaeser, R. T. Littleton, and T. M. Tritt, 2000. High figure of merit in partially filled ytterbium skutterudite materials. *Appl. Phys. Lett.* **77**, 1855.

46. L. D. Chen, T. Kawahara, X. F. Tang, T. Goto, T. Hirai, J. S. Dyck, W. Chen, and C. Uher, 2001. Anomalous barium filling fraction and n-type thermoelectric performance of $Ba_yCo_4Sb_{12}$. *J. Appl. Phys.* **90**, 1864.

47. W. Y. Zhao, P. Wei, Q. J. Zhang, C. L. Dong, L. S. Liu, and X. F. Tang, 2009. Enhanced thermoelectric performance in barium and indium double-filled skutterudite bulk materials via orbital hybridization induced by indium filler. *J. Am. Chem. Soc.* **131**, 3713.

48. B. C. Sales, B. C. Chakoumakos, and D. Mandrus, 2000. Thermoelectric properties of thallium-filled skutterudites. *Phys. Rev. B* **61**, 2475.

49. W. Kim, J. Zide, A. Gossard, D. Klenov, S. Stemmer, A. Shakouri, and A. Majumdar, 2006. Thermal conductivity reduction and thermoelectric figure of merit increase by embedding nanoparticles in crystalline semiconductors. *Phys. Rev. Lett.* **96**, 045901.

50. C. J. Vineis, A. Shakouri, A. Majumdar, and M. G. Kanatzidis, 2010. Nanostructured thermoelectrics: Big efficiency gains from small features. *Adv. Mater.* **22**, 3970.

51. M. Zebarjadi, K. Esfarjani, Z. Bian, and A. Shakouri, 2010. Low-temperature thermoelectric power factor enhancement by controlling nanoparticle size distribution. *Nano Lett.* **11**, 225.

52. M. Zebarjadi, K. Esfarjani, A. Shakouri, J. H. Bahk, Z. X. Bian, G. Zeng, J. Bowers, H. Lu, J. Zide, and A. Gossard, 2009. Effect of nanoparticle scattering on thermoelectric power factor. *Appl. Phys. Lett.* **94**, 202105.

53. X. Zhen and C. Li-Dong, 2010. Recent progress of thermoelectric nano-composites. *J. Inorg. Mater.* **25**, 561.

54. J. He, J. R. Sootsman, S. N. Girard, J.-C. Zheng, J. Wen, Y. Zhu, M. G. Kanatzidis, and V. P. Dravid, 2010. On the origin of increased phonon scattering in nanostructured PbTe based thermoelectric materials. *J. Am. Chem. Soc.* **132**, 8669.

55. C. W. Nan, X. P. Li, and R. Birringer, 2000. Inverse problem for composites with imperfect interface: Determination of interfacial thermal resistance, thermal conductivity of constituents, and micro-structural parameters. *J. Am. Ceram. Soc.* **83**, 848.

56. W. Kim, S. Singer, A. Majumdar, J. Zide, A. Gossard, and A. Shakouri, 2005. Role of nanostructures in reducing thermal conductivity below alloy limit in crystalline solids. *ICT: 2005 24th International Conference on Thermoelectrics*, Clemson, US, 9-12533.

57. D. Vashaee and A. Shakouri, 2004. Improved thermoelectric power factor in metal-based super-lattices. *Phys. Rev. Lett.* **92**, 106103.

58. C. J. Vineis, T. C. Harman, S. D. Calawa, M. P. Walsh, R. E. Reeder, R. Singh, and A. Shakouri, 2008. Carrier concentration and temperature dependence of the electronic transport properties of epitaxial PbTe and PbTe/PbSe nanodot superlattices. *Phys. Rev. B* **77**, 235202.

59. G. P. Meisner, D. T. Morelli, S. Hu, J. Yang, and C. Uher, 1998. Structure and lattice thermal conductivity of fractionally filled skutterudites: Solid solutions of fully filled and unfilled end members. *Phys. Rev. Lett.* **80**, 3551.

60. M. M. Koza, M. R. Johnson, R. Viennois, H. Mutka, L. Girard, and D. Ravot, 2008. Breakdown of phonon glass paradigm in La- and Ce-filled Fe_4Sb_{12} skutterudites. *Nat. Mater.* **7**, 805.

61. J. R. Sootsman, R. J. Pcionek, H. J. Kong, C. Uher, and M. G. Kanatzidis, 2006. Strong reduction of thermal conductivity in nanostructured PbTe prepared by matrix encapsulation. *Chem. Mater.* **18**, 4993.

62. Q. Zhang, J. He, T. J. Zhu, S. N. Zhang, X. B. Zhao, and T. M. Tritt, 2008. High figures of merit and natural nanostructures in $Mg_2Si_{0.4}Sn_{0.6}$ based thermoelectric materials. *Appl. Phys. Lett.* **93**, 102109.

63. J. H. Bahk, Z. Bian, M. Zebarjadi, J. M. O. Zide, H. Lu, D. Xu, J. P. Feser et al. 2010. Thermoelectric figure of merit of $(In_{0.53}Ga_{0.47}As)_{0.8}(In_{0.52}Al_{0.48}As)_{0.2}$ III-V semiconductor alloys. *Phys. Rev. B* **81**, 235209.

64. G. A. Lamberton, R. H. Tedstrom, T. M. Tritt, and G. S. Nolas, 2005. Thermoelectric properties of Yb-filled Ge-compensated $CoSb_3$ skutterudite materials. *J. Appl. Phys.* **97**, 113715.

65. Xiong Z, J. D. X. Chen, X. Huang, H. Gu, L. Chen, W. Zhang, unpublished.

66. A. F. May, E. S. Toberer, A. Saramat, and G. J. Snyder, 2009. Characterization and analysis of thermo-electric transport in n-type $Ba_8Ga_{16-x}Ge_{30+x}$. *Phys. Rev. B* 80, 125205.

67. A. F. May, J.-P. Fleurial, and G. J. Snyder, 2010. Optimizing thermoelectric efficiency in $La_{3-x}Te_4$ via Yb substitution. *Chem. Mater.* 22, 2995.

68. S. Q. Bai, Y. Z. Pei, L. D. Chen, W. Q. Zhang, X. Y. Zhao, and J. Yang, 2009. Enhanced thermoelectric performance of dual-element-filled skutterudites $Ba_xCe_yCo_4Sb_{12}$. *Acta Mater.* **57**, 3135.

69. J. R. Sootsman, H. Kong, C. Uher, J. J. D'Angelo, C. I. Wu, T. P. Hogan, T. Caillat, and M. G. Kanatzidis, 2008. Large enhancements in the thermoelectric power factor of bulk PbTe at high temperature by synergistic nanostructuring. *Angew Chem Int Edit* **47**, 8618.

70. J. Ding, L. D, Chen, Hui Gu, unpublished.

71. L. D. Hicks, T. C. Harman, and M. S. Dresselhaus, 1993. Use of quantum-well superlattices to obtain a high figure of merit from nonconventional thermoelectric-materials. *Appl. Phys. Lett.* **63**, 3230.

72. L. D. Hicks and M. S. Dresselhaus, 1993. Thermoelectric figure of merit of a one-dimensional conductor. *Phys. Rev. B* **47**, 16631.

4

C.L. Wan
Nagoya University
CREST

Y.F. Wang
Nagoya University

N. Wang
Nagoya University

Y.E. Putri
Nagoya University

W. Norimatsu
Nagoya University

M. Kusunoki
Nagoya University

K. Koumoto
Nagoya University
CREST

Layer-Structured Metal Sulfides as Novel Thermoelectric Materials

4.1 Introduction

Thermoelectric materials have been considered to give an effective solution for the increasing energy crisis nowadays.[1] By taking advantage of the Seebeck effect, thermoelectric materials can generate electricity from waste heat that widely exists in automobile exhaust, various industrial processes, and even renewable energy, such as solar heat. The figure of merit of thermoelectric materials is defined as follows: $ZT = S^2\sigma T/k$, where S, σ, and k represent Seebeck coefficient, electrical conductivity, and thermal conductivity, respectively. Since the ZT values of the current materials are too low for cost-effective applications, various efforts have been made to improve them. The concept of phonon-glass electron-crystal (PGEC) was proposed and has become a general guideline for developing new thermoelectric materials.[2] In order to obtain a PGEC material, the idea of complex structure was put forward which imagines a material with distinct regions providing different functions.[1] It is believed that the ideal thermoelectric material would have regions of the structure composed of a high-mobility semiconductor that provides the electron-crystal electronic structure, interwoven with a phonon glass. The phonon-glass region would be ideal for housing dopants and disordered structures without disrupting the carrier mobility in the electron-crystal region.[1]

Based on the above idea, we intercalate a layer of MS (M = Pb, Bi, Sn) into the van der Waals gap of layered TiS_2, forming a series of misfit layer compounds $(MS)_{1+x}(TiS_2)_2$.[3–5] They consist of an alternative stacking of CdI_2-type TiS_2 trigonal antiprismatic layers and rock-salt-type MS slabs, which could be viewed as a natural superlattice.[6] The TiS_2 layer can provide thermopower as well as electron pathway according to Imai's research on TiS_2 single crystal.[7] The MS layer was intercalated into the gap of the TiS_2 layers to form a modulated structure which would suppress the transport of phonons by the interaction between the MS layer and TiS_2 layer and/or disruption of the periodicity of TiS_2 in the direction perpendicular to the layers.

Moreover, the structure of these natural superlattice materials can be varied to some extent, including the species of the host material, the intercalated material as well as the ratio of these two components, thus constituting the large family of chalcogenide misfit layer compounds, $(MX)_{1+x}(TX_2)_n$ (M = Pb, Bi, Sn, Sb, rare earth elements; T = Ti, V, Cr, Nb, Ta; X = S, Se; n = 1, 2, 3).[6] The structure and physical properties of misfit layer compounds have been intensively investigated in the 1990s.[6,8,9] By now, only a few studies have been performed on the thermoelectric properties of these compounds. Miyazaki et al.[10] prepared polycrystalline samples of $(Yb_{1.90}S_2)_{0.62}NbS_2$ and obtained a ZT value of 0.1 at 300 K. A modulated elemental reactants method was developed to make thin films of the misfit layer compounds with designed composition.[11–13] Low thermal conductivities were found and an ability of tuning the carrier concentration was also demonstrated, but a ZT below 0.02 was reported for the $(PbSe)_{0.99}(WSe_2)_x$ superlattice thin films because of the low in-plane electrical conductivity.[14]

4.2 Microstructure and Thermoelectric Properties of $(MS)_{1+x}(TiS_2)_2$

4.2.1 Crystal Structure

The crystal structure of $(SnS)_{1.2}(TiS_2)_2$ is analyzed as an example. The stage-2 compound $(SnS)_{1.2}(TiS_2)_2$ consists of alternative stacks of one SnS layer and paired TiS_2 layers. Its crystal structure can be analyzed by analogy with the available data of stage-1 compound $(SnS)_{1.2}TiS_2$, because the subsystem of SnS and TiS_2 are more or less identical in these two types of compositions.[15,16] Both of the two subsystems, SnS and TiS_2, of $(SnS)_{1.2}(TiS_2)_2$ are triclinic. The space groups are C-1 for SnS and F-1 for TiS_2. The SnS layer consists of deformed slices of SnS with a thickness of half the cell edge of the distorted NaCl-type SnS. Each Sn atom is coordinated by five S atoms within the same layer and two or three S atoms of the TiS_2 layers. The Sn atoms are slightly pushed out of the $a-b$ plane with respect to the sulfur atoms in the same plane. The structure of the TiS_2 part is hardly distorted compared to the crystal structure of $1T$-TiS_2, in which Ti is octahedrally coordinated. The Ti atoms have a trigonally antiprismatic coordination by six S atoms. In $(SnS)_{1.2}(TiS_2)_2$, the van der Waals gap still exists between the paired TiS_2 layers as in $1T$-TiS_2. The SnS layer and the paired TiS_2 layers stack in the direction of the c axis, although the two c axes of the two subsystems slightly diverge. Inside the layers, the b axes are parallel and of equal length while the a axes, also being parallel, have different lengths and the ratio is about 3:5. Therefore, the value of 1.2 in the chemical formula of $(SnS)_{1.2}(TiS_2)_2$ is obtained by taking account of the ratio of lattice parameters and number of atoms contained in one unit cell.

The crystal structure of $(SnS)_{1.2}(TiS_2)_2$ cannot be refined by an XRD pattern here, due to the presence of translational disorder, which will be shown below. The atomic structure along the incommensurate axis ([100] zone axis) is directly observed by HRTEM, as shown in Figure 4.1. The layered structure is shown, though detailed composition cannot be determined immediately. We simulated the patterns of separated SnS and TiS_2 layers using the available structure data of $(SnS)_{1.2}TiS_2$, which agrees well with the observed pattern. (See the square with white frame in Figure 4.1.) It then allows reconstruction of the whole crystal structure of $(SnS)_{1.2}(TiS_2)_2$ along the a axis by "translating" the observed pattern, which is shown on the right side of Figure 4.1. It can be seen that the SnS layer and the paired TiS_2 layers stack alternatively as expected. The paired TiS_2 layers, separated by a van der Waals gap, almost stack in the same way as that in pure TiS_2. In contrast to the consistency of TiS_2 layers stacking in the whole crystal, the relative position between the SnS layer and TiS_2 layer varies. The SnS layers deposited on both sides of the same paired TiS_2 reference layers occupy two different relative positions. The Sn atoms of the SnS layer either reside in the middle of the two sulfur atom rows of the neighboring TiS_2 layer or stay exactly vertical to one of them by a translational displacement of $b/4$ along the b-axis. In the direction of the commensurate b axis, the lattice parameters of SnS and TiS_2 are equal and the first case is reported in almost all the previous structural study of misfit layer compounds.[6, 17–19] The second case has rarely been

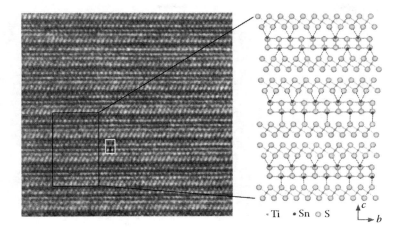

FIGURE 4.1 HRTEM image and simulated crystal structure of $(SnS)_{1.2}(TiS_2)_2$ along the [100] zone axis.

observed, and we anticipate that strain may play a role and the translational disorder may have been formed during pressure sintering.

4.2.2 Electrical Properties

Both TiS_2 and $(MS)_{1+x}(TiS_2)_2$ show a metallic electrical conductivity in the "in-plane" direction as shown in Figure 4.2. They also show anisotropic behavior and the "in-plane" values are much higher than the "cross-plane" values. The anisotropy of TiS_2 and $(MS)_{1+x}(TiS_2)_2$ is much lower than their respective single crystals, because the $(00l)$ planes of the polycrystalline samples are not perfectly oriented and any deflection can decrease this anisotropy. However, the degrees of orientation of the $(00l)$ planes in TiS_2 and $(MS)_{1+x}(TiS_2)_2$ are almost the same according to the rocking curve measurement, thereby enabling reasonable comparison of transport properties in the same direction for these compositions. In the "in-plane" direction, all the $(MS)_{1+x}(TiS_2)_2$ compounds show higher electrical conductivity than TiS_2 and it increases in the sequence of Sn, Pb, Bi in the whole temperature range.

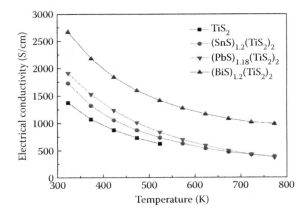

FIGURE 4.2 Electrical conductivities of TiS_2 and $(MS)_{1+x}(TiS_2)_2$.

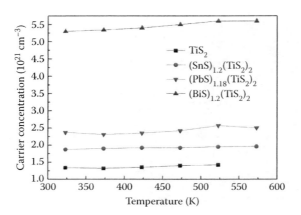

FIGURE 4.3 Carrier concentrations of TiS_2 and $(MS)_{1+x}(TiS_2)_2$.

The electrical conductivity of materials is determined by the carrier concentration and mobility. Hall measurement was performed to analyze the electron transport properties in these misfit layer compounds. The Hall coefficients are all negative, showing that the dominant carriers in these compounds are electrons. As shown in Figure 4.3, all the compositions show high carrier concentrations which are almost temperature independent, supporting the metallic conduction mechanism. All the $(MS)_{1+x}(TiS_2)_2$ compositions have higher carrier concentration than that of TiS_2. A general viewpoint has been accepted that there exists electron transfer from the MX layer to the TX_2 layers in the misfit layer compounds $(MX)_{1+x}(TX_2)_n$, which also accounts for its stability.[6] In this case, there must be electron transfer from the MS layer to the TiS_2 layer, resulting in an increase in carrier concentration. From the carrier concentrations and the lattice parameters, we can estimate the number of electrons received per Ti atom for $(BiS)_{1.2}(TiS_2)_2$, $(SnS)_{1.2}(TiS_2)_2$, and $(PbS)_{1.18}(TiS_2)_2$ is 0.45, 0.16, and 0.2, respectively. Much more electron transfer takes place in $(BiS)_{1.2}(TiS_2)_2$ than the other two compositions, because the nominal valence of bismuth is 3+ here and one can easily deduce that one electron can be transferred from one BiS layer to two TiS_2 layers, and hence each Ti atom receives 0.6 electron, which is in reasonable agreement with the above estimation. The carrier concentration of TiS_2 is higher than that of stoichiometric single crystal TiS_2,[7] due to either excess Ti atoms into the van der Waals gap or sulfur vacancies, which can hardly be avoided in the synthesis process of TiS_2.

The mobilities for both TiS_2 and $(SnS)_{1.2}(TiS_2)_2$ show temperature dependency proportional to $T^{-1.5}$, showing that the electrons are mainly scattered by acoustic phonons (Figure 4.4). It is noticed that in the in-plane direction, TiS_2 has a mobility of 6.8 $cm^2 V^{-1} s^{-2}$ at around 295 K, which is close to the corresponding value of 7.7 $cm^2 V^{-1} s^{-2}$ found in TiS_2 single crystal with the same carrier concentration.[20] This demonstrates that the $(00l)$ planes of the polycrystalline samples here are highly oriented so that the in-plane electron transport properties are close to those of a single crystal. In the in-plane direction, the electron mobility of TiS_2 is almost maintained after the intercalation of SnS or PbS layers. However, the BiS intercalation has much lower mobility. The electron transfer from the MS layers to the TiS_2 layers may also change the effective mass, resulting in different mobilities. An estimation of the effective mass will be shown below.

The Seebeck coefficients of $(MS)_{1+x}(TiS_2)_2$ are decreased compared with that of TiS_2 (Figure 4.5). It has generally been accepted that $1T$-TiS_2 is a semiconductor with an indirect gap of 0.2–0.3 eV, with its conduction and valence bands consisting of Ti $3d$ and S $3p$ states, respectively.[21,22] In stoichiometric TiS_2, the Fermi level lies in the bottom of Ti $3d$ band, making multivalley structure with six small electron pockets around the L-point in the hexagonal Brillouin zone.[7] Large Seebeck coefficient was observed due to high density of states just above the Fermi level as well as phonon-mediated inter-valley scattering of

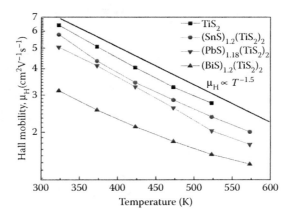

FIGURE 4.4 Hall mobilities of TiS_2 and $(MS)_{1+x}(TiS_2)_2$.

conduction electrons. In the nonstiochiometric TiS_2, excess Ti atoms residing in the van der Waals gap or sulfur vacancies become electron donors, resulting in the shift of Fermi level into the 3d band from the bottom and a decrease in Seebeck coefficient. In $(MS)_{1+x}(TiS_2)_2$, a rigid band model can hold and the band structure is a superposition of those of MS and TiS_2 subsystem, as in the case of $(SnS)_{1.2}TiS_2$.[15,23,24] It is realized that the d orbitals of Ti plays an important role in determining the physical properties of TiS_2-based materials and the degree of band filling, their energy levels and the width of the d-band significantly affect their thermoelectric properties.[23] In $(MS)_{1+x}(TiS_2)_2$, the position of the Fermi level in the 3d orbital of Ti atom can be shifted upward due to band filling by those electrons transferred from the MS layer. The density of states effective mass $m^\star$ values for $(BiS)_{1.2}(TiS_2)_2$, $(SnS)_{1.2}(TiS_2)_2$, and $(PbS)_{1.18}(TiS_2)_2$, were calculated to be $6.3m_0$, $4.8m_0$, and $4.5m_0$, respectively, where m_0 is the bare electron mass. It can be seen that $(BiS)_{1.2}(TiS_2)_2$ has the highest effective mass, resulting in the lowest mobility as shown in Figure 4.4.

As shown in Figure 4.6, the power factors of the $(MS)_{1+x}(TiS_2)_2$ compositions are lower than that of TiS_2. At lower temperatures, the power factors almost increase in the order of Bi < Pb < Sn, indicative of increased carrier concentration. Although the carrier concentration in $(MS)_{1+x}(TiS_2)_2$ is not yet optimized, it can be expected that further reduction in carrier concentration would increase the power

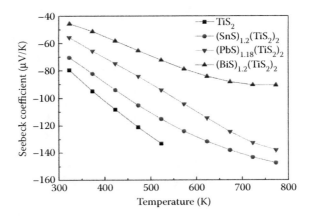

FIGURE 4.5 Seebeck coefficients of TiS_2 and $(MS)_{1+x}(TiS_2)_2$.

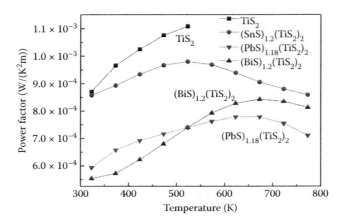

FIGURE 4.6   Power factors of TiS_2 and $(MS)_{1+x}(TiS_2)_2$.

factor. In fact, an optimum carrier concentration for the highest power factor of a typical thermoelectric material is of the order of 10^{20} cm^{-3} [1], which is one order of magnitude lower than the corresponding values for the $(MS)_{1+x}(TiS_2)_2$ compositions ($>10^{21}$ cm^{-3}).

4.2.3 Thermal Conductivity

The thermal conductivities of TiS_2 and $(MS)_{1+x}(TiS_2)_2$ are shown in Figure 4.7. All the $(MS)_{1+x}(TiS_2)_2$ compositions have lower thermal conductivities than TiS_2 in the in-plane directions in the whole temperature range. Since the thermal conductivity comes from two sources: (1) electrons and holes transporting heat (k_e) and (2) phonons travelling through the lattice (k_l), the electronic thermal conductivity (k_e) is directly related to the electrical conductivity through the Wiedemann–Franz law: $k_e = L_0 T\sigma$, where the Lorentz number, L_0, is 2.44×10^{-8} W S^{-1} K^{-2}. The values of k_e of these $(MS)_{1+x}(TiS_2)_2$ compositions were calculated and plotted in Figure 4.7. It can be seen that the k_e in $(MS)_{1+x}(TiS_2)_2$ is close to or even higher than that of TiS_2. The main reason of the reduction of the thermal conductivity is the reduction of lattice thermal conductivity.

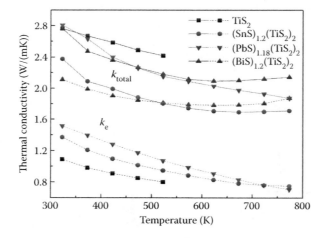

FIGURE 4.7   Total thermal conductivities (k_{total}, solid line) and electron thermal conductivities (k_e, dashed line) of TiS_2 and $(MS)_{1+x}(TiS_2)_2$.

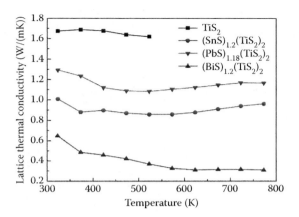

FIGURE 4.8 Lattice thermal conductivities of TiS$_2$ and $(MS)_{1+x}(TiS_2)_2$.

It can be noticed in Figure 4.8 that the $(MS)_{1+x}(TiS_2)_2$ has much lower lattice thermal conductivities than TiS$_2$ in the in-plane directions. To analyze the low thermal conductivities, the kinetic theory of thermal conductivity was used.

$$k = 1/3C_v Vl \qquad (4.1)$$

where C_v, V, and l represent the heat capacity, sound velocity, and phonon mean free path, respectively. The heat capacity makes limited contribution to the low thermal conductivity, as the heat capacity approaches $3k_B$ per atom at temperatures higher than the Debye temperature, according to the Dulong–Petit law. Therefore, the reduced thermal conductivity mainly arises from reduction in sound velocity or phonon mean free path.

The sound velocity is determined by the density and the elastic constants of a solid. As shown in Figure 4.9, the sound velocity has three polarization modes, including one longitudinal mode and two transverse modes. The two transverse velocities are equal in the cross-plane direction because of nearly isotropic structure inside the layers and random orientation of the a–b planes of different grains. In contrast, in the in-plane direction, the two transverse velocities are not equal because of the layered structure. A pulse–echo method was used to measure these sound velocities with a 30 MHZ longitudinal transducer and a 20 MHZ transverse transducer. The measured values are listed in Table 4.1.

Compared with pure TiS$_2$, the longitudinal velocities of the misfit layer compounds are a little decreased, which can be attributed to the increase in density. In contrast, the transverse sound velocities,

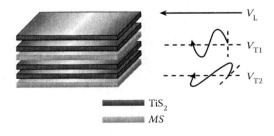

FIGURE 4.9 Schematic illustration of the longitudinal and transverse sound velocities of the layered $(MS)_{1+x}(TiS_2)_2$ compounds.

TABLE 4.1   Densities, Longitudinal and Transverse Sound Velocities, and Shear Moduli of TiS_2, and $(MS)_{1+x}(TiS_2)_2$

Material	ρ (g/cm³)	V_L (m/s)	V_{T1} (m/s)	V_{T2} (m/s)	G_1 (GPa)	G_2 (GPa)
TiS_2	3.21	5284	2799	3295	25.0	34.7
$(BiS)_{1.2}(TiS_2)_2$	4.57	3662	1350	1688	8.3	13.0
$(PbS)_{1.18}(TiS_2)_2$	4.69	3834	1120	1837	5.9	15.8
$(SnS)_{1.2}(TiS_2)_2$	3.87	4111	1578	2352	9.6	21.4

especially V_{T1}, apparently decreased, which arise from the softening of atomic bonding. The transverse polarization is a kind of shear movement, and the velocity is determined by shear modulus as follows:

$$V_T = \sqrt{\frac{G}{\rho}} \tag{4.2}$$

where G is the shear modulus and ρ is the density. The shear modulus is calculated by the above equation and shown in Table 4.1. The shear moduli of the misfit layer compounds are much lower than those of pure TiS_2 due to the intercalation of the MS layers into the TiS_2 layers. It can also be seen that the velocities of the two transverse waves (V_{T1} and V_{T2}) are different, as V_{T1} is mainly determined by the interlayer bonding while V_{T2} is determined by the intralayer bonding. For V_{T1}, the weak interlayer bonding between the MS layer and TiS_2 layer arises either from the electrostatic interaction due to the electron transfer between these layers or weak covalent bonds between the M atom and the sulfur atoms in the TiS_2 layers.[15,25] For V_{T2}, the intralayer bonding is weakened by intercalating MS layers, possibly due to the incommensurate structure or disruption of periodicity of TiS_2 layers in the direction perpendicular to the layers.

Besides the softening of transverse sound velocities, additional investigation is required to understand the difference in the lattice thermal conductivity of the $(MS)_{1+x}(TiS_2)_2$ compounds.[26] It is found that $(PbS)_{1.18}(TiS_2)_2$ has an ordered structure and each Pb atom is coordinated by two sulfur atoms in the neighboring TiS_2 layer along the [100] direction. Some SnS layers in $(SnS)_{1.2}(TiS_2)_2$ show translational displacement of $b/4$ along the b-axis, resulting in a single bonding configuration for the Sn atom with the sulfur atoms in the neighboring TiS_2 layer. In $(BiS)_{1.2}(TiS_2)_2$, stacking faults appear because of the coexistence of $(BiS)_{1.2}TiS_2$ and $(BiS)_{1.2}(TiS_2)_2$ phases. Besides, the atomic planes are even found to be bent or distorted along the [310] direction in $(BiS)_{1.2}(TiS_2)_2$. The SAED patterns also support the increasing lattice disorder in the sequence of Pb, Sn, Bi for $(MS)_{1+x}(TiS_2)_2$, as the electron patterns become more and more diffuse (Figure 4.10). Finally, the lattice thermal conductivity decreased significantly as the disorder increases. It is even more interesting to find that the relaxation time of the electrons is almost

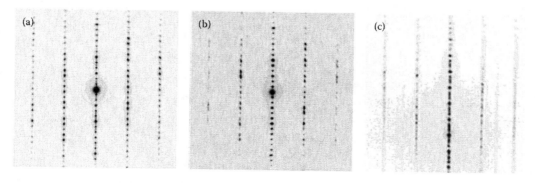

FIGURE 4.10   SAED patterns of (a) $(PbS)_{1.18}(TiS_2)_2$, (b) $(SnS)_{1.2}(TiS_2)_2$, and (c) $(BiS)_{1.2}(TiS_2)_2$ along the [100] zone axis.

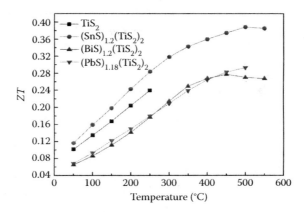

FIGURE 4.11 Electrical conductivities of TiS_2 and $(MS)_{1+x}(TiS_2)_2$.

insensitive against these planar defects, suggesting that the electron transport is unaffected while the phonons are strongly scattered.

4.2.4 *ZT* Value

It has been found that all the misfit layer compounds show relatively high *ZT* values in the in-plane direction (Figure 4.11). However, compared with pure TiS_2, only $(SnS)_{1.2}(TiS_2)_2$ shows a higher *ZT* value. The intercalation of *MS* layer does reduce the lattice thermal conductivity significantly, but the electron transfer effect in $(PbS)_{1.18}(TiS_2)_2$ and $(BiS)_{1.2}(TiS_2)_2$ is too strong, leading to a reduction in the power factor and an increase in the electronic thermal conductivity. It is required to reduce the carrier concentrations of the misfit layer compounds to optimize the thermoelectric performances.

4.3 Concluding Remarks

Thermoelectric properties of a series of misfit layer compounds $(MS)_{1+x}(TiS_2)_2$ (M = Pb, Bi, Sn) are studied, which appear to be promising for medium-temperature *n*-type thermoelectric materials. This naturally modulated structure shows a low lattice thermal conductivity. Measurement of sound velocities shows that the ultralow thermal conductivity partially originates from the softening of the transverse sound velocities due to weak interlayer bonding. Various planar defects including translational displacement and stacking faults are found in those misfit layer compounds and further reduce the lattice thermal conductivity. Meanwhile, electron transfer from the *MS* layer to the TiS_2 layer deteriorates the thermoelectric performance by reducing the power factor and increasing the electronic thermal conductivity. The SnS intercalation compound $(SnS)_{1.2}(TiS_2)_2$ shows the least electron transfer and the *ZT* value reaches 0.37 at 700 K. Reduction in the carrier concentration in these misfit layer compounds is required to achieve higher *ZT* value.

Moreover, we propose a large family of misfit layer compounds $(MX)_{1+x}(TX_2)_n$ (M = Pb, Bi, Sn, Sb, rare earth elements; T = Ti, V, Cr, Nb, Ta; X = S, Se; n = 1, 2, 3) with natural superlattice structures for possible candidates for high-performance thermoelectric materials, including both *n*- and *p*-type.

References

1. Snyder, G. J., Toberer, E. S. 2008. Complex thermoelectric materials. *Nat. Mater.* 7:105–14.
2. G.A., Slack. 1995. *CRC Handbook of Thermoelectrics*. Boca Raton, FL: CRC.

3. Wan, C., Wang, Y., Wang, N., Koumoto, K. 2010. Low-thermal-conductivity $(MS)_{1+x}(TiS_2)_2$ (M = Pb, Bi, Sn) misfit layer compounds for bulk thermoelectric materials. *Materials* 3:2606–17.

4. Wan, C. et al. 2010. Development of novel thermoelectric materials by reduction of lattice thermal conductivity. *Sci. Tech. Adv. Mater.* 11:044306.

5. Wan, C., Wang, Y., Wang, N., Norimatsu, W., Kusunoki, M., Koumoto, K. 2011. Intercalation: Building a natural superlattice for better thermoelectric performance in layered chalcogenides. *J. Electron. Mater.* 40:1271–80.

6. Wiegers, G. A. 1996. Misfit layer compounds: Structures and physical properties. *Prog. Solid State Chem.* 24:1–139.

7. Imai, H., Shimakawa, Y., Kubo, Y. 2001. Large thermoelectric power factor in TiS_2 crystal with nearly stoichiometric composition. *Phys. Rev. B* 64:241104.

8. Wiegers, G. A., Meerschaut, A. 1992. Structures of misfit layer compounds $(MS)_n TS_2$ (M = Sn, Pb, Bi, rare earth metals; T = Nb, Ta, Ti, V, Cr; $1.08 < n < 1.23$). *J. Alloy Compd.* 178:351–68.

9. Rouxel, J., Meerschaut, A., Wiegers, G. A. 1995. Chalcogenide misfit layer compounds. *J. Alloy Compd.* 229:144–57.

10. Miyazaki, Y., Ogawa, H., Kajitani, T. 2004. Preparation and thermoelectric properties of misfit-layered sulfide $[Yb_{1.90}S_2]_{(0.62)}NbS_2$. *J. Appl. Phys.* 43:L1202–L04.

11. Chiritescu, C., Cahill, D. G., Heideman, C., Lin, Q. Y., Mortensen, C., Nguyen, N. T., Johnson, D., Rostek, R., Bottner, H. 2008. Low thermal conductivity in nanoscale layered materials synthesized by the method of modulated elemental reactants. *J. Appl. Phys.* 104:033533.

12. Heideman, C., Nyugen, N., Hanni, J., Lin, Q., Duncombe, S., Johnson, D. C., Zschack, P. 2008. The synthesis and characterization of new $[(BiSe)_{(1.10)}]_{(m)}[NbSe_2]_{(n)}$, $[(PbSe)_{(1.10)}]_{(m)}[NbSe_2]_{(n)}$, $[(CeSe)_{(1.14)}]_{(m)}[NbSe_2]_{(n)}$ and $[(PbSe)_{(1.12)}]_{(m)}[TaSe_2]_{(n)}$ misfit layered compounds. *J. Solid State Chem.* 181:1701–06.

13. Lin, Q., Heideman, C. L., Nguyen, N., Zschack, P., Chiritescu, C., Cahill, D. G., Johnson, D. C. 2008. Designed synthesis of families of misfit-layered compounds. *Eur. J. Solid State Inorg. Chem.* 2008:2382–85.

14. Mavrokefalos, A., Lin, Q. Y., Beekman, M., Seol, J. H., Lee, Y. J., Kong, H. J., Pettes, M. T., Johnson, D. C., Shi, L. 2010. In-plane thermal and thermoelectric properties of misfit-layered $[(PbSe)_{(0.99)}]_{(x)}$ $(WSe_2)_{(x)}$ superlattice thin films. *Appl. Phys. Lett.* 96:181908.

15. Fang, C. M., deGroot, R. A., Wiegers, G. A., Haas, C. 1996. Electronic structure of the misfit layer compound $(SnS)_{(1.20)}TiS_2$: Band structure calculations and photoelectron spectra. *J. Phys.-Condens. Mat.* 8:1663–76.

16. Wiegers, G. A., Meetsma, A., Deboer, J. L., Vansmaalen, S., Haange, R. J. 1991. X-ray crystal-structure determination of the triclinic misfit layer compound $(SnS)_{1.20}TiS_2$. *J. Phys.-Condens. Mat.* 3:2603–12.

17. Auriel, C., Meerschaut, A., Roesky, R., Rouxel, J. 1992. Crystal-structure determination and transport-properties of a new misfit layer compound $(Pbse)_{1.12}(NbSe_2)_2$. *Eur. J. Solid State Inorg. Chem.* 29:1079–91.

18. Meerschaut, A., Auriel, C., Rouxel, J. 1992. Structure determination of a new misfit layer compound $(PbS)_{1.18}(TiS_2)_2$. *J. Alloy Compd.* 183:129–37.

19. Meerschaut, A., Guemas, L., Auriel, C., Rouxel, J. 1990. Preparation, structure determination and transport-properties of a new misfit layer compound—$(PbS)_{1.14}(NbS_2)_2$. *Eur. J. Solid State Inorg. Chem.* 27:557–70.

20. Klipstein, P. C., Bagnall, A. G., Liang, W. Y., Marseglia, E. A., Friend, R. H. 1981. Stoichiometry dependence of the transport-properties of TiS_2. *J. Phys. C Solid State* 14:4067–81.

21. Wilson, J. A. 1978. Modeling contrasting semimetallic characters of TiS_2 and $TiSe_2$. *Phys. Status Solidi B* 86:11–36.

22. Barry, J. J., Hughes, H. P., Klipstein, P. C., Friend, R. H. 1983. Stoichiometry effects in angle-resolved photoemission and transport studies of $Ti_{1+x}S_2$. *J. Phys. C Solid State* 16:393–402.

23. Meerschaut, A. 1996. Misfit layer compounds. *Curr. Opin. Solid State Mater. Sci.* 1:250–59.

24. Martinez, H., Auriel, C., Gonbeau, D., Pfister-Guillouzo, G., Meerschaut, A. 1998. Electronic structure of two misfit layer compounds: $(PbS)_{(1.18)}(TiS_2)$ and $(PbS)_{(1.18)}(TiS_2)_{(2)}$. *J. Electron. Spectrosc. Relat. Phenom.* 95:145–58.

25. Ohno, Y. 1991. Electronic-structure of the misfit-layer compounds $PbTiS_3$ and $SnNbS_3$. *Phys. Rev. B* 44:1281–91.

26. Wan, C. 2010. Unpublished data.

5

Thermoelectric Properties of Quantum Wires within Chrysotile Asbestos Nanotubes

M.V. Vedernikov
Ioffe Physical-Technical Institute of the Russian Academy of Sciences

Y.V. Ivanov
Ioffe Physical-Technical Institute of the Russian Academy of Sciences

O.N. Uryupin
Ioffe Physical-Technical Institute of the Russian Academy of Sciences

Y.A. Kumzerov
Ioffe Physical-Technical Institute of the Russian Academy of Sciences

5.1 Introduction

Thermoelectric properties of low-dimensional semiconductors have attracted much attention in recent years. Considerable success has been achieved in the preparation of superlattices [1] and the quantum-dot structures [2,3] with the dimensionless thermoelectric figure of merit $ZT > 2$. However, the prospects for thermoelectric applications of quasi-one-dimensional systems are not evident. For nanowires, the calculations give both very high values of the thermoelectric figure of merit [4–7] and the more conservative results [8]. The experimental values of ZT measured on different quasi-one-dimensional systems do not exceed unity [9–12] in spite of a dramatic decrease in the thermal conductivity.

We would like to stress the two important features of present-day thermoelectric investigations of quantum wires. First, diameters of Bi, $Bi_{1-x}Sb_x$, InSb, Bi_xTe_{1-x}, and Si nanowires, whose thermoelectric properties have been experimentally studied [9–16], are relatively large (>10 nm). But according to theoretical estimates [4–7], only nanowires with diameters less than 10 nm can have values of ZT exceeding unity by several times. Second, all calculations of thermoelectric figure of merit of nanowires are based on the Fermi gas (Fermi liquid) model and essentially use quasiclassic Boltzmann equation formalism describing transport of charged quasiparticles. However, it is well known that, in one-dimensional conductors, the Coulomb interaction cannot be considered as a small perturbation. In a one-dimensional wire, even weak electron–electron interaction leads to formation of the special state known as a Luttinger liquid [17–20]. The main feature of this state is the absence of individual excitations similar to the Fermi liquid quasiparticles. Only the collective phonon-like excitations exist in the electronic Luttinger liquid (LL). Moreover, the charge and spin excitations are independent of each other (spin–charge separation). A LL has a rather specific combination of transport properties. Its electrical conductance [21,22] and the thermopower [23–26] grow simultaneously with increasing temperature. Therefore, calculations of the thermoelectric figure of merit of quantum wires based on the Fermi liquid model are inapplicable to ultrathin nanowires.

In this chapter, we will discuss some experimental data on the thermoelectric properties of ultrathin nanowires with diameters smaller than 10 nm and will show that the majority of those can be understood within the framework of an LL theory. The properties of quasi-one-dimensional InSb, Bi, and $Bi_{1-x}Sb_x$ alloy nanowires prepared by a high-pressure injection of molten semiconductors into chrysotile asbestos matrices [27,28] will be mainly considered. Of course, these nanostructures do not exhaust all one-dimensional conductors under study. Ultrathin MoSe [29], $Mo_6S_3I_6$ [30], In_2O_3 [31], and polymer [32,33] nanowires, whose electronic systems exhibit the LL behavior, are described in the literature. However, as far as we know, the thermopower of the indicated wires was not studied and these materials are not interesting for thermoelectric applications. Of special note are carbon nanotubes, which also exhibit the LL behavior [34–36], and bismuth networks embedded in porous host materials (Vycor glass, porous alumina, silica gel) with average sizes of pores from 4 to 15 nm [37–39]. Here we will not discuss these nanostructures because the former have specific two-dimensional crystal and electronic structures and the latter are three-dimensional networks, in which a Fermi liquid rather than an LL is formed.

5.2 Chrysotile Asbestos

Chrysotile, $Mg_3Si_2O_5(OH)_4$, is the most commonly encountered form of asbestos [27]. The natural mineral chrysotile is mined in large quantities. It can also be synthesized by hydrothermal reactions [40] in the systems SiO_2–MgO, SiO_2–$MgCl_2$ or SiO_2–$Mg(OH)_2$. Only the synthetic chrysotile can have the stoichiometric composition [41], the composition of natural asbestos depends on a mineral deposit.

Chrysotile asbestos has a layered structure [40–43]. Each layer is composed of a sheet of Si-centered oxygen tetrahedra joined to a sheet of $MgO_2(OH)_4$ octahedra. The characteristic feature of this structure is a mismatch of lateral dimensions of the sheets. The mismatch is compensated by curving of the layers and formation of tubes with concentric or spiral multilayer walls. Macroscopic bundles of densely packed parallel tubes form chrysotile asbestos (Figure 5.1). The inner and outer diameters of tubes usually fall in the range $1 < d < 10$ nm and $10 < D < 100$ nm, respectively [27,41]. The length of the tubes may be as much as 1 cm. The dimensions of tubes depend on a mineral deposit or conditions of preparation of synthetic chrysotile. However, each sample shows a sharp distribution of inner diameters of nanotubes.

In some respects, chrysotile asbestos nanotubes are similar to carbon nanotubes. On the other hand, there are differences important for practical applications. They are dielectric, uncapped, and much

FIGURE 5.1 Schematic illustration of a close-packed array of nanotubes in chrysotile asbestos.

longer than carbon nanotubes. These properties make it possible to use chrysotile asbestos samples as matrices for a preparation of conductive ultrathin nanowires.

Different methods may be used for a fabrication of nanowires within channels of asbestos nanotubes. For example, InP nanowires have been synthesized [44] by the metal-organic chemical vapor deposition (MOCVD). However, the majority of nanowires are produced by filling of channels in chrysotile matrices with molten semiconductors or metals under a high hydrostatic pressure [27]. InSb, Bi, $Bi_{1-x}Sb_x$, In, Sn, and Te wires were prepared by this method. Chysotile samples with channel diameters of about 5 nm were taken for a fabrication of ultrathin nanowires discussed in Sections 5.4 and 5.5.

5.3 Electronic Transport in a Luttinger Liquid

In three-dimensional metals the Coulomb interaction leads to the formation of a Fermi liquid [45]. In many respects, a Fermi liquid is similar to a Fermi gas because fermionic elementary excitations (quasiparticles) of the former are in one-to-one correspondence with electrons of the latter. In a one-dimensional space, two charged classical particles cannot change the mutual arrangement. Otherwise, they would collide in some point, where their potential energy would be infinite. Therefore, in one-dimensional conductors with a small electron density, the special state of an electronic system known as the Wigner crystal is formed at low temperatures [46]. Electrons occupy the sites of a periodic lattice and only collective bosonic elementary excitations similar to phonons in solids are possible. If the electron density in a one-dimensional conductor is sufficiently large (more than the inverse effective Bohr's radius of the material) and the Coulomb interaction is screened by a metal gate, the kinetic energy of electrons dominates over the potential energy [46]. In this case, an LL is formed [17–20,45] instead of a Wigner crystal. A LL retains the short-range spatial correlations between electrons and the collective phonon-like elementary excitations.

Physical properties of an LL differ markedly from those of a Fermi liquid. For thermoelectric applications, the transport properties of a LL are of crucial importance. Unfortunately, a rigorous theory of electronic transport in an LL is absent at present. The simplest version of the LL theory describes spinless electrons, which are scattered by a single potential barrier [21,47]. The limit of a large barrier is important for practical purposes because strong defects are responsible predominantly for the resistance of nanowires. Recall briefly some results describing this limiting case.

The distinguishing feature of an LL is a power-law energy dependence of the tunneling density of states of physical electrons [19,21,47]

$$\rho(\varepsilon) \propto \left|\varepsilon - \varepsilon_F\right|^\gamma, \tag{5.1}$$

where ε_F is the Fermi energy and the exponent γ (≥ 0) is governed by both a strength of the electron–electron interaction and a location of the tunneling contact, that is, values of the exponent are different for tunneling of electrons into the middle and into the end of a one-dimensional wire (note that only the latter case is important for barrier scattering). One can see that, unlike a Fermi liquid, the tunneling density of states of an LL vanishes at $\varepsilon = \varepsilon_F$. Therefore, tunneling of electrons through the barrier is impossible at $T = 0$. As a result, the linear conductance of an LL vanishes at this temperature.

A current in an LL arises only at finite temperature and/or a nonzero voltage drop across the barrier. According to the theory [21,47], the electrical current $I \propto VT^\alpha$ and the conductance $G \propto T^\alpha$ at low bias voltage ($eV \ll k_B T$). Here V is the voltage drop across the single barrier, $-e$ is the electron charge, and k_B is the Boltzmann constant. The exponent $\alpha = 2\gamma$, if the potential barrier is located in a bulk of a wire (i.e., between two semi-infinite LLs), and $\alpha = \gamma$, if the barrier is located between a semi-infinite LL and a noninteracting lead with a Fermi liquid. In this limit, current–voltage characteristics of an LL are linear and the conductance is described by a power-law function of temperature. At high bias ($eV \gg k_B T$), the current $I \propto V|V|^\beta$ and the differential conductance $G \propto |V|^\beta$, where $\beta = \alpha$ in the framework of the LL theory. In other words, the current–voltage characteristics are nonlinear, while the differential conductance is independent of temperature in this limit. It must be emphasized that the exponent α depends on strength of the electron–electron interaction and a barrier location. In addition, it is different for the spinless and spin-$\frac{1}{2}$ models. In the spinless case under consideration,

$$\alpha_{LL-LL} = \frac{2}{g} - 2, \quad \alpha_{FL-LL} = \frac{1}{g} - 1, \tag{5.2}$$

where g is the Luttinger parameter, which is determined by a strength of the electron–electron interaction. For the repulsive interaction $g < 1$. If the interaction is absent, $g = 1$. In Equation 5.2, the former defines α in the case, when the potential barrier is located between two LLs, the latter corresponds to the barrier location between an LL and a Fermi liquid of a lead.

All foregoing holds in the limit of a large barrier. If the barrier is small, the conductance of an LL attached to noninteracting leads approaches $e^2/2\pi\hbar$ with increasing temperature [48]. This value coincides with the conductance (per spin orientation) of a perfect noninteracting wire.

The power-law dependences indicated above may be used also for description of electron transport in nanowires with a small number of independent defects (impurities, constrictions, etc.). However in the case of N identical barriers, the voltage drop across the single barrier $V = V_{tot}/N$, where V_{tot} is the total applied bias. In the opposite case of large density of defects, the nanowire can be modeled by a chain of quantum dots divided by potential barriers [49]. This approach takes into account the electron–electron interaction in the framework of Coulomb blockade theory. In this model, the electrical current $I(T,V)$ and the conductance $G(T,V)$ are also described by the same power-law dependences, which have been presented above, but the exponents α and β depend on degree of disorder and satisfy the inequality $\alpha \gg \beta$. Much less is known about one-dimensional interacting conductors with an arbitrary disorder. It is hoped that the transport in these wires will also be described by the power laws. However, the exponents α and β should depend on both the interaction and the disorder [50,51].

In an LL, the thermoelectric power induced by a nonlinearity of the electron spectrum is equal to zero [52]. The backscattering of electrons by a potential barrier leads to the finite thermopower, which proves to be a linear function of temperature [23–26], as in bulk metals. Moreover, the estimates show [24,25] that, in an LL with strong electron–electron interaction ($g \ll 1$), the impurity-induced thermopower $S \approx S_0/g^2$, where S_0 is the thermopower of the same nanowire but without the interaction. Since the typical values of $g \sim 0.3$, the thermopower can be strongly enhanced by the interaction due to an appearance of a pseudogap in the tunneling density of states (3.1) near the Fermi level.

The results of calculations of the thermal conductance and the Lorenz number of an LL are varied. In the simplest case of a spinless pure (defect-free) LL connected to noninteracting leads, the thermal conductance is suppressed by an electron–electron interaction [53–55] because the heat in a LL is transported by bosonic excitations, which, in contrast to electrons, are scattered by the contacts between the nanowire and the noninteracting leads. In this case, the Lorenz number of an LL is considerably less, than the universal value $L_0 = \pi^2 k_B^2/3e^2$, and decreases with an increase of temperature and interaction strength. In other more complicated cases taking into account effects of disorder, spin degrees of freedom, and so on, values of the Lorenz number can be both greater than and less than L_0 [23,53,56,57]. For example, in some conditions, a huge violation of the Wiedemann–Franz law, when the Lorenz number of an LL differs from L_0 by thousands times, is possible [57].

It is generally believed that the electron–phonon interaction has a minor effect on transport properties of a LL [58]. The phonon backscattering of electrons is rather weak due to a small number of phonons with the component of the wave vector $q \approx 2k_F$ directed along the nanowire. At the same time, the electron-impurity backscattering is enhanced by the electron interaction because of a small tunneling density of states (1) in the vicinity of the Fermi level. However, in some cases, an influence of the electron–phonon scattering on transport properties of an LL can be considered [58,59]. Effects of electron–phonon coupling should be more noticeable in nanowires, which have several quantum conduction channels (modes) and small Debye temperatures.

5.4 InSb Nanowires

Bulk indium antimonide does not fall into the category of the best thermoelectric materials. However, some estimates [6] show that the thermoelectric figure of merit of InSb quantum wires drastically increases as their thickness decreases. Besides, the unique electronic properties allow one to consider these nanowires as the model to study an influence of strong electron correlations on thermoelectric properties of nanostructures.

The extremum of the conduction band of bulk InSb is located at the Γ-point of the Brillouin zone. An electronic spectrum of the band near this point is isotropic, nonparabolic and may be written in the form coinciding with the dispersion relation for a relativistic particle [60]

$$\varepsilon(k) = \sqrt{m^2 s^4 + \hbar^2 s^2 k^2} - ms^2, \tag{5.3}$$

where $s = \sqrt{\varepsilon_g/2m} \approx 10^6$ m/s, ε_g is the band gap, and $m = 0.013\, m_0$ is the electron effective mass at the bottom of the conduction band (at 300 K). Using this dispersion relation, it is easy to show that energy spacing between the first and second size-quantized subbands of 5 nm InSb nanowire is $\sim\pi\hbar s/d \approx 400$ meV. Hence, at not-too-heavy doping, only the first subband is occupied at room temperature and electrons have only one orbital degree of freedom, that is, the electronic system of the nanowire is one dimensional.

A small value of the electron mass and a large value of the velocity s lead to a small density of states in InSb and, as a result, to strong degeneracy of the electron system at relatively small impurity concentrations and at relatively high temperatures. For example, in bulk InSb, the Fermi energy of 400 meV is observed [61] at donor concentration of $\sim 10^{19}$ cm^{-3}. Recall that an increase in the density of states with decreasing nanowire diameter reduces the Fermi energy at a fixed electron concentration.

Finally, the electron spectrum (5.3) in a wide region near the Fermi energy may be well approximated by a linear dispersion relation due to its relativistic form. All listed properties of indium antimonide ensure a fulfilment of the necessary conditions, restricting an applicability of the LL model to a description of electronic transport in nanowires.

Before proceeding further, it is reasonable to discuss a question, whether nanowires embedded in chrysotile asbestos nanotubes have a stoichiometric composition and a crystalline structure. The x-ray diffraction pattern of InSb nanowires in asbestos matrix [62] is shown in Figure 5.2. The positions of the marked peaks are close to those of standard polycrystalline bulk InSb. The crystalline phase of wires has the lattice constant $a = 6.482(2)$ Å and matches well with the zinc-blende structure of bulk InSb with the lattice constant $a = 6.4782$ Å. Hence, the polycrystalline InSb nanowires are really formed in channels of asbestos nanotubes.

Consider now the thermoelectric properties of 5 nm InSb nanowires embedded in an asbestos matrix. Figure 5.3 shows [28] the temperature dependences of the normalized conductance of a few samples in a double-logarithmic scale. The dependences can be fitted well by power-law functions in accordance with the LL theory. Values of the exponent α for different samples range from 2 to 7. It seems likely that this scatter in the data is caused by different disorder in nanowires, the different distribution of wires over the diameters in samples, and variations of the Fermi energy. The $G(T)$ dependences of nanowires drastically differ from those of bulk InSb and of a three-dimensional InSb network within Vycor glass also shown in Figure 5.3. It must be emphasized that pores in a Vycor glass have approximately the same size as the channels in chrysotile nanotubes. Therefore, the quantized electron spectrums in the both nanostructures approximately coincide with each other. We believe that the single

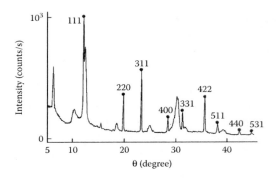

FIGURE 5.2 X-ray diffraction pattern of the composite consisting of InSb nanowires within a chrysotile asbestos matrix. The diffraction peaks denoted by dots relate to InSb. (From Kumzerov, Y.A. et al., *Fiz. Tverd. Tela*, **48**, 1498–1503, 2006 (*Phys. Solid State*, **48**, 1584–1590, 2006). With permission.)

essential difference is the three-dimensional nature of the InSb network within Vycor glass. As a result, the Fermi liquid rather than the LL is formed in the InSb network.

The current–voltage characteristics of one sample with InSb nanowires [28] are plotted on a double-logarithmic scale for different temperatures in Figure 5.4. One can see that the characteristics are strongly nonlinear at low temperatures. Moreover, in accordance with LL theory, the low-temperature curves can be fitted by the linear function at small bias voltages and approach the power law $I \propto V|V|^{\beta}$ with $\beta \approx 3.4$ at large voltages. It is possible to suppose that these two ranges of voltages correspond to the above-mentioned limits $eV \ll k_{B}T$ and $eV \gg k_{B}T$, respectively (recall that V is the voltage drop across a single potential barrier). If this assumption is correct and all barriers have the comparable transmission coefficients, the density of strong defects in a nanowire may be easily estimated. For example, using

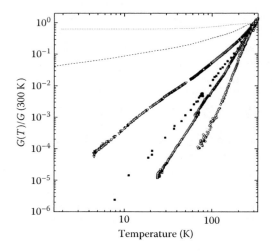

FIGURE 5.3 Temperature dependences of the normalized linear conductance of the different InSb samples in a double-logarithmic scale. The four lower curves correspond to 5 nm InSb nanowires embedded in chrysotile matrices. The dashed curve shows the conductance of bulk InSb extracted from a crack in asbestos. The dotted curve represents the conductance of InSb in Vycor glass. (From Zaitsev-Zotov, S.V. et al., Luttinger-liquid-like transport in long InSb nanowires, *J. Phys.: Condens. Matter*, **12**, L303–L309, 2000. Copyright Institute of Physics and IOP Publishing. With permission.)

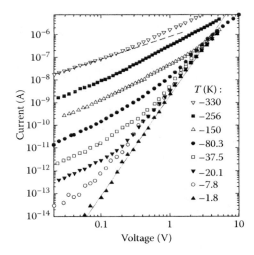

FIGURE 5.4 Current–voltage characteristics of a one sample of 5 nm InSb nanowires within a chrysotile matrix at different temperatures in a double logarithmic scale. The dashed and solid lines are the dependences $I \propto V$ and $I \propto V^{4.4}$, respectively. (From Zaitsev-Zotov, S.V. et al., Luttinger-liquid-like transport in long InSb nanowires, *J. Phys.: Condens. Matter*, **12**, L303–L309, 2000. Copyright Institute of Physics and IOP Publishing. With permission.)

the current–voltage characteristic for $T = 20.1$ K, one finds that a change of the regimes of small and large voltage drop across a barrier occurs at bias $V_{tot} = NV \sim 0.2$ V, where N is the number of strong defects within a nanowire. Therefore, the substitution of this value into the relation $eV = eV_{tot}/N \sim k_B T$ gives immediately $N \sim 100$. Taking into account that nanowires in the sample had the length ~0.2 mm [28], one obtains the defect density of ~500 mm^{-1}. It is a reasonable value.

The LL theory describes a low-energy physics of one-dimensional conductors. It means that an energy of excitations should be much less than the Fermi energy. Therefore, an applicability of the LL model is restricted by temperatures $k_B T \ll \varepsilon_F$. The Fermi energy in InSb cannot be very high. For example, the limiting electron concentration in heavily doped bulk InSb does not exceed 10^{19} cm^{-3} that corresponds to the Fermi energy of ~0.4 eV [61]. The 41.5 nm InSb nanowire unintentionally doped with accidental impurities [63] had the electron concentration of ~10^{18} cm^{-3} and the Fermi energy of 0.2–0.3 eV. It seems likely that ultrathin nanowires under study have still less Fermi energy. In any case, the applicability of the LL model to InSb nanostructures should be restricted by temperature not exceeding a few hundreds of degrees. Therefore, it would be interesting to find this limiting temperature for nanowires under consideration.

The temperature dependences of the conductance and the thermopower of InSb nanowires embedded in an asbestos matrix in the temperature range from 80 to 400 K are shown in Figures 5.5 and 5.6, respectively [64]. At temperatures less than about 300 K, the $G(T)$ dependence is well fitted by the power-law function $G \propto T^{2.2}$ and the thermopower is well described by the linear function $S \propto T$ in accordance with the LL theory. But the behavior of the dependences changes considerably at temperatures higher than about 300 K. The rate of a variation of the conductance with temperature increases. Simultaneously, the rate of a variation of the thermopower decreases. These features cannot be explained in the framework of the LL theory. It is also difficult to explain them only by a violation of the constraint $k_B T \ll \varepsilon_F$. In this case, a decrease in the rate of the conductance growth would be more reasonable [65]. One possible explanation of such behavior is the occupation of other subbands or bands by charge carriers at high temperatures. However, at the moment, an unambiguous interpretation of such behavior is absent.

Let us discuss more carefully a behavior of the thermopower of 5 nm InSb nanowires at $T < 300$ K. On the one hand, the $S(T)$ dependence is linear as is seen from Figure 5.6. It is typical for strongly

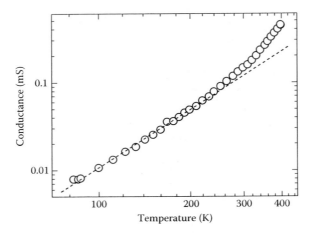

FIGURE 5.5 Temperature dependence of the linear conductance of 5 nm InSb nanowires within a chrysotile matrix in a double-logarithmic scale. The dashed line is the dependence $G \propto T^{2.2}$. (Adapted from Uryupin, O.N. et al., *J. Electron. Mater.*, **38**, 990–993, 2009.)

degenerate electron systems including both Fermi and Luttinger liquids. On the other hand, values of the thermopower (exceeding 200 μV/K at 300 K) are very high for strongly degenerate semiconductors. For comparison, the temperature dependence of the thermopower of the 41.5 nm InSb nanowire with strongly degenerate Fermi gas [14,63] is also shown in Figure 5.6. One of possible explanations for this effect is an enhancement of the thermopower by the electron–electron interaction in an LL [24,25]. Using the found value of the exponent $\alpha = 2.2$ and the results obtained in Ref. [24], it is easy to estimate the ratio of the thermopower of an LL to the thermopower of the same quasi-one-dimensional nanowire but without an electron–electron interaction. This enhancement coefficient is equal to ~4.5 for both the spinless and spin-$\frac{1}{2}$ models of an LL. Note that the thermopowers of the nanostructures, whose $S(T)$ dependences are shown in Figure 5.6, differ from each other by the coefficient ~7. Although a direct comparison of these values of the enhancement coefficient is impossible because the 41.5 nm InSb

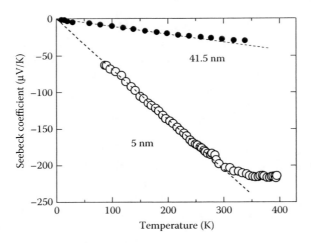

FIGURE 5.6 Seebeck coefficient as a function of temperature for 5 nm InSb nanowires within a chrysotile matrix (open circles) and the individual 41.5 nm InSb nanowire (filled circles). The dashed lines are the dependences $S \propto T$. (Adapted from Uryupin, O.N. et al., *J. Electron. Mater.*, **38**, 990–993, 2009; Zhou, F. et al., *J. Phys.: Condens. Matter*, **18**, 9651–9657, 2006; Seol, J.H. et al., *J. Appl. Phys.*, **101**, 023706, 2007.)

nanowire has a large number of conduction channels (~10) and the different Fermi energy, an agreement in order of magnitude between the experimental and calculated values speaks in favor of the interaction-enhanced thermopower of an LL.

Thermal properties of ultrathin InSb nanowires are also unusual. In Figure 5.7, the temperature dependences of the thermal conductivity of chrysotile asbestos along its nanotubes and of the composite, consisting of 5 nm InSb nanowires embedded in an asbestos matrix, are shown [62]. Using these data, the authors of the indicated work had estimated the effective thermal conductivity κ of InSb nanowire arrays (different from that of a free-standing nanowire [66]), whose temperature dependence is presented in Figure 5.8, curve 1. In the same figure, the corresponding dependencies for three bulk InSb samples with different concentrations of charge carriers and defect structures [67] are shown for comparison. One can see that the low-temperature thermal conductivity of the nanowire array is a few orders of magnitude less than those of bulk samples. This decrease is typical for all nanostructures and is commonly accounted for by the additional phonon boundary scattering. At the temperature of ~30 K, the $\kappa(T)$ dependence passes through a maximum. Moreover, the left and right tails of the peak have asymptotic behavior close to T^3 and T^{-1}, respectively. These features are characteristic for the bulk lattice thermal conductivity. However, the fast growth of the $\kappa(T)$ dependence at $T > 100$ K cannot be explained in the framework of a lattice dynamics. Of course, the procedure of a determination of the effective thermal conductivity of nanowires is rather artificial and is not very exact. Nevertheless, an increase of the nanowire contribution to the thermal conductivity of the composite with increasing temperature is revealed in Figure 5.7 too.

Note that similar temperature dependences of the thermal conductivity of the quasi-one-dimensional organic Bechgaard salts have been obtained in Ref. [68]. Measuring the electrical conductivity of these salts, the authors estimated the electronic contribution to the thermal conductivity in the framework of the Fermi liquid theory. It proved to be very small and could not be responsible for the growth of κ with increasing temperature. On the base of the obtained data, the authors of Refs. [62,68] have concluded that the revealed high-temperature behavior of the thermal conductivity of InSb nanowires and Bechgaard salts is evidence in favor of a formation of an LL in these structures. More precisely, it

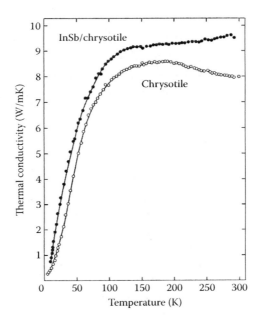

FIGURE 5.7 Temperature dependences of the thermal conductivity of chrysotile asbestos (open circles) and InSb nanowires within a chrysotile matrix (filled circles). (From Kumzerov, Y.A. et al., *Fiz. Tverd. Tela*, **48**, 1498–1503, 2006 (*Phys. Solid State*, **48**, 1584–1590, 2006). With permission.)

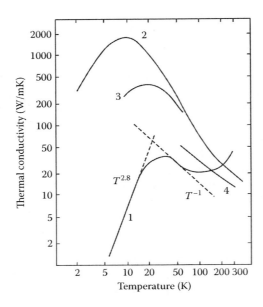

FIGURE 5.8 Temperature dependences of the thermal conductivity of (1) InSb nanowires embedded in a chryso-tile matrix and (2–4) bulk n- and p-type InSb samples: (2) the sample with $n \sim 7 \times 10^{13}$ cm^{-3}, (3) the sample irradi-ated by 2-MeV electrons at 429 K, and (4) the sample with $p \sim 10^{20}$ cm^{-3}. (From Kumzerov, Y.A. et al., *Fiz. Tverd. Tela*, **48**, 1498–1503, 2006 (*Phys. Solid State*, **48**, 1584–1590, 2006); *Thermal Conductivity of Solids: Handbook*, edited by A.S. Okhotin, Energoizdat, Mascow, 1984 [in Russian]. With permission.)

was suggested that an additional thermal flow is transported by spin excitations arising due to the spin–charge separation in an LL.

5.5 Bi and Bi$_{1-x}$Sb$_x$ Nanowires

In contrast to indium antimonide, bulk bismuth has the complex anisotropic band structure. Besides, in bismuth films and wires with transverse sizes less than about 100 nm, surface states have a deter-mining effect on electronic properties of the nanostructures (see Refs. [69,70] and references therein). Therefore, in addition to size-quantized subbands formed from the bulk band structure, the subbands originating from surface states exist in bismuth nanowires and films. Moreover, numerous quantized one-dimensional subbands originating from surface states can cross the Fermi level, since energy spac-ing between them is small because of large carrier masses ($\sim 0.3\ m_0$ [69,70]). Thus, several quantum conduction channels with unknown electron and hole spectra should exist in Bi nanowires embedded in an asbestos matrix.

The interchannel interaction of charge carriers [71,72] in a Bi nanowire decreases values of the expo-nents α defining transport properties of an LL. The large dielectric constant of bismuth (~ 100 at 300 K [73]) also results in a decrease of the electron–electron interaction and α. In other words, values of the exponents α should be small compared with InSb nanowires and distinctive features of an LL can be manifested weakly in Bi nanowires. In addition, a large density of states of surface electron and hole pockets restricts the possibility of increasing the Fermi energy by means of doping. As a result, the LL theory is applicable to Bi nanowires at smaller temperatures than it does in the case of InSb nanowires. At last, bismuth has the small bulk Debye temperature (112 K at $T = 300$ K [74]) and even lower surface Debye temperature [69]. Therefore, the electron–phonon backscattering [59] can have a noticeable effect on transport properties of Bi nanowires at $T \sim 100$ K due to a large number of phonons with the wave vector $q \geq 2k_F$ and the possibility of interband and intersubband scattering. So, the traditional approaches used in an LL theory are likely inapplicable to Bi nanowires but electron correlations still should manifest themselves in some transport properties.

Due to surface states, the ultrathin Bi nanowire is a much better metal than the bulk material [69]. Therefore, in the framework of the Fermi liquid theory, it was to be expected that the electrical conductance of Bi nanowires embedded in asbestos nanotubes should decrease with increasing temperature because of electron–phonon scattering. However, the electrical conductance of this nanostructure only slightly depends on temperature [75] in the range from 100 to 270 K (see Figure 5.11). One way to explain such behavior is to take into account the backscattering of electrons by potential barriers, which leads to a power-low increase in the conductance of an LL with increasing temperature as was shown in the previous sections. A competition between a scattering of electrons by phonons and defects can lead to the weak dependence of the conductance from temperature.

The other argument in favor of strong electron correlations in nanowires under study is the nonlinearity of their current–voltage characteristics. As in the case of InSb nanowires, these characteristics are practically linear at $T \sim 300$ K but lowering temperature changes their behavior [76]. For example, the I–V dependence at $T = 83$ K is shown in Figure 5.9. One can see that the initial portion of the curve is linear but the nonlinearity appears at relatively high voltages. The two regions of the current–voltage characteristic can correspond to the limiting cases of electron tunneling in an LL, which are specified by the inequalities $eV \ll k_B T$ and $eV \gg k_B T$ and were mentioned in Section 5.3. The nonlinear portion of the characteristic may be fitted by the power-law function with the exponent $\beta \approx 0.4$, that is the electron–electron interaction in Bi nanowires is weaker than in InSb wires. However, the estimate is rough because it is fulfilled for the narrow range of biases.

Much attention has been given to the study of the thermoelectric power of Bi nanowires. However, an understanding of a thermopower behavior, when the nanowire thickness decreases, is still absent. On the one hand, according to the Fermi gas theory, a size quantization of the electron spectrum increases the density of states in the vicinity of the L-point subband edge and decreases the overlap of the L- and T-point extrema as a diameter of a Bi nanowire decreases [4,5]. These effects should reduce the Fermi energy (measured from the edge of a lowest L-point subband) and increase the thermopower in comparison with that of bulk bismuth. Moreover, the formation of an LL in ultrathin nanowires can also enhance the thermopower as was indicated in Section 5.3. On the other hand, it is known that the concentration of charge carriers sharply increases with decreasing transverse sizes of films and

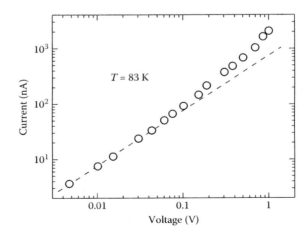

FIGURE 5.9 Current–voltage characteristic of 5 nm Bi nanowires within a chrysotile matrix at 83 K in a double-logarithmic scale. The dashed line is the dependence $I \propto V$. (Adapted from Vedernikov, M.V. et al., *Proc. 22 Int. Conf. On Thermoelectrics*, La Grande-Motte, 2003, pp. 355–358.)

wires [69,70,77] due to a growing contribution of surface band carriers. This effect can prevent the reduction of Fermi energy and stabilize or even decrease the thermopower of a nanowire.

Figure 5.10 (open symbols) shows the temperature dependences of the thermopower of 65 and 40 nm nanowires embedded in porous anodic alumina [13]. These nanostructures have the comparable Seebeck coefficients. It seems likely that such relatively thick nanowires can be described by the Fermi gas theory. The calculation, taking into account only the L- and T-point carrier pockets, predicts a more sharp dependence of the thermopower on a wire diameter [13] than the presented experimental results exhibit. The discrepancy can be explained by a large difference of dopant concentrations in the samples. The other reason of the discrepancy is a large contribution of surface-band carriers to the thermopower. This contribution should increase with decreasing nanowire thickness and it may be supposed that the ultrathin Bi nanowires with an intrinsic carrier concentration will have a less thermopower than the "thick" nanowires. However, as it will be shown now, this prediction does not hold for ultrathin Bi nanowires.

We will not discuss here about the unusual results obtained in Ref. [38], where, in particular, the absolute values of the thermopower of ~100 mV/K have been measured at temperatures of about 100 K in the composite consisting of bismuth nanowires with diameters of about 9 nm embedded in porous alumina. These measurements have never been repeated, as far as we know. Let us consider the temperature dependence of the thermopower of 5 nm nanowires embedded in the chrysotile asbestos matrix [76,78], which is shown in Figure 5.10 (lower curve). In contrast to InSb nanowires, this dependence differs from the linear one but this feature is not surprising in view of the special properties of Bi nanowires indicated above. Absolute values of the thermopower exceed considerably those of ~50 nm Bi nanowires. Of course, an increase in the Seebeck coefficient may be caused by different reasons, one of which is the enhancement of the thermopower by electron–electron interaction.

It is useful to adduce an argument in favor of the last assumption. In Figure 5.10 (upper curve), the temperature dependence of the thermopower of the composite, containing a Bi network embedded in the porous Vycor glass, is shown [78]. The similar dependence has been obtained in Ref. [37] too. The network consists of short nanowires (their length is comparable with the thickness), whose diameters are ~7 nm and almost coincide with thicknesses of nanowires within the asbestos matrix. Besides,

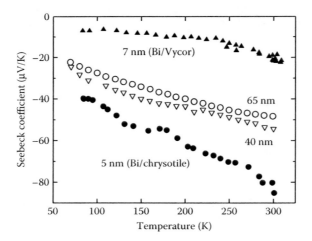

FIGURE 5.10   Seebeck coefficient as a function of temperature for 5 nm bismuth nanowires within a chrysotile matrix (filled circles), 7 nm bismuth network within Vycor glass (filled triangles), and 40 and 65 nm Bi nanowires embedded in anodic alumina matrix (open symbols). (Adapted from Vedernikov, M.V. et al., *Proc. 22 Int. Conf. On Thermoelectrics*, La Grande-Motte, 2003, pp. 355–358; Ivanov, Y.V. et al, in *Proceedings of the XI International Workshop "Thermoelectrics and Their Applications,"* St. Petersburg, Russia, 2008, edited by M.V. Vedernikov (St. Petersburg, 2008), pp. 44–47 [in Russian]; Lin, Y.-M. et al, *Appl. Phys. Lett.*, 81, 2403–2405, 2002.)

pores in the both nanostructures were filled with molten bismuth of the same purity using the same preparation technique. The measurements of the thermopower were performed by means of the one setup. The principal difference of the nanostructures is the three-dimensional nature of the Bi network in a Vycor glass. Therefore, the Fermi liquid should be formed in the Bi network instead of the LL. We believe that it is the main reason of the dramatic difference between thermopowers of the nanostructures under consideration. In other words, the formation of an LL in Bi nanowires within the asbestos matrix enhances the thermopower by a factor of 4–5 in comparison with the Bi network embedded in the porous glass. Note that the Seebeck coefficient of the composite on the base of a Vycor glass is much less than the thermopower of 40 and 65 nm Bi nanowires. This result is in good agreement with the assumption made above, which states that the contribution of surface carriers reduces the thermopower of nanostructures.

For thermoelectric applications, $Bi_{1-x}Sb_x$ alloy nanowires are particularly promising. The calculation [79] shows that the dimensionless thermoelectric figure of merit of n-type nanowires with a composition $x \approx 0.25$ and a diameter of about 30 nm is 2.5 at 77 K. In addition, the figure of merit increases as the diameter is decreased. This evaluation does not take into account the formation of an LL in nanowires and effects of surface electronic states. Therefore, the measured figure of merit of ultrathin $Bi_{1-x}Sb_x$ nanowires may be both smaller and larger than the predicted one.

At present, it is impossible to identify unambiguously electronic processes responsible for a behavior of ultrathin $Bi_{1-x}Sb_x$ nanowires. There are little data on electronic properties of these nanostructures. Note only that the Fermi energies for some extrema of the surface band structure of bulk antimony are rather large (~100 meV [80]). If the tendency to an increase of the Fermi energy with increasing x is pronounced with $Bi_{1-x}Sb_x$ nanowires, then LL features can manifest themselves in a wider temperature range than in the case of bismuth wires. In other words, the larger an antimony content the higher the temperatures, at which the LL behavior should be distinguishable in $Bi_{1-x}Sb_x$ nanowires. In addition, the Debye temperature of $Bi_{1-x}Sb_x$ alloys rises [81] as x increases. Therefore, the effect of an electron–phonon scattering on transport properties of the nanostructures under consideration should decrease with increasing antimony content.

Figure 5.11 shows [75,82,83] the temperature dependences of the electrical conductance (normalized to $G(80\ K)$) of 5 nm $Bi_{1-x}Sb_x$ nanowires within asbestos matrix with $x = 0, 0.15, 0.25$. Figure 5.12 presents

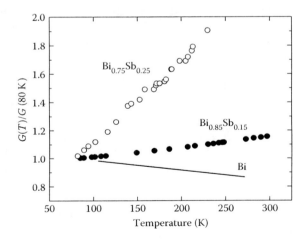

FIGURE 5.11 Temperature dependences of the normalized conductance of 5 nm $Bi_{1-x}Sb_x$ nanowires ($x = 0, 0.15$, 0.25) embedded in a chrysotile matrix. (Adapted from Ivanova, M.S. et al., *Microporous Mater.*, **4**, 319–322, 1995; Vedernikov, M.V. et al., in *Proceedings of the X International Workshop "Thermoelectrics and Their Applications,"* *St. Petersburg, Russia, 2006*, edited by M.V. Vedernikov (St. Petersburg, 2006), pp. 210–223 [in Russian]; Uryupin, O.N. et al., *Proc. 3rd Europ. Conf. on Thermoelectrics*, Nancy, France, 2005, c.35–38.)

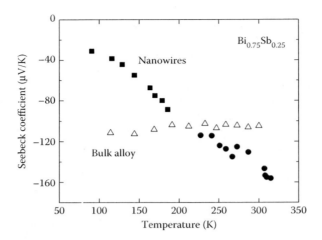

FIGURE 5.12   Seebeck coefficient as a function of temperature for 5 nm $Bi_{0.75}Sb_{0.25}$ nanowires within a chrysotile matrix (filled symbols) and the same bulk alloy obtained in the course of a preparation of the nanowires (open triangles). Two sorts of the filled symbols correspond to measurements before and after an irreversible jump of the sample resistance. (Adapted from Vedernikov, M.V. et al., in *Proceedings of the X International Workshop "Thermoelectrics and Their Applications," St. Petersburg, Russia, 2006*, edited by M.V. Vedernikov (St. Petersburg, 2006), pp. 210–223 [in Russian]; Uryupin, O.N. et al., *Proc. 3rd Europ. Conf. on Thermoelectrics*, Nancy, France, 2005, c. 35–38.)

the temperature dependences of the thermopower of 5 nm $Bi_{0.75}Sb_{0.25}$ nanowires and bulk alloy of the same composition [82,83], which had been formed in the course of the nanowire preparation. Unfortunately, in $Bi_{0.75}Sb_{0.25}$ samples, some nanowires are frequently broken up at cooling and heating. Therefore, the conductance has been measured over a narrow temperature range. The thermopower depends weakly on a number of wires in a sample and the two pieces of the $S(T)$ dependence (before and after of an irreversible jump of the resistance) are shown in Figure 5.12.

As expected, the behavior of the presented $G(T)$ and $S(T)$ dependences of $Bi_{1-x}Sb_x$ nanowires is in better agreement with the LL theory than that of the bismuth nanostructures. In contrast to Bi nanowires, the conductance of $Bi_{1-x}Sb_x$ wires increases monotonically with the increase of temperature. Besides, the rate of change of the conductance grows as the antimony content increases. The $G(T)$ dependence for $x = 0.25$ is adequately described by a power law with the exponent $\alpha \approx 0.6$. Hence, the electron–electron interaction in these wires is stronger than in bismuth nanowires (in the latter $\beta \approx 0.4$) within the accuracy of the analysis. The temperature dependence of the Seebeck coefficient of $Bi_{0.75}Sb_{0.25}$ nanowires is close to a linear function. Moreover, the thermopower of wires is much more "metallic" than that of the bulk alloy having a comparable impurity concentration (compare the two dependences in Figure 5.12). It seems reasonable to assume that a carrier concentration in nanowires is larger than in the bulk alloy and is dominated by surface states. The absolute value of the room-temperature thermopower of $Bi_{0.75}Sb_{0.25}$ nanowires is well above the corresponding value for bismuth wires (see Figure 5.10). The enhancement of the thermopower may be caused by an increase of the electron–electron interaction.

5.6 Concluding Remarks

In this chapter, we have summarized the thermoelectric properties of ultrathin nanowires embedded in asbestos nanotubes, taking into account the electron–electron interaction. It has been shown that, in most cases, transport properties of InSb nanowires may be explained in the framework of the LL theory. An interpretation of properties of Bi and $Bi_{1-x}Sb_x$ alloy nanowires is ambiguous because of the special features of these materials. However, a number of phenomena in these nanostructures (e.g., a strong dependence of the thermopower on the topology of the nanowire ensemble) also may be understood, if electron correlations are taken into account.

We did not analyze here an applicability of Fermi-liquid approaches to a description of electronic transport in ultrathin nanowires. Such analysis can be found, for example, in Refs. [28,29,33]. The main purpose of this chapter was to demonstrate the necessity of considering strong electron correlations in the course of an investigation of thermoelectric transport in quasi-one-dimensional systems.

A number of characteristic properties of ultrathin nanowires may be used for a production of promising thermoelectric materials. Among these are the large thermopower of strong-degenerated quasi-one-dimensional semiconductors enhanced by the electron–electron interaction, a small electronic contribution to the thermal conductance arising under appropriate conditions and leading to the small Lorenz number, and the reduced lattice thermal conductivity resulting from the boundary phonon scattering. At the same time, the electrical conductance of an LL induced by a disorder is suppressed by interaction and the Lorenz number can exceed the value L_0 in some cases. Therefore, the use of ultrathin nanowires alone does not ensure the high thermoelectric figure of merit of materials produced on their base. The optimization of properties and operating conditions of nanowires is required for thermoelectric applications.

References

1. Venkatasubramanian, R., Siivola, E., Colpitts, T., and O'Quinn, B., *Nature*, **413**, 597–602, 2001.
2. Hsu, K.F., Loo, S., Guo, F., Chen, W., Dyck, J.S., Uher, C., Hogan, T., Polychroniadis, E.K., and Kanatzidis, M.G., *Science*, **303**, 818–821, 2004.
3. Harman, T.C., Walsh, M.P., Laforge, B.E., and Turner, G.W., *J. Electron. Mater.*, **34**, L19–L22, 2005.
4. Hicks, L.D. and Dresselhaus, M.S., *Phys. Rev. B*, **47**, 16631–16634, 1993.
5. Lin, Y.-M., Sun, X., and Dresselhaus, M.S., *Phys. Rev. B*, **62**, 4610–4623, 2000.
6. Mingo, N., *Appl. Phys. Lett.*, **84**, 2652–2654, 2004 (Erratum: *Appl. Phys. Lett.*, **88**, 149902, 2006).
7. Bejenari, I., Kantser, V., and Balandin A.A., *Phys. Rev. B*, **81**, 075316, 2010.
8. Broido, D.A. and Reinecke, T.L., *Phys. Rev. B*, **64**, 045324, 2001.
9. Hochbaum, A.I., Chen, R., Delgado, R.D., Liang, W., Garnett, E.C., Najarian, M., Majumdar, A., and Yang, P., *Nature*, **451**, 163–167, 2008.
10. Boukai, A.I., Bunimovich, Y., Tahir-Kheli, J., Yu, J.-K., Goddard, W.A., and Heath, J.R., *Nature*, **451**, 168–171, 2008.
11. Zhou, J., Jin, C., Seol, J.H., Li, X., and Shi, L., *Appl. Phys. Lett.*, **87**, 133109, 2005.
12. Zhou, F., Moore, A.L., Pettes, M.T., Lee, Y., Seol, J.H., Ye, Q.L., Rabenberg, L., and Shi, L., *J. Phys. D: Appl. Phys.*, **43**, 025406, 2010.
13. Lin, Y.-M., Rabin, O., Cronin, S.B., Ying, J.Y., and Dresselhaus, M.S., *Appl. Phys. Lett.*, **81**, 2403–2405, 2002.
14. Zhou, F., Seol, J.H., Moore, A.L., Shi, L., Ye, Q.L., and Scheffler, R., *J. Phys.: Condens. Matter*, **18**, 9651–9657, 2006.
15. Huber, T.E., Nikolaeva, A.A., Gitsu, D.V., Konopko, L.A., and Graf, M.J., *Proc. XXV Int. Conf. on Thermoelectrics*, Vienna, 2006, pp. 224–227.
16. Nikolaeva, A., Huber, T.E., Gitsu, D., and Konopko, L., *Phys. Rev. B*, **77**, 035422, 2008.
17. Haldane, F.D.M., *J. Phys. C: Solid State Phys.*, **14**, 2585–2609, 1981.
18. Voit, J., *Rep. Prog. Phys.*, **58**, 977–1116, 1995.
19. Giamarchi, T., *Quantum Physics in One Dimension*, Oxford University Press, Oxford, 2003.
20. Deshpande, V.V., Bockrath, M., Glazman, L.I., and Yacoby, A., *Nature*, **464**, 209–216, 2010.
21. Kane, C.L. and Fisher, M.P.A., *Phys. Rev. B*, **46**, 15233–15262, 1992.
22. Furusaki, A. and Nagaosa, N., *Phys. Rev. B*, **47**, 4631–4643, 1993.
23. Kane, C.L. and Fisher, M.P.A., *Phys. Rev. Lett.*, **76**, 3192–3195, 1996.

24. Krive, I.V., Bogachek, E.N., Scherbakov, A.G., and Landman, U., *Phys. Rev. B*, **63**, 113101, 2001.
25. Krive, I.V., Romanovsky, I.A., Bogachek, E.N., Scherbakov, A.G., and Landman, U., *Low Temp. Phys.* **27**, 821, 2001.
26. Romanovsky, I.A., Krive, I.V., Bogachek, E.N., and Landman, U., *Phys. Rev. B*, **65**, 075115, 2002.
27. Kumzerov, Y. and Vakhrushev, S., Nanostructures within porous materials, in *Encyclopedia of Nanoscience and Nanotechnology*, edited by H.S. Nalwa, American Scientific Publishers, Los Angeles, 2004, Vol. 7, pp. 811–849.
28. Zaitsev-Zotov, S.V., Kumzerov, Y.A., Firsov, Y.A., and Monceau, P., Luttinger-liquid-like transport in long InSb nanowires, *J. Phys.: Condens. Matter*, **12**, L303–L309, 2000.
29. Venkataraman, L., Hong, Y.S., and Kim, P., *Phys. Rev. Lett.* **96**, 076601, 2006.
30. Uplaznik, M., Bercic, B., Remskar, M., and Mihailovic, D., *Phys. Rev. B*, **80**, 085402, 2009.
31. Liu, F., Bao, M., Wang, K.L., Li, C., Lei, B., and Zhou, C., *Appl. Phys. Lett.*, **86**, 213101, 2005.
32. Aleshin, A.N., Lee, H.J., Park, Y.W., and Akagi, K., *Phys. Rev. Lett.* **93**, 196601, 2004.
33. Aleshin, A.N., *Phys. Solid State*, **49**, 2015–2033, 2007.
34. Bockrath, M., Cobden, D.H., Lu, J., Rinzler, A.G., Smalley, R.E., Balents, L., and McEuen, P.L., *Nature*, **397**, 598, 1999.
35. Yao, Z., Postma, H.W.Ch., Balents, L., and Dekker, C., *Nature*, **402**, 273, 1999.
36. Kong, W.J., Lu, L., Zhu, H.W., Wei, B.Q., and Wu, D.H., *J. Phys.: Condens. Matter*, **17**, 1923, 2005.
37. Huber, T., Nikolaeva, A., Gitsu, A., Konopko, D., Graf, M.J., and Huang, J., arXiv:cond-mat/0311112.
38. Heremans, J.P., Thrush, C.M., Morelli, D.T., and Wu, M.-C., *Phys. Rev. Lett.*, **88**, 216801, 2002.
39. Dresselhaus, M.S. and Heremans J.P., in *Thermoelectrics Handbook: Macro to Nano*, edited by D.M. Rowe, CRC Press, Taylor & Francis Group, Boca Raton, FL, 2006, pp. 39-1–39-20.
40. Jancar, B. and Suvorov, D., *Nanotechnology*, **17**, 25–29, 2006.
41. Falini, G., Foresti, E., Gazzano, M., Gualtieri, A.F., Leoni, M., Lesci, I.G., and Roveri, N., *Chem. Eur. J.*, **10**, 3043–3049, 2004.
42. Foresti, E., Hochella, M.F., Jr., Kornishi, H., Lesci, I.G., Madden, A.S., Roveri, N., and Xu, H., *Adv. Funct. Mater.*, **15**, 1009–1016, 2005.
43. Balan, E., Mauri, F., Lemaire, C., Brouder, C., Guyot, F., Saitta, A.M., and Devouard, B., *Phys. Rev. Lett.*, **89**, 177401, 2002.
44. Romanov, S.G., Butko, V.Y., Kumzerov, Y.A., Yates, N.M., Pemble, M.I., Agger, J.R., Anderson, M.W., and Torres, C.M.S., *Phys. Solid State*, **39**, 641–648, 1997.
45. Giuliani, G. and Vignale, G., *Quantum Theory of the Electron Liquid*, Cambridge University Press, Cambridge, 2005.
46. Meyer, J.S. and Matveev, K.A., *J. Phys.: Condens. Matter*, **21**, 023203, 2009.
47. Fisher, M.P.A. and Glazman, L.I., in *Mesoscopic Electron Transport* (NATO ASI Series E, Vol. 345) edited by L.L. Sohn, L.P. Kouwenhoven, G. Schön, Kluwer Academic Publishers, Dordrecht, 1997, pp. 331–374 (*preprint* arXiv:cond-mat/9610037).
48. Furusaki, A. and Nagaosa, N., *Phys. Rev. B*, **54**, 5239–5242, 1996.
49. Fogler, M.M., Malinin, S.V., and Nattermann, T., *Phys. Rev. Lett.*, **97**, 096601, 2006.
50. Renn. S.R. and Arovas, D.P., *Phys. Rev. B*, **51**, 16832–16839, 1995.
51. Zhou, Z., Xiao, K., Jin, R., Mandrus, D., Tao. J., Geohegan, D.B., and Pennycook, S., *Appl. Phys. Lett.*, **90**, 193115, 2007.
52. Ivanov, Y.V., *J. Phys.: Condens. Matter*, **22**, 245602, 2010.
53. Fazio, R., Hekking, F.W.J., and Khmelnitskii, D.E., *Phys. Rev. Lett.*, **80**, 5611–5614, 1998.
54. Krive, I.V., *Low Temp. Phys.*, **24**, 377–379, 1998.
55. Gutman, D.B., Gefen, Y., and Mirlin, A.D., *Phys. Rev. B*, **80**, 045106, 2009.
56. Li, M.-R. and Orignac, E., *Europhys. Lett.*, **60**, 432–438, 2002.
57. Gard, A., Rasch, D., Shimshoni, E., and Rosch, A., *Phys. Rev. Lett.*, **103**, 096402, 2009.
58. Komnik, A. and Egger, R., arXiv:cond-mat/9906150.
59. Seelig, G., Matveev, K.A., and Andreev, A.V., *Phys. Rev. Lett.*, **94**, 066802, 2005.

60. Askerov, B.M., *Electron Transport Phenomena in Semiconductors*, World Scientific, Singapore, 1994.

61. Filipchenko, A.S. and Nasledov, D.N., *Phys. Stat. Sol. (a)* **27**, 11, 1975.

62. Kumzerov, Y.A., Smirnov, I.A., Firsov, Y.A., Parfen'eva, L.S., Misiorek, H., Mucha, J., and Jezowski, A., *Fiz. Tverd. Tela*, **48**, 1498–1503, 2006 (*Phys. Solid State*, **48**, 1584–1590, 2006).

63. Seol, J.H., Moore, A.L., Saha, S.K., Zhou, F., and Shi, L., *J. Appl. Phys.*, **101**, 023706, 2007.

64. Uryupin, O.N., Vedernikov, M.V., Shabaldin, A.A., Ivanov, Y.V., Kumzerov, Y.A., and Fokin, A.V., *J. Electron. Mater.*, **38**, 990–993, 2009.

65. Yue, D., Glazman, L.I., and Matveev, K.A., *Phys. Rev. B*, **49**, 1966–1975, 1994.

66. Prasher, R., *J. Appl. Phys.*, **100**, 034307, 2006.

67. Okhotin, A.S., Borovikova, R.P., Nechaeva, T.V., and Pushkarskiy, A.S., *Thermal Conductivity of Solids: Handbook*, Energoizdat, Moscow, 1984 [in Russian].

68. Lorenz, T., Hofmann, M., Grüninger, M., Frelmuth, A., Uhrig, G.S., Dumm, M., and Dressel, M., *Nature*, **418**, 614–617, 2002.

69. Hofmann, P., *Prog. Surf. Sci.*, **81**, 191–254, 2006.

70. Huber, T.E., Nikolaeva, A., Konopko, L., and Graft, M.J., *Phys. Rev. B*, **79**, 201304, 2009.

71. Matveev, K.A. and Glazman, L.I., *Phys. Rev. Lett.*, **70**, 990–993, 1993.

72. Egger, R., *Phys. Rev. Lett.*, **83**, 5547–5550, 1999.

73. Gerlach, E., Grosse, P., Rautenberg, M., and Senske, W., *Phys. Stat. Sol. (b)*, **75**, 553–558, 1976.

74. Fischer, P., Sosnowska, I., and Szymanski, M., *J. Phys. C: Solid State Phys.*, **11**, 1043, 1978.

75. Ivanova, M.S., Kumzerov, Y.A., Poborchii, V.V., Ulashkevich, Y.V., and Zhuravlev, V.V., *Microporous Mater.*, **4**, 319–322, 1995.

76. Vedernikov, M.V., Uryupin, O.N., Ivanov, Y.V., and Kumzerov, Y.A., *Proc. 22 Int. Conf. on Thermoelectrics*, La Grande-Motte, 2003, pp. 355–358.

77. Komnik, Y. F. and Andrievskii, V.V., *Fiz. Nizk. Temp.*, **1,** 104, 1975 [*Sov. J. Low Temp. Phys.*, **1**, 51, 1975].

78. Ivanov, Y.V., Uryupin, O.N., Vedernikov, M.V., Kumzerov, Y.A., and Fokin, A.V., in *Proceedings of the XI International Workshop "Thermoelectrics and Their Applications,"* St. Petersburg, Russia, 2008, edited by M.V. Vedernikov (St. Petersburg, 2008), pp. 44–47 [in Russian].

79. Rabin, O., Lin, Y.-M., and Dresselhaus, M.S., *Appl. Phys. Lett.*, **79**, 81–83, 2001.

80. Höchst, H. and Ast, C.R., *J. Electron Spectrosc. Related Phenomena*, **137–140**, 441–444, 2004.

81. Gopinathan, K.K. and Padmini, A.R.K.L., *J. Phys. D: Appl. Phys.*, **7**, 32–40, 1974.

82. Vedernikov, M.V., Uryupin, O.N., Ivanov, Y.V., Kumzerov, Y.A., and Fokin, A.V., in *Proceedings of the X International Workshop "Thermoelectrics and Their Applications,"* St. Petersburg, Russia, 2006, edited by M.V. Vedernikov (St. Petersburg, 2006), pp. 210–223 [in Russian].

83. Uryupin, O.N., Ivanov, Y.V., Vedernikov, M.V., Kumzerov, Y.A., and Fokin, A.V., *Proc. 3rd Europ. Conf. on Thermoelectrics*, Nancy, France, 2005, c. 35–38.

6

Bismuth Telluride Alloys for Waste Energy Harvesting and Cooling Applications

H. Scherrer
Ecole des Mines

S. Scherrer
Ecole des Mines

6.1 Introduction

Studies and measurements of the electric properties of bismuth telluride and other compounds in the series were done as early as 1910. These materials were used to produce thermoelectric refrigeration about room temperature in the early 1950s. Since extensive works were performed on a worldwide basis on the Bi_2Te_3–Sb_2Te_3–Bi_2Se_3–Sb_2Se_3 alloy systems. An effort was also directed toward developing a Bi_2Te_3 thermoelectric material for power generation used at hot sides of about 300°C. The research results presented in the literature [1–8] show that, at room temperature, the best thermoelectric materials are solid solutions based on bismuth telluride. As a consequence of the results, the research for a value of the dimensionless figure of merit $ZT \leq 1$ can be established for these materials. In most of the investigated samples, it is difficult to relate the thermoelectric properties to the thermodynamic basic state of the material.

Our present objective is to clarify in the phase diagram the solid line limiting the region of existence of the chalcogenide compounds, we present the obtained results.

6.2 General Properties of Bi_2Te_3, Sb_2Te_3 and Their Solid Solutions

6.2.1 Chemical Properties

First, we shall review the physico-chemical properties of bismuth telluride and solid solutions based on it. These materials have been studied, for many years, but a number of essential questions have not been conclusively answered. Some thermodynamics data and the density of Bi_2Te_3 are given in Table 6.1 [9–17].

TABLE 6.1   Melting Temperature T_m, Latent Heat of Fusion ΔH_m, Debye Temperature Θ_D, Density, Specific Heat C_p of Bi_2Te_3, Sb_2Te_3, Bi_2Se_3

Compounds	T_m(°C)	ΔH_m (kcal/mol)	θ_D(°K)	d (g/cm³)	C_p (cal deg^{-1} mol^{-1}) $T < 550$°C
Bi_2Te_3	585 [9–10]	29.0 [14]	155 [15] 164.9 [16]	7.8587 [17]	$36.0 + 1.30 \cdot 10^{-2}$ T $- 3.11 \cdot 10^{-5}$ T² [14]
Sb_2Te_3	618.5 [11] 621.6 [12]	23.6 [11]	—	6.57 [13]	—
Bi_2Se_3	706 [13]	—	—	7.308 [13]	—

Russian experiments and thermodynamic calculations [18] have shown that the vaporization of Bi_2Te_3 in the temperatures interval 700–1000 K results from the following equation:

$$Bi_2Te_3 \Leftarrow 2\,BiTe(gaz) + 1/2\,Te_2(gaz) \tag{6.1}$$

The temperature dependence of the total pressure of the saturated vapors P_{tot} above the solid bismuth telluride in the range 720–850 K is described by

$$\log P_t = -A/T + B \tag{6.2}$$

The temperature behavior of the different pressures above the solid antimony telluride in the range 716–825 K is described by the same type of equations. The coefficients of Equation 6.2 are given in Table 6.2.

Besides, Brebrick [8], Boncheva et al. [19] proposed the same temperature dependence of the P_{Te2} above solid Bi_2Te_3 in the range 719–827 K by the following equation:

$$\log P_{Te_2} = 10.3 - \frac{10.700}{T}. \tag{6.3}$$

The vaporization of the solid Bi_2Se_3 proceeds by a reaction analogous to Equation 6.1.

6.2.2 Structure

Bismuth selenide and antimony telluride have a similar crystallographic structure to bismuth telluride. Many authors [20–25] have described these compounds as a rhombohedral unit cell consisting of five atoms per cell (two Bi, three chalcogen) with the point group R $\bar{3}$ m. It is often more convenient to observe the structure by a hexagonal primitive cell.

For example, the hexagonal cell of Bi_2Te_3 is formed by a set of layers perpendicular to the third-order axis of symmetry (C-axis). The atoms of separate layers are identical and form a plane hexagonal lattice.

TABLE 6.2   Coefficients in Equation 6.2 for Bi_2Te_3 and Sb_2Te_3

	Coefficient	P_{tot}	P_{BiTe}	P_{Te_2}
Bi_2Te_3	A	10.443	10.443	10.443
	B	11.054	10.831	9.445
Sb_2Te_3	A	10.003	10.003	10.003
	B	10.925	10.573	10.573

TABLE 6.3   Lengths of Bonds and Angles between Them in Bi_2Te_3

Bonds	Bond Lengths (Å)	Angles
Bi–Te$^{(2)}$	3.22	85° 30′
Bi–Te$^{(1)}$	3.42	89° 20′
Te$^{(1)}$–Te$^{(1)}$	3.57	85° 42′

The layers alternate following the sequence:

$$-Te^{(1)}-Bi-Te^{(2)}-Bi-Te^{(1)}-$$

This sequence is called a quintet and the hexagonal cell is formed by three quintets. The lamellar structure of Bi_2Te_3 and the weakness of Te$^{(1)}$–Te$^{(1)}$ bonds between two quintets are responsible for the easy cleavage along the planes perpendicular to the *C*-axis.

Because of the presence of a layered structure, the Bi_2Te_3 properties are very anisotropic. The Te$^{(1)}$–Te$^{(1)}$ bonds are considered to be of van der Waals type whereas the Te$^{(1)}$–Bi and Bi–Te$^{(2)}$ are of ionic–covalent type [26]. The lengths of bonds and angle between them are given in Table 6.3.

Taking into account the crystallographic structure of Bi_2Te_3, each atom has three neighbors in the above plane and three neighbors in the below one as the hexagonal compact lattice or cubic compact lattice in a direction perpendicular to the [111] direction. So we can represent the hexagonal cell by the stacking of deformed cubes along their diagonal directions [111] (Figure 6.1).

This type of representation is interesting to study some physical properties, for instance, heterodiffusion experiments or point defects analysis [27].

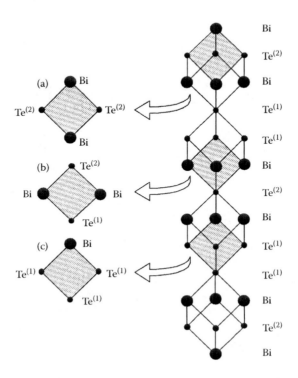

FIGURE 6.1   Representation of the crystal structure of Bi_2Te_3 by the stacking of cubic cells along their [111] directions. On the left part, the faces of the cube giving the first neighbors of Bi atoms are represented.

TABLE 6.4   Crystallographic Data for Bismuth Telluride and Bismuth Selenide

Symbol	Magnitude Bi_2Te_3	Bi_2Se_3	Units of b	Definition
a_0	10.418 Å	9.841 Å		Rhombohedral vector at 0 K
α	24° 12′ 40″	24° 16′	1.000	Rhombohedral angle at 0 K
b_0	1.6731 Å^{-1}			Reciprocal lattice vector
β	61° 30′ 37″			Rhombohedral angle for reciprocal lattice
V	169.11 Å			Unit cell volume
ΓA	0.8366 Å^{-1}		0.5000	1/2 (100)
ΓD	0.8556 Å^{-1}		0.5114	1/2 (100)
ΓZ	0.3108 Å^{-1}		0.1858	1/2 (111)
θ_1	7° 6′ 50″			Angle between ΓA and ΓY
θ_2	14° 0′ 50″			Angle between ΓD and ΓY

Source: Lovett, D.R. 1977. *Semimetals and Narrow Band Gap Semiconductors*, Pion Limited, London.

Crystallographic data for the unit cells and Brillouin zone available for bismuth telluride and bismuth selenide are given in Table 6.4 [28]. The Brillouin zone for each is similar to that for bismuth but is strongly narrowed because the unit cell has a large dimension in the trigonal direction (Figure 6.2).

The compounds Bi_2Te_3, Bi_2Se_3, and Sb_2Te_3 form isomorphous solid solutions, because they have the same class of symmetry, similar lattice dimensions and chemical properties (Table 6.5).

The covalent radii for Se, Bi, Te, and Sb are respectively equal to 1.16, 1.46, 1.36, 1.40 Å. So in solid solutions of Bi_2Te_3–Sb_2Te_3, the Sb atoms substitute in Bi sites as the following:

$$-Te^{(1)}-Sb-Te^{(2)}-Bi-Te^{(1)}$$

In Bi_2Te_3–Bi_2Se_3 solutions, the same atoms may occupy $Te^{(1)}$ or $Te^{(2)}$ sites giving the sequences:

$$-Te^{(1)}-Bi-Se-Bi-Te^{(1)}$$
$$-Se-Bi-Te^{(2)}-Bi-Te^{(1)}$$

It is admitted that the Se atoms, being more electronegative, take the place of the $Te^{(2)}$ atoms and increase the ionic bonds [32,33]. When all the $Te^{(2)}$ sites are filled, $Te^{(1)}$ sites become occupied at random by Se atoms.

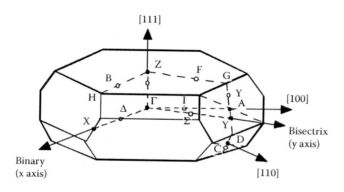

FIGURE 6.2   Brillouin zone for bismuth telluride and bismuth selenide with lengths in the z-direction exaggerated.

TABLE 6.5   Space Group and Hexagonal Lattice Parameters for Bi_2Te_3, Sb_2Te_3, Bi_2Se_3 at 300 K

Compounds	Space Group	Lattice Parameters (Å)	
		A	c
Bi_2Te_3 [29]	R $\bar{3}$ m	4.3835	30.360
Sb_2Te_3 [30]	R $\bar{3}$ m	4.275	30.490
Bi_2Se_3 [31]	R $\bar{3}$ m	4.934	28.546

The lattice parameters of Bi_2Te_3–Sb_2Te_3 were measured in Refs. [34–36] and for Bi_2Te_3–Bi_2Se_3 in Refs. [31,34,36,37]. The parameter a in both cases obeys the Vegard's law. It decreases during the transition from Bi_2Te_3 to Sb_2Te_3. The change in parameter c with respect to the composition of the solid solution is more complicated. This parameter c decreases linearly with 30 mol% Bi_2Se_3 in Bi_2Te_3 (Bi_2Te_2Se compound); then deviations from Vegard's law are observed, due to the substitution of the $Te^{(1)}$ atoms by Se atoms.

6.2.3 Phase Diagrams of Bi–Te, Sb–Te, and Bi–Se

6.2.3.1 Deviations from Stoichiometry and Point Defects

The phase diagrams temperature composition of these systems are typical and present a single congruently melting compound whose ideal crystal structure corresponds to the formulae Bi_2Te_3 or Sb_2Te_3 and Bi_2Se_3. The compounds exhibit such small deviations from stoichiometry that on the scale of conventional phase diagrams are solely a line phase.

As shown in Figure 6.3, the initial components may be dissolved in the chemical compound to give homogeneous solid solutions (shaded zone). The atoms of the dissolved component form either substitutional solid solutions (antisite defects) or solid solution with vacancies in the sites of atoms of the other component or interstitial solid solutions.

The limit of concentration of the dissolved components depends on the temperature and corresponds to the solidus line. This solidus line presents a retrograde solubility. Consequently, if the compound, grown from the melt, is slowly cooled, the solid solution dissociates and it appears in the crystal grains or the grain boundaries. During rapid cooling a supersaturated solid solution is formed.

In all cases, the concentration of the excess component may be changed by annealing the as-grown compound at different temperatures. With the assumption that the resulting point defects are electrically active, the concentration of the current carriers is also changed.

For Bi_2Te_3 the solidus line spread out on both the sides of the stoichiometric composition [7,8] (within the limit 59.8–60, 2 at.% Te, in the temperature range 460–520°C). The maximum on the liquidus line

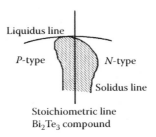

FIGURE 6.3   Phase diagram around Bi_2Te_3 compound with exaggerated abscissa.

has a composition of 40.065 at.% Bi and 59.935 at.% Te. Ingots of Bi_2Te_3 obtained by crystallization from a melt of stoichiometric composition have a conduction of p-type. Deviations from stoichiometry were also observed in the compounds Bi_2Se_3 and Sb_2Te_3 [38]. At the maximum on the liquidus line, the solids have the composition:

$$Sb_2Te_3 \begin{pmatrix} Sb\ 40.4\ atom\% \\ Te\ 59.6\ atom\% \end{pmatrix} \quad Bi_2Se_3 \begin{pmatrix} Bi\ 40.02\ atom\% \\ Se\ 59.98\ atom\% \end{pmatrix}$$

Thus, the three compounds present an excess of the more electronegative component. This defect may lead to the formation of donor or acceptor centers. Tight-binding studies of crystal stability and defects in Bi_2Te_3 were published [39]. The calculations are made from a basic cubic cell stabilized by the formation of five layers of planes along the [111] axis of the cubic cell (Figure 6.1). Antisite defects appear as single donors (Te_{Bi}) or single acceptors (Bi_{Te}). Vacancies are strong perturbations: V_{Te} are double donors and V_{Bi} are triple acceptors.

Previous accurate measurements of the density of Bi_2Te_3 samples [40] and heterodiffusion studies of Se atoms in Bi_2Te_3 [27] showed that defects of antisite type should be predominant for samples with excess of Bi. For the samples with an excess of Te, the vacancies V_{Bi} and the antisite-type Te_{Bi} are both possible.

For the compound Sb_2Te_3, as in Bi_2Te_3, the excess of Sb atoms forms antisite defects [38]. Therefore, ingots grown by a Bridgman method have always a p-type conduction. Concerning the solid solutions Bi_2Te_3–Sb_2Te_3, if the content of Sb_2Te_3 increases in Bi_2Te_3, the antisite defects Bi_{Te} are gradually replaced by the same type of defects Sb_{Te}. The solid solutions always have the p-type conduction.

Defects have been less well studied in Bi_2Se_3, but we know that the excess of Bi atoms would preferentially occupy interstitial positions. Therefore, Bi_2Se_3 has always an n-type conduction.

6.2.3.2 Doping in Bi_2Te_3 and Solid Solutions

Usually, in bismuth telluride and its solid solutions, the wanted concentration of charge carriers is obtained by deviation of stoichiometry introducing an excess of bismuth or tellurium atoms into the primary melt or by dopant impurities.

In solid solutions Bi_2Te_3–Bi_2Se_3 used as n-type materials, the carrier concentration is adjusted by doping with halogens. In p-type materials, the holes concentration due to an excess of Sb atoms must be lowered by introducing an excess of Te atoms into the native melt. Halogens show a donor action because they replace Te atoms in the lattice. They may give one electron to conduction band. Using as an approximation the hydrogen-like model, it appears that the ionization energy of a halogen atom is very low and therefore this atom in Bi_2Te_3 is almost full ionized. From a review of literature data [41], several groups of donors may be considered below (Table 6.6).

Impurities are chosen taking into account their solubility in the host material, their electrical activity and the stability of the doped material at high temperatures. However, the doping action of impurities must be determined from homogeneous single crystals or crystals with large grains, without microheterogeneities. It is important to note that solid solutions with the best thermoelectric efficiency are doped with $SbBr_3$ [42]. The electrical activity of Br is equal to 1 [43]. Otherwise, solid solutions doped with copper give a high thermoelectric efficiency but their properties are temporally unstable.

TABLE 6.6 Donors in Bi_2T_3 and Its Solid Solutions

Atoms	Halogen atoms	I, Cl, Br
	Metal atoms	Cu, Ag, Au, Zn, Cd
Compounds fully or partially dissociated	Metal halide	CuBr, $CuBr_2$, AgCl, AgI, Cn, $CdBr_2$, $ZnCl_2$, $CdCl_2$, $HgCl_2$
	Sb, Bi trihalides	SbI_3, BiI_3, $SbCl_3$, $BiCl_3$, $SbBr_3$, $BiBr_3$

Tin, lead, germanium (group IV) and arsenic, antimony, bismuth (group V) act as acceptor atoms in Bi_2Te_3 and its alloys [41]. The electrical activity of these elements depends on the composition of the alloys, the amount and the distribution of the doping introduced in the host material [44–46].

6.3 Preparation of Standard Binary Compounds Bi_2Te_3, Sb_2Te_3 and Their Solid Solutions

6.3.1 Introduction

The characterization of the thermoelectric properties of cooling materials suitable for use at 300 K as bismuth telluride, antimony telluride, and their solid solution is supported currently by a great number of published results. In most of the studied samples, it is difficult to connect the thermoelectric properties to the basic state of the material. The intrinsic properties are hidden by effects of doping or excess in one of components. We want to show that high-quality single crystals must be prepared under well-defined thermodynamical conditions.

6.3.2 Binary Compounds: Bi_2Te_3–Sb_2Te_3

The main problem in preparing chalcogenide-based thermoelectric materials is the high partial pressure of tellurium or selenium. Preparation must be carried out in a quartz ampoule sealed under vacuum. In a general way, the phase rule ($F = C - P + 2$) therefore indicates the degrees of freedom of the thermodynamic system. For example, in the case of binary compounds (Bi_2Te_3 or Sb_2Te_3) the number of component C is 2, the number of phases P is 3 (solid, liquid, and vapor) then the thermodynamic system will be monovariant. Figure 6.3 displays that at a temperature T the liquidus and solidus compositions are fixed. This property can be used in the preparation of single crystals by THM [47].

To avoid the preparation of many single crystals at different temperatures, that is, different maximum stoichiometric deviations, the solidus line can be explored by a saturation annealing technique [48].

The solidus line limits the field of existence of the chalcogenide compounds at high temperatures. The resulting deviation from stoichiometry can be expressed in terms of native defects, which are electrically active and related to the thermoelectric properties. The concentration of charge carriers is measured by the Hall effect [49,50].

In Figure 6.4, the number of excess carriers for Bi_2Te_3 are displayed as a function of annealing temperatures in the range from 570°C to the melting point 585.5°C for the bismuth-rich side, and from the melting point to 560°C for the tellurium-rich side. It should be noted that the carrier concentrations on the p-type side are greater than on the n-type side. The authors observed that the change in type occurs at 583.5°C.

Figure 6.5 shows the number of excess carriers for Sb_2Te_3 as a function of annealing temperatures between 558°C and the melting point on the Sb-rich side, and between the melting point and 500°C on the Te-rich side [47]. The behavior of Sb_2Te_3 is strongly p-type with carrier concentration ranging between 10^{20} and 10^{21} cm^{-3}. The melting point of the compound is incongruent and a retrograde solubility is observed on the side of the lower concentrations [51,52]. As seen later, Sb_2Te_3 is uninteresting as a thermo element but its behavior is important in obtaining knowledge of Bi–Sb–Te ternary solid solutions.

6.3.3 Ternary Solid Solutions: (Bi,Te,Se)–(Bi,Sb,Te)

The figure of merit of these materials could be increased by forming solid solutions between isomorphous crystals. It has been argued that the short-range disorder would decrease the lattice thermal conductivity.

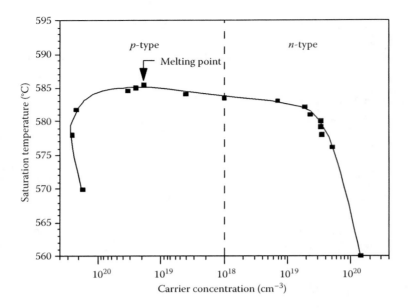

FIGURE 6.4   Solidus of Bi_2Te_3 for different saturation temperatures plotted as a function of carrier concentration.

The best positive (*p*-type) composition appears to be close to $(Sb_2Te_3)_{75}(Bi_2Te_3)_{25}$, while the best negative (*n*-type) composition lies near $(Bi_2Te_3)_{95}$ $(Bi_2Se_3)_5$. The preparation of ternary solid solutions is more complex. In the application of the phase rule, the number of components will be three, the thermodynamical system becomes bivariant and two parameters are necessary to fix the equilibrium of the system. Knowledge of the liquidus compositions in equilibrium with the solidus compositions (isoconcentration line) yields the bivariant system in a monovariant system. Then the temperature is always the parameter for preparation by THM and of the saturation annealing technique [48].

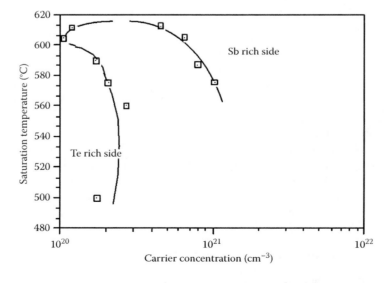

FIGURE 6.5   Solidus of Sb_2Te_3 for different saturation temperatures plotted as a function of carrier concentration.

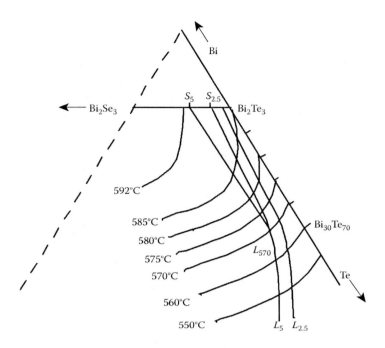

FIGURE 6.6   Liquidus and isoconcentration lines for $Bi_2(Te_{1-x}Se_x)_3$ solid solutions with $x = 0.025$ and 0.05.

Figure 6.6 represents an interesting region of the (Bi,Te,Se) phase diagram. On this scale the stoichiometric deviations are not apparent and S represents the point on the solidus of the solid solutions containing different concentrations of Bi_2Se_3 ($S_5 = 5\%$, $S_{2.5} = 2.5\%$ of Bi_2Se_3 in Bi_2Te_3), L is the corresponding experimental isoconcentration line. Different experimental isothermal curves of the liquidus surface are plotted. The intersection between an isothermal curve and an isoconcentration curve gives the composition of the liquidus in equilibrium with the corresponding solidus at the desired temperature (tie-line) [53].

Likewise, Figure 6.7 represents the (Bi,Sb,Te) ternary phase diagram on the tellurium-rich side from the pseudo binary cross section [54]. On this diagram the experimental isothermal curves of the liquidus at different temperatures, and the experimental isoconcentration lines for two concentrations S_1, S_2, 20% and 25% of Bi_2Te_3 in Sb_2Te_3, respectively are drawn.

With knowledge of these ternary diagrams it is now possible to prepare single crystals under defined thermodynamical conditions. Then, the solidus line of different solid solution compositions corresponding to the maximum stoichiometric deviations can be explored. Figure 6.8 displays the number of excess carriers as a function of annealing temperatures on the Te-rich side for $S_{2.5}$, S_5 (Bi,Te,Se) solid solutions and Bi_2Te_3 compounds.

It is supposed that the native major defects responsible for stoichiometric deviations were antisite defects. The total number of defects ($Te_{Bi} + Se_{Bi}$), like the carrier concentration, decreases with the increasing content of selenium, probably connected with the formation energy of the Se_{Bi} defect larger than that of the Te_{Bi} [53].

Figure 6.9 shows the number of excess carriers as a function of annealing temperature on the Te-rich side, for (S_1), (S_2) (Bi,Sb,Te) solid solutions, and Sb_2Te_3 compounds [54–56]. The rapid decrease of Hall carrier concentration with the increasing mole contents of Bi_2Te_3 is related to the decrease of (Sb_{Te}) antisite defects. Also the retrograde solubility of Sb_2Te_3 becomes less important, which permits easier preparation of the solid solutions.

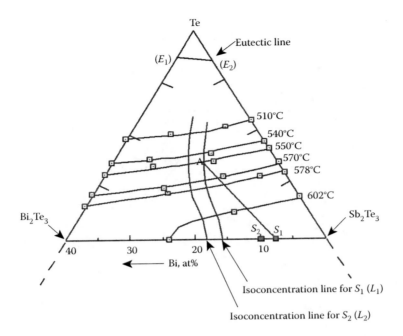

FIGURE 6.7 Liquidus temperatures and isoconcentration lines for $Bi_8Sb_{32}Te_{60}$ and $Bi_{10}Sb_{30}Te_{60}$ solid solutions for the Te-rich field.

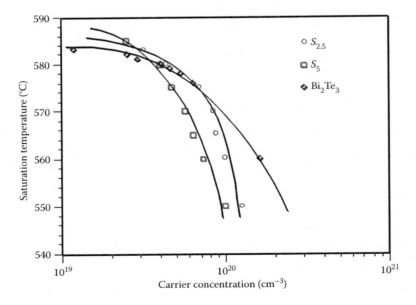

FIGURE 6.8 Hall carrier concentration as a function of saturation temperature for $S_{2.5}$, S_5, and Bi_2Te_3 on the Te-rich side.

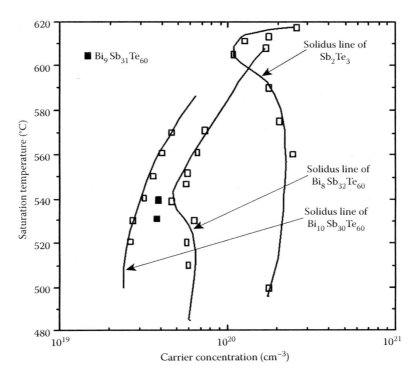

FIGURE 6.9 Hall carrier concentration as a function of saturation temperature for S_1, S_2 solid solutions, and Sb_2Te_3 on the Te-rich side.

6.4 Thermoelectric Properties: Figure of Merit

The following results were obtained for single crystals of *p*- and *n*-type compounds and solid solutions. These were characterized at 300 K by Hall effect, Seebeck coefficient, thermal and electrical conductivity measurements.

All the data have been reported in previous publications [49–60]. The figure of merit, *Z*, is obtained from the experimental measurements of α, σ, and λ with

$$Z = \frac{\alpha^2 \sigma}{\lambda}$$

α is the Seebeck coefficient, σ, λ are the electrical and thermal conductivities, respectively.

Due to the strong anisotropic behavior of the thermal and electrical conductivities the figure of merit of these materials is higher in a direction parallel (Z_{11}) rather than perpendicular (Z_{33}) to the cleavage planes, except for Sb_2Te_3 compound.

The figure of merit Z_{11} for Bi_2Te_3 is plotted vs. the liquidus composition in Figure 6.10. The best values are obtained for *n*-type material. It can be seen that in *n*-type material, Z_{11} has a maximum value of 2.9×10^{-3} K^{-1} at about 64% of Te, corresponding to a charge carrier concentration of 2×10^{19} cm^{-3}.

Because of the low value of the energy gap (0.16 eV) the material is partially degenerate, and the corresponding Fermi level should not be far from the bottom of the conduction band at room temperature. For small stoichiometric deviations, conduction by minority carriers cannot be neglected. The experimental results of Bi_2Te_3 at 300 K show a significant effect of two band conduction upon all the thermoelectric parameters [49,50].

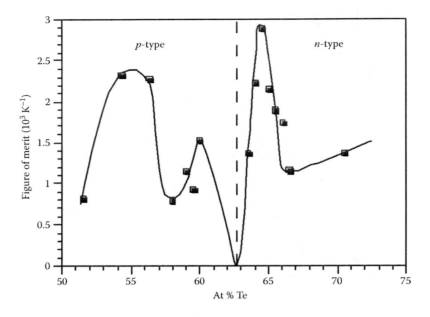

FIGURE 6.10   Figure of merit vs the liquidus composition for both *n*- and *p*- type Bi$_2$Te$_3$.

For the *p*-type Sb$_2$Te$_3$ compound the estimated figure of merit Z_{33} is higher than Z_{11}, (Figure 6.11)

$$\begin{cases} Z_{33} = 1.6 \times 10^{-3} \text{ K}^{-1} \text{ with } p = 2.1 \times 10^{20} \text{ cm}^{-3} \\ Z_{11} = 0.65 \times 10^{-3} \text{ K}^{-1} \text{ with } p = 1 \times 10^{20} \text{ cm}^{-3} \end{cases}.$$

The modeling of the transport properties in Sb$_2$Te$_3$ leads to find an optimal figure of merit Z equal to 3×10^{-3} K^{-1} for a carrier concentration of about 3×10^{19} cm^{-3}. Any carrier concentration so low has been measured in Sb$_2$Te$_3$. To decrease the carrier concentration in Sb$_2$Te$_3$, (Bi,Sb,Te) ternary solid solutions

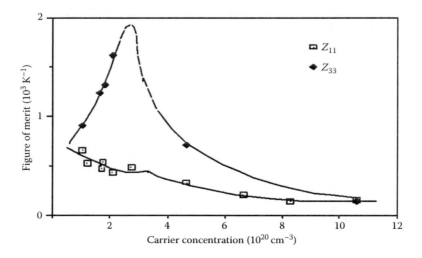

FIGURE 6.11   Figure of merit as a function of carrier concentration for Sb$_2$Te$_3$.

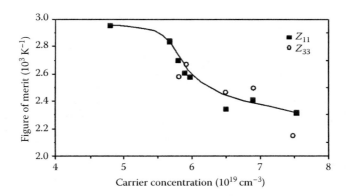

FIGURE 6.12 Figure of merit as a function of carrier concentration for S_1 solid solution.

must be taken into account. So the maximum value Z_{11} obtained for (S_1) solid solution (20% Bi_2Te_3, 80% Sb_2Te_3) was 2.95×10^{-3} K^{-1} corresponding to a carrier concentration of 4.8×10^{19} cm^{-3}. It should be noticed that the figure of merit is isotropic (Figure 6.12). This composition is interesting to be considered for sintered materials.

A different behavior for (S_2) solid solution (25% Bi_2Te_3, 75% Sb_2Te_3) was observed and the anisotropic figure of merit Z_{11} has a maximum value of 2.9×10^{-3} K^{-1} with $p = 3.4 \times 10^{19}$ cm^{-3} (Figure 6.13). It is interesting to study the thermoelectric properties of a solid solution (S_3) with a composition (22.5% Bi_2Te_3, 77.5% Sb_2Te_3) between (S_1) and (S_2). The maximum value of the figure of merit is 3.2×10^{-3} K^{-1} for the sample saturated at 530°C with the carrier concentration $p = 3.9 \times 10^{19}$ cm^{-3}.

For *n*-type $S_{2.5}$ and S_5 solid solutions, the results of the figure of merit Z_{11} as a function of carrier concentration are plotted in Figure 6.14. The best values are about $Z_{11} = 2.9 \times 10^{-3}$ K^{-1} with $n = 4 \times 10^{19}$ cm^3. It is noted that Z_{11} is higher than 2.5×10^{-3} K^{-1} over a larger range of carrier concentration than for the Bi_2Te_3 compound. Consequently, it is easier to optimize Z_{11} for the solid solutions than for the Bi_2Te_3 compound.

The typical results for all the thermoelectric parameters at 300 K obtained for the optimized materials are compiled in Table 6.7.

It is possible to obtain *n*-type ternary solid solutions from the ternary phase diagram (Bi,Sb,Te). The composition such as $(Bi_xSb_{1-x})_2Te_3$ with $0 \le x \le 0.7$ should be investigated [61].

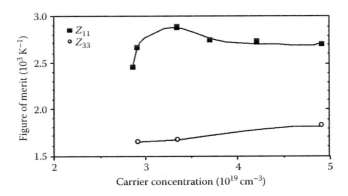

FIGURE 6.13 Figure of merit as a function of carrier concentration for S_2 solid solution.

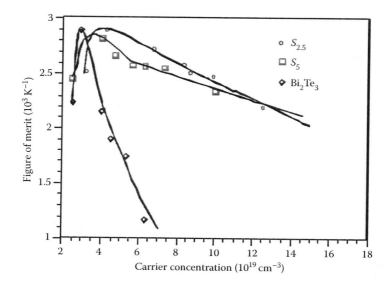

FIGURE 6.14 Figure of merit as a function of carrier concentration for $S_{2.5}$, S_5 solid solutions, and Bi_2Te_3 on the Te-rich side.

TABLE 6.7 Experimental Results for Single Crystals

Column	n–p	T (°C)	C_p	ρ	μ_H	α	λ	Z
Bi_2Te_3	n	582	2.3	10.0	212	240	2.0	2.9
$Sb_2Te_3 \perp$	p	575	21.0	3.2	244	92	1.6	1.6
$Sb_2Te_3 {}_{/\!/}$	p	605	10.0	1.9	313	83.0	5.6	0.6
$Bi_8Sb_{32}Te_{60}$	p	540	4.8	10.0	176	194	1.3	3.0
$Bi_{10}Sb_{30}Te_{60}$	p	540	3.5	13	177	225	1.4	2.9
$Bi_9Sb_{31}Te_{60}$	p	530	3.9	8.9	190	206	1.5	3.2
$Bi_{40}Sb_{57}Te_3$	n	580	4.0	11.0	140	223	1.6	2.8
$Bi_{40}Te_{58.5}Se_{1.5}$	n	580	4.3	11.0	150	230	1.7	2.9

Note: Column n–p: type of conduction; T (°C): saturation temperature in °C; C_p: carrier concentration $\times 10^{19}$ cm^{-3}; ρ: resistivity $\mu\Omega$ m; μ_H: mobility of carriers cm^2 v^{-1} s^{-1}; α: absolute Seebeck coefficient μV K^{-1}; λ: thermal conductivity W m^{-1} K^{-1}; Z: figure of merit 10^{-3} K^{-1}.

6.5 Discussion

In conclusion, a method has been indicated to prepare single crystals of high quality in a reproducible way. The conditions of preparation are well defined of a thermodynamical point of view from the phase diagram studies.

It appears from our results that an optimal Seebeck coefficient (or an optimal carrier concentration) exists at room temperature, leading to the best dimensionless figure of merit ZT for a given material. For n-type materials our results show that the ternary solid solutions do not give ZT better than 0.9, value obtained also for the Bi_2Te_3 compound which is always a good thermo-element. For p-type materials, the Sb_2Te_3 compound is not suitable, but the (Bi,Sb,Te) solid solutions are a way to control the carrier concentration and to achieve the corresponding high value $ZT \sim 1$. It is important to note that the crystalline quality of the optimized single crystals is associated with a very good homogeneity and a low density of dislocations [62].

Generally to produce a large quantity of material for commercial applications a powder metallurgy process is used. Numerous papers have been published on the subject. A good analysis of the influence of the composition and the texture on the thermoelectric and mechanical properties is necessary to optimize the alloy's performance [63].

Thermoelectric materials can operate only around an optimum working temperature. If we consider the figure of merit ZT as a function of temperature of p- and n-type bulk thermoelectric materials, bismuth telluride and alloys are always among the best for low temperatures (<250°C). Practically, the thermal–electrical conversion is quite low (<8%). Some new interest was generated when it was proposed that the use of nanostructures, such as quantum wells and nanowires could improve the efficiencies of the conventional thermoelectric materials [64–68]). For example, compared to the experimental values of ZT for bulk Bi_2Te_3 ($ZT \sim 1$) it was predicted that the confinement effects of the carriers leads to $ZT > 2$ for the structures smaller than 20 Å.

The experimental efforts of the research community in finding new structures give rise to a range of controversial but interesting results.

So, nanostructures with confirmed high ZT values are few. For example, some systems can be cited: bulk nanocomposites of $(Bi_xSb_{1-x})_2 Te_3$ [69,70], resonant-doped Bi_2Te_3 and PbTe [71–73] and other systems [74–80]. We observe that the improvements in ZT seem to have arisen mainly from a reduced lattice thermal conductivity. So, higher values of ZT have not yet been achieved because it is very difficult to control the charge carrier concentration in nanostructures. The presence of impurities and other defects of the structure reduces the electronic mobility. Then, for multilayer thin films, and superlattices mainly, interfaces and boundaries lead to some thermal and electrical problems of contact to solve. Many of these structures are not thermally stable and present some fragility for commercial applications.

As we have seen, our results show the existence of an optimal Seebeck coefficient α_{opt} at 300 K. The problem is to know if it is always true for nanostructures. To improve the figure of merit, an increase in the power factor ($\alpha^2\sigma$) is desirable .The increase in the power factor can be obtained by a change in the magnitude of the density of states near the Fermi level E_F.

Through a local peak due to resonant impurity levels [72]. This effect induces an enhancement of the Seebeck coefficient α. However, the efforts must be focused on the increase of the electrical conductivity σ for an optimal Seebeck coefficient α_{opt}. Through the Mott expression

$$\alpha = \frac{\pi^2}{3}\frac{k_B^2}{e}\ T\left(\frac{d\ln\sigma(E)}{dE}\right)_{E=EF}$$

the scattering mechanisms, via the relaxation time dependent of the electron energy must be considered. For example, nanoscale inclusions enhance the thermoelectric properties of bulk materials [81,82]. The grain boundary potential barrier in bulk nanocomposite alloys of (Bi,Sb,Te) can filter the electronic energy distribution and influence the power factor. Nanoscale events must be optimized in size and composition. This new approach provides means to obtain efficient materials.

References

1. Rosi, F.D., Abeles, B., and Jensen, R.V., *J. Phys. Chem. Solids*, 1959, 10, 191.
2. Rosi, F.D. and Ramberg, F.G. 1960. In *Thermoelectricity*, P.M. Egli, ed., Wiley, New York.
3. Goldsmid, H.J. 1986. *Thermoelectric Refrigeration*, Pion Limited, London.
4. Heikes, R.R. and Ure, R.W. Jr., 1961. *Thermoelectricity: Science and Engineering*, Interscience, New York.
5. Rowe, D.M. and Bhandari, C.M. 1983. *Modern Thermoelectrics*, Holt Rinehart and Winston, London.
6. Rowe, D.M., ed. 1995. *Handbook of Thermoelectrics*, CRC Press, Boca Raton.
7. Satterthwaite, C.D. and Ure, R.W., *Phys. Rev.*, 1957, 108, 1164.
8. Brebrick, R.F., *J. Phys. Chem. Solids*, 1969, 30, 719.
9. Abrikosov, N.Kh. and Bankina, V.F., *ZhNKh*, 1958, 3, 659.

10. Zhukov, A.A. et al., *Izvest. Akad. Nank. S.S.S.R., Metally*, 1983, 1, 183.
11. Hawlett, B., Miara, S., and Bever, M., *Trans. AIME*, 1964, 230, 1367.
12. Abrikosov, N.Kh., Poretskaya, I.V., and Ivanova, I.P., *ZhNKh*, 1959, 4, 2525.
13. Abrikosov, N.Kh. et al. *Semiconductor Compounds and Its Properties*, Nauka, Moscow, 1967.
14. Boiling, G.J. *Chem. Phys.*, 1960, 33, 305.
15. Itskevitch, E.C., *Sov. Phys. J.E.T.P.*, 1960, 11, 255.
16. Jenkins, J.O., Rayne, J.A., and Ure, R.W., *Phys. Rev. B*, 1972, 5, 3171.
17. Goldsmid, H.J. 1960. *Proceedings of the 5th Conference on Physics of Semiconductors*, p. 1015, Prague.
18. Gol'tsman, B.M. et al.,1972. *Thermoelectric Semiconductors Materials Based on Bi2Te3*, Nauka, Moscow.
19. Boncheva, Z., Pashinkin, A.S., and Novoselova, A.V., *Neorg. Mater.*, 1968, 4, 291.
20. Lange, P.W., *Naturwissenschaften*, 1939, 27, 133.
21. Francombe, M.H., *Br. J. Appl.*, 1958, 9, 415.
22. Brebrick, R.F., *J. Appl. Crystallogr.*, 1968, 1, 241.
23. Mallinson, R.B., Rayne, J.R., and Ure, R.W. Jr., *Phys. Rev.*, 1968, 17–5, 1049.
24. Wyckoff, R.W.G. 1964. *Crystal Structures*, Interscience, New York.
25. Semiletov, S.A., *Kristallographiya*, 1956, 1, 403.
26. Drabble, J.R. and Goodmann, C.H.I., *J. Phys. Chem. Solids*, 1958, 5, 142.
27. Chitroub, M., Scherrer, S., and Scherrer, H., *J. Phys. Chem. Solids*, 2000, 61, 1693.
28. Lovett, D.R. 1977. *Semimetals and Narrow Band Gap Semiconductors*, Pion Limited, London.
29. Kullmann, W., Geurts, J., Richter, W. et al. *Phys. Status Solidi (b)*, 1984, 125, 131.
30. Smith, M.J., Knight, R.J., and Spencer, C.W., *J. Appl. Phys.*, 1962, 33, 2186.
31. Miller, G.R., Che-Yu, Li., and Spencer, C.W., *J. Appl. Phys.*, 1973, 34, 1398.
32. Stary, Z., Horak, J., Stordeur, M., and Stölzer, M., *J. Phys. Chem. Solids*, 1988, 49, 97.
33. Kutasov, V.A., Svechnikova, T.E., and Chizhevskaya, S.N., *Sov. Phys. Solid State*, 1987, 10, 1724.
34. Birkholz, H., *Z. Naturforsch*, 1958, 13a, 780.
35. Stavosa, M.M. and Abrikosov, N.Kh., *Jzv. An. S.S.S.R. Neorg. Mater.*, 1970, 6, 1090.
36. Nakajima, S.*J. Phys. Chem. Solids*, 1963, 24, 479.
37. Wiese, J.R. and Muldawer, L. *J. Phys. Chem. Solids*, 1960, 15, 13.
38. Offergeld, G. and Van Caken Berghe, *J. Nat.*, 1959, 10, 185.
39. Pecheur, P. and Toussaint, G., *J. Phys. Chem. Solids*, 1994, 55, 327.
40. Miller, G.R. and Che-Yu, Li., *J. Phys. Chem. Solids*, 1965, 28, 173.
41. Przyluski, J. and Borkowski, K. 1982. *Proceedings of 4th Int. Conf. on Thermoelectric Energy Conversion*, Arlington.
42. Sotirova, M., Sotirov, St., and Andreev, A., *Freiberg Forschung. B*, 1975, 43, 175.
43. Perrin, D., Chitroub, M., Scherrer, S., and Scherrer, H. *J. Phys. Chem. Solids*, 2000, 61, 1687.
44. Abrikosov, N.C. et al., *Zhi. Neorg. Mater.*, 1981, 14, 428.
45. Tamura, H., *Jpn. J. Appl. Phys.*, 1966, 7, 593.
46. Süssmann, H., Priemuth, A., and Proehl, U. *Phys. State Solids*, 1984, 82, 561.
47. Borshchevsky, A. 1995. *Handbook of Thermoelectrics*, D.M. Rowe ed., p. 83.
48. Scherrer, H., Weber, S., and Scherrer, S. *Phys. Lett. A*, 1980, 77, 189.
49. Fleurial, J.-P., Gailliard, L., Triboulet, R., Scherrer, H., and Scherrer, S., *J. Phys. Chem. Solids*, 1988, 49, 1237.
50. Fleurial, J.-P., Gailliard, L., Triboulet, R., Scherrer, H., and Scherrer, S., *J. Phys. Chem. Solids*, 1988, 49, 1249.
51. Gailliard, L., Caillat, T., Scherrer, H., and Scherrer, S. 1989, *Proceedings of 8th Int. Conf. on Thermoelectric Energy Conversion*, H. Scherrer and S. Scherrer, eds., p. 12, Nancy.
52. Caillat, T. 1991. Doctoral thesis, INPL Nancy.
53. Carle, M. 1992. Doctoral thesis, INPL Nancy.
54. Caillat, T., Carle, M., Scherrer, H., and Scherrer, S., *J. Phys. Chem. Solids*, 1992, 53, 227.
55. Caillat, T., Carle, M., Pierrat, P., Scherrer, H., and Scherrer, S., *J. Phys. Chem. Solids*, 1992, 53, 1121.

56. Caillat, T., Gailliard, L., Scherrer, H., and Scherrer, S., *J. Phys. Chem. Solids*, 1993, 54, 575.

57. Carle, M., Caillat, T., Lahalle-Gravier, C., Scherrer, S., and Scherrer, H., *J. Phys. Chem. Solids*, 1995, 56, 195.

58. Carle, M., Caillat, T., Lahalle-Gravier, C., Scherrer, S., and Scherrer, H., *J. Phys. Chem. Solids*, 1995, 56, 201.

59. Lahalle-Gravier, C., Scherrer, S., and Scherrer, H., *J. Phys. Chem. Solids*, 1996, 57, 1713.

60. Scherrer, H. and Scherrer, S. In *Handbook of Thermoelectrics*, D.M. Rowe, ed., 1994, pp. 211–337. CRC Press, Boca Raton, FL, chap. 9.

61. Martin-Lopez, R., Lenoir, B., Dauscher, A., Scherrer, H., and Scherrer, S., *Solid State Commun.*, 1998, 108, 285–288.

62. Scherrer, H., Chiroub, M., Roche, G., and Scherrer, S. *Proceedings of 17th International Conference on Thermoelectrics*, 1998, p. 115, Nagoya.

63. Simard, J.M., Vasilevskiy, D., and Turenne, S. *Proceedings of 22nd International Conference on Thermoelectrics*, 2003, pp. 13–18. La Grande Motte, France.

64. Dresselhaus, M.S., Lin, Y.M., Cronin, S.B., Rabin, O., Black, M.R., Dresselhaus, G., and Koga, T., Quantum wells and quantum wires for potential thermoelectric applications. *Recent Trends in Thermoelectric Materials Research III*, 2001, 71, 1–121.

65. Dresselhaus, M.S., Chen, G., Tang, M.Y., Yang, R.G., Lee, H., Wang, D.Z., Ren, Z.F., Fleurial, J.P., and Gogna, P., New directions for low-dimensional thermoelectric materials. *Adv. Mater.* 2007, 19(8), 1043–1053.

66. Dresselhaus, M.S., Chen, G., Ren, Z.F., Dresselhaus, G., Henry, A., and Fleurial, J.P., New composite thermoelectric materials for energy harvesting applications. *JOM*, 2009, 61(4), 86–90.

67. Hicks, L.D. and Dresselhaus, M.S., Effect of quantum-well structures on the thermoelectric figure of merit. *Phys. Rev. B*, 1993, 47(19), 12727–12731.

68. Hicks, L.D. and Dresselhaus, M.S., Thermoelectric figure of merit of a one-dimensional conductor. *Phys. Rev. B*, 1993, 47, 16631–16634.

69. Lan, Y., Poudel, B., Ma, Y., Wang, D., Dresselhaus, M.S., Chen, G., and Ren, Z., Structure study of bulk nanograined thermoelectric bismuth antimony telluride. *Nano Letters*, 2009, 9(4), 1419–1422.

70. Poudel, B., Hao, Q., Ma, Y., Lan, Y., Minnich, A., Yu, B., Yan, X. et al. High-thermoelectric performance of nanostructured bismuth antimony telluride bulk alloys. *Science*, 2008, 320(5876), 634–638.

71. Heremans, J.P., Jovovic, V., Toberer, E.S., Saramat, A., Kurosaki, K., Charoenphakdee, A., Yamanaka, S., and Snyder, G.J., Enhancement of thermoelectric efficiency in PbTe by distortion of the electronic density of states. *Science*, 2008, 321(5888), 554–557.

72. Jaworski, C.M., Kulbachinskii, V., and Heremans, J.P., Resonant level formed by tin in Bi_2Te_3 and the enhancement of room-temperature thermoelectric power. *Phys. Rev. B*, 2009, 80(23), 233201.

73. Jaworski, C.M., Tobola, J., Levin, E.M., Schmidt-Rohr, K., and Heremans, J.P., Antimony as an amphoteric dopant in lead telluride. *Phys. Rev. B*, 2009, 80(12), 125208.

74. Harman, T.C., Taylor, P.J., Spears, D.L., and Walsh, M.P., Thermoelectric quantum-dot superlattices with high ZT. *J. Electron. Mater.* 2000, 29(1), L1–L4.

75. Harman, T.C., Taylor, P.J., Walsh, M.P., and LaForge, B.E., Quantum dot superlattice thermoelectric materials and devices. *Science* 2002, 297(5590), 2229–2232.

76. Harman, T.C., Walsh, M.P., Laforge, B.E., and Turner, G.W., Nanostructured thermoelectric materials. *J. Electron. Mater.*, 2005, 34(5), L19–L22.

77. Harman, T.C., Spears, D.L., and Manfra, M.J., High thermoelectric figures of merit in PbTe quantum wells. *J. Electron. Mater.*, 1996, 25(7), 1121–1127.

78. Venkatasubramanian, R., Siivola, E., Colpitts, T., and O'Quinn, B., Thin-film thermoelectric devices with high room-temperature figures of merit. *Nature*, 2001, 413(6856), 597–602.

79. Chowdhury, I., Prasher, R., Lofgreen, K., Chrysler, G., Narasimhan, S., Mahajan, R., Koester, D., Alley, R., and Venkatasubramanian, R., On-chip cooling by superlattice-based thin film thermoelectrics. *Nature Nanotechnol.*, 2009, 4(4), 235–238.

80. Snyder, G.J. and Toberer, E.S., Complex thermoelectric materials. *Nature Mater.*, 2008, 7(2), 105–114.

81. Faleev, S.V. and Leonard, F. Theory of enhancement of thermoelectric properties of materials with nanoinclusions. *Phys. Rev. B*, 2008, **77**, 214304, 2008.

82. Kishimoto, K. Yamamoto, K., and Koyanagi, T. Influences of potential barrier scattering on the thermoelectric properties of sintered n-type PbTe with a small grain size, *Jpn. J. Appl. Phy.*, 2003, 42, 501.

7

Optimization of Solid Solutions Based on Bismuth and Antimony Chalcogenides above Room Temperature

L.N. Lukyanova
Ioffe Physical-Technical Institute of the Russian Academy of Sciences

V.A. Kutasov
Ioffe Physical-Technical Institute of the Russian Academy of Sciences

P.P. Konstantinov
Ioffe Physical-Technical Institute of the Russian Academy of Sciences

V.V. Popov
Ioffe Physical-Technical Institute of the Russian Academy of Sciences

7.1 Introduction

Solid solutions based on bismuth and antimony chalcogenides with atomic substitutions in Bi and Te sublattices are known to be high-performance thermoelectrics for use in thermoelectric modules for operation above room temperature at optimal compositions and charge carrier concentrations. The bulk Bi_2Te_3-based thermoelectric materials remain essential for industrial application[1-4] in spite of significant achievements over the last few years in the development of new nanobulk materials.[5-8] In Chapter 7, thermoelectric properties of n- and p-type Bi_2Te_3-based solid solutions with substitutions Bi → Sb and Te → Se, S are considered through the temperature interval 300–550 K. An increase in the figure-of-merit Z is shown to be observed for the compositions with optimum relations between the effective density-of-states mass (m/m_0), the carrier mobility with account of degeneracy (μ_0), and the lattice thermal conductivity (κ_L) as a function of temperature, composition, and carrier concentration.[9–14] Changes of scattering mechanism of charge carriers due to substitutions of atoms in the Bi and Te sublattices are taken into account for calculations of the m/m_0, μ_0 and κ_L values[15–17] in terms of the parabolic model of the energy spectrum for isotropic scattering mechanism. The materials under consideration have a complex band structure described by a many-valley model of energy spectrum.[18–25] Changes of the parameters of ellipsoidal constant-energy surface and scattering of charge carriers affect the thermoelectric and galvanomagnetic properties of the solid solutions. Therefore, the features of the figure of merit for solid solutions were analyzed and optimized simultaneously using thermoelectric and galvanomagnetic properties.

7.2 Review of the Figure-of-Merit Features of Thermoelectrics Based on Bismuth and Antimony Chalcogenides

High-performance thermoelectrics based on bismuth and antimony chalcogenides with optimum properties above room temperature are actively studied because they are widely used for design of thermoelectric modules operating above room temperature. Some of the results on studies of the figure-of-merit features for n-[26–28] and p-type[29–32] solid solutions in dependence on composition and temperature are given in Figure 7.1. An increase in the figure-of-merit Z is observed in the compositions p-$(Bi_2Te_3)_x(Sb_2Te_3)_{1-x}$ at $x = 0.21$ and 0.20,[31] and p-$(Bi_{0.2}Sb_{0.8})_2Te_3$,[32] which are prepared by the extrusion technique (Figure 7.1, curves 1, 2) and optimized in the temperature interval 300–370 K. Single crystals of the p- and n-type solid solutions have been grown by a melt-feed Czochralski technique[26,28–30] (Figure 7.1, curves 4–6). An increase in Z values was found in the temperature interval 370–420 K for the composition p-$Bi_{0.5}Sb_{1.5}Te_3$ doping with 4 mol% Bi_2Se_3[29] (Figure 7.1, curve 4) for the value of the Seebeck coefficient $\alpha = 160$ μV K^{-1} at room temperature. The compositions $Bi_{0.6}Sb_{1.4}Te_3$ and $Bi_{0.4}Sb_{1.6}Te_3$ are effective at higher temperatures up to 700 K[30] (Figure 7.1, curves 5, 6) with increase in the carrier concentration at $\alpha = 170$ μV K^{-1} and 140 μV K^{-1} at $T = 300$ K, respectively.

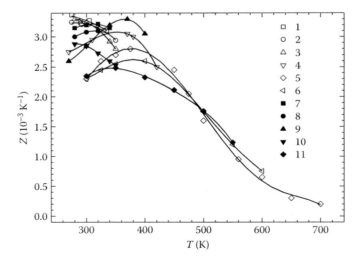

FIGURE 7.1 Temperature dependence of the figure-of-merit for solid solutions based on bismuth and antimony chalcogenides. *P*-type: 1—$(Bi_2Te_3)_x(Sb_2Te_3)_{1-x}$, $x = 0.21$,[31] 2—$(Bi_2Te_3)_x(Sb_2Te_3)_{1-x}$, $x = 0.20$,[31] 3—$Bi_{0.5}Sb_{1.5}Te_3 + 4$ mol% Bi_2Se_3,[29] 4—$Bi_{0.4}Sb_{1.6}Te_3$,[30] 5—$Bi_{0.4}Sb_{1.6}Te_3$,[30] 6—$Bi_{0.1}Sb_{1.6}Te_3$.[30] *N*-type: 7—$Bi_2Te_{2.85}Se_{0.15} + 0.05$ at% Cu,[28] 8—$Bi_2Te_{2.85}Se_{0.15} + 0.5$ at% CdTe,[28] 9—$Bi_{1.996}Sn_{0.004}Te_{0.15}Se_{0.15}$,[26] 10—$(Bi_{0.95}Sb_{0.05})_2(Te_{0.95}Se_{0.05})_3$,[32] 11—$Bi_2Te_{2.4}Se_{0.6}$.[27]

Single crystals of the *n*-type solid solutions $Bi_2Te_{2.85}Se_{0.15}$ doped with 0.05 mol% Cu and 0.5 mol% CdTe (Figure 7.1, curves 7–8),[28] have been optimized near room temperature. These compositions have high values of Z at temperatures up to 350 K. For the multicomponent single-crystal $Bi_{1.996}Sn_{0.004}Te_{0.15}Se_{0.15}$ containing tin,[26] the higher Z value was found in the interval 350–420 K in comparison with the composition $(Bi_{0.95}Sb_{0.05})_2(Te_{0.95}Se_{0.05})_3$ (Figure 7.1, curve 9, 10).[32] Besides, the Seebeck coefficient, which is sensitive to fluctuations of the carrier concentration, shows the high homogeneity of the thermoelectric properties on the length and cross section of the single crystals doped with tin.[26] The solid solution $Bi_2Te_{2.4}Se_{0.6}$ was prepared by directed crystallization method.[27] The higher figure of merit[27] at optimum charge carrier concentration at $T > 450$ K is related to increase of the band gap E_g with growth of the Se content in solid solution (Figure 7.1, curve 11)[33] because the band gap (E_g) increasing permits shifting the intrinsic conductivity contribution to higher temperatures.

The Bi_2Te_3-based solid solutions, like its parent compound, is a narrow-gap indirect semiconductor with $E_g = 0.14$–0.15 eV in accordance with the recent calculations of the electronic structure.[34–36] This value is in a good agreement with optical measurements carried out in Refs. 37,38. However, according to direct measurements by tunneling experiments the band gap E_g for Bi_2Te_3 is equal approximately $E_g = 0.20$ eV.[39] For Sb_2Te_3 the optical band gap is equal to $E_g = 0.21$ eV[40] and according to tunneling experiments $E_g = 0.25$ eV.[39] The $E_g = 0.25$ eV for Sb_2Te_3 is in a good agreement from optical studies of *p*-type Bi_2Te_3–Sb_2Te_3 alloys at room temperature.[40]

But studies of the optical absorption spectra of the *n*-$Bi_2Te_{2.7}Se_{0.3}$ and the multicomponent solid solutions *n*-$Bi_{1.8}Sb_{0.2}Te_{2.7}Se_{0.15}S_{0.15}$ and *p*-$Bi_{0.75}Sb_{1.25}Te_{2.91}Se_{0.09}$ at $T = 300$ K showed that the interband electronic transitions are direct ones, and the optical band gap values are close for the studied solid solutions.[41,42] The E_g values change from 0.23 eV to (0.238 ± 0.015) eV,[41,42] and that agree with the data measured earlier for the composition *n*-$Bi_2Te_{3-y}Se_y$.[33] The maximum value of $E_g = 0.29$ eV for system of the solid solution $Bi_2Te_{3-y}Se_y$ was found near to the composition $Bi_2Te_2Se_1$.[33] The analysis of the band gap shows that the increase of E_g is observed with increase in the Sb content in the *p*-type solid solutions $Bi_{2-x}Sb_xTe_3$ in spite of some distinctions of the E_g values measured by different authors. An increase in the E_g for the multicomponent solid solutions of *n*-type $Bi_{2-x}Sb_xTe_{3-y-z}Se_yS_z$ was not observed in comparison with the system $Bi_2Te_{3-y}Se_y$.[41–43]

7.3 Thermoelectric Properties

7.3.1 Samples for Studies

Solid solutions based on bismuth and antimony chalcogenides with substitutions of atoms in both sublattices of bismuth telluride Bi → Sb and Te → Se, S and with atomic substitutions in bismuth or in tellurium sublattices (Te → Se, S and Te → Se) were studied. The solid solutions were grown by the method of directed crystallization using precision stabilization of temperature at the crystallization front for obtaining homogeneous ingots. Additional homogeneity of the ingots was achieved by reversive rotation of quartz ampoules with synthesized materials. Such technique of formation of multicomponent solid solutions allows one to obtain textured ingots, in which single-crystal grains are mainly oriented along the axis of growth, which is perpendicular to the axis of third-order C_3. Clevage planes (0001) in single-crystal grains are parallel to the growth axis, which is a direction of maximum figure-of-merit value. The alloying impurity $CdCl_2$ and excess of Te in comparison with stoichiometric composition were introduced in the solid solutions for optimization of charge carrier concentration.

7.3.2 Power Factor

Temperature dependence of the power factor $\alpha^2\sigma$ for *n*-type $Bi_{2-x}Sb_xTe_{3-y-z}Se_yS_z$, and for *p*-type $Bi_{2-x}Sb_xTe_{3-y}Se_y$ solid solutions were obtained from the Seebeck coefficient α and electrical conductivity σ reported in the Refs. 9–14 (Figure 7.2). An increase in the carrier concentration and the content of the substituted atoms in the solid solutions leads shifting the maximum of the power factor

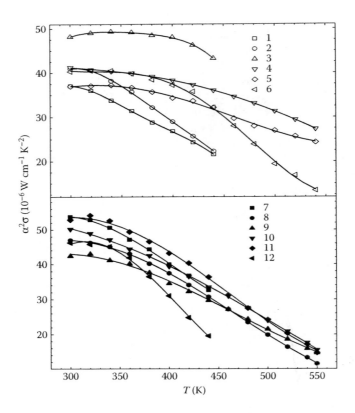

FIGURE 7.2 Temperature dependence of power factor $\alpha^2\sigma$ for n-$Bi_{2-x}Sb_xTe_{3-y-z}Se_yS_z$ (1–6), and p-$Bi_{2-x}Sb_xTe_{3-y}Se_y$ (7–12) solid solutions. The curve notation is the same as in Figure 7.2. 1—($x = 0.16$, $y = z = 0.12$), 2—($x = 0.16$, $y = z = 0.06$), 3—($x = 0$, $y = z = 0.09$), 4—($x = 0$, $y = 0.6$), 5—($x = z = 0$, $y = 0.9$), 6—($x = 0$, $y = z = 0.15$), 7, 9—($x = 1.55$, $y = 0$), 8—($x = 1.6$, $y = 0.06$), 10, 11—($x = 1.6$, $y = 0$), 12—($x = 1.3$, $y = 0.06$).

temperature dependence to high temperature (Figure 7.2) due to growth of the energy gap E_g. The power factor increases with increasing E_g^{33} for the n-type solid solutions with atomic substitutions Te → Se for high content of Se atoms at $y = 0.6$ and 0.9 in comparison with the substitutions T → Se and S at $y = z = 0.15$ (Figure 7.2, curves 4, 5, and 6).

High values of $\alpha^2\sigma$ in the range of temperatures 300–450 K are found. The changes of charge carrier scattering are taken into account due to the influence of the substitutions of atoms in the Bi and Te sublattices and filling additional bands in the conduction and the valence bands of Bi_2Te_3-based solid solutions.

The power factor increases in the p-type composition with increasing Sb content from $x = 1.55$ to 1.6 and increases with the carrier concentration (Figure 7.2, curves 7, 11). The power factor $\alpha^2\sigma$ sharply decreases with rise in the temperature for n- and p-type solid solutions with substitutions of bismuth telluride atoms in both sublattices due to the additional charge carrier scattering on these atoms (Figure 7.2, curves 1, 2, and 8, 12) and decrease in the carrier mobility.

7.3.3 Thermal Conductivity

Thermal conductivity κ considerably decreases in n- and p-type solid solutions with substitutions on both bismuth telluride sublattices as a result of the additional distortions of the crystal lattice comparing with atomic substitutions on the Bi or Te sublattices only. A reduction of κ is observed in the n-type compositions at $x = 0.16$, $y = z = 0.12$, and 0.06 (Figure 7.3, curves 1, 2), but in p-type compositions the

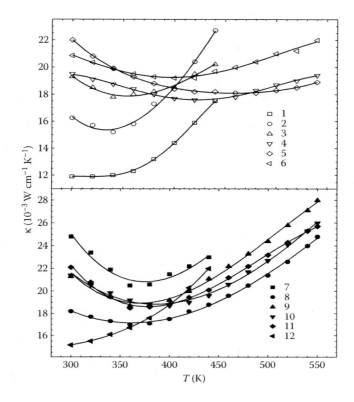

FIGURE 7.3 Temperature dependence of the thermal conductivity κ for n-type solid solutions $Bi_{2-x}Sb_xTe_{3-y-z}Se_yS_z$ (1–6), and for p-type solid solutions $Bi_{2-x}Sb_xTe_{3-y}Se_y$ (7–12). The curve notation is the same as in Figure 7.2.

thermal conductivity κ decreases at $x = 1.6$ and 1.3, $y = 0.06$ (Figure 7.3, curves 8, 12). An increase of electron concentration and the substituted Se atoms in the solid solutions enable to avoid an appreciable growth of the thermal conductivity with the temperature rises up to 400–550 K due to weak influence of the intrinsic conductivity for n-type compositions at $x = z = 0$, $y = 0.6$ and 0.9 (Figure 7.3, curves 4, 5).

For p-type solid solution at $x = 1.55$, $y = 0$, an increase of the carrier concentration leads to decrease in the κ value (Figure 7.3, curves 7, 9). With an increase of Sb atoms in the solid solutions from $x = 1.55$ to 1.6, the thermal conductivity decreases for the samples with close carrier concentrations at the temperatures more than 350 K (Figure 7.3, curves 9, 11).

7.3.4 Figure of Merit

An increase of the figure-of-merit Z in the n-type composition with substitutions Bi $\rightarrow$ Sb and Te $\rightarrow$ Se, S at $x = 0.16$, $y = z = 0.12$ at the temperatures up to 350 K is defined by the low thermal conductivity, in spite of low value of the power factor for the sample with optimum electron concentration for specified temperatures (Figures 7.2 through 7.4, curves 1).

In the temperature interval 350–450 K, the figure of merit Z increases in n-type solid solution with small atomic substitutions only on tellurium sublattice at $x = 0$, $y = z = 0.09$ due to a significant increase of the power factor, while the thermal conductivity increases (Figures 7.2 through 7.4, curves 3). At temperatures more than 450 K, the value of Z increases in the sample with higher electron concentration at Te $\rightarrow$ Se ($x = z = 0$, $y = 0.6$) atomic substitutions in comparison with the composition with more content of Se atoms at $x = z = 0$, $y = 0.6$ owing to an increase in the power factor and low thermal conductivity (Figure 7.4, curves 4, 5).

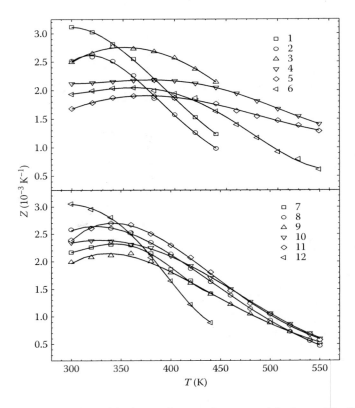

FIGURE 7.4 Temperature dependence of the figure-of-merit Z for n-type solid solutions $Bi_{2-x}Sb_xTe_{3-y-z}Se_yS_z$ (1–6), and for p-type solid solutions $Bi_{2-x}Sb_xTe_{3-y}Se_y$ (7–12). The curve notation is the same as in Figure 7.2.

An increase in the figure of merit Z for p-type solid solution is observed in the composition with substitutions of atoms in both sublattices of Bi_2Te_3 ($x = 1.3$ and $y = 0.06$) in the temperature interval from 300 to 370 K for optimal carrier concentration due to low thermal conductivity (Figures 7.3, 7.4, curves 12). In the temperature interval from 370 to 550 K, an increase in Z for the p-type solid solution with the content of Sb at $x = 1.6$ $y = 0$ is determined by the high-power factor $\alpha^2\sigma$ for the sample (Figures 7.2, 7.4, curves 11) with more high-charge carrier concentration.

7.4 Effective Mass, Charge Carrier Mobility, and Lattice Thermal Conductivity

7.4.1 Effective Mass

The values of m/m_0, μ_0, and κ_L were calculated in terms of the parabolic model of the energy spectrum for isotropic scattering mechanism of charge carriers. In the calculation of m/m_0, μ_0, and κ_L, the effective scattering parameter r_{eff} was used.[15–17] The substitutions of atoms on bismuth telluride sublattices influence on scattering processes in comparison with the widely used acoustic scattering mechanism, and it leads to change in the scattering parameter. The additional conduction and valence bands in the solid solutions also influence on the parameter r_{eff}. The value of the r_{eff} parameter was calculated from experimental the Seebeck coefficient and galvanomagnetic properties treated in the framework of the many-valley energy spectrum model.[15–17] The calculated r_{eff} values changes from (−0.7) to (−0.8) in dependence on the composition and carrier concentration of the solid solutions optimized above room temperature.

The effective mass m/m_0 grows with increasing carrier concentration in *n*- and *p*-type solid solutions (Figure 7.5). The temperature dependences of m/m_0 weaken with increasing carrier concentration owing to the influence of the intrinsic conductivity at higher temperatures in *n*-type solid solutions with atomic substitutions of Te sublattices (Figure 7.5, curves 3–6) in comparison with the compositions at $x = 0.16$ and $y = z = 0.12$ and 0.06 (Figure 7.5, curves 1, 2) at lower carrier concentration. In *p*-type compositions, weakening the temperature dependence of m/m_0 is observed with increase in the carrier concentration compared to the samples at $x = 1.3$ and $y = 0.06$ at lower carrier concentration (Figure 7.5, curve 12).

The effective mass m/m_0 increases in the solid solutions containing a higher content of substituted Se atoms in the *n*-type compositions at $y = 0.6$ and 0.9, and $x = z = 0$ compared to the substitutions Te → Se + S ($y = z = 0.15$) (Figure 7.5, curves 4–6). The m/m_0 value also increases with increasing Te content in the compositions with the substitutions of Bi → Sb и Te → Se, S atoms at $x = 0.16$ $y = z = 006$, and 0.12 (Figure 7.5, curves 1, 2). In *p*-type solid solutions the effective mass m/m_0 slightly increases with growth of Sb content from 1.55 to 1.6 (Figure 7.5, curves 3 and 4, 5) in the samples with close carrier concentration.

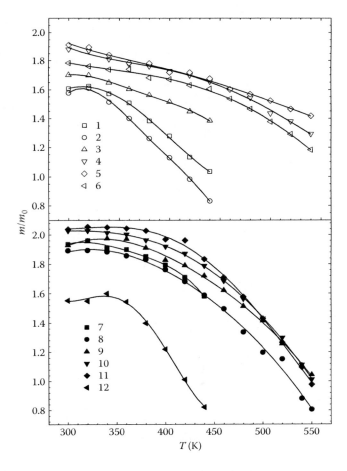

FIGURE 7.5 Temperature dependence of the density-of-states effective mass m/m_0 for *n*-type solid solutions $Bi_{2-x}Sb_xTe_{3-y-z}Se_yS_z$ (1–6), and for *p*-type solid solutions $Bi_{2-x}Sb_xTe_{3-y}Se_y$ (7–12). The curve and sample notation is the same as in Figure 7.2.

7.4.2 Charge Carrier Mobility

The carrier mobility μ_0 decreases, and the temperature dependence of μ_0 weakens with growth in charge carrier concentration and atomic substitutions in the solid solutions owing to scattering on neutral atoms of additional components and scattering on atoms of alloying impurity (Figure 7.6). In *n*-type solid solutions the carrier mobility μ_0 is higher in the composition with smaller content of Se and S atoms at $y = z = 0.09$ and $x = 0$ (Figure 7.6, curve 3) in comparison with the substitutions at $x = 0.16$ and $y = z = 0.12$ (Figure 7.6, curve 1) in spite of higher electron concentration in the sample with atomic substitutions Te → Se, S. The carrier mobility decreases in the *n*-type solid solutions with increase in the content of Se atoms from $y = 0.6$ to 0.9 and with increase in electron concentration (Figure 7.6, curves 4, 5). The mobility also decreases in the composition containing selenium and sulfur atoms at $y = z = 0.15$ in comparison with substitutions Te → Se (Figure 7.6, curves 4–6) due to additional electron scattering on sulfur atoms.

The mobility μ_0 decreases in *p*-type solid solutions with growth of carrier concentration (Figure 7.6, curves 7–12). The slopes of the $\mu_0(T)$ dependence increase from $|1.75|$ to $|1.9|$ with increase in the content of Sb atoms from $x = 1.55$ to 1.6 for the samples with close carrier concentrations (Figure 7.6, curves 9, 11). In the composition with substitutions of atoms in both sublattices of bismuth telluride Bi → Sb and Te → Se at $x = 1.3$ $y = 0.06$, the slope of the $\mu_0(T)$ dependence decreases up to $|1.6|$. However, the slope of the $\mu_0(T)$ dependence is high in the solid solution with higher Sb content ($x = 1.6$, $y = 0.06$), and the slope

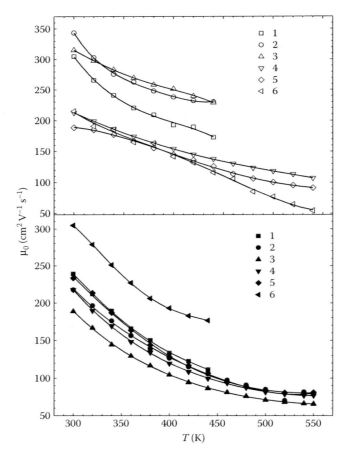

FIGURE 7.6 Temperature dependence of the carrier mobility μ_0 for *n*-type solid solutions $Bi_{2-x}Sb_xTe_{3-y-z}Se_yS_z$ (1–6), and for *p*-type solid solutions $Bi_{2-x}Sb_xTe_{3-y}Se_y$ (7–12). The curve and sample notation is the same as in Figure 7.2.

is equal to $|1.8|$ (Figure 7.6, curves 8, 12). It is mentioned that the changes of the effective mass and the carrier mobility are associated with changes of the anisotropy of the constant energy surface depending on composition and electron concentration in the solid solutions.[14,17,21–25]

7.4.3 Lattice Thermal Conductivity

The distortions of the crystal lattice grow with increase in carrier concentration and content of atomic substitutions in the solid solutions that leads to increase of phonon scattering and a decrease in the lattice thermal conductivity κ_L (Figure 7.7). The expression for thermal conductivity κ is given as

$$\kappa = \kappa_L + \kappa_e + \kappa_{np}, \tag{7.1}$$

where κ_L is the lattice thermal conductivity, $\kappa_e = L(r, \eta)\,\sigma T$ is the electronic thermal conductivity, κ_{np} is the contribution of the intrinsic conductivity, which rises with higher temperature, η is the reduced Fermi level. The calculations of the Lorentz number $L(r_{eff}, \eta)$ were carried out with a view to the effective scattering parameter r_{eff}.[9,12,15–17]

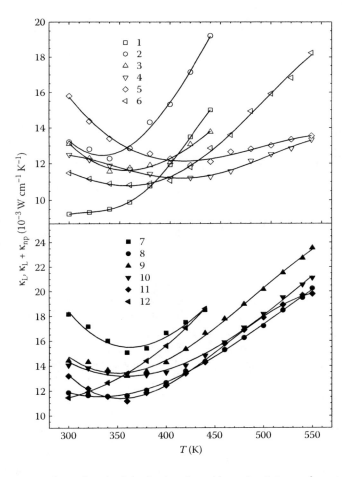

FIGURE 7.7 Temperature dependence of the lattice thermal conductivity κ_L for *n*-type solid solutions $Bi_{2-x}Sb_xTe_{3-y-z}Se_yS_z$ (1–6), and for *p*-type solid solutions $Bi_{2-x}Sb_xTe_{3-y}Se_y$ (7–12). The curve and sample notation is the same as in Figure 7.2.

Growth of crystal lattice distortions with an increase in the contents of the atomic substitutions Bi → Sb and Te → Se, S ($x = 0.16$ and $y = z = 0.12$) leads to decrease in the lattice thermal conductivity κ_L at the temperatures below 350 K comparing with the composition with low contents of Se and S atoms at $y = z = 0.06$ (Figure 7.7, curves 1, 2) for n-type solid solutions. With the temperature rise above 350 K, an appreciable contribution of the intrinsic conductivity κ_{np} is observed in the samples 1–3 in Figure 7.7 at the temperatures above 350 K. In the samples with higher carrier concentrations, the influence of the intrinsic conductivity rises at $T > 420$ K in composition with Te → Se, S ($y = z = 0.15$) and Te → Se ($y = 0.6$) atomic substitutions (Figure 7.7, curves 5–7). An increase in the content of substituted atoms on the bismuth sublattice from 1.55 to 1.6 (Figure 7.7, curves 6, 8 and 10, 11) and the substitutions on both sublattices of bismuth telluride at $x = 1.3$ and 1.6 and $y = 0.06$ brings a decrease in the lattice thermal conductivity κ_L for p-type solid solutions.

The dependences of the density-of-states effective mass m/m_0, the carrier mobility with account of degeneracy μ_0, and the lattice thermal conductivity κ_L on composition, temperature, and charge carrier concentration were used for analysis and optimization of the figure-of-merit Z in the solid solutions for temperatures above 300 K.

7.5 Figure-of-Merit Optimization

The figure-of-merit Z is determined by the ratios between values of the effective mass, the carrier mobility and the lattice thermal conductivity (Figure 7.4):

$$ZT \sim \text{const} \left(\frac{m}{m_0} \right)^{3/2} \mu_0 \, T^{5/2} \kappa_L^{-1} \tag{7.2}$$

In the n-type solid solutions n-Bi$_{2-x}$Sb$_x$Te$_{3-y-z}$Se$_y$S$_z$, the figure-of merit-Z increases in the composition with atomic substitutions Bi → Sb and Te → Se, S at $x = 0.16$, $y = z = 0.12$. Increasing Z values occurs due to growth of the density-of-states effective mass m/m_0, which accompanied by decreasing the lattice thermal conductivity κ_L (Figures 7.5, 7.7, curves 1) for optimized carrier concentrations at the temperature interval 300–350 K. With growth in temperature more than 350 K, an increase in Z is observed in n-type solid solutions with substitutions of atoms only in tellurium sublattice in the samples with higher electron concentration, an influence of the intrinsic conductivity being shifted to high temperatures. The value of Z increasing in the temperature interval 350–450 K for the composition with small contents of the substituted atoms at $x = 0$, $y = z = 0.09$ is related to increase in the carrier mobility μ_0 (Figures 7.5, 7.6, curves 3).

At the temperatures 450–550 K the increase in the figure-of-merit Z is found in the composition at $x = z = 0$, $y = 0.6$. An increase in Z values is explained by low lattice thermal conductivity κ_L and increase in the carrier mobility μ_0 in comparison with the composition with more contents of Se atoms ($y = 0.9$), and with simultaneous substitutions of Te by Se and S atoms ($x = 0$, $y = z = 0.15$) (Figures 7.6, 7.7, curves 4–6). The optimization of composition and carrier concentration for the n-Bi$_{2-x}$Sb$_x$Te$_{3-y-z}$Se$_y$S$_z$ solid solutions permits to achieve the average value of <ZT> parameter close to 0.95 in the temperature interval 300–450 K, and to 0.85 in the interval 450–550 K.

For the p-type solid solutions p-Bi$_{2-x}$Sb$_x$Te$_{3-y}$Se$_y$ at $x = 1.3$ and $y = 0.06$, the figure-of-merit Z increases due to high carrier mobility μ_0 and the low lattice thermal conductivity κ_L at the temperatures from 300 to 370 K (Figures 7.4, 7.6, 7.7, curves 12). The value of Z increases in the composition with more content of Sb atoms at $x = 1.6$, $y = 0$ for optimal carrier concentration within the temperature interval 370–550 K as a combined result increasing of the effective mass m/m_0 and slopes of the temperature dependences of the mobility μ_0, and the low lattice thermal conductivity κ_L (Figures 7.4 through 7.7, curves 11). The average value of <ZT> is close to 0.95 in the temperature interval 300–370 K, and to 0.7 in the interval 370–550 K for p-type solid solutions.

7.6 Review of Galvanomagnetic Properties of Thermoelectrics Based on Bismuth and Antimony Chalcogenides

The constant energy surfaces of valence and conductivity bands of the materials under consideration are described by six-valley model with isotropic scattering of charge carriers[18–20,23,24] near room temperature and with anisotropic scattering mechanism at low temperatures.[23–25,44–47] For the first time, galvano-magnetic properties for Bi$_2$Te$_3$ have been studied by Drabble with coauthors.[18,19] Galvanomagnetic coefficients (GMC) such as the resistivity, Hall effect, and magnetoresistivity tensor components ρ_{ij}, ρ_{ijk}, and ρ_{ijkl}, respectively, were measured for *n*- and *p*-type Bi$_2$Te$_3$ in low magnetic fields. The ratios of the effective-mass tensor components m_i/m_j and the degeneracy parameter $\beta_d(\eta)$ were calculated in terms of the many-valley model of energy spectrum. The relaxation time was approximated by a power-law function:

$$\tau = \tau_0 E^r, \tag{7.3}$$

where τ_0 is an energy-independent factor, r is the scattering parameter, for acoustic scattering mechanism $r = -0.5$.

GMC for n-Bi$_2$Te$_3$ also studied in a wide interval of carrier concentrations in weak magnetic fields.[20] The arbitrary scattering parameter r was not set in advance in calculating the degeneracy parameter $\beta_d(\eta)$ together with the reduced Fermi level in Ref. 20. The degeneracy parameter $\beta_d(\eta)$[20] occurs lower than similar value at $r = -0.5$ calculating. A partial degeneration of the multicomponent solid solutions with various atomic substitutions in the sublattices of bismuth telluride was also found from fundamental absorption band studies of the n-Bi$_{2-x}$Sb$_x$Te$_{3-y-z}$Se$_y$S$_z$ and p-Bi$_{2-x}$Sb$_x$Te$_{3-y}$Se$_y$ solid solutions.[41,42] Therefore, calculations of the scattering parameter r are required for analysis and optimization of the thermoelectric properties of the Bi$_2$Te$_3$-based solid solutions.

We calculated the effective scattering parameter together with the reduced Fermi level from the experimental data on galvanomagnetic and thermoelectric properties for various compositions of bismuth telluride-based materials in the wide interval of charge carrier concentrations.[15–17] Changing the scattering parameter r_{eff} occurs due to filling additional bands in the conduction and the valence bands and scattering of charge carrier on neutral atoms of the solid solution components, and on atoms of alloying impurity. The effective scattering parameter changes from (-0.7) to (-0.8) with increase in carrier concentration for solid solutions optimized in the temperature range above room temperature. The value of r_{eff} modifies to (-0.35) at $T = 80$ K in the solid solution optimized for low-temperature range owing to an impurity scattering.

The similar values of r parameter from (-0.6) to (-0.7) were reported for the Bi$_2$Te$_3$ and Bi$_2$Te$_{3-x}$Se$_x$ solid solutions in Refs. 48,49. The scattering parameter r[50] was also calculated from the Shubnikov–de Haas effect using Fermi energy E_F.[50] The parameter r is not equal to -0.5 and depends on carrier concentration. The value of r for solid solution p-Sb$_2$Te$_x$ <Ag,Sn> is equal to (-0.6).[50]

Studies of Shubnikov–de Haas effect showed that the anomalous temperature and magnetic field behavior of the Hall coefficient and resistivity for high magnetic fields may be explained quantitatively using two-valence (upper light and lower heavy hole) band structure model of Bi$_2$Te$_3$,[51] Sb$_2$Te$_3$[50] and p-(Bi$_{1-x}$Sb$_x$)$_2$Te$_3$,[52] and n-Bi$_2$(Se$_x$Te$_{1-x}$)$_3$[53] solid solutions. The two valence and conduction bands structure are discussed for Sb$_{2-x}$In$_x$Te$_3$ and Sb$_2$Te$_{3-y}$Se$_y$,[54] p-(Bi$_{1-x}$Sb$_x$)$_2$Te$_3$ doped with Sn[55] and Ag, Sn, Ga[50] from the measurements of the Shubnikov–de Haas effect. Two-valence band with a single and six valley structure is developed for the interpretation of galvanomagnetic effects in p-type systems (Bi$_x$Sb$_{1-x}$)$_2$Te$_3$ and Sb$_2$Te$_3$ under the assumption of anisotropic scattering mechanism in each band.[45] The consideration of two-band effects to explain the peculiarities of the calculated lattice thermal resistivity for p-(Bi$_{1-x}$Sb$_x$)$_2$Te$_3$[56] is discussed.

7.7 Many-Valley Model of Energy Spectrum

The magnetoresistivity ρ_{ijlk}, Hall effect ρ_{123}, ρ_{312}, and electrical resistivity ρ_{11} tensor components are used to calculate of the parameters u, v, w, which define in the terms of many-valley model the shape of the constant-energy ellipsoids, and the degeneracy parameter β_d governing the scattering mechanism of charge carriers:

$$\frac{\rho_{312}}{\rho_{123}} = y_1(u,v,w) \tag{7.4}$$

$$\frac{\rho_{11}\rho_{ijkl}}{\rho_{123}^2} = y_m(u,v,w,\beta_d), \quad m = 1,2,3 \tag{7.5}$$

where ρ_{ijkl} are the transverse components of the magnetoresistivity tensor ρ_{1133} and ρ_{1122}; and ρ_{1111} is the longitudinal component.

The parameters u, v, w, β_d are determined from solution of the equation system (4–5) by minimizing the objective function $\chi(u, v, w, \beta_d)$, which is sum of the squared deviations between the experimental GMC ratios y_i^e and the calculated values of the function $y_i(u, v, w, \beta_d)$[21–25]:

$$\chi(u,v,w,\beta_d) = \sum_{i=1}^{4}\left(\frac{(y_i^e - y_i)(u,v,w,\beta_d)}{y_i^e}\right)^2 \tag{7.6}$$

The components of the effective-mass tensor using the u, v, w parameters and the tilt angle (Θ) can be expressed in the form

$$\frac{m_1}{m_3} = \frac{\dfrac{w}{v}}{vs^2 + 2cs\sqrt{v - w} + c^2} \tag{7.7}$$

$$\frac{m_3}{m_2} = \frac{s^2v + 2cs\sqrt{v - w} + c^2}{s^2 - 2cs\sqrt{v - w} + c^2v} \tag{7.8}$$

$$tg2\Theta = \frac{2\sqrt{v - w}}{v - 1} \tag{7.9}$$

where $s = \sin \Theta$, $c = \cos \Theta$, and Θ is the tilt angle between the principal axes of the constant-energy ellipsoids and the crystallographic axes.

Next, the scattering parameter r and the reduced Fermi level η were calculated from the expression for the degeneracy parameter β_d determining for isotropic relaxation time as

$$\beta_d(r,\eta) = \frac{(2r + 3/2)^2 F_{2r+1/2}^2(\eta)}{(r + 3/2)(3r + 3/2)F_{r+1/2}(\eta)F_{3r+1/2}(\eta)}, \tag{7.10}$$

where $F_s(\eta)$ are Fermi integrals of the type:

$$F_s(\eta) = \int_0^{\infty} \frac{x^s}{e^{x-\eta} + 1}dx, \tag{7.11}$$

and from the expression of the Seebeck coefficient α

$$\alpha \frac{k}{e}\left[\frac{(r + 5/2)F_{r+3/2}(\eta)}{(r + 3/2)F_{r+1/2}(\eta)} - \eta\right] \quad (7.12)$$

The objective function $R(r, \eta)$ for calculation r and η has the form

$$R(r,\eta) = \frac{(\alpha_c(r,\eta) - (\alpha_e(r,\eta))^2 + (\beta_c(r,\eta) - (\beta_e(r,\eta))^2}{\alpha_c^2(r,\eta) + \alpha_e^2(r,\eta)}, \quad (7.13)$$

where $\alpha_e(r, \eta)$ and $\alpha_c(r, \eta)$ are the experimental and calculated values of the Seebeck coefficient from Equation 7.12; $\beta_e(r, \eta)$ is the degeneracy parameter derived from Equations 7.4, 7.5; and $\beta_c(r, \eta)$ is the value calculated from Equation 7.10. The r values, which named the effective scattering parameter r_{eff},[15–17] were used for calculations of the m/m_0, μ_0, and κ_L values in the parabolic model of the energy spectrum for isotropic scattering mechanism of charge carriers for studied solid solutions.

7.8 Galvanomagnetic Properties

GMC tensor components were measured by DC method for n- and p-type Bi_2Te_3-based solid solutions in the range of the magnetic fields at $H = 25$–28 kOe, where the variations of GMC from H value are weakened. Tables 7.1 and 7.2 list the ratios of the GMC tensor components. The parameters of constant-energy surface u, v, w, degeneracy parameter β_d, principal components of the effective-mass tensor (m_1, m_2, m_3), and the tilt angle Θ between the principal axes of the constant-energy ellipsoids and the crystallographic axes were calculated in accordance with expressions 7.4, 7.5, 7.8 through 7.11, (Tables 7.3, 7.4). Figure 7.8 schematically shows one of the six equivalents of the constant-energy ellipsoids for the conduction and valence bands for the solid solutions with high figure-of-merit Z above room temperatures.

The constant-energy ellipsoids are more strongly compressed along the binary (m_1, X') and the bisector axes (m_2, Y') for the compositions of n- and p-type solid solutions with high the figure-of-merit Z and ZT- parameter at optimal carrier concentration above room temperature. In n-type solid solutions the

TABLE 7.1 Ratios of GMC for n-$Bi_{2-x}Sb_xTe_{3-y-z}Se_yS_z$ Solid Solutions

N	x, y, z	ρ_{312}/ρ_{123}	$\rho_{11}\rho_{1133}/\rho_{123}^2$	$\rho_{11}\rho_{1122}/\rho_{123}^2$	$\rho_{11}\rho_{1111}/\rho_{123}^2$
1	0.16, 0.12, 0.12	1.51	6.56	3.85	12.9
2	0, 0.09, 0,09	1.0	1.16	3.87	4.5
3	0, 0.6, 0	1.24	2.81	7.07	8.43
4	0, 0.15, 0.15	1.03	1.35	0.73	1.82

TABLE 7.2 Ratios of GMC for p-$Bi_{2-x}Sb_xTe_{3-y}Se_y$ Solid Solutions

N	x, y	ρ_{312}/ρ_{123}	$\rho_{11}\rho_{1133}/\rho_{123}^2$	$\rho_{11}\rho_{1122}/\rho_{123}^2$	$\rho_{11}\rho_{1111}/\rho_{123}^2$
1	1.6	1.46	0.65	1.3	1.51
2	1.55	1.54	0.51	1.29	1.49
3	1.5	1.0	1.16	3.87	4.5
4	1.3, 0.06	2.29	4.06	4.0	4.87
5	1.5, 0.06	1.25	1.0	3.34	2.38

TABLE 7.3   Calculated Parameters of Constant-Energy Surface u, v, w, Degeneracy Parameter β_d, Components of the Effective-Mass Tensor, and the Tilt Angle Θ for n-Bi$_{2-x}$Sb$_x$Te$_{3-y-z}$Se$_y$S$_z$ Solid Solutions

N	x, y, z	u	v	w	β_d	$\chi^2\, 10^{-4}$	m_3	m_1	m_2	Θ
1	0.16, 0.12, 0.12	12.2	0.52	3.45	0.26	0.7	0.15	0.06	0.04	42
2	0, 0.09, 0.09	4.99	0.41	2.11	0.5	0.5	0.28	0.06	0.11	
3	0, 0.6, 0	3.38	1.23	0.55	0.2	1.23	0.24	0.02	0.06	40
4	0, 0.15, 0.15	6.95	0.58	1.96	0.8	1.99	0.22	0.03	0.05	41

TABLE 7.4   Calculated Parameters of Constant-Energy Surface u, v, w, Degeneracy Parameter β_d, Components of the Effective-Mass Tensor, and the Tilt Angle Θ for p-Bi$_{2-x}$Sb$_x$Te$_{3-y}$Se$_y$ Solid Solutions

N	x, y	u	v	w	β_d	$\chi^2,\, 10^{-4}$	m_3	m_1	m_2	Θ
1	1.6, 0	3.97	1.4	0.97	0.62	2.23	0.22	0.03	0.05	39
2	1.55, 0	4.04	1.48	1.32	0.63	1.6	0.24	0.05	0.17	33
3	1.5, 0	2.85	1.26	0.94	0.24	0.26	0.28	0.06	0.11	41
4	1.3, 0.06	10.2	1.39	2.59	0.52	1.02	0.24	0.02	0.06	42
5	1.5, 0.06	2.97	2.02	1.58	0.39	1.55	0.22	0.07	0.17	29

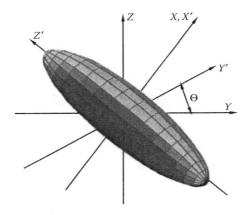

FIGURE 7.8   One of equivalent of the constant-energy ellipsoids for the conduction and valence bands of the solid solution Bi$_2$Te$_3$ based.

ellipsoids are more strongly compressed along the binary and the bisector axes at $x = 0.16$, $y = z = 0.12$; $x = 0$, $y = z = 0.09$, and $x = z = 0$, $y = 0.6$ compositions with high Z and ZT values in the different range of the temperature interval 300–550 K (Figure 7.4, curves 1, 3, 4, Table 7.3) comparing with the composition at $x = 0$, $y = z = 0.15$ with lower Z (Figure 7.4, curve 6, Table 7.3).

The ellipsoid compression grows along the binary and the bisector axes for the compositions of p-type at $x = 1.3$, $y = 0.06$ with high figure-of-merit near room temperature up to 370 K comparing with in the composition at $x = 1.6$, $y = 0.06$ (Figure 7.4, curves 12 and 8, Table 7.4). But for the temperatures more than 370 K the compression increases in the composition with more content of Sb atoms at $x = 1.6$ with high figure-of-merit Z in comparison with the composition at $x = 1.55$ (Figure 7.4, curves 11 and 9, Table 7.4). The compression of the constant-energy ellipsoids is accompanied by the turn on the tilt angle Θ of principal axes of the constant-energy ellipsoids with respect to the crystallographic axes.

7.9 Conclusions

Thus, the figure-of-merit Z increases for n- and p-type solid solutions based on bismuth and antimony chalcogenides at optimum charge carrier concentration above room temperature for optimum relations between the values of the effective mass (m/m_0), the charge carrier mobility (μ_0), and the lattice thermal conductivity (κ_L). These values were determined in the parabolic model of the energy spectrum for isotropic scattering mechanism of charge carriers and the changes of charge carrier scattering were taken into account. In n- and p-type solid solutions with atomic substitutions in both sublattices of bismuth telluride, the value of Z increases due to decreasing the lattice thermal conductivity and increasing the density-of-states effective mass. An increase in the substituted atoms in tellurium sublattices for n-type and in bismuth sublattices for p-type compositions is also associated with growth of the effective mass and reduction of the lattice thermal conductivity. For n-type solid solutions with small content of the substituted atoms in tellurium sublattices, the increase in Z is determined with an increase in the carrier mobility. The relations between the anisotropy of the constant-energy surface and the values of the figure of merit Z are defined by variations of the shape and orientation of the principal axes of the constant-energy ellipsoids (m_1. m_2, m_3), which turn on the tilt angle (Θ) between the principal axes of the ellipsoids and the crystallographic axes.

References

1. Fleurial J.-P., G.J. Snyder, J. Patel et al. 2001, Solid-state power generation and cooling micro/nanodevice for distributed system architectures. *Proceedings of the 20th International Conference on Thermoelectrics*, Beijing, China, 24–9.
2. Rowe D. M. 2007, Economic thermoelectric recovery of low temperature heat. *Proceedings of the V European Conference on Thermoelectrics*, Odessa, Ukraine, 11–8.
3. Baukin V.E., A.P. Vyalov, I.A. Gershberg, G.K. Muranov, O.G. Sokolov, and Ph.J. Takhistov. 2002, Constraction optimization of high-power thermoelectric generators. *Proceedings of the VIII Interstate Seminar on Thermoelectrics and their Applications*, St. Petersburg, Russia, 411–16.
4. Pustovalov A.A. 2007, Role and prospects of application of RTG on base of plutonioum-238 for planetary exploration. *Proceedings of the V European Conference on Thermoelectrics*, Odessa, Ukraine, 24–8.
5. Minnich A.J., M.S. Dresselhaus, Z.F. Ren, and G. Chen. 2009, Bulk nanostructured thermoelectric materials: Current research and future prospects. *Energy Environ. Sci.*, 2:466–79.
6. Vasilevskiy D., S. Turenne, and R. Masur. 2008, Thermoelectric extruded alloys for module manufactoring: 10 Years of development at École Polytechnique de Montréal. *Proceedings of the V European Conference on Thermoelectrics* Paris, France, I-04-I04-6.
7. Winkler M., J.D. Köning, S. Buller et al. 2010, Bi_2Te_3, Sb_2Te_3 and Bi_2Te_3/Sb_2Te_3-superlattices created using the nanoalloing approach. *Proceedings of the 8th European Conference on Thermoelectrics*, Como, Italy, 19–24.
8. Böttner H., D.G. Ebling, A. Jacquot, U. Kühn, and J. Schmidt. 2007, Melt spinning of bismuth and partially alloying with IV–VI compounds for thermoelectric application. *Proceedings of the V European Conference on Thermoelectrics*, Odessa, Ukraine, pp. 104–107.
9. Lukyanova L.N., V.A. Kutasov, P.P. Konstantinov, and V.V. Popov. 2010, Features of the behavior of figure-of-merit for p-type solid solutions based on bismuth and antimony chalcogenides. *J. Electron. Mater.*, 39:2070–73.
10. Luk'yanova L.N., V.A. Kutasov, P.P. Konstantinov, and V.V. Popov. 2010, Thermoelectric figure-of-merit in p-type bismuth and antimony-chalcogenide-based solid solutions above room temperature. *Phys. Solid State*, 52:1599–605 [2010, *Fiz. Tverd. Tela*, 52:1492–97].
11. Luk'yanova L.N., V.A. Kutasov, P.P. Konstantinov, and V.V. Popov. 2006, Specific features in the behavior of the effective mass and mobility in n-$(Bi, Sb)_2(Te, Se, S)_3$ solid solutions. *Phys. Solid State*, 48:1856–62 [2006, *Fiz. Tverd. Tela*, 48:1751–56].

12. Luk'yanova L.N., V.A. Kutasov, and P.P. Konstantinov. 2008, Multicomponent n-$(Bi,Sb)_2(Te,Se,S)_3$ solid Solutions with different atomic substitutions in the Bi and Te sublattices. *Phys. Solid State*, 50:2237–44 [2008, *Fiz. Tverd. Tela*, 50:2143–49].

13. Luk'yanova L.N., V.A. Kutasov, and P.P. Konstantinov. 2007, Solid solutions based on bismuth and antimony chalcogenides above room temperatures. *Proceedings of the V European Conference on Thermoelectrics*. Odessa, Ukraine, pp. 172–77.

14. Luk'yanova L.N., V.A. Kutasov, P.P. Konstantinov, V.V. Popov, and G.N. Isachenko. 2007, Optimization of multicomponent thermoelectrics based on Bi_2Te_3 in the range below and above room temperature. *Proceedings of the XXVI International Conference on Thermoelectrics*, Jeju Island, Korea, pp. 67–72.

15. Kutasov V.A. and L.N. Lukyanova. 1984, Carrier density dependence of the effective scattering parameter of solid solutions based on bismuth telluride. *Sov. Phys. Solid State*, 26:1515–16 [1984, *Fiz. Tverd. Tela*, 26:2501–04].

16. Kutasov V.A. and L.N. Luk'yanova. 1986, Carrier density dependence of the anisotropy parameter of n-type Bi_2Te_3 and solid solutions derived from it. *Phys. Solid State*, 28:502–03 [1986, *Fiz. Tverd. Tela*. 28:899–902].

17. Luk'yanova L.N., V.A. Kutasov, V.V. Popov, and P.P. Konstantinov. 2004, Galvanomagnetic and thermoelectric properties of p-$Bi_{2-x}Sb_xTe_{3-y}Se_y$ solid solutions at low temperatures (<220 K). *Fiz. Tverd. Tela*, 46:1366–71 [2004 *Phys. Solid State*, 46:1404–09].

18. Drabble J.R., R.D. Groves, and R. Wolfe. 1958, Galvanomagnetic effects in n-type bismuth telluride. *Proc. Phys. Soc.*, 71:430–43.

19. Drabble, J.R. 1958, Galvanomagnetic effects in p-type bismuth telluride. *Proc. Phys. Soc.*, 72:380–90.

20. Caywood, L.P. and G.R. Miller. 1970, Anisotropy of the constant-energy surfaces in n-type Bi_2Te_3 and Bi_2Se_3 from galvanomagnetic coefficients. *Phys. Rev.*, 2:3210–20.

21. Kutasov V.A., L.N. Luk'yanova, and P.P. Konstantinov. 2000, An analysis of the thermoelectric efficiency of n-$(Bi,Sb)_2(Te,Se,S)_3$ solid solutions within an isotropic scattering model. *Phys. Solid State*, 42:2039–46 [2000, *Fiz. Tverd. Tela*, 42:1985–91].

22. Luk'yanova L.N., V.A. Kutasov, and P.P. Konstantinov. 2004, Effective mass and mobility in p-$Bi_{2-x}Sb_xTe_{3-y}Se_y$ solid solutions below room temperature. *Proceedings of the 2nd European Conference on Thermoelectrics*, Krakow, Poland, pp. 149–152.

23. Lukyanova L.N., V.A. Kutasov, V.V. Popov, and P.P. Konstantinov. 2005, Galvanomagnetic properties of multicomponent solid solutions based on Bi and Sb chalcogenides. *Proceedings of the XXIV International Conference on Thermoelectrics*. Clemson University, SC, USA, pp. 426–29.

24. Lukyanova L.N., V.A. Kutasov, V.V. Popov, P.P. Konstantinov, and M.I. Fedorov. 2006, Anisotropic scattering in the $(Bi, Sb)_2(Te, Se, S)_3$ solid solutions. *Proceedings of the XXV International Conference on Thermoelectrics*. Vienna, Austria, pp. 496–99.

25. Lukyanova L.N., V.A. Kutasov, P.P. Konstantinov, and V.V. Popov. 2010, Improved thermoelectrics based on bismuth and antimony chalcogenides for temperatures below 240 K. *J. Adv. Sci. Technol.*, 74:77–82.

26. Svechnikova T.E., V.S. Zemskov, M.K. Zhitinskaya et al. 2006, Properties of Sn-doped $Bi_2Te_{3-x}Se_x$ single crystals. *Inorg. Mater.*, 42:135–42.

27. Gol'tsman B.M., V.A. Kudinov, and I.A. Smirnov. 1972. *Semiconducting Thermoelectric Materials Based on Bi2Te3*, Nauka, Moscow, p. 320.

28. Svechnikova T.E. and Konstantinov P.P. 2000, Thermoelectric properties of $Bi_2Te_{2.85}Se_{0.15}$. Solid solutions with various concentrations of carriers. *Proceeding of the VII Interstate Seminar on Thermoelectrics and Their Applications*, St. Petersburg, Russia, pp. 24–9.

29. Ivanova L.D., L.I. Petrova, Yu.V. Granatkina, and V.S. Zemskov. 2007, Thermoelectric materials based on Sb_2Te_3-B_2Te_3 solid solutions with optimal performance in the range 100–400 K. *Inorg. Mater.*, 43:1044–48.

30. Ivanova L.D. and Yu.V. Granatkina. 2000, Thermoelectric materials based on $Sb_3Te_3-B_2Te_3$ single crystals in the range 100–700 K. *Inorg. Mater.*, 36:810–16.

31. Sokolov O.V., S.Ya. Skipidarov, N.I. Duvankov, and G.G. Shabunina. 2007, Physical characterization of the processes occurring in extrusion and their effect on phase diagram and thermoelectric properties of the materials in $Bi_2Te_3-Bi_2Se_3$ and $Sb_2Te_3-Bi_2Te_3$ systems. *Proceedings of the V European Conference on Thermoelectrics*, Odessa, Ukraine, pp. 140–3.

32. Vasilevskiy D.V., N. Kukhar, S. Turenne, and R.A. Masut. 2007, Hot extruded $(Bi, Sb)_2(Te, Se)_3$ alloys for advanced thermoelectric modules. *Proceedings of the V European Conference on Thermoelectrics*. Odessa, Ukraine, pp. 64–7.

33. Greenway D.L. and G. Harbeke. 1965, Band structure of bismuth telluride, bithmuth selenide, and their respective alloys. *J. Phys. Chem. Sol.*, 26:1585–04.

34. Pecheur P. and G. Toussaint. 1989, Electronic structure and bonding in bismuth telluride. *Phys. Lett. A*. 135:223–5.

35. Oleshko E.V. and Korylishin V.N. 1985, Quasirelativistic band structure of Bi_2Te_3, *Sov. Phys. Sol. State*, 27:1723–26 [1985, *Fiz. Tverd. Tela*, 2856–9].

36. Larson P., S.D. Mahanti, and M.G. Kanatzidis, 2000, Electronic structure and transport of Bi_2Te_3 and $BaBiTe_3$. *Phys. Rev. B*, 61:8162–71.

37. Austin J.G. and A. Sheard. 1957, Some optical properties of $Bi_2Te_3-Bi_2Se_3$ alloys. *J. Electr. Control*, 3:236–37.

38. Thomas G.A., D.H. Rapkine, R.B. Van Dover et al. 1992, Large electronic-density increase on cooling a layered metal: Doped Bi_2Te_3. *Phys. Rev. B*, 46:1553–56.

39. Funagai K., Y. Miyahara, H. Ozaki, and V.A. Kulbachinskii. 1996, Tunneling spectroscopy of band edge structures of Bi_2Te_3 and Sb_2Te_3. *Proceedings of the International Conference on Thermoelectrics*, Pasadena, USA, pp. 408–11.

40. Sehr R and L. R. Testardi. 1962, The optical properties of p-type $Bi_2Te_3-Sb_2Te_3$ alloys between 2–15 microns. *J. Phys. Chem. Solids*, 23:1219–24.

41. Veis A.N., L.N. Lukyanova, and V.A. Kutasov. 2010, Study of absorption coefficient in solid solutions of bismuth-telluride-based near the edge of fundamental band. *Proceedings of the XII Interstate Seminar on Thermoelectrics and Their Applications*, Saint-Petersburg, Russia, pp. 130–5.

42. Veis A.N., L.N. Lukyanova, and V.A. Kutasov. 2010, Optical absorption in solid solutions $Bi_{2-x}Sb_xTe_{3-y}Se_{y/2}S_{y/2}$ in the range of fundamental band edge. *Proceedings of the XII Interstate Seminar on Thermoelectrics and Their Applications*, Saint-Petersburg, Russia, pp. 136–41.

43. Bekdurdyev Ch.L., B.M. Goltsman, V.A. Kutasov, and A.V. Petrov. 1974, Forbidden gap width of $Bi_2Te_{2.4}Se_{0.3}S_{0.3}$ solid solution. *Fiz. Tverd. Tela*, 16:2121–22.

44. Asworth N.A., J.A. Rayne, and R.W. Ure. 1971, Transport properties of Bi_2Te_3. *Phys Rev. B*, 3:2646–61.

45. Simon, G. and Eichler, W. 1981, Investigations on a two-valance band model for Sb_2Te_3. *Phys. Stat. Sol.*, 107:201–06.

46. Efimova B.A., I.Ia. Korenblit, V.I. Novikov, and A.G. Ostroumov. 1961, Anisotropy of galvanomagnetic properties in p-Bi_2Te_3. *Fiz. Tverd. Tela*, 3:2746–60.

47. Efimova B.A., V.I. Novikov, and A.G. Ostroumov. 1962, Anisotropy of galvanomagnetic properties in n-Bi_2Te_3. *Fiz. Tverd. Tela*, 4:302–04.

48. Kaibe H., Y. Tanaka, M. Sakata, and I. Nishida. 1989, Anisotropic galvanomagnetic and thermoelectric properties of n-type Bi_2Te_3 single crystal with the composition of a useful thermoelectric cooling material. *J. Phys. Chem. Solids*, 50:945–50.

49. Ohsigi I. J., T. Kojima, H. Kaibe, and I. Nishida. 1989, Analysis of the anisotropic resistivities and hall coefficients of sintered n-type Bi_2Te_3 Single. *Proceedings of the VIII International Conference on Thermoelectrics*, Nancy, France, pp. 195–99.

50. Kulbachinskii V.A., A.V.G. Kytin, and P.M. Tarasov. 2006, Fermi surface and thermoelectric power of $(Bi_{1-x}Sb_x)_2Te_3$ single crystals doped by Ag, Sn, Ga. *Proceedings of the XXV International Conference on Thermoelectrics*. Austria, Vienna, pp. 459–64.

51. Köhler, H. 1976, Non-parabolic E(k) relation of the lowest conduction band of Bi_2Te_3, *Phys. Stat. Sol.* (b), 73(1):95–104.
52. Köhler H., W. Haigis, and A. Middendorff. 1976, Shubnikov-de haas investigations on *n*-type $Bi_2(Se_xTe_{1-x})_3$. *Phys. Stat. Sol.* (b), 78:637–42.
53. Kulbachinskii V.A., Z.M. Dashevskii, M. Inoue et al. 1995, Valence-band changes in $Sb_{2-x}In_xTe_3$ and $Sb_2Te_{3-y}Se_y$ by transport and Shubnikov de Haas effect measurements. *Phys. Rev.* B, 512:10915–22.
54. Köhler H. and A. Freudenberger. 1977, Investigations of the highest valance band in $(Bi_{1-x}Sb_x)_2Te_3$ crystals. *Phys. Stat. Sol* (b), 84:195–203.
55. Kulbachinskii V.A., A.Yu. Kaminskii, R.A. Lunin et al. 2002, Quantum oscillations of Hall resistance, magnetoresistance in a magnetic field up to 54 T and the energy spectrum of Sn doped layered semiconductors p-$(Bi_{1-x}Sb_x)_2Te_3$. *Semicond. Sci. Technol.*, 17:1133–40.
56. Müller E. and Süßmann H. 1998, Lattice thermal resistivity and two valence band model—A new understanding of transport properties of p-$(Bi_{1-x}Sb_x)_2Te_3$. *Proceedings of the XVII International Conference on Thermoelectrics*, Nagoya, Japan, pp. 85–88.

8

Effect of Vacancy Distribution on the Thermoelectric Properties of Gallium and Indium Chalcogenides

Ken Kurosaki
Osaka University

Shinsuke Yamanaka
Osaka University

8.1 Introduction

Ga and In chalcogenides such as Ga_2Te_3, In_2Te_3, and Ga_2Se_3 share the same crystal structure, namely a defect zinc-blende cubic crystal (space group: $F\text{-}43m$). Due to the valence mismatch between the cation and the anion, a third of the cation sites are structural vacancies; that is, the chemical formula A_2B_3 (A = Ga, In; B = Te, Se) can be written as $A_2VA_1B_3$, where VA indicates a vacancy. These vacancies are thought to be distributed with various states. They are also expected to affect the thermoelectric properties of Ga and In chalcogenides. By forming such vacancies in a crystal, a significant reduction in the lattice thermal conductivity due to strong phonon–vacancy scattering was recently observed in InSb–In_2Te_3 solid solutions.[1] The thermoelectric properties of Ga and In chalcogenides, including the electrical resistivity, the Seebeck coefficient, and the mobility of In_2Te_3, have been investigated.[2–4] The mobility of In_2Te_3 was found to be much smaller than those of isoelectronic binary compounds such as InSb and to be independent of temperature, which is probably due to the predominance of electron–vacancy scattering.[2] Furthermore, the electrical conductivities of Zn and Cu-doped Ga_2Te_3 and In_2Te_3 were found to be almost independent of impurity concentrations.[5] In addition to the thermoelectric properties, the optical properties of Ga_2Te_3 and In_2Te_3 single crystals[6] and the dielectric and ac conductivity properties of as-grown Ga_2Te_3 crystals[7] have been examined.

As mentioned above, the electrical properties of Ga and In chalcogenides have been considerably investigated. However, little research has been done on the thermal conductivities of these compounds. Our group has been investigating the effect of the vacancy distribution on the thermal conductivities (κ)

of Ga_2Te_3, In_2Te_3, Ga_2Te_3–In_2Te_3 solid solutions, and Ga_2Se_3. In this chapter, we review recent research on the thermal conductivities of these Ga and In chalcogenides and discuss the relationship between the vacancy distribution and the thermal conductivities.

8.2 Thermoelectric Properties of Ga_2Te_3, In_2Te_3, and $(Ga, In)_2Te_3$ Solid Solutions

Ga_2Te_3 and In_2Te_3 have different vacancy distributions. Desai et al. demonstrated that In_2Te_3 has two components: a low-temperature α-phase with regularly ordered vacancies and a high-temperature β-phase with fully disordered vacancies; the former transforms to the latter at about 893 K.[8] α-In_2Te_3 has a face-centered cubic lattice with $a = 1.850$ nm, which is approximately three times larger than the lattice parameter of β-In_2Te_3 ($a = 0.616$ nm).[9] On the other hand, Ga_2Te_3 has a mesoscopic super-structured phase with periodic vacancy planes,[10] as observed in Ga_2Se_3.[11]

Figure 8.1 shows high-resolution transmission electron microscopy (TEM) images, electron diffraction patterns, and x-ray diffraction (XRD) patterns obtained from a single grain of a sintered pellet of $(Ga_{1-x}In_x)_2Te_3$ ($x = 0$, 0.25, 0.5, 0.75, and 1).[12] The TEM micrographs and electron diffraction patterns were obtained with the electron beam aligned in the [1 $\overline{1}$0] direction. In the electron diffraction pattern of Ga_2Te_3, superlattice spots exist at the ~1/10 positions between two neighboring fundamental spots in the [111] direction, in addition to the fundamental Bragg reflections due to the zinc-blende structure. The high-resolution TEM image of Ga_2Te_3 reveals plane defects due to the vacancy-rich plane that are two-dimensionally arranged throughout the whole sample. These plane defects appear in the (111) plane at approximately 10-lattice (equivalent to 3.5 nm) intervals,[13] which is consistent with the electron diffraction pattern. Based on *ab initio* calculations, Ishikawa and Nakayama[14] suggested that the small difference in the negativities of Ga and Te is important in the formation of the vacancy planes. On the other hand, In_2Te_3 possesses a triple periodicity in the [111] direction, as shown in the high-resolution TEM image. This is highly consistent with the electron diffraction pattern, in which the superlattice spots are located at the one-third positions. In the high-resolution TEM images of the $(Ga, In)_2Te_3$ solid solutions, $(Ga_{1-x}In_x)_2Te_3$ ($x = 0.25$, 0.5, and 0.75), there are two-dimensional (2D) vacancy planes in the (111) plane (like Ga_2Te_3); however, the interval periodicity is random. In particular, the vacancy planes are located with approximately 3–15 lattice intervals in the solid solution samples. These TEM results are very consistent with the electron diffraction patterns, which have diffuse superlattice reflections and a streak.

In the XRD patterns, almost all the peaks can be indexed by the zinc-blende structure. In the XRD pattern of Ga_2Te_3, a few peaks are observed before and after the (111) peak, suggesting that the vacancies may be concentrated in the (111) plane. Assuming that peaks around the (111) peak in the XRD pattern are due to satellite reflections from the 2D vacancy planes, the periodicity of the planes calculated from the XRD peaks is approximately 3.5 nm, which is consistent with the TEM results. On the other hand, In_2Te_3 can be indexed as $3 \times 3 \times 3$ of the zinc-blende unit cell, which is highly consistent with the reported crystal structure of α-In_2Te_3 in which vacancies exist regularly in the $3 \times 3 \times 3$ of the zinc-blende unit cell.[9,15] Peaks derived from the ordered vacancy distribution in the In_2Te_3 crystal are indicated by asterisks. Satellite peaks are also observed before and after the (111) peak in the XRD patterns of the $(Ga, In)_2Te_3$ solid solutions, but they are broader. This peak broadening is most probably due to the random periodicity of the vacancy plane intervals.

Figure 8.2a shows the temperature dependence of κ of the fully dense bulk samples of Ga_2Te_3 and α-In_2Te_3.[16] κ of α-In_2Te_3 decreases with increasing temperature approximately according to a T^{-1} relation, whereas κ of Ga_2Te_3 exhibits rather flat temperature dependence. The κ values of Ga_2Te_3 are unexpectedly low, being about 0.5 W m^{-1} K^{-1}. The molecular weight of α-In_2Te_3 (612.4) is greater than that of Ga_2Te_3 (522.24); as expected, α-In_2Te_3 (159 K) has a lower Debye temperature (θ) evaluated from the sound velocities than Ga_2Te_3 (245 K). Based on these characteristics, α-In_2Te_3 is predicted to have a lower κ than Ga_2Te_3; however, the opposite result is observed experimentally. This possibly indicates

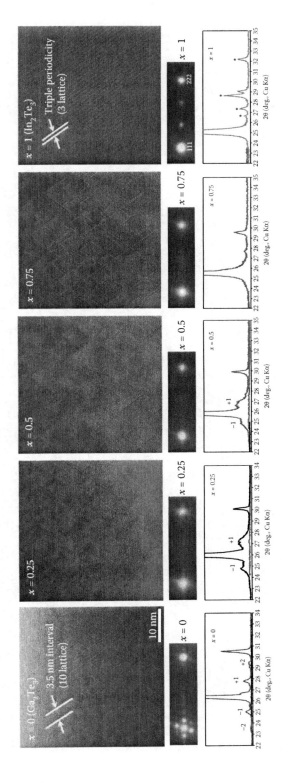

FIGURE 8.1 High-resolution TEM images, electron diffraction patterns, and XRD patterns taken from single grains of sintered pellets of $(Ga_{1-x}In_x)_2Te_3$ ($x = 0, 0.25, 0.5, 0.75,$ and 1). (Adapted from Yamanaka, S. et al. 2009. *Journal of Electronic Materials* 38:1392.)

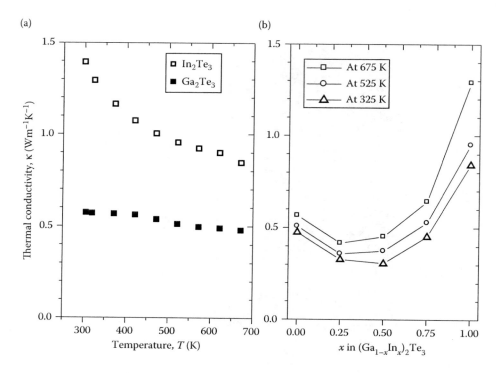

FIGURE 8.2　Thermal conductivity (κ) of Ga_2Te_3 and α-In_2Te_3. (a) Temperature dependence of the κ; (b) relationship between the κ and x in $(Ga_{1-x}In_x)_2Te_3$. (Adapted from Kurosaki, K. 2008. *Applied Physics Letters* 93:012101-1.)

highly effective phonon scattering in Ga_2Te_3. As Figure 8.2b shows, the $(Ga, In)_2Te_3$ solid solutions exhibit a lower κ than those of Ga_2Te_3 and In_2Te_3. The solid solutions (especially those with compositions having $x = 0.25$ and 0.5 for which $\kappa = 0.25$ to 0.5 W m^{-1} K^{-1}, respectively) have very low values of κ (even lower than that of Ga_2Te_3). The principal causes for this are probably the alloying effect between Ga and In and the random periodicity of the vacancy plane intervals.

Figure 8.3 shows the temperature dependences of the electrical resistivity (ρ), the Seebeck coefficient (S), and the power factor (S^2/ρ) of $(Ga_{1-x}In_x)_2Te_3$ ($x = 0, 0.25, 0.5, 0.75,$ and 1).[12] S of all samples have large positive values and a negative temperature dependence, whereas ρ has relatively high values and a negative temperature dependence. ρ systematically increases with x, whereas there is no clear relationship between the magnitude of S and x. All the solid solutions have lower power factors than Ga_2Te_3. No improvement in the power factor can be achieved in the solid solution system. Although its electrical performance is not exceptional, Ga_2Te_3 exhibits a relatively high ZT value (approximately 0.16 at 850 K), primarily due to its very low κ. This means that it is possible to further improve ZT; namely, ZT could be enhanced after optimizing the carrier concentration. However, doping and the deviation from the stoichiometric composition are expected to affect the formation of vacancy planes and consequently the crystal symmetry. Therefore, it is important to account for the symmetry breaking effect when optimizing the carrier concentration.

8.3　Effect of Vacancy Distribution on the Thermal Conductivities of Ga_2Te_3 and Ga_2Se_3

2D vacancy planes with approximately 3.5 nm intervals are present in Ga_2Te_3. They scatter phonons efficiently, resulting in a very low κ. Due to this abnormally low κ, Ga_2Te_3 and related materials are very promising for thermoelectric applications. However, it is unclear as to what effect the size and periodicity of the 2D vacancy planes have on κ. In addition, it is unclear as to whether only 2D vacancy planes reduce

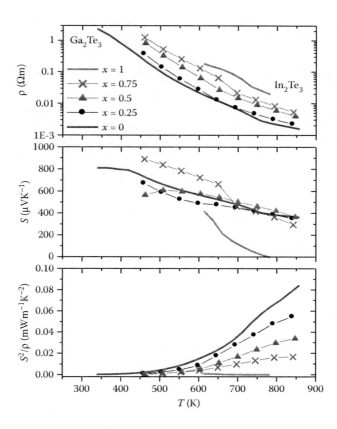

FIGURE 8.3 Temperature dependence of the electrical resistivity (ρ), the Seebeck coefficient (S), and the power factor (S^2/ρ) of $(Ga_{1-x}In_x)_2Te_3$ ($x = 0$, 0.25, 0.5, 0.75, and 1). (Adapted from Yamanaka, S. et al. 2009. *Journal of Electronic Materials* 38:1392.)

κ or whether point vacancies can also reduce κ. Here, we present recent research results on the effect of the vacancy distribution on κ of Ga chalcogenides.

The effect of periodicity of the 2D vacancy planes on κ has been investigated[17] using two kinds of Ga_2Te_3 (see Table 8.1). One is Ga_2Te_3 with randomly arranged vacancy planes; it was prepared by annealing at 673 K for 14 days and then furnace cooling (referred to below as GT-LT: Ga_2Te_3 annealed at a low temperature). The other is Ga_2Te_3 with regularly arranged vacancy planes (approximately

TABLE 8.1 Annealing Conditions and Vacancy Distributions of Ga_2Te_3 and Ga_2Se_3

Sample Name	GT-LT (Ga_2Te_3 Annealed at Low Temperature)	GT-HT (Ga_2Te_3 Annealed at High Temperature)	GS-LT (Ga_2Se_3 Annealed at Low Temperature)	GS-HT (Ga_2Se_3 Annealed at High Temperature)
Annealing temperature (K)	673	973	873	1173
Annealing time (days)	14	14	30	7
Cooling method	Furnace cooled	Quenching	Quenching	Quenching
Vacancy distribution	In-plane with random periodicity	In-plane with regular periodicity (approximately 3.5 nm intervals)	Point	In-plane with random periodicity

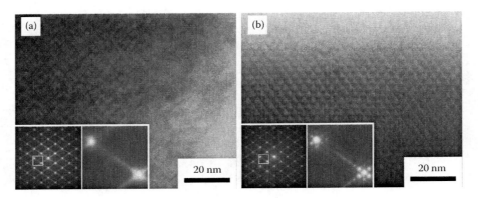

FIGURE 8.4 High-resolution TEM images and electron diffraction patterns taken from single grains of sintered pellets of Ga_2Te_3 annealed under different conditions. (a) Data for GT-LT: Ga_2Te_3 annealed at a low temperature (in-plane vacancies with random periodicity); (b) data for GT-HT: Ga_2Te_3 annealed at a high temperature (in-plane type vacancies with regular periodicity). (Adapted from Kim, C. E. et al. 2009. *Physica Status Solidi RRL* 3:221.)

3.5 nm intervals); it was prepared by annealing at 973 K for 14 days and then water quenching (referred to below as GT-HT: Ga_2Te_3 annealed at a high temperature).

Figure 8.4 shows high-resolution TEM images and electron diffraction patterns taken from a single grain of a sintered pellet of Ga_2Te_3 annealed under different conditions.[17] In the high-resolution TEM image of the GT-HT sample (Figure 8.4b), plane defects due to the vacancy-rich plane are two-dimensionally distributed throughout the whole sample. These plane defects appear in the (111) plane with approximately 10 lattice (equivalent to 3.5 nm) intervals. Similar two-dimensional (2D) vacancy planes are also observed in the GT-LT sample, but their periodicity is random (Figure 8.4a). These high-resolution TEM results are highly consistent with the electron diffraction patterns. That is, the electron diffraction pattern of the GT-HT sample contains superlattice spots at the ~1/10 positions between two neighboring fundamental spots in the [111] direction as well as the fundamental Bragg reflections due to the zinc-blende structure. The electron diffraction pattern of the GT-LT sample contains diffuse superlattice reflections and a streak.

The effect of the vacancy distribution (plane or point) on κ has been investigated using two kinds of Ga_2Se_3 (see Table 8.1). One is Ga_2Se_3 with ordered vacancies as point defects, which was prepared by annealing at 873 K for 30 days and then water quenching (referred to below as GS-LT: Ga_2Se_3 annealed at a low temperature). The other is Ga_2Se_3 with randomly arranged vacancy planes, which was prepared by annealing at 1173 K for 7 days and then water quenching (referred to below as GS-HT: Ga_2Se_3 annealed at a high temperature).

Figure 8.5 shows high-resolution TEM images and electron diffraction patterns taken from single grains of sintered pellets of Ga_2Se_3 annealed under different conditions.[18] The diffraction pattern of Figure 8.5a is consistent with the (001) reciprocal lattice plane of Ga_2Se_3 (space group: Cc (No. 9), $a = 0.6608$ nm, $b = 1.16516$ nm, $c = 0.66491$ nm, $\alpha = 90°$, $\beta = 108.84°$, $\gamma = 90°$) in which vacancies induced by the valence mismatch between the cation and the anion are regularly arranged in the Ga sublattice.[19,20] The high-resolution TEM image of Figure 8.5a shows that no vacancy planes are formed, suggesting that the vacancies in the GS-LT sample exist as point defects. On the other hand, in the high-resolution TEM image of the GS-HT sample (Figure 8.5b), there are 2D vacancy planes in the (111) plane with random periodicity similar to the GT-LT sample (Figure 8.4a). In addition, a streak between the fundamental Bragg reflections due to the zinc-blende structure is observed in the electron diffraction pattern of the GS-HT sample.

Figure 8.6a and b shows the temperature deendences of κ of Ga_2Te_3 and Ga_2Se_3 with various vacancy distributions, respectively.[18] Here, κ can be considered to correspond to κ_{lat} because these Ga chalcogenides have quite low σ values (e.g., 0.42 Ω^{-1} m^{-1} at 338 K for Ga_2Te_3).[16] As shown in Figure 8.6a, the κ

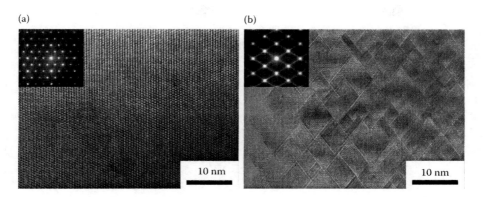

FIGURE 8.5 High-resolution TEM images and electron diffraction patterns taken from single grains of sintered pellets of Ga_2Se_3 annealed under different conditions. (a) Data for GS-LT: Ga_2Se_3 annealed at a low temperature (point vacancies); (b) data for GS-HT: Ga_2Se_3 annealed at a high temperature (in-plane vacancies with random periodicity). (Adapted from Kim, C.E. et al. 2011. *Journal of Electronic Materials* 40:999.)

values of both the GT-HT and GT-LT samples are exceptionally low, being about 0.5 W m⁻¹ K⁻¹. This possibly indicates highly effective phonon scattering due to the 2D vacancy plane structures. There is little difference between the κ values of the GT-LT and GT-HT samples. This implies that the presence of 2D vacancy planes significantly reduces κ, whereas the periodicity of the vacancy planes has little effect on κ. On the other hand, κ of the GS-LT sample with point vacancies is clearly higher than that of the GS-HT sample with in-plane vacancies (see Figure 8.6b). In addition, the GS-LT and GS-HT samples clearly have different temperature dependences of κ; that is, κ of the GS-LT sample with point vacancies

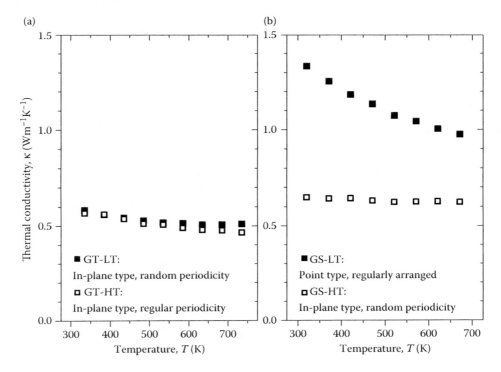

FIGURE 8.6 Temperature dependences of the thermal conductivities (κ) of (a) Ga_2Te_3 and (b) Ga_2Se_3 with various vacancy distributions. (Adapted from Kim, C.E. et al. 2011. *Journal of Electronic Materials* 40:999.)

decreases with increasing temperature approximately according to a T^{-1} relation, whereas that of the GS-HT sample with in-plane vacancies exhibits rather flat temperature dependence.

From these experimental results, the following conclusions are obtained:

1. The presence of vacancies alone does not result in effective phonon scattering.
2. Vacancies should form an in-plane defect structure to achieve effective phonon scattering.
3. The periodicity of the vacancy planes (i.e., regular or random) has little effect on phonon scattering.

8.4 Summary and Conclusions

The Ga and In chalcogenides, Ga_2Te_3, In_2Te_3, and Ga_2Se_3, have a zinc-blende structure except that a third of the cation sites are structural vacancies due to the valence mismatch between the cation and the anion. In Ga_2Te_3, the vacancies are concentrated in the (111) plane and form 2D vacancy planes, whereas in In_2Te_3, the vacancies are present as point defects. Ga_2Te_3 exhibits unexpectedly low κ, which is lower than that of In_2Te_3; this is most probably due to highly effective phonon scattering by the 2D vacancy planes. In addition, two kinds of Ga_2Te_3 can be obtained by controlling the annealing conditions: one has randomly arranged 2D vacancy planes and the other has regularly arranged 2D vacancy planes (at 3.5 nm intervals). Although both forms of Ga_2Te_3 exhibit relatively low κ with a rather flat temperature dependences, there is little difference between the values of κ for samples with different vacancy plane distributions. On the other hand, there are two forms of Ga_2Se_3: one includes 2D vacancy planes, whereas the other contains vacancies as point defects. Although these two forms of Ga_2Se_3 have the same structures with the same chemical composition and the same vacancy density, their κ values are quite different. Ga_2Se_3, which has point vacancies, exhibits a relatively high κ with a T^{-1} temperature dependence, whereas that with in-plane vacancies exhibits quite a low κ with a rather flat temperature dependence, like that of Ga_2Te_3.

Based on the κ measurement results for Ga and In chalcogenides with various vacancy distributions, it can be concluded that phonons are more effectively scattered by vacancies when they are present as an in-plane defect than when they are distributed as point defects. Introducing structural vacancies with in-plane type defect structures is a promising new method for reducing κ of thermoelectric materials and increasing *ZT*.

Acknowledgments

This work was supported in part by a Grant-in-Aid for Scientific Research (No. 21760519 & 23686091) from the Ministry of Education, Culture, Sports, Science and Technology of Japan. The TEM observations were carried out by Dr. Manabu Ishimaru of the Institute of Scientific and Industrial Research, Osaka University, Japan.

References

1. Pei, Y., and Morelli, D. T. 2009. Vacancy phonon scattering in thermoelectric In_2Te_3-InSb solid solutions. *Applied Physics Letters* 94:122112-1.
2. Zhuze, V. P., Sergeeva, V. M., and Shelykh, A. I. 1961. Electrical properties of semiconducting In_2Te_3 with defective structure. *Soviet Physics – Solid State* 2:2545.
3. Nagat, A. T., Nassary, M. M., and El-Shaikh, H. A. 1991. Investigation of thermoelectric power of indium sesquitelluride monocrystals. *Semiconductor Science and Technology* 6:979.
4. Lakshminarayana, D., Patel, P. B., Desai, R. R., and Panchal. C. J. 2002. Investigation of thermoelectric power in indium sesquitelluride(In_2Te_3) thin films. *Journal of Materials Science: Materials in Electronics* 13:27.

5. Koshkin, V. M., Gal'chinetskii, L. P., and Korin, A. I. 1972. Electrical conductivity of heavily doped $B_2^{III}C_3^{VI}$ semiconductors. *Soviet Physics – Semiconductors* 5:1718.

6. Sen, S., and Bose, D. N. 1984. Electrical and optical properties of single crystal In_2Te_3 and Ga_2Te_3. *Solid State Communications* 50:39.

7. Seki, Y., Kusakabe, M., and Kashida, S. 2010. Dielectric and ac conductivity studies in as-grown Ga_2Te_3 crystals with the defect zinc-blende structure. *Physica Status Solidi A* 207:203.

8. Desai, R. R., Lakshminarayana, D., Patel, P. B., Patel, P. K., and Panchal, C. J. 2005. Growth and structural properties of indium sesquitelluride (In_2Te_3) thin films. *Materials Chemistry and Physics* 94:308.

9. Woolley, J. C., and Pamplin, B. R. 1959. The ordered crystal structure of In_2Te_3. *Journal of the Less-Common Metals* 1:362.

10. Guymont, M., Tomas, A., and Guittard, M. 1992. The structure of Ga_2Te_3 an x-ray and high-resolution electron microscopy study. *Philosophical Magazine A* 66:133.

11. Teraguchi, N., Kato, F., Konagai, M., Takahashi, K., Nakamura, Y., and Otsuka, N. 1991. Vacancy ordering of Ga_2Se_3 films by molecular beam epitaxy. *Applied Physics Letters* 59:567.

12. Yamanaka, S., Ishimaru, M., Charoenphakdee, A., Matsumoto, H., and Kurosaki, K. 2009. Thermoelectric characterization of $(Ga,In)_2Te_3$ with self-assembled two-dimensional vacancy planes. *Journal of Electronic Materials* 38:1392.

13. Nakamura, Y., and Hanada, T. 1997. Order-mesoscopic transition in compound semiconductor containing structural vacancies. *Electron Microsc.* 32:150.

14. Ishikawa, M., and Nakayama, T. 1997. Electronic structures of vacancy-plane-superstructured Ga_2Te_3 and Ga_2Se_3. *Phys. Low-Dimens. Semicond. Struct.* 11/12:95.

15. Zaslavskii, A. I., and Sergeeva, V. M. 1961. The polymorphism of In_2Te_3. *Sov. Phys. Solid State* 2:2556.

16. Kurosaki, K., Matsumoto, H., Charoenphakdee, A., Yamanaka, S., Ishimaru, M., and Hirotsu, Y. 2008. Unexpectedly low thermal conductivity in natural nanostructured bulk Ga_2Te_3. *Applied Physics Letters* 93:012101-1.

17. Kim, C. E., Kurosaki, K., Ishimaru, M., Jung, D. Y., Muta, H., and Yamanaka, S. 2009. Effect of periodicity of the two-dimensional vacancy planes on the thermal conductivity of bulk Ga_2Te_3. *Physica Status Solidi RRL* 3:221.

18. Kim, C.E., Kurosaki, K., Ishimaru, M., Muta, H., and Yamanaka, S. 2011. Effect of vacancy distribution on the thermal conductivity of Ga_2Te_3 and Ga_2Se_3. *Journal of Electronic Materials* 40: 999–1004.

19. Villars, P., and L.D. Calvert. 1991. *Pearson's Handbook of Crystallographic Data for Intermetallic Phases.* 2nd ed., Materials Park, OH: ASM International.

20. Lubbers, D., and Leute, V. 1982. The crystal structure of β-Ga_2Se_3. *Journal of Solid State Chemistry* 43:339.

9

Thermoelectric Inverse Clathrates

Matthias Falmbigl
University of Vienna

Peter F. Rogl
University of Vienna

9.1 Introduction

Clathrates are periodic solids in which tetrahedrally bonded atoms form a three-dimensional framework of cages that enclose guest atoms, which do not engage in covalent bonds with the host lattice. The clathrate type-I structure consists of a host lattice of 24-vertex tetrakaidecahedra $[5^{12}6^2]$ (12 pentagonal and 2 hexagonal faces), which share hexagonal faces and form nonintersecting channels enclosing isolated 20-vertex pentagon dodecahedra $[5^{12}]$. Structures of type-I crystallize in the centrosymmetric space group $Pm\bar{3}n$ (No. 223; Pearson symbol cP54). This clathrate type-I framework contains 46 atoms E distributed over three sites[*]: 6d, 16i, and 24k. Filler (guest) atoms G are incorporated (clathrated) in two small cavities (site 2a) at the centers of pentagon dodecahedra and in six larger cavities (site 6c) at the centers of tetrakaidecahedra, resulting in a general formula $G_{2+6}[E_{46}]$ (for details see Ref. [1]).

Besides the general class of "intermetallic" clathrates, which consists of cationic guest atoms in the voids of a covalently bonded framework (incorporating anionic elements in order to balance the guest charges) [1], there exists a group of so-called "inverse clathrates" with reversed polarity of the guest–host relation [2]. The voids in these "polycationic" compounds are occupied by anionic halogen or tellurium ions and electrons are transferred from the framework to the guest atoms. Following the Zintl concept for semiconductors, Menke and von Schnering [3] were the first to synthesize polycationic clathrates of type-I, $Ge_{38}A_8X_8$, A = P, As, Sb; X = Cl, Br, I. The growing interest in thermoelectric applications caused a significantly increased interest in Si-, Ge-, and Sn/P-based inverse clathrates and there have been

---

[*] Wyckoff sites refer to a standardized setting of the crystal structure. According to this standardization Wyckoff sites 6c and 6d interchange.

successful efforts to substitute the group 14 elements (Si, Ge, Sn) partially by P, As, Sb, Ga, Te, I, In, Zn, Cd, Cu [4]. Furthermore, Sn-based inverse clathrates were shown to derive from an ideal composition $Sn_{24}A_{22-y}\square_yX_8$ (A = P, As, Sb; X = Te, Cl, Br, I), which on Sn/M substitution (M is Cu, Zn, In etc.) may develop a varying amount of vacancies in the framework [5–7].

As only some of the hitherto synthesized inverse clathrates are stable at elevated temperatures, thermoelectric properties (S, λ, ρ) are still lacking for many compounds. Table 9.1 is a comprehensive compilation of crystal structures and thermoelectric parameters for all inverse clathrates hitherto reported in the literature.

TABLE 9.1   Overview on Inverse Clathrate Compounds, Structural and Physical Property Data

Compound	Structure Type / Space Group	Lattice Parameter (Å)	Seebeck Coefficient (S) (µV/K) (300 K)	Electrical Cond. (σ) (S/m) (300 K)	Thermal Cond. (κ) (W/mK) (300 K)	ZT	Ref.
			Inverse Clathrates of Type-I				
Si-based:							
$Si_{46-x}P_xTe_8$ $x =$ 11			60	50,000			[8]
12			90	30,000			[8]
13			135	6000	3.9	0.45 (900 K)	[8]
14			210	1000			[8]
15			220	300			[8]
16	$Pm\bar{3}n$	9.96457 (9)	–10	0.158	4.4	0.06 (900 K)	[8]
17			250 (500 K)	0.0158		0.06 (900 K)	[8]
$Si_{46-x}P_xTe_y$ $x =$ $y =$	$Pm\bar{3}$	(9.9789 (2)) 6.6 ≤ y ≤ 7.5 (9.9724 (1)); $x \sim 2y$; 1100°C					[9]
13.0 6.88	$Pm\bar{3}$	9.9808 (2)					[10]
13.3 6.70	$Pm\bar{3}$		5556				[9]
13.6 6.98	$Pm\bar{3}$	9.9794 (2)					[10]
14.1 7.25	$Pm\bar{3}$	9.9750 (1)	1000				[10,9]
14.7 7.35	$Pm\bar{3}$	9.9702 (3)					[10]
15.1 7.50	$Pm\bar{3}$	9.9724 (1)	16.6 (100 K)				[9]
15.2 7.55	$Pm\bar{3}$		4808				[9]
$Si_{38}Te_8Te_8$	$P\bar{4}3n$						[11,12]
Both are high-pressure forms (5 GPa, 1200°C)	$R3c$	10.457 (1) $a_R = 10.465$ (1); $\alpha_R = 89.88$ (1) $a_h = 14.79$; $c_h = 18.11$					[11,12]
$Si_{40}P_6I_{6.5}$	$Pm\bar{3}n$	10.130 (1)					[13]
$Si_{46-x}I_xI_8$ (5 GPa, 700°C)	$Pm\bar{3}n$	10.4195 (7) ($x = 1.5$ (5))					[14]
Ge-based:							
$Ge_{38}P_8X_8$ X= Cl		10.3514 (3)					[3]
Br	$P\bar{4}3n$	10.4074 (4)					[3]
I	$P\bar{4}3n$	10.5067 (6)	1.11–6.25				[3]
$Ge_{38}P_8Br_{8-x}I_x$	$P\bar{4}3n$	10.33 ($x = 2$)					[3]
$Ge_{30.1}P_{15.9}Te_{7.4}$	$Pm\bar{3}n$	10.3376 (2)	750	3.16	0.9	0.0006 (300 K)	[15]
$Ge_{32}Ga_3As_{11}I_8$	$P\bar{4}3n$	10.616					[3]
		10.507 (2)					[16]

TABLE 9.1 (continued) Overview on Inverse Clathrate Compounds, Structural and Physical Property Data

Compound	Structure Type / Space Group	Lattice Parameter (Å)	Seebeck Coefficient (S) (μV/K) (300 K)	Electrical Cond. (σ) (S/m) (300 K)	Thermal Cond. (κ) (W/mK) (300 K)	ZT	Ref.
$Ge_{14}Ga_{12}Sb_{20}I_8$	$P\bar{4}3n$	11.273 (2)					[3,16]
$Ge_{38}As_8X_8$ X= Br		10.5161 (5)					[3]
I	$Pm\bar{3}n$	10.625		100			[17]
I	$P\bar{4}3n$	10.6158 (4)		0.02–0.008			[3]
$Ge_{25}As_{21}I_8$	$Pm\bar{3}n$	10.5963 (6)					[18]
$Ge_{38}Sb_8X_8$ X= Br		10.7893 (8)					[3]
I		10.8697 (6)		$1\text{–}2 \times 10^{-5}$			[3]
I	$Pm\bar{3}n$	10.8892 (2)	−800	0.001	1.15		[19]
$Ge_{38}Sb_8I_{7.81}$	$Pm\bar{3}n$						[20]
$Ge_{38+x}Sb_{8-x}I_{8-x}$	$(0 < x \leq 0.86)$ 10.850 (3) $(x = 0.86)$						[20]
$Ge_{38-y/4}\square_{y/4}Sb_{8+y}I_8$	$(0 < y \leq 6.83)$ 10.946 (4) $(y = 6.83)$						[20]
$Ge_{36}Sb_{10}I_8$	$Pm\bar{3}n$	10.8907 (2)					[21]
$Ge_{40.0}Te_{5.3}\square_{0.7}I_8$	$Pm\bar{3}n$ or $P23$	10.815 (1)					[22]
$Ge_{43.3}I_{2.7}I_8$	$Pm\bar{3}n$	10.814 (3)					[23]
Sn-based:							
$Sn_{24}P_{19.3}\square_{2.7}Cl_yI_{8-y}$ y = 0	$Pm\bar{3}n$	10.954 (1)	80	1500	1.8	0.02 (300 K)	[24,25]
$0 \leq y \leq 0.8$, 400°C 0.25	$Pm\bar{3}n$	10.948 (1)					[25]
0.49	$Pm\bar{3}n$	10.9408 (7)					[25]
0.8	$Pm\bar{3}n$	10.9331 (8)					[25]
$Sn_{24}P_{19.3}\square_{2.7}Br_xI_{8-x}$ x = 2	$Pm\bar{3}n$			1340	0.5	0.02 (300 K)	[25]
$0 \leq x \leq 8$, 400°C 2.35	$Pm\bar{3}n$	10.9200 (10)					[26]
3	$Pm\bar{3}n$			400			[25]
3.14	$Pm\bar{3}n$	10.9140 (10)					[26]
4.62	$Pm\bar{3}n$	10.8860 (10)					[26]
5	$Pm\bar{3}n$			340			[25]
6.1	$Pm\bar{3}n$	10.8440 (10)					[26]
7	$Pm\bar{3}n$			405			[25]
8	$Pm\bar{3}n$	10.8142 (7)	180	334			[25,26]
8	$Pm\bar{3}n$		145	7330	1.75 (140 K)	0.03 (300 K)	[27]
$Sn_{24}P_{19.3-x}As_x\square_{2.7}I_8$	$Pm\bar{3}n$	$(a = 10.9358)$ $0 \leq x \leq 15.75$ (11.1495); 450°C					[28]
$Sn_{24}P_{19.3}\square_{2.7}I_8$ + 6.5 at%Ni	$Pm\bar{3}n$	10.9550 (2)	131	6897	2	0.02 (300 K)	[7]
$Sn_{19.3}Cu_{4.7}P_{22}I_8$	$Pm\bar{3}n$	10.847 (1)	600	0.2			[7,29]
$Sn_{19.3}Cu_{4.7}As_{22}I_8$	$Pm\bar{3}n$	11.1736 (3)					[30]
$Sn_{19.3}Cu_{3.7}Zn_1P_{21.2}\square_{0.8}I_8$	$Pm\bar{3}n$	10.8773 (4)	620	0.83			[7]
$Sn_{19.3}Cu_{2.7}Zn_2P_{20.9}\square_{1.1}I_8$	$Pm\bar{3}n$	10.8876 (3)	525	3.6	0.66	0.0006 (300 K)	[7]
$Sn_{19.3}Cu_{1.7}Zn_3P_{19.9}\square_{2.1}I_8$	$Pm\bar{3}n$	10.8915 (3)	433	7.9	0.84	0.0005 (300 K)	[7]

continued

TABLE 9.1   (**continued**) Overview on Inverse Clathrate Compounds, Structural and Physical Property Data

Compound	Structure Type Space Group	Lattice Parameter (Å)	Seebeck Coefficient (S) (µV/K) (300 K)	Electrical Cond. (σ) (S/m) (300 K)	Thermal Cond. (κ) (W/mK) (300 K)	ZT	Ref.
$Sn_{24-x}Zn_xP_{22-y}\square_yBr_8$	$y = (14 - 2x)/5; \ 0 \leq x \leq 7; \ 300°C$ slowly cooled						[6]
$Sn_{16.8}Zn_{7.2}P_{22}Br_8$	$Pm\bar{3}n$	10.7254 (2)					[6]
$Sn_{17}Zn_7P_{22}Br_8$	$Pm\bar{3}n$	10.7449 (2)		500			[6]
$Sn_{17}Zn_7P_{22}Br_8$	$Pm\bar{3}n$		40	0.3	1.55 (140 K)		[27]
$Sn_{24-x}Zn_xP_{22-y}\square_yI_8$	$y = (14 - 2x)/5; \ 0 \leq x \leq 7; \ 300°C$ slowly cooled						[6]
$Sn_{20}Zn_4P_{20.8}\square_{1.2}I_8$	$Pm\bar{3}n$	10.881 (1) (SC), 10.883 (2) (XPD)					[6]
$Sn_{17.2}Zn_{6.8}P_{22}I_8$	$Pm\bar{3}n$	10.8425 (6) (SC)					[6]
$Sn_{17}Zn_7P_{22}I_8$	$Pm\bar{3}n$	10.8458 (3) (XPD)		0.25			[6]
$Sn_{18}Zn_6P_{21}\square_1I_8$	$Pm\bar{3}n$	10.8420 (3)	314	0.02			[31]
$Sn_{21}Zn_3P_{20}\square_2I_8$	$Pm\bar{3}n$	10.8901 (6)	242	204			[31]
$Sn_{22}Zn_2P_{19.5}\square_{2.5}I_8$	$Pm\bar{3}n$	10.9299 (4)	256	482			[31]
$Sn_{24}As_{19.3}\square_{2.7}I_8$	Unknown	22.179 (1)					[24]
$(Sn_{20.5}\square_{3.5})As_{22}I_8$	$Pm\bar{3}n$	11.092 (1)	−180	0.95	0.5	2 × 10⁻⁵ (300 K)	[32]
	$F23$ or $Fm\bar{3}$	22.1837 (4)					[32]
$Sn_{14}In_{10}P_{21.2}\square_{0.8}I_8$	$P4_2/m$	$a = 24.745$ (3); $c = 11.067$ (1)					[5]
$Sn_{10}In_{14}P_{22}I_8$	$Pm\bar{3}n$	11.0450 (7)					[5]
$Sn_{38}Sb_8I_8$	$Pm\bar{3}n$	12.0447 (3)	−600	0.1	0.7		[19]
Inverse Clathrates of Type-III							
$Si_{172-x}P_xTe_y, \ x = 2y$		$a = 19.2632(3); \ c = 10.0706(2)$					[33]
$20 \leq y \leq 22, 1150°C$	$P4_2/mnm$	$a = 19.2573(6); \ c = 10.0525(7)$ (XPD; $y = 21.2$)		2000			[33]
Inverse Clathrates Related to Type-VII							
$Te_{7+x}Si_{20-x}; \ x \sim 2.5$	$Fd\bar{3}c$	$a = 21.136(2)$					[34]

Note: $\square$ is a symbol for vacancies.

9.2 Structural Chemistry and Crystallographic Relations of Inverse Clathrates

Among the manifold of clathrate structures, nine true hydrate clathrate structure types have been identified, and seven types of which have so far been observed with isopointal polyanionic inter-metallide element combinations [1]. Among inverse clathrates, besides the type-I structure, hitherto only one compound $Si_{172-x}P_xTe_y$ ($x = 2y$, $y \sim 21$; Te-atoms are guest atoms) has been reported to crys-tallize in the clathrate type-III structure [33], and the structure of the high-pressure phase $Si_{20-x}Te_{7+x}$ ($x \sim 2.5$) with truncated Te-octahedra (each enclosing a Te-centered (Si,Te)-pentagon dodecahe-dron) is related to clathrate type-VII [34]. Although the early structure determination for $Ge_{38}A_8X_8$ (A = P, As, Sb and X = Cl, Br, I) promoted a noncentrosymmetric space group $P\bar{4}3n$ [3], a redeter-mination of the structure of $Ge_{38}Sb_8I_8$ unambiguously revealed centro-symmetry ($Pm\bar{3}n$) and thus isotypism with clathrate type-I [35]. Similarly, a neutron diffraction study of $Ge_{38}As_8I_8$ revealed $Pm\bar{3}n$ as the true space group [17].

Albeit the majority of the known inverse clathrates adopts the type I structure with space group *Pm* $\bar{3}$ *n*, atom site preference or atom/vacancy ordering has proven to be a driving force toward the formation of superstructures and/or low-symmetry structure variants. Full or partial atom order, such as claimed for $Ge_{38}A_8X_8$ [3], $Si_{38}Te_8Te_8$ [11], or for the substitution variant $Ge_{14}Ga_{12}Sb_{20}I_8$ [16], was undoubtedly observed for $Ge_{40}Te_{5.3}\square_{0.7}I_8$ [22], $Si_{46-x}P_xTe_y$ [9,10], $(Sn_{20.5}\square_{3.5})As_{22}I_8$ [32], $Sn_{24}As_{19.3}\square_{2.7}I_8$ [24], and (polyanionic) $Ba_8Cu_{16}P_{30}$ [36], the latter crystallizing in an orthorhombic superstructure of the type-I clathrate adopting space group *Pbcn*. Defect ordering plays a significant role in binary $Ba_8Ge_{43}\square_3$ [37] enlarging the unit cell 8-fold ($a = 2a_0$, $Ia\bar{3}d$) with respect to the parent type-I as well as in $Sn_{14}In_{10}P_{21.2}\square_{0.8}I_8$ [5], the latter enlarging the volume under reduction of symmetry to a tetragonal space group $P4_2/m$ ($a = a_0\sqrt{5}$, $c = c_0$). Figure 9.1 summarizes for all inverse clathrate type-I structures reported in the literature the crystallographic group–subgroup relationships among the aristo-type and the lower symmetry structures derived from it. As a consequence of vacancy formation, atom positions neighboring the vacancies usually are described by crystallographic split positions taking care of the fact that the presence of an atom or of a void in a site defines a slightly different bonding distance to the surrounding framework atoms.

In most clathrate type I structures the 6d site is preferred by vacancies irrespective of anionic or cationic clathrate compounds [1,2]. Accordingly, the adjacent 24k site appears in the form of a threefold split site (where only one of the sites is occupied at a time) vanishing in correlated manner with the disappearance of the void in 6d. Due to the different atomic radii in randomly mixed neighboring atom sites, splitting of crystallographic positions was also observed for vacancy-free inverse clathrates, for example, $Sn_{17}Zn_7P_{22}I_8$, $Sn_{17}Zn_7P_{22}Br_8$ [6] and $Sn_{19.3}Cu_{4.7}As_{22}I_8$ [30]. Although there is a considerable difference in the volumes of the 24- and 20-vertex cages, only a slight preference of the smaller cage for the smaller Br- and Cl-atom was found in $Sn_{24}P_{19.3}\square_{2.7}Br_xI_{8-x}$ [25,26] and $Sn_{24}P_{19.3}\square_{2.7}Cl_xI_{8-x}$ [25]. An interesting case is encountered with $(Sn_{20.5}\square_{3.5})As_{22}I_8$ [32], where defects arise on the split Sn-positions. Although there is no sign for off-center motion on both iodine sites, the As-atoms deviate from the center of the 6d and 16i positions to $1/4 \times As1$ in 24k and $1/3 \times As2$ in 48l sites, respectively [32]. Electron diffraction, furthermore, suggests Sn-atom/vacancy ordering in a face-centered $2a \times 2a \times 2a$ supercell (see Figure 9.1); however, with

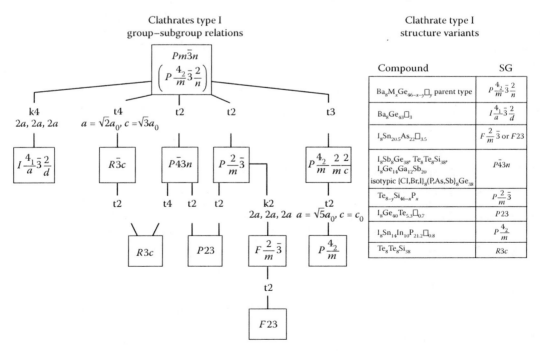

FIGURE 9.1 Crystallographic group–subgroup relationships among clathrate type-I structures.

intergrowth of randomly distributed small domains with different type of ordering [32]. Whereas As/P shows a random distribution on the 16i and 6d sites of $Sn_{24}As_xP_{19.3-x}\square_{2.7}I_8$ ($0 \leq x \leq 16$; ~450°C) with a small preference of As for the 6d site, there is no P-solubility in $(Sn_{20.5}\square_{3.5})As_{22}I_8$ [28].

Atom ordering or strong and directed ionocovalent interactions may constitute the reason for symmetry reductions observed in the low- and high-temperature, high-pressure forms of $Si_{38}Te_{16}$ (clathrate type-I derivatives, [11,12,34]) with Te-atoms exclusively sharing framework sites with Si. Tellurium atoms in $Si_{20-x}Te_{7+x}$ ($x \sim 2.5$) [11,12,34], however, adopt a dual function as guest and host atoms in a structure that may be seen at the verge from a three-dimensional (3D) Si-host lattice to a 3D Te-framework. Confirmation of the puzzling bonding situation in these compounds by DFT calculations is still missing. Te adopts the guest sites in $Si_{46-x}P_xTe_y$ (S.G. $Pm\bar{3}$) [9,10] but prefers distinctly the sites with higher concentration of Si-atoms in its coordination sphere. Te-site preference and Te/vacancy segregation in the partially filled pentagon dodecahedra (essentially confirmed by neutron powder diffraction [9]) were supposed to be the origin of the symmetry reduction. Deviating from the general behavior, vacancies in $Ge_{40}Te_{5.3}\square_{0.7}I_8$ [22] are located in the 24k Ge-site, tellurium being distributed in the framework sites 6d and 16i.

DFT calculations [38] and experimental observations [16] confirmed the site preference of elements such as Sb or Ga, which try to avoid direct Sb–Sb or Ga–Ga contacts in the framework for instance in $Ge_{14}Ga_{12}Sb_{20}I_8$.

Vacancies and mixed occupation of sites may alter the next nearest coordination shell of atoms. For the ideal type-I clathrates the tetrahedrally coordinated next-nearest neighbors of framework atoms are: 6d[24k$_4$], 16i[24k$_3$16i$_1$], 24k[6d$_1$16i$_2$24k$_1$]. Judging from a Mulliken population analysis of Si-based clathrates under the principle of topological charge stabilization [39], electronegative atoms were said to prefer sites accumulating the highest Mulliken population among the cage-forming atoms, whereas electropositive elements tend to accumulate in the lowest Mulliken population [40]. As one can also conceive the clathrate type-I structure as a set of pentagonal dodecahedra bridged by additional atoms in site 6d, these sites 6d with the lowest Mulliken atomic populations are the most likely candidates for vacancy formation. As discussed above, vacancies in the 6d site provoke a splitting of neighboring 24k sites resulting in three-coordinated 24k atoms.

In many cases, ^{119}Sn Mößbauer spectroscopy was used to either confirm the Sn-site occupation, or different local environment and bonding for Sn-atoms, for instance, in $Sn_{24}P_{19.3}\square_{2.7}I_8$ [24], $(Sn_{20.5}\square_{3.5})$ $As_{22}I_8$ [32], $Sn_{14}In_{10}P_{21.2}\square_{0.8}I_8$, $Sn_{10}In_{14}P_{22}I_8$, [5], $Sn_{24-x}Zn_xP_{22-y}\square_yX_8$; $1 < x < 7$; X = Br, I [6], or $Sn_{19.3}Cu_{4.7}As_{22}I_8$ [30]. In all these cases the isomer chemical shift and quadrupole splitting found in the Mößbauer spectra are in accordance with the results from single-crystal x-ray diffraction and the expectations from chemical bonding. ^{31}P-NMR [6,9,33] and ^{119}Sn-NMR [6] spectroscopy revealed small signal shifts typically observed in nonconducting diamagnetic materials. Broadening of the NMR signal width supported the type of atom disorder and vacancy formation in these clathrate compounds.

9.3 Synthesis, Thermal Stability, and Phase Relations of Inverse Clathrates

Several techniques have so far been employed to successfully synthesize inverse clathrates. The commonly used method relies on intimate mixing of stoichiometric amounts of the elements or halides (SnI_4, TeI_4, etc.) in a ceramic mortar followed by compacting the powder blend in a steel die. The synthesis reaction is then carried out in evacuated quartz ampoules with slow heating to the desired temperature taking care of the high vapor pressure of phosphorous (triple point of red P at 589.5°C, 4.4 MPa [41]). On some occasions a two-step sintering process at various temperatures and time durations was reported [5,6,10,22,24–26,28–30,32]. The temperature used for the synthesis is governed by the group-IV element of the compound, increasing from Sn, where temperatures around 500°C were used, to Ge (around 700°C [3]) and Si (1100°C [10]). As halogens can act as transport agents, the ampoules were sometimes mounted in a temperature gradient to obtain single crystals from transport reactions. For the synthesis of Si-based clathrates $Si_{46-x}I_xI_8$ and $Si_{38}Te_{16}$ a high-pressure–high-temperature technique was employed. The reaction of stoichiometric amounts of the elements contained in a h-BN cell was carried out under a pressure of 5 GPa either in a cubic multi-anvil press at 700°C ($Si_{46-x}I_xI_8$) [14] or in a belt-type press at 1200°C ($Si_{38}Te_{16}$) [11]. In both cases pyrophyllite was used as pressure transmitting medium. More recently, Si- [8], Ge- [15,19] as well as Sn- [19] and Sn/P-based [7] clathrates were produced via spark plasma sintering or uniaxial hot pressing of finely ground (ball milled) starting material (elements and/ or master alloys). After milling the samples were partially or fully mechanically alloyed [19].

Data on the thermal stability of inverse clathrates are listed in Table 9.2.

A detailed investigation on the compounds $Ge_{38}\{P,As,Sb\}_8\{Cl,Br,I\}_8$ [3] revealed that all samples decompose in one step with pure Ge as the final residue. A single-step decomposition was also confirmed by thermogravimetry for ternary type-I Si–P–Te clathrates [9,42]. In general, Si-based compounds show high-temperature stability even above about 700°C, whereas Ge- and Sn-based compounds decompose in the temperature range of 500–700°C. $Ge_{43.3}I_{2.7}I_8$ [23] was said to decompose above 510°C into Ge, GeI_2, and I_2. The highest thermal stability was reported for the type-III compound $Si_{130}P_{42}Te_{21}$. Here an incongruent decomposition at ~1240°C was observed resulting in Si and clathrate-I (S.G. *Pm* 3) [33]. Incongruent melting at ~1200°C was reported for $Si_{46-x}P_xTe_y$ ($x \sim 2y$) losing Te and P on heating [9]. All these data refer to decomposition in vacuum [9,33].

Few data concern the stability of cationic clathrates in air: type-III $Si_{130}P_{42}Te_2$ was claimed to be oxidation-resistant in air up to ~1200°C [33]. Although crystals of $Si_{46-2x}P_{2x}Te_x$ were said to decompose in air within a few days [42], $Si_{46-x}P_xTe_y$ was mentioned to be stable against oxidation up to 1020°C [9]. Chemical resistance of $Ge_{38}\{P,As,Sb\}_8\{Cl,Br,I\}_8$ against oxidation on air, moisture as well as against dilute acids and bases was reported in Ref. [3]. Powders of some Sn-based compounds were described to be air- and moisture-stable: $Sn_{24}P_{19.3}\square_{2.7}Br_xI_{8-x}$; $0 < x < 8$ [26], $Sn_{24-x}Zn_xP_{22-y}\square_yX_8$; $1 < x \leq 7$; X = Br, I [6], $Sn_{19.3}Cu_{4.7}As_{22}I_8$ [30] and $Sn_{24}As_{19.3}\square_{2.7}I_8$ [24].

Information on phase equilibria involving cationic clathrate compounds is still limited and essentially concerns the system Si–P–Te [8–10,33,42]. The successful synthesis of $Si_{46-2x}P_{2x}Te_x$ via chemical vapor transport ($TeCl_4$) was based on a thermodynamic modeling of solid-gas phase equilibria [42] revealing as a by-product the phase diagram of the ternary system Si–P–Te (see Figure 9.2).

TABLE 9.2 Decomposition Temperature of Inverse Clathrates

Compound	Decomposition T (°C)	Conditions	Ref.
$Si_{30}P_{16}Te_8$	950	TG	[42]
$Si_{32}P_{14}Te_7$	1100	TG	[42]
$Si_{46-x}P_xTe_y$	1020	On air	[9]
	1170 ($y = 7.4$)	TG, high vacuum	[9]
	1220 ($y = 6.7$)	TG, high vacuum	[9]
$Si_{130}P_{42}Te_{21}$	1240	TG, high vacuum	[33]
	~1200	On air	[33]
$Ge_{38}P_8Cl_8$	630–760	TG, high vacuum	[3]
$Ge_{38}P_8Br_8$	690–775	TG, high vacuum	[3]
$Ge_{38}P_8I_8$	600–750	TG, high vacuum	[3]
$Ge_{38}As_8Br_8$	670–810	TG, high vacuum	[3]
$Ge_{38}As_8I_8$	670–800	TG, high vacuum	[3]
$Ge_{38}Sb_8I_8$	515–675	TG, high vacuum	[3]
$Ge_{43.3}I_{2.7}I_8$	510	—	[23]
$Sn_{19.3}Cu_{4.7}As_{22}I_8$	575	—	[33]
$Sn_{20.5}\square_{3.5}As_{22}I_8$	525	—	[33]

Note: TG, Thermogravimetry.

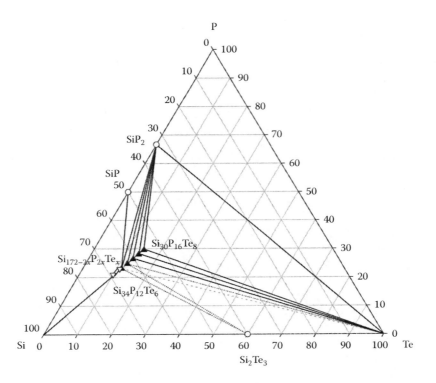

FIGURE 9.2 Phase diagram of the Si–P–Te system based on the thermodynamically modeled phase equilibria up to 973 K. (Adapted from Philipp, F., and P. Schmidt. 2008. *J. Crystal Growth* 310: 5402–5408.) The coexistence areas between the clathrate-phase $Si_{46-x}P_xTe_y$ ($x \sim 2y$; black triangles), Si_2Te_3 and Te below 700°C are plotted with dark-gray dotted lines and above 973 K with light-gray dashed lines. The homogeneity region of the clathrate type-III, $Si_{172-2x}P_{2x}Te_x$ ($20 \leq x \leq 22$) at a temperature of 1425 K is indicated with unfilled triangles. (Adapted from Zaikina, J. V. et al. 2008. *Chem. Eur. J.* 14: 5414–5422.)

For the modeling of the $Si_{46-2x}P_{2x}Te_x$ phase the authors assumed a nonideal solid solution in the range of $6 \leq x \leq 8$ and adjusted the thermodynamic parameters (see Table 9.3; for further details see Ref. [42]) according to the experimentally observed stability of clathrate-I up to 950°C (from thermogravimetry and vapor transport). Similarly, tie lines and coexistence areas were matched with the results from solid-state reactions and residuals from transport experiments. It shall be noted, that for the calculated phase diagram [42] the clathrate-III compound $Si_{172-2x}P_{2x}Te_x$ [33] was not taken into account and the coexistence area of the Si-poor clathrate $Si_{46-2x}P_{2x}Te_x$ with SiP_2 and Te is different from the experimental results

TABLE 9.3 Specific Heat and Heat and Entropy of Formation Data for Inverse Clathrates at 300 K

Compound	C_p (J K^{-1} mol^{-1})	ΔH^0_{298} (k J mol^{-1})	S^0_{298} (J K^{-1} mol^{-1})	Ref.
$Si_{30}P_{16}Te_8$	1212[a]	−2030.1	1254.2	[42]
$Si_{34}P_{12}Te_6$	1155[b]	−1904.5	1157.2	[42]
$Ge_{38}Sb_8I_8$	1330			[19]
$Sn_{38}Sb_8I_8$	1365			[19]
$Sn_{24}P_{19.6}\square_{2.4}I_8$	1295			[26]
$Sn_{17}Zn_7P_{22}I_8$	1300			[6]

[a] $C_p = 1102.8 + 362.6 \times 10^{-3} \cdot T$.
[b] $C_p = 1050.5 + 347.5 \times 10^{-3} \cdot T$.

found at higher temperatures [9]. Furthermore, data on structure and homogeneity region of the phase $Si_{46-x}P_xTe_y$ ($x \sim 2y$) are controversial: whereas Kishimoto et al. [8] claimed a solid solution $Si_{46-x}P_xTe_8$ with clathrate type I (S.G. $Pm\bar{3}n$) extending at 850–1000°C for $11 \leq x \leq 17$, Zaikina et al. [9,10] reported a much smaller homogeneity region at 1100°C for the solid solution $Si_{46-x}P_xTe_y$, $6.6 \leq y \leq 7.5$, $x \sim 2y$, with a low-symmetry version (S.G. $Pm\bar{3}$) of clathrate type-I (see Figure 9.1). Albeit weak reflections (hkl) for $l \neq 2n$ were reported to accompany the symmetry reduction [9], they may have been left unnoticed in [8], where only the Rietveld analysis of $Si_{30}P_{16}Te_8$ has been reported. It should be noted that the lattice parameter for $Si_{30}P_{16}Te_8$ ($a = 9.96457(9)$ Å; [8]) fits well to the essentially linear lattice parameter dependency for $Si_{46-x}P_xTe_y$ [9] extrapolated to $y = 8$, $x = 2y$. It should also be mentioned, that according to Kishimoto et al. [8] secondary phases (Si) + Si_2Te_3 were observed after sintering at 850–1000°C in all samples $Si_{46-x}P_xTe_8$ with $11 < x < 17$ except for the Zintl-compound $Si_{30}P_{16}Te_8$ (samples were prepared by ball milling and SPS at 40 MPa). According to the investigation of the homogeneity range of the two clathrates at 1100°C by Zaikina et al. [9,33], coexistence of $Si_{46-x}P_xTe_y$ with clathrate-III was reported for $y < 6.6$, whereas for $y > 7.5$ SiP_2 and Si_2Te_3 were identified as secondary phases. The homogeneity range of $Si_{172-2x}P_{2x}Te_x$ at 1150°C was derived as $20 \leq x \leq 22$ [33]. In samples with $x < 20$ pure Si was found in equilibrium but at higher P- and Te-contents only $Si_{46-x}P_xTe_y$ was forming.

The existence of independently prepared $Ge_{38+x}Sb_{8-x}(I_{8-x}\square_x)$ [20] and $Ge_{36}Sb_{10}I_8$ [21] indicates a homogeneity region not only as a function of Sb/Ge exchange but also as a function of iodine deficiency.

9.4 Are Inverse Clathrates Zintl Compounds?

Most inverse clathrate compounds are semiconductors and show a balanced valence with the octet rule satisfied for both electropositive and electronegative components. These are the typical requirements to be considered as Zintl phases (for details see Ref. [40]). The relatively simple rules of valence counting using the Zintl–Klemm approach allow in a variety of complex structures like clathrates or skutterudites a fast but rough estimation of the transport properties. For good thermoelectric properties a slight valence imbalance is favorable to create a heavily doped semiconductor with the Fermi level residing near a band gap but inside a band. The general valence rule (described in details in Ref. [43]) can as well be applied to characterize inverse clathrate compounds. In contrast to the polyanionic clathrates, where the guest atoms donate electrons to the framework, which can be compensated by the formation of vacancies and/or substitution, the guest atoms in inverse clathrates attract electrons from the host lattice. Thus, in a simple approach halogen guest atoms are considered as $[X^{-1}]$ and tellurium as $[Te^{-2}]$ using the anion formal valence whereas the framework atoms are described by applying the formal valence formula for cations, for example, the compound $Si_{30}P_{16}Te_8$, which is exactly fulfilling the Zintl count, is written in terms of this concept as $[Si^0]_{30}[P^{+1}]_{16}[Te^{-2}]_8$. Also vacancy-containing compounds like $Sn_{14}In_{10}P_{21.2}\square_{0.8}I_8$ can be described: $[Sn^0]_{14}[In^{-1}]_{10}[P^{+1}]_{21.2}[\square^{-4}]_{0.8}[I^{-1}]_8$. For both examples the formal anionic and cationic valences are compensating each other and thus those compounds are supposed to exhibit semiconducting properties.

Applying this concept to the compounds listed in Table 9.1, most inverse clathrates are perfectly in line with the Zintl formalism. Only a few exceptions with a valence imbalance have been reported, mainly for Si-based compounds. The rather large solubility range observed for the ternary clathrate-I compound $Si_{46-x}P_xTe_y$ [8–10] allows tuning transport properties [8]. It was reported that p-type doping starting from the valence balanced sample with $x = 16$ (exhibiting an electrical conductivity typical for an intrinsic semiconductor) going down to $x = 11$ decreases the electrical resistivity significantly. Recently, also Ge-based inverse clathrate compounds have been described [18,21], which show a charge imbalance ($Ge_{25}As_{21}I_8$ [13 excess electrons] and $Ge_{36}Sb_{10}I_8$ [2 excess electrons]). For charge balanced $Ge_{38}Sb_8I_8$ [20] the solubility of Sb and Ge was investigated and an increasing Sb-content was found to be compensated by vacancy formation. An interesting example is the subiodide $Ge_{4.1}I \equiv Ge_{43.3}I_{2.7}I_8$ [23], which encompasses iodine atoms in two oxidation states (confirmed by XPS-experiments): I^- as filler atoms in 2a and 6c but I^{3+} sharing the 6d sites with Ge atoms (2.67 I + 3.33 Ge). It should be noted that for isostructural $Si_{44.5}I_{1.5}I_8$ the Zintl balance is not fully achieved and furthermore $I_{1.5}$ was claimed to enter the 16i Si-positions [14]. More puzzling cases are encountered with $Si_{38}Te_{16}$ and $Si_{20-x}Te_{7+x}$ ($x \sim 2.5$) [11,12,34] where Te–Si distances indicate the Te-atoms in various bonding states.

For the Sn/pnicogen-based inverse clathrates no deviations from the Zintl–Klemm concept have been observed so far; however, all attempts to create sufficient p- or n-type doping failed due to limited solubility for the substituting elements or vacancy formation compensating for additional electrons. By varying the composition in $Sn_{19.3}Cu_{4.7-x}Zn_xP_{22-y}\square_yI_8$ ($x = 1, 2, 3$) a change in the charge carrier concentration (from 1.3×10^{16} cm^{-3} for $x = 0$ to 6.2×10^{17} cm^{-3} for $x = 3$) and a lowering in the band gap can be achieved, but the electrical conductivity increases insufficiently due to vacancy formation at the phosphorus 6d-site [7].

9.5 Physical, Thermoelectric, and Mechanical Properties of Inverse Clathrates

9.5.1 Specific Heat, Off-Center Motion and Rattling of Guest Atoms

Although many Zintl compounds with large unit cells and complex crystal structures exhibit very low lattice thermal conductivities (close to κ_{min}) without evidence for (i) rattling atoms with large atom displacement parameters or (ii) nanostructuring, low κ_{phonon} values in clathrates seem to indicate increased scattering of the heat-carrying phonons due to tunneling and/or rattling of the guest atoms in the large 24-vertex cages [$5^{12}6^2$]. Recent studies have shown that the host structure plays an important role for the disorder and dynamics of the guest atom. Sales et al. [44] suggested that the guest atom in the large clathrate cage can be regarded as an Einstein oscillator relying on a weak bonding between the host structure and the guest atom further assuming that the motion of the guest atom is to a large extent independent of the collective motion of the host framework. Guest atoms with low Einstein energies are expected to result in lower κ_{phonon}, when guest atom phonons hybridize with the acoustic phonons of the host. It was shown for anionic clathrates that for the combination of the two modes an avoided crossing may occur and the acoustic phonon dispersion branch reaches the Brillouin zone boundary at reduced energy [45]. The Einstein temperatures (θ_E) can either be extracted from x-ray or neutron diffraction data via the temperature-dependent atom displacement parameters (ADPs) of guest atoms or from low-temperature specific heat data. Alternatively, inelastic neutron or Raman scattering can elucidate the phonon modes directly.

The majority of specific heat data available in the literature for inverse clathrates were obtained using the Dulong–Petit law [19,26] or estimated according to Neumann–Kopp's rule [42] in order to provide C_V for the lattice thermal conductivity. Data are listed in Table 9.3. Only for $Sn_{17}Zn_7P_{22}I_8$ temperature-dependent heat capacity was measured from 6 to 400 K. The result agrees well with the Dulong–Petit law at high temperatures and shows that no low-temperature phase transition occurs [6]. In order to gain information on the lattice dynamics, a quantitative description was attempted by the authors [31] following the approach of Junod et al. [46,47], that is, modeling the specific heat data as presented in Ref. [6] in the form of $(Cp - \gamma T)/T^3$ versus $\ln T$ to reveal deviations from the simple Debye model, which at low temperature causes a T^3 dependence (for details see Ref. [48]). A fairly good description of the temperature-dependent specific heat is obtained yielding a Debye temperature $\theta_D = 240$ K and two Einstein-like contributions, $\theta_{EL1} = 48$ and $\theta_{EL2} = 88$ K, with corresponding spectral widths of 0.3 and 12.6 K, respectively. The results are shown in Figure 9.3. Although the low-temperature data (<15 K) did not allow an accurate fitting to the model resulting in large error bars for the Sommerfeld value $\gamma \sim 0.075$ J mol^{-1} K^{-2}, Debye- and Einstein temperatures are in good agreement with results from other measurement techniques (see below).

Alike for polyanionic clathrates, also in cationic clathrates enhanced ADPs were observed from x-ray diffraction data for guest atoms located in the 6c-position within the large tetrakaidecahedral cages. Whereas information mainly covers data from room temperature investigations, temperature-dependent ADPs can hitherto only be retrieved from single-crystal x-ray data published for $Sn_{24}P_{19.3}\square_{2.7}Br_8$

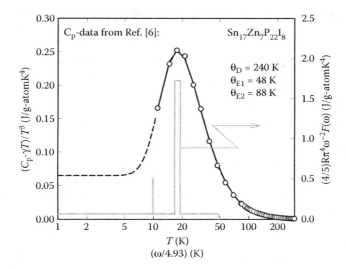

FIGURE 9.3 Temperature-dependent specific heat of $Sn_{17}Zn_7P_{22}I_8$ plotted as $(C_p-\gamma T)/T^3$ vs ln T. (Adapted from Kovnir, K. A. et al. 2005. *Solid State Sciences* 7: 957–968.) The black line indicates the fit to the model of Junod. (Adapted from Junod, A., D. Bichsel, and J. Muller. 1979. *Helvet. Phys. Acta* 52: 580-596; Junod, A., T. Jarlborg, and J. Muller. 1983. *Phys. Rev. B* 27: 1568–1585.) The spectral function $F(\omega)$ (shown in gray) was calculated using θ_D, θ_{E1}, θ_{E2} with a spectral width for the Einstein functions of 0.3 and 11.6 K.

for which a Debye temperature, $\theta_D = 220$ K, was calculated from an average value of all framework positions [26]. Figure 9.4 shows a fit to the data for $Sn_{24}P_{19.3}\square_{2.7}Br_8$ [26] applying the Debye model to the framework atoms and the Einstein model to the Br2 atoms located in the large cages [31]. The 16i-position was selected for the evaluation of the Debye temperature because the ADPs at this crystallographic position are usually unaffected by contributions from vacancy formation or splitting.

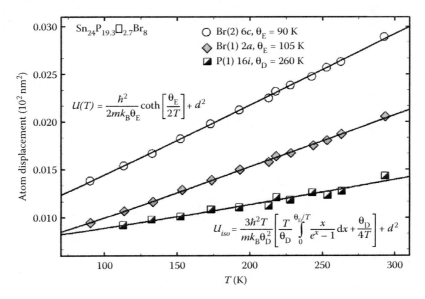

FIGURE 9.4 Atomic displacement parameters extracted from the data reported for $Sn_{24}P_{19.3}\square_{2.7}Br_8$. (Adapted from Kovnir, K. A. et al. 2004. *Inorg. Chem.* 43:3230–3236.) The solid lines show fits to the Debye- and Einstein model, using the formulae given in the graph.

The resulting value of $\theta_D = 260$ K is slightly higher than that extracted earlier [26], whereas the Einstein temperature of $\theta_E(Br_2) = 90$ K compares well with the results from specific heat data [31]. As the temperature dependence of the Br1-ADPs is slightly steeper than for the framework atoms, the Einstein model was tentatively applied for the Br1-atoms located in the pentagon dodecahedra resulting in a $\theta_E(Br1) = 105$ K. In contrast to most published data on Einstein temperatures here only isotropic instead of anisotropic displacement parameters had been available for the calculation. ADPs from x-ray powder data also served to define $\theta_D = 159$ K, $\theta_E = 52$ K for $Sn_{38}Sb_8I_8$ and $\theta_D = 185$ K, $\theta_E = 78$ K for $Ge_{38}Sb_8I_8$, respectively [19].

Although details on off-center motion in cationic clathrates have been scarce, a possible off-center position of Br/I-atoms in $Sn_{24}P_{19.3}\square_{2.7}Br_xI_{8-x}$ was excluded from x-ray single-crystal intensity data analyses [26]. Similarly, no sign for off-center motion was encountered for both iodine sites in $(Sn_{20.5}\square_{3.5})As_{22}I_8$; however, the As-atoms deviate from the center of the 6d and 16i positions and are located in $1/4 \times As1$ in 24k and $1/3 \times As2$ in 48l sites, respectively [32]. Unfortunately no specific heat curve is available to analyze the influence of the Sn-defects and the deflection of As-atoms from their center sites.

Low-frequency Raman peaks for $Si_{44}I_2I_8$ (75 and 101 cm^{-1}) [49], $Ge_{38}Sb_8I_8$ (53 and 58 cm^{-1}) and $Sn_{38}Sb_8I_8$ (29 and 35 cm^{-1}) [50] were attributed to iodine rattling modes in the large and smaller cages, whereas frequency peaks at 120–500, 75–273, and 40–195 cm^{-1}, respectively, were identified as framework vibrations. In contrast to the results by Raman scattering the phonon density of states (PDOS) obtained from neutron scattering experiments [51] did not identify any peaks coming from the vibration of iodine guests. The frequencies found for the rattlers in $Ge_{38}Sb_8I_8$ are in good agreement with the Einstein temperature of 78 K (=54.2 cm^{-1}; $\theta_D = 185$ K) derived from isotropic atomic displacement parameters (Rietveld refinement) [19]. The iodine Raman frequencies of 29 and 35 cm^{-1} for $Sn_{38}Sb_8I_8$ [50] are lower than those for $Ge_{38}Sb_8I_8$ indicating a weaker guest–host interaction in the larger Sn-cages.

Debye temperatures, Einstein temperatures and Raman data for anionic clathrates-I seem to compare well with those typical for germanium-based cationic clathrates with θ_E values of about 60 K, which for silicon-based clathrates range around 70 K. It should be noted that ADPs for the guest atoms in the smaller cages (pentagon dodecahedra) are practically identical with those of the host atoms. Although off-center motion of guest atoms have so far not been encountered in cationic clathrates, large ADPs of guest atoms in tetrakaidecahedra seem to contribute via phonon interaction with the rattling modes to a reduction of thermal conductivity.

9.5.2 Electrical Conductivity

Naturally, the electrical conductivity of all inverse clathrates hitherto measured exhibits semiconducting behavior (room temperature data are listed in Table 9.1). In most cases the temperature dependence can be described using the thermal activation model typical for intrinsic semiconductors (Equation 9.1),

$$\rho(T) = \rho_0 + C \exp \frac{\Delta E_{gap}}{2k_B T}$$

where ρ_0 is the resistivity at infinitely high temperature, ΔE_{gap} is the band gap energy and C is a constant. Plots of $\ln \sigma$ (electrical conductivity) versus T^{-1} in different temperature ranges, which are shown in Figure 9.5 (160–1000 K) and Figure 9.6 (10 to ~200 K), essentially confirm for most inverse clathrate compounds the expected linear temperature dependence. However, a few exceptions have been reported so far. In samples $Si_{46-x}P_xTe_y$ it was found that with higher P-content the character of conductivity changes from thermal activation type to the behavior of heavily doped semiconductors [9,8]. In the same system the Zintl-balanced clathrate-III compound $Si_{130}P_{42}Te_{21}$, shows the expected thermally activated semiconducting behavior [33]. Substitution of the iodine guest atoms in $Sn_{24}P_{19.3}\square_{2.7}Br_xI_{8-x}$ ($x = 0$–8) and $Sn_{24}P_{19.3}\square_{2.7}Cl_yI_{8-y}$ ($y \leq 0.8$) always leads to an increase of the band

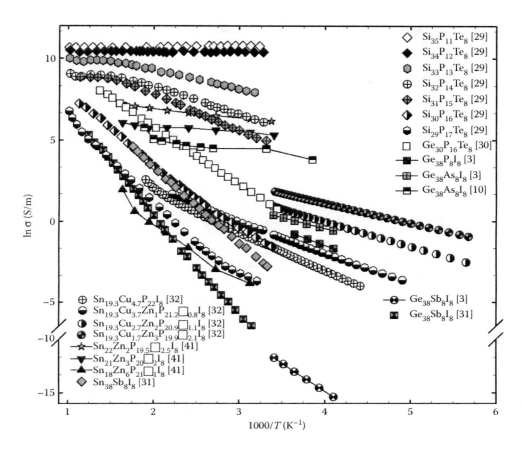

FIGURE 9.5 Electrical conductivity as a function of T^{-1} of inverse clathrates in a temperature range of 160–1000 K.

gap. In the case of the Br-substitution a linear change of the band gap with increasing Br-content from 0.03 to 0.14 eV was found [25]. Also the Cl-content increased ΔE slightly to 0.05 eV. Substitution of Br for I significantly reduced the electrical conductivity, a fact that was attributed to the chemical compression and its influence on the Sn-states near E_F [25]. Quantum mechanical calculations [5] demonstrated that for the hypothetical composition $Sn_{24}P_{19}\square_3I_8$ the main contribution to the density of states just below the Fermi-level originates from the 3 + 3 coordinated Sn-atoms in a split 24k site, which are generally observed in case of vacancy formation at the 6d-position. The increase of the band gap was explained by the shrinking of the unit cell, which changes the chemical environment of the aforementioned Sn-atoms [25]. In contrast to this, Kovnir et al. [6] found for the vacancy-free Zn-containing clathrate-I compounds $Sn_{17}Zn_7P_{22}Br_8$ (0.15 eV [6] or 0.11 eV [27]) and $Sn_{17}Zn_7P_{22}I_8$ (0.25 eV [6]) a much higher electrical conductivity for the Br-containing sample. Temperature and frequency dependences of the complex impedance in samples $Sn_{24}P_{19.3}\square_{2.7}Br_xI_{8-x}$ ($x = 0-8$) [52] revealed dielectric anomalies, which were attributed to a substantial contribution from the grain boundaries. A detailed investigation of the transport properties of $Sn_{24}P_{19.3}\square_{2.7}Br_8$ and $Sn_{17}Zn_7P_{22}Br_8$ at low temperatures also revealed significant differences in the temperature-dependent characteristics of electrical conductivity attributed to the diversity in the crystal structures [27], namely the presence of 3 + 3 coordinated Sn-atoms in $Sn_{24}P_{19.3}\square_{2.7}Br_8$. The rather dramatic change in the electrical conductivity in this compound (see Figure 9.6) was attributed to a change in the balance of charge carriers and is consistent with a sharp decrease in the Seebeck coefficient at 125 K (see Figure 9.9).

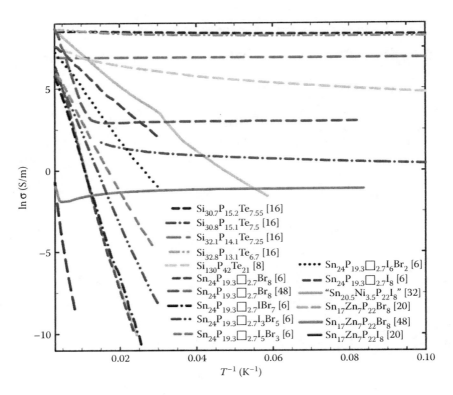

FIGURE 9.6 Electrical conductivity as a function of T^{-1} of inverse clathrates in a temperature range of 10 to ~200 K.

A more detailed investigation of the system $Sn_{24-x}Zn_xP_{22-y}\square_yI_8$ [31] above room temperature also revealed that a higher content of vacancies inside the framework improves electrical conductivity (Figure 9.7). Although the electrical resistivity as well as the band gap changes dramatically, the temperature dependency for all samples still shows the typical slope of intrinsic semiconductors. Similarly, in $Sn_{19.3}Cu_{4.7-x}Zn_xP_{22-y}\square_yI_8$ ($x = 1, 2, 3$) a change in the charge carrier concentration (from 1.3×10^{16} cm^{-3} for $x = 0$ to 6.2×10^{17} cm^{-3} for $x = 3$) and a decrease of the band gap can be achieved [7]. Charge carrier concentrations, estimated to be in the range of 10^{14}–10^{15} cm^{-3} for $Ge_{38}Sb_8I_8$ and $Sn_{38}Sb_8I_8$ [19] and 5×10^{17} cm^{-3} as experimentally verified for $Ge_{38}As_8I_8$ [17], indicate semiconducting behavior. Although the real mechanism of Ni-incorporation in "$Sn_{20.5}Ni_{3.5}P_{22}I_8$" remains still unclear, the charge carrier density calculated using the slope of $S(T)$ is in the order of 10^{20} cm^{-3}, which is already promising for thermoelectrics [31]. For the compound $(Sn_{20.5}\square_{3.5})As_{22}I_8$, which is the only one reported so far to form vacancies at the Sn-sites and thus crystallizes in an 8 times larger face-centered unit cell (see Figure 9.1), a rather large band gap of 0.45 eV was reported [32]. $Ge_{40.0}Te_{5.3}\square_{0.7}I_8$, which adopts a subgroup of $Pm\bar{3}n$, shows a thermal activation type electrical conductivity with an even larger ΔE_{gap} of 0.78 eV. For the electron-balanced compounds $Ge_{38}Sb_8I_8$, $Sn_{38}Sb_8I_8$, and $Ge_{30}P_{16}Te_8$ Kishimoto et al. [15,19] found activation energies of $E_a = (1/2)\Delta E_{gap} = 0.48, 0.40,$ and 0.31 eV, respectively. For equivalent substitutions the Sn-based clathrate $Sn_{38}Sb_8I_8$ exhibits a larger thermal conductivity than the Ge-based clathrate. According to the results of Menke and von Schnering [3], the electrical conductivity for $Ge_{38}A_8I_8$, with A = P, As and Sb decreases significantly from P to Sb. For $Si_{29}P_{17}Te_8$ a band gap of 1.24 eV was estimated from the temperature-dependent conductivity [8]. Addition of 6.5 at% of Ni in the compound "$Sn_{20.5}Ni_{3.5}P_{22}I_8$" [7] revealed a beneficial influence on the improvement of the electrical conductivity.

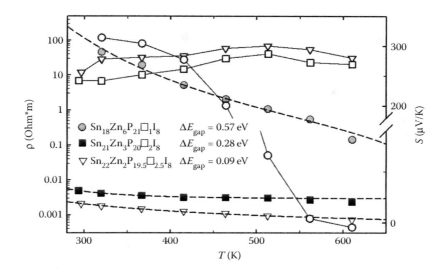

FIGURE 9.7 Electrical resistivity and Seebeck coefficient for samples $Sn_{24-x}Zn_xP_{22-y}\square_yI_8$. Filled symbols correspond to the left axis (electrical resistivity) and open symbols correspond to the right axis (Seebeck coefficient). The dashed lines represent fits to Equation 9.1. Black lines are guides to the eyes.

9.5.3 Thermal Conductivity

Figure 9.8 summarizes all data on thermal conductivity $\kappa(T)$ hitherto reported in the literature. Semiconducting clathrates generally exhibit a rather low thermal conductivity due to their rather complex crystal structure and the ability of phonon scattering via rattling of the guest atoms in the large voids. From Figure 9.8 it is obvious that the mean atomic mass of the compounds governs thermal conductivity. The particularly low thermal conductivity of $(Sn_{20.5}\square_{3.5})As_{22}I_8$ reaching the theoretical minimum is attributed to (i) a high degree of structural disorder by vacancy formation in random distribution, (ii) splitting of tin-atom positions as well as (iii) the shift of arsenic atoms from their ideal position in a face-centered $2 \times 2 \times 2$ superstructure of the clathrate type-I, and (iv) a random intergrowth of differently ordered structure domains [32]. As has been discussed in Section 5.1, both anionic and cationic clathrates show similar features in the ADPs for host and guest atoms. Consequently, large ADPs also in cationic clathrates refer to guest atoms rattling in the larger 24-vertex cages $[5^{12}6^2]$. A detailed evaluation in the system $Sn_{24}P_{19.3}\square_{2.7}Br_xI_{8-x}$ with $x = 0, 2, 4$ [25] showed that a random distribution of different guest atoms in the cages lowers thermal conductivity (see also Figure 9.8). Furthermore, substitutional atom disorder in the framework was shown to lead to a reduction of thermal conductivity from modeling of the thermal conductivity in terms of the Debye model for $Sn_{24}P_{19.3}\square_{2.7}Br_8$ and $Sn_{17}Zn_7P_{22}Br_8$ [27] taking into account scattering mechanisms (grain boundaries, point defects, Umklapp processes, and resonant scattering). Interestingly, and in contrast to $Sn_{17}Zn_7P_{22}Br_8$, almost no "rattling" effect was encountered for $Sn_{24}P_{19.3}\square_{2.7}Br_8$. This was attributed to the shrinking of the $[5^{12}6^2]$ cages in this compound caused by the vacancies [27]. Except for $Sn_{24}P_{19.3}\square_{2.7}Br_8$ and $Sn_{17}Zn_7P_{22}Br_8$ [27], a common feature of most cationic clathrates is the absence of a pronounced low-temperature maximum in $\kappa(T)$, characterizing, in general, simple metals and insulators of sufficiently low defect/impurity concentrations. Scattering of both, electrons and phonons on impurities, vacancies, grain boundaries, or other static defects significantly suppresses the initial rise of $\kappa(T)$, resulting in a practically featureless temperature dependency.

9.5.4 Seebeck Coefficients and *ZT*

As most inverse clathrates exhibit hole-dominated conductivity, efforts were made to synthesize corresponding materials with negative Seebeck coefficients. A compilation of all temperature-dependent

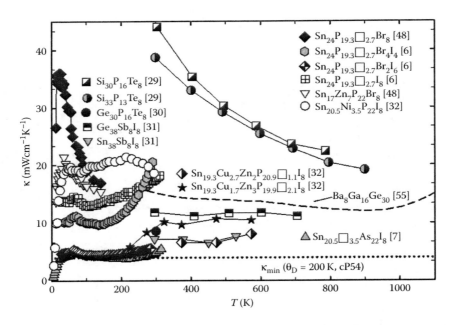

FIGURE 9.8 Thermal conductivity of inverse clathrates. For comparison the thermal conductivity of polycrystalline $Ba_8Ga_{16}Ge_{30}$ is given as a dashed line. (Adapted from Toberer, E. S. et al. 2008. *Phys. Rev.* B 77: 075203.) The theoretical minimum according to the expression by Cahill and Pohl [54] is indicated as a dotted line. Whereas a T^3-correction accounting for radiation losses was applied to the low-temperature data of "$Sn_{20.5}Ni_{3.5}P_{22}I_8$" [7]. Such a correction seems to be missing for $Sn_{24}P_{19.3}\square_{2.7}Br_4I_4$[25].

data on Seebeck coefficients (S) of inverse clathrate compounds hitherto available in the literature is presented in Figure 9.9. Room-temperature data are listed in Table 9.1. All compounds exhibiting a rather large electrical resistivity show a decrease of the Seebeck coefficient with increasing temperature. This can be attributed to the rise of a temperature-dependent charge carrier population in the conduction band. In $Si_{46-x}P_xTe_8$ another scenario can be identified for samples deviating substantially from the Zintl concept. These more metallic clathrates show an almost linear increase of S with temperature due to a nearly constant charge carrier concentration [8]. For three compounds, namely $Ge_{38}Sb_8I_8$, $Sn_{38}Sb_8I_8$ [19], and $Sn_{20.5}\square_{3.5}As_{22}I_8$ [32], n-type conductivity was reported. In agreement with the electrical conductivity data the Seebeck coefficient for $Ge_{38}Sb_8I_8$ is higher than for $Sn_{38}Sb_8I_8$. An increase of the Seebeck coefficient from 80 ($x = 0$) to 180 ($x = 8$) μV/K was achieved by substitution of iodine by bromine in $Sn_{24}P_{19.3}\square_{2.7}Br_xI_{8-x}$. Stefanoski et al. [27] found a sharp decrease of S at 125 K going to lower temperatures for $Sn_{24}P_{19.3}\square_{2.7}Br_8$. This is accompanied by a change in the balance of charge carriers in this temperature regime as already mentioned in Section 5.2.

The thermoelectric figure-of-merit ZT was reported in most cases only at a specific temperature and is listed in Table 9.1. ZT as a function of temperature for inverse clathrates is presented in Figure 9.10. Despite a high Seebeck coefficient as well as low thermal conductivity can be easily inferred, the usually low value of ZT for inverse clathrates can generally be attributed to their poor electrical conductivity [7,15,19]. Consequently for "$Sn_{20.5}Ni_{3.5}P_{22}I_8$" the improvement of ZT compared to the other Sn/P-based clathrates was achieved by reducing the electrical resistivity. For the compounds $Sn_{24}P_{19.3}\square_{2.7}Br_xI_{8-x}$ and $Sn_{20.5}\square_{3.5}As_{22}I_8$ moderate values of 80–180 and −180 μV/K diminish the ZT-value [25,32]. Although a later study reported for $Sn_{24}P_{19.3}\square_{2.7}Br_8$ an acceptable room-temperature power factor ($S^2\sigma$) of 130 W/mK^2, the ZT-values do not exceed 0.03 at 300 K [27]. The highest ZT-value hitherto reported for inverse

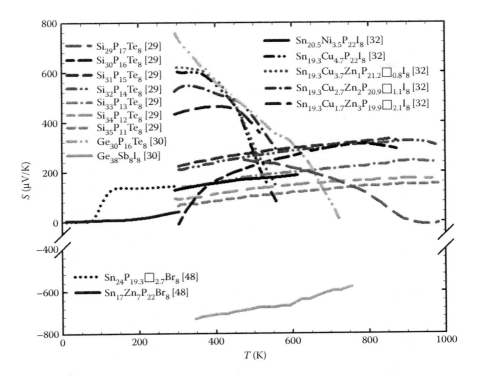

FIGURE 9.9 Temperature dependence of the Seebeck coefficient for inverse clathrates.

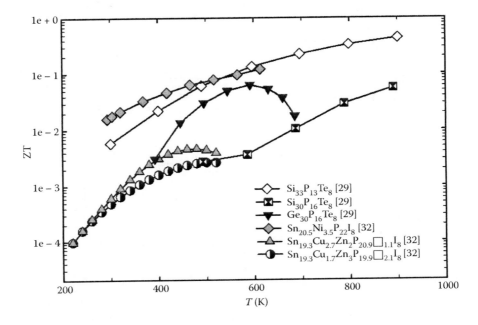

FIGURE 9.10 Thermoelectric figure of merit *ZT* of inverse clathrates.

clathrates was obtained for $Si_{33}P_{13}Te_8$ (0.45 at 900 K) [8]. In this system the rather high thermal conductivity impedes a further improvement of ZT to a value interesting for thermoelectric applications.

9.5.5 Mechanical Stability

Mechanical properties have been explored for a few inverse clathrates. Thermal expansion data are only available for two Sn/P-based samples: in both cases the thermal expansion coefficients of 10×10^{-6} K^{-1} (capacitance dilatometer) for $Sn_{19.3}Cu_{1.7}Zn_3P_{19.9}\square_{2.1}I_8$ and of 11×10^{-6} K^{-1} (single-crystal x-ray lattice parameters) for $Sn_{24}P_{19.3}\square_{2.7}Br_8$ compare well with the data compiled for anionic Sn-based clathrates such as $Cs_8Sn_{42}Zn_4$ [55] and $Cs_8Sn_{44}\square_2$ [55] (see Figure 9.11) [7].

Resonant ultrasound spectroscopy measurements (at RT) for $Sn_{19.3}Cu_{4.7}P_{22}I_8$ [31] revealed a Young's modulus of 84 GPa and a bulk modulus of 57 GPa. The isothermal bulk modulus $B_0 = 64.7$ GPa ($B' = 4.5$) for $Ge_{38}Sb_8I_8$ was derived from XRD measurements at high pressure (diamond anvil cell) by applying the Murnaghan equation of state [50]. The bulk modulus $B_0 = 95(5)$ GPa of $Si_{44.5}I_{1.5}I_8$ (Murnaghan fit to volume vs. pressure) [56] is in line with data gained from *ab initio* calculations (91 GPa for $Si_{46}I_8$ and 87 GPa for unstuffed Si_{46}) [57,58]. The kink at 17 GPa in the volume–pressure curve for $Si_{44.5}I_{1.5}I_8$ corresponds to a Si–Si distance of 0.230 nm and was taken as a stability limit for Si-cage structures [56]. The small volume reduction at 17 GPa together with practically unchanged x-ray spectra, however, was not indicative of a phase transition [56]. Bulk moduli were also calculated for hypothetical Te_8Si_{46} ($B_0 = 95$ GPa), for $Si_{46}Br_8$ ($B_0 = 92$ GPa), for $Si_{46}Sn_8$ and $Si_{46}Ge_8$ (both $B_0 = 97$ GPa) as well as for hypothetical type-II clathrate $Si_{136}I_{24}$ ($B_0 = 90$ GPa); the latter appears consistent with $B_0 = 86$ GPa for the empty cage Si_{136} [57]. The comparison of the bulk moduli of 52.8 GPa ($B' = 3.4$) for $Ba_8Ge_{43}\square_3$ and 64.7 GPa ($B' = 4.5$) for $Ge_{38}Sb_8I_8$ suggests that vacancies make the structure more compressible [50]. Raman scattering at high pressure showed that rattling frequencies in $Ge_{38}Sb_8I_8$ undergo frequency softening above 16 GPa due to weakening of the guest–host interaction [50]. At pressures above 36 GPa

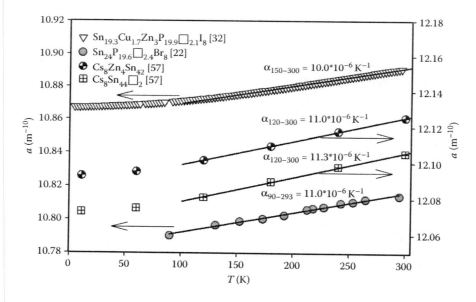

FIGURE 9.11 Thermal expansion of the two inverse clathrate compounds $Sn_{24}P_{19.6}\square_{2.7}Br_8$ and $Sn_{24}Cu_{1.7}Zn_3$ $P_{19.9}\square_{2.1}I_8$. For comparison temperature-dependent lattice parameters of two Sn-based polyanionic clathrates are plotted. (Adapted from Nolas, G. S. et al. 2000. *Chem. Mater.* 12: 1947–1953.) The black solid lines represent linear fits in the temperature range, where the thermal expansion was calculated.

amorphization starts and is completed at about 50 GPa for $Ge_{38}Sb_8I_8$ [50], at 47 GPa for $Si_{44}I_2I_8$ and at 40 GPa for Ba_8Si_{46}, respectively [59].

Although only a few data are available hitherto on the mechanical properties, in general the order of magnitude of elastic property data is the same for both polycationic as well as polyanionic clathrates.

9.5.6 Magnetic Properties

The magnetic susceptibility of several inverse clathrate compounds was measured in the temperature range from 4.2 to ~400 K [6,9,22,25,26,30,32,33] using a SQUID magnetometer. In most cases [6,22,25,26,30,32,33] a diamagnetic behavior as expected for Zintl phases was found. It was pointed out that a dominating core contribution of iodine to the diamagnetic susceptibility exists for $Sn_{17}Zn_7P_{22}I_8$ [6]. From a detailed analysis of the data [9,25], a significant diamagnetic structural contribution (χ_{struct}) appears, also reported for polyanionic clathrates [60], which was explained by molecular ring currents around the large cages in the clathrate-I structure. Low-temperature upturns in the magnetic susceptibilities for $Si_{46-x}P_xTe_y$ were attributed to paramagnetic impurities or paramagnetic point defects. Moreover, rather small values for the structural increment were found, which hint that other intrinsic paramagnetic contributions to χ_{struct} may play an important role. All together the measurement of the magnetic susceptibility confirmed that $Si_{46-x}P_xTe_y$ does not behave like a typical Zintl phase [9].

9.6 Theoretical Calculations

Band structure calculations for $Sn_{24}P_{22}I_8$, $Sn_{24}P_{19}\square_3I_8$, and $Sn_{10}In_{14}P_{22}I_8$ [5] showed that $Sn_{24}P_{22}I_8$ is a metal (with the Fermi-level E_F in the conduction band beyond a gap of about 0.1 eV), while $Sn_{24}P_{19}\square_3I_8$ and $Sn_{10}In_{14}P_{22}I_8$ are essentially semiconductors with E_F at the top of the valence band with a gap of about 0.01 eV for $Sn_{24}P_{19}\square_3I_8$ and about 0.05 eV for $Sn_{10}In_{14}P_{22}I_8$. The main contribution to the bands just below E_F mainly originates from the metal orbitals. As already mentioned in Chapter 5.2 the 3 + 3 coordinated Sn-atoms in the vacancy containing compound $Sn_{24}P_{19}\square_3I_8$ give rise to states just below E_F; the two sharp peaks directly below the Fermi level in $Sn_{10}In_{14}P_{22}I_8$ could be attributed to In [5]. Extended Hückel calculations for $Sn_{19.3}Cu_{4.7}P_{22}I_8$ suggested that besides p- and s-orbitals also the d-orbitals of Cu are contributing to form a tetrahedral bonding to the three neighboring P-atoms and one Cu-atom [29].

As a typical feature, ionic guest–host bonding was encountered due to weak interaction between the framework and iodine–guest atoms, which in the tetrakaidecahedral cages can carry a charge up to –0.95 [5]. Inelastic neutron scattering data for $Si_{46}M_8$ (M = Na, K, Ba) and $Si_{46-x}I_xI_8$ at 300 K reveal low-frequency molecular modes (fingerprints of rattling ions in the host cages) [51]. Whereas all guest atoms are strongly coupled to the host in the smaller cages, low-frequency molecular modes for the larger cages are gradually disappearing for the sequence M = Na, K, Ba in $Si_{46}M_8$ and $Si_{46-x}I_xI_8$ [51]. The guest–host coupling strength, calculated from ab initio (LDA) calculations increases for the sequence M = Na, K, Ba in $Si_{46}M_8$, however, was not evaluated for $Si_{46-x}I_xI_8$ [51].

The calculated band gaps in $Sn_{24}P_{19}\square_3I_8$ (0.01 eV) and $Ge_{30}P_{16}Te_8$ (0.29–0.48 eV) [5,15] compare well with the values gained from the temperature-dependent electrical conductivity measurements. An *ab initio* study (DFT-LDA) of Si_{46} and (hypothetical) $Si_{46}I_8$ clearly revealed the stabilization effect of the iodine atom at the cage center (by ~0.65 eV; no iodine dimers I_2) [57] and showed $Si_{46}I_8$ and $Si_{46}Te_8$ to be p-type semiconductors with a band gap of 1.6 eV and a Fermi level located at 0.26 and 0.68 eV, respectively, below the top of the valence band, which exhibits a substantial contribution from the 5p-guest atom orbitals. The cohesive energy of iodine inside the framework was evaluated to be 3.1 eV for $Si_{46}Te_8$ [57], 1.7 eV for $Si_{46}I_8$ [57], but only 0.7 eV for $Si_{46}Br_8$ [57]. Increasing hybridization of conduction band states (predominately from the bottom of the conduction bands) and the guest atom orbitals leads to an increase of the band gap reaching ~2 eV for the *p*-doped semiconductor $Si_{46}Br_8$ [57]. Real stoichiometry $Si_{46-x}I_xI_8$, however, may shift the Fermi level to the top of the valence band (insulating state from conductivity measurements) [51,57]. For hypothetical type-II clathrate $Si_{136}I_{24}$ the binding energy was calculated to be 1.65 eV (*p*-type semiconductor) [57].

9.7 Conclusion

Although several inverse clathrate compounds exhibit a remarkably high thermal stability, such as both clathrate-I and clathrate-III solutions in the system Si–P–Te [9,10,33,42], most inverse clathrates undergo thermal decomposition already above 500–600°C with high partial vapor pressures of halogens and/or pnictogens at even lower temperatures. The proximity to Zintl-based compositions renders inverse clathrates to be semiconductors with high Seebeck coefficients but accordingly high electrical resistivities. Similar to anionic clathrates, the complex crystal structure of inverse clathrates, stabilized by random framework atom substitution and/or vacancy formation with split sites on the next-nearest-neighbor coordination, results in generally low thermal conductivities. Although off-center motion of filler atoms have rarely been encountered, rattling of guest atoms, weakly bonded in the cages, supports the low thermal conductivity. As thermoelectric applications request $ZT > 1$, optimization procedures will have to focus on achieving high electrical conductivities either via proper chemical substitutions or via suitable secondary grain boundary phases at sufficiently high bulk Seebeck coefficients and low thermal conductivities (via nanostructuring).

References

1. Rogl, P. 2006. Formation and crystal chemistry of clathrates. In *CRC Handbook of Thermoelectrics*, ed. D.M. Rowe, pp. 32-1–32-24. CRC Press, Boca Raton, FL.
2. Shevelkov, A.V. 2008. Chemical aspects of the design of thermoelectric materials. *Russian Chemical Reviews* 77(1): 1–19.
3. Menke, H., and H. G. von Schnering. 1973. The cage compounds $Ge_{38}A_8X_8$ with A = P, As, Sb and X = Cl, Br, J. *Z. Anorg. Allg. Chem.* 395: 223–238.see also: 3a. Menke, H., and H. G. von Schnering. 1972. New clathrates of the type $Ge_{38}A_8X_8$ with A = P, As, Sb and X = Cl, Br, I. *Naturwiss* 59(9): 420. 3b. von Schnering, H. G., and H. Menke. 1972. $Ge_{38}P_8I_8$ and $Ge_{38}As_8I_8$, a new class of compounds with clathrate structure. *Angew. Chemie* 84: 30–31.
4. Kovnir, K. A., and A. V. Shevelkov. 2004. Semiconducting clathrates: Synthesis, structure and properties. *Russian Chemical Reviews* 73(9): 923–938.
5. Shatruk M. M., K. A. Kovnir, M. Lindsjoe, I. A. Presniakov, L. A. Kloo, and A. V. Shevelkov. 2001. Novel compounds $Sn_{10}In_{14}P_{22}I_8$ and $Sn_{14}In_{10}P_{21.2}I_8$ with clathrate I structure: Synthesis and crystal and electronic structure. *J. Solid State Chem.* 161: 233–242.
6. Kovnir, K. A., M. M. Shatruk, L. N. Reshetova et al. 2005. Novel compounds $Sn_{20}Zn_4P_{22-v}I_8$ ($v = 1.2$), $Sn_{17}Zn_7P_{22}I_8$, and $Sn_{17}Zn_7P_{22}Br_8$: Synthesis, properties, and spectral features of their clathrate-like crystal structures. *Solid State Sci.* 7: 957–968.
7. Falmbigl, M., P. Rogl, E. Bauer, M. Kriegisch, H. Müller, and S. Paschen. 2009. On the thermoelectric potential of inverse clathrates. *MRS Symposium Proceedings* 1106-N06-03.
8. Kishimoto, K., T. Koyanagi, K. Akai, and M. Matsuura. 2007. Synthesis and thermoelectric properties of type-I clathrate compounds $Si_{46-x}P_xTe_8$. *Jpn. J. Appl. Phys.* 46: L746–L748.
9. Zaikina, J. V., K. A. Kovnir, U. Burkhardt et al. 2009. Cationic clathrate I $Si_{46-x}P_xTe_y$ (6.6(1) $\leq y \leq 7.5(1)$, $x \leq 2y$): Crystal structure, homogeneity range, and physical properties. *Inorg. Chem.* 48(8): 3720–3730.
10. Zaikina, J. V., K. A. Kovnir, U. Schwarz, H. Borrmann, and A. V. Shevelkov. 2007. Crystal structure of silicon phosphorus telluride, $Si_{46-x}P_xTe_y$ ($y = 7.35, 6.98, 6.88$; $x \leq 2y$), a cationic clathrate-I. *Z. Kristallogr. NCS* 222: 177–179.
11. Jaussaud, N., P. Toulemonde, M. Pouchard et al. 2004. High pressure synthesis and crystal structure of two forms of a new tellurium–silicon clathrate related to the classical type I. *Solid State Sci.* 6:401–411.
12. Jaussaud, N., M. Pouchard, P. Gravereau, S. Pechev, G. Goglio, and C. Cros. 2005. Structural trends and chemical bonding in Te-doped silicon clathrates. *Inorg. Chem.* 44: 2210–2214.

13. Kovnir, K. A., A. N. Uglov, J. V. Zaikina, and A. V. Shevelkov. 2004. New cationic clathrate: Synthesis and structure of $[Si_{40}P_6]I_{6.5}$. *Mendeleev Commun.* 14(4): 135–136.

14. Reny, E., S. Yamanaka, C. Cros, and M. Pouchard. 2001. A new silicon clathrate compound: $I_8Si_{46-x}I_x$. *AIP Conference Proceedings* (*Nanonetwork Materials*, ed. by S. Saito et al.) 590: 499–502.

15. Kishimoto, K., K. Akai, N. Muraoka, T. Koyanagi, and M. Matsuura. 2006. Synthesis and thermoelectric properties of type-I clathrate $Ge_{30}P_{16}Te_8$. *Appl. Phys. Lett.* 89: 172106.

16. Menke, H., and H. G. von Schnering. 1976. The partial substitution of Ge with GaAs and GaSb in the cage compounds $Ge_{38}As_8I_8$ and $Ge_{38}Sb_8I_8$. *Z. Anorg. Allg. Chem.* 424: 108–114.

17. Chu, T. L., S. S. Chu, and R. L. Ray. 1982. Germanium arsenide iodide: A clathrate semiconductor. *J. Appl. Phys.* 53(10): 7102–7103.

18. Ayouz, K., M. Kars, A. Rebbah, and H. Rebbah. 2009. $I_8As_{21}Ge_{25}$. *Acta Crystallogr.* E 65: i15.

19. Kishimoto, K., S. Arimura, and T. Koyanagi. 2006. Preparation and thermoelectric properties of sintered iodine-containing clathrate compounds $Ge_{38}Sb_8I_8$ and $Sn_{38}Sb_8I_8$. *Appl. Phys. Lett.* 88: 222115.

20. Jin, Z., Z. Tang, A. Litvinchuk, and A. M. Guloy. 2008. Inverse clathrate with a significant nonstoichiometric compositional range. *Abstracts of Papers, 235th ACS National Meeting*, New Orleans, LA, USA, April 6–10.

21. Kars, M., T. Roisnel, V. Dorcet, A. Rebbah, and L. C. Otero-Diaz. 2010. $I_8Sb_{10}Ge_{36}$. *Acta Crystallogr.* E 66: i47.

22. Kovnir, K. A., N. S. Abramchuk, J. V. Zaikina et al. 2006. $Ge_{40.0}Te_{5.3}I_8$: Synthesis, crystal structure, and properties of a new clathrate-I compound. *Z. Kristallogr.* 221: 527–532.

23. Nesper, R., J. Curda, and H. G. von Schnering. 1986. $Ge_{4.06}I$, an unexpected germaniumsubiodide— A tetragermanioiodonium(III)-iodide with clathrate structure $[Ge_{46-x}I_x]I_8$, $x = 8/3$. *Angew. Chemie* 25: 369.

24. Shatruk, M. M., K. A. Kovnir, A. V. Shevelkov, I. A. Presniakov, and B. A. Popovkin. 1999. First tin pnictide halides $Sn_{24}P_{19.3}I_8$ and $Sn_{24}As_{19.3}I_8$: Synthesis and the clathrate-I type of the crystal structure. *Inorg. Chem.* 38: 3455–3457.

25. Zaikina, J. V., W. Schnelle, K. A. Kovnir, A. V. Olenev, Y. Grin, and A. V. Shevelkov. 2007. Crystal structure, thermoelectric and magnetic properties of the type-I clathrate solid solutions $Sn_{24}P_{19.3(2)}$ Br_xI_{8-x} ($0 \le x \le 8$) and $Sn_{24}P_{19.3(2)}Cl_yI_{8-y}$ ($y \le 0.8$). *Solid State Sci.* 9: 664–671.

26. Kovnir, K. A., J. V. Zaikina, L. N. Reshetova, A. V. Olenev, E. V. Dikarev, and A. V. Shevelkov. 2004. Unusually high chemical compressibility of normally rigid type-I clathrate framework: Synthesis and structural study of $Sn_{24}P_{19.3}Br_xI_{8-x}$ solid solution, the prospective thermoelectric material. *Inorg. Chem.* 43:3230–3236.

27. Stefanoski, S., L. N. Reshetova, A. V. Shevelkov, and G. S. Nolas. 2009. Low-temperature transport properties of $Sn_{24}P_{19.3}Br_8$ and $Sn_{17}Zn_7P_{22}Br_8$. *J. Electron. Mat.* 38(7): 985–989.

28. Kelm, E. A., J. V. Zaikina, E. V. Dikarev, and A. V. Shevelkov. 2009. Distribution of phosphorus and arsenic atoms in the solid solution $Sn_{24}As_xP_{19.3-x}I_8$ with the structure of clathrate-I. *Russ. Chem. Bull. Int. Ed.* 58: 746–750.

29. Shatruk, M. M., K. A. Kovnir, A. V. Shevelkov, and B. A. Popovkin. 2000. A new zintl phase $Sn_{19.3}Cu_{4.7}P_{22}I_8$ with a structure of the clathrate-I type: Target-directed synthesis and structure. *Russ. J. Inorg. Chem.* 45(2): 153–159. Translated from *Zh. Neorganicheskoi Khimii* 45(2): 203–209.

30. Kovnir, K. A., A. V. Sobolev, I. A. Presniakov et al. 2005. $Sn_{19.3}Cu_{4.7}As_{22}I_8$: A new clathrate-I compound with transition-metal atoms in the cationic framework. *Inorg. Chem.* 44: 8786–8793.

31. Falmbigl, M. 2010. PhD thesis, University of Vienna.

32. Zaikina, J. V., K. A. Kovnir, A. V. Sobolev et al. 2007. $Sn_{20.5}\square_{3.5}As_{22}I_8$: A largely disordered cationic clathrate with a new type of superstructure and abnormally low thermal conductivity. *Chem. Eur. J.* 13: 5090–5099.

33. Zaikina, J. V., K. A. Kovnir, F. Haarmann et al. 2008. The first silicon-based cationic clathrate III with high thermal stability: $Si_{172-x}P_xTe_y$ ($x = 2y$, $y > 20$). *Chem. Eur. J.* 14: 5414–5422.

34. Jaussaud, N., M. Pouchard, G. Goglio et al. 2003. High pressure synthesis and structure of a novel clathrate-type compound: $Te_{7+x}Si_{20-x}$ ($x \sim 2.5$). *Solid State Sci.* 5: 1193–1200.

35. Mudryk, Ya., P. Rogl, C. Paul et al. 2002. Thermoelectricity of clathrate I Si and Ge phases. *J. Phys.: Condens. Matter* 14: 7991–8004.

36. Dünner, J., and A. Mewis. 1995. $Ba_8Cu_{16}P_{30}$-a new ternary variant of the clathrate I-type structure. *Z. Anorg. Allg. Chem.* 621: 191–196.

37. Carrillo-Cabrera, W., S. Budnyk, Y. Prots, and Y. Grin. 2004. Ba_8Ge_{43} revisited: A $2a' \times 2a' \times 2a'$ superstructure of the clathrate-I type with full vacancy ordering. *Z. Anorg. Allg. Chem.* 630: 2267–2276.

38. Blake, N. P., D. Bryan, S. Lattumer, L. Mollnitz, G. D. Stucky, H. Metiu. 2001. Structure and stability of the clathrates $Ba_8Ga_{16}Ge_{30}$, $Sr_8Ga_{16}Ge_{30}$, $Ba_8Ga_{16}Si_{30}$, and $Ba_8In_{16}Sn_{30}$. *J. Chem. Phys.* 114(22): 10063–10074.

39. Gimarc, B.M. 1983. Topological charge stabilisation. *J. Am. Chem. Soc.* 105: 1979–1984.

40. Miller, G. J. 1996. Structure and bonding at the Zintl border. In *Chemistry, Structure and Bonding of Zintl Phases and Ions*, ed. S. M. Kauzlarich, pp. 1–59. Wiley-VCH Publishers Inc., NY.

41. Smits, A., and S. C. Brockhorst. 1916. The system phosphorus from the point of view of the theory of allotropy. *Z. Physik. Chem.* 91: 249–312.

42. Philipp, F., and P. Schmidt. 2008. The cationic clathrate $Si_{46-2x}P_{2x}Te_x$ crystal growth by chemical vapour transport. *J. Crystal Growth* 310: 5402–5408.

43. Toberer, E. S., A. F. May, and G. J. Snyder. 2010. Zintl chemistry for designing high efficiency thermoelectric materials. *Chem. Mater.* 22: 624–634.

44. Sales, B. C., B. C. Chakoumakos, and D. Mandrus. 1999. Atomic displacement parameters and the lattice thermal conductivity of clathrate-like thermoelectric compounds. *J. Solid State Chem.* 146: 528–532.

45. Christensen, M., S. Johnsen, and B. B. Iversen. 2010. Thermoelectric clathrates of type I. *Dalton Trans.* 39: 978–992.

46. Junod, A., D. Bichsel, and J. Muller. 1979. Modification of the acoustic phonon spectra as a function of electronic state density in the A 15 type superconductor. *Helvet. Phys. Acta* 52: 580–596.

47. Junod, A., T. Jarlborg, and J. Muller. 1983. Heat-capacity analysis of a large number of A15-type compounds. *Phys. Rev. B* 27: 1568–1585.

48. Melnychenko-Koblyuk, N., A. Grytsiv, L. Fornasari et al. 2007. Ternary clathrates Ba–Zn–Ge: Phase equilibria, crystal chemistry and physical properties. *J. Phys.: Condens. Matter* 19: 216223.

49. Shimizu, H., T. Kume, T. Kuroda, S. Sasaki, H. Fukuoka, and S. Yamanaka. 2003. High-pressure Raman study of the iodine-doped silicon clathrate $I_8Si_{44}I_2$. *Phys. Rev. B* 68: 212102.

50. Shimizu, H., R. Oe, S. Ohno et al. 2009. Raman and x-ray diffraction studies of cationic type-I clathrate $I_8Sb_8Ge_{38}$: Pressure-induced phase transitions and amorphization. *J. Appl. Phys.* 105: 043522.

51. Reny, E., A. San Miguel, Y. Guyot et al. 2002. Vibrational modes in silicon clathrate compounds: A key to understanding superconductivity. *Phys. Rev. B* 66: 014532.

52. Yakimchuk, A.V., J. V. Zaikina, L. N. Reshetova, L. I. Ryabova, D. R. Khokhlov, and A. V. Shevelkov. 2007. Impedance of $Sn_{24}P_{19.3}Br_xI_{8-x}$ semiconducting clathrates. *Low Temp. Phys.* 33(2–3): 276–279.

53. Toberer, E. S., M. Christensen, B. B. Iversen, and G. J. Snyder. 2008. High temperature thermoelectric efficiency in $Ba_8Ga_{16}Ge_{30}$. *Phys. Rev. B* 77: 075203.

54. Cahill, D. G., and R. O. Pohl. 1989. Heat flow and lattice vibrations in glasses. *Solid State Commun.* 70:927–930.

55. Nolas, G. S., B. C. Chakoumakos, B. Mahieu, G. J. Long, and T. J. R. Weakley. 2000. Structural characterization and thermal conductivity of type-I tin clathrates. *Chem. Mater.* 12: 1947–1953.

56. San Miguel, A., P. Melinon, D. Connetable et al. 2002. Pressure stability and low compressibility of intercalated cagelike materials: The case of silicon clathrates. *Phys. Rev. B* 65: 054109.

57. Connetable, D. 2007. Structural and electronic properties of p-doped silicon clathrates. *Phys. Rev. B* 75: 125202.

58. Connetable, D., V. Timoshevskii, E. Artacho, and X. Blase. 2001. Tailoring band gap and hardness by intercalation: An *ab initio* study of I8@Si-46 and related doped clathrates. *Phys. Rev. Letters* 87(20): 206405.

59. San Miguel, A., A. Merlen, P. Toulemonde et al. 2005. Pressure-induced homothetic volume collapse in silicon clathrates. *Europhys. Lett.* 69(4): 556–562.

60. Paschen, S., V. H. Tran, M. Baenitz, W. Carrillo-Cabrera, Yu. Grin, and F. Steglich. 2002. Clathrate Ba_6Ge_{25}: Thermodynamic, magnetic, and transport properties. *Phys. Rev. B* 65: 134435.

10

Recent Advances in the Development of Efficient N-Type Skutterudites

Ctirad Uher
University of Michigan

10.1 Introduction

Skutterudites are binary compounds with the chemical formula MX_3, where M = Co, Rh, Ir and the pnicogen atom X = P, As, Sb. Since the mid-1990s, they have been intensively studied as one of the novel families of prospective thermoelectric materials. It is now more than 5 years since the last *CRC Handbook on Thermoelectric Materials* [1] was published in which I reported on the progress and perspectives of skutterudites as power-generating thermoelectric materials for the intermediate range of temperatures 500–900 K. Since that time, major advances have been made in the development of n-type skutterudites with the dimensionless figure of merit now approaching and, perhaps even exceeding, values as high as 1.5. The progress made is a reflection of the world-wide interest in this family of thermoelectrics and the firm belief that skutterudites have a great prospect to make a major impact in the area of waste industrial heat recovery and specifically as it applies to automotive operations. Now that reliable measurements of mechanical properties are also at hand, engineers and scientists recognize the advantages skutterudites offer in terms of their robust structure that bodes well for the fabrication of modules expected to withstand the harsh environment of high temperatures and mechanical stresses.

In this chapter I attempt to update the reader about the advances made in understanding the filling process of structural voids in the skutterudite lattice that led to rational choices of filler species including advantages of filling with more than one kind of a filler, the importance of nanostructuring in the fabrication of efficient, skutterudite-based thermoelectric materials, and the development of innovative synthesis routes such as melt-spinning to prepare bulk skutterudites in processing times an order of magnitude shorter than those required in the traditional synthesis relying on melting and long-term annealing. I will also discuss interesting alternative approaches such as partial compensation and compensating double-doping on the pnicogen rings, both of which yielded ZT values in excess of unity, and that should be considered as part of the grand strategy of developing outstanding skutterudite materials.

10.2 Structure, Nature of Filling, and Thermal Transport

Skutterudites crystallize with the body-centered-cubic structure in the space group Im3. Invariably, we shall be concerned only with cobalt triantimonide-based skutterudites ($CoSb_3$) since of all possible skutterudite compounds they have not only the best thermoelectric properties but their chemical constituents are also the least expensive, readily available, and environmentally benign. In the skutterudite structure the Co atoms occupy the so-called c-sides while antimony is located at the g-sites of the structure, (see Figure 10.1). The most characteristic and unique feature of the skutterudite lattice are the near-square planar rings of Sb atoms depicted in Figure 10.1. The location of the Sb atom is given by two positional parameters that, together with the lattice parameter, fully specify the skutterudite structure [2]. In Figure 10.1, we note the presence of six such antimony rings in a large square divided into eight smaller squares. This implies that there are two small squares, one at the upper left front and another at the lower right back, which do not contain a ring of Sb atoms. In other words, the skutterudite lattice contains two structural voids, often called cages. Thus, instead of $CoSb_3$, one can equivalently describe cobalt triantimonide as $\square_2Co_8Sb_{24} = 2\square Co_4[Sb_4]_3$ where the symbol $\square$ represents the void and $[Sb_4]$ highlights the presence of four-member Sb rings. It is customary to take only one-half of the unit cell including a single void and designate the skutterudite structure as $\square Co_4Sb_{12}$. Since each Co atom contributes nine electrons and each Sb atom contributes three electrons, this designation leads to the valence electron count of 72 (or 144 for the entire unit cell) which reflects the truly diamagnetic and semiconducting character of the binary skutterudite structure.

The bonding arrangement is appreciated by inspecting Figure 10.2 which depicts the true unit cell of the skutterudite structure centered at the position of the filler atom at (000) (position 2a). The figure highlights the octahedral bonding of the Co atom and the formation of tilted $CoSb_6$ octahedrons sharing corners with the neighboring octahedrons. The planar Sb_4 ring structure of antimony atoms is a consequence of the tilt of these octahedrons. Details concerning bonding can be found in the original paper by Dudkin [3] and elsewhere [1,2].

As was first shown by Jeitschko and Brown [4], the voids can be filled with foreign species that occupy the body-centered sites of the cubic lattice resulting in compounds called filled skutterudites* and designated as

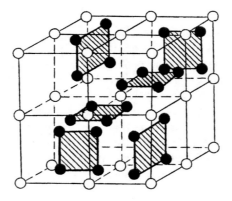

FIGURE 10.1 The unit cell of $CoSb_3$ shifted by one-quarter distance along the body diagonal. (Reprinted from Uher, C., Skutterudites: Promising power conversion thermoelectric. In *Proceedings of the 22nd International Conference on Thermoelectrics*, pp. 42–47. IEEE Catalog Number 03TH8726, Piscataway, NJ, 2003. Copyright 2003 by IEEE. With permission.)

---

* Strictly speaking, it is not the neutral $[Co_4Sb_{12}]$ complex but the charged $[T_4Sb_{12}]^{4-}$ complex where T is the group 8 metal such as Fe rather than the group 9 Co that forms the basis of the filled skutterudites. The semiconducting and diamagnetic form of the filled skutterudite $R^{4+}[T_4Sb_{12}]^{4-}$ with the valence electron count of 72 then requires a tetravalent electropositive filler R^{4+}. Detailed discussion can be found, e.g., in Refs. [1] and [2].

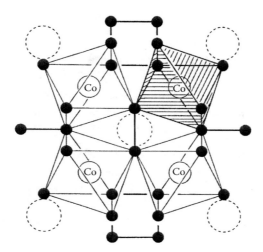

FIGURE 10.2 The unit cell of the skutterudite structure depicting the octahedral coordination of Co by Sb atoms with one octahedron highlighted. Dashed circles represent the filler positions. (Reprinted from Uher, C., Skutterudites: Promising power conversion thermoelectric. In *Proceedings of the 22nd International Conference on Thermoelectrics*, pp. 42–47. IEEE Catalog Number 03TH8726, Piscataway, NJ, 2003. Copyright 2003 by IEEE. With permission.)

$R_yCo_4Sb_{12}$ where R stands for the filler atom and y indicates the fractional occupancy of the available void sites. With typical fillers such as rare earths or alkaline earths, where the fillers enter the lattice not as neutral atoms but as ions, the structure becomes rapidly saturated with electrons and this leads to limitations on the void occupancy called filling fraction limits (FFL). In pure $CoSb_3$, the FFL can be as low as 10% in the case of trivalent Ce [5] and as high as 45% in the case of divalent Ba [6]. Using high-pressure synthesis, it has been reported that Sn can attain 100% occupancy of the void sites in $CoSb_3$ [7]. Filling the voids in $CoSb_3$ is thus an exceptionally effective doping mechanism whereby the density of electrons can be enhanced by 2–4 orders of magnitude depending on the degree of filling. If filling is the only structural modification, such filled skutterudites become strongly n-type conductors with carrier densities in the range 10^{20}–10^{22} cm^{-3}. One can reasonably ask how it is that the materials with such near-metallic carrier densities can be considered good thermoelectrics? The answer lies in the existence of flat bands at the bottom of the conduction band derived from the filler (particularly in the case of rare-earths fillers) that result in heavy effective masses that in turn support a large Seebeck coefficient [8].

The filling fraction limits have been established for most of the fillers of interest by painstaking experimental work during the past dozen or so years (see Ref. [1]). While the empirical evidence indicated that the charge state and the size of the filler are among the important factors determining FFL, until recently, no theoretical underpinning supporting the findings existed. Nor was it clear whether the maximum FFL is an intrinsic property of the filler or is influenced by external factors such as a particular synthesis process. Recently, using the density functional methods, Shi et al. [9,10] carried out detailed analysis of FFL for various impurities in the $CoSb_3$ matrix. Making a reasonable assumption that the FFL will be reached when the formation enthalpy of the filler in $CoSb_3$ turns positive, that is, when the filled $CoSb_3$ becomes unstable, the calculations revealed that the formation enthalpies remain negative even at full filling, in contrast to the experimental data. This led to a conclusion that factors other than just the interaction of the filler with the host material are influential in setting up the FFL. Among the key factors, the formation of secondary phases (e.g., $CoSb_2$, $RCoSb_2$) between the filler and the host atoms seemed most plausible. Proceeding to set up the respective formation enthalpies and Gibbs free energies, Shi et al. eventually established the FFL by minimizing the Gibbs free energy with respect to the filling fraction y and came to an important conclusion: an impurity atom R will be able to fill the void of the $CoSb_3$ structure if its Pauling electronegativity with respect to that of the electronegativity of Sb satisfies the following simple inequality:

$$X_{Sb} - X_R > 0.80. \tag{10.1}$$

The authors compared their theoretically estimated filling fraction limits with those determined by experiment and found very good agreement across all rare-earth and alkaline-earth filler species. The above simple criterion thus provides an excellent and powerful guiding principle to determine which species are likely to fill structural voids in the antimony-based skutterudite lattice. It is also interesting to note that electronegativity of constituent elements once again enters (the reader may recall a comment by G. A. Slack [11] concerning the important role of small electronegativity difference plays in forming a good thermoelectric material) as the key ingredient in understanding thermoelectric properties of solids even if, in this case, the inequality is reversed.

Equation 10.1 should govern any element intended to fill the voids in $CoSb_3$ even if more than one species is often desirable. The idea behind multifilling is to scatter as broad a spectrum of normal phonon modes as possible. In this respect, a question what are the best combinations of fillers has been addressed by the theoretical work of Yang et al. [12], who used a lattice-dynamical model based on density-functional calculations to obtain resonant frequencies and spring constants of a large number of filler species including rare earths, alkaline earths, and alkali elements. The results are summarized in Table 10.1 and have served well as a guiding principle during the past couple of years in the development of highly efficient, multifilled, *n*-type skutterudites [13,14].

It has been known since the mid-1950s [15] that binary skutterudites possess very high carrier mobilities, particularly as it concerns holes. Coupled with rather large effective masses, the thermoelectric power factor of binary skutterudites looked quite impressive and presented a tantalizing opportunity to consider binary skutterudites as potentially prospective thermoelectric materials. The initial enthusiasm, however, was tuned down quickly when the data on the thermal conductivity became available and indicated very high, phonon-dominated thermal conductivity of the order of 10 W/m-K or more [16,17]. Viewing skutterudites as an example of this Phonon–Glass–Electron–Crystal (PGEC) paradigm, Slack suggested [11] that filler species in the skutterudite lattice might act as independent, Einstein-like oscillators with large atomic displacements ("rattlers") that should scatter heat carrying acoustic phonon modes and thus reduce the lattice thermal conductivity. Without this idea, the development of skutterudites as novel thermoelectric materials would have likely been arrested or at least much delayed. The corner stone of the PGEC

TABLE 10.1 Spring Constant k and Resonance Frequency ω_o in the [111] and [100] Directions for a Number of Fillers in Antimonide Skutterudite

R	Mass (10^{-26} kg)	[111]		[100]	
		k (N/m)	ω_o (cm^{-1})	k (N/m)	ω_o (cm^{-1})
La	23.07	36.10	66	37.42	68
Ce	23.27	23.72	54	25.18	55
Eu	25.34	30.16	58	31.37	59
Yb	28.74	18.04	42	18.18	43
Ba	22.81	69.60	93	70.85	94
Sr	14.55	41.62	90	42.56	91
Na	3.819	16.87	112	17.18	113
K	6.495	46.04	141	46.70	142

Source: Reprinted with permission from Yang, J. et al., Dual-frequency resonant phonon scattering in $Ba_xR_yCo_4Sb_{12}$ (R = La, Ce, and Sr), *Appl. Phys. Lett.*, 90, 192111-1–192111-3. Copyright (2007), American Institute of Physics.

paradigm, as originally assumed to apply to skutterudites and other cage-like structures such as clathrates, was the notion of fillers strongly interacting with the acoustic phonons while having little impact on the electronic transport. As even a casual look at the effect of fillers on the carrier transport immediately reveals, the damage to the mobility of charged carriers is significant and skutterudites should not be viewed as a good example of the PGEC concept. Nevertheless, the idea of fillers acting as rattlers took hold and has been the driving force for the development of skutterudites as novel thermoelectric materials ever since. In fact, several early experiments provided an undeniable support for the rattling theory. Foremost among them were thermal transport measurements [18–21] that clearly demonstrated an order of magnitude reduction in the thermal conductivity upon filling the skutterudite lattice. Other studies [22–24] also provided strong supporting evidence for fillers in the skutterudite structure acting as rattlers. Inelastic neutron scattering [24] as well as nuclear inelastic scattering [25] identified the localized vibration modes in filled skutterudites and depicted them as essentially harmonic in nature. While the question as to how a harmonic mode of the filler can interact with the acoustic phonons and reduce the lattice thermal conductivity has been intriguing and not yet fully answered, a possible scenario was identified. It postulated inelastic resonant scattering [26] in which a small amount of energy of the interacting acoustic phonon is trapped in the excited state of a rattler which subsequently decays into phonons with wave vectors incoherent with respect to the wave vector of the absorbed phonon. In this picture, it is assumed that only acoustic phonons with frequencies near the frequency of the localized rattler can interact strongly with the localized mode. Thus, the scenario of fillers acting as independent, uncorrelated oscillators seemed plausible.

In 2006, during the 25th International Conference on Thermoelectrics in Vienna, Koza et al. [27] presented an invited talk on the dynamics of the filler species in the fully filled $LaFe_4Sb_{12}$ and $CeFe_4Sb_{12}$ based on their careful high-resolution inelastic neutron scattering experiments that seriously questioned the idea of fillers acting as localized rattlers. In this presentation and the subsequent paper [28] the authors argued that no direct microscopic evidence is available showing that the motion of the fillers is decoupled from that of the acoustic phonons and the purported experimental evidence for rattlers is based on measurements that do not address the key issues such as phase coherence, collectivity and coupling of modes. According to Koza et al., the earlier low-resolution density-of-states data reflect chiefly the strong contributions from van Hove singularities, that is, optical phonons and zone boundary phonons. Using inelastic neutron scattering facilities at the European neutron source Institute Laue Langevin in Grenoble, Koza et al. performed time-of-fight measurements and augmented their experiments with *ab initio* calculations of vibrational properties. Based on their results they concluded that their studies provide unequivocal evidence of quasi-harmonic, coherently coupled motions of the filler with the framework atoms which is in contrast to the notion of independently rattling filler species.

It is somewhat unfortunate that the skutterudite compounds in Refs. [27] and [28] were fully filled Fe-based antimonide skutterudites rather than partially filled compounds with a random spatial distribution of fillers. Neither are fully filled structures the compositions used in the thermoelectric forms of skutterudites, nor is the implicitly strong coupling between the filler species of fully filled skutterudites reflective of weak coupling in only partially occupied voids of $CoSb_3$ subjected to FFL. Therefore, the findings of Koza et al. need not necessarily be relevant to $R_yCo_4Sb_{12}$ ($y < 1$) and one should be careful before dismissing the rattling picture outright. Indeed, the most recent studies by Shi et al. [14] on triple-filled $CoSb_3$ with the overall void occupancy not exceeding 20% that used a very similar technique of powder-averaged inelastic neutron scattering spectra gathered with the time-of-flight spectrometer at the NIST Center for Neutron Research are in sharp contrast with the findings of Koza et al. The results unambiguously indicate the phase incoherence of the filler vibrations that follows from the quadratic dependence of the difference scattering function $\Delta S(Q,E)$ on the wave vector transfer Q, one of the key criterions for the filler vibrations being incoherent [28,29]. Shi et al. put forward a suggestion that there is a characteristic length scale corresponding to the average nearest inter-filler distance that corresponds to the effective screening length of 12.6 Å [9] which in terms of the fractional occupancy corresponds to 29% filling. For filler occupancies less than 29% (inter-filler separations greater than 12.6 Å), interactions are weak giving rise to phase incoherence in the motion of the fillers. If, on the other hand, the distance between the fillers is smaller than 12.6 Å, the vibrations develop phase coherence. Clearly, more work is needed to settle the issue whether the fillers do or do not behave as independent vibrational entities and ascertain the boundary conditions that delineate the two distinctly different modes of behavior.

One should also note that very recent *ab initio* calculations coupled with molecular dynamics simulations [30] of the phonon conductivities in $CoSb_3$ and $Ba_xCo_4Sb_{12}$ indicated significant changes in the bond lengths upon filling with the most notable being the increase in the short Sb–Sb bond on the Sb_4 ring. Such bond softening caused by the presence of the filler has a strong influence on the phonon density of states and the overall vibration spectrum, and leads to flattening of the acoustic branches and a reduction in the cutoff frequency. This, in turn, results in around 30% decrease in the group velocity of acoustic branches. Moreover, the frequency and group velocity of optical branches are also reduced. As a consequence, bond softening on account of filling the skutterudite voids may be a contributing factor to the overall decrease in the lattice thermal conductivity observed in filled skutterudites.

10.3 Nanostructuring

As is well known, a good thermoelectric material should have high electrical conductivity, large Seebeck coefficient, and poor ability to conduct heat. Detailed discussions concerning what kind of materials, what type of the band structure, and what scattering mechanisms are favorable to yield a high-performing thermoelectric can be found elsewhere [31–34]. Here we are interested in skutterudites as power-generating thermoelectric materials and thus we are concerned about their properties at temperatures above the ambient, that is, above their Debye temperature θ_D (~320 K). In comparison with low temperatures where minute structural changes turn into spectacular changes in the transport behavior, the regime of high temperatures ($T > \theta_D$) is relatively insensitive and the options are limited. Apart from the general recipe recommending maximum disruption to the phonon system consistent with the minimal impact on the mobility of charge carriers and making use of doping to tune the electronic properties, until recently, there was not much one could do to improve thermoelectric properties.

In 1993, Hicks and Dresselhaus [35,36] considered the role of dimensionality on the thermoelectric properties and showed that lower-dimensional structures can have significantly enhanced figures of merit in comparison with the usual bulk thermoelectrics. When one or more spatial dimensions are comparable to the characteristic length scale of electrons and phonons, new physical phenomena emerge that are uncommon or difficult to realize in the 3-D world, and they offer great opportunities for improving thermoelectric properties. As far as the charge carrier system is concerned, phenomena such as carrier confinement and energy filtering [37,38] may alter band structure, energy dependence of the density of states, or preferably filter out the more energetic charge carriers and thus play a positive role in enhancing the Seebeck coefficient. At the same time, high surface-to-volume ratio associated with nanostructural features and numerous interfaces that inevitably arise in structures built up of nanometer-scale structural units greatly enhance opportunities for phonon scattering with a consequent reduction in the parasitic influence of the lattice thermal conductivity.

It is unrealistic to think that we can make practical power-generating thermoelectric modules based on bunches of thin wires or thin-film superlattice structures. The cost would be prohibitive and the stability of such devices at high temperatures is highly questionable. Moreover, some of the spectacular values of *ZT* reported on various MBE-grown quantum dot structures [39] have since been "renormalized" to more mundane values [40,41]. On the other hand, if we could import some of the clearly beneficial features of the lower-dimensional structures into bulk thermoelectrics and demonstrate dramatic improvements in the figure of merit, this would represent a definite milestone and open thermoelectric technology to a variety of industrial applications. It is in this context that the word "nanostructuring" (meaning the incorporation of nanometer-scale structural features in the otherwise bulk matrix) has become firmly established in the literature and scientific presentations during the past half-a-dozen years or so.

Just like everything else, nanostructuring is not the universal panacea that will work under all circumstances. Yes, the transport of phonons is likely to be adversely affected with any kind of a nanostructure, but one must always keep in mind that benefits gained by lowering the thermal conductivity must not be squandered by even greater degradation of the electronic properties. Nanometer-scale features lacking coherence with the host matrix and, even worse, inclusions having ionic character are likely to degrade carrier mobility to a far greater extent than any reduction in the mean-free path of phonons and such nanostructures will not be beneficial but rather undesirable.

When considering nanostructuring, it is useful to keep the perspective of what group of phonons we are trying to target. Phonon spectra have a broad frequency range (from zero to several THz) and because phonons of different frequencies interact differently with entities constituting a solid, not all phonons are equal as far as their ability to conduct heat is concerned. At the temperature range of interest (300–900 K) for thermoelectric power generation, the dominant scattering mechanism of phonons are intrinsic Umklapp processes leading to the T^{-1} temperature variation of the lattice thermal conductivity [42,43]. The question arises, what structural disorder can one introduce to degrade the flow of heat further? Starting with the smallest possible length scale it is the atomic disorder (point defects) such as atomic substitutions, doping, vacancies, interstitials, and alloying. Defects of this kind lead to the well-known Rayleigh scattering [44] with the relaxation rate $\tau^{-1} \sim \omega^4$. At very low temperatures, the dominant phonon wavelength is large and, rather than point defects imposing a limit on the thermal conductivity, it is the crystal boundaries (microns to mm length scale) that constrain the mean-free path of phonons leading to the T^3 variation of the thermal conductivity. At higher temperatures where the phonon wavelength is more compatible with the scale of point defects, the Rayleigh scattering is an important scattering mechanism of phonons. The alloy limit (the lowest value of the lattice thermal conductivity resulting when alloying component A with component B) was viewed as the lowest value of thermal conductivity one can achieve in a given material system. Indeed, all successful thermoelectric materials used for cooling applications are isostructural alloys relying on this mechanism [31] to bring down the lattice thermal conductivity. However, the strong frequency dependence of Rayleigh scattering implies that high-frequency phonons are scattered by point defects far more effectively than the low-frequency end of the spectrum. Thus, what carries heat in a solid are low-frequency (long-wavelength) phonons, longitudinal and transverse acoustic branches occasionally assisted by a small contribution from optical modes in more complex structures. To disrupt heat conduction, one must therefore effectively scatter these kinds of phonons. Until recently, fabrication routes of bulk thermoelectric materials yielded either single crystals with the typical sample size of several millimeters or polycrystalline structures with the grain size of several to tens of microns. Either length scale is too large in comparison with the dominant phonon wavelength above ambient temperatures (~ a few nm). With structural features on the scale of nanometers, one might possibly affect a somewhat broader range of phonon frequencies and further reduce the mean-free path of phonons so that the thermal conductivity may fall below the alloy limit. It is here where innovative synthesis methods leading to the formation of nanometer-scale structural features in the bulk matrix have had a profound impact on heat transport in general and thermoelectricity in particular. Canonical examples of the use of nanostructuring in reducing the lattice thermal conductivity and improving thermoelectric performance are the so-called LAST materials [45] and ErAs nanoparticles embedded in the InGaAs matrix [46]. In the former case, compositional modulations on the nanometer-scale in modified chalcogenides of the form $AgPb_mSb_nTe_{2+m}$ resulted in strongly suppressed lattice thermal conductivities leading to greatly enhanced values of ZT reaching 1.7 at 700 K. In the latter case, epitaxially grown but randomly distributed ErAs nanoparticles were shown to scatter effectively mid-wavelength phonons that could not be affected by alloy scattering in the InGaAs matrix and thus the thermal conductivity of this nanocomposite material was reduced below the alloy limit.

In general, bulk nanostructured materials can be fabricated by two principal approaches: (1) using intrinsic (thermodynamically driven) phase segregation/reformulation phenomena during cooling of the melt such as spinodal decomposition [47], nucleation and growth [48], matrix encapsulations [49], and compound formation [50]; (2) relying on extrinsic processes of preparing nanosized powders by grinding, milling, or wet chemistry routes that are then compacted into ingots/pellets by pressing [51,52]. Both approaches have proved successful in terms of enhancing the figure of merit although, in my opinion, the first approach is far superior when it comes to fabrication of bulk nanocomposites for power generation applications where extended exposure to a high-temperature environment is inevitable. The key issue here is the long-term stability of the nanocomposite. While the materials prepared by intrinsic processes are thermodynamically stabilized and the nanometer-scale features are unlikely to grow or dissolve unless one comes close to the melting point, the nanostructured compacts suffer from the presence of oxides and organic binders on the surfaces of particles and, more important, the nanograins readily coalesce and the long-term exposure to even moderate temperatures is likely to anneal out the benefits of nanostructuring. It is, however, possible that such nanostructured materials might find use in cooling applications where the material is not exposed to elevated temperatures.

In the context of skutterudites, early effort to control the size of microcrystallites in $CoSb_3$ focused on the use of ball milling followed by hot pressing [53] and on annealing the hot-pressed samples at various temperatures [54]. The first attempt to prepare a skutterudite matrix with submicron grains was made by chemical precipitation of nanopowders of $Co_{1-x}Ni_xSb_3$ ($0 < x < 0.275$) from aqueous solutions that were subsequently compacted into pellets by uniaxial pressure under various conditions [55]. While the size of the precipitated powders was small (about 40 nm), SEM images clearly indicated that the grain size of the compacted pellets is nowhere near this length scale and approaches 1 μm. Although the thermal conductivity of high-density samples was reduced to about one-half of the value typical of samples fabricated by the traditional long-term annealing of quenched melts (~10 W/m-K for grain sizes a few tens of microns), the aggressive grain growth clearly indicated that attempting to prepare skutterudites by compacting nanometer-size powders is unlikely to result in a viable nanostructured material. The subsequent more detailed paper by the same team [56] provided data on the grain size as a function of annealing conditions and further attested to the fact that a prolonged exposure of compacted skutterudite nanopowders to temperatures where they would be expected to operate as active elements in power-generating modules would completely obliterate the nanostructure length scale.

Similarly unfortunate fate would also inevitably meet nanostructured skutterudites prepared by mechanical alloying using planetary ball mills and sintered by any of the established techniques. As shown by Liu et al. [57], fine-grained $CoSb_3$ with the average grain size down to 50 nm can be prepared by mechanical alloying but only if the SPS processing temperature is kept at not more than 300°C. As the sintering temperature is increased to 600°C and held there for merely 5 min, the grains grow rapidly to an average size of 300 nm. One can only extrapolate what would happen if such $CoSb_3$ was exposed to the operational temperature range of skutterudites for an extended period of time! The same research team used the same mechanical alloying synthesis followed by SPS also with Te-doped $CoSb_3$ [58] and actually reported an impressively high $ZT = 0.93$ at 820 K with the main gain originating in a much reduced thermal conductivity due to the average grain size of 160 nm. Unfortunately, the authors did not enquire what would happen to the grain size and consequently ZT if the sample was kept at the measuring temperature for any reasonable length of time.

A very different situation is encountered with nanostructures in the skutterudite matrix that are driven by thermodynamic processes such as phase segregation. This was demonstrated in the work with Yb-filled antimonide skutterudite with an intentional Sb overstoichiometry fabricated by the melt-spinning process [59,60]. This recently developed synthesis procedure [61,62] that incorporates ultra-fast solidification of the melt by ejecting the molten charge on a cold rotating drum of copper yields materials with a very fine grain structure. The technique has generated much excitement as it seems to be broadly applicable to a variety of thermoelectric materials. In the case of skutterudites, it replaces very time-consuming traditional synthesis routes that require annealing times on the order of a week or so in order to overcome very slow kinetics and allow completion of the peritectic reaction. The very fine grain structure of the melt-spun ribbons requires only a few minutes of processing using spark plasma sintering (SPS) to complete the skutterudite phase. Since this Handbook contains a separate chapter on melt spinning processing of thermoelectric materials, I will say no more about this technique except that it has also a great potential for scaling up the production of skutterudites to quantities needed in manufacturing thermoelectric modules. Figure 10.3 illustrates the influence of melt spinning

FIGURE 10.3 Field emission scanning electron micrographs of (a) a polycrystalline $Yb_{0.2}Co_4Sb_{12}$ prepared by the traditional synthesis followed by SPS compaction. The grain size here is a few microns. (b) The same composition processed by melt spinning followed by SPS. Melt spinning yields a considerably smaller grain size of typically 300 nm. (c) $Yb_{0.2}Co_4Sb_{12+0.6}$ with the excess of Sb prepared by melt spinning followed by SPS. The excess of Sb segregates on the crystal boundaries in the form of a fine nanostructure of ~30 nm extent. (Reprinted with permission from Li, H. et al., Rapid preparation method of bulk nanostructured $Yb_{0.3}Co_4Sb_{12+y}$ compounds and their improved thermoelectric performance, *Appl. Phys. Lett.*, 93, 252109-1–252109-3. Copyright 2008, American Institute of Physics.)

on the grain size of the skutterudite matrix. Panel (a) depicts the microstructure (a few micrometers) in a sample synthesized by the traditional melting and long-term annealing method with the final compaction made by SPS. In contrast, panels (b) and (c) show a much finer grain structure (~300 nm) achieved in ingots processed by melt spinning with the copper drum rotating with the linear speed of 10 m/s and followed by SPS consolidation. Panel (c) represents a skutterudite with excess of Sb that is segregated on crystal boundaries in the form of a fine (~30 nm) nanostructure.

Using melt spinning in the preparation of double-filled skutterudites where one of the filler species was indium, Li et al. [50] made an interesting and important discovery that the resulting filled skutterudite contained a finely dispersed InSb nanophase (see Figure 10.4). Samples with this InSb nanophase displayed very low lattice thermal conductivity that helped to enhance the dimensionless figure of merit to values over 1.4 at 800 K. According to the theory of filling fraction limit [9], indium should not be able to fill structural voids in the skutterudite lattice because the electronegativity difference between Sb ($X_{Sb} = 2.05$) and In ($X_{In} = 1.78$) does not come close to satisfying Equation 10.1. Rather than filling the voids, indium seems to bond with Sb and form an *in situ* InSb nanophase. Thus, this compound formation process is another possible route how to form a nanostructure in the skutterudite matrix. It should be pointed out that there are at least three reports in the literature describing indium as a filler element in $CoSb_3$ and discussing its influence on the transport properties [63–65]. However, none of these studies presented detailed structural information and it is thus not clear that In indeed occupied the void sites. We have recently tried to double-fill $CoSb_3$ with In and Yb in my laboratory using the classical synthesis consisting of melting followed by prolonged annealing and detected the presence of InSb with both XRD and high-resolution transmission electron microscopy. Consequently, nanostructuring via compound formation is not unique to melt-spinning processing but, rather, reflects a thermodynamic barrier the skutterudite lattice poses to certain species trying to enter the void sites. In the case of indium, it seems energetically more favorable to form InSb than to fill the voids. Detailed modeling and analysis of the relevant thermodynamic processes would help to shed more light on the role of indium in the skutterudite lattice.

In a subsequent study, Li et al. [66] prepared a broader range of skutterudite compositions with In and Ce and they noted that the density of the InSb nanophase increases with the increasing content of In. Some of these nanocomposite samples reached *ZT* values greater than 1.5 at 800 K. Moreover, since there must be some Sb deficiency in stoichiometrically prepared skutterudites upon In bonding with Sb to form InSb, the same skutterudite compounds were also prepared with a slight excess of Sb and their lattice thermal conductivities compared with the nominally stoichiometric sample. Interestingly, the thermal conductivity

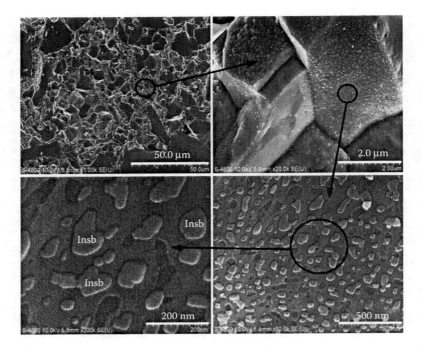

FIGURE 10.4   Field emission scanning electron micrograph of the bulk $Ce_{0.15}Co_4Sb_{12}$/InSb nanocomposite. (Reprinted with permission from Li, H. et al., High performance $In_xCe_yCo_4Sb_{12}$ thermoelectric materials with in-situ forming nanostructured InSb phase, *Appl. Phys. Lett.*, 94, 102114-1–102114-3. Copyright 2009, American Institute of Physics.)

of the nominally stoichiometric compound had a lower lattice thermal conductivity, possibly suggesting the presence of vacancies on the Sb sublattice. Although such point defects scatter preferentially high-frequency phonons, by eliminating their contribution, the thermal conductivity might be decreased [67].

Since the lowest values of the lattice thermal conductivity of skutterudites achieved so far are still a factor of 2 to 3 larger than the minimum thermal conductivity estimated for the skutterudite lattice [68], there is a growing interest in identifying processes that lead to the formation of stable nanostructures able to degrade the flow of phonons yet having a minimal impact on the charge carrier mobility. The above examples illustrate the progress made during the past couple of years and, undoubtedly, new and exciting mechanisms of nanostructure formation in the skutterudite matrix will emerge in the future.

10.4 Partial Charge Compensation

It is well known that the structure of $CoSb_3$ tolerates up to about 25% substitution of Fe for Co [69]. The trivalent state Fe^{3+} exactly matches the size of Co^{3+} and provides the right number of electrons for bonding. However, this leaves only five electrons in the d-shell and it is thus impossible to pair all spins. When Fe substitutes for Co one therefore expects the skutterudite to become a paramagnet. Concentrations of Fe higher than 25% result in the formation of secondary phases. With the content of Fe kept below 25%, the thermal conductivity of $Co_{1-x}Fe_xSb_3$ is dramatically reduced [68,70,71] compared to the parent $CoSb_3$. At the first glance, this is surprising since the masses as well as the ionic radii of Co and Fe differ little and thus the alloy scattering should not be significant. Similarly strong influence of the presence of Fe at the sites of Co was noticed earlier [5,72] in the thermal conductivity of partially compensated filled skutterudites of the form $R_yFe_xCo_{4-x}Sb_{12}$ that were originally explored as prospective *p*-type forms of filled skutterudites [73,74]. It was eventually rationalized [75] that such partially filled skutterudites can be viewed as solid solutions of the fully filled and nearly empty end members of the form $(CeFe_4Sb_{12})_\alpha(\square Co_4Sb_{12})_{1-\alpha}$

TABLE 10.2 Room Temperature Values of Transport Parameters

Actual (EMPA) Content	n (10^{20} cm^{-3})	σ (S/cm)	μ (cm^2/V-s)	S (μV/k)	κ (W/m-K)	ZT
$Ba_{0.08}Yb_{0.09}Co_4Sb_{12}$	4.04	2711	346	−109.9	3.93	0.25
$Ba_{0.06}Yb_{0.09}Co_{3.95}Fe_{0.05}Sb_{12}$	3.61	1742	27.1	−135.6	3.16	0.30
$Ba_{0.06}Yb_{0.22}Co_{3.77}Fe_{0.23}Sb_{12}$	2.97	1116	21.3	−154.3	2.44	0.33

and thus the reduced thermal conductivity has its origin not in the difference between Fe and Co, but in the difference between a compound with a fully filled cage and a compound with a completely empty cage, that is, the mass difference of 100%. Of course, when substituting Fe for Co in either the binary or filled skutterudite lattice, one introduces a positive hole into the carrier spectrum and drives the system toward p-type conduction. But what if the content of Fe is kept very low and the skutterudite stays well within the n-type domain of conduction? Could the benefit of a reduced thermal conductivity exceed the penalty one expects to pay for stronger charge carrier scattering with the net enhancement in the figure of merit? This question was posed by the author recently [76] and affirmed by the experimental results. Specifically, a series of double-filled and partially compensated skutterudite samples $Ba_{0.1}Yb_yCo_{4-x}Fe_xSb_{12}$ with $x = 0.05$ and 0.23 was synthesized, including a reference sample with no Fe content, and their transport properties were studied up to 800 K. All samples were single-phase compounds prepared by the traditional melting and long-term annealing method and compacted by spark plasma sintering (SPS). The key room-temperature transport parameters are given in Table 10.2.

As expected, with the increasing content of Fe, the density of electrons decreases and Fe has a detrimental influence on the electrical conductivity also via strong scattering which is reflected in an order of magnitude lower carrier mobility. A somewhat surprising benefit comes from a significantly enhanced Seebeck coefficient in the Fe-containing samples and, of course, an expected reduction in the thermal conductivity. Overall, the dimensionless figure of merit improves when a small amount of Fe substitutes for Co. This general trend extends to the entire temperature range covered and the ZT behavior as a function of temperature is depicted in Figure 10.5. The figure of merit of the two Fe-containing

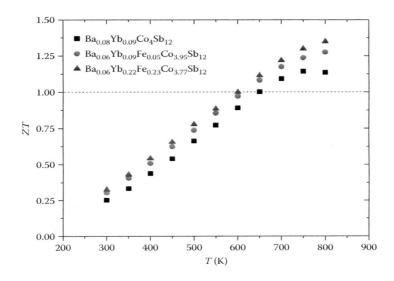

FIGURE 10.5 Dimensionless figure of merit of the partially compensated double-filled skutterudites. Actual (EPMA) compositions of the samples are given. Samples containing Fe have higher values of ZT in comparison with the Fe-free sample in the entire temperature range. (Data from Uher, C., Li, C.-P., and Ballikaya, S., *J. Electron. Mater.*, DOI:10.1007/s11664-009-0990-1, 2009.)

skutterudites is higher than that of the double-filled skutterudite sample free of Fe and the sample with a higher concentration of Fe shows a marginally higher value of ZT in excess of 1.3. Moreover, unlike the Fe-free sample, the figure of merit of the two samples containing Fe does not saturate by 800 K.

Although much more work is needed to establish the optimal concentration of Fe as well as the choice and concentration of the filler species, the preliminary data indicate that the concept of partial compensation might be an additional route how to improve the thermoelectric performance of n-type-filled skutterudites. It would be of interest to make use of this approach in the melt-spinning synthesis and combine it with the beneficial features of nanostructuring.

10.5 Compensating Double-Doping of CoSb₃

As already noted, stoichiometric $CoSb_3$ is a true diamagnetic semiconductor. To increase the carrier density one can dope the structure on either the cation or anion sublattice, replacing a certain number of Co or Sb atoms with elements from the neighboring columns of the periodic table. Specifically, to increase the density of electrons, one can either replace some Co atoms with Ni or, alternatively, some Sb atoms with elements of the group VIB such as Te. Conversely, to enhance the density of holes, one would replace Co with Fe or Sb atoms with the group IVB elements such as Ge or Sn. Typically, there are solubility limits imposed on such atomic substitutions stemming from the requirement that the substituting atom must be able to form the same kind of bond as the atom it replaces and that the ionic sizes of the guest–host pair must be comparable. Exceeding the solubility limits leads to the formation of multiphase structures. In the case of $CoSb_3$, the cationic sublattice can accommodate up to 10% of Ni atoms [77] and detailed studies of the effect of Ni on the transport properties of $CoSb_3$ were carried out for both high [78] and low [79] doping levels. The anionic sublattice is much less tolerant and accepts not more than 5% of Te ($x = 0.15$ in the formula $CoSb_{3-x}Te_x$) [80] and about 2.4% of Sn [81]. Doping was extensively explored as a means of tuning the electronic properties as well as reducing the thermal conductivity. In the case of atomic substitutions on the anionic sublattice it seemed a particularly reasonable approach to reduce the lattice thermal conductivity [82–85] because it focused on the disruption of pnicogen rings the vibrations of which dominate the heat-carrying phonons. In spite of valiant efforts, the degradation of the heat transport never reached the level necessary to make doped $CoSb_3$ a viable thermoelectric material. Experience gained during the past decade pointed to the inevitable conclusion that to make skutterudites with high ZT, one had to fill the voids of the structure.

Recently, experiments with double-doping on the anionic sublattice of $CoSb_3$ [86,87] have led to surprisingly high values of the figure of merit that call for reexamination of benefits of doping and their potential incorporation in the grand strategy of synthesizing skutterudite compounds with exceptionally high thermoelectric performance. In their work with mechanically alloyed $CoSb_{3-x}Te_x$, Liu et al. [58] observed that the highest ZT value of 0.93 at 820 K was obtained at $x = 0.15$, the highest level of Te doping the $CoSb_3$ structure can accommodate. The authors surmised that a higher Te content might yield even higher ZT. But how can one achieve such high doping levels without forming secondary phases? One possibility is to compensate electrons donated by Te with holes coming from the group IVB elements such as Si, Ge, Sn, or Pb. Trying all four elements, Liu et al. [86] found Sn as the most effective and synthesized a series of double-doped skutterudites $CoSb_{2.75}Sn_{0.25-x}Te_x$ with $x = 0.125$, 0.150, 0.175, and 0.200 using mechanical alloying followed by SPS. The average grain size of the *as-prepared* material was 140 nm, comparable to other binary skutterudites the authors have fabricated using planetary milling and SPS sintering and this length scale already assures a low thermal conductivity structure. Moreover, the presence of the compensating Sn on the Sb sublattice effectively extended the solubility range of Te from $x = 0.15$ to 0.20 and this, together with the modified vibration spectrum of the Sb_4 rings (detected by shifts in the Raman spectra) resulted in even greater suppression of the lattice thermal conductivity on account of strong point-defect scattering. Consequently, ZT values of 1.1 at 820 K were recorded, representing exceptionally high values of the figure of merit for an unfilled skutterudite! The authors also observed a few "black" regions resembling nanodots in high-resolution TEM images taken on these

alloys and reasoned that such nanodots might facilitate enhanced phonon scattering and thus further degrade the already low lattice thermal conductivity. Since the samples were fabricated by compacting ball-milled powders—the same technique the team used in the synthesis of $CoSb_3$ described in Section 10.3—the initial grain structure of the skutterudite matrix (~140 nm) would not survive extended exposure to elevated temperatures and the values of *ZT* would inevitably deteriorate. On the other hand, it would be of interest to look at what the nanodots do when subjected to prolonged annealing.

The possibility of forming a nanostructure by double-doping on the pnicogen rings is an intriguing idea taken up by Su et al. [87]. To circumvent the problem with grain growth, the authors used the traditional synthesis based on melting and long-term annealing followed by SPS sintering that yields materials with the grains on the scale of 10 μm. Using this technique, they prepared a series of double-doped skutterudites compounds with compositions $CoSb_{2.75}Ge_{0.25-x}Te_x$ ($x = 0.125–0.20$). The use of Ge instead of Sn was to provide a greater mass contrast with Sb. Just as in the case of Sn, the presence of Ge assures a greater solubility of Te and all compounds are single-phase skutterudites with no traces of impurity phases. Structural analysis using high-resolution transmission electron microscopy (HRTEM) revealed the presence of finely dispersed nanostructure in the form of circular nanodots of about 30 nm diameter embedded in the matrix, Figure 10.6. Energy-dispersive spectroscopy (EDS) yielded the normalized chemical composition of the matrix and the nanodots as $CoSb_{2.90}Ge_{0.02}Te_{0.135}$ and $CoSb_{2.65}Ge_{0.125}Te_{0.312}$, respectively. The nanodots are thus rich in Ge and Te while the matrix is poor in these elements. Referring to Figure 10.7, the formation of the nanostructure can be rationalized as follows: when a Te atom (group VIB element) substitutes for Sb, it creates a local charge and significantly disturbs delicate bonding within the Sb_4 ring. When a group IVB element such as Ge occupies the same ring where Te already sits, the charge is effectively neutralized, the bonding is "healed," and this situation represents an energetically more favorable configuration than when Te is the sole occupant of the site. One might even argue that Te and Ge prefer to sit next to each other rather than being in the diagonal positions. However, such speculations should wait for detailed all energy calculations.

The mechanism of nanostructure formation by compensated double-doping on the pnicogen rings of the skutterudite structure very much resembles the formation of nanostructure in the case of the LAST

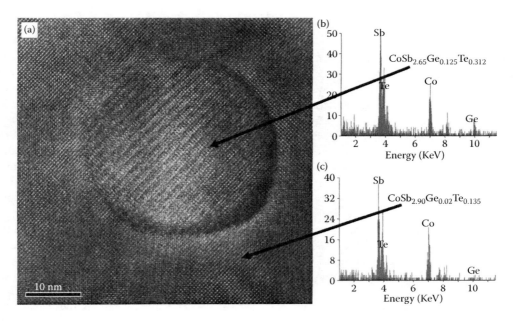

FIGURE 10.6 (a) High-resolution TEM image of $CoSb_{2.75}Ge_{0.05}Te_{0.20}$. (b) Energy-dispersive spectroscopy (EDS) of the nanostructured region. (c) EDS of the matrix.

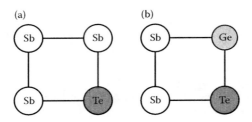

FIGURE 10.7   Schematic representation of the pnicogen ring with (a) Te substituting for Sb; and (b) both Te and Ge substituting on the same ring.

materials, the modified chalcogenides of composition $AgPb_mSbTe_{2+m}$ [45]. In that case it is the average rock salt structure of the LAST system where the 2^+ charge of Pb is supplied by a close proximity of Ag^{1+} and Sb^{3+} which gives rise to nanoprecipitates rich in Ag and Sb and the matrix relatively poor in these elements.

In contrast to the work with mechanically alloyed double-doped $CoSb_{2.75}Sn_{0.25-x}Te_x$ skutterudites [86] where high ZT values reported were obtained on samples with the initially 140 nm grain structure of the matrix, the equally impressive ZT values [87] obtained on $CoSb_{2.75}Ge_{0.25-x}Te_x$ ($x = 0.125$–0.20) compounds with the matrix grains on the scale of 10 µm are undoubtedly the result of the presence of nanometer-scale compositional variations (nanodots) that provide strong scattering for phonons and consequently very low lattice thermal conductivity. One should note that values of $ZT = 1.1$ achieved with both double-doped binary skutterudite systems above rival the values of ZT obtained on filled skutterudites. This in itself is a remarkable development. One might argue that such modified binary skutterudites are more appealing from the perspective of practical applications than are filled skutterudites since there is no need for any of the highly reactive elements such as rare earths or alkaline earths that often serve as preferred filler species. In any case, the compensating double doping of binary skutterudites should be tried in combination with other effective synthesis approaches in an attempt to fabricate high-performance skutterudite-based thermoelectric materials.

10.6 Conclusions

During the last half-a-dozen years we have witnessed intensified worldwide competition to secure access to fossil fuels, gradual depletion of the existing oil fields, and enthusiasm for more risky deep water oil drilling being tempered by major environmental disasters such activity can generate. The need for the development of alternative energy sources has never been greater. While the objective is clear, the road-map of how to achieve the goal is not well defined as none of the alternative energy-producing pathways is free of either the technological or societal problems. Thermoelectricity—a process of energy conversion by purely solid state means—has a potential to transform from a merely niche area of technology to one of the forefront industrial methods of providing environmentally friendly cooling/air conditioning and harvesting and conversion of waste industrial heat into electricity.

To achieve such a lofty goal, one needs new thermoelectric materials which are mechanically robust, plentiful in supply, economically affordable and, above all, having conversion efficiencies far superior to the current state-of-the-art thermoelectric materials. Among several new prospective thermoelectrics, skutterudite compounds based on $CoSb_3$ have generated much excitement and their intensive development during the past decade has resulted in spectacular improvements in their figure of merit that now reach (and perhaps even exceed) values of 1.5 at 800 K. The progress with n-type skutterudites was enabled by a combination of insightful theoretical input and broad-based experimental efforts that contributed to the understanding of the important issues such as filling fraction limitations, the pivotal role nanostructuring plays in enhancing scattering of heat-conducting phonons, as well as the development of new synthesis routes that yield not just high figures of merit but are also scalable and compatible with the production of industrial quantities of thermoelectric materials. Several independent approaches that have been identified to lead to improvements in the figure of merit of *n*-type $CoSb_3$-based skutterudites should now be consolidated in one grand strategy to prepare a superior thermoelectric material that aims for the new horizon of $ZT \sim 2$.

While the development of *p*-type forms of skutterudites has, so far, lagged much behind the development of their *n*-type cousins, the latest encouraging results [88] hold the promise that, even in this area, the progress is at hand or soon will be realized. Coupled with their excellent mechanical properties [89,90], skutterudite compounds are perhaps the best candidate materials for power generation in the intermediate range of temperatures between 500 and 900 K which covers a vast amount of waste heat generated in industrial processes including transportation.

The time is ripe for taking a leap and investing in a large-scale production of modules based on skutterudites and putting them to the test.

Acknowledgment

The author gratefully acknowledges support from the Center for Solar and Thermal Energy Conversion, an Energy Frontier Research Center funded by the U.S. Department of Energy, Office of Basic Energy Sciences under Award Number DE-SC0000957.

References

1. Uher, C., Skutterudite-based thermoelectrics. In *Handbook of Thermoelectricity: From Macro to Nano*, D.M. Rowe, ed., Ch. 34, CRC, Taylor & Francis, Boca Raton, FL, 2006.
2. Uher, C., Skutterudites: Prospective novel thermoelectrics. In *Semiconductors and Semimetals*, Vol. 69, T.M. Tritt, ed., pp. 139–253, Academic Press, New York, 2001.
3. Dudkin, L.D., The chemical bond in semiconducting cobalt triantimonide, *Sov. Phys.—Tech. Phys.*, 3, 216–220, 1958.
4. Jeitschko, W. and Brown, D.J., $LaFe_4P_{12}$ with filled $CoAs_3$-type structure and isotypic lanthanoid-transition metal polyphosphides, *Acta Crystallogr. B*, 33, 3401–3406, 1977.
5. Morelli, D.T., Meisner, G.P., Chen, B., Hu, S., and Uher, C., Cerium filling and doping of cobalt triantimonide, *Phys. Rev. B*, 56, 7376–7383, 1997.
6. Chen, L.D., Kawahara, T., Tang, X.F., Goto, T., Hirai, T., Dyck, J.S., Chen, W., and Uher, C., Anomalous barium filling fraction and *n*-type thermoelectric performance of $Ba_yCo_4Sb_{12}$, *J. Appl. Phys.*, 90, 1864–1868, 2001.
7. Takizawa, H., Okazaki, K., Uheda, K., Endo, T., and Nolas, G.S., High pressure synthesis of new filled skutterudites, *Mat. Res. Soc. Symp. Proc.*, 691, 3747, 2002.
8. Nordström, L. and Singh, D.J., Electronic structure of Ce-filled skutterudites, *Phys. Rev. B*, 53, 1103–1108, 1996.
9. Shi, X., Zhang, W., Chen, L.D., and Yang, J., Filling fraction limits for intrinsic voids in crystals: Doping in skutterudites, *Phys. Rev. Lett.*, 95, 185503-1–185503-4, 2005.
10. Shi, X., Zhang, W., Chen, L.D., Yang, J., and Uher, C., Theoretical study of the filling fraction limits for impurities in $CoSb_3$, *Phys. Rev. B*, 75, 235208-1–235208-9, 2007.
11. Slack, G.A., New materials and performance limits for thermoelectric cooling. In *CRC Handbook of Thermoelectrics*, D.M. Rowe, ed., Ch. 34, pp. 407–440, CRC Press, Boca Raton, FL, 1995.
12. Yang, J., Zhang, W., Bai, S.Q., Mei, Z., and Chen, L.D., Dual-frequency resonant phonon scattering in $Ba_xR_yCo_4Sb_{12}$ (R = La, Ce, and Sr), *Appl. Phys. Lett.*, 90, 192111-1–192111-3, 2007.
13. Shi, X., Kong, H., Li, C.-P., Uher, C., Yang, J., Salvador, J. R., Wang, H., Chen, L.D., and Zhang, W., Low thermal conductivity and high thermoelectric figure of merit in *n*-type $Ba_xYb_yCo_4Sb_{12}$ double-filled skutterudites, *Appl. Phys. Lett.*, 92, 182101-1–182101-3, 2008.

14. Shi, X., Yang, J., Salvador, J.R., Chi, M., Cho, J.Y., Wang, H., Bai, S., Yang, J., Zhang, W., and Chen, L. D., Multiple-filled skutterudites: High thermoelectric figure of merit through separately optimizing electrical and thermal transports, *J. Am. Chem. Soc.* 133, 7837–7846, 2011.

15. Dudkin, L.D. and Abrikosov, N.Kh., A physicochemical investigation of cobalt antimonides, *Zh. Neorg. Khimie*, 1, 2096–2105, 1956.

16. Slack, G.A. and Tsoukala, V.G., Some properties of semiconducting $IrSb_3$, *J. Appl. Phys.*, 76, 1665–1671, 1994.

17. Morelli, D.T., Caillat, T., Fleurial, J.-P., Borshchevsky, A., Vandersande, J., Chen, B., and Uher, C., Low temperature transport properties of p-type $CoSb_3$, *Phys. Rev. B*, 51, 9622–9628, 1995.

18. Morelli, D.T. and Meisner, G.P., Low temperature properties of the filled skutterudite $CeFe_4Sb_{12}$, *J. Appl. Phys.*, 77, 3777–3781, 1995.

19. Nolas, G.S., Slack, G.A., Morelli, D.T., Tritt, T.M., and Ehrlich, A.C., The effect of rare-earth filling on the lattice thermal conductivity of skutterudites, *J. Appl. Phys.*, 79, 4002–4008, 1996.

20. Chen, B., Uher, C., Morelli, D.T., Meisner, G.P., Fleurial, J.-P., Caillat, T., and Borshchevsky, A., Low temperature transport properties of the filled skutterudite $CeFe_{4-x}Co_xSb_{12}$, *Phys. Rev. B*, 55, 1476–1480, 1997.

21. Nolas, G.S., Cohn, J.L., and Slack, G.A., Effect of partial filling on the lattice thermal conductivity of skutterudites, *Phys. Rev. B*, 58, 164–170, 1998.

22. Sales, B.C., Mandrus, D., and Williams, R.K., Filled skutterudite antimonides: A new class of thermoelectric materials, *Science*, 272, 1325–1328, 1996.

23. Sales, B.C., Mandrus, D., Chakoumakos, B.C., Keppens, V., and Thompson, J.R., Filled skutterudites: Electron crystal and phonon glass, *Phys. Rev. B*, 56, 15081–15089, 1997.

24. Keppens, V., Mandrus, D., Sales, B.C., Chakoumakos, B.C., Dai, P., Coldea, R., Maple, M.B., Gajewski, D.A., Freeman, E.J., and Bennington, S., Localized vibrational modes in metallic solids, *Nature*, 395, 876–878, 1998.

25. Long, G.J., Hermann, R.P., Grandjean, F., Alp, E.E., Sturhahn, W., Johnson, C.E., Brown, D.E., Leupold, O., and Rüffer, R., Strongly decoupled europium and iron vibrational modes in filled skutterudites, *Phys. Rev. B*, 71, 140302-1–140302-4, 2005.

26. Grannan, E.R., Randeria, M., and Sethna, J.P., Low-temperature properties of a model glass. II. Specific heat and thermal transport, *Phys. Rev. B*, 41, 7799–7821, 1990.

27. Koza, M.M., Johnson, M.R., Viennois, R., Mutka, H., Girard, L., and Ravot, D., Dynamics of La and Ce filled iron skutterudite structures, *Proc. 25th Int. Conf. on Thermoelectrics*, IEEE Catalog Number 06TH8931, pp. 70–73, 2006.

28. Koza, M.M., Johnson, M.R., Viennois, R., Mutka, H., Girard, L., and Ravot, D., Breakdown of phonon glass paradigm in La- and Ce-filled Fe_4Sb_{12} skutterudites, *Nature of Materials*, 7, 805–810, 2008.

29. Squires, G.L., in *Introduction to the Theory of Thermal Neutron Scattering*, Cambridge University Press, Cambridge, UK, 1978.

30. Huang, B. and Kaviany, M., Filler-reduced phonon conductivity of thermoelectric skutterudites: *Ab initio* calculations and molecular dynamics simulations, *Acta Materialia*, 58, 4516–4526, 2010.

31. Ioffe, A.F., *Semiconductor Thermoelements and Thermoelectric Cooling*, Infosearch, London, 1957.

32. Rowe, D.M. and Bhangari, C.M., *Modern Thermoelectrics*, Holt, Rinehart and Winston, London, 1983.

33. Goldsmid, H.J., *Electronic Refrigeration*, Pion, London, 1986.

34. Nolas, G.S., J. Sharp, and Goldsmid, H.J., *Thermoelectrics: Basic Principles and New Materials Development*, Springer-Verlag, Berlin, 2001.

35. Hicks, L.D. and Dresselhaus, M.S., Effect of quantum well structures on the thermoelectric figure of merit, *Phys. Rev. B*, 47, 12727–12731, 1993.

36. Hicks, L.D. and Dresselhaus, M.S., Thermoelectric figure of merit of a one-dimensional conductor, *Phys. Rev. B*, 47, 16631–16634, 1993.

37. Ravich, Y.I., Efimova, B.A., and Tamarchenko, V.I., Scattering of current carriers and transport phenomena in lead chalcogenides, *Phys. Stat. Solidi (b)*, 43, 453–469, 1971.
38. Vashaee, D. and Shakouri, A., Improved thermoelectric power factor in metal-based superlattices, *Phys. Rev. Lett.*, 92, 106103-1–106103-4, 2004.
39. Harman, T.C., Taylor, P.J., Walsh, M.P., and Laforge, B.E., Quantum dot superlattice thermoelectric materials and devices, *Science*, 297, 2229–2232, 2002.
40. Vineis, C.J., Harman, T.C., Calawa, S.D., Walsh, M.P., Reeder, R.E., Singh, R., and Shakouri, A., Carrier concentration and temperature dependence of the electronic transport properties of epitaxial PbTe and PbTe/PbSe nanodot superlattices, *Phys. Rev. B*, 77, 235202-1–235202-14, 2008.
41. Koh, Y.K., Vineis, C.J., Calawa, S.D., Walsh, M.P., and Cahill, D.G., Lattice thermal conductivity of nanostructured thermoelectric materials based on PbTe, *Appl. Phys. Lett.*, 94, 153101-1–153101-3, 2009.
42. Peierls, R.E., *Quantum Theory of Solids*, Clarendon Press, Oxford, 1955.
43. Berman, R., *Thermal Conduction in Solids*, Clarendon Press, Oxford, 1979.
44. Rayleigh, Lord, *Theory of Sound* (2nd edn.), Vol. 2, Dover Publications, NY, 1945.
45. Hsu, K.F., Loo, S., Guo, F., Chen, W., Dyck, J.S., Uher, C., Hogan, T., Polychroniadis, E.K., and Kanatzidis, M.G., Cubic $AgPb_mSbTe_{2+m}$: Bulk thermoelectric materials with high figure of merit, *Science*, 303, 818–821, 2004.
46. Kim, W., Zide, J., Gossard, A., Klenov, D., Stemmer, S., Shakouri, A., and Majumdar, A., Thermal conductivity reduction and thermoelectric figure of merit increase by embedding nanoparticles in crystalline semiconductors, *Phys. Rev. Lett.*, 96, 045901-1–045901-4, 2006.
47. Androulakis, J., Lin, C.-H., Kong, H., Uher, C., Wu, C.-I., Hogan, T., Cook, B.A., Caillat, T., Paraskevopoulos, K.M., and Kanatzidis, M.G., Spinodal decomposition and nucleation and growth as a means to bulk nanostructured thermoelectric, *J. Am. Chem. Soc.*, 129, 9780–9788, 2007.
48. Heremans, J.P., Thrush, C.M., and Morelli, D.T. Thermopower enhancement in PbTe with Pb precipitates, *J. Appl. Phys.*, 98, 063703-1–063703- 6, 2005.
49. Sootsman, J.R., Pcionek, R.J., Kong, H., Uher, C., and Kanatzidis, M.G., Strong reduction of thermal conductivity in nanostructured PbTe prepared by matrix encapsulation, *Chem. Mater.*, 18, 4993–4995, 2006.
50. Li, H., Tang, X.F., Zhang, Q., and Uher, C., High performance $In_xCe_yCo_4Sb_{12}$ thermoelectric materials with *in-situ* forming nanostructured InSb phase, *Appl. Phys. Lett.*, 94, 102114-1–102114-3, 2009.
51. Poudel, B., Hao, Q., Ma, Y., Lan, Y., Minnich, A., Yu, C., Yan, X. et al., High thermoelectric performance of nanostructured bismuth antimony telluride bulk alloys, *Science*, 320, 634–638, 2008.
52. Liu, W.-S., Zhang, B.-P., Li, J.-F., Zhang, H.-L., and Zhao, L.-D., Enhanced thermoelectric properties in $CoSb_{3-x}Te_x$ alloys prepared by mechanical alloying and spark plasma sintering, *J. Appl. Phys.*, 102, 103717-1–103717-7, 2007.
53. Nakagawa, H., Tanaka, H., Kasama, A., Anno, H., and Matsubara, K., Grain size effects on thermoelectric properties of hot-pressed $CoSb_3$, *Proc. 16th Int. Conf. on Thermoelectrics*, IEEE Catalog Number 97TH8291, Piscataway, NJ, pp. 351–355, 1997.
54. Anno, H., Hatada, K., Shimizu, H., Matsubara, K., Notohara, Y., Sakakibara, T., Tashiro, H., and Motoya, K., Structural and electronic transport properties of polycrystalline p-type $CoSb_3$, *J. Appl. Phys.*, 83, 5270–5276, 1998.
55. Bertini, L., Stiewe, Ch., Toprak, M., Williams, S., Platzek, D., Mrotzek, A., Zhang, Y. et al., Nanostructured $Co_{1-x}Ni_xSb_3$ skutterudites: Synthesis, thermoelectric properties, and theoretical modeling, *J. Appl. Phys.*, 93, 438–447, 2003.
56. Toprak, M.S., Stiewe, Ch., Platzek, D., Williams, S., Bertini, L., Müller, E., Gatti, C., Zhang, Y., Rowe, M., and Muhammed, M., The impact of nanostructuring on the thermal conductivity of thermoelectric $CoSb_3$, *Adv. Funct. Mater.*, 14, 1189–1196, 2004.

57. Liu, W.-S., Zhang, B.-P., Li, J.-F., and Zhao, L.-D., Thermoelectric property of fine-grained $CoSb_3$ skutterudite compound fabricated by mechanical alloying and spark plasma sintering, *J. Phys. D: Appl. Phys.*, 40, 566–572, 2007.

58. Liu, W.-S., Zhang, B.-P., Li, J.-F., Zhang, H.-L., and Zhao, L.-D., Enhanced thermoelectric properties in $CoSb_{3-x}Te_x$ alloys prepared by mechanical alloying and spark plasma sintering, *J. Appl. Phys.*, 102, 103717-1–103717-7, 2007.

59. Li, H., Tang, X., Zhang, Q., and Uher, C., Rapid preparation method of bulk nanostructured $Yb_{0.3}Co_4Sb_{12+y}$ compounds and their improved thermoelectric performance, *Appl. Phys. Lett.*, 93, 252109-1–252109-3, 2008.

60. Li, H., Tang, X., Su, X.L., Zhang, Q., and Uher, C., Nanostructured bulk $Yb_xCo_4Sb_{12}$ with the high thermoelectric performance prepared by the rapid solidification method, *J. Phys. D: Appl. Phys.*, 42, 145409-1–145409-9, 2009.

61. Li, H., Tang, X., Su, X.L., and Zhang, Q., Preparations and thermoelectric properties of high performance Sb additional $Yb_{0.2}Co_4Sb_{12+y}$ bulk materials with nanostructure, *Appl. Phys. Lett.*, 92, 202114-1–202114-3, 2008.

62. Li, Q., Lin, Z.W., and Zhou, J., Thermoelectric materials with potential high power factors for electricity generation, *J. Electr. Mater.*, 38, 1268–1272, 2009.

63. He, T., Chen, J.Z., Rosenfeld, H.D., and Subramanian, M.A., Thermoelectric properties of indium-filled skutterudites, *Chem. Mater.*, 18, 759–762, 2006.

64. Jung, J.Y., Ur, S.C., and Kim, I.H., Thermoelectric properties of $In_zCo_4Sb_{12-y}Te_y$ skutterudites, *Mater. Chem. Phys.*, 108, 431–434, 2008.

65. Peng, J.Y., Alboni, P.N., He, J., Zhang, B., Su, Z., Holgate, T., Gothard, N., and Tritt, T.M., Thermoelectric properties of (In,Yb) double-filled $CoSb_3$ skutterudite, *J. Appl. Phys.*, 104, 053710-1–053710-5, 2008.

66. Li, H., Tang, X., Su, X., Zhang, Q., and Uher, C., Rapid synthesis and enhanced thermoelectric performance of $In_xCe_yCo_4Sb_{12+z}$ with *in-situ* formed InSb nanophase, submitted for publication.

67. Pei Y. and Morelli, D.T., Vacancy phonon scattering in thermoelectric In_2Te_3–InSb solid solutions, *Appl. Phys. Lett.*, 94, 122112-1–122112-3, 2009.

68. Nolas, G.S., Cohn, J.L., and Slack, G.A., Effect of partial void filling on the lattice thermal conductivity of skutterudites, *Phys. Rev. B*, 58, 164–170, 1998.

69. Dudkin L.D. and Abrikosov, N.Kh., On the doping of the semiconductor compound $CoSb_3$, *Sov. Phys.-Solid State*, 1, 126–133, 1959.

70. Katsuyama, S., Shichijo, Y., Ito, M., Majima, K., and Nagai, H., Thermoelectric properties of the skutterudite $Co_{1-x}Fe_xSb_3$ system, *J. Appl. Phys.*, 84, 6708–6712, 1998.

71. Yang, J., Meisner, G.P., Morelli, D.T., and Uher, C., Iron valence in skutterudites: Transport and magnetic properties of $Co_{1-x}Fe_xSb_3$, *Phys. Rev. B*, 63, 014410-1—014410-11, 2000.

72. Sharp, J.W., Jones, E.C., Williams, R.K., Martin, P.M., and Sales, B.C., Thermoelectric properties of $CoSb_3$ and related alloys, *J. Appl. Phys.*, 78, 1013–1018, 1995.

73. Fleurial, J.-P., Borshchevsky, A., Caillat, T., Morelli, D.T., and Meisner, G.P., High figure of merit in Ce-filled skutterudites, *Prof. 15th Int. Conf. on Thermoelectrics*, IEEE Catalog 96TH8169, Piscataway, NJ, pp. 91–96, 1996.

74. Sales, B.C., Mandrus, D., and Williams, R.K., Filled skutterudite antimonides: A new class of thermoelectric materials, *Science*, 272, 1325–1328, 1996.

75. Meisner, G.P., Morelli, D.T., Hu, S., Yang, J., and Uher, C., Structure and lattice thermal conductivity of fractionally filled skutterudites: Solid solutions of fully filled and unfilled end members, *Phys. Rev. Lett.*, 80, 3551–3554, 1998.

76. Uher, C., Li, C.-P., and Ballikaya, S., Charge-compensated *n*-type skutterudites, *J. Electron. Mater.*, 39, 2122–2126, 2010.

77. Dudkin L.D. and N. K. Abrikosov, N. Kh., Effect of Ni on the properties of the semiconducting compound $CoSb_3$, *Zh. Neorg. Khim.*, 2, 212–221, 1957.

78. Anno, H., Matsubara, K., Notohara, Y., Sakakibara, T., and Tashiro, H., Effect of doping on the transport properties of $CoSb_3$, *J. Appl. Phys.*, 86, 3780–3786, 1999.

79. Dyck, J.S., Chen, W., Yang, J., Meisner, G.P., and Uher, C., Effect of Ni on the transport and magnetic properties of $Co_{1-x}Ni_xSb_3$, *Phys. Rev. B*, 65, 115204-1–115204-9, 2002.

80. Zobrina B.N. and Dudkin, L.D., Investigation of the thermoelectric properties of $CoSb_3$ with Sn, Te, and Ni impurities, *Sov. Phys.-Solid State*, 1, 1668–1674, 1960.

81. Koyanagi, T., Tsubouchi, T., Ohtani, M., Kishimoto, K., Anno, H., and Matsubara, K., Thermoelectric properties of $Co(M_xSb_{1-x})_3$ (M = Ge, Sn, Pb) compounds, *Proc. 15th Int. Conf. on Thermoelectrics*, IEEE Catalog 96TH8169, Piscataway, NJ, pp. 107–111, 1996.

82. Y. Nagamoto, Y., Tanaka, K., and Koyanagi, T., Transport properties of heavily doped n-type $CoSb_3$, *Proc. 17th Int. Conf. on Thermoelectrics*, IEEE Catalog Number 98TH8365, Piscataway, NJ, pp. 302–305, 1998.

83. Yang, J., Morelli, D.T., Meisner, G.P., Chen, W., Dyck, J.S., and Uher, C., Effect of Sn substituting for Sb on the low-temperature transport properties of ytterbium-filled skutterudites, *Phys. Rev. B*, 67, 165207-1–165207-6, 2003.

84. Li, X.Y., Chen, L.D., Fan, J.F., Zhang, W.B., Kawahara, T., and Hirai, T., Thermoelectric properties of Te-doped $CoSb_3$ by spark plasma sintering, *J. Appl. Phys.*, 98, 083702-1–083702-6, 2005.

85. Wojciechowski, K.T., Tobola, J., and Leszczynski, J., Thermoelectric properties and electronic structure of $CoSb_3$ doped with Se and Te, *J. Alloys and Comp.*, 361, 19–27, 2003.

86. Liu, W.-S., Zhang, B.-P., Zhao, L.-D., and Li, J.-F., Improvement of thermoelectric performance of $CoSb_{3-x}Te_x$ skutterudite compounds by additional substitution of IVB-group elements for Sb, *Chem. Mater.*, 20, 7526–7531, 2008.

87. Su, X., Li, H., Wang, G., Chi, H., Zhou, X., Tang, X.F., Zhang, Q., and Uher, C., Structure and transport properties of double-doped $CoSb_{2.75}Ge_{0.25-x}Te$, (x = 0.125–0.20) with *in-situ* nanostructure, *Chem. Mater.*, 23, 2948–2955, 2011.

88. Rogl, G., Grytsiv, A., Bauer, E., Rogl, P., and Zehetbauer, M., Thermoelectric properties of novel skutterudites with didymium: $DD_y(Fe_{1-x}Co_x)_4Sb_{12}$ and $DD_y(Fe_{1-x}Ni_x)_4Sb_{12}$, *Intermetallics*, 18, 57–64, 2010.

89. Salvador, J.R., Yang, J., Shi, X., Wang, H., Wereszczak, A.A., Kong, H., and Uher, C., Transport and mechanical properties of Yb-filled skutterudites, *Philos. Mag.*, 89, 1517–1534, 2009.

90. Zhang, L., Rogl, G., Grytsiv, A., Ruchegger, S., Koppensteiner, J., Spieckermann, F., Kabelka, H. et al., Mechanical properties of filled antimonide skutterudites, *Mater. Sci. Eng. B*, 170, 26–31, 2010.

91. Uher, C., Skutterudites: Promising power conversion thermoelectric. In *Proceedings of the 22nd International Conference on Thermoelectrics*, pp. 42–47. IEEE Catalog Number 03TH8726, Piscataway, NJ, 2003.

Silicide Thermoelectrics: State of the Art and Prospects

M.I. Fedorov
Ioffe Physical-Technical Institute of the Russian Academy of Sciences

V.K. Zaitsev
Ioffe Physical-Technical Institute of the Russian Academy of Sciences

11.1 Introduction

Many silicides are cheap, ecologically friendly, accessible, mechanically and chemically strong materials. Higher silicides of transition metals (Cr, Mn, Fe, Ru, and Re), magnesium silicide (Mg_2Si), and its solid solutions with magnesium stannide and magnesium germanide are the most interesting. Silicides of transition metals as thermoelectrics were proposed by E.N. Nikitin's in his paper in 1958.[1] The main advantage of these thermoelectrics was high density of states, but the current carrier mobility was too low. Later Nikitin and coworkers showed a favorable set of properties in magnesium compounds with the elements of the 4th group.[2] These materials had been studied intensively in the 1960s, but later the interest decreased. The new interest was warmed by the information about high figure of merit achieved in the Mg_2Si–Mg_2Sn solid solutions[3] and growing interest in thermoelectric method of energy conversion.

11.2 Higher Silicides of Transition Metals

Melting points and crystal structure parameters of silicides are shown in Table 11.1. As shown in Table 11.1, all thermoelectrics have a symmetry worse than cubic. Although they have different crystal structures, all these structures (except that of $CrSi_2$) could be considered as slightly deformed tetragonal structures. All unit cells have many atoms that results in high density of states ($d \sim 10^{21}$ cm^{-3}), and low current carrier mobility ($u \sim 10$ cm^2 V^{-1} s^{-1}). Two of these silicides: $FeSi_2$ and Ru_2Si_3 have a phase transition. The phase transition from low-temperature orthorhombic to tetragonal phase takes place when the

TABLE 11.1   Crystal Structures of Silicides, Which Could be Used in Thermoelectric Devices

Material	Melting Point, K	Syngony	a, nm	b, nm	c, nm	$\alpha,^{\circ}$	Reference	Phase Transition[a]
$CrSi_2$	1763	Hexagonal	0.4431	—	0.6364	—	[4,5]	
Mn_4Si_7	1430	Tetragonal	0.5525	—	1.7463	—	[6]	1240 K[b]
$FeSi_2$	1490	Orthorhombic	0.9863	0.7791	0.7833	—	[7]	1210 K (O/T)
Ru_2Si_3	1970	Orthorhombic	1.1074	0.8957	0.5533	—	[8]	1250 K (O/T)
$ReSi_{1.75}$	2213	Triclinic	0.3138	0.3120	0.7670	89.9	[9]	

[a] In this column the lowest temperature of the phase transition and its type are shown. (O/T) means orthorhombic to tetragonal.

[b] The nature of phase transition is not determined yet.

temperature becomes higher than ~1200 K. There were indications that phase transition could take place in higher manganese silicide (HMS).

Differential thermal analysis of some HMS samples made recently[10] showed that at approximately 1240 K calorification or heat absorption takes place when heating or cooling. This effect is reversible and small hysteresis takes place. It shows the possibility of a phase transition at this temperature, which should be confirmed by other experiments.

Crystal structure of these materials has preferential direction of higher symmetry, as a result of which anisotropy of transport properties in these materials could take place. Really, all compounds have anisotropy of all thermoelectric parameters. Besides iron disilicide all compounds allow producing large-enough single crystals, so it is not difficult to measure the anisotropy of their transport properties. In iron disilicide because of the phase transition, it is hardly possible to produce a single crystal from the melt. The chemical transport and high-temperature solution methods are the ways to produce single crystals at the temperature below the phase transition point. For the first time anisotropy of Seebeck coefficient in β-$FeSi_2$ was measured by Japanese researchers.[11] They used their method for the determination of the anisotropy of doped iron disilicide crystals.[12] Last year, using a statistic development of the measurements results,[13] we confirmed significant Seebeck coefficient anisotropy in β-$FeSi_2$.[13] This method is useful and for other silicides.

In Ru_2Si_3 and HMS layered inclusions take place in the crystals in some cases. The influence of these inclusions on thermoelectric properties is under investigation. It is necessary to understand whether the inclusions determine the anisotropy of thermoelectric properties of HMS single crystals. We produce single crystals of HMS by chemical transport[13] and high-temperature solution.[14] In both cases, no inclusions were found in these crystals. Using the above-mentioned method we have shown that Seebeck coefficient anisotropy of these crystals is practically the same as in directionally crystallized crystals. So, one could say that the anisotropy in HMS is determined by HMS itself and not by the inclusions.

In Table 11.2, the updated data for thermoelectric properties of higher silicides of transition metals have been shown. Here ZT_{max}—is the highest dimensionless figure of merit achieved in optimized material; ΔS—is the absolute value of the difference of Seebeck coefficient measured along the highest symmetry axis and in orthogonal direction.

TABLE 11.2   Some Parameters of Higher Silicides of Transition Metals

Material	Type	ZT_{max}	ΔS, μV/K ($T = 300$ K)
$CrSi_2$	p	0.25[15]	70[14]
$MnSi_{1.7}$	p	0.9	99[16]
$FeSi_2$	n, p	0.4(n)[17] 0.3(p)[18,19]	100[11], 150[13]
Ru_2Si_3	n, p	0.4(n)[20]; 0.3(p)[21] 0.8(n)[20]	~25[22]
$ReSi_{1.75}$	p	0.8[23]	~450[24]

It is necessary to say that the properties of a material could be determined well if a few researchers form various laboratories showed similar results. If some result is much better than the other, either it could be wrong, or it could be a new step in the material study. For example, the highest *ZT* value for n-FeSi$_2$ was reported by Hesse[17] in 1964, and then the majority of researcher obtained *ZT* values for β-FeSi$_2$ in the range about 0.2–0.3. In this situation, for example, the value *ZT* = 1 for n-FeSi$_2$[25] should be repeated by other researcher, before one can refer this value as the β-FeSi$_2$ parameter. The similar position was used for some other results showing very high *ZT* values for other silicide materials.

11.3 Mg$_2$Si and Its Solid Solutions

Recently, Mg$_2$Si and the solid solutions between the compounds Mg$_2$Si–Mg$_2$Ge–Mg$_2$Sn attract special attention. These materials show high figure of merit,[3,26] have no toxic components and they have the lowest density among all efficient thermoelectrics.[27]

One can outline three main directions of the study of these materials. These are the study of Mg$_2$Si, Mg$_2$Si–Mg$_2$Ge solid solutions and Mg$_2$Si–Mg$_2$Sn solid solutions. Let us discuss all these directions separately.

11.3.1 Magnesium Silicide

Magnesium silicide has high enough thermal conductivity of crystal lattice ($\kappa_p \sim 9$ W·m^{-1} K^{-1} at RT). It does not allow producing efficient thermoelectric for the room-temperature region. Because lattice thermal conductivity decreases inversely to absolute temperature, at high temperature the figure of merit could be high enough. Another useful feature of this material is strong dependence of lattice thermal conductivity on impurity amount.

Figure 11.1 shows the temperature dependencies of the product $\kappa_p T$ for Mg$_2$Si-based materials. The highest doping level is 1–2 at.% of impurity (samples 4, 7, and 9). Samples 3, 6, and 8 have intermediate doping level. One can see that there is a strong dependence of lattice thermal conductivity on doping level independently of the impurity used. There are some papers where the other impurities (P[28], Al[29], and La[30]) were used for *n*-type material production, but antimony and bismuth allowed obtaining the highest figure of merit.

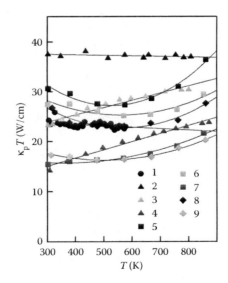

FIGURE 11.1 Temperature dependence of the product $\kappa_p T$ for Mg$_2$Si: nondoped (1, 2, 5) and doped by Sb (3, 4, 6, 7) and Bi (8, 9). 1—[31]; 2–4—[32]; 5–7—[33]; 8, 9—[34].

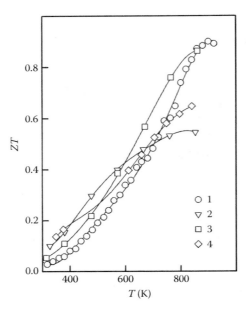

FIGURE 11.2 Temperature dependence of the dimensionless figure of merit for optimum-doped Mg_2Si. 1—[35]; 2—[33]; 3—[34]; 4—[36].

Figure 11.2 shows dimensionless figure of merit for the best samples studied by various researchers. One can see that at room temperature $ZT < 0.1$ whereas it achieves $ZT \approx 0.9$ near 900 K. This material hardly can be used in two-stage module together only with $Bi_2(Te,Se)_3$. Intermediate stage or intermediate material for complex leg is necessary for the temperature range 550–700 K.

11.3.2 $Mg_2Si–Mg_2Ge$ Solid Solutions

These components create continuous solid solutions, which can be produced easily because of low difference between solidus and liquidus temperatures.[37] The first disadvantage of this material is using very expensive germanium and second is relatively high lattice thermal conductivity because the masses of Ge and Si atoms differ not very much.

Figure 11.3 shows dimensionless figure of merit for the best samples of $Mg_2Si–Mg_2Ge$ solid solutions studied by various researchers. Two compositions were studied $Mg_2Ge_{0.3}Si_{0.7}$ and $Mg_2Ge_{0.4}Si_{0.6}$. The highest figure of merit ($ZT \approx 1$) was obtained for the $Mg_2Ge_{0.4}Si_{0.6}$ solid solution,[38] whereas at room temperature the $Mg_2Ge_{0.3}Si_{0.7}$ is better(O. Sh. Gogishvili, private communication). Most probably, that unity is the highest possible figure of merit for these solid solutions because of relatively high lattice thermal conductivity.

11.3.3 $Mg_2Si–Mg_2Sn$ Solid Solutions

The highest figure of merit both at high temperature and at room temperature is achieved in $Mg_2Si–Mg_2Sn$ solid solutions. These solid solutions are the cheapest among the solid solutions based on Mg_2X compounds. They also have the lowest lattice thermal conductivity among them.[27]

These $Mg_2Si–Mg_2Sn$ solid solutions are more difficult in production because of large difference between solidus and liquidus temperatures. Phase diagram of the system $Mg_2Si–Mg_2Sn$[27] shows that there is no solid solution in the range between 40 and 60 mol% of Mg_2Si. In 2004, Japanese researchers reported that they synthesized the $Mg_2Sn_{0.5}Si_{0.5}$ solid solution.[39] Therefore, the phase diagram should be studied additionally.

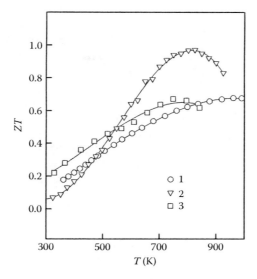

FIGURE 11.3 Temperature dependence of the dimensionless figure of merit for optimum-doped Mg_2Si–Mg_2Ge solid solutions. 1—$Mg_2Ge_{0.3}Si_{0.7}$ (O.Sh. Gogishvili, private communication); 2—$Mg_2Ge_{0.4}Si_{0.6}$[35]; 3—$Mg_2Ge_{0.4}Si_{0.6}$.[38]

In comparison with the Mg_2Si–Mg_2Ge solid solutions, the Mg_2Si–Mg_2Sn solid solutions additionally have a possibility to move subbands of conduction band. It allows increasing power factor at practically the same lattice thermal conductivity.[40]

Figure 11.4 shows dimensionless figure of merit for the best samples of Mg_2Si–Mg_2Sn solid solutions studied by various researchers. The obtained results reliably show the maximum figure of merit higher than unity.

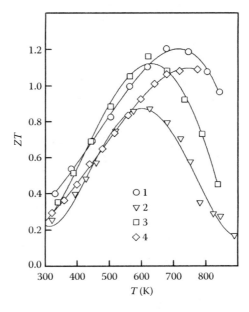

FIGURE 11.4 Temperature dependence of the dimensionless figure of merit for optimum-doped Mg_2Si–Mg_2Sn solid solutions. 1—$Mg_2Si_{0.4}Sn_{0.6}$[3]; 2—$Mg_2Si_{0.5}Sn_{0.5}$[41]; 3—$Mg_2Si_{0.5}Sn_{0.5}$[42]; 4—$Mg_2Si_{0.4}Sn_{0.6}$.[43]

11.4 Nanostructuring

The Mg_2X solid solutions are promising materials. Because they have a cubic structure, they could be prepared in polycrystalline form. Therefore, the bulk material prepared from nanopowder could have significantly higher figure of merit similarly to the result obtained for Bi_2Te_3-based nanocomposites.[44]

Nevertheless, it is hardly possible to accept the explanation of high figure of merit in these materials by the existence of natural nanostructures.[43] It was shown earlier that both lattice thermal conductivity and power factor are in good agreement with the existing theory of solid state. In the framework of Klemens' theory,[45] lattice thermal conductivity of all solid solutions in the three binary systems could be described very well.[46] The increase of power factor of $Mg_2Si_{0.4}Sn_{0.6}$ in comparison with other solid solutions is described by optimum positions of conduction band subbands[40] and by the absence of interband scattering in these subbands.[47] Therefore, if the found natural nanostructures increase thermoelectric figure of merit there are other factors, which decrease it, because the properties of such samples are practically the same as in usual materials. Destroying this unknown factor one can increase the figure of merit of these materials additionally.

11.5 *P*-Type Materials

When presenting a new *n*-type material, the first question the presenter should answer is the question about a possibility to produce similar *p*-type material. In the solid solution systems under discussion, all good results have been obtained for *n*-type material. If the electron energy spectrum is favorable for creation of *n*-type material, the hole energy spectrum is not favorable. Nevertheless, the attempts to develop *p*-type material take place and some first results have been obtained.

Figure 11.5 shows dimensionless figure of some *p*-type materials in Mg_2Si–Mg_2Sn–Mg_2Ge system.

The obtained results show that now figure of merit of *p*-type material is too low to compete with the best *p*-type materials.

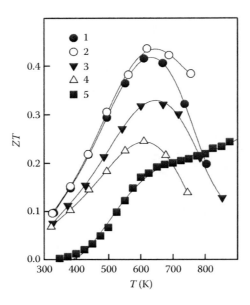

FIGURE 11.5 Temperature dependence of the dimensionless figure of merit for *p*-type Mg_2Si–Mg_2Sn–Mg_2Ge solid solutions. $1-Mg_2Si_{0.4}Sn_{0.6}$; $2-Mg_2Si_{0.3}Sn_{0.7}$; $3-Mg_2Ge_{0.4}Sn_{0.6}$; $4-Mg_2Ge_{0.3}Sn_{0.7}$; $5-Mg_2Ge_{0.4}Si_{0.6}$. $1-4-[48]$, $5-[35]$.

11.6 Devices

Thermoelements on the base of silicon compounds have been produced for long time. As much as we know the first reliably working thermoelectric generator with p-leg of higher manganese silicide and n-leg of $Mg_2Ge_{0.3}Si_{0.7}$ was developed in the USSR and had been worked for 5000 h at a maximum working temperature of 920 K.

Recently, Komatsu company developed thermoelectric modules where the Mg_2Si–Mg_2Sn solid solution was used as n-leg and higher manganese silicide as p-leg. It allowed achieving a very high coefficient of performance for double-stage module (12%).[49] The low-temperature stage was made of the materials based on Bi_2Te_3–Bi_2Se_3–Sb_2Te_3 solid solutions.

These results show that the industrial application of these materials could be achieved soon.

11.7 Discussion

There are two principal ways for the development of efficient silicide materials. The first one is the traditional way to find new impurities and/or the method of material production able to produce more homogeneous material with lower amount of defects. This way, probably, could give essential *ZT* increase for some materials. This way is the preferable one for the materials studied insufficiently such as chromium disilicide or higher silicides of rhenium and ruthenium. For other materials, where the majority of possible impurities have been tested, only the creation of some structures could give essential increase of the figure of merit. In this case, the nanostructure creation is of the highest interest. It is necessary to discuss separately the nanostructure creation in magnesium-based compounds and higher silicides of transition metals. Magnesium-based compounds have a cubic structure and it is possible to produce polycrystalline material with practically the same properties as in a single crystal or even with better properties, if boundary scattering gives useful effect for *ZT* increase. Probably, some increase of *ZT* could be achieved when using nanosize powder for material preparation.

Another situation takes place for higher silicides of transition metals. All these materials are strongly anisotropic and their polycrystalline form has a lower figure of merit, as a rule, than that in some direction of a single crystal. It is connected with eddy currents arising in the polycrystalline medium. A good example of such a situation is bismuth telluride, having the direction, where the maximum figure of merit takes place. The anisotropy is stronger for *n*-type material than for *p*-type. Therefore, it was predictable that figure of merit increase caused by nanostructuring was achieved in *p*-type material[44] and no increase was achieved yet for *n*-type material. Nevertheless, in pure higher manganese silicide some effect of "positive" averaging takes place,[50] when the figure of merit of pressed material is higher than that in any direction of single crystal, although no positive averaging takes place in heavily doped material. It has the highest figure of merit only in the textured crystal where all grains are oriented practically in the same direction. The study of the role of milling parameters on the thermoelectric properties of mechanically alloyed chromium silicide showed additional danger in nanograin production—contaminating the material by mill materials.[51] Therefore, only an experimental study could show the possibility of figure of merit increase due to using the nanosize grains.

There is one more way for the improvement of material thermoelectric properties. It is the creation of multiphase structure. If scattering conditions for electrons and phonons are favorable, *ZT* can be increased. Such a system has to be thermodynamically stable at working temperature to be used for long time. The study of such a system, based on higher manganese silicide, has been carried out in 1983.[52] There was no essential *ZT* increase in this system. There are similar works on improvement of higher manganese silicide properties by adding $CeSi_2$ micrograins.[53] Unfortunately, micrograins of $CeSi_2$ gave no positive effect to power factor also. This discussion shows that although there was no new step in the improvement of thermoelectric properties, nobody showed the impossibility of the increase of figure of merit using one of these methods.

11.8 Conclusion

Silicides are promising thermoelectrics, some of them show high thermoelectric figure of merit. Higher transition metal silicides show unusual transport properties, and it can be useful in the search of new thermoelectrics. All these materials probably can be used for the creation of anisotropic thermoelements. Following study of the physics of silicides could open new ways for thermoelectrics improvement. The materials based on magnesium silicide even now have high figure of merit and are very close to industrial application.

References

1. Nikitin E.N. *Zhurnal Tekhnicheskoj Fiziki* 28, 23, 1958 (in Russian).
2. Nikitin E.N., Bazanov V.G., Tarasov V.I. *Sov. Phys. Solid State* 3(12), 2650–2651, 1961.
3. Zaitsev V.K., Fedorov M.I., Gurieva E.A., Eremin I.S., Konstantinov P.P., Samunin A.Yu., Vedernikov M.V. *Phys. Rev. B* 74, 045207, 2006.
4. Maex K. van Rossum M. (Ed.), *Properties of Metal Silicides*, INSPEC, London, 1995. See articles by M. Ostling and C. Zaring (pp. 15 and 31).
5. Samsonov G.V., Vinetskii I.M. *Handbook of Refractory Compounds*, Plenum Press, New York, 1980.
6. Karpinskij O.G., Evseev V.A. *Izv. AN SSSR (Neorg. Mater.)* 5, 525, 1969 (in Russian).
7. Wandji R., Dusausoy Y., Protas J., Roques B. *C.R. Acad. Sci. C.* 267, 1587, 1968.
8. Poutcharovsky D.J., Parthe E. *Acta Crystallogr. B* 30, 2692, 1974.
9. Gottlieb U., Lambert-Andron B., Nava F., Affronte M., Laborde O., Rouault A., Madar R. *J. Appl. Phys.* 78, 3902, 1995.
10. Fedorov M.I., Germanovich O.S., Zaitsev V.K. *Thermoelectrics and Their Applications,* Ioffe Physical-Technical Institute, St. Petersburg, p. 269, 2008 (in Russian).
11. Kuramitsu M., Yoshio M., Takeda M., Ohsugi I.J. 23rd International Conference on Thermoelectrics, Adelaide, Australia, *Proceedings ICT'2004.* IEEE, #125, 2005.
12. Ishida Y., Kayamura K., Takeda M., Ohsugi I.J. XXV International Conference on Thermoelectrics, Vienna, Austria, *Proceedings ICT'2006*, IEEE, 566, 2006.
13. Zaitsev V.K., Andreev A.A., Ivanov Yu.V., Fedorov M.I. *Thermoelectrics and Their Applications*, Ioffe Physical-Technical Institute, St. Petersburg, 226, 2008 (in Russian).
14. Solomkin F.Yu., Zaitsev V.K., Kartenko N.F., Kolosova A.S. *Thermoelectrics and Their Applications*, Ioffe Physical-Technical Institute, St. Petersburg, 257, 2008 (in Russian).
15. Voronov B.K., Dudkin L.D., Trusova N.N. *Khimicheskaya svyaz v poluprovodnikah*, Minsk, "Nauka i Tekhnika," 291, 1969 (in Russian).
16. Zaitsev V.K. *CRC Handbook of Thermoelectrics*, ed. Rowe D.M., CRC Press, New York, p. 299, 1995.
17. Hesse J. *Z. Angew, Phys.*, 28, 133, 1969.
18. Birkholz U., Gross E., Stoehrer U. Polycrystalline, in *CRC Handbook of Thermoelectrics*, ed. by Rowe D.M., CRC Press, New York, p. 287, 1995.
19. Chen H.Y., Zhao X.B., Stiewe C., Platzek D., Mueller E. *J. Alloys Comp.*, 433, 338, 2007.
20. Arita Y., Mitsuda S., Nishi Y., Matsui T., Nagasaki T. *J. Nucl. Mater.* 294, 202, 2001.
21. Ivanenko L., Filonov A., Shaposhnikov V., Krivosheev A., Behr G., Souptel D., Schumann J. et al. *Twenty-Second International Conference on Thermoelectrics*, La Grande Motte, France, *Proceedings of ICT'03*, IEEE, p. 157, 2003.
22. Simkin B.A., Hayashi Y., Inui H. *Intermetallics* 13, 1225, 2005.
23. Sakamaki Y., Kuwabara K., Jiajun G., Inui H., Yamaguchi M., Yamamoto A., Obara H. *Mater. Sci. Forum* 426–432, 1777, 2003.
24. Gu J.-J., Oh M.-W., Inui H., Zhang D. *Phys. Rev. B* 71, 113201, 2005.
25. Sugihara, S., Morikawa, K., Igarashi, Y., Nishiyama, K. XXV International Conference on Thermoelectrics, Vienna, Austria, *Proceedings ICT'06*, IEEE, p. 711, 2006.

26. Fedorov M.I., Zaitsev V.K., Vedernikov M.V. XXV International Conference on Thermoelectrics, Vienna, Austria, *Proceedings ICT'06*, IEEE, p. 111, 2006.
27. Zaitsev Z.V.K., Fedorov M.I., Eremin I.S., Gurieva E.A., *Thermoelectrics Handbook: Macro to Nano*, CRC Press, New York, Chapter 29, 2006.
28. Tani J., Kido H. *Jap. J. Appl. Phys.* 46, 3309, 2007.
29. Tani J., Kido H. *J. Alloys Compounds* 466, 335, 2008.
30. Zhang Q., He J., Zhao H.B., Zhang S.N., Zhu T.J., Yin H., Tritt T.M. *J. Phys. D* 41, 185103, 2008.
31. Labotz R.J., Mason D.R. *J. Electrochem. Soc.* 110, 120, 1963.
32. Zaitsev V.K. *Thermoelectric Properties and Thermal Conductivity of Some Solid Solution Based on Magnesium Stannide*, PhD thesis, Leningrad, 1969.
33. Tani J., Kido H. *Intermetallics* 15, 1202, 2007.
34. Tani J.-I., Kido H. *Physica B* 364, 218, 2005.
35. Mars K., Ihou-Mouko H., Pont G., Tobola J., Scherrer H. *J. Electron. Mater.* 38, 1360, 2009.
36. Akasaka M., Iida T., Matsumoto A., Yamanaka K., Takanashi Y., Imai T., Hamada N. *J. Appl. Phys.* 104, 013703, 2008.
37. Labotz R.J., Mason D.R., O'Kane D.F. *J. Electrochem. Soc.* 110, 127, 1963.
38. Akasaka M., Iida T., Nishio K., Takanashi Y. *Thin Solid Films* 515, 8237, 2007.
39. Isoda Y., Shioda N., Fujiu H., Imai Y., Shinohara Y. *23rd International Conference on Thermoelectrics*, Adelaide, Australia, *Proceedings ICT'2004*, #124.pdf, 2004.
40. Pshenay-Severin D.A., Fedorov M.I. *Phys. Sol. St.* 49, 1633, 2007.
41. Isoda Y., Nagai T., Fujiu H., Imai Y., Shinohara Y. XXVI International Conference on Thermoelectrics, Jeju, Korea, *Proceedings ICT'07*, IEEE, 251, 2007.
42. Isoda Y., Nagai T., Fuziu H., Imai Y., Shinohara Y. XXV International Conference on Thermoelectrics, Vienna, Austria, *Proceedings ICT'06*, IEEE, p. 406, 2006.
43. Zhang Q., He J., Zhu T.J., Zhang S.N., Zhao X.B., Tritt T.M. *Appl. Phys. Lett.* 93, 102109, 2008.
44. Poudel B., Hao Q., Ma Y., Lan Y., Minnich A., Yu B., Yan X. et al. *Science* 320, 634, 2008.
45. Klemens P.G., *Phys. Rev.* 119, 507, 1960.
46. Zaitsev V.K., Tkalenko E.N., Nikitin E.N. *Sov. Phys. Solid State* 11, 274, 1969.
47. Fedorov M.I., Pshenay-Severin D.A., Zaitsev V.K., Sano S., Vedernikov M.V. *Twenty-Second International Conference on Thermoelectrics*, La Grande Motte, France, *Proceedings of ICT'03*, IEEE, p. 142, 2003.
48. Fedorov M. I., Zaitsev V. K., Isachenko G. N., *Solid State Phenom.*, 170, 286, 2011.
49. Kaibe H., Aoyama I., Mukoujima M., Kanda T., Fujimoto S., Kurosawa T., Ishimabushi H., Ishida K., Rauscher L., Sano S. XXIV International Conference on Thermoelectrics, Clemson, SG, USA, *Proceedings ICT'05*, IEEE, p. 227, 2005.
50. Zaitsev V.K., Engalychev A.E., Ktitorov S.A., Petrov Yu.V., Rakhimov K.A. Incommensurate structures and transport properties of higher manganese silicide and some materials on its basis. *Preprint of Ioffe Physical-Technical Institute*, 1983, Leningrad, 46pp.
51. Dasgupta T., Umarji A.M. *J. Alloys Comp.* 461, 292, 2008.
52. Zaitsev V.K., Rakhimov K.A., Engalychev A.E. *Appl. Solar Energy* (English translation of *Geliotekhnika*) 25, 15, 1989.
53. Zhou A.J., Zhu T.J., Ni H.L., Zhang Q., Zhao X.B. *J. Alloys and Comp.* 455, 255, 2008.

Thermoelectric Properties of Intermetallic Hybridization Gap and Pseudo-Gap Systems: Fe$_2$VAl and CoSi

Donald T. Morelli
Michigan State University

12.1 Introduction

Many scientific approaches, both experimental and theoretical/computational, have been brought to bear on the problem of thermoelectric materials science. The ultimate goal of these studies is, of course, to maximize the thermoelectric figure of merit Z of a material. Large Z paves the way for many applications in both power generation and solid state heating and cooling. The figure of merit itself is comprised of the three fundamental transport parameters of Seebeck coefficient (S), electrical conductivity (σ), and thermal conductivity (κ) and is given by the equation

$$Z = \frac{S^2 \sigma}{\kappa} \tag{12.1}$$

Even a precursory examination of Equation 12.1 reveals the conundrum of thermoelectricity: this particular combination of these three parameters is not conducive to producing high figure of merit. The tendency in nature is that, for instance, high electrical conductivity σ, as in a good metal like copper or silver, is accompanied by minute Seebeck coefficient, and vice versa; and similarly high electrical conductivity also implies high thermal conductivity. These "contraindicated" properties have frustrated the development of materials with high figure of merit for decades if not centuries. Noting that the dimensions of Z are inverse temperature, it is traditional to classify thermoelectric materials by the dimensionless quantity ZT, where T is the absolute temperature. Despite intense experimental and theoretical scrutiny of this problem, few thermoelectric materials have been discovered with ZT values exceeding unity. New approaches continue to be brought to bear on this issue in the hopes that materials with much higher figure of merit can be realized.

A bit more insight into this intriguing and knotty problem can be attained by noting that the thermal conductivity is the sum of electronic (κ_E) and lattice (k_L) parts:

$$ Z = \frac{S^2\sigma}{\kappa} = \frac{S^2 ne\mu}{\kappa_E + \kappa_L} = \frac{S^2\sigma}{LT\sigma + \kappa_L} \tag{12.2} $$

and writing the electronic part in terms of the electrical conductivity via the Wiedemann–Franz relation. In the limit of very high σ (a metal), the dimensionless figure-of-merit approaches S^2/L. If we allow the Lorenz number to assume its Sommerfeld value, this implies that for ZT to exceed unity, S must be larger for 156 uV K^{-1}. Such large values of Seebeck coefficient for metallic conductors are not known. In the limit of small electronic thermal conductivity, ZT approaches $S^2\sigma T/\kappa_L$, and high Z can be achieved either by maximizing the numerator $S^2\sigma$, know as the power factor P, or by minimizing the lattice thermal conductivity.

The development of new thermoelectric materials in the nearly two centuries since the discovery of these effects has largely focused on identifying semiconductors for which the power factor is optimized, and then applying various techniques, amongst these solid solution formation and grain size reduction, to reduce the lattice thermal conductivity. This approach was largely successful as it led ultimately to the development of polycrystalline semiconductors such as $(Bi,Sb)_2(Te,Se)_3$, $(Pb,Sn)(Te,Se)$, and (Si,Ge). However, it should be pointed out that one cannot expect to increase ZT to arbitrarily high values since the lattice thermal conductivity cannot be reduced below a minimum value corresponding to a phonon mean free path of one (or one half) interatomic spacing. Significantly larger values of ZT will be achieved only in materials which have high power factor P.

In this chapter we focus on a class of intermetallic compounds consisting of alloys between transition metal elements with partially filled d-shells, principally Fe and Co, and elements from the main group of the periodic table, mainly Si and Al. In order to focus the discussion we consider the cases of Fe_2VAl and $CoSi$. Interestingly, these alloy systems possess some of the largest known thermoelectric power factors in the near-room temperature range.

12.2 Electronic Band Structure of Transition-Metal Aluminides and Silicides

Most alloys between metallic elements are metals themselves. In some special cases, notably those involving transition metals and aluminum or silicon, both theoretical predictions and experiments suggest the existence of a gap in the band structure. Some classic examples of this behavior are provided by $FeSi$[1,2] and $RuAl_2$.[3–5] It has long been suggested that such behavior is due to hybridization effects between transition-metal d-electrons and (s,p) electrons originating from Al (or Si).

Using first principles electronic band structure methods, Weinert and Watson studied theoretically the occurrence of such behavior in some 150 different binary and ternary alloys based on transition metals and aluminum.[6] They showed that in some circumstances a true gap is formed while in others only a deep "well" occurs in the density of states at the Fermi level. The former situation gives rise to semiconductor-like behavior whereas the latter implies a semimetallic state. Two examples of this type of behavior are given in Figure 12.1, which shows the electronic density of states versus energy for $RuAl_2$ and Fe_2VAl. In this figure the zero in energy corresponds to the Fermi energy. It is clearly seen (Figure 12.1a) that for $RuAl_2$ their calculations predict a gap in the DOS into which the Fermi level falls; this system should be a semiconductor, and indeed infrared absorption measurements detect such a state. For Fe_2VAl,

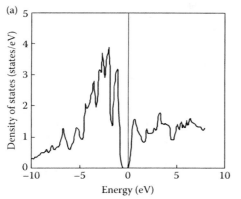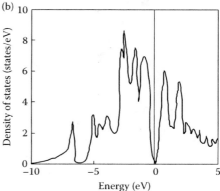

FIGURE 12.1 Electron density of states (DOS) for (a) $RuAl_2$ and (b) Fe_2VAl. In the former case, a true gap forms near the Fermi energy ($E = 0$) while in the latter a deep well, or "pseudo-gap," forms, leading to a semimetallic state. (Adapted from M. Weinert and R.E. Watson, *Physical Review B* **58**, 9732, 1998.)

on the other hand, the strong dip in the density of states near the Fermi energy does not extend all the way down to the abscissa; instead there is a deep "well" in the DOS near the Fermi energy. This means that for this system both electrons and holes are available for transport, and we have a semimetal.

A similar situation can occur with alloys of transition metals and silicides, the previously mentioned case of FeSi being one example. Here we focus on the alloy CoSi.

Pan et al., performed first principles band structure calculations of this alloy and showed that, like Fe_2VAl, the band structure features a minimum, but not a gap, very close to the Fermi energy.[7] These calculations confirmed and extended some studies by earlier workers.[8–10] Once again this behavior can be attributed to hybridization effects involving the d-electrons of Co with the (s,p) electrons of Si. The situation with regard to electron DOS for Fe_2VAl and CoSi is quite different from that expected for a bulk three-dimensional solid in the free electron approximation, for which one expects N(E) to vary as $E^{1/2}$ up to the Fermi energy.

12.3 Implications for Thermoelectricity

The existence of gaps or deep wells in the density of states of transition metal aluminides and silicides can have important consequences for their thermoelectric properties, the Seebeck coefficient in particular. To see why this is so, we can write the conductivity and Seebeck coefficient using the following expressions[11]:

$$\sigma(E) = ne\mu \propto g(E)\mu(E) \tag{12.3}$$

$$S = \frac{\pi^2}{3}\frac{k_B}{e}\left(k_B T\right)\frac{1}{\sigma}\left(\frac{d\sigma(E)}{dE}\right)_{E=E_F} = \frac{\pi^2}{3}\frac{k_B}{e}\left(k_B T\right)\left(\frac{1}{n}\frac{dn}{dE} + \frac{1}{\mu}\frac{d\mu}{dE}\right) \tag{12.4}$$

where n is the carrier concentration and μ the carrier mobility. It is clear that the conductivity, for a given carrier mobility, increases with increasing density of states $g(E)$, as this implies an increasing carrier concentration. The Seebeck coefficient, on the other hand, depends on the logarithmic energy derivative of the conductivity, and thus on the sum of the logarithmic energy derivatives of the carrier concentration and the electron mobility. Thus one might expect that any increase of the density of states near the Fermi energy can give rise to an increase in Seebeck coefficient. The sharp features in the DOS for Fe_2VAl and $CoSi$, as well as other transition metal alloys, are one example of this situation. In the next sections we will summarize theoretical and experimental research that has been done on these two classes of materials.

12.4 Full Heusler Alloys Based on Fe_2VAl

The full Heusler alloy Fe_2VAl was first studied within the context of the strengthening of the intermetallic alloy Fe_3Al by transition metal substitutions on the Fe site. Nishino et al.,[12] reported that the substitutional alloys $(Fe_{1-x}V_x)_3Al$ remained single phased and isostructural (D03 structure) out to $x = 0.4$. They observed that the lattice parameter first decreased, reaching a minimum near $x = 0.33$ before increasing again. The value of $x = 0.33$ corresponds to the Heusler alloy Fe_2VAl. The substitution of V for Fe in this system is very similar to that which occurs in Fe_3Si.[13] A very surprising finding was the appearance of a negative temperature coefficient of resistivity (TCR), even at small values of vanadium substitution.[14] For x values less than 1/3, a crossover from a metallic resistivity at low temperature to a region of negative TCR occurred close to the Curie temperature of the alloy. At $x = 1/3$ (Fe_2VAl), the resistivity remained semiconductor-like down to 2 K and reached values approaching 4000 μΩ cm. Photoemission experiments by the same authors, however, clearly demonstrated the absence of an energy gap in the electronic structure.

Some understanding of the behavior of Fe_2VAl was provided by Weinert and Watson[6] who showed that, in this compound as well as several other transition metal aluminides, hybridization between Al and transition metal atoms results in a deep well or pseudogap in the electronic band structure near the Fermi level. In some cases, such as $RuAl_2$, a true gap is formed in the density of states and the material is a true semiconductor. In others, such as Fe_2VAl, a true gap does not form but the density of states exhibits a deep well, characteristic of a semimetallic state. Several other band structure calculations have confirmed the semimetal character of these latter alloys.[15–17] In either case, however, the rapidly varying energy dependence of the density of states near the Fermi level is expected to give rise to large Seebeck coefficients.

Nishino et al.[18] first reported the effect of off-stoichiometry on the thermoelectric properties of Fe_2VAl. He showed that the stoichiometric compound exhibited semiconductor-like behavior with modest (~ + 40 μV K^{-1} at 300 K) Seebeck coefficient. However, in slightly Al-deficient samples, the resistivity at low temperature assumed a metallic behavior and the Seebeck coefficient evolved to large (~ −140 μV K^{-1} at 250 K) negative values. This was consistent with the Fermi level moving away from the center of the pseudogap and close to the edge of the conduction band. By combining their resistivity and Seebeck results, one calculates a thermoelectric power factor of approximately 30 μW cm^{-1} K^{-2}, comparable to state-of-the-art thermoelectric alloys such as Bi_2Te_3 near and below room temperature. Large power factors are a hallmark of the pseudogap state, arising from a combination of a nearly metallic resistivity and large Seebeck coefficient.

Vasundhara et al.[19] explored the effect on the thermoelectric properties of Si doping on the Al site in Fe_2VAl alloys. These authors found that, similar to Al-deficiency, Si doping gives rise to a large negative Seebeck coefficient and a metallic resistivity. At a doping level of 3% Si the Seebeck coefficient exhibits a negative minimum value of −230 μV K^{-1} at 200 K. In spite of the high Seebeck coefficient, the resistivity of the doped alloys is less than 1 mΩ cm, the prototypical value for a thermoelectric material. The highest power factor occurs for Si content of 6% and exceeds 80 μW cm^{-1} K^{-2} at 150 K, a factor of 2 larger than Bi_2Te_3.

Nishino et al.[20] prepared Ge-doped samples with results similar to those for Si-doped samples. The largest negative Seebeck coefficient in this system was −140 μV K^{-1} at 250 K for 5% doping. The resistivity of this sample was on the order of 0.3 mΩ cm, yielding a power factor of approximately 65 μW cm^{-1} K^{-2}.

Skoug et al.[21] confirmed the large Seebeck coefficient and power factor in $Fe_2VAl_{1-x}Si_x$ and also found similar results for Sn-doped samples. The maximum power factors in their samples were 50 μW cm^{-1} K^{-2} near 250 K and 45 μW cm^{-1} K^{-2} near 300 K in 10% Si- and Sn-doped samples, respectively. Lue et al.[23] also studied $Fe_2VAl_{1-x}Si_x$ system and reported that the largest Seebeck coefficient (~ −130 μV K^{-1}) occurred near 300 K for a sample with $x = 0.05$.

Lue and Kuo[24] investigated the thermoelectric properties of samples containing excess V. They found that while the Seebeck coefficient exhibited a trend similar to that found with Al deficiency and Group-IV element doping, the resistivity increased rather than decreased. They took this as evidence that the band overlap was decreased with increasing vanadium concentration. Some of the results on the temperature dependence of the Seebeck coefficient in optimized samples are shown in Figure 12.2.

Figure 12.3 is a compilation of room temperature Seebeck data as a function of the concentration x of the substitutional atom in $Fe_2VAl_{1-x}M_x$ (M = Si, Ge, or Sn). For all these systems we see clearly that the Seebeck coefficient reaches an absolute maximum in the vicinity of $x = 0.1$ before rapidly decaying back toward more metallic values for x approaching unity (Fe_2VSi). At this composition, the Fermi level moves close to the conduction band edge and maximum Seebeck coefficient is achieved.

In spite of the observation of large thermoelectric power factor in Fe_2VAl-based alloys, in some cases twice that of Bi_2Te_3, none of these alloys exhibit high thermoelectric figure of merit because the thermal conductivity is too large. Large thermal conductivity is a combination of a large electronic component (on the order of 3–4 W m^{-1} K^{-1}) and a substantial lattice thermal conductivity (on the order of 20 W m^{-1} K^{-1}). With a total thermal conductivity on the order of 25 W m^{-1} K^{-1} the dimensionless figure of merit is only on the order of 0.1, even for samples with the highest power factor. Some work has demonstrated that the thermal conductivity can be reduced by impurity substitution. In the work by Lue et al. on $Fe_2VAl_{1-x}Si_x$, for instance,[23] the thermal conductivity for $x = 0$ was found to be reduced by approximately a factor of 4 from the peak value near 50 K and by a factor of 2 at room temperature. Most of this reduction was due to a decrease in lattice thermal conductivity. The strong effect of Si substitution is surprising, since the mass difference between Al and Si is quite small. The authors attribute their observations to the appearance of another type of defect, such as a vacancy, upon substitution on the Al site. In their work on $Fe_2VAl_{1-x}Ge_x$, Nishino et al.[20] found a room temperature reduction in thermal conductivity of closer to a factor of 3, consistent with the larger mass difference between Ge and Al.

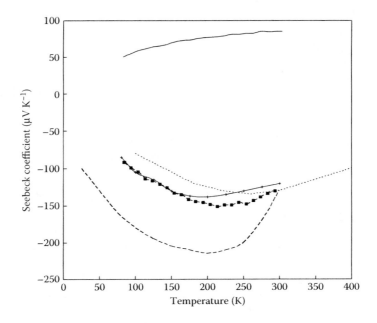

FIGURE 12.2 Seebeck coefficient below room temperature for some $Fe_2VAl_{1-x}M_x$ samples. Solid line: Fe_2VAl,[21] solid squares: M = Si, $x = 0.06$,[21] dotted line: M = Ge, $x = 0.05$,[20] crosses: M = Sn, $x = 0.05$,[21] dashed line: M = Si, $x = 0.06$.[19]

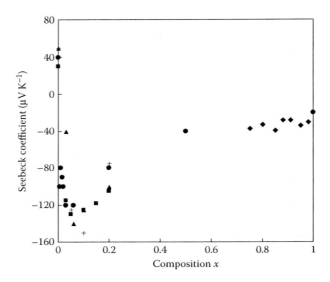

FIGURE 12.3   Seebeck coefficient as a composition for $Fe_2VAl_{1-x}M_x$. The Seebeck coefficient reaches an absolute maximum near $x = 0.1$, where the Fermi energy moves close to the conduction band edge. Sample designation: circles: M = Si,[19] squares: M = Ge,[20] triangles: M = Si,[21] crosses: M = Si,[21] and diamonds: M = Si.[22]

In spite of these efforts on reducing lattice thermal conductivity, Fe_2VAl-based alloys do not have ZT values exceeding about 0.1. In order to achieve higher values of ZT, further reductions in lattice thermal conductivity are necessary, together with reduction in electronic thermal conductivity. The latter might be achieved by, for instance, finding alloys of the same or similar crystal structure with true gaps in their electron density of states, thus allowing for smaller carrier concentrations and higher Seebeck coefficients.

12.5 Alloys Based on CoSi

The intermetallic alloy CoSi also exhibits a band structure featuring a deep well in the density of states. The crystal structure of this alloy is cubic, space group $P2_13$ (no. 198).[25] Single crystals have been successfully synthesized using both the Bridgman[8] and floating zone techniques.[26] Single-phase polycrystalline samples can be formed by the arc-melting process.[27]

The electronic properties of these alloys were first studied many years ago by Asanabe et al.[28] as part of a study of FeSi. Pure CoSi samples were found to be n-type with mobility near 100 $cm^2 V^{-1} s^{-1}$ at room temperature and approaching 1000 $cm^2 V^{-1} s^{-1}$ at cryogenic temperatures. With addition of Fe both the Hall coefficient and Seebeck coefficient eventually cross over to positive values. The authors concluded that CoSi is a semimetal with a small (20 meV) band overlap and heavy effective masses (2 m_e for electrons and 4–10 m_e for holes, depending on Fe content).

The semimetallic nature of CoSi was confirmed by the recent band structure calculations of Pan et al.[7] The hole mass is heavier than that of the electron mass, but the volume of the electron pockets is larger than that of the holes; both of these theoretical results are consistent with the experimental observations. While earlier work found that the Seebeck coefficient of CoSi was only about −50 uV/K, more recently, it has been demonstrated that this system can display Seebeck coefficient close to −100 uV/K. Fedorov and Zaitsev[29] have pointed out that the Seebeck coefficient is reduced in samples that deviate from the stoichiometric 1:1 ratio. Perhaps most surprisingly, however, is that this system achieves quite large Seebeck coefficient in the presence of high electrical conductivity. This has been demonstrated in several recent studies of doped CoSi which we describe presently.

Ren et al.[26] studied the influence of both B and Ge substitution for Si on CoSi. Interestingly, they found that upon substitution of 2% Ge and 0.5% B, the electrical resistivity and lattice thermal conductivity were both decreased while the absolute magnitude of the Seebeck coefficient increased. This resulted in room temperature power factors in the range of 35–50 μW cm^{-1} K^{-2}. Kuo et al.[30] investigated CoSi$_{1-x}$Ge$_x$ and found that while undoped CoSi had a resistivity near 2 mΩ cm at 300 K, samples doped with x in the range of 0.02–0.15 had resistivity an order of magnitude lower. In spite of this large drop in resistivity, the Seebeck maintained values in the range of 60–80 μV/K, once again leading to substantial thermoelectric power factor. A third study on Ge-doped CoSi was carried out by Skoug et al.,[31] who found absolute Seebeck coefficients in the range of 70–100 μV/K, resistivity near 0.2 mΩ cm, and power factor as high as 60 μW cm^{-1} K^{-2}. These authors extended the range of Ge substitution to all the way to $x = 0.5$, consistent with the earlier work of Wald and Michalik[32] showed that the solubility limit of Ge in CoSi was as high as $x = 0.67$. These Ge-substituted results are interesting because in a rigid band model one would not expect doping to occur under Ge substitution. It is possible that the large decrease in resistivity is not an intrinsic effect but rather is related to an improvement in microstructure, as several workers note that while pure CoSi is very brittle, samples with substituted Ge and B seem to be more mechanically robust. Clearly more detailed work on the microstructure in these systems would be very desirable.

Interestingly, although one might expect B to be a p-type dopant in CoSi, the results of Ren et al.[26] show that, at least for the one sample studied with a doping level of $x = 0.005$, the Seebeck coefficient remained negative in the temperature range 300–900 K. It is not clear from this study, however, whether the B is actually substituted for Si. This is not the case for Al doping, which shows a clear conversion to p-type behavior. Lue et al.[27] prepared CoSi$_{1-x}$Al$_x$ samples by arc melting with x ranging from 0 to 0.15 and found that for all samples with $x > 0$ the Seebeck coefficient became positive, at least for some temperatures in the range of 4–400 K. For $x = 0.02$ the Seebeck coefficient went through a maximum near 50 K and crossed over to negative values above about 100 K. For higher doping levels, the peak temperature moved to higher temperature. This behavior can be understood in terms of a two-band model with the Seebeck coefficient determined by the partial Seebeck coefficient of electrons and holes weighted by their respective partial conductivities. In spite of a decrease in resistivity with Al doping, the power factors for these alloys were moderate due to the small value of Seebeck coefficient arising from the compensation of electrons and holes.

A second study concerning Al doping was performed by Li et al.[33] on single crystals of CoSi synthesized using a floating-zone technique. From x-ray and SEM studies, these authors found a clear indication that Al can substitute for Si up to at least $x = 0.2$, with the maximum solubility somewhere in the range $0.2 < x < 0.3$. These authors also identified a clear hole doping effect from Hall measurements. Interestingly, the resistivity strongly increased with Al doping, a trend exactly opposite of that observed by Lue et al.[27] As Li et al. point out, however, the earlier study was on arc-melted samples, which displayed an order of magnitude higher resistivity, even for $x = 0$. Again this suggests that it is likely that defects play a crucial role in determining the electrical transport in polycrystalline CoSi. In spite of this difference in resistivity, the results of Li et al. and Lue et al. on Seebeck coefficient are largely in agreement, with positive values exhibited at low temperature and a crossover to negative values at high temperature.

Several measurements of the thermal conductivity of CoSi and various doped samples and solid solutions have been reported. Figure 12.4 shows a collection of data on pure CoSi. The data of Ren et al.[26] were taken on a single crystal while the other specimens shown here were polycrystals synthesized by arc melting. The thermal conductivity shows a peak near 50 K with a value near 50 W m^{-1} K^{-1}. Room temperature values range between 13 and 16 W m^{-1} K^{-1}, and the thermal conductivity is practically constant from room temperature to 1000 K. Part of the slight variation between the measurements can be accounted for by different amounts of electronic contribution to the thermal conductivity, which is on the order of 25% at room temperature.

The moderately large lattice thermal conductivity of CoSi limits ZT ~ 0.1 at 300 K in spite of the large power factor. Recognizing this, several workers have explored reducing the thermal conductivity using mass-difference scattering by doping or solid solution formation. Some of the results are summarized in

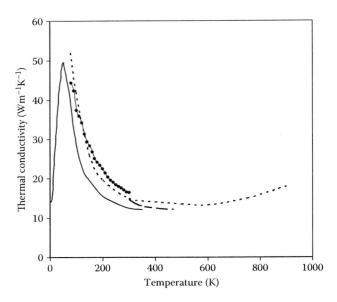

FIGURE 12.4 Thermal conductivity of pure CoSi, as measured by various workers. Solid line: Ref. 30; dotted line: Ref. 29; dashed line: Ref. 26; solid circles: this author (unpublished).

Figure 12.5. Kuo et al.[30] studied the effect of Ge substitution for silicon up to $x = 0.15$ and found the peak value of thermal conductivity at 50 K was suppressed by more than a factor of 2 but at room temperature the diminution was a modest 10%. Very similar results were obtained by Lue et al. for Al substitution,[27] even though the mass difference between Al and Si is much less than that between Ge and Si. The authors suggest that, similar to the findings in $Fe_2VAl_{1-x}Si_x$, this strong effect is not directly due to Al substitution but has its origin in the appearance of some other type of defect such as a vacancy in the doped samples.

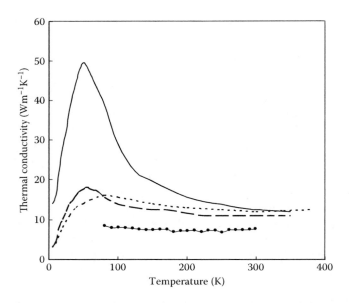

FIGURE 12.5 Thermal conductivity of some of $CoSi_{1-x}M_x$ samples. Solid line: pure CoSi, Ref. 30; dashed line: M = Ge, $x = 0.15$, Ref. 30; dotted line: M = Al, $x = 0.15$, Ref. 27; filled circles: M = Ge, $x = 0.5$, this author (unpublished).

12.6 Summary

Intermetallic alloys of transition metals and aluminum or silicon display unusual electronic and thermal transport properties due to sharp features in the electron density of states arising from band hybridization effects. Two prototype families displaying these traits are the Heusler alloys based on Fe_2VAl and the CoSi family. A characteristic attribute of these materials is the coexistence of large Seebeck coefficient ($S >$ ~100 $\mu V\ K^{-1}$) and metallic-like resistivity ($\rho <$ ~500 $\mu\Omega$ cm). This behavior can be exploited for thermoelectric applications, as it gives rise to thermoelectric power factor (S^2/ρ) in some cases a factor of 2 larger than state-of-the-art thermoelectric materials in the room temperature range. High thermal conductivity, a combination of moderately large lattice thermal conductivity and modest electronic thermal conductivity, limit thermoelectric figure of merit ZT to values on the order of 0.1. Further work in thermal conductivity reduction and exploration of related alloys is necessary to determine whether these limitations can be overcome.

References

1. L.F. Mattheiss and D.R. Hamann, Band structure and semiconducting properties of FeSi, *Physical Review B* **47**, 13114, 1993.

2. C.-H. Park, Z.-X. Shen, A.G. Loeser, D.S. Dessau, D.G. Mandrus, A. Migliori, J. Sarrao, and Z. Fisk, Direct observation of a narrow band near the gap edge of FeSi, *Physical Review B* **52**, R16981, 1995.

3. D. Nguyen Manh, G. Trambly de Laissardiere, J.P. Julien, D. Mayou, and F. Cyrot-Lackmann, Electronic structure and hybridization effects in the compounds Al_2Ru and Ga_2Ru, *Solid State Communications* **82**, 329, 1992.

4. S.E. Burkov and S.N. Rashkeev, On optical properties of quasicrystals and approximants, *Solid State Communications* **92**, 525, 1994.

5. D. Mandrus, V. Keppens, B.C. Sales, and J.L. Sarrao, Unusual transport and large diamagnetism in the intermetallic semiconductor $RuAl_2$, *Physical Review B* **58**, 3712, 1998.

6. M. Weinert and R.E. Watson, Hybridization-induced band gaps in transition-metal aluminides, *Physical Review B* **58**, 9732, 1998.

7. Z.J. Pan, L.T. Zhang, and J.S. Wu, Electronic Structure and transport properties of $Co(Si_{1-x}M_x)$ (M = Al, P): First-principles study, *Computational Materials Science* **39**, 752, 2007.

8. E.N. Nikitin, P.V. Tamarin, and V.I. Tarasov, Thermal and electrical properties of cobalt monosilicide between 4.2 and 1600 K, *Soviet Physics Solid State* **11**, 2002, 1970.

9. Y. Imai, M. Mukaida, K. Kobayashi, and T. Tsunoda, Calculation of the density of states of transition metal monosilicides by a first-principle pseudopotential method using plane-wave basis, *Intermetallics* **9**, 261, 2001.

10. J. Guevara, V. Vildosola, J. Milano, and A.M. Llois, Half-metallic character and electronic properties of inverse magnetoresistant $Fe_{1-x}Co_xSi$ alloys, *Physical Review B* **69**, 184422, 2002.

11. J.P. Heremans, V. Jovovic, E.S. Toberer, A. Saramat, K. Kurosaki, A. Charoenphakdee, S. Yamanaka, and G.J. Snyder, Enhancement of thermoelectric efficiency in PbTe by distortion of the electronic density of states, *Science* **321**, 554, 2008.

12. Y. Nishino, C. Kumada, and S. Asano, Phase stability of Fe_3Al with addition of 3d transition elements, *Scripta Materialia* **36**, 461, 1997.

13. V.A. Niculescu, T.J. Burch, and J.I. Budnick, A local environment description of hyperfine fields and atomic moments in $Fe_{3-x}T_xSi$ alloys, *Journal of Magnetism and Magnetic Materials* **39**, 223, 1983.

14. Y. Nishino, M. Kato, S. Asano, K. Soda, M. Hayasaki, and U. Mizutani, Semiconductorlike behavior of electrical resistivity in Heusler-type Fe_2VAl compound, *Physical Review Letters* **79**, 1909, 1997.

15. D.J. Singh and I.I. Mazin, Electronic structure, local moments, and transport in Fe_2VAl, *Physical Review B* **57**, 14352, 1998.

16. G.Y. Guo, G.A. Botton, and Y.J. Nishino, Electronic structure of possible 3d heavy-fermion compound, *Journal of Physics: Condensed Matter* **10**, L119, 1998.

17. R. Weht and W.E. Pickett, Excitonic correlations in the intermetallic Fe_2VAl, *Physical Review B* **58**, 6855, 1998.

18. Y. Nishino, H. Kato, M. Kato, and U. Mizutani, Effect of off-stoichiometry on the transport properties of the Heusler-type Fe_2VAl compound, *Physical Review B* **63**, 233303, 2001.

19. M. Vasundhara, V. Srinivas, and V.V. Rao, Low temperature electrical transport in Heusler-type $Fe_2V(AlSi)$ alloys, *Journal of Physics Condensed Matter* **17**, 6025, 2005.

20. Y. Nishino, S. Deguchi, and U. Mizutani, Thermal and transport properties of the Heusler-type $Fe_2VAl_{1-x}Ge_x$ ($0 \leq x \leq 0.20$) alloys: Effect of doping on lattice thermal conductivity, electrical resistivity, and Seebeck coefficient, *Physical Review B* **74**, 115115, 2006.

21. E.J. Skoug, C. Zhou, Y. Pei, and D.T. Morelli, High thermoelectric power factor near room temperature in full Heusler alloys, *Journal of Electronic Materials* **38**, 1221, 2009.

22. C.S. Lue, Y.K. Kuo, S.N. Horng, S.Y. Peng, and C. Chent, Structural, thermal, and electronic properties of $Fe_2VSi_{1-x}Al_x$, *Physical Review B* **71**, 064202, 2005.

23. C.S. Lue, C.F. Chen, J.Y. Lin, Y.T. Yu, and Y.K. Kuo, Thermoelectric properties of quaternary Heusler alloys $Fe_2VAl_{1-x}Si_x$, *Physical Review B* **75**, 064204, 2007.

24. C.S. Lue and Y.K. Kuo, Thermoelectric properties of the semimetallic Heusler compounds $Fe_{2-x}V_{1+x}M$ (M = Al, Ga), *Physical Review B* **66**, 085121, 2002.

25. G. Ghosh, G.V. Narasimha Roa, V.S. Sastry, A. Bharathi, Y. Hariharan, and T.S. Radhakrishnan, X-ray powder diffraction data of CoSi, *Powder Diffraction* **12**, 252, 1997.

26. W.L. Ren, C.C. Li, L.T. Zhang, K. Ito, and J.S. Wu, Effects of Ge and B substitution on thermoelectric properties of CoSi, *Journal of Alloys and Compounds* **392**, 50, 2005.

27. C.S. Lue, Y.K. Kuo, C.L. Huang, and W.J. Lai, Hole-doping effect on the thermoelectric properties and electronic structure of CoSi, *Physical Review B* **69**, 125111, 2004.

28. S. Asanabe, D. Shinoda, and Y. Sasaki, Semimetallic properties of $Co_{1-x}Fe_xSi$ solid solutions, *Physical Review* **134**, A774, 1964.

29. M.I. Fedorov and V.K. Zaitsev, Semimetals as materials for thermoelectric generators, in *CRC Handbook of Thermoelectrics*, ed. D.M. Rowe (CRC Press, Boca Raton, 1995), p. 321.

30. Y.K. Kuo, K.M. Sivakumar, S.J. Huang, and C.S. Lue, Thermoelectric properties of $CoSi_{1-x}Ge_x$ alloys, *Journal of Applied Physics* **98**, 123510, 2005.

31. E. Skoug, C. Zhou, Y. Pei, and D.T. Morelli, High thermoelectric power factor in alloys based on CoSi, *Applied Physics Letters* **94**, 022115, 2009.

32. F. Wald and S.J. Michalik, The ternary system cobalt-germanium-silicon, *Journal of the Less-Common Metals* **24**, 277, 1971.

33. C.C. Li, W.L. Ren, L.T. Zhang, K. Ito, and J.W. Su, Effects of Al doping on the thermoelectric performance of CoSi single crystal, *Journal of Applied Physics* **98**, 063706, 2005.

13

Novel Power Factor of Si–Ge System

Yoichi Okamoto
National Defense Academy

Hiroaki Takiguchi
National Defense Academy

13.1 Introduction

It is well known that conventional SiGe alloy crystal materials have good thermoelectric performance in a higher-temperature region. The thermoelectric power of typical SiGe alloy crystals is around 200 µV/K.[1] However, there are a few papers which report 2–100 times higher thermoelectric power than conventional SiGe alloy crystal materials.[2–7]

Most of these *non-ordinary* SiGe system has lower reproducibility and shorter life-time compared with conventional SiGe alloy crystals. Therefore, there is a high possibility that these *non-ordinary* SiGe system are in a metastable and/or nonequilibrium phase.

It is possible to research the mechanism of the unstable superior thermoelectric properties of these SiGe systems by actualized sample preparation condition using molecular beam epitaxy (MBE). To clarify the real nature of these *non-ordinary* SiGe system, a wide range of measurements, such as XRD, Raman scattering, photo acoustic spectroscopy, XRD, SEM, transport properties, and so on, were performed.

13.2 Overviews of Some Systems with Extremely High Thermoelectric Power

The extraordinary high thermoelectric power of amorphous Si/(Ge + Au) superlattice thin films have been reported. But there are other Si-related systems with extraordinary high thermoelectric power, such as Si/(Ge + B) superlattice, porous Si, and so on.

Figures 13.1 through 13.7 show the thermoelectric properties of some samples which have superior thermoelectric properties compared to conventional bulk SiGe materials. Figure 13.1 shows the temperature dependence of thermoelectric power for Si/(Ge + Au) amorphous superlattice thin films which are the first reported samples.[2,3] This system has the highest thermoelectric power in samples with conventional electrical resistivity. These samples are being continually researched in detail, and are samples that become central to this chapter. Figure 13.2 shows the temperature dependence of thermoelectric power for Si/(Ge + B) amorphous superlattice thin films which are fabricated by ion beam sputtering method.[4] Figure 13.3 shows layer thickness dependence of electrical conductivity, Seebeck coefficient, and power factor for nano porous Si.[5] Figure 13.4 also shows density dependence of Seebeck coefficient of nano porous Si.[6] Figure 13.5 displays results on the SiGe thin film. However, it changed the way of thinking a little. Electrical resistivity is decreased and thermoelectric power is maintained to the same value, instead of attempting an increase of thermoelectric power. (It has increased power factor.[7]) In these publications, there is large difference in the improvement of thermoelectric property. However, the common characteristic is that there is a microstructure in the samples.

Figures 13.6 and 13.7 show the two examples that we made in bulk form.[8,9] Thin films have a restricted field of application. Therefore, to expand application fields, attempts were made to make bulk samples while keeping superior thermoelectric properties.

Figure 13.6 shows the temperature dependence of thermoelectric properties (thermoelectric power, electrical resistivity, and power factor) for samples prepared by the melt-spinning method. In the calculation of power factor, thermal conductivity is assumed as same as conventional polycrystalline SiGe. Figure 13.7 shows the temperature dependence of thermoelectric power for the sample prepared by the powder-molded method from ultra-fine powders crushed with planetary ball mill. However, for every temperature dependence measurement (it has "annealing effect"), sintering and/or recrystallization, thermoelectric properties changed. From 1st to 10th measurement the thermoelectric power is lower than conventional bulk SiGe crystal. At 16th measurement, the highest thermoelectric power is observed and decreases at the 18th measurement.

To reveal the true character of these SiGe system material with extraordinary high thermoelectric properties, many kinds of measurements such as XRD, Raman scattering, photo acoustic spectroscopy, TEM, SEM, transport properties and so on were performed on SiGe amorphous super lattice thin films. From the results of XRD, Raman scattering, TEM, and SEM, SiGe samples have super lattice

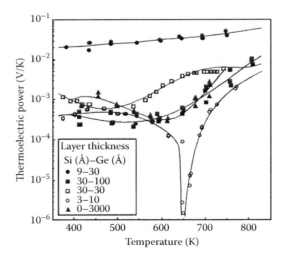

FIGURE 13.1   The temperature dependence of thermoelectric power for Si/(Ge + Au) amorphous superlattice thin films. All samples have around 300 nm total thickness with various artificial intervals. (Adapted from Okamoto, Y. et al., 1999. *Jpn. J. Appl. Phys.* **38**: L945–L947.)

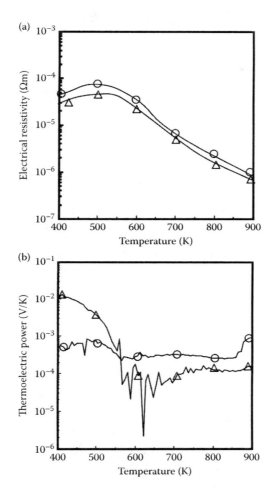

FIGURE 13.2 The temperature dependence of electrical resistivity in (a) thermoelectric power (b) for Si/(Ge + B) amorphous superlattice thin films, which has artificial superlattice structure of [Si(3 nm)/Ge(1 nm)] × 20 periods, and was deposited at 623 K by the ion beam sputtering method. (Adapted from Kawahara, T. et al., 2002. *Jpn. J. Appl. Phys.* **41**: L949–L951.)

structure with each layer in an amorphous phase as deposited. After moderate annealing the super lattice structure defuses and gradually recrystallized. It is evident that the superior thermoelectric properties are measured only on the way to the materials recrystallization.

At the earlier research stage it was presumed that there are metastable amorphous phase with superior thermoelectric properties in many amorphous phases.[2,3,10,11] Later it was found that when the recrystallization progressed and microcrystals exist in thin-film, superior thermoelectric properties appear. Therefore, it was concluded that the main origin of superior thermoelectric properties is in a mixture state of amorphous and nanosized micro crystals.[12,13]

13.3 Sample Preparations

13.3.1 Vacuum Deposition

The Si/(Ge + Au) superlattice thin films were deposited in a ultrahigh vacuum molecular beam epitaxial chamber which possessed the required accuracy and degree of control.

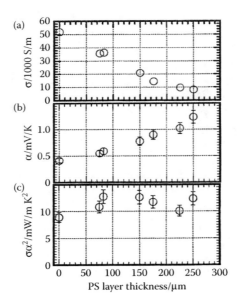

FIGURE 13.3 The layer thickness dependence of electrical conductivity in (a) Seebeck coefficient (b) and power factor (c) for heavily doped nanoporous Si. (Adapted from Yamamoto, A. et al., 1998. *Proc. of 17th Int. Conf. Thermoelectrics*, pp. 198–201.)

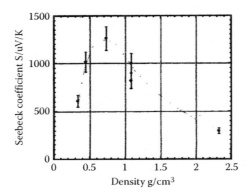

FIGURE 13.4 The density dependence of Seebeck coefficient of heavily doped nanoporous Si. (Adapted from Yamamoto, A., Takazawa, H., and Ohta, T. 1999. *Proc. of 18th Int. Conf. Thermoelectrics*: pp. 428–431.)

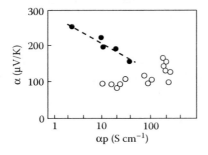

FIGURE 13.5 The relation of electrical conductivity and thermoelectric power for heavily doped Si–Ge alloy thin film. (Adapted from Kodato, S. 1985. *J. Non-Cryst. Sol.* **77&78**: 893–896.)

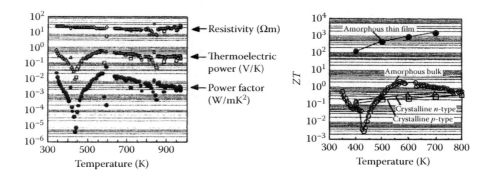

FIGURE 13.6 The temperature dependence of thermoelectric properties (thermoelectric power, electrical resistivity, and power factor) for samples by the melt-spinning method. (Left side figure noted as Figure 13.6 in the original paper). The temperature dependence of nondimensional figure-of-merit *ZT* for this sample, amorphous Si–Ge–Au superlattice thin film, and conventional Si–Ge alloy crystal. (The right-side figure is noted as Figure 13.7 in the original paper.) (Adapted from Lee, S. M. et al., 2002. *Mat. Res. Soc. Proc.* **691**: 215–220.)

The sample preparation, constituents, and structural analysis followed procedures described in previous papers.[2,3,10–15] Figure 13.8 is a schematic of the UHV–MBE system (Eiko; EV-10 & PV-100 mod.). Figure 13.9 shows the UHV–MBE system that has two main chambers (growth room-A and growth room-B) and two load lock chambers. Growth room-A is equipped with four evaporation sources (three electron beam guns and one K-cell; not drawn in Figure 13.8) and variable temperature sample holder (RT ~ 1200 K and at 77 K). Growth room-B equipped with three evaporation sources (two electron beam guns and one ion beam sputter gun) and variable temperature sample holder (77K–800 K). The substrate temperature was monitored by a thermocouple. To enable the sample transportation all chambers are connected via vacuum tunnels.

Six of the seven evaporation source systems (except for K-cell) were monitored and controlled from evaporation rate monitors (INFICON; XTC) and PC (NEC; PC-9801). The thickness monitors were calibrated using a surface morphology micrometer (Veeco; Dektak-3030) and low-angle XRD (x-ray

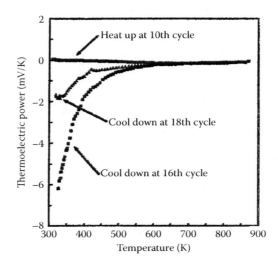

FIGURE 13.7 The temperature dependence of thermoelectric power for the sample which is made by powder-molded method from ultra-fine powders crushed with planetary ball mill. (Adapted from Abe, M. et al., 2009. *Mat. Res. Soc. Proc.* **1145**: 1145-MM04–07.)

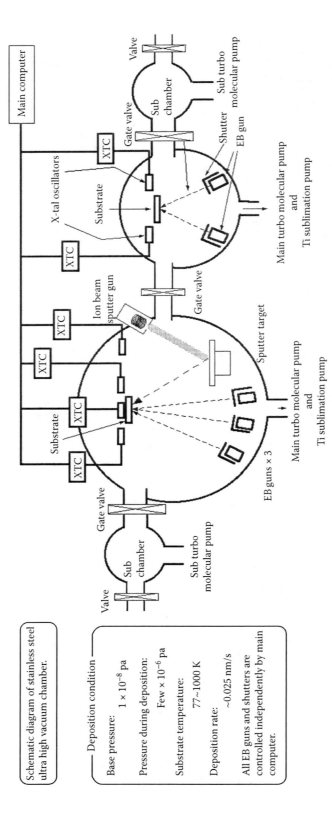

FIGURE 13.8   Schematic representation of the UHV–MBE system.

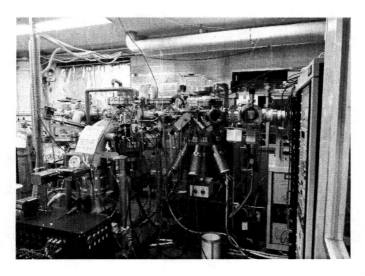

FIGURE 13.9 The photograph of UHV–MBE.

diffraction). Deposition rates from all electron beam guns and ion beam sputter gun are easily controlled in the range of 0.02–0.5 and 0.0001–0.05 nm/s, respectively .

As is well known when fabricating amorphous thin film it is very important that the evaporated atoms are quickly quenched on the substrate. Therefore, lower-temperature evaporation source and substrate are preferable.

The base pressure in the stainless-steel chamber with TMP main pumps and liquid N_2 shrouds was around 1×10^{-8} Pa, but rose to 1–5×10^{-6} Pa during operation of the evaporation sources.

13.3.2 Ion Beam Sputtering

The Si/(Ge + B) superlattice thin films were deposited by using ion beam sputtering gun in an ultrahigh vacuum chamber.[16] This ion beam sputtering method is one of the physical deposition methods. This method was adopted to achieve the nonequilibrium quenched phase.

In the electron beam heating method and electrical resistance heating method, source materials are heated up to the vapor phase and transported to the substrate. On the substrate surface, transported vapor loses its thermal energy and changes to a solid-phase thin film. On the one hand, in the ion beam sputtering method, source atoms are sputtered from source material target by the impact of the collision of high-speed ions, and are deposited on the substrate. There are some advantages in employing the ion beam sputtering method. Because the sample does not evaporate, film maintaining the source material composition can be made easily. It is easy to make a nonequilibrium phase. On the other hand, making a single crystal-like thin film is not so good.

Figure 13.10 is a schematic of the ion beam sputtering deposition system.

13.3.3 Melt Spinning

Although successful in making thin-film samples with superior thermoelectric power regrettably, the reproducibility of the sample is poor. Attempts were made to prepare the sample with higher thermoelectric properties in the state of bulk. There were two reasons for this. The first is to increase the range of applications and second to facilitate discussing the mechanism responsible for generating high thermoelectric power.

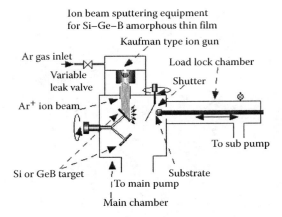

FIGURE 13.10　Schematic representation of the ion beam sputtering deposition system.

At an early phase of research, it was discussed that some metastable amorphous phase was responsible for superior thermoelectric properties.[2,3] The melt-spinning method, which makes an amorphous phase by quenching from melted liquid, was thought to be one of the best methods, because the control of the quenching rate was large. This melt-spinning method is a general industrial technique for the production of a quenched phase or amorphous materials. The schematic model chart and photograph of a small device are shown in Figures 13.11 and 13.12, respectively. In Figure 13.12, part of copper roller and spray nozzle are as seen through a round-shaped window. In the melt-spinning method the heated and melted raw material is sprayed on the copper roller at a high speed. The sprayed sample rapidly quenched from the liquid phase to the solid phase, because the copper roller rotates at high speed, has large thermal capacity, and good thermal conductivity. In this melt-spinning method, over several thousands of foil-shaped samples can be made.[8]

Figure 13.13 shows the conventional optical photograph (a) and SEM photograph (b). As seen in Figure 13.13a, a large amount of the foil-shaped and/or tape-shaped samples were generated at one time during sample fabrication. Figure 13.13b is an SEM photograph of a sample surface. It was presumed that the morphology of the surface was transcribed from the surface of the copper roller.

Many of the samples produced could not be measured because of too small size or high electrically resistivity. Consequently, only several tens of samples were measured. Very high thermoelectric power

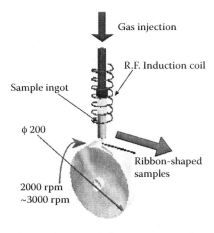

FIGURE 13.11　The schematic model chart of a small-sized melt-spinning method machine.

FIGURE 13.12   The photograph of a small-sized melt-spinning method machine.

FIGURE 13.13   The conventional optical photograph (a) and SEM photograph (b) of samples made by the melt-spinning method. (Adapted from Lee, S. M. et al., 2002. *Mat. Res. Soc. Proc.* **691**: 215–220.)

($\sim 10^{-1}$V K^{-1}), shown in Figure 13.6, was found in only several samples. However, the electrical resistivity is very high ($\sim 10^{+1}$ Ωm), the power factor is improved several times compared with conventional crystalline SiGe. It is concluded that reproducibility is unacceptable as only several samples showed high thermoelectric properties out of the several thousands fabricated.

13.3.4 Mechanical Milling

Another technique used to make bulk sample as well as the melt-spinning method is "Mechanical milling technique (planetary ball milling)." This method is different from other methods. The other methods produce amorphous phase by quenching the nonordered state materials (liquid phase or vapor phase). This method adds kinetic energy to the ordered state materials (crystal of the solid phase), and results in fine-powdered strained materials, amorphous phase materials, and so on. This technique is to make submicron size powders by crushing and grinding from conventional size bulk alloy ingot, followed by using the submicron size powders to form bulk size samples. The raw material and agate balls in the mill are subjected to planetary movement, which is a combination of the orbital and rotational

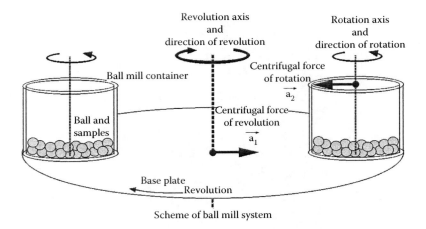

FIGURE 13.14   The schematic diagram of the mechanical milling (planetary ball milling) system.

movement. The raw materials are crushed into an amorphous phase. Figures 13.14 and 13.15 show the schematic diagram and photograph of the mechanical milling (planetary ball milling) system. Figure 13.16[17] shows the SEM photograph of the fine powder samples that have superior thermoelectric power (~4 mV K^{-1}) as shown in Figure 13.7.[9] Obviously, there were many particles smaller than 100 nm in diameter. Figure 13.17[17] shows the distribution of particle diameter. The sample shown in Figures 13.7 and Figure 13.16 is designated No. 23 in Figure 13.17. In the mechanical milling technique, the processed samples were pulverized rather than transformed into an amorphous phase. Only when a lot of very fine particles existed in the fabricated sample was a high thermoelectric power observed. This fact indicates that the quantum effect of a nanocrystal is responsible for the high thermoelectric characteristic of SiGeAu thin films.

FIGURE 13.15   The photograph of the mechanical milling (planetary ball milling) system.

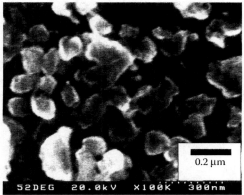

SEM image of sample no. 23

FIGURE 13.16 The SEM photograph of the fine powder samples made by mechanical milling technique (planetary ball milling).

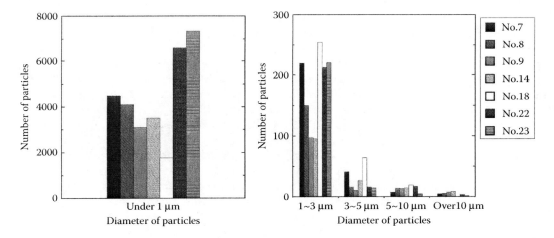

FIGURE 13.17 The distribution of particle diameter made by mechanical milling technique (planetary ball milling). (Adapted from Abe, M. 2009. Mechanical fabrication of Si–Ge bulk samples with superior thermoelectric power. Master's thesis, National Defense Academy, Kanagawa in Japanese.)

13.4 Properties Change and Recrystallization by Annealing

13.4.1 Structural Analysis by Using XRD, RAMAN, TEM, and SEM

Sample structure was analyzed by using XRD, RAMAN, TEM, and SEM. From these results, as-deposited condition, almost all samples have amorphous superlattice structure with intended spacings.[3] (Some samples deposited on heated substrate have crystal superlattice structure.[15]) Following several annealing cycles, artificial spacing disappears and recrystallization and constituent segregation takes place. Depending on the constituent and annealing conditions, a mixture form of nanosized crystal and amorphous phase can be achieved.[12,13]

The collapse of artificial lattice structure and recrystallization starts at almost the same time as annealing. The collapse of artificial lattice structure requires only one cycle of annealing.[3] On the other hand, recrystallization requires 5–20 annealing cycles. Of course, it depends on the heat treatment condition. Usually, during one cycle of annealing the temperature rises from room temperature to 1000 K

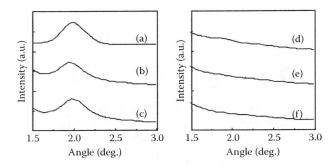

FIGURE 13.18 The annealing temperature dependence of low-angle region of XRD profiles of samples that have superlattice structure of Si layer: 0.9 nm, Ge layer: 3.0 nm, total repeated interval 300 pairs and were deposited sapphire substrate of ~300 K. (a) As deposited, (b) annealed at 350 K, (c) 450 K, (d) 500 K, (e) 550 K, and (f) 650 K. (Adapted from Uchino, H. et al., 2000. *Jpn. J. Appl. Phys.* **39**: 1675–1677.)

at a rate of 10 K/min, maintained for 10 min then cooled down at a rate of −10 K/min to room temperature. The change of thermoelectric properties progresses gradually with annealing cycles as described later.[10,18,19]

Figures 13.18 and 13.19 show the low- and high-angle region of XRD profiles, respectively.[3] In this case, a single deposited sample was divided into eight pieces, and then processed nonannealed, and annealed at 350, 450, 500, 550, and 650 K, respectively. The annealing time was for 15 min. The purpose of measurement was to clarify at which temperature the collapse of artificial lattice structure and recrystallization starts. It was understood that the collapse of artificial lattice structure and recrystallization started at the same temperature. The samples have Si layer: 0.9 nm, Ge layer: 3.0 nm, total repeated interval 300 pairs and were deposited on a sapphire substrate at ~300 K.

Figure 13.20a and b show cross-sectional TEM photographs of the sample (Si: 3.0 nm/Ge + Au: 3.0 nm/50 pairs) as deposited and after 623 K, 10 min annealing, respectively.[13] Figure 13.20a shows a typical stripe pattern of artificial interval which correspond to intended interval of 6.0 nm. On the other hand, Figure 13.20b shows distributed small particles of about 10 nm or less.

Figure 13.21a and b shows plan view TEM photographs of the same condition sample (Si: 3.0 nm/ Ge + Au: 3.0 nm/50 pairs) as deposited and after 623 K 10 min annealing, respectively.[20] In Figure 13.21a, no feature structure is seen, so it is in the amorphous phase. On the other hand, Figure 13.21b shows distributed small particles as shown in Figure 13.20b. These photographs show as-deposited samples having artificial lattice structures consisting of amorphous individual layers. After suitable condition of annealing, samples changed into the state of distributed microcrystal with a few nanometers in the matrix of the amorphous phase. To complete the recrystallization, annealing was excessively repeated.

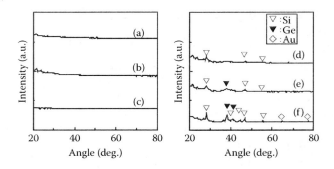

FIGURE 13.19 The annealing temperature dependence of a high-angle region of XRD profiles of samples shown in Figure 13.17. (Adapted from Uchino, H. et al., 2000. *Jpn. J. Appl. Phys.* **39**: 1675–1677.)

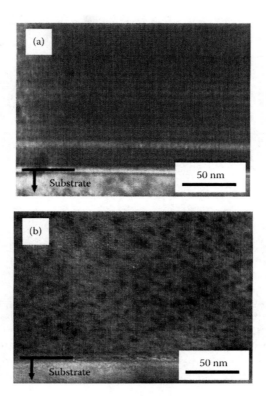

FIGURE 13.20 The cross-sectional TEM photographs of the sample (Si: 3.0 nm/Ge + Au: 3.0 nm/50 pairs), as deposited (a) and after 623 K 10 min annealing (b). (Adapted from Takiguchi, H., Fukui, K., and Okamoto, Y. 2010. *Jpn. J. Appl. Phys.* **49**: 115602–115607.)

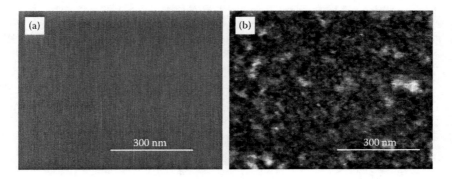

FIGURE 13.21 The plane view TEM micrographs of the same condition sample as in Figure 13.20; (a) as deposited and (b) after 623 K 10 min annealing. (Adapted from Takiguchi, H. et al., 2010. *J. Elec. Mater.* **39**: 1627–1633.)

13.4.2 Thermoelectric Properties

Thermoelectric properties (electrical resistivity and thermoelectric power) changed with thermal annealing. Some as-deposited samples showed relatively lower thermoelectric properties. Thermoelectric properties increased with annealing cycles. On the other hand, other samples showed higher thermoelectric properties as-deposited condition.[17]

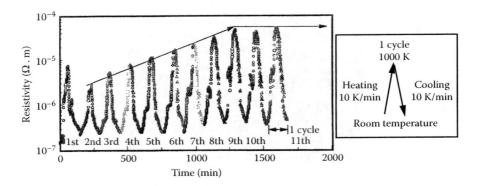

FIGURE 13.22 The electrical resistivity as a function of annealing cycles. The sample was (Si: 3.0 nm/Ge + Au: 3.0 nm/50 pairs) deposited on sapphire (~77 K during deposition). The one annealing procedure was heating with rate 10 K/min, keep annealing temperature 10 min, and cooling with rate of −10 K/min. (Adapted from Lee, S. M. et al., 2002. *Proc. of 20th Int. Conf. Thermoelectrics*, Beijing, pp. 348–351.)

Figures 13.22 and 13.23 display the electrical resistivity and power factor as a function of annealing cycles. The sample in both figures was (Si: 3.0 nm/Ge + Au: 3.0 nm/50 pairs) deposited on sapphire (~77 K during deposition). The annealing procedure was heating with rate 10 K/min, maintaining the annealing temperature for 10 min, and cooling at a rate of −10 K/min. The abscissas are elapsed time and the temperature with elapsed time. The measurements are displayed only from 1st to 11th annealing cycle, although they were continued until the 15th annealing cycle. The thermoelectric properties hardly changed from the 9th to the 15th annealing cycle. Obviously, the electrical resistivity has gradually changed up to the 9th annealing cycle. Moreover, also the power factor gradually increases to the 7th annealing cycle. The maximum value power factor exceeded 10^2 W m^{-1} K^{-2}. However, a significant change cannot be seen in conventional XRD pattern in this region of the annealing cycle. Figure 13.24 shows the power factor measured at 1st and 10th annealing cycle as a function of temperature. The plotted data of 1st annealing cycle ("as deposited" in figure) were not "temperature dependence" in a strict meaning, because the sample structure changed (collapse of artificial lattice structure and progress of recrystallization) during the measurement. Both the dependency of the temperature and the structural change are superimposed on these data. It was thought that the influence of the structural change was

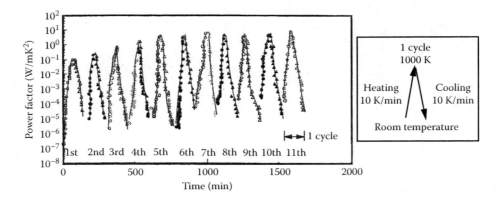

FIGURE 13.23 The power factor as a function of annealing cycles. The sample and annealing procedure are the same as in Figure 13.22. (Adapted from Lee, S. M. et al., 2002. *Proc. of 20th Int. Conf. Thermoelectrics*, Beijing, pp. 348–351.)

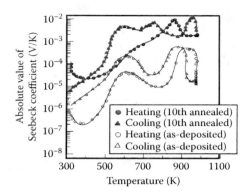

FIGURE 13.24 The power factor measured at 1st and 10th annealing cycle as a function of temperature. (Adapted from Lee, S. M. et al., 2002. *Proc. of 20th Int. Conf. Thermoelectrics*, Beijing, pp. 348–351.)

not comparatively large on the data of 10th annealing cycle, because the structural change (recrystallization) had progressed considerably.

In any event, at the 1st annealing cycle, thermoelectric power was around 10^{-6} V K^{-1} at room temperature and around 10^{-4} V K^{-1} at 900 K, respectively. At the 10th annealing cycle, it improved to around 10^{-5} at RT and around 10^{-3} V K^{-1} at 900 K, respectively. From the analysis of the structure and the measurement of a thermoelectric characteristic it is concluded that the collapse of superlattice structure and recrystallization starts almost at the same time. However, several annealing cycles are required for the completion of recrystallization, while the collapse of the artificial lattice is completed after one cycle of annealing.

13.4.3 Electronic Band Structure by PAS

To discuss the change mechanism of thermoelectric properties, an analysis on electric band structure is important. The samples were examined using photo acoustic spectroscopy (PAS). PAS measures the light absorption directly and provides information on the electronic band structure. From the results of PAS, changes in the electronic band structure were correlated with annealing cycles and sample constitutions.

Figure 13.25 shows the (1-transmission) spectrum for some samples. The sample information in Figure 13.25a indicates the Si layer thickness nm/Ge layer thickness nm/number of total intervals/substrate temperature/times of annealing.[21] The absorption edge shifted to the long-wavelength side (i.e., lower energy side) as the substrate temperature increased. Figure 13.26 shows the (1-transmission) spectrum for sample of (Si = 3.0 nm/Ge = 3.0 nm/50 pairs) measured at room temperature. Annealing temperature and duration time were 673 K and 10 min, respectively.[22] Clearly, the change depending on the number of annealing cycle was not too large, although the spectrum has changed after only one annealing cycle. Figure 13.27 shows the absorption spectrum for sample-B of (Si = 3.0 nm/Ge = 5.0 nm/38 pairs) measured at room temperature with number of annealing cycle as a parameter. Spectrum information on this figure, such as "Series-B-5," means the spectrum of 5-times-annealed sample-B. Annealing temperature and duration time are same to those in Figure 13.26 (673 K and 10 min). The word "sample-B" is the sample notation in Reference 22.

The number of annealing cycle dependence of the peak position and the valley position is shown in Figure 13.28. It changed greatly after the first time annealing cycle, with a systematic change after second time annealing cycle. However, it could not explain the mechanism of the directionality of the change. At any rate, an optical characteristic (electronic-banded structure-related characteristic) changes with an annealing cycle.

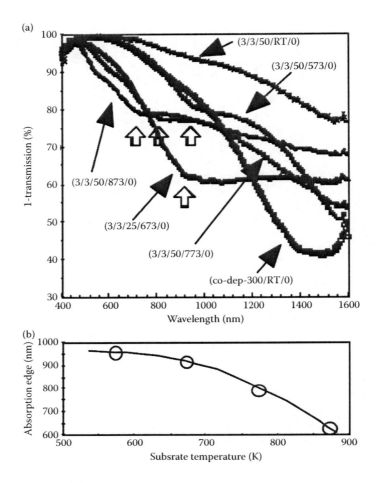

FIGURE 13.25 The (a); (1-transmission) spectrum and (b); substrate temperature dependence of absorption edge for some samples. The sample marks in figure (a) mean the (Si layer thickness nm/Ge layer thickness nm/number of total intervals/substrate temperature/times of annealing). (Adapted from Okamoto, Y. et al., 2004. *Proc. of 23rd Int. Conf. Thermoelectrics,* Adelaide: #074.pdf.)

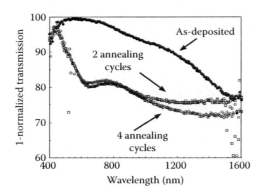

FIGURE 13.26 The (1-transmission) spectrum for a sample of (Si = 3.0 nm/Ge = 3.0 nm/50 pairs) measured at RT. Annealing temperature and duration time were 673 K and 10 min, respectively. (Adapted from Takiguchi, H., Okamoto, Y., and Morimoto, J. 2007. *Jpn. J. Appl. Phys.* **46**: 4622–4625.)

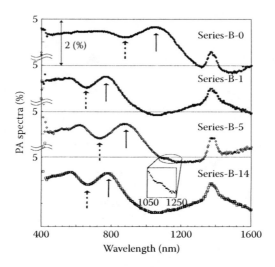

FIGURE 13.27 The absorption spectrum of PAS measurement for a sample of (Si = 3.0 nm/Ge = 5.0 nm/38 pairs) measured at RT with number of annealing cycle as a parameter. (Adapted from Takiguchi, H., Okamoto, Y., and Morimoto, J. 2007. *Jpn. J. Appl. Phys.* **46**: 4622–4625.)

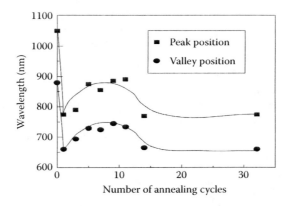

FIGURE 13.28 The peak position and the valley position of absorption spectrum as a function of number of annealing cycle. (Adapted from Takiguchi, H., Okamoto, Y., and Morimoto, J. 2007. *Jpn. J. Appl. Phys.* **46**: 4622–4625.)

13.5 Mechanism of Recrystallization

There are some indications of the mechanism responsible for superior thermoelectric properties and are as follows:

1. Samples have an amorphous artificial lattice structure in the as-deposited condition.
2. Recrystallization starts at lower temperature than the original temperature of recrystallization which is reported in conventional amorphous SiGe.
3. Recrystallization is not completed after only one annealing cycle and progressed gradually with further annealing cycles.
4. The artificial superlattice structure collapsed with the start of recrystallization.
5. After the artificial lattice structure collapsed, superior thermoelectric properties were obtained.

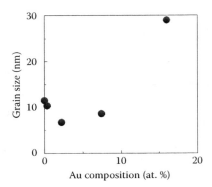

FIGURE 13.29 The Au concentration of the microcrystal size calculated from FWHM value of XRD and TEM micrograph. (Adapted from Takiguchi, H., Aono, M., and Okamoto, Y. 2011. *Jpn. J. Appl. Phys.* **50**: 041301-1–041301-5.)

6. After annealing, nanostructure (nano-crystals) were confirmed in the thin film.
7. The superior thermoelectric properties were demonstrated with the sample that was made mechanically by the milling technique.

From the above-mentioned facts, it is apparent that both artificial superlattice structure and amorphous phase do not have an important role. While nanostructure (nano-crystals) does have a major role for the origin of superior thermoelectric properties.[14,22] How is the nanostructure (nanocrystals) generated in amorphous superlattice thin film? It is deduced that metal-induced crystallization (MIC)[23–27] has an important role on the mechanism of recrystallization, especially in the early stage of recrystallization, MIC is dominant. In MIC mechanism, dopant metallic atoms have very important roles. Dopant metallic atoms assist to create crystal nucleation even at a lower temperature than conventional recrystallization resulting in many small-sized crystal nucleation occurring. In the other words, small-sized crystal generation nucleuses are nanocrystals. When the anneal temperature increased over conventional recrystallization temperature, nanocrystals grow to conventional size crystals. So the number of microcrystals decreases and their sizes increase. In the later stage of recrystallization, local area liquid-phase recrystallization is important. In this stage, crystal growth is dominant. The number of microcrystals decrease and their sizes increase.

Figure 13.29 shows the Au concentration dependence of nanocrystal size.[14] The ordinate represents the diameter of a nanocrystal which was calculated from full-width at half-maximum (FWHM) of XRD profile using Schrerr's equation and it was confirmed with TEM observation. Clearly, it has the minimum value (5–6 nm) at around 2.5 at.% of Au concentration.

13.6 Relationship of Extremely High Thermoelectric Power and Structure

As-deposited samples with artificial superlattice structure do not have superior thermoelectric properties. Fully recrystallized samples also do not have superior thermoelectric properties. Only those having appropriate composition and after moderately annealed with distributed nanocrystal structure have superior thermoelectric properties. It is deduced that the existence of nanosized crystals is necessary.

In Figure 13.29, the size of the microcrystals have a minimum value at a Au concentration of 3–4 at.%. This mechanism follows the following model.[14] With appropriate temperature annealing, the samples start to recrystallize due to the MIC effect. When the added metal (here, Au) concentration is low, the density of the crystal nucleation for recrystallization is low. The recrystallized area is stable at the annealing temperature. Therefore, the crystalline growth will stop at some size. The density of the

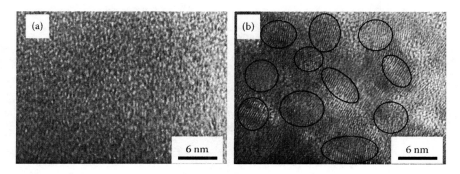

FIGURE 13.30 The surface TEM photograph (plane view not cross-sectional view) of (a) as-deposited sample of the sample and (b) annealed with optimum conditions (Au; 2.2 at.%, 600 K, 10 min). (Adapted from Takiguchi, H., Aono, M., and Okamoto, Y. 2011. *Jpn. J. Appl. Phys.* **50**: 041301-1–041301-5.)

crystal nucleation for recrystallization increases, when added metal concentration is increased a little. The conventional liquid-phase crystal growth commences when the added metal concentration is increased further, and the size of the generated microcrystal increases.

Figure 13.30 shows a surface TEM photograph (not cross sectional) of (a) as-deposited sample of the sample and (b) annealed under optimum conditions (Au; 2.2 at.%, 600 K, 10min).[14] Obviously, as-deposited sample is in a typical amorphous phase and no microstructure is observed. On the other hand, after anneal, the structure is that of a single crystal of roughly 5–10 nm. The ovals in figure show single crystal for a comparison. From the experimental results, the following is concluded. The Si/ (Ge + Au) artificial superlattice thin-film could be changed from the amorphous artificial superlattice structure into the structure with the nanocrystal distributed in the amorphous matrix by the control of added Au concentration and annealing conditions. In addition, the size of the nanocrystal can be uniquely controlled by the added Au concentration under the best annealing condition for the added Au concentration.

Figure 13.31a and b shows the added Au concentration dependence of electrical resistivity and absolute value of thermoelectric power, respectively.[14] When Figures 13.29 and 13.31 are compared, superior thermoelectric power (10^{-3} V K^{-1} class) is obviously obtained only in a specific added Au concentration area, where the size of the generated nanocrystal is minimized. Therefore, superior thermoelectric properties are presumed to originate from nanocrystal, that is, quantum size effect. If this presumption is confirmed then the superior thermoelectric power is generated when nanocrystal are formed. Figure 13.32 shows the annealing temperature dependence of thermoelectric power of SiGeB thin with boron rather than Au as an additive.[12] These samples were made on the same condition as shown in Figure 13.2, only annealing temperatures were different. Evidently superior thermoelectric power (>1 mV K^{-1}) is observed only for annealing temperature of about 900°C. Figure 13.33 shows the annealing temperature dependence of the grain size which is estimated from (111) plane peak of XRD measurement.[12] Obviously, superior thermoelectric power was observed only in the area where the size of the nanocrystal is small enough. Therefore, it is concluded that there was a very high possibility that the quantum effect of nanocrystal is responsible for the extraordinary high thermoelectric power of SiGe system thin films.

13.7 Conclusion

Comparison of annealing cycle dependence of thermoelectric properties and structure indicates that a mixture of nanosized crystal and amorphous phase is most important for extraordinary high thermoelectric power. However, there are many factors which have effects on the thermoelectric properties. It is concluded that the most important factor is quantum size effect in SiGe nanocrystals (nanodots).

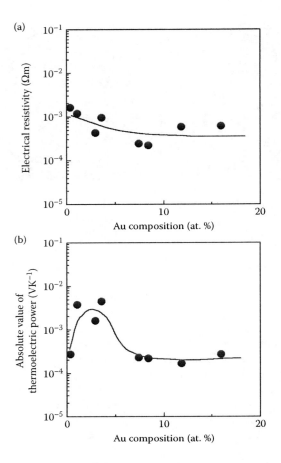

FIGURE 13.31 (a) Electrical resistivity and (b) absolute value of thermoelectric power as a function of added Au concentration. (Adapted from Takiguchi, H., Aono, M., and Okamoto, Y. 2011. *Jpn. J. Appl. Phys.* **50**: 041301-1–041301-5.)

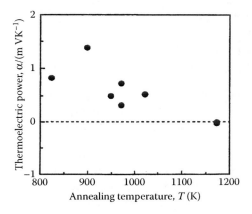

FIGURE 13.32 The annealing temperature dependence of thermoelectric power of SiGeB thin film. These samples were made on the same condition as shown in Figure 13.2. (Adapted from Takiguchi, H. et al., 2000. *Mater. Trans.* **51**: 878–881.)

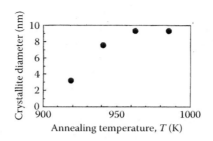

FIGURE 13.33 The grain size of SiGeB thin film estimated from (1 1 1) plane peak of XRD measurement as a function of annealing temperature. (Adapted from Takiguchi, H. et al., 2000. *Mater. Trans.* **51**: 878–881.)

References

1. Vining, C. B. 1955. *CRC Handbook of Thermoelectrics*, edited by D. M. Rowe, pp. 329–337. Boca Raton: CRC Press.
2. Okamoto, Y., Uchino, H., Kawahara, T., and Morimoto, J. 1999. Anomalous large thermoelectric power of the Si and Au doped Ge superlattice thin film. *Jpn. J. Appl. Phys.* **38**: L945–L947.
3. Uchino, H., Okamoto, Y., Kawahara, T., and Morimoto, J. 2000. The study of the origin of the anomalous large thermoelectric power of the Si and Au doped Ge superlattice thin film. *Jpn. J. Appl. Phys.* **39**: 1675–1677.
4. Kawahara, T., Lee, S. M., Okamoto, Y., Morimoto, J., Sasaki, K., and Hata, T. 2002. Anomalous large thermoelectric power on heavily B-doped SiGe thin films. *Jpn. J. Appl. Phys.* **41**: L949–L951.
5. Yamamoto, A., Takimoto, M., Ohta, T., Whitlow, L., Miki, K., Sakamoto, K., and Kamisako, K. 1998. Two dimensional quantum net of heavily doped porous silicon. *Proc. of 17th Int. Conf. Thermoelectrics*, Nagoya, pp. 198–201.
6. Yamamoto, A., Takazawa, H., and Ohta, T. 1999. Thermoelectric transport properties of porous silicon nanostructure. *Proc. of 18th Int. Conf. Thermoelectrics*, Baltimore, pp. 428–431.
7. Kodato, S. 1985. Si-Gealloy film with very high electrical conductivity and thermoelectric power. *J. Non-Cryst. Sol.* **77&78**: 893–896.
8. Lee, S. M., Okamoto, Y., Kawahara, T., and Morimoto, J. 2002. The fabrication and thermoelectric properties of amorphous Si–Ge–Au bulk samples. *Mat. Res. Soc. Proc.* **691**: 215–220.
9. Abe, M, Takiguchi, H., Okamoto, Y., Miyazaki, H., and Morimoto, J. 2009. Mechanical fabrication of Si–Ge–B bulk samples with superior thermoelectric power. *Mat. Res. Soc. Proc.* **1145**: 1145-MM04-07.
10. Lee, S. M., Okamoto, Y., Kawahara, T., and Morimoto, J., 2002. Annealing temperature dependence of the amorphous structure of amorphous Si–Ge–Au thin films. *Jpn. J. Appl. Phys.* **41**: 5336–5337.
11. Okamoto, Y., Miyata, A., Sato, Y., Takiguchi, H., Kawahara, T., and Morimoto, J. 2003. The measurement of annealing cycle effects of Si–Ge–Au amorphous thin film with anomalously large thermoelectric power by using photoacoustic spectroscopy. *Jpn. J. Appl. Phys.* **42**: 3048–3051.
12. Takiguchi, H., Matoba, A., Sasaki, K., Okamoto, Y., Miyazaki, H., and Morimoto, J. 2000. Structural properties of heavily B-doped SiGe thin films for high thermoelectric power. *Mater. Trans.* **51**: 878–881.
13. Takiguchi, H., Fukui, K., and Okamoto, Y. 2010. Annealing temperature dependence of crystallization process of SiGeAu thin film. *Jpn. J. Appl. Phys.* **49**: 115602–115607.
14. Takiguchi, H., Aono, M., and Okamoto, Y. 2011. Nano-structural and thermoelectric properties of SiGeAu thin films. *Jpn. J. Appl. Phys.* **50**: 041301-1–041301-5.
15. Sasaki, K., Nabetani, Y., Miyashita, H., and Hata, T. 2000. Heteroepitaxial growth of SiGe films and heavy B doping by ion-beam sputtering. *Thin Solid Films* **369**: 171–174.

16. Okamoto, Y., Saeki, J., Ohstuki, T., and Takiguchi, H. 2010. Structural effects on thermal conductivity of SiGeAu superlattice thin films. *Jpn. J. Appl. Phys.* **49**: 85801–85805.

17. Abe, M. 2009. Mechanical fabrication of Si–Ge bulk samples with superior thermoelectric power. Master's thesis, National Defense Academy, Kanagawa in Japanese.

18. Lee, S. M., Okamoto, Y., Kawahara, T., and Morimoto, J. 2002. Effects of substrate temperature on thermoelectric properties of amorphous Si/Ge thin films. *Proc. of 20th Int. Conf. Thermoelectrics*, Beijing, pp. 348–351.

19. Lee, S. M., Okamoto, Y., Kawahara, T., and Morimoto, J. 2003. The study for the mechanism on the noble thermoelectric properties of the amorphous Si-Ge-Au thin films. *Proc. of 21th Int. Conf. Thermoelectrics*, Long Beach, pp. 126–129.

20. Takiguchi, H., Yoshikawa, Z., Miyazaki, H., Morimoto, J., and Okamoto Y. 2010. The role of Au on thermoelectric properties of amorphous Ge/Au and Si/Au thin films. *J. Elec. Mater.* **39**: 1627–1633.

21. Okamoto, Y., Miyata, A., Inoue, Y., Kawahara, T., and Morimoto, J. 2004. Substrate temperature dependence of recrystallization process of Si–Ge–Au amorphous thin films evaluated by using PAS. *Proc. of 23rd Int. Conf. Thermoelectrics*, Adelaide: #074.pdf.

22. Takiguchi, H., Okamoto, Y., and Morimoto, J. 2007. Direct measurement of optical absorption for Si–Ge–Au amorphous thin films by using photoacoustic spectroscopy. *Jpn. J. Appl. Phys.* **46**: 4622–4625.

23. Oki, F., Ogawa, Y., and Fujiki, Y. 1969. Effect of deposited metals on the crystallization temperature of amorphous germanium film. *Jpn. J. Appl. Phys.* **8**: 1056.

24. Radnoczi, G., Robertsson, A., Hentzell, H. T. G., Gong, S. F., and Hasan, M.-A. 1991. Al induced crystallization of a-Si. *J. Appl. Phys.* **69**: 6394–6399.

25. Tan, Z. and Heald, S. M. 1992. Gold-induced germanium crystallization *Phys. Rev. B* **46**: 9505–9510.

26. Choi, J. H., Kim, D. Y., Kim, S. S., Park, S. J., and Jang, J. 2004. Polycrystalline silicon obtained by metal induced crystallization using different metals. *Thin Solid Films* **451–452**: 334–339.

27. Aoki, T., Kanno, H., Kenjo, A., Sadoh, T., and Miyao, M. 2006. Au-induced lateral crystallization of a-Si1-xGex (x: 0–1) at low temperature. *Thin Solid Films* **508**: 44–47.

14

Boride Thermoelectrics: High-Temperature Thermoelectric Materials

Takao Mori
National Institute for
Materials Science

14.1 Introduction

For possible waste heat applications, there is a large incentive to develop thermoelectric materials that can function at high temperatures. Boron cluster compounds are attractive as materials because of their excellent stability under high temperatures (typical melting points >2000°C) and "unfriendly" (e.g., acidic, corrosive) conditions. Importantly, they have also been found to typically possess intrinsic low thermal conductivity [1–4].

As a representative of boron compounds, boron carbide is one of the few thermoelectric materials that actually has a history of being commercialized. Modules containing boron carbide were previously sold by the Hi-Z Technology Inc., although they do not seem to be sold now. One of the obstacles for wide application of boron carbide is that although it had been searched for extensively for more than 20 years, a viable *n*-type counterpart could not be found. Higher borides containing the boron icosahedra are predominantly *p*-type due to the intrinsically electron-deficient nature of boron [5].

As an exciting development, novel boride compounds exhibiting intrinsic *n*-type behavior have recently been discovered [6,7]. Such *n*-type behavior was previously thought to be not possible for higher borides [8]. Advancements in the development of this promising series of compound as viable high-temperature thermoelectric materials, together with those of both well known and other novel borides will be described in this chapter. The origins of the intrinsic low thermal conductivity in boron cluster compounds will also be discussed, since they may yield effective principles to design intrinsically low thermal conductivity materials not limited to borides.

In the previous *CRC Handbook on Thermoelectrics* [9], it was reported that modules containing quantum well structures of boron carbide exhibited $ZT \sim 4$, which is a world record. The second stage of development of boron-rich compounds as good high-temperature thermoelectric materials can be considered to be just underway with the novel developments to be described in this chapter, and such development should be vigorously pursued further.

14.2 Comments on Particular Features of Boron

Boron is an interesting element, tending to form two-dimensional atomic nets and clusters in compounds. In this sense, it is similar to carbon, which has been much more extensively studied for such materials as fullerenes, nanotubes, and graphite-related materials. However, since boron has only three valence electrons available, there tends to be an electron deficiency in forming a three-dimensional network of conventional covalent bonds. This particular circumstance gives rise to the fascinating variety of clusters and structures that are found in boron-rich compounds. For example, B_{12} boron icosahedra bonded inside compounds lack two electrons. This is because 38 electrons are required to satisfy the bonding; 26 electrons for 13 bonding molecular orbitals for intracluster bonding, and 12 electrons for intercluster bonding [10]. However, since boron has three electrons in its outer shell, there are only $12 \times 3 = 36$ electrons available, leading to a deficiency of two electrons. This is why B_{12} can often be found in states like $[B_{12}H_{12}]^{2-}$. The combination of electron-deficient boron with rare earth atoms is a good one, since localized rare earth atoms can occupy voids among the boron atomic networks while supplying electrons to stabilize and form intriguing novel structures. The shell of f-electrons further can supply or enable tuning of attractive electronic properties. As a synthesis method it has also been found that the addition of small amounts of third elements like carbon, nitrogen, and silicon can function as bridging sites and result in the formation of novel and varied rare earth boron cluster structures. Many novel rare earth borides have been discovered within the last decade [5].

14.3 Synthesis Methods

Boride compounds, which can melt, can usually easily be obtained by arc melting the elements or by crystal growth with various techniques (floating zone, Czochralski, flux growth, etc.). A setup of a typical floating zone (FZ) crystal growth furnace is shown in Figure 14.1a. This is a crucible-free method that enables the growth of high-quality and large-sized single crystals of borides or other compounds with melting points up to 2900 K.

Although they are high-temperature materials, many of the novel borides discovered in the past decade [5] will not melt stably and therefore need to be synthesized by solid-state reactions at high temperatures. One useful method for heating to such temperatures is radio-frequency (rf) induction heating in BN crucibles wrapped around with susceptors like graphite wool (Figure 14.2). This method enables good control of synthesis at temperatures up to 2300 K, which is usually far beyond the reach of conventional electrical furnaces.

Synthesis of borides containing rare earth can be carried out in the following way. Although the amounts of rare earth content in higher borides are not high compared to more metal-rich metallic borides, a further cost-saving method is to start from relatively inexpensive rare earth oxides. Powders of REB_n (RE = rare earth) can be prepared with the borothermal reduction method by heating above 1600 K,

$$RE_2O_3 + (2n + 3)B \rightarrow 2REB_n + 3BO. \tag{14.1}$$

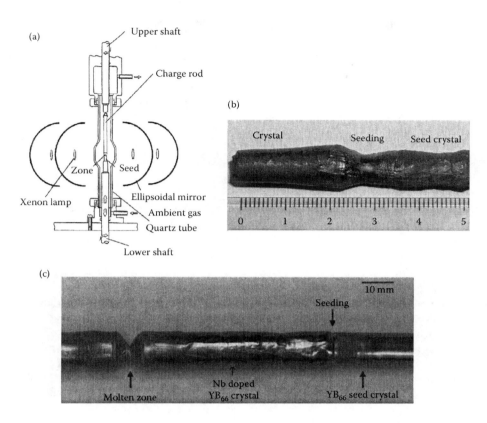

FIGURE 14.1 (a) Setup of a floating zone (FZ) crystal growth furnace and examples of grown crystals of (b) $TmB_{44}Si_2$ and (c) Nb-doped YB_{66}.

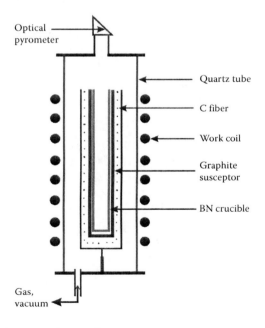

FIGURE 14.2 Setup of an rf induction heating furnace.

In the case of binary borides such as REB_2, REB_4, REB_6, REB_{12}, REB_{25}, REB_{50}, REB_{66}, and so on, this can be a one-shot method to obtain pure phase compounds, by heating at the temperature necessary for compound formation. For ternary or quartenary compounds like the rare earth boron carbonitrides, $REB_{17}CN$, $REB_{22}C_2N$, and $REB_{28.5}C_4$, and borosilicides $REB_{44}Si_2$, the desired amounts of boron, carbon, boron nitride, or silicon can be further added to the REB_n powder, and heated and reacted again to form the compounds. Among the recently discovered higher borides, large crystals of $REB_{44}Si_2$ can be grown by the FZ method as shown in Figure 14.1b.

14.4 Well-Known (Classical) Boron Compounds

14.4.1 Boron Carbide

Regarding the thermoelectric properties of well-known boron compounds, the thermoelectric properties of boron carbide have been investigated for over half a century [11], and have been extensively studied from the early 1980s with increasing understanding and improvement of properties [12,13]. As mentioned above, it is one of the very few thermoelectric materials that actually has a history of commercialization. Since many reviews have been written previously on boron carbide [14,15], I will just cover the basics, and some recent results.

The structure of boron carbide is depicted in Figure 14.3. It is rhombohedral and composed of B_{12} icosahedra layers and C–B–C chains. However, as can be surmized, there is a wide homogeneity range, with boron and carbon atoms replacing each other. Carbon is reported to enter the B_{12} icosahedra and form $B_{11}C$, while the C–B–C chain can also be composed of different combinations [16,17]. Boron carbide has been known to take a composition of $B_{12+x}C_{3-x}$ ($0.06 \leqq x \leqq 1.7$). In other words, while being a crystalline, nonalloy material, it can take a very wide range of composition from $B_{4.1}C$ to $B_{10.5}C$. It should be noted that the "ideal" composition of B_4C (with $B_{11}C$ and C-B-C chains) does not appear to exist, and compounds synthesized successfully without any free carbon impurities are always more boron-rich. However, the notation B_4C had traditionally been used as the chemical formula of boron carbide (despite such a carbon-rich compound not forming), and therefore, the notation is still used with parentheses like "B_4C." The flexibility (i.e., wide homogeneity region) of the compound has enabled easy and effective tuning of the thermoelectric properties.

Boron carbide is a *p*-type material and exhibits large Seebeck coefficients while having high carrier densities and relatively small electrical resistivities. This has been attributed by Emin and coworkers to a small bipolaron hopping mechanism in which hole charge carriers generated by the replacement of carbon atom sites with boron are self-trapped [12–15,18]. A softening mechanism can explain the large

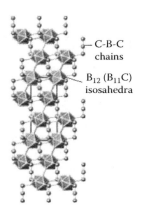

FIGURE 14.3   Structure of boron carbide.

Seebeck coefficients. There has also been a claim by Werheit that the bipolaron picture is incorrect [19]. However, the fact remains that the bipolaron picture gives an intuitive explanation of boron carbide's combination of large Seebeck coefficients and high electrical conductivity (due to a high number of carriers despite a low mobility). However, the band picture may give an alternate view of this compound [19] and these matters should be investigated further.

Since boron carbide was demonstrated to be an exemplary high-temperature thermoelectric material, extensive studies on doping foreign elements into it have been carried out. For example, a ZT of ~0.6 has been reported for Si-doped boron carbide [20] and TiB_2-doped boron carbide [21]. However, despite many efforts, n-type behavior was not realized except for weakly and in very limited temperature regions like that observed for Ni-doped boron carbide [22]. Hydrogenized boron carbide was reported to have n-type behavior [23], but the detailed composition or thermal stability of the material at high temperatures is not clear.

Compared to Bi_2Te_3 and some other compounds recently being studied, the ZT values given above are not particularly high. However, it should be stressed that the ZT of boron carbide (and some other borides that will be introduced in this chapter) sharply increases with increasing temperature and the material is stable at high temperatures (1500 K for example) where other materials would be easily destroyed or have their thermoelectric properties severely degraded. Thus, it is an attractive option for high-temperature thermoelectric applications for utilizing waste heat in factories and incinerators, for example. Furthermore, the ZT ~4 reported by Hi-Z Technology Inc. for quantum well modules containing boron carbide remains a world record [9], and illustrates the high potential of boron carbide as a thermoelectric material.

14.4.2 β-Boron

β-boron has also been extensively studied although the thermoelectric properties are not as excellent as boron carbide. Undoped β-boron is an insulator that can take large Seebeck coefficients close to 1 mV/K at room temperature and larger than 400 μV/K at 1000 K [24,25]. However, since the electrical conductivity is low, maximum ZT is around 0.01.

Slack et al. first reported a comprehensive series of doping of metal atoms (Ce, Ni, Co, Fe, Cu, Cr, Mn, V, Al, Ti, Ta, Nb, Si, Ge, Mg, Zr, Hf, Sc) into β-boron [24,26]. One attractive feature of the higher borides in general is that since they are cluster compounds, there are voids among the clusters into which it is possible to readily dope foreign elements. β-Boron has four well-known sites; A1, A2, D (sometimes split), E sites, into which metal atoms can be doped [26]. Regarding notable thermoelectric properties, generally a very large increase in the electrical conductivity has been achieved through metal doping [24]. Cu-doped β-boron was reported to attain a ZT around 0.5, while heavy vanadium doping of β-boron results in n-type behavior [24]. However, the vanadium doping is a simple metallization, and with the Seebeck coefficients trending small (or even positive) at higher temperatures, it is not a viable n-type counterpart to boron carbide.

Werheit et al., Nakayama et al., and others have also investigated the thermoelectric properties of various transition metal doping of β-boron [27,28]. However, neither a ZT value larger than 0.5 nor n-type behavior in a wide temperature range except for vanadium doping like Slack's result has been reported up to now.

To summarize, metal doping into β-boron has been shown to be a powerful method for modifying the thermoelectric properties, despite the thermoelectric figure of merit not attaining the level of the highest values reported for boron carbide.

14.4.3 Others

Asides from the heavily vanadium-doped β-boron among the boron icosahedra cluster compounds [24], we note that divalent alkaline-earth hexaborides like CaB_6 have also been found to exhibit n-type behavior as first shown by the three different groups of Paderno [29], Etourneau [30], and Shvartsman

[31]. Recently, Yagasaki et al. and Takeda et al. have also reported this behavior and optimized the properties significantly [32–34]. However, the hexaborides are a different class of compounds compared to the boron icosahedra cluster compounds like boron carbide, since they are basically dirty metals/small gap semiconductors with relatively large thermal conductivity. The *n*-type behavior is not unexpected, especially for inadvertently doped samples.

A viable *n*-type counterpart to boron carbide was recently discovered among the novel borides and will be described in detail in the next section.

Regarding other well-known boride systems, REB_{66} exhibits Seebeck coefficients larger than 700 µV/K at room temperature [35], but similar to undoped β-boron, the electrical conductivity is poor and the thermoelectric figure of merit is small. The crystal structure of REB_{66} is cubic (space group *Fm3c*) with a large unit cell of *a* = 23.44 Å, for example, for YB_{66} [36]. As schematically depicted in Figure 14.4, the structure is basically formed from super-icosahedra $B_{12}(B_{12})_{12}$, where a central icosahedron is bonded to and surrounded by 12 icosahedra. The super-icosahedra have two different orientations and form two cubic sublattices. B_{80} clusters and rare earth pair sites occupy the holes created by the arrangement of the super-icosahedra. Usually, only one of the rare earth sites is occupied (i.e., a partial occupancy of the rare earth site of 50%), although a small homogeneity region has been found [37]. An attempt to improve the thermoelectric properties through transition metal doping was made [38,39]. Nb-doping was attempted first and an example of a single crystal grown using an FZ furnace is shown in Figure 14.1c. Nb replaces the boron dumbbell site, which is located inside the B_{80} cluster as depicted in Figure 14.5 and occupancy values from 89% to 97% were tried. The Seebeck coefficients are shown in Figure 14.6. A factor 3 increase of the power factor at room temperature was obtained through the doping that reduces the electrical resistivity. However, in the important high-temperature region (*T* > 550 K) the undoped sample actually exhibited the highest power factor due to its large characteristic temperature T_0, which causes a steeper reduction of resistivity as temperature increases. The electrical conductivity of REB_{66} and most of the higher borides can be described by Mott's variable range hopping for three-dimensional systems [40,41];

$$\sigma = \sigma_0 exp\left[-\left(T_0/T\right)^{0.25}\right]. \tag{14.2}$$

This result is interesting as in this case it counter-intuitively advocates manufacturing a high T_0, despite an actual higher initial resistivity, when considering extremely high-temperature properties. However,

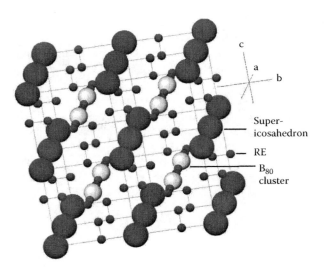

FIGURE 14.4   Schematic structure of REB_{66}.

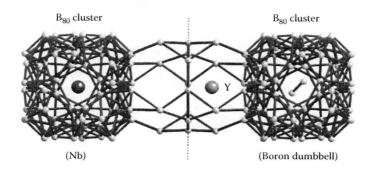

FIGURE 14.5 Structure of Nb-doped YB_{66} close to the doping site inside the B_{80} cluster.

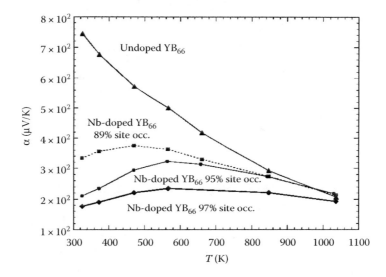

FIGURE 14.6 Temperature dependence of the Seebeck coefficient α of undoped and Nb-doped YB_{66}.

this is not a path to lead to excellent thermoelectric efficiency, and the thermoelectric properties of the REB_{66} system do not appear to be so promising (maximum of $ZT \sim 0.01$ for samples studied so far [39]).

Despite this, an interesting new aspect of the thermal conductivity of REB_{66} was discovered and will be described in more detail in the last section.

To summarize, boron carbide appears to be the most promising system among the well-known borides for thermoelectric application, and it has actually been commercialized in the past. Boron carbide exhibits p-type, and despite various rigorous efforts over the years, it was not possible to obtain a viable n-type counterpart based on the well-known borides.

14.5 Novel Borides

14.5.1 Rare Earth Boron Carbonitrides; $REB_{17}CN$, $REB_{22}C_2N$, and $REB_{28.5}C_4$

14.5.1.1 Intrinsic *N*-Type Counterpart to Boron Carbide

As noted in the Introduction, in a striking new development, n-type behavior was observed in a homologous series of rare earth borocarbonitrides [6,7]. Starting with boron carbide, all the boron icosahedra compounds discovered up to now have shown p-type behavior. A viable n-type counterpart to boron

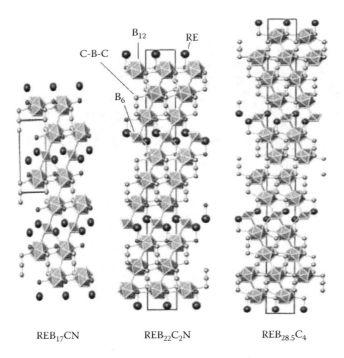

FIGURE 14.7 Structure of the homologous rare earth borocarbonitrides, $REB_{17}CN$, $REB_{22}C_2N$, and $REB_{28.5}C_4$.

carbide had been searched for over 20 years, but a good option was not found. As we have noted above, only after very heavy vanadium doping could n-type behavior be forcibly induced in β-boron. However, this phenomenon was metallization, with the absolute value of Seebeck coefficients trending very small as temperature was increased toward the important high-temperature region. The discovery in the rare earth boron carbonitrides was the first example of intrinsic n-type behavior ever being observed for a higher boride. Developing an n-type counterpart to boron carbide or $REB_{44}Si_2$ is extremely important in terms of the potential development of this class of compounds as viable thermoelectric materials since reasonable application will require p- and n-type legs.

The structures of the whole homologous series of rare earth boron carbonitrides, $REB_{17}CN$, $REB_{22}C_2N$, and $REB_{28.5}C_4$, are depicted in Figure 14.7. The structure is trigonal (space group *P3m1*) for $REB_{17}CN$ and rhombohedral (space group *R3m*) for $REB_{22}C_2N$ and $REB_{28.5}C_4$. For example, the lattice constants are $a = 5.588$ Å, $b = 10.878$ Å, $a = 5.614$ Å, $b = 44.625$ Å, and $a = 5.638$ Å, $b = 56.881$ Å for $HoB_{17}CN$, $HoB_{22}C_2N$, and $HoB_{28.5}C_4$, respectively.

The compounds have a layered structure along the c-axis with B_{12} icosahedra and C–B–C chain layers residing in between B_6 octahedral and rare earth atomic layers. The B_{12} icosahedra and C–B–C chain layers increase successfully along the series of $REB_{17}CN$, $REB_{22}C_2N$, and $REB_{28.5}C_4$. In the limit of the boron icosahedra layers going to infinity, the compound is actually analogous to boron carbide.

14.5.1.2 Densification Methods and Doping

Unlike the $REB_{44}Si_2$ compounds, $REB_{17}CN$, $REB_{22}C_2N$, and $REB_{28.5}C_4$ will not melt stably and therefore, large crystals cannot be grown. Therefore, sintered powders are compacted in various ways for measurements. A challenge facing these materials for application is that densification processes have not been fully developed yet. Conventional hot press methods and cold pressing and annealing methods yield samples with densities of only around 50% of the theoretical value [6]. Straightforwardly applying spark plasma sintering (SPS) has been found to increase the density to ~70%, which is still too low [7].

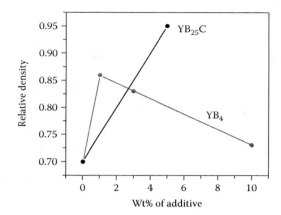

FIGURE 14.8 Dependence of densification of $YB_{22}C_2N$ on additive concentration.

Even with the densification by SPS being as it is, there is large improvement in electrical conductivity, with resistivities decreasing by close to 2 orders of magnitude compared to conventionally hot-pressed samples. Accompanying the decrease in resistivity, the absolute values of the Seebeck coefficients also show small decreases, but the effect is much smaller than the resistivity improvement. In fact, the power factor at 1000 K increases by close to 2 orders with the density change from 50% to 70% [7].

It was found that certain sintering additives in the SPS synthesis can be quite effective for densification of the rare earth borocarbonitrides [42,43]. As shown in Figure 14.8, the addition of YB_4 has been found to have some effect, yielding samples with higher densities of ~85%. Interestingly, the addition of $YB_{25}C$, which is not a common compound, resulted in a high densification of 93%. The reason for this densification was found to be that with the addition of $YB_{25}C$, the starting temperature of shrinkage decreases as shown in Figure 14.9, and this is quite effective for getting higher-density samples. The explicit mechanism of the decrease in the shrinkage temperature is still not clear.

However, one detrimental feature of the $YB_{25}C$ addition is that it appears to promote growth of boron carbide impurities in the sample. Boron carbide impurities have been shown to be strongly detrimental to the *n*-type characteristics. Initial samples of $REB_{28.5}C_4$ samples exhibited *p*-type behavior [6]. It was

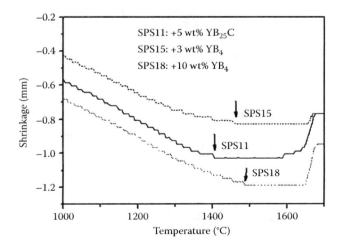

FIGURE 14.9 Dependence of shrinkage temperature of $YB_{22}C_2N$ on additive.

later revealed by SEM analysis that small inclusions of boron carbide, difficult to detect by powder x-ray diffraction due to overlap of peaks, were present. Inclusion-free $REB_{28.5}C_4$ samples were successfully synthesized and shown to possess *n*-type behavior similar to $REB_{22}C_2N$ and $REB_{28.5}C_4$ [44], illustrating the powerful effect of boron carbide impurities. Further systematic work on the densification processes of the rare earth borocarbonitrides should be carried out to find additives that are beneficial for densification and at the same time beneficial as a dopant for the thermoelectric properties. This can be said to be an important topic not just for the rare earth borocarbonitrides, but also for other sintered bodied materials, and a systematic analysis should be valuable.

Transition metal doping of $YB_{22}C_2N$ as a representative of the rare earth borocarbonitrides has also been carried out in an attempt to improve the properties [45,46]. As mentioned in Section 14.4, transition metal doping of the voids in boron cluster compounds like β-boron, has been found to be an effective way to control the electronic properties. Co and V doping of $YB_{22}C_2N$ appear to be the most promising, with the absolute values of the Seebeck coefficient increasing by up to 180%, while simultaneously a significant reduction up to two orders in electrical resistivity was obtained [45,46]. The reduction in resistivity is considered mainly due to the seeding effect observed earlier, where doping/seeding of the sintered bodies with highly electrically conductive metallic boride powders like REB_4 and REB_6, decreased the electrical resistivity dramatically. The Seebeck coefficients and thermal conductivities do not show a large change, so this seeding method is revealed to be a highly effective way to increase the figure of merit for such sintered bodies [7].

14.5.2 Rare Earth Borosilicides, $REB_{44}Si_2$

A view of the crystal structure of the $REB_{44}Si_2$ compounds is shown in Figure 14.10. The rare earth atoms have a single crystallographic site and form infinite ladders in the direction of the *c*-axis while there are 5 crystallographically independent B_{12} icosahedra and a $B_{12}Si_3$ polyhedron [47]. The structure is orthorhombic with space group *Pbam*. Lattice parameters are $a = 16.651$ Å, $b = 17.661$ Å, $c = 9.500$ Å for $TbB_{44}Si_2$, for example [48].

Regarding the thermoelectric properties of "as is" $REB_{44}Si_2$ compounds, the Seebeck coefficients α increase monotonically and take values exceeding 200 μV/K at 1000 K (Figure 14.11) [49]. As a representative of the series, $ErB_{44}Si_2$ has been found to exhibit a low thermal conductivity of <0.02 W cm^{-1} K^{-1} at high temperatures [3]. $REB_{44}Si_2$ compounds have melting points of around 2300 K and a previous extrapolation of the temperature dependence of the figure of merit to 1500 K yielded an estimated *ZT*

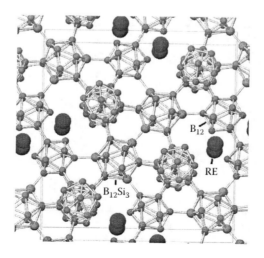

FIGURE 14.10   Structure of $REB_{44}Si_2$.

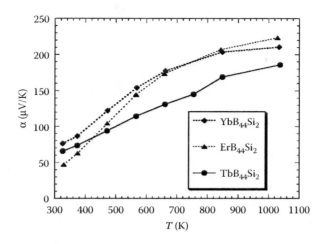

FIGURE 14.11 Temperature dependence of the Seebeck coefficient α of $REB_{44}Si_2$.

for $ErB_{44}Si_2$ of 0.12 [3]. With improved crystals the extrapolated ZT value of nondoped $REB_{44}Si_2$ is ~0.2 (Mori T., unpublished). Considering that these results are for "as is," that is, nondoped, noncomposition optimized compounds, this can be considered as a good starting point to further develop them as high-temperature thermoelectric materials.

The thermoelectric figure of merit of REB_{66} is largely inferior to $REB_{44}Si_2$. Therefore, in the new icosahedral compound $REB_{44}Si_2$, is found a compound that retains the low thermal conductivity of the B_{12} icosahedral borides but shows promising thermoelectric properties for extremely high temperatures [49].

In terms of morphology, $REB_{44}Si_2$ compounds can be melted and therefore, it is possible to grow large crystals like that shown in Figure 14.1b. However, it has been found difficult to grow a high-quality single crystal of $REB_{44}Si_2$ and attempts are still underway. Arc melting of the elements is also an easy way to obtain these compounds. Since the $REB_{44}Si_2$ compounds can be more easily melted than boron carbide, processing them may be easier, and therefore work should continue to modify/dope the $REB_{44}Si_2$ compounds to improve their properties as a possible p-type alternative to boron carbide.

14.6 Control of Morphology: Zinc Doping Effect

Recently, a zinc doping method was discovered to control the morphology (i.e., crystallinity) of the borosilicide compound and it also has the potential to be an easy, inexpensive method to apply to other high-temperature materials [50]. As a result of the modification of morphology, the power factor of the borosilicide could be improved significantly and will be described.

A series of transition metal doping (Mo, Mn, Fe, Rh, Ti, Cu, Zn) was systematically carried out on $YB_{44}Si_2$. The samples were prepared by arc melting and were dense bodies, compared to the nondense sintered bodies of rare earth borocarbonitrides. A 30% increase in the power factor was obtained for zinc-doped $YB_{44}Si_2$ (Figure 14.12). However, surprisingly, chemical analysis revealed that zinc is actually not retained in the doped sample product. Zinc doping was found to change the morphology of the sample, namely, improving the crystallinity (Figure 14.13). Borosilicides and silicides tend to have a problem with Si aggregations (silicon "threads") forming easily in the materials, and the improved crystallinity led to a significant increase in the power factor.

Such improvement of the crystallinity is beneficial when the system possesses intrinsic low thermal conductivity like the boron cluster compounds. The doping is quite unobtrusive and secondary effects

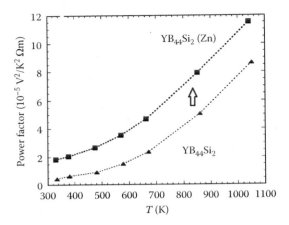

FIGURE 14.12 Temperature dependence of the power factor for undoped and Zn-doped (Zn does not remain in the final product compound) $YB_{44}Si_2$.

FIGURE 14.13 SEM pictures of the morphology for undoped and Zn-doped (Zn does not remain in the final product compound) of $YB_{44}Si_2$.

from electronic doping do not have to be worried about. This zinc doping can potentially be a wide ranging and an easy, inexpensive method to control the morphology of many high-temperature materials, not just limited to borosilicides, silicides, and borides [50].

14.7 Origins of Intrinsic Low Thermal Conductivity

14.7.1 Introduction

An interesting phenomenon is that boron cluster compounds are typically observed to exhibit low thermal conductivity [1–4], despite being hard compounds with high sound velocities. Several striking origins of this behavior can be proposed and I will review them here, since such aspects might be utilized as effective principles to develop other materials with intrinsically low thermal conductivity.

We note that there is a huge boom now of ball milling powders and sintering them in small grains to lower the thermal conductivity [51]. However, this is quite an extrinsic phenomenon. Such processing can readily be done later for materials that possess more intrinsic mechanisms to lower the thermal conductivity. Such intrinsic low thermal conductivity can be a "built-in" advantage in the development of thermoelectric materials, and it is important to try to develop such possible mechanisms further.

The possible mechanisms are: (A) high unit cell atomic density (or "crystal complexity" or "amorphous concept"), (B) "rattling" in voids in cluster compounds, (C) "symmetry mismatch effect," (D) disorder, and (E) particular features of the network-like crystal structure.

14.7.2 High Unit Cell Atomic Density (or "Crystal Complexity" or "Amorphous Concept")

Golikova previously proposed the so-called "amorphous concept" to interpret the basic physical properties of complex borides [35]. Namely, the number of atoms in the unit cell, N, is taken as a critical parameter to determine whether amorphous behavior is observed or not. For example, α-rhombohedral boron ($N = 12$) is a typical crystalline semiconductor, while β-boron ($N = 105$) has VRH electrical conductivity and crystalline thermal conductivity. REB_{66} ($N = 1600$), of which the crystal structure we redraw from the schematic Figure 14.4 by depicting all the atoms in Figure 14.14, exhibits VRH electrical conductivity and amorphous-like thermal conductivity.

On the other hand, Slack made a more quantitative treatment of "crystal complexity" [52]. Assuming that the optic phonons do not carry heat, an approximation of thermal conductivity can be given by

$$\kappa \approx BM\delta(\theta)^3 n^{-2/3} T^{-1} \gamma^{-2}, \tag{14.3}$$

where B is a constant, M the mean atomic weight, δ a length parameter in which δ^3 is the volume of the primitive cell, θ the Debye temperature, n the number of atoms in the primitive cell, and γ the Gruneisen constant. There is a dependence of κ on n as $n^{-2/3}$. Namely, the greater the crystal complexity, the lower the thermal conductivity, and Slack proposed a threshold of $n \approx 3 \times 10^4$ above which a compound will behave like a glass.

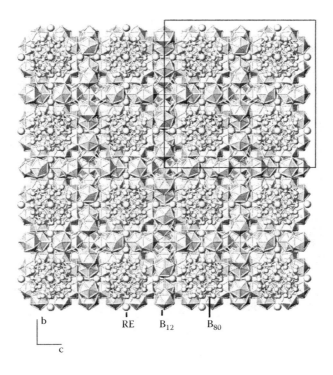

b
c
RE B_{12} B_{80}

FIGURE 14.14 Detailed structure of REB_{66}.

14.7.3 "Rattling" in Voids in Cluster Compounds

"Rattling" has been proposed to occur in cage compounds like skutterudites or clathrates where heavy metal atoms are located inside cage-like structures, in which they "rattle" like Einstein oscillators and scatter acoustic phonons to reduce thermal conductivity [53–55]. We have observed a trend for the thermal conductivity to be lower for rare earth higher borides, where rare earth atoms reside in the voids among the boron cluster network structure, compared to compounds like beta-boron and boron carbide. "Rattling" may also occur in the voids among such cluster compounds and should be elucidated further.

14.7.4 "Symmetry Mismatch Effect"

The basic building block of higher borides is the B_{12} icosahedra cluster. A prominent symmetry of this cluster is the fivefold symmetry (Figure 14.15), which obviously does not match the symmetry of any crystal structure. This is quite original, but it can be proposed that there is some kind of mechanism to lower the thermal conductivity when there exists a large mismatch of the crystal structure symmetry with that of the basic building blocks. The explicit mechanism of such an effect, if it exists, is not clear and should be investigated further [56].

14.7.5 Disorder

A way to more quantitatively evaluate and have a measure of the disorder in hopping systems has been proposed by using the localization length as a measure [4]. The localization length ξ can be given by [41]

$$\xi = [18.1/(k_B T_0 D(E_F))]^{1/3},$$

where T_0 is the characteristic temperature defined in Equation 14.2 and $D(E_F)$ is the density of states at the Fermi energy. Localization occurs when carriers cannot propagate throughout the compound because their wave functions are localized by the disorder/randomness. The localization length can be

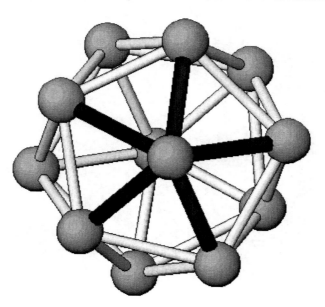

FIGURE 14.15　View of the B_{12} icosahedron. The fivefold symmetry is in bold.

considered to be a quantitative measure of disorder in similar systems since strong disorder will directly cause shorter localization lengths. This can be especially useful to actually gauge the effects that doping has on disorder.

14.7.6 Particular Features Embedded in the Crystal Structure (e.g., Boron Dumbbell)

These are features that may not be so readily inducible to artificial control, however, are still important to take note of and understand. An example given here is the boron dumbbell that can be observed in some borides. In the case of REB_{66}, a boron dumbbell is located in the void in the B_{80} cluster (Figure 14.5). As described in Section 14.4.3, regarding the transition metal doping of REB_{66}, the transition metal atoms can replace the boron dumbbell, as depicted in Figure 14.5. While such doping can be considered to increase the disorder, a striking effect was observed wherein the thermal conductivity increased for the doped samples. While this doping effect itself is bad for thermoelectric prospective, it indicates that the boron dumbbell was functioning to lower the thermal conductivity [38,39]. It can be envisaged that the boron dumbbell has some degrees of freedom/motion in the B_{80} cluster and this might be functioning to scatter phonons. It is an intriguing phenomenon and should be investigated further.

14.7.7 Comparison of Two Borides

The thermal conductivity of two boride systems $REB_{44}Si_2$ and REB_{66} was compared and analyzed in a study to gauge the efficaciousness of the possible mechanisms for intrinsic low thermal conductivity given above.

$YbB_{44}Si_2$ exhibits crystalline behavior [4] while YB_{66} exhibits glass-like thermal conductivity [1,2] (Figure 14.16). A quantitative treatment of (A), (B), (D) was carried out and it was indicated that in this case, disorder is the dominating driving force for the differences in the two compounds. The rare earth sites of $REB_{44}Si_2$ have full occupancy while those of REB_{66} have a partial occupancy of only around 50%, not to mention the low occupancy of the B_{80} cluster, and it is indicated that the disorder from such partial occupancy has a large effect [4].

Boron dumbbells, described in (E) above, can also be considered to be contributing to the low thermal conductivity of REB_{66} versus $REB_{44}Si_2$, however, the large difference indicates (D), disorder, to be a dominating driving force in this case.

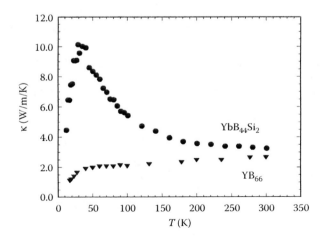

FIGURE 14.16 Temperature dependence of the thermal conductivity κ of $YbB_{44}Si_2$ and YB_{66}.

To summarize, we have described several interesting mechanisms for intrinsic low thermal conductivity. Further understanding of these mechanisms and attempted implementation through synthesis and modification (large unit cell, "rattlers" in cluster compounds, symmetry mismatch, quantitative control and monitoring of disorder, boron dumbbells, etc.) may yield routes for developing novel ultralow thermal conductivity materials.

14.8 Conclusions

Boron compounds are shown to be promising systems for their high-temperature thermoelectric properties. A brief review on some well-known boron compounds was given together with exciting new developments in some novel compounds. $REB_{44}Si_2$ is a promising p-type high-temperature system that retains the low thermal conductivity exhibited by compounds such as REB_{66} while improving the electrical properties and thermoelectric figure of merit significantly. The rare earth boron carbonitrides, $REB_{17}CN$, $REB_{22}C_2N$, and $REB_{28.5}C_4$, represent the first instance of intrinsic n-type behavior in boron icosahedra compounds, and are interesting as a potential counterpart to p-type boron carbide, which is one of the few thermoelectric materials that has actually been commercialized in the past.

Furthermore, despite being hard materials with high sound velocities, boron icosahedra compounds exhibit intrinsic low thermal conductivity. This can be considered as one of the "built-in" advantages of these compounds as viable thermoelectric materials. Several possible origins of the low thermal conductivity were reviewed, since they can offer good principles to design ultralow thermal conductivity materials.

Acknowledgments

The author gratefully acknowledges support from AOARD (AOARD 104144) and the Scientific Research on Priority Areas of New Materials Science Using Regulated Nano Spaces, the Ministry of Education, Science, Sports and Culture, Grant-in-Aid, Japan.

References

1. Slack G.A., Oliver D.W., and Horn F.H., *Phys. Rev. B*, 4, 1714–1720, 1971.
2. Cahill D.G., Fischer H.E., Watson S.K., Pohl R.O., and Slack G.A., *Phys. Rev. B*, 40, 3254–3260, 1989.
3. Mori T., *Physica B*, 383, 120–121, 2006.
4. Mori T., Martin J., and Nolas G., *J. Appl. Phys.*, 102, 073510, 2007.
5. Mori T., Higher borides, in Gschneidner Jr. K.A., Bunzli J.C., Pecharsky V., editors. *Handbook on the Physics and Chemistry of Rare Earths*, Vol. 38, Elsevier, Amsterdam, North-Holland, 2008, 105–173.
6. Mori T. and Nishimura T., *J. Solid State Chem.*, 179, 2908, 2006.
7. Mori T., Nishimura T., Yamaura K., and Takayama-Muromachi E., *J. Appl. Phys.*, 101, 093714, 2007.
8. Emin D., *J. Solid State Chem.*, 177, 1619, 2004.
9. Ghamaty S., Bass J.C., and Elsner N.B., in, Rowe D.M., editor. *Thermoelectrics Handbook, Micro to Nano*, Boca Raton, CRC Press, 2006, 57.
10. Longuet H.C. and Roberts M. DeV., *Proc. Roy. Soc. A*, 230, 110, 1955.
11. Samsonov G.V., Makarenko G.N., and Tsebulya G. G., *Izv. Akad. Nauk SSSR, Otd. Tekhn. Nauk*, 4, 1960.
12. Wood C. and Emin D., *Phys. Rev. B*, 29, 4582–4587, 1984.
13. Aselage T.L., Emin D., McCready S.S., and Duncan R.V., *Phys. Rev. Lett.*, 81, 2316–2319, 1998.
14. Aselage T.L. and Emin D., in, Kanatzidis, M.G., editor. *Chemistry, Physics and Materials Science of Thermoelectric Materials, Beyond Bismuth Telluride*. Dordrecht, Kluwer Academic, 2003, 55.
15. Aselage T.L. and Emin D., in Rowe D.M., editor. *Thermoelectrics Handbook*, Boca Raton, CRC Press, 2002, 31.

16. Emin D., *Phys. Rev. B*, 38, 6041, 1988.
17. Werheit, H., Rotter, H.W., Meyer, F.D., Hillebrecht, H., Shalamberidze, S.O., Abzianidze, T.G., and Esadze, G.G., *J. Solid State Chem.*, 177, 569–574, 2004.
18. Emin D., *Phys. Today*, 40, 55, 1987.
19. Werheit, H., *J. Phys., Condens. Matter*, 18, 10655–10662, 2006.
20. Cai K.F., Nan C.W., and Min X.M., *Mater. Res. Soc. Symp. Proc.*, 545, 131, 1999.
21. Goto T., *Kinzoku* (in Japanese), 68, 1086, 1998.
22. Liu C.H., *Mater. Lett.*, 49, 308–312, 2001.
23. Lunca-Popa P., Brand J.I., Balaz S., Rosa L.G., Boag N.M., Bai M., Robertson B.W., and Dowben P.A., *J. Phys. D*, 38, 1248, 2005.
24. Slack G.A., Rosolowski J.H., Hejna C., Garbauskas M., and Kasper J.S., *Proc. 9th Int. Symp. Boron, Borides and Related Compounds*, Duisberg, 1987, 132.
25. Golikova O.A., Amandzhanov N., Kazanin M.M., Klimashin G.M., and Kutasov V.V., *Phys. Stat. Sol.*, 121, 579, 1990.
26. Slack G.A., Hejna C.I., Garbauskas M.F., and Kasper J.S., *J. Solid State Chem.*, 76, 52, 1988.
27. Werheit H., Schmechel R., Kueffel V., and Lundström T., *J. Alloys. Compd.*, 262–263, 372, 1997.
28. Nakayama T., Shimizu J., and Kimura K., *J. Solid State Chem.*, 154, 13, 2000.
29. Paderno Yu.B., *Poroshk. Metallurgija*, 11, 70–73, 1969; Paderno Yu.B., *Electron. Technol.*, 3, 175, 1970.
30. Etourneau J., *J. Solid State Chem.*, 2, 332–342, 1970.
31. Avetisyan A.O., Goryachev Yu.M., Kovenskaya B.A., and Shvartsman E.I., *Izv. Akad. Nauk SSSR Neorg. Mater.*, 15, 663–666, 1979.
32. Yagasaki K., Notsu S., Shimoji Y., Nakama T., Kaji R., Yokoo T., Akimitsu J., Hedo M., and Uwatoko Y., *Physica B*, 329–333, 1259–1260, 2003.
33. Takeda M., Fukuda T., Domingo F., and Miura T., *J. Solid State Chem.*, 177, 471–475, 2004.
34. Takeda M., Domingo F., Miura T., and Fukuda T., *Mater. Res. Soc. Symp. Proc.*, 691, 209–214, 2002.
35. Golikova O.A., *Phys. Stat. Sol. A*, 101, 277, 1987.
36. Richards S.M. and Kasper J.S., *Acta Cryst. B*, 25, 237, 1969.
37. Higashi I., Kobayashi K., Tanaka T., and Ishizawa Y., *J. Solid State Chem.*, 133, 16, 1997.
38. Tanaka T., Kamiya K., Numazawa T., Sato A., and Takenouchi S., *Z. Kristallogr.*, 221, 472, 2006.
39. Mori T. and Tanaka T., *J. Solid State Chem.*, 179, 2889–2894, 2006.
40. Mott N.F., *J. Non-Cryst. Solids*, 1, 1, 1968.
41. Efros A.L. and Shklovskii B.I., in, Efros A.L., and Pollak M., editor. *Electron–Electron Interactions in Disordered Systems*, North-Holland, Amsterdam, 1985, 409–482.
42. Berthebaud D., Nishimura T., and Mori T., *J. Mater. Res.*, 25, 665–669, 2010.
43. Berthebaud D., Nishimura T., and Mori T., *J. Electron. Mat.*, 40, 682–686, 2011.
44. Mori T., Burkhardt U., Schnelle W., and Grin Y., Origin of the n-type behavior in rare earth boron carbonitrides, to be submitted.
45. Prytuliak A. and Mori T., *J. Electron. Mat.*, 40, 920–925, 2011.
46. Prytuliak A. and Mori T., Vanadium seeding effect on $YB_{22}C_2N$, to be submitted.
47. Higashi I., Tanaka T., Kobayashi K., Ishizawa Y., and Takami M., *J. Solid State Chem.*, 133, 11, 1997.
48. Mori T. and Tanaka T., *J. Phys. Soc. Jpn.*, 68, 2033–2039, 1999.
49. Mori T., *J. Appl. Phys.*, 97, 093703, 2005.
50. Mori T., Berthebaud D., Nishimura T., Nomura A., Shishido T., and Nakajima K., *Dalton Trans.*, 39, 1027–1030, 2010.
51. Poudel B., Hao Q., Ma Y., Lan Y., Minnich A., Yu B. Yan X. et al., *Science*, 320, 634, 2008.
52. Slack G.A., in Seitz F., Turnbull D., and Ehrenreich H. editors. *Semiconductors and Semimetals*, Vol. 34, ed., Academic Press, New York, 1979, 1.

53. Uher C., in Tritt T.M. editor. *Semiconductors and Semimetals*, Vol. 69, Academic Press, New York, 2000, 139, and references therein.

54. Nolas G.S., Morelli D.T., and Tritt T.M., *Annu. Rev. Mater. Sci.*, 29, 89, 1999, and references therein.

55. Nolas G.S., Slack G.A., Morelli D.T., Tritt T.M., and Ehrlich A.C., *J. Appl. Phys.*, 79, 4002, 1996.

56. Mori T., *J. Soc. Inorg. Mater. Jpn.*, 17, 428, 2010.

15

Polymer Thermoelectric Materials

Yoshikazu
Shinohara
*National Institute for
Materials Science*

15.1 Introduction

Research and development of thermoelectric materials were intensified by the arms race between the United States and the Soviet Union in the 1950s. At the time, many telluride compounds, including BiTe and PbTe, as well as SiGe, were found, and the fundamentals of the study of thermoelectric materials were established. After the end of the Cold War, research into consumer applications advanced mainly using BiTe.

One typical example is semiconductor laser packages for optical communications. Highly responsive performance, precise regulation, and compactness for thermoelectric devices fit the needs of the packages well. Refrigerators used in hospitals and hotels and on-board cool boxes are other example applications utilizing the compactness and noiselessness of thermoelectric devices.

Thermoelectric watches are an example of a familiar consumer application of these devices (Figure 15.1). The watch works using the temperature difference of 1–2°C between the human body and environmental air. This is an example of application of thermoelectric devices as ubiquitous power sources.

It has not been long since polymer materials were first focused on as thermoelectric materials, and Japan currently leads research and development in this field. In this report, we will introduce the current state, problems, expectations, and perspectives for thermoelectric conjugated polymer materials.

15.2 Expectations of Conjugated Polymer Materials

Currently, what is most hoped for thermoelectric conversion is the recovery of exhaust heat. Approximately 90% of greenhouse gas (GHG) discharged in Japan is carbon dioxide produced from the combustion of fossil fuels such as petroleum oil, coal, and fuel gas. Considering the international commitment to a 50% reduction of GHG by 2050, a drastic correspondence is needed. For the GHG reduction, not only the conversion to renewable energy but also the development of effective technologies for utilizing energy is strongly required.

FIGURE 15.1 Thermoelectric wrist watch.

In exhaust heat recovery, it is important to find how to recover heat which is distributed thinly and widely. Because effective recovery of low-temperature waste-heat (below 200°C) using hot water is limited, needs for new heat recovery methods are particularly high in various industries. Thus, applications of thermoelectric conversion are promising. For low-temperature waste heat, as well as solar batteries, the device area is more important than the temperature difference to gain output power.

The inorganic materials whose working temperature region is responsive to the applications mentioned above are BiTe materials, but they have many problems with regard to applications as follows:

1. Their principal elements are heavy metals.
2. Resources are few, and the countries producing them are limited.
3. They are fragile and have only a few limited forms.
4. It is difficult to make highly integrated and large area devices from them.
5. Much energy (emission of CO_2) is required to produce thermoelectric devices.

It was reported that artificial superlattice multilayered films synthesized by the molecular beam epitaxy (MBE) have performance more than two times that of bulk materials,[1] but the MBE process is very difficult to utilize for consumer applications.

Thus, thermoelectric conjugated polymer materials are currently attracting attention. The advantages of polymer materials are as follows:

1. Abundant resources (petroleum: fuels → materials)
2. Lightweight
3. Flexible
4. High yield rate of process
5. Integration technology (screen printing, ink jet printing)

The problems of BiTe materials are conversely the advantages of polymer materials. In other words, the only inconvenient issue appearing in consumer applications is the thermoelectric characteristics of polymer materials.

15.3 Current State of Thermoelectricity Studies

15.3.1 Conjugated Polymers

There are some conductive polymers applied to electric parts and antistatic agents. Table 15.1 shows the chemical structures of typical conductive polymers. The characteristic is p-conjugated double bonds, on which electric carriers can move easily. The physical characteristics are shown in Table 15.2.[2] Polyacetylene has a high conductivity of approximately 10^5 S/cm, but is unstable in air. Polypyrrole, polyaniline, and polythiophene are comparatively stable in air, but their champion data of conductivity show a small value of approximately 10^3 S/cm.

For inorganic materials, doping is conducted by the incorporation of additives into crystal structures. However, for polymer materials, doping is done by the ionization of additives in the vicinity of the polymer main chains, and thus most of the additives are not stably fixed around the main chains. For this reason, the stably realized conductivity is 1/5–1/10 times the champion data shown in table. The practically reproducible conductivity corresponds to at most 1/2 the values of BiTe materials.

TABLE 15.1 Typical Conductive Polymers

- Polyacetylene
- Polypyrrole
- Poly(phenylene vinylene)
- Polyaniline
- Polythiophene

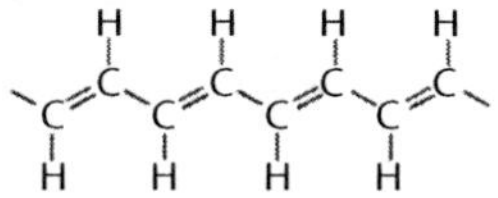

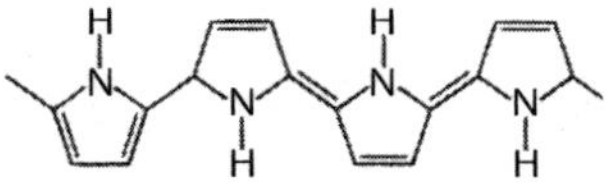

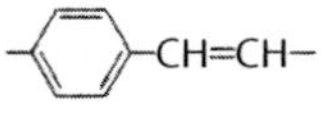

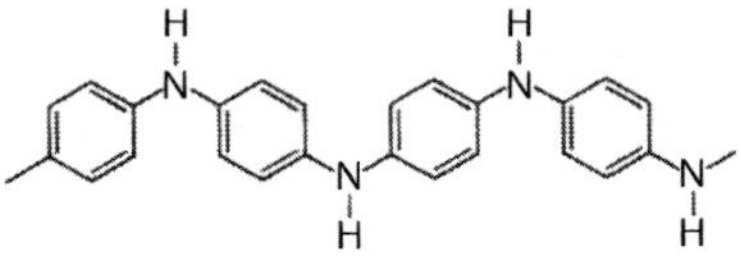

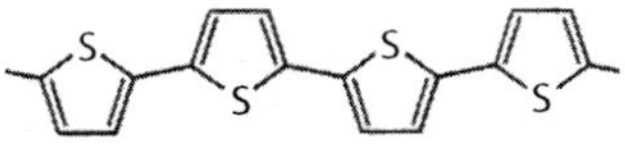

TABLE 15.2 Physical Properties of Typical Conductive Polymers

	Maximum Electrical Conductivity (S/cm)	Carrier Mobility (m^2/V s)	Bandgap (eV)
Polyacetylene	1.7×10^5 (I$_2$)	~1	1.4–1.5
Polypyrrol	1.5×10^3 (ClO$_4^-$)	~10^{-2}	3.2
Polyaniline	1.5×10^2 (I$_2$)	~10^{-2}	3.3
Poly(phenylene vinylene)	1.4×10^4 (H$_2$SO$_4$)	~10^{-3}	3
Polythiophene	5.5×10^3 (I$_2$)	~10^{-2}	2.2
Bi$_2$Te$_3$	9×10^2	5×10^2	0.15

The carrier mobility of conducting polymers is approximately 1 cm^2/Vs, 2–4 orders smaller than that of BiTe materials. To obtain conductivities comparable with those of BiTe materials, the carrier densities of the polymer systems must be 2–4 orders larger than those of BiTe materials. The larger carrier density results in the smaller Seebeck coefficient, because the coefficient is inversely proportionate to the square root of the carrier density.

In addition, as shown in the table, band gaps of conducting polymers are 10 times larger than those for BiTe materials. For inorganic materials, the narrow band gap causes a steep increase of state density distribution near the Fermi level, leading to the increase of the thermoelectric power.

For conducting polymers, details of the carrier conduction mechanism have not been clarified yet. Thus, although the results obtained for inorganic materials are not always applied to polymer materials, the low carrier mobility and large band gap of the polymer materials are problems in enhancing their thermoelectricity performance.

Figure 15.2 shows the relationship between the Seebeck coefficient and conductivity of a *p*-type polymer material at room temperature,[3] together with the data of a representative *p*-type inorganic material. Seebeck coefficients and conductivities of the polymer materials are pretty smaller than those of the inorganic materials. Even for polyacetylene, which has a comparatively high performance, the power factor ($S^2\sigma$, S: Seebeck coefficient, and σ: conductivity) is approximately 1/1000 times that of BiTe compounds.

It is a common trend between inorganic and polymer materials that the higher the conductivity or carrier density, the smaller the Seebeck coefficient; compared with the Seebeck coefficients of the inorganic materials, those of polymer materials are small because of high carrier densities. For polymer materials, the improvement for enhancing the Seebeck coefficients and the development of new materials with high carrier mobility and narrow band gaps are expected.

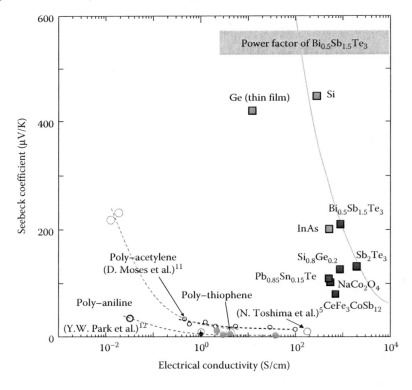

FIGURE 15.2 The relationship between Seebeck coefficients and electrical conductivities of conductive polymers. (Adapted from N. Toshima, *Material Stage*, 10(1), 42, 2010.)

15.3.2 Control of Thermoelectric Effect

The thermoelectric effect is related with the mobility of carriers, the effective mass, and the state density distributions. The figure of merit of thermoelectric materials, Z, is

$$Z = \frac{S^2\sigma}{\kappa} \propto m^{*3/2}\left(\frac{\mu}{\kappa_L}\right) \tag{15.1}$$

where S is the Seebeck coefficient, σ is the conductivity, κ is the thermal conductivity, m^* is the effective mass of carriers, μ is the mobility of carriers, and κ_L is the lattice component of thermal conductivity. The FoM, Z, indicates the energy conversion efficiency when a temperature difference of 1°C is applied, and the power factor, $S^2\sigma$, mentioned above, corresponds to the numerator of Equation 15.1.

For inorganic materials, the main control parameters were the mobility and the lattice components of thermal conductivities, but by utilizing computer science technologies the state density distribution has now been added to the control parameters. For polymer materials, the lattice components of thermal conductivities are small (e.g., approximately 0.1 W/Km for commercial conducting polyaniline, while 1–10 kW/Km for most of the inorganic materials), and thus the investigation concerning the mobility is currently being focused on.

The carrier transfer pathways in polymers are between and within the main chains. To increase the carrier mobility, it is important

1. To make the carrier transfer between the main chains easy, and
2. To make the carrier migration distance short in materials by enhancing the degree of main-chain orientation.

By studying the relationship between the side-chain size and Seebeck coefficient or conductivity of polyalkylthiophene, it has been found that (1) smaller side chains induce higher thermoelectric properties, (2) polythiophene, which has no side chains, shows an unprecedented high figure of merit, and (3) the degree of main-chain orientation is an important factor.[3] These results are shown in Figure 15.3.[3,4] The reduction of the distance between main chains due to shortening of the side-chain size was achieved in polyalkylchiophene, and the reduction of the mean distance between main chains was realized by increasing the main-chain orientation degree of polythiophene, which has no side chains. Both reductions enhance the carrier mobility between the main chains of polythiophene.

Figure 15.4 shows the relationship between the electrical conductivity and the figure of merit of polythiophene films estimated from Equation 15.1 by assuming that the thermal conductivity of the films, κ, is 0.1 W/Km.[3,4] The figure of merit increases by two orders by shortening the side-chain length, that is, from a polyalkylthiophene to finally polythiophene. Moreover, by enhancing the degree of orientation of polythiophene molecules, the figure of merit increases by two orders, achieving a value of 1.5×10^{-4}. This value corresponds to 1/20 times that of BiTe series materials used practically.

Particularly using polyaniline series materials, Toshima et al. have systematically investigated the relationship between their degree of orientation and thermoelectric properties.[5–8] Experimental results of spin coating and self-standing films suggest the significance of the polymer main-chain orientation. Figure 15.5 shows the extension effect to the thermoelectricity of polyaniline films. In Figure 15.5, the left axis indicates the power factor $S^2\sigma$, and the right axis indicates a dimensionless performance index ZT, where ■ represents the data of the extension direction for the stretched films, and □ represents the data of the vertical direction (with respect to extension direction) for the stretched films, and ○ represents the data for the unstretched films. By stretching the films, the figure of merit of the extension direction increases five times, and the figure of merit of the vertical direction increases two times.

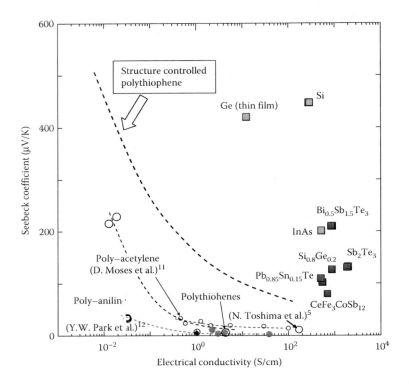

FIGURE 15.3   Change in thermoelectric properties of polythiophene series by controlling molecular structures.

As mentioned above, it is important to make the carrier transfer between the main chains easy for increase in the electrical conductivity of polymer materials. Organic–inorganic hybrid materials are an interesting idea to increase the electrical conductivity as shown in Figure 15.6.[9] It is reported that addition of platinum or gold nanoparticles by 1 wt% enhanced the electrical conductivities of poly-aniline without deterioration of the Seebeck coefficents.[10] Metal nanoparticles play a supplementary role of bridging the main chains to make the carrier transfer easy.

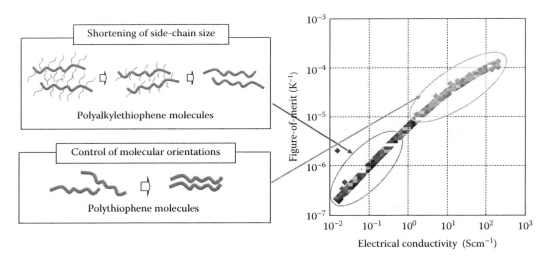

FIGURE 15.4   The variations of figure of merit of polythiophene series with the molecular structures.

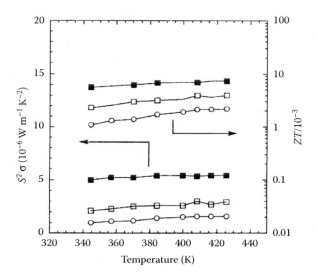

FIGURE 15.5 The effect of extension on thermoelectric properties of polyaniline.

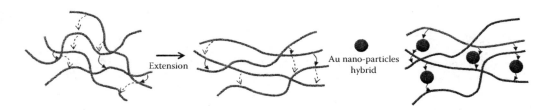

FIGURE 15.6 Schematic view of enhancement of electrical conductivity of conductive polymers by extension and hybrid with Au nanoparticles.

15.4 Problems and Perspective

15.4.1 Problems for Conjugated Polymers

The problems for conjugated polymers are summarized as follows:

1. The Seebeck coefficients are smaller by one order than those of inorganic materials.
2. There are no *n*-type, high-conductivity materials.
3. High conductivity is not stably reproducible.

However, it was reported that the effective mass of carriers for conjugated polymers are larger than that of inorganic materials,[2] indicating that the polymers have a potential for high Seebeck coefficients. It is important whether the excellent transfer characteristics of the polymer main chains are effectively brought out. Most of the conducting polymers are the *p*-type, and do not stably reproduce conductivities larger than 100 S/cm. The stabilization and high dispersivity of additives are also important.

15.4.2 Perspective

It is no exaggeration that inherent functions of elements themselves are found to be utilized in traditional material development. However, as the use of practical materials increases, that of the elements increases, leading to resource problems. Considering this situation, it is necessary to change to material development independent of the inherent functions of the elements themselves.

Materials development, which aims at function occurrence due to the organization and structuring of molecules, but is independent of the inherent characteristics of elements, is just sustainable. Currently, based on the function occurrence mechanism clarified, the direction for exploring new materials is gradually being discussed. However, in the current state, the function occurrence for most of the inorganic thermoelectric materials, such as BiTe systems, depends on the inherent characteristics of elements, for example, heavy elements, narrow band gaps, intermetallic compounds, and large lattice constants.

For organization and restructuring of inorganic materials, molecular beam methods have been used, and it will take many years to develop new materials for practical use. However, for polymer materials, it is comparatively easy to control the molecular structures and alignments, and in addition, their variations are enormous. Thus, polymer materials are best suitable for creating new functional materials.

Considering the thermoelectricity occurrence mechanisms found for inorganic materials, the main stream of thermoelectric material development in the twenty-first century will be generating thermoelectricity functions with polymer materials in which new structures are built. In other words, familiar consumer applications of thermoelectric materials can be realized by the development of thermoelectric polymer materials. However, even for polymer materials, the development of thermoelectric materials still depends on the inherent characteristics of the elements. Thus, we expect that the occurrence of thermoelectric functions due to novel polymer structures will be realized by activating exchanges with the inorganic material researchers.

References

1. R. Venkatasubramanian, E. Siivola, T. Colpitts, B. O'Quinn, Thin-film thermoelectric devices with high room-temperature figures of merit, *Nature*, 413, 597, 2001.
2. K. Tanaka, K. Akagi, Evolution of conductive polymers, In *Dr. Hideki Shirakawa and Conductive Polymers*, editors, K. Akagi, K. Tanaka, Tokyo, Kagaku-Dojin, Publishing Co., Inc., p. 53, 2002.
3. Y. Shinohara et al., Power evaluation of PbTe with continuous carrier concentration gradient, *Materials Science Forum*, 492–493, 141, 2005.
4. Y. Shinohara et al., A new challenge of polymer thermoelectric materials as ecomaterials, *Materials Science Forum*, 539–543, 2329, 2007.
5. N. Toshima, H. Yan, T. Ohta, Electrically conductive polyaniline films as organic thermoelectric materials, *Proc. 19th Int. Conf. Thermoelectrics*, Cardiff, United Kingdom, p. 214, 2000.
6. H. Yan, T. Ohta, N. Toshima, Stretched polyaniline films doped by (±)-10-camphorsulfonic acid: Anisotropy and improvement of thermoelectric properties, *Macromol. Mater. Eng.*, 286, 139, 2001.
7. N. Toshima, H. Yan, M. Kajita, Thermoelectric properties of spin-coated polyaniline films, *Proc. 21st Int. Conf. Thermoelectrics*, Longbeach, p. 147, 2002.
8. H. Yan, T. Isida, N. Toshima, Thermoelectric properties of electrically conductive polypyrrole film, *Proc. 20th Int. Conf. Thermoelectrics*, Beijing, p. 310, 2001.
9. N. Toshima, Conductive polymers as a candidate of organic thermoelectric materials, *Material Stage*, 10(1), 42, 2010.
10. N. Toshima, Organic-inorganic hybrid, In *Handbook of Thermoelectric Energy Conversions*, editor, T. Kajikawa, Tokyo, NTS, p. 320, 2009.
11. D. Moses, A. Denenstein, J. Chen, A. J. Heeger, P. McAndrew, T. Woerner, A. G. MacDiarmid, Y. W. Park, Effect of nonuniform doping on electrical transport in *trans*-$(CH)_x$: Studies of the semiconductor-metal transition, *Physical Review B*, 25(12), 7652, 1982.
12. Y. W. Park, Y. S. Lee, C. Park, L. W. Shacklette, R. H. Baughman, Thermopower and conductivity of metallic polyaniline, *Solid State Communications*, 63(11), 1063, 1987.

16

Thermomechanical Properties of Thermoelectric Materials

E.D. Case
Michigan State University

16.1 Introduction

In this chapter, we shall first discuss the nature of thermal fatigue and the intimate linkages between thermal fatigue and waste heat recovery applications of thermoelectric materials. The thermoelastic and the energy balance models for thermal shock will be reviewed, along with the roles material parameters and microstructure play in these models. Numerical examples of the thermal fatigue resistance parameters will illustrate differences among current thermoelectric materials.

After initially considering thermal shock, we will shift the focus to a discussion of damage evolution during thermal fatigue. Key differences between thermal fatigue and thermal shock will be highlighted, especially in terms of the role of crack toughening mechanisms and microstructure. Finally, three examples of nonthermoelectric, brittle material systems that are highly resistant to thermal fatigue damage will be considered and in each of these examples the role of crack-microstructure interaction is highlighted. These thermal-fatigue-resistant systems provide case studies in the application of the principles discussed in the chapter as well as pointing to practical ways to potentially improve the mechanical integrity thermoelectric materials for energy-harvesting applications, especially with regard to resisting thermal fatigue damage.

Since many of those engaged in thermoelectric research deal with the electrical and thermal properties of thermoelectric instead of mechanical properties, this chapter includes a review of the pertinent aspects of the thermal shock and thermal fatigue of brittle materials.

16.1.1 Mechanical and Thermal Properties Critical for Thermoelectric Applications

The mechanical properties such as the elastic moduli (Young's modulus, E, shear modulus, G, and Poisson's ratio, v) are needed to describe a solid's response to imposed mechanical or thermal stresses. For example, finite element modeling of the stress–strain behavior of a solid requires the construction of a stiffness matrix which is populated by Young's modulus and Poisson's ratio data (Kaliakan 2002). If thermal stresses are involved, then data for the thermal expansion coefficient α is also required. In addition, if a material fractures in response to the imposed stresses, then the fracture strength, σ_f, the effective fracture surface energy, γ_f, and/or the fracture toughness, K_C, are important parameters. If the stresses are imposed via thermal gradients or transients, then knowledge of the material's thermal expansion coefficient and the thermal conductivity are required to characterize the stress state. Additional thermal parameters that can be important include the surface heat transfer coefficient, h, and the thermal conductivity, k. Discussions of the thermoelastic and energy balance models of thermal shock will deal with the role of the mechanical properties E, v, σ_f, γ_f, K_C and the thermal properties α, h, and k in thermal shock damage in brittle materials.

Candidate thermoelectric materials for waste heat recovery applications typically include heavily doped semiconductors (Hogan et al. 2007) and a variety of ceramic oxides such as the cobalt oxide materials $Ca_3Co_4O_{9+\delta}$ (Wang et al. 2010a) and $Na_xCo_2O_4$ (Ito and Sumiyoshi 2010) as well as number of other oxides including ZnO-based materials (Park et al. 2007), $CuAlO_2$ (Liou et al. 2008) as well as Gd-doped $CaMnO_3$ (Lan et al. 2010), Dy-doped $La_xSr_{1-x}TiO_3$ (Wang et al. 2010b) and La-doped $BaSnO_3$ (Yasukawa et al. 2010). Both the heavily doped semiconductors and the ceramic oxide are brittle materials, thus this chapter focuses on the mechanical properties of brittle materials.

The recovery of waste heat from industrial and automotive sources usually involves repetitive heating and cooling. For example, in waste heat recovery from automobile or truck exhaust, thermoelectric generators capturing waste heat must endure thousand or tens of thousands of cold start-ups and hot shut-downs during the lifetime of an engine. Thus a key issue of applying thermoelectric materials to waste heat recovery is that of how well a brittle thermoelectric can perform in a thermal fatigue environment, where by "fatigue" we mean the accumulation of damage that occurs in the material on a cycle-by-cycle basis and in particular the damage that we are most concerned with is macrocrack growth as that leads directly to catastrophic failure of the body, say a thermoelectric leg in a module within a thermoelectric generator. (In fact, as we shall see later in this chapter, if we can arrange for many small cracks to grow simultaneously rather than one large macrocrack, the resistance to thermal fatigue damage can be greatly enhanced.) While a few decades ago the prevailing view was that brittle materials did not undergo fatigue (since brittle materials by definition have limited dislocation motion), it is now widely recognized that brittle materials do exhibit both mechanical and thermal fatigue under repetitive loading (Bhowmick et al. 2007; Ritchie 2009; Wachtman et al. 2009).

The mechanical properties of thermoelectric materials, as is the case with other brittle materials, are functions of temperature (Ravi et al. 2008; Ren et al. 2008a, 2009; Salvador et al. 2009) and composition (Ren et al. 2007, 2008b), although explicit dependencies on temperature and composition of mechanical properties will not be treated as a separate topic in this chapter.

The microstructure of thermoelectric materials, including grain size (Ni et al. 2010) and porosity (Ni et al. 2009) impacts both mechanical property values and the nature of crack growth in brittle materials. We shall emphasize the impact microstructure upon crack growth, especially crack growth driven by thermal shock and thermal fatigue. Also, since (i) most mechanical and thermal properties are functions of the number density of microcracks within the specimen and (ii) thermal fatigue can generate significant microcracking in a brittle material, the effect of microcracking on mechanical properties will be briefly reviewed later in this chapter.

The hardness, *H*, is a mechanical property often measured for thermoelectrics and other brittle materials via nanoindentation or microindentation. Hardness testing is relatively quick and straightforward and requires only a planar specimen surface that can be finely polished, as opposed to the standard 3 mm × 4 mm × 45 mm parallelepiped specimen needed for conventional fracture strength measurements (ASTM 2008). For example, using a Vickers microindentation tip, a specimen on the order of several millimeters on a side can easily accommodate an array of 10–20 indentations. (For a Vickers indentation diagonal length of 2a, the indentation impression centers should be spaced at least 10a apart from neighboring indentations as well as at least 10–20a from the nearest specimen edge or other elastic discontinuity in order to avoid interference effects.) As is the case for other brittle materials, *H* in thermoelectric materials is a function of composition (Ren et al. 2008b), applied load and grain size (Ni et al. 2010). *H* is tangentially linked to thermal fatigue, since is it a measure of scratch resistance during specimen handling and surface scratches in brittle materials can act as strength controlling flaws. Thus, while *H* is an important mechanical parameter of brittle materials, since this chapter focuses on mechanical and thermal parameters important to thermal fatigue, *H* will not be discussed further in this chapter and the interested reader is referred to the recent, comprehensive text by Gilman on hardness and the relationship of hardness to other physical properties (Gilman 2009).

16.1.2 Thermal Shock and Thermal Fatigue

Thermal fatigue occurs when microcrack and macrocrack damage accumulates in body subjected to repeated heating and cooling. If, for example, the temperature excursion is sufficiently small or the heating/cooling rate is low, then thermal fatigue does not induce damage. In waste heat recovery applications of thermoelectric materials, however, the in-service environment may be such that thermal fatigue damage will accumulate in the thermoelectric material.

Unfortunately, the literature contains few studies of thermal fatigue (or even mechanical fatigue) of brittle materials. Also, there are currently no thermal fatigue studies of thermoelectric materials in the literature and even the database on the mechanical properties of thermoelectric materials is quite limited.

In contrast to thermal fatigue, there is an extensive literature on the thermal shock of brittle materials, where thermal shock refers to a single thermal cycle. However, for thermoelectric materials, even thermal shock studies are absent.

Most thermal shock studies have been carried out by measuring the residual strength (the fracture strength following the thermal shock) as a function of ΔT, the quench temperature difference. In this type of study, one can define three temperature regimes in terms of the damage induced by thermal shock (Hasselman 1963, 1969):

Regime I. $0 < \Delta T < \Delta T_C$: The quench temperature difference ΔT is sufficiently small that the shock does not result in a measureable change in the physical property being monitored in property *P*.

Regime II. $\Delta T_C \leq \Delta T < \Delta T_P$: Beginning with the critical quench temperature difference, ΔT_C, there can be an abrupt decrease in property *P*. However as noted by Bradt and coworkers (Ashizuka et al. 1983; Ohira and Bradt 1988; Bradt et al. 1993), if a sufficient number of specimens are thermally shocked (say, roughly on the order of 30–50), then rather than a step-function change in property *P*, the decrease in *P* is sigmoidal in shape and the fracture data can be analyzed in terms of Weibull statistics (Bradt et al. 1993). As ΔT is increased above ΔT_C, often a plateau in property P is observed such that P does not change appreciably until ΔT reaches ΔT_P.

Regime III. $\Delta T > \Delta T_P$: Quench temperature differences $\Delta T > \Delta T_P$ often result in a gradual, quasi-static decrease in property P.

If the material undergoing thermal shock has a sufficient population of pre-existing flaws, Regime II may be diminished or absent altogether. These three thermal shock regimes can be understood in terms of Hasselman's energy balance model of thermal shock, which will be discussed later in this chapter (Hasselman 1963, 1969).

16.2 Fracture Toughness K_C, the Critical Flaw Length a_c, and the Grain Size GS

For brittle materials, the fracture strength, σ_f, typically depends on the preexisting flaw length such that $\sigma_f \sim 1/a_c^{0.5}$ where a_c = length of largest flaw in the specimen (the critical flaw). Thus, if a brittle solid, say, has a fracture strength of 60 MPa and a maximum flaw size of 30 μm, if we machine the specimen and cause the maximal flaw length to extend from 30 to 300 μm, fracture strength will drop to about 20 MPa. The magnitude of the strength drop will in fact also depend on a number of other factors, including the details of how the load is applied and the location and shape of the flaw. This simple example highlights the fact that fracture strength is only in part dependant on the material but is also sensitive to the detailed specimen size and geometry along with the flaw size, shape and location within the specimen.

The concept of the stress intensity factor, K, for brittle linearly elastic materials, K can be written as

$$K = Y\sigma_{app}(\pi a)^{0.5} \tag{16.1a}$$

where σ_{app} is the applied stress, a is the crack length and Y is a unitless geometric factor that is a function of specimen geometry, details of ap$_\parallel$ lied load, crack shape and location (Hertzberg 1983). For specimens loaded in bend, Y for small surface cracks is typically on the order of 1.1 (Hertzberg 1983).

The "linearly elastic" nature of K in Equation 16.1a refers to the concept that for an ideally brittle material, the slope of the stress–strain curve is constant until fracture. Thus the K in Equation 16.1a embodies a core concept of linear elastic fracture mechanics (LEFM). However, as microcracks nucleate and/or grow in a brittle material under stress, the stress–strain curve can become nonlinear and elastically nonlinear problems can be treated by the more generalized concept of the J-integral (Kanninen and Popelar 1985). Fracture in refractory materials has been treated using the J-integral approach (Homeny et al. 1980; Bradt 2004). However, since the LEFM approach and the thermal shock resistance parameters, R^i (which will be discussed in the following section) are typically used in the literature to treat thermal shock problems, the J-integral approach will not be discussed further in this chapter.

When the applied load is increased to the point that the specimen fractures, then by definition $\sigma_{app} = \sigma_f$, $a = a_c$, and $K = K_C$ where σ_f, a_c, and K_C are the fracture strength, the critical flaw length and the critical stress intensity factor such that

$$K_C = Y\sigma_f(\pi a_c)^{0.5} \tag{16.1b}$$

Thus, the condition $K = K_C$ gives a criterion for when a crack will grow in a brittle material.

The critical stress intensity factor, K_C, is also called the fracture toughness of the material. Since flaw size in polycrystalline materials tend to scale with grain size, GS, then fracture strength $\sigma_f \sim 1/GS^{0.5}$ (Barsoum 1997; Rice 2000).

As K_C increases, the "flaw tolerance" increases, thus K_C increases (for a fixed value of the specimen and flaw geometric factor Y), then the a_c required for fracture increases. Therefore, increasing K_C directly increases flaw tolerance since it takes a larger flaw to generate fracture for a given applied stress.

The concept of K_C is also related to the fracture surface energy, γ_f, and the strain energy release rate, G. The fracture surface energy measures the energy required to form a crack and includes not only the breaking of atomic bonds but contributions from the acoustic energy radiated by the growing crack (evidenced by the audible noise that accompanies the fast propagation of large cracks) as well as limited plastic deformation processes if the material is not ideally brittle (Wachtman 2009). The strain energy release rate refers to the strain energy per unit area involved in formation of the crack surfaces. K_C, γ_f, and G are related by

$$2\gamma_f = G = \frac{K_C^2}{E'} \tag{16.1c}$$

where E' is a function of Young's modulus E and Poisson's ratio ν that depends on whether the loading is in plane strain or plane stress (Wachtman 2009). The factor of 2 pre-multiplying γ_f in Equation 16.1c reflects the formation of two crack surfaces by the propagating crack.

In the following sections, we begin our discussion with thermal shock, with refers to a single thermal cycle ($N = 1$). Of course for thermoelectric materials applications, thermal fatigue ($N > 1$) is much more germane than thermal shock, however much of the literature on the effect of thermal transients on brittle materials deals exclusively with thermal shock rather than thermal fatigue.

16.3 Thermal Shock Characterization of Brittle Materials

A typical thermal shock test consists of (i) heating a body, (ii) rapidly immersing the body in a quenching fluid, and (iii) measuring the residual fracture strength, where residual strength refers to the fracture strength retained by the specimen following the shock (Hasselman 1969; Gupta 1972; Chen et al. 2009). Likely far more than 99% of literature describing thermal transient testing and analysis of materials deals with thermal shock rather than thermal fatigue.

Why is such a preponderance of thermal shock studies compared to thermal fatigue studies? One reason is that the typical residual strength thermal shock test cannot be easily extended to thermal fatigue measurements. For example, for thermal shock testing, the stochastic nature of strength dictates that a statistically adequate number of specimens be fractured at each ΔT value (say a minimum of 5 but preferably 10–30 specimens for each thermal shock condition). Of course, thermal shock testing, by definition, applies to a single thermal shock cycle. If one adds a "third dimension," namely the number of thermal fatigue cyclic to our residual strength versus ΔT analysis and we wish to characterize the thermal fatigue damage for a range of number of N (number of thermal cycle) values, say, 1, 3, 5, 10, 15, 20, 30 thermal cycles or more and at the various N values we gather data at different ΔT values, then it becomes unworkable to extend the classical retained strength protocol from thermal shock to thermal fatigue.

In contrast to residual strength versus measurements, several researchers have explored thermal fatigue damage nondestructively in terms of, say, monitoring the elastic modulus, thermal conductivity or the dielectric constant of brittle specimens as a function of the number of cumulative number of thermal fatigue cycles (Lee et al. 1989; Lee 1900; Lee and Case 1992; Case et al. 1993; Kumakawa and Niino 1993; Sinha et al. 1994; Lisjak et al. 1997; Rendtorff and Aglietti 2010). Nevertheless, the number of thermal fatigue studies of brittle materials in the literature is quite small and currently there are no thermal fatigue studies of thermoelectric materials in the open literature.

16.3.1 Thermal Shock Resistance Parameters, R^i

The various expressions for the thermal shock resistance parameters R^i give at best a relative ranking of the resistance to thermal shock failure for a group of materials. There are two distinct "families" or types of thermal shock resistance parameters, with one family based on thermoelastic theory and the other family on energy balance considerations. We shall briefly outline both approaches obtaining R. These discussions will highlight both the background and assumptions involved in the thermal shock resistance parameters.

16.3.1.1 Thermoelastic Model for Stress Resistance Parameters

We shall begin our discussion of the thermoelastic model with the solution of Fourier law of heat conduction for an infinite plate of half thickness H for the spatial coordinate z, where the z axis is normal of the surface of the plate, $T =$ plate temperature at coordinate z and time t, and $\kappa_z =$ thermal diffusivity of the plate along the z direction (Cheng 1951; Manson 1966; Zhao et al. 2000).

$$\frac{\partial^2 T}{\partial z^2} = \frac{1}{\kappa_z}\frac{\partial T}{\partial t}, \quad |z| \leq H \tag{16.2}$$

Typically the solution is performed by separation of variables for convective heat transfer conditions such that at the top and bottom surfaces of the plate (Cheng 1951; Manson 1966; Zhao et al. 2000) with the initial and boundary conditions

$$T = T_i \quad \text{at } t = 0 \tag{16.3a}$$

$$k\frac{\partial T(z,t)}{\partial z} = -h(T - T_\infty) \quad \text{for } z = \pm H \text{ (the plate's outer surfaces)} \tag{16.3b}$$

$$\frac{\partial T(z,t)}{\partial z} = 0 \quad \text{at } z = 0 \text{ (the mid-plane of the plate)} \tag{16.3c}$$

where k is the thermal conductivity of the plate, h is the surface heat transfer coefficient, T_i is the initial temperature of the plate, and T_∞ is the ambient temperature (the initial temperature of the convective medium surrounding the plate). In terms of the physics of the problem, it is important to note that the surface heat transfer coefficient h is a measure of the rate of heat energy transfer across the interface between the plate and the ambient medium.

For Equation 16.3b, the heat transfer condition is typically considered in terms of the surface heat transfer coefficient, $h = h_C$ for convective heat transfer, but the form of Equation 16.3b is identical for the case of radiative heat transfer, where the heat transfer coefficient is given by $h = h_R$. However, for combined radiative and convective heat transfer, one can define an effective heat transfer coefficient $h_{eff} = h_C + h_R$ (Kreith 1973), thus Equation 16.3b and the solutions of Equation 16.2 that follow apply to convective, radiative, or combined heat transfer conditions. Rather than distinguish among h_{eff}, h_C, and h_R, the solution below employs h to denote convective, radiative, or combined heat transfer.

As is typically the case for second-order partial differential equations, the general solution to Equation 16.2 can be written in terms of the infinite series which in this case is given by (Zhao et al. 2000)

$$\frac{T(z,t) - T_i}{T_i - T_\infty} = -1 + 2\sum_{n=1}^{\infty} \frac{\sin\beta_n \cos(\beta_n z/H)}{\beta_n + \sin\beta_n \cos\beta_n}\exp\left(-\beta_n^2\frac{\kappa_z t}{H^2}\right) \tag{16.4}$$

The set of coefficients β_n that appear within the infinite series represent the roots of the transcendental equation $\beta \tan \beta = Bi$, where Bi is the Biot modulus (Manson 1966; Zhao et al. 2000). The Biot modulus is a nondimensional parameter defined by $Bi = ah/k$, where a is a characteristic specimen dimension, h is the surface heat transfer coefficient, and k is the thermal conductivity of the thermally shocked solid.

As described by Zhao et al. (2000), from the temperature distribution given by Equation 16.4 one can in turn evaluate the normalized stress $\bar{\sigma}(z,t)$ distribution in the plate such that

$$\bar{\sigma}(z,t) = 2\sum_{n=1}^{\infty} \frac{\sin\beta_n}{\beta_n + \sin\beta_n \cos\beta_n} \left\{\cos\left(\beta_n \frac{z}{H}\right) - \frac{\sin\beta_n}{\beta_n}\right\} \exp\left(-\beta_n^2 \frac{\kappa_z t}{H^2}\right) \tag{16.5}$$

The argument of the exponential function includes as a factor the quantity $\kappa t/H^2$, which is the Fourier number, a normalized, nondimensional time variable designated by F. The surface stress begins at zero at $F = 0$, rises to a maximum within milliseconds or tens or milliseconds for fluid quenching media such as water or silicone oil (Lee and Case 1992), with a much slower drop-off in stress as a function of time (Zhao et al. 2000; Case 2002).

The relationship for the maximum surface stress, σ_{max} can be approximated in terms of a simple closed-form function of Bi (Cheng 1951; Manson 1966; Zhao et al. 2000), written as f(Bi), namely (Manson 1966; Zhao et al. 2000)

$$\sigma_{max} = \frac{E(T)}{1 - \upsilon(T)} \alpha(T) \cdot (T_i - T_\infty) \cdot f(\text{Bi}) \tag{16.6}$$

where $E(T)$ is the Young's modulus, $\upsilon(T)$ is Poisson's ratio, $\alpha(T)$ is the linear coefficient of thermal expansion, α. The quench temperature difference, $\Delta T = (T_i - T_\infty)$, where T_i is the initial specimen temperature and T_∞ is the temperature of the quenching medium.

An often-used form of the approximation for f(Bi) was presented by Manson, which is labeled here as f_{Manson}(Bi) (Manson 1966) is given by

$$f_{\text{Manson}}(\text{Bi}) = (1.5 + 3.25/\text{Bi} - 0.5\exp(-16/\text{Bi}))^{-1} \tag{16.7}$$

The expression of the maximum surface stress given by Equation 16.6 is often used in the literature to approximate the maximum stress developed during thermal shock, even when the component does not approximate the infinite plate geometry.

Let us now consider two physically and conceptually important asymptotes for f_{Manson}(Bi) as a function of Bi, namely those for Bi → 0 and Bi → ∞. An important physical insight that helps us to interpret Equations 16.6 and 16.7 is that h is an extremely sensitive function of the time rate of change of temperature, dT/dt (Manson 1966; Zhao et al. 2000), such that a given quenching medium, the magnitude of h can change by several orders of magnitude. Thus, for a rapid quench, h (as well as Bi since Bi $= ha/k$) is very large and as the rate of quench decreases, h decreases dramatically. As an example, quenching solid specimen from, say, 600 K into a room temperature water bath approximates an infinite quench relatively well. As can be seen by inspection of Equation 16.7, as Bi → ∞ (a very rapid quench) then f_{Manson}(Bi) → 1. Likewise, as Bi → 0 (an extremely mild quench), then f_{Manson}(Bi) approaches Bi/3.25. The function f_{Manson}(Bi) is a nonlinear, monotonic increasing function of Bi (Manson 1966; Case et al. 1993).

In addition to expressing the response to a thermal shock in terms of σ_{max}, we can characterize the crack tip stresses in terms K_{imax} the maximum stress intensity factor developed during thermal shock, which for in an infinite flat plate of thickness L was given by Noda (1991) as

$$K_{imax} = \frac{E\alpha\Delta T\sqrt{a}}{1 - \nu} f_{\text{Noda}}(\text{Bi}) \tag{16.8}$$

where E, α, ΔT, and ν are as defined above. The function f_{Noda}(Bi) is in turn given by

$$f_{\text{Noda}}(\text{Bi}) = \{(3.04\exp(-1.6\gamma) + 0.76\exp(3.6\gamma)) - 0.21\exp(-(15 + 10^4\gamma)/\text{Bi})\}^{-1} \qquad (16.9)$$

where $\gamma = a/L$ and a is the crack length and L is the plate thickness. Thus the magnitude of $f_{\text{Noda}}(\text{Bi})$ and hence K_{imax} depend on both the material parameters and the crack length normalized by the plate thickness. In addition to using expressions such as those for σ_{max} and K_{imax} to characterize the stresses or stress intensity factors to characterize the potential for thermal fatigue damage for a particular quench conditions, a set of thermal shock resistance parameters, most commonly including R, R', R''', R'''', and R_{ST} are frequently used to rank a material's susceptibility to thermal shock damage. The first two of those parameters, R and R' are based directly on the thermoelastic model described above while the factors R''', R'''' and R_{ST} are based on an energy balance model, which will be discussed in beginning of Section 16.3.2.3.

16.3.1.2 Parameters R and R' and the Thermoelastic Model

Knehans and Steinbrech (1982) provided direct experimental evidence of crack wake mechanisms operating during crack growth that exhibited R-curve behavior. Based on the thermoelastic approach to thermal shock, Kingery (Kingery 1955), developed the thermal shock resistance parameters R and R' by first considering the relationship $\Delta T_C = RS =$ critical quench temperature difference = minimum temperature change that induces fracture where R is the thermal shock resistance parameter. The unitless parameter S is a shape factor which depends on the geometry of the thermally shocked body.

Although the factor S is called a "shape factor," thermal shock and fatigue are functions of the specimen size even for a fixed geometry, for example, for plate-shaped specimens or for cylindrical specimens (Becher 1980, 1993; Lutz 1995; Damani et al. 2000). Becher and coworkers have done extensive studies of the size effect for thermal shock of brittle materials (Becher 1980, 1993). In a thermal shock study of alumina, zirconia, boron carbide, a glass ceramic and silicon nitride, plots of the critical quench temperature difference, ΔT_C, versus specimen thickness displayed a roughly bilinear behavior, with an initial "low thickness" branch for which ΔT_C dropped rapidly with increasing specimen thickness followed by a second, also roughly linear branch for which ΔT_C was approximately constant with respect to further increases in specimen thickness (Becher 1993). For example, for silicon nitride bars quenched into 22°C water, ΔT_C was ~1100°C for 1 mm thick specimens while for silicon nitride specimens between 5 and 10 mm in thickness, ΔT_C was ~400°C. Also, for 1 mm thick zirconia specimens, ΔT_C was ~500°C while for specimens thicker than ~3 mm, ΔT_C was nearly constant at about 240°C. Thus, for a fixed brittle material (where the specimens were nominally identical except for thickness), the thermal shock resistance (as gauged by ΔT_C) was dramatically higher for 1 mm thick specimens as opposed to specimens that were several mm thick. The trends observed by Becher and coworkers are consistent with the dimensional dependence of the Biot modulus, Bi, and its role in the generation of thermal stresses during thermal shock, as discussed in Section 16.3.1.1.

These "size effect" results (Becher 1980, 1993; Lutz 1995) underscore the fact that for the relationship $\Delta T_C = RS$, the specimen geometry including specimen dimension, can have a dramatic effect on thermal shock resistance. Of course, for specimens with complicated geometrical shapes, the S factor will become far more complicated than for simple bar or cylinder shapes.

In terms of the thermal shock resistance parameters R and R', for the maximum surface transient thermal stresses, σ_{max}, when the quench temperature $\Delta T = \Delta T_C$, then by definition $\sigma_{\text{max}} = \sigma_{\text{f}}$. Rearranging our σ_{max} expression (Equation 10.6) without writing the explicit T dependence of the mechanical and thermal parameters, we obtain

$$\Delta T_C = \frac{\sigma_{\text{f}}(1 - v)}{f(\text{Bi})E\alpha} \qquad (16.10)$$

If we use the Manson (1966) form of $f(\text{Bi}) = f_{\text{Manson}}$ (Equation 16.7), for very small values of Bi, the function $f(\text{Bi})$ may be approximated as $f(\text{Bi}) \sim \text{Bi}/3.25$ and the ΔT_C expression becomes

$$\Delta T_C = \frac{3.25\sigma_f(1 - \nu)}{(\text{ha})E\alpha}k \tag{16.11}$$

If we now neglect the numerical factor of 3.25 in the numerator and the factor (ha) in the denominator and recall that $\Delta T_C = R'S$, then we obtain Kingery's (1955) expression for the thermal shock damage parameter R', namely

$$R' = \frac{\sigma_f k(1 - \upsilon)}{E\alpha} = Rk \tag{16.12}$$

The R' expression is thus only a very rough description of the resistance to thermal shock damage since (i) $f(\text{Bi}) \sim \text{Bi}/3.25$ is a very poor approximation to $f(\text{Bi})$ except for when $\text{Bi} \to 0$ and (ii) h (surface heat transfer coefficient), was neglected where h is in fact a very sensitive function of temperature and the quench medium (ambient environment of the shocked body). Thus, h includes the information about the quench medium and as h changes, the rate of heating/cooling changes. If in Equation 16.11, for ΔT_C we set $f(\text{Bi}) = 1$ (corresponding $\text{Bi} \to \infty$ and an infinity rapid cooling rate), then we obtain the thermal shock parameter R, where $Rk = R'$.

Note that from the relationship $\Delta T_C = RS$, if one neglects (or sets equal to unity) the shape factor S, then one has $R = \Delta T_C$. However, earlier in this section, Becher's experimental results on the size effect of thermal shock show that in general S is not unity. Thus although R has units of temperature and is related to ΔT_C, R is typically a very poor estimate of ΔT_C. This is underscored by the original intent of the R and R' factors (Kingery 1955), which was to be a ranking system of thermal shock resistance among materials, not absolute quantities. Likewise, in the following section, the factors R''', R'''', and R_{ST} were intended to serve to a set of materials' resistance to thermal shock (Hasselman 1963, 1969).

It is important that the basic "philosophy" represented by R and R' is that one attempts to avoid crack nucleation by making the strength high, the thermal conductivity high and the Young's modulus and thermal expansion low.

16.3.1.3 The Energy Balance Approach

Using an energy balance approach, Hasselman (1963, 1969) assumed a body with multiple cracks and the cracks propagate under thermal stress. For the thermal shock resistance parameters, R and R' the concept was to inhibit crack nucleation. In contrast, with the energy balance approach, with the thermal shock resistance parameters R''' and R'''' one seeks to minimize crack propagation, that is, one "allows" multiple short cracks to propagate rather than a single catastrophic crack.

Hasselman (1963) formulated expressions for the thermal shock resistance parameters R''' and R'''' by considering a sphere of radius b that is thermal shocked. Based on relationships from Boley and Weiner (1960) and Timoshenko and Goodier (1951), Hasselman (1963) obtained the following relationship for the total elastic stored strain energy, W:

$$W = \frac{4\pi b^3 \sigma_f^2(1 - \nu)}{7E} \tag{16.13}$$

Assuming the total stored elastic energy to be consumed by the formation of N cracks, each of the average area A and assuming an effective fracture surface energy γ_f, then

$$W = 2A\gamma_f N \tag{16.14}$$

where the factor of 2 accounts for the fracture surface energy on each of the two opposing faces of each crack. Then, substituting the right-hand side of Equation 16.14 for W in Equation 16.13 yields

$$A = \frac{2\pi b^3 \sigma_f^2 (1 - v)}{7 E N \gamma_f} \tag{16.15}$$

In order to minimize the mean crack area A, we can invert both sides of the above equation. By omitting the factors including the sphere volume, the fracture surface energy, γ_f, and numeric prefactors in the expression for A (Equation 16.15), Hasselman (1963) obtained the thermal shock resistance parameter R''' given by

$$R''' = \frac{E}{\sigma_f^2 (1 - v)} \tag{16.16}$$

(Note that the factors in the expression for A that are not included in R''' are equivalent to a numerical factor multiplied by the inverse of the crack number density for the sphere.)

Hasselman introduced a second thermal shock resistance parameter by then including a multiplicative factor of γ_f, such that (Hasselman 1963)

$$R'''' = \frac{E \gamma_f}{\sigma_f^2 (1 - v)} \tag{16.17}$$

where Hasselman noted that R''' could be used to rank the thermal shock damage potential in materials with similar values of γ_f and R'''' could be used to compare materials with significantly different values of γ_f.

However, one can also view R''' as the minimum stored elastic strain energy at fracture available for crack growth and R'''' as the minimum distance that a thermal-shock-induced crack propagates after nucleation (Kingery et al. 1976; Wang and Singh 1994). Thus, R''' and R'''' apply to kinetic crack propagation. For bodies with high fracture strength, σ_f, the stored elastic energy is high and both R''' and R'''' are small such that cracks propagate kinetically, that is, in addition to the energy needed to form the crack faces the crack grows rapidly with kinetic energy that causes catastrophic failure. (E.g., numerous filmstrips show brittle, high strength materials loaded under tension or compression that literally explode immediately after the onset of crack growth.)

After introducing R''' and R'''' in 1963, Hasselman developed an additional thermal shock resistance parameter in 1969 (Hasselman 1969), again using an energy balance technique. Assumed a body contains N circular, planar microcracks per unit volume, with radius c (Hasselman 1969), Hasselman obtained a new total elastic energy expression, W_{total} per unit volume, as the sum of two terms such that the first term was

$$W_{crack} = (\text{crack area}) * (\text{crack surface energy}) = 2\pi N c^2 G \tag{16.18}$$

where G is the strain energy release rate (Equation 16.1c). The second term contributing to W_{total} is given by

$$W_{elastic} = \frac{3(\alpha \Delta T)^2 E_0}{2(1 - 2v)} \left[1 + \frac{16(1 - v^2) N c^3}{9(1 - 2v)} \right]^{-1} \tag{16.19}$$

where E_0 is the elastic modulus in the uncracked state and $W_{total} = W_{crack} + W_{elastic}$. Using the Griffith criterion for crack growth (Griffith 1920), Hasselman evaluated

$$\frac{\partial W_{total}}{\partial c} = \frac{\partial (W_{crack} + W_{elastic})}{\partial c} = 0 \tag{16.20}$$

to first determine the condition for $\Delta T = \Delta T_C$ (the critical quench temperature difference that corresponds to the onset of crack propagation) and then solved for ΔT_C to obtain

$$\Delta T_C = \left[\frac{\pi G E_0 (1 - 2\nu)^2}{2 E_0 \alpha^2 (1 - \nu^2)} \right]^{1/2} \left[1 + \frac{16(1 - \nu^2)Nc^3}{9(1 - 2\nu)} \right]^{-1} (c)^{-1/2}. \tag{16.21}$$

From the second term in square brackets in Equation 16.21, if one lets

$$B = \frac{16(1 - \nu^2)Nc^3}{9(1 - 2\nu)} \tag{16.22}$$

then for short cracks $B \ll 1$ and (from Equation 16.21) ΔT_C can be approximated as

$$\Delta T_C = \left[\frac{\pi G (1 - 2\nu)^2}{2 E_0 \alpha^2 (1 - \nu^2)c} \right]^{1/2}. \tag{16.23}$$

From Equation 16.23, Hasselman re-derived his 1963 expression for R'''' (Hasselman 1963), the kinetic growth conditions under which short cracks growth rapidly. However, in the 1969 paper (Hasselman 1969), Hasselman also considers long cracks, where for long cracks, $B + 1 \rightarrow B$, thus Equation 16.21 becomes

$$\Delta T_C = \left[\frac{\pi G (1 - 2\nu)^2}{2 E_0 \alpha^2 (1 - \nu^2)} \right]^{1/2} \left[\frac{9(1 - 2\nu)}{16(1 - \nu^2)Nc^3} \right] (c)^{-1/2} \tag{16.24}$$

which Hasselman rewrote as

$$\Delta T_C = \left[\frac{128 \pi G (1 - \nu^2)N^2 c^5}{81 \alpha^2 E_0} \right]^{1/2}. \tag{16.25}$$

From Equation 16.25 Hasselman extracted the crack damage parameter R_{ST} to describe the "thermal stress crack stability" parameter, R_{ST}, such that (Hasselman 1969)

$$R_{ST} = \left[\frac{G}{\alpha^2 E_0} \right]^{1/2} \tag{16.26a}$$

where R_{ST} is a measure of the resistance to quasistatic crack growth for long, preexisting cracks in a thermally shocked body. Based on the relationships in Equation 16.1c, some authors express R_{ST} as

$$R_{ST} = \frac{K_C (1 - \nu^2)^{0.5}}{\alpha E} \tag{16.26b}$$

16.3.1.4 Comparison of the Thermoelastic Approach and the Energy Balance Model

Hasselman's thermal shock model (Hasselman 1969) includes several key assumptions concerning the quenched body, including: (1) no body forces (e.g., external mechanical stresses) act on the body; (2) the entire body experiences a uniform cooling (or heating) temperature change, ΔT; (3) the body is subject to "fixed grip" conditions, that is, the exterior surfaces of the body are constrained from displacement in response to ΔT; (4) the body is ideally brittle so that no stress relaxation occurs via plastic deformation mechanics such as dislocation motion or viscous flow; and (5) the homogeneously distributed microcracks (N microcracks per unit volume) propagate simultaneously when the body is heated or cooled.

Especially in terms of assumptions (2) and (3) above, Hasselman notes that his model "…represents the worst possible condition of thermal stress as the body as a whole is stressed to the maximum value of thermal stress. As a result, the calculated values of the extent of crack propagation are expected to be greater than the extent of crack propagation for most cases of thermal stress failure" (Hasselman 1969, p. 601).

It should also be noted that the five thermal shock resistance parameters discussed here cannot be straightforwardly intercompared since each represents difference measures of the resistance to crack growth in response to a thermal shock. The impossibility to directly compare the various R factors is further underscored by the fact that each parameter has different physical units: (i) temperature for R, (ii) power/length for R', (iii) reciprocal stress for R''', (iv) length for R'''' and the product of temperature and length squared R_{ST}. Thus, the five parameters represent five separate ranking systems for the tendency of a material to resist thermal shock damage and the use of each factor depends on whether one wishes to rank materials based on the desire to (1) avoid crack nucleation or (2) minimize crack propagation.

The material parameters pertinent to R factor calculation are given in Table 16.1 for a number of brittle materials, including several materials considered as candidate materials for laser windows (phosphate glass, Al_2O_3, MgO, $BeAl_2O_4$, YAG, MgF_2, LiF, KCl, and NaCl) (Krupke et al. 1986), as well as the material properties for α-SiC (Andersson and Rowcliffe 1998), which is highly resistant to thermal shock damage. Data for two thermoelectric materials, Yb-filled skutterudite and PbTe (Salvador et al. 2009) is also included (Table 16.1). Since K_C data for Yb-filled skutterudite and PbTe were not available in the work by Salvador et al., K_C values for a n-type $CoSb_3$ skutterudite (Ravi et al. 2008) and an n-type PbTe-based material (LAST, a lead–antimony–silver–tin compound) (Ren et al. 2008c) were

TABLE 16.1 The Fracture Toughness, K_C, Fracture Strength, σ_f, Young's Modulus, E, Poisson's Ratio, ν, Thermal Expansion, α, and Thermal Conductivity, k, for Selected Brittle Materials

Material	K_C (MPa$\sqrt{m}$)	σ_f (MPa)	E (GPa)	ν	$\alpha \times 10^{-6}$/K	k (W/m K)	References[*]
Phosphate glass	0.45	90	52	0.26	11	0.6	(a)
Al_2O_3	2.20	444	405	0.25	7	28.0	(a)
MgO	0.77	154	261	0.18	14	59.0	(a)
$BeAl_2O_4$	2.60	520	446	0.30	8	23.0	(a)
$MgAl_2O_4$	1.20	240	171	0.30	9	12.5	(a)
YAG	1.40	280	282	0.28	7	10.0	(a)
MgF_2	0.90	180	138	0.27	15	21.0	(a)
LiF	0.36	72	91	0.30	37	11.0	(a)
KCl	0.14	28	39	0.30	40	9.2	(a)
NaCl	0.16	32	43	0.30	41	9.2	(a)
α-SiC	4.0	612	427	0.14	5	110.0	(b)
Yb-filled skutterudite	1.7(d)	111	135	0.20	11	3.7	(c)
PbTe	0.35(e)	50	70	0.22	20	2.8	(c)

[*] The references listed as (a)–(e) are given in the footnotes. The data listed in this table is used to calculate the thermal shock resistance factors R, R', R''', R'''', and R_{ST} in Table 16.2. (a) (Krupke et al. 1986), (b) (Andersson and Rowcliffe 1998), (c) (Salvador et al. 2009), (d) (Ravi et al. 2008), (e) (Ren et al. 2008c).

TABLE 16.2 The Thermal Shock Resistance Factors R, R', R''', R'''', and R_{ST}, Where the R and R' Are Computed from Equation 16.12

Material	R (°C)	R' (W/m)	R''' $(MPa)^{-1} \times 10^{-6}$	R'''' (cm)	R_{ST} (°C m^2) $\times 10^{-2}$
Phosphate glass	112	70	8.7	1.5	76
Al$_2$O$_3$	123	3436	2.7	7.3	81
MgO	36	2115	13.4	2.3	22
BeAl$_2$O$_4$	102	2346	2.4	9.3	73
MgAl$_2$O$_4$	109	1365	4.2	4.3	78
YAG	107	1067	5.0	4.9	74
MgF$_2$	63	1333	5.8	3.1	43
LiF	15	165	25.1	1.3	11
KCl	13	116	71.1	0.5	9
NaCl	13	117	60.0	0.6	9
α-SiC	274	30,130	1.3	15.2	208
Yb-filled skutterudite	61	227	13.7	14.4	110.8
PbTe	28	77	35.9	1.6	25

Note: The factors R''', R'''', and R_{ST} were computed from Equations 16.16, 16.17, and 16.26b, respectively. Physical property data for the R factor computation was taken from Table 16.1. Ideally, for each of the R factors, the resistance to thermal shock damage increases as the numerical value of that R factor increases.

used to estimate to the estimate K_C data for Yb-filled skutterudite and PbTe, respectively. The set of materials listed in Table 16.1 represent a wide range of material properties, especially with respect to K_c, σ_f, α, and k.

Using the physical property data listed in Table 16.1, values of the five thermal shock resistance parameters discussed in this chapter, namely R, R', R''', R'''', and R_{ST} were computed (Table 16.2), The values of R and R' are computed from Equation 16.12 and the values of factors R''', R'''', and R_{ST} were computed from Equations 16.16, 16.17, and 16.26b, respectively.

The R-factor concept states that for each of the individual R-factors, R^i, the thermal shock resistance increases as the value of R^i increases. Recall that each of the R^i factors is a physically different interpretation of factors that contribute to good thermal shock resistance and in fact the physical units are different for each R factor (Table 16.2). Thus, while one can compare materials within a given R^i-factor description, it makes on physical sense to compare materials among one or more R^i categories.

As one would expect, the R^i values for the alkali halides KCl and NaCl are uniformly high and the R^i values for α-SiC are uniformly low (Table 16.2). For Yb-filled skutterudite and PbTe, the two thermoelectric materials included in Table 16.2, the higher fracture strength and thermal conductivity values for the skutterudite leads to higher R and R' values (Equation 16.12). PbTe has the higher R''' value of the two thermoelectrics since PbTe has the lower fracture strength (Equation 16.16), but the higher K_C and fracture surface energy (Equation 16.1c) values for the skutterudite lead to higher R'''' (Equation 16.17) and R_{ST} (Equation 16.26b) values than the PbTe.

16.4 Thermal Fatigue Damage

In terms of the literature on the thermal fatigue of brittle materials, including the TE literature, on those rare occasions when thermal transient damage is analyzed, typically R factor analysis is done. Also, for TE materials and the literature in general, most often either R or R' is evaluated.

In addition, too often, R factors are treated as absolute quantities rather than the ranking system that they are intended to be. Researchers often take the numerical values of R factors much too seriously. Are R factors of any use for thermoelectric materials research for in the direct conversion of waste heat to electricity, since the energy conversion will typically involve thermal fatigue? The answer is likely "yes," since the R factors indicate general trends but one should keep in mind that R factors arise from thermal shock ($N = 1$) studies. R factors do not fully address thermal fatigue and as noted above, the R factors do not even fully address thermal shock.

16.4.1 Microcracking and Thermal Fatigue

Accumulated microcrack damage induced by thermal fatigue damage can degrade mechanical, thermal, and electrical properties (Ainsworth and Herron 1974; Lee and Case 1989; Kim et al. 1990; Zhang et al. 1997; Case 1999; Ivon et al. 1999). What can be done to inhibit the accumulation of damage that results from the repeated thermal cycling? Make the material tougher, that is, more resistant to crack growth, or in other words, boost fracture toughness, K_C. However, first we will consider the nature of microcracking in brittle materials, and then we will discuss mechanisms that are available to us for enhancing the toughness of brittle materials. As we shall see, some of the important toughening mechanisms for brittle materials become inoperative during even mechanical or thermal fatigue.

16.4.2 Microcrack-Induced Physical Property Changes

Several researchers have developed models for microcrack-induced changes in physical properties such as the elastic moduli (Budiansky and O'Connell 1976; Laws and Brockenbrough 1987), electrical conductivity σ_e (Hoening 1978/79) and thermal conductivity, k (Hoening 1984). The theoretical dependencies of the Young's modulus E along with σ_e, and k on microcracking damage can be summarized (Case 2002) as

$$E = E_0(1 - f_E(\nu)\varepsilon) \tag{16.27a}$$

$$\sigma_e = \sigma_{e0}(1 - g\varepsilon) \tag{16.27b}$$

$$k = k_0(1 - h\varepsilon) \tag{16.27c}$$

where E_0, σ_{e0}, and k_0 represent the microcrack-free values of Young's modulus, electrical conductivity and thermal conductivity, respectively. The unitless quantity ε is the microcrack damage parameter. For the case of circular cracks of uniform radius a, then ε is $a^3\Lambda$, where Λ is the volumetric number density of microcracks (the number of microcracks per unit volume) (Budiansky and O'Connell 1976, Laws and Brockenbrough 1987). The function $f_E(\nu)$ depends on (i) the relative orientation of the microcracks and (ii) ν, the Poisson's ratio of the microcracked solid (Laws and Brockenbrough 1987). In contrast g and h are not dependent on material properties but are only functions of the relative microcrack orientation.

In addition to the theoretical work, there is considerable experimental evidence for microcracked-induced changes in a broad range of thermal, electrical, and mechanical properties (Rice 2000, Case 2002). Microcracking generated by thermal fatigue in particular has been observed to lower the Young's modulus (Lee and Case 1989, 1990, Chiu and Case 1992, Case et al. 1993, Rendtorff 2010), fracture strength (Ainsworth and Herron 1974, Chiu and Case 1992, Katigiri et al. 1994), thermal conductivity (Kumakawa and Niino 1993, Zhang et al. 1997) and dielectric constant (Ivon et al. 1999) of brittle materials.

16.4.3 Physical Property Changes Induced by Microcracking

As reviewed by Case (2002), if the thermal cycling is performed with a constant quench temperature difference, ΔT, and in a fixed quenching medium, then the evolution of thermal fatigue damage typically results in (i) an initial, exponential drop in the measured property P as the number of thermal fatigue cycles, N, increases and (ii) a subsequent steady state or saturation damage regime. The initial drop followed by a state–state region has been observed by a number of researchers for physical properties P that include fracture strength, elastic modulus, thermal conductivity, dielectric strength, and hardness such that (Case 2002)

$$P(N) = P_0 \pm (P_0 - P_{SAT})\{1 - \exp(-\delta N)\} \tag{16.28}$$

where

P_0 = the value of property P for the undamaged body
P_{SAT} = saturation or steady-state value of P
δ = rate constant, describes change in P as a function of N
N = number of thermal fatigue cycles.

Microcracking due to thermal fatigue has several important implications for the in-service life of thermoelectric materials used for energy harvesting in terms of both the thermoelectric properties and the mechanical integrity of the thermoelectric materials. First, the unitless figure of merit, ZT, for thermoelectric materials is defined by

$$ZT = S^2\sigma_e T/k \tag{16.29}$$

Since increasing microcrack number densities and or microcrack size diminishes both the electrical conductivity σ_e and thermal conductivity k (Case 2002), thus the property σ_e/k may remain relatively constant with the increase in microcrack damage, analogous to what has been observed in terms of the affect of the porosity on σ_e and k. In addition, some researchers have observed that the Seebeck coefficient is relatively insensitive of volume fraction porosity, P.

Thus, microcracking in brittle materials can occur via thermal fatigue, that microcracking can induce significant changes in mechanical, thermal and electrical properties. However, a key question is, how does microcracking affect the resistance to crack growth, that is, does microcracking affect the toughness of a brittle material?

16.5 Toughening Mechanisms

For the mechanical properties such as the elastic modulus and fracture strength, the damage saturation behavior observed in many brittle materials provides that possibility that if the thermoelectric element survives the first "few" thermal cycles (where "few" is a relative term that will depend on the empirical rate constant δ), then the thermoelectric material may withstand many more thermal cycles with a limited further decrease in mechanical properties.

This assumes that environmentally induced crack growth (Wachtman et al. 2009) does not significantly degrade the material over time. However, preliminary experiences on the stress corrosion behavior in the thermoelectric chalcogenide materials LAST (lead–antimony–silver–tellurium) and LASTT (lead–antimony–silver–tellurium–tin) indicate that the environmentally assisted slow crack growth is extremely limited, if it exists at all (Hall unpublished work). At present, there is no work in the literature concerning either the presence or absence of slow crack growth in other thermoelectric material systems. Of course, if the thermal fatigue is sufficiently severe, the body can fracture catastrophically on the first cycle or early in the thermal fatigue process.

Thus, the Kingery approach (the thermoeleastic model) to thermal shock damage resistance is to prevent macrocracks nucleation while the Hasselman (energy balance) approach advocates having an ensemble of preexisting microcracks that extend under thermal shock but do not lead to catastrophic failure. In the thermoelastic model (R and R'), a high fracture strength and thermal conductivity and a low elastic modulus and thermal expansion coefficient are beneficial to avoiding crack nucleation and for the energy balance model (R''', R'''') high values of elastic modulus and fracture surface energy and low strength values favor the resistance to kinetic crack growth.

16.5.1 Intrinsic and Extrinsic Crack Toughening Mechanisms

A number of toughening mechanisms have been identified for brittle materials and these mechanisms can be divided into two categories, namely (i) intrinsic mechanisms that involve microstructural features in the specimen itself and (ii) extrinsic mechanism in which the growing macrocrack interacts with microstructural features in the microstructure (Ritchie 1999, 2009).

16.5.2 Extrinsic Crack Toughening Mechanisms

The relationship between fracture toughness and crack length is called an "R-curve" or a crack-growth resistance curve (Ritchie 1999, 2009; Wachtman et al. 2009), where the R designation for the crack growth resistance curve is not directly related to the R parameters (R, R', R''', R'''', and R_{ST}) that are commonly used to estimate the resistance to thermal shock damage. We will explore each of these toughening mechanisms and comment on whether or not a given mechanism may be effective for toughening of TE materials with respect to the thermal fatigue conditions generated by energy-harvesting applications.

16.5.2.1 Transformation Toughening

Transformation toughening occurs when a crystallographic phase transformation, such as martensitic transformation (rapid distortion of lattice accompanied by volume change) leads to a compressive stress near the tip of a macrocrack (Evans and Cannon 1986). The imposed compressive stress inhibits crack growth and thus by definition, increase the fracture toughness (Wachtman et al. 2009). Transformation toughening is observed in steel, zirconia and other materials that exhibit martensitic transformation, but no martensitic transformations have been documented in TE materials, thus toughening due to phase transformation is unavailable to TE materials.

16.5.2.2 Grain Size and Crystal Symmetry Effects on Toughening

For polycrystalline ceramics, K_C is independent of grain size for materials with cubic symmetry (Monroe and Smyth 1978; Rice 1981) but K_C is a function of grain size for ceramics with lower symmetry than cubic. According to Rice (1981, 2000), this dichotomy between cubic and noncubic ceramics is related to thermal expansion anisotropy (TEA). The thermal expansion coefficient is isotropic in a cubic material but anisotropic in noncubic crystalline materials. In general, physical properties such as thermal expansion, optical index of refraction, thermal diffusivity, thermal conductivity, and mass diffusivity that are described by second-rank tensors are isotropic in cubic materials and anisotropic in noncubic crystalline materials (Nye 1985). (Physical properties described by higher rank tensors, such as the elastic constants, as anisotropic even for materials with cubic symmetry (Nye 1985).)

For materials with a lower crystal symmetry than cubic, a maximum in K_C versus grain size, GS, has been be observed near Gs_{crit}, the critical grain for microcracking due to TEA. (GS_{crit} refers to the critical grain size for TEA-induced microcracking (Cleveland and Bradt 1978). GS_{crit} is a function of a function of $1/\Delta\alpha_{max}$, where $\Delta\alpha_{max}$ = the maximum difference in the thermal expansion coefficients along the crystallographic axes. GS_{crit} can vary from about 1 μm for aluminum titanate (which has a very degree high thermal expansion anisotropy) to roughly 100 μm for alumina (which is nearly isotropic).

The rise in toughness with increasing grain size for noncubic materials with GS < Gs_{crit} was interpreted by Bennison and Lawn as being related to TEA-generated stresses leading to an increase in the clamping force on grains during grain bridging of the larger grains (Bennison and Lawn 1989). For GS > Gs_{crit} the fall in toughness with further increases in grain has been attributed to microcrack link-up.

Since many thermoelectric materials that are possible candidates for energy-harvesting applications, including chalcogenide-based thermoelectric materials, skutterudites, TAGS and Mg_2Si are cubic, tailoring the grain size will do little to improve K_C. Furthermore, if the enhanced fracture toughness for noncubic materials with GS < Gs_{crit} is indeed linked to grain bridging, then toughening via this mechanism will not be effective for materials undergoing fatigue, since essentially all crack bridging mechanisms degrade during cyclic loading, as will be discussed in the next section.

It is important not to confuse grain size with toughness in this context. The length of the largest flaw, a_c, in a specimen scales with GS, thus reducing GS boosts fracture strength by lowering a_c. Nevertheless for cubic materials, K_C is unchanged with GS.

16.5.2.3 Crack Bridging (in the Wake of a Growing Macrocrack)

Toughening by crack bridging (also called crack wake toughening) subsumes several individual types of crack toughening mechanisms, each of which involves the interaction of crack surfaces across the adjacent faces of a macrocrack. For crack wake toughening, the fracture toughness, K_C, of a specimen increases as the crack length increases.

"Grain-bridging" is a particular crack bridging mechanism that involves the interaction of polycrystalline grain faces across a crack. Originally the term "grain bridging" referred the situation in which a macrocrack had branched around a grain (Figure 16.1), such that macrocrack wake included a grain or grains bridging across the crack faces of the microcrack (Swanson et al. 1987). More recently, crack bridging is often used to indicate any situation in grains on opposing macrocrack faces interaction via a frictional process. For grain bridging to be effective, generally mesoscale grain dimensions on the order of tens of microns across are needed or grains must be elongated in shape (Ritchie 1999, 2009). A broader term, "crack bridging" refers to asperities that can interact across the opposing faces of a crack and such asperities can be in the form of grains, whiskers, fibers, or whiskers that bridge crack faces.

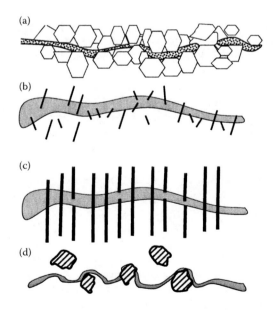

FIGURE 16.1 Schematic drawings of extrinsic (crack wake) toughening mechanisms of (a) grain bridging, (b) whisker bridging, (c) fiber bridging, and (d) inclusion bridging. In (b), "whiskers" refer to single crystal materials of a high aspect ratio. For example, SiC whiskers are available having lengths of roughly 5 to 10 microns and diameters of a few tenths of a micron. Fiber reinforcements (c) in brittle materials are generally polycrystalline or amorphous with diameters often in the range from about 10 to 20 µm.

In the late 1980s and early 1990s, it was discovered that the crack wake mechanisms for toughening literally "break down" during fatigue (Mai and Lawn 1987; Hay and White 1993; Ritchie 1999, 2009). When a cyclic load is applied to a crack, the asperities responsible for crack wake toughening interact and degrade. For example, Ritchie and coworkers have found that grain-bridging in silicon carbide ceramic is consistent with concept that under cyclic mechanical fatigue conditions, "... the crack advance mechanism does not change and that the influence of cyclic loads is solely to progressively degrade crack bridging, i.e., the *local* crack-tip stress intensity factor remains at K_0" (p. 71) (Gilbert and Ritchie 1998; Ritchie 1999), where K_0 is the crack's stress intensity factor before the cracks propagates.

16.5.3 Intrinsic Toughening Mechanisms (Ahead of the Growing Macrocrack)

16.5.3.1 Crack Deflection

Faber and Evans analyzed the crack deflections induced when a growing crack impinges on inclusions with idealized geometries, namely spheres, disks and rods. Some key results of the Faber and Evans (1983a,b) analysis were that (i) the crack deflection should be a function of the volume fraction and shape of inclusions, but not the inclusion size, (ii) the "scattering strength" of inclusion shapes was ranked as rod > disc > sphere, and (iii) the likelihood crack deflection should increase as the volume fraction of inclusions increase, but reaching a maximum at about 20 vol% of inclusions. While the Faber and Evans analysis did indicate some general trends, experimental work found that the rod shape was not as effective as predicted in deflecting cracks and that the maximum toughening in many composite systems corresponds to a volume fraction of a few percent rather than 20%.

There are a number of types of crack–microstructure interactions that can lead to crack deflection. For example, Blanks et al. observed crack deflection at the interfaces between SiC–SiC laminates, where the individual layers were alternating dense and porous SiC layers (Blanks et al. 1998).

16.5.3.2 Crack Bowing and Crack Branching

Crack bowing (also called "crack pinning") is another intrinsic toughening mechanism. Lange (1970) and later Green (Green et al. 1979) modeled the crack bowing mechanism after the concepts of "line tension" and bowing of a dislocation held by pinning points in a crystalline lattice (Hirth and Lothe 1982). More recently, Wang et al. (2001) using plexiglass (PMMA) plates with arrays of drilled holes as model specimens to experimentally investigate crack bowing in advanced macrocracks in brittle materials. Also, the intrinsic toughening via crack bowing, crack branching and crack deflection has been by Ni-alloyed $NbCr_2$ (Nie et al. 2009), La_2O_3 particles in a $MoSi_2$ (Zhang et al. 2003), $TiB_2 + 2.5CrSi_2$ composites (Murthy et al. 2010), n-type $CoSb_3$ based nanocomposites (Mi et al. 2007). Also, crack bowing by rigid inclusions has been identified as a toughening mechanism in brittle (glassy polymers) (Norman and Robertson 2003).

If roughly equiaxed particles (inclusions) are present (or are added) then crack bowing (also called crack pinning) can occur. If high aspect ratio particles are added, then filamentary crack bridging is more likely (Wachtman et al. 2009).

16.5.3.3 What Microstructures Promote Intrinsic Crack Toughening?

For crack deflection, crack bowing (crack pinning), and crack branching (Figure 16.2), the material needs to be inhomogeneous, that is, the material needs to include second phases, particles or filaments inclusions, whiskers, fibers, or pores (Ritchie 2009; Wachtman et al. 2009). In this case, pores can also act as "inclusions" (Rice 2000). In the absence of macrocracks that grow very rapidly, crack branching also depends on a dispersion of second-phase particles. Addition of micro- or nano-particles can be effective in toughening TE materials by crack bowing/pinning.

Separately from an interest in nano- or micro-particle additions for fatigue resistance, there has been considerable interest recently in nanocomposite thermoelectric systems that seek to boost the

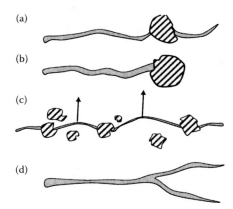

FIGURE 16.2 Schematic illustrations of intrinsic (crack progress zone) toughening mechanisms for (a) crack deflection via an inclusion, (b) crack blunting by impinging on a pore, (c) crack bowing (also called crack pinning), and (d) crack branching. For each of the schematics except (c), the cracks are depicted from an edge view. For (c), the crack bowing mechanism is illustrated from a vantage point normal to the plane of the growing macrocrack, where the two arrows depict the propagation direction of the macrocrack.

dimensionless figure of merit ZT (Equation 16.29), where nanoparticles are either (i) added to the thermoelectric matrix and or (ii) precipitated from the matrix via chemical and processing control (Dresselhaus et al. 2007; Kanatzidis 2010). For example, recent research on nanoparticle additions to thermoelectric materials includes work on $CoSb_3$ nanoscale powder additions microscale $CoSb_3$ powders prior to hot pressing (Mi et al. 2007), silicides in SiGe (Mingo et al. 2009), p-type $Bi_{0.4}Sb_{1.6}Te_3$ nanocomposites (Fan et al. 2010), PbTe-Si eutectic composites (Sootsman et al. 2010), Ag_2Te nano-scale precipitates in $AgSbTe_2$-based compounds (Zang et al. 2010), and $Zn_4Sb/Bi_{0.5}Sb_{1.5}Te_3$ nanocomposites (Sun et al. 2010). The goal of adding these nanoparticles to thermoelectric materials has been largely to decrease the thermal conductivity of the material via phonon scattering. Pores as a mechanism to possibly enhance ZT has also been reported (Lee and Grossman 2009; Lee et al. 2010). For example, a 30-fold enhancement of the ZT of nanoporous Ge over dense, bulk Ge has been reported (Lee and Grossman 2009), but in the ZT of nanoporous SiGe decreased compared to dense SiGe (Lee et al. 2010). However, Lee et al. (2010) note that their analysis, for materials with low effective electron mass and low charge carrier relaxation time, the addition of nanopores may increase ZT. Thus, for some materials, pores and other inclusions may possibly serve the dual purposes of enhancing both thermal fatigue resistance and ZT.

In terms of the potential processing of thermoelectric materials for waste heat applications, whiskers or fibers added to the materials would likely be (i) ineffective as toughening agents for extrinsic toughing with respect to thermal fatigue loading as discussed above and (ii) inhibit sintering or powder specimens because such high-aspect ratio particles inhibit particle packing the green state, prior to sintering (Rice 2003).

In contrast, dispersing relatively equiaxed nano- or micro-particles in thermoelectrics would interfere much less with densification and would afford the potential for intrinsic crack-toughening mechanisms (Figure 16.2), which in general can be effective in reducing the growth of macrocracks via fatigue. A consequence of adding inclusions to a thermoelectric material can also be the generation of microcracks within the thermoelectric.

In addition to interacting with a growing macrocrack (as an intrinsic toughening mechanism), inclusions generally have thermal expansion coefficients that are different than the matrix, thus the thermal expansion mismatch among phases in a multiphase body can lead to local stresses that can in turn generate microcracks (Davidge and Green 1968; Luo and Stevens 1993; Luo 1996; Sridhar et al. 1996; Litovsky et al. 1999) in the absence of external body forces. Also, even if the CTE mismatches alone do not generate microcracks, the CTE mismatch still generates stresses on the grain scale which when superimposed on stresses arising from other mechanisms such as thermal transients and/or mechanical stresses, can ultimately lead to microcracking (Wachtman et al. 2009). Of course, there is a critical size, GS_{CRIT}, above which microcracking due to thermal expansion mismatch occurs (Davidge and Green 1968).

Also, crystallographic phase changes in inclusions (if the phase change is accompanied by a volume change and/or a shear of the atomic lattice) can induce microcracking in the matrix material (Fu et al. 1984; Yang and Chen 2000; Tiefenbach et al. 2000). Analogous to the cases of microcracking due to thermal expansion mismatch and thermal expansion anisotropy, there is a critical particle (inclusion) size for microcracking due to a crystallographic phase transformation. Thus, inclusions of materials that potentially undergo phase transformations (such as hafnia and zirconia (Evans and Cannon 1986)) could be added to the thermoelectric, not for transformation toughening in the crack wake, but to promote the formation of microcracks in the matrix of the thermoelectric material.

Microcracking can enhance the thermal shock resistance of ceramics by lowering the Young's modulus of the microcracked body and thus decreasing the stored elastic strain energy in the body (Swain 1990).

Also, even for highly anisotropic materials, the grain size needed for thermal expansion anisotropy-induced cracking is typically at least a few microns across. Thus, when nano-sized particles are added, they are likely not to induce microcracking in the matrix material.

In addition, nanoparticle and/or micro-particle additions also can inhibit grain growth at high temperature, and thus possibly retain nano-size grains and thermoelectric and strength advantages due to small grain size.

16.6 Examples of Material Systems with Excellent Thermal Shock Resistance

In regard to developing materials with good thermal fatigue properties for thermoelectric materials, it is worthwhile to consider examples of materials systems that show exceptional resistance to thermal fatigue damage. The three examples considered here are: (1) thermal spray coatings and bulk materials, which have been developed over the last few decades, (2) refractory materials for furnace linings which in their present form have been developed over about the past 100 years or more, and (3) pottery cookware, which has been developed over the past several thousand years.

16.6.1 Thermal Sprayed Coatings and Bulk Specimens

The thermal spray process refers to the melting and subsequent freezing of powders as they pass quickly through a heat source (a flame or plasma, for example) and then impinge on a solid surface. The nature of the thermal spray process itself leads to a porosity and microcracks between the individual "splat droplets." Thermal spray processes are used to produce thermal barrier coatings (TBCs) that are currently being used for thermal barrier coatings that protect aircraft and industrial turbine engines from high temperatures and corrosion (Padture et al. 2002). The temperature drop across the low thermal conductivity TBC materials in turn lowers the operating temperature of the superalloys that make up the turbine components, allowing the superalloys to function at significantly higher temperatures than would otherwise be possible.

Thermal barrier coatings are typically made up of three layers: (1) a ceramic outer layer roughly 50–100 μm thick often made of yttria-stabilized-zirconia (YSZ), lanthanum zirconate or rare earth oxides and next (2) a thermally grown oxide layer (TGO) below the ceramic layer, and finally (3) a bond coat of NiCrAlY, NiCoCrAlY, or an aluminide roughly 75 to 150 microns thick. The bond coat anchors the TBC to a metallic substrate, which is often a nickel- or nickel–cobalt-based superalloy (Padture et al. 2002; Wright and Evans 1999). In order to maintain a sharp thermal gradient across the ceramic outer layer of the TBC, materials with a low thermal conductivity are best. The thermal conductivity of yttria-stabilized-zirconia is roughly 1 W/mK (Wright and Evans 1999), which is roughly on par with a number of thermoelectric materials (Kanatzidis 2010). For thermal barrier coatings, Wright and Evans (1999) state that "The TBC itself must be 'strain tolerant' to avoid instantaneous delamination. This is achieved by incorporating either microcracks or aligned porosity..." (p. 255).

There are interesting parallels between thermal barrier coatings (TBC) and thermoelectric materials. For example, both thermoelectrics and the superalloy substrates for TBC are cooled, establishing a very large thermal gradient across the materials. Also, in order to function in the desired manner, both thermal barrier coatings and thermoelectrics must develop substantial thermal gradients along the direction of heat flow and to accomplish that, their thermal conductivity must be low. For example, for TBCs, the thermal conductivity, k, of the YSZ top coat ceramic materials themselves is roughly 2.3 W/mK, but k for YSZ in the thermal-sprayed condition (porous and microcracked) is in the neighborhood of 0.8–1.7 W/mK (Padture et al. 2002). If one then refers to the thermoelectrics' literature, the k values of the TBC's ceramic top coat are comparable to those of very good thermoelectric materials (Kanatzidis 2010). Thus, in terms of the thermal environment and the thermal properties of the materials, TBC's and thermoelectrics for waste heat recovery have several things in common. However, over the past two decades or more, the nature of the microstructure (cracks and pores between the splats) has been optimized to allow TBC's to have phenomenal levels of thermal shock/thermal fatigue resistance.

In addition to coatings such as thermal barrier coatings, bulk, free standing specimens can be fabricated by thermal spray techniques. Damani et al. (2000) note that in bulk plasma-sprayed alumina specimens, the interfaces are weak between the columnar gamma- and alpha-alumina grains that, along with porosity, make up much of the microstructure. As a consequence, the bulk plasma sprayed alumina specimens have low fracture strength and low fracture toughness, but the resistance to thermal shock damage is excellent for these materials. Once again, preexisting defects such as pores and microcracks tend to reduce the Young's modulus which reduces the thermal stresses induced at a given value of thermal strain. In addition, as outlined in Hasselman's energy balance model (Hasselman 1963, 1969), energy is likely dissipated by the simultaneous growth of an ensemble of microcracks.

16.6.2 Refractory Materials

In the sense used here, refractories are thermally insulating furnace-lining materials that also generally require resistance to erosion and chemical attack. Refractory materials are often classified as either "shaped" or "unshaped," where shaped refers to refractory bricks and geometric castings and unshaped are monolithic ceramics (Banerjee 2004). Monolithic refractories are typically fabricated in-place with various casting or gunning techniques (Banerjee 2004). A common characteristic of most refractory bricks and castables is the good (often exceptional) resistance to thermal shock and thermal fatigue.

In terms of the thermal shock of refractory materials, Bradt refers to Hasselman's energy balance model (Hasselman 1963, 1969) in noting that "The thermal stress damage resistance decreases as the strength increases and does so as the square of the strength. High strength refractories are much more susceptible to thermal shock damage" (Bradt 2004, p. 18). Furthermore in regard to adding hard inclusions to refractories, Bradt further observes that "Refractory manufacturers have known this for some time as their ceramic engineers increased the thermal shock resistance of their fireclays by adding crushed 'bats' to their refractory mix" (Bradt 2004, p. 18). The role of such inclusions, according to Bradt, is to generate cracks in the specimen (Bradt 2004) either during processing or during mechanical and/or thermal loading of the refractories. These cracks in turn can extend under thermal loading.

16.6.3 Cookware (Pottery)

Archeologists have identified the widespread use of temper in the fabrication of ceramic cooking ware (Kilikoglou et al. 1998; Tite et al. 2001; Montana et al. 2007). In the context of pottery, "temper" refers to inclusions such as quartz grains, limestone particles and shells added to the clay prior to firing. In a review of the strength, fracture toughness, and thermal shock resistance of ancient ceramic cookware, Tite et al. noted that "to produce pottery with high strength requires high firing temperatures and low temper concentrations. Conversely, to produce pottery with high toughness and thermal shock resistance requires low firing temperature and high temper concentrations" (Tite et al. 2001, p. 301). Furthermore, Tite et al. state that "... the routine use of high temper concentrations and low firing temperature in the production of cooking ware suggests that the high thermal shock resistance was a factor that at least influenced technological choice in this case" (p. 301).

The low firing temperature has the effect that the porosity of the fired body will be lower as the process of vitrification interlocking the grains within the clay will be less complete at lower firing temperatures (Kingery et al. 1976). Tite also points out that the use of low firing temperatures and high concentrations of temper in cooking pots has been identified in Europe, Asia, and North American going back to Neolithic times and continuing up to much more recent times (Tite et al. 2001). Furthermore, laboratory tests done on clay-based ceramics and on ancient ceramics indicate that the toughening effect increases up a volume fraction loading of about 20% of inclusions. As the inclusion volume fraction increase above 20%, the fracture toughness is relatively constant but the strength continues to decrease as more inclusions are added. It is interesting to note that in their analysis of crack deflection by inclusions, Faber and Evans (1983a, 1983b) came to a similar conclusion, namely that fracture toughness would increase for increasing volume fractions of inclusions up to a loading of about 20%.

Thus, hundreds and even thousands of years before his birth, Hasselman's concepts of energy balance, kinetic and quasistatic crack propagation were being put to use by ancient potters in Rome (Montana et al. 2007), Greece (Tite et al. 2001), and North America (Tite et al. 2001), with presumably the rather straightforward motivation of not wanted to have a good dinner spoiled by the failure of a ceramic cooking pot.

16.7 Conclusions and Strategies for Reducing Thermal Fatigue Damage in TE Materials for Energy-Harvesting Applications

The examples in Section 16.6 show that one approach to making brittle materials resistant to thermal fatigue is to make the materials with a relatively low strength and a significant volume fraction of inclusions, either "hard" inclusions or pores. Of course, typically strength decreases as consequence to adding inclusions. "Low strength" in this case is a balance between (i) having sufficient strength to survive material processing, such as the cutting of legs and assembling modules for thermoelectric generators and (ii) not having strength so high that a growing crack acquires so much kinetic energy that it fractures the specimen. Thus, regardless of the theoretical constructions we use to interpret the results given in Section 16.6, it is clear that the microstructure–crack interactions employed in those material systems can enhance thermal fatigue in brittle materials. Similar strategies could be applied to thermoelectric materials used in the harvesting of waste heat.

The thermoelastic and energy balance approaches to analyzing thermal shock were discussed, but it is worth repeating that neither of these approaches deals with specimen size and/or shape effects. As discussed in Section 16.3.2.2, the thermal shock work of Becher et al. (Becher and Warwick 1993) yielded a ΔT_C (onset temperature of thermal shock cracking) of about 1100°C for 1 mm thick Si_3N_4 specimens shocked into a 22°C water bath and a ΔT_C of 400°C for 5 mm thick specimens of the same material shocked into the same water bath. This tremendous change in ΔT_C is not accounted for by any of the thermal shock resistance parameters, R^i, because from the outset the R^i factors were meant to serve as a ranking system, independent of the size and shape of the thermally shocked body.

In addition, none of the R factors reviewed here explicitly include the Biot modulus, Bi, thus the R factors do not include the important information (for convectively and radiatively heated/cooled systems) concerning the surface heat transfer coefficient, h, and thus does not account for differing heat transfer characteristics of the surrounding medium. Nevertheless, differences in the quench medium can lead to pronounced differences in thermal shock and thermal fatigue. Since R and R' are based on thermoelastic analysis, they could be extended to include h, and R' does include the thermal conductivity, k, of the quenched material. Of course, R''', R'''' and R_{ST} do not include the Biot modulus, h, or k since Hasselman-based R''', R'''' and R_{ST} directly on stored elastic strain energy and crack surface energies expressions of Timoshenko and Goodier (1951), Walsh (1965), and Sack (1946).

In addition to the R factors, expressions for approximating the surface maximum stress (Equation 16.6) or approximating the stress intensity factor K (Equation 16.8) are available for simple geometries such as plates or cylinders. For complicated shapes, one should likely turn to numerical analysis, for example, a finite element analysis.

In regard to thermal fatigue, there is also the overall caution that all of the R factors were intended to rank thermal shock resistance, rather than thermal fatigue resistance. Thus R factors serve at best as rough guideline in thermal fatigue problems.

For thermoelectric materials, during thermal fatigue, the microcrack damage accumulation can degrade transport properties as well as some of the mechanical properties (damage accumulation lowers the elastic moduli and fracture strength), but the accumulation of microcrack damage can enhance the resistance to thermal fatigue damage likely in part by lowering the Young's modulus of the micro-cracked body and thus decreasing the stored elastic strain energy in the body (Swain 1990).

Summarizing from the discussions in this chapter, possible mechanisms to enhance fracture toughness and reduce thermal fatigue damage in the form of macrocrack propagation include:

- The transformation toughening mechanism, but it is not available for TE materials.
- Grain size engineering, which is likely of limited use, especially for cubic TE materials. Also, if Bennison and Lawn (1989) are correct that the toughness enhancement for GS < Gscrit is linked to grain bridging, the grain size engineering will be ineffective for toughening of all TE materials (both cubic and noncubic) with regard to thermal fatigue.
- Likewise, the entire menagerie of crack bridging, extrinsic toughening mechanisms is thought to be defeated by fatigue.

Thus crack wake mechanisms (the extrinsic toughening mechanisms shown in Figure 16.1) keenly are important for thermal shock and noncyclic mechanical loading, crack wake mechanisms are unlikely to produce significant toughening under the thermal fatigue conditions encountered in energy-harvesting applications. However, intrinsic toughening mechanisms such as crack pinning, crack bowing, crack deflection, and crack branching may help toughen TE materials (Figure 16.2) against thermal fatigue. In addition, as suggested by Swain (1990), the reduction in the local elastic modulus, E, of a quenched body due to microcrack extension is likely very significant in the thermal fatigue resistance of brittle materials. This local reduction is E likely also plays an important role in the thermal fatigue damage saturation behavior discussed by Case (2002).

The temporal nature of the loading is a fundamental difference between fatigue conditions and the quasi-static loading that is typical of most mechanical test techniques. During fatigue, the maximum stress may only be applied for very brief time so that, in contrast to the case for quasi-static loading, the frontal crack process zone may not fully develop during one or even several fatigue cycles. Thus, mechanical testing involving quasi-static loading may be of limited use in characterizing the thermal fatigue damage inherent in waste heat recovery conditions for thermoelectrics. Also, the evolution of the frontal process zone during thermal fatigue may be an interesting topic to explore.

To activate the intrinsic crack-toughening mechanisms, one can (i) add micro- or nano-particles to powder processed materials and/or (ii) employ chemical techniques to process the materials so that nanoparticles or micro-inclusions are precipitated from the TE material matrix. Advantages of particulate additions are that typically only a small volume fraction of the appropriate powder is needed for mechanical property enhancement (a few volume percent, sometimes less). Also, at low volume fraction levels, there is little increased cost to the thermoelectric material unless the powder additions are extremely expensive. If the nano- or micro-inclusions are relatively equiaxed, the particle additions should not interfere with densification. Also, nanoscale particles should not induce microcracking due to thermal expansion mismatch or TEA.

Nano- and micro-particles in thermoelectric matrices have been widely shown to enhance thermo-electric properties, especially to boost ZT by lowering the thermal conductivity via phonon scattering. Thus, the addition of inclusions to engineer the thermal fatigue resistance could proceed in tandem with the efforts to tailor thermoelectric properties and long as the inclusions added to enhance one property did not act to degrade the other. It may even be found that inclusions of a particular size and composition (or combination of sizes and compositions) could simultaneously improve ZT and enhance thermal fatigue resistance.

Acknowledgments

The author acknowledges the financial support of the Department of Energy through the grant "Revolutionary Materials for Solid State Energy Conversion" Center, an Energy Frontier Research Center funded by the U.S. Department of Energy, Office of Science, Office of Basic Energy Sciences under Award Number DE-SC0001054.

References

Ainsworth, J.H. and Herron, R.H., Thermal damage resistance of refractories, *Am. Ceram. Soc. Bull.*, 53:533–538, 1974.

Andersson, T. and Rowcliffe, D.J., Thermal cycling of indented ceramic materials, *J. Eur. Ceramic Soc.*, 18:2065–2071, 1998.

Ashizuka, M., Easler, T.E. and Bradt, R.C., Statistical study of thermal shock damage of a borosilicate glass, *J. Am. Ceram. Soc.*, 66:542–550, 1983.

ASTM C 1161-02c, Standard test method for flexural strength of advanced ceramics at ambient temperatures, 2008.

Banerjee, S., Properties of refractories, pp. 11–38 in *Refractories Handbook*, edited by C. Schacht, Marcel Dekker, New York, 2004.

Barsoum, M.W., Chapter 11 in *Fundamentals of Ceramics*, McGraw-Hill, New York, 1997.

Becher, P.F. and Warwick, W.H., Factor influencing the thermal shock behavior of ceramics, pp. 37–48, in *Thermal Shock and Thermal Fatigue Behavior of Advanced Ceramics*, eds. G.A. Schneider and G. Petzow, Kluwer Academic Publishers, Dordrecht, The Netherlands, 1993.

Becher, P.F., Lewis III, D., Karmen, K.C., and Gonzalez, A.C., Thermal shock resistance of ceramics: Size and geometry effects in quench tests, *Am. Ceram. Soc. Bull.*, 59:542–548, 1980.

Bennison, S.J. and Lawn, B.R., Role of interfacial grain-bridging sliding friction in the crack-resistance and strength properties of non-transforming ceramics, *Acta Metall.*, 37:2659–2671, 1989.

Bhowmick, S., Melendez-Martinez, J.J. and Lawn, B.R., Bulk is silicon is susceptible to fatigue, *Appl. Phys. Lett.*, 91:201902, 2007.

Blanks, K.S., Kristoffersson, A., Carlstrom, E., and Clegg, W. J., Crack-deflection in ceramic laminates using porous interlayers, *J. Europ. Ceram. Soc.*, 18:1945–1951, 1998.

Boley, B.A. and Weiner, J.H., *Theory of Thermal Stresses*, Wiley, New York, 1960.

Bradt, R.C., Ashizuka, M., Easler, T.E., and Ohira, H., Statistical aspects of the thermal shock damage and the quench-strengthening of ceramics, pp. 447–458, in *Thermal Shock Behavior and Thermal Fatigue of Advanced Ceramics*, eds. G.A. Schneider and G. Petzow, Kluwer Academic Publishers, the Netherlands, 1993.

Bradt, R.C., Fracture of refractories, pp. 1–10 in *Refractories Handbook*, edited by C. Schacht, Marcel Dekker, New York, 2004.

Budiansky, B. and O'Connell, R.J., Elastic moduli of a cracked solid, *Int. J. Solids Structures*, 12:81–97, 1976.

Case, E.D., Kim, Y. and Lee, W.J., Cyclic thermal shock in SiC whisker reinforced alumina and in other ceramic systems, pp. 393–406, *Thermal Shock Behavior and Thermal Fatigue of Advanced Ceramics*, eds. G.A. Schneider and G. Petzow, Kluwer Academic Publishers, the Netherlands, 1993.

Case, E.D., Relationships among changes in electrical conductivity, thermal conductivity, thermal diffusivity, and elastic modulus for microcracked materials, pp. 401–409 in *Ceramic Eng. and Sci. Proc.*, Volume 20, American Ceramic Society, Inc., Westerville, OH, 1999.

Case, E.D., The saturation of thermomechanical fatigue in brittle materials, pp. 137–208, a chapter in *Thermo-Mechanical Fatigue and Fracture* (part of a series on Fracture), editor M. H. Alibadi, WIT Press, Southampton, UK, 2002.

Chen, J.K., Tang, K.L. and Chang, J.T., Effects of zinc oxide on thermal shock behavior of zinc sulfide-silicon dioxide ceramics, *Ceram. Int.*, 35:2999–3004, 2009.

Cheng, C.M., Resistance to thermal shock, *J. Am. Rocket Soc.*, 21:147–153, 1951.

Chiu, C.C. and Case, E.D., Influence of quenching on fracture strength, elastic modulus, and internal friction of glass plates, *J. Mater. Sci.*, 27:2353–2362, 1992.

Cleveland, J.J. and Bradt, R.C., Grain size dependence of spontaneous cracking in ceramics, *J. Am. Ceram. Soc.*, 61:478–481, 1978.

Damani, R.J., Rubesla, D. and Danzer, R., Fracture toughness, strength and thermal shock behavior of bulk plasma sprayed alumina effects of heat treatment, *J. Eur. Ceram. Soc.*, 20:1439–1452, 2000.

Davidge, R.W. and Green, D.J., The strength of two-phase ceramic/glass materials, *J. Mater. Sci.*, 3:629–634, 1968.

Dresselhaus, M.S., Chen, G., Tang, M.Y., Yang, R., Lee, H., Wang, D., Ren, Z., Fleurial, J.P. and Gogna, P., New directions for low-dimensional thermoelectric materials, *Adv. Mater.*, 19:1043–1053, 2007.

Evans, A.G. and Cannon R.M., Toughening of brittle solids by martensitic transformations, *Acta Metall.*, 34:761–800, 1986.

Faber, K.T. and Evans A.G., Crack deflection processes—I. Theory, *Act Metall.*, 31:565–576, 1983a.

Faber, K.T. and Evans A.G., Crack deflection processes—II. Experiment, *Act Metall.*, 31:577–584, 1983b.

Fan, S., Zhao, J., Guo, J., Yan, Q., Ma, J. and Hng, H.H., P-type Bi0.4Sb1.6Te3 nanocomposites with enhanced figure of merit, *Appl. Phys. Lett.*, 96:182104-1, 2010.

Fu, Y., Evans, A.G. and Kriven, W.M., Microcrack nucleation in ceramics subject to a phase transformation, *J. Am. Ceram. Soc.*, 67:626–630, 1984.

Gilbert, C.J. and Ritchie, R.O., On the quantification of bridging tractions during subcritical crack growth under monotonic and cyclic loading in a grain-bridging silicon carbide ceramic, *Acta Mater.*, 46:609–616, 1998.

Gilman, J.J., *Chemistry and Physics of Mechanical Hardness*, Wiley, Hoboken, New Jersey, 2009.

Green, D.J., Nicholson, P.S. and Embury, G.H., Fracture of a brittle particulate composite -Part 2 theoretical aspects, *J. Mater. Sci.*, 14:1657–1661, 1979.

Griffith, A.A., The phenomena of rupture and flow in solids, *Philos. Trans. R. Soc. Lond.*, A221:163–198, 1920.

Gupta, T.K., Strength degradation and crack propagation in thermally-shocked Al_2O_3, *J. Am. Ceram. Soc.*, 55:249–253, 1972.

Hall B.D., Anderson C. and Case E. D., unpublished work.

Hasselman, D.P.H., Elastic energy at fracture and surface energy as design criteria for thermal shock, *J. Am. Ceram. Soc.*, 46:535–540, 1963.

Hasselman, D.P.H., Unified theory of thermal shock fracture initiation and crack propagation in brittle ceramics, *J. Am. Ceram. Soc.*, 52:600–604, 1969.

Hay, J.C. and White, K.W., Grain-bridging mechanisms in monolithic alumina and spinel, *J. Am. Ceram. Soc.*, 76:1849–1854, 1993.

Hertzberg, R.W., Chapter 8 in *Deformation and Fracture Mechanics of Engineering Materials*, Second Edition, John Wiley and Sons, New York, 1983.

Hirth J.P. and Lothe, J. *Theory of Dis,ocation*, 2nd Edition, Wiley, New York, 1982.

Hoening, A., Electrical conductivity of a cracked solid, *Pure and Applied Geophysics*, 117:690–710, 1978/79.

Hoening, A., Thermal conductivity of a cracked solid, pp. 281–287 in *Environment Effects on Composite Materials*, ed. G. S. Springer, Technomic Pub. Co., Lancaster, PA, 1984.

Hogan, T.P, Downey, A., Short, J., D'Angelo, J., Wu, C.I., Quarez, E., Androulakis et al., Nanostructured thermoelectric materials and high efficiency power generation modules, *J. Electron. Mater.*, 36(7):704–710, 2007.

Homeny, J., Darroudi, T. and Bradt, R.C., J-integral measurements of the fracture of 50% alumina refractories, *J. Am. Ceram. Soc.* 63:326–331, 1980.

Ito, M. and Sumiyoshi, J., Enhancement of thermoelectric performance of $Na_xCo_2O_4$ with Ag dispersion by precipitation from Ag^+ aqueous solution, *J. Sol-Gel Sci. Technol.*, 55:354–359, 2010.

Ivon, A.I., Kolbunov, V.R. and Chernenko, I.M., Stability of electrical properties of vanadium dioxide based ceramics, *J. Eur. Ceram. Soc.*, 19:1883–1888, 1999.

Kaliakan, V.N., *Introduction to Approximate Solution Techniques, Numerical Modeling and Finite Element Methods*, Marcel Dekker Inc., New York, 2002.

Kanatzidis, M.G., Nanostructured themoelectrics: The new paradigm?, *Chem. Mater.*, 22:648–659, 2010.

Kanninen, M.F. and Popelar, C.H., *Advanced Fracture Mechanics*, Oxford University Press, New York, 1985.

Katigiri, N., Hattori, Y., Ota, T. and Yamai, I., Grain size dependence of thermal shock resistance in $KZr_2(PO_4)_3$ ceramic, *J. Ceramic Soc. Japan*, 102:715–718, 1994.

Kingery, W.D., Factors affecting thermal stress resistance of ceramic materials, *J. Am. Ceram. Soc.*, 38:3–15, 1955.

Kingery, W.D., Bowen, H.K. and Uhlmann, D.R., *Introduction to Ceramics*, 2nd Edition, Wiley, New York, 1976.

Kilikoglou, V., Vekinis, G., Maniatis, Y. and Day, P. M., 1998, Mechanical performance of quartz tempered ceramics: Part 1, strength and toughness, *Archaeometry*, 40:261–279, 1998.

Kim, Y.M., Lee, W.J. and Case, E.D, Thermal fatigue behavior of ceramic matrix composites: a comparison among fiber reinforced, whisker reinforced, particulate reinforced and monolithic ceramics, pp. 871–881, *Proceedings of the Fifth Annual Meeting of the American Society for Composites*, Technomic Publications, Lancaster, PA, 1990.

Knehans, R. and Steinbrech, R., Memory effect of crack resistance during slow crack growth in notched aluminum oxide bend specimens, *J. Mater. Sci. Lett.*, 1:327–329, 1982.

Kreith, F., *Principles of Heat Transfer*, 3rd Edition, Harper & Row, New York, 1973.

Krupke, W.F., Shinn, M.D., Marion, J.E., Caird, J.A. and Stokowski, S.E., Spectroscopic, optical, and thermomechanical properties of neodymium- and chromium-doped gadolinium scandium gallium garnet, *J. Opt. Soc. Am. B*, 3(1), 102–114, 1986.

Kumakawa, A. and Niino, M., Thermal fatigue characteristics of functionally gradient materials for aerospace applications, pp. 393–406, *Thermal Shock Behavior and Thermal Fatigue of Advanced Ceramics*, eds. G.A. Schneider and G. Petzow, Kluwer Academic Publishers, the Netherlands, 1993.

Lan, J., Lin, Y.-H., Fang, H., Mei, A., Nan, C.-W., Liu, Y., Xu, S., and Peters, M., High-temperature thermoelectric behaviors of fine-grained Gd-doped $CaMnO_3$ ceramics, *J. Am. Ceram. Soc.*, 93:2121–2124, 2010.

Lange, F.F., The interaction of crack-front with second phase dispersion, *Philos. Mag.*, 22:983–992, 1970.

Laws, N. and Brockenbrough, J.R., The effect of micro-crack systems on the loss of stiffness of brittle solids, *Int. J. Solids Structures*, 23(9):1247–1268, 1987.

Lee J.H. and Grossman, J.C., Thermoelectric properties of nanoporous Ge, *Appl. Phys. Lett.*, 95:013106, 2009.

Lee, H., Vashaee, D., Wang, Z., Dresselhaus, M.S., Ren, Z.F. and Chen G., Effects of nanoscale porosity on thermoelectric properties of SiGe, *Appl. Phys. Lett.*, 107, 094308, 2010.

Lee, W.J. and Case, E.D., Cyclic thermal shock in sic whisker reinforced alumina composites, *Mater. Sci. Eng.*, A119:113–126, 1989.

Lee, W.J. and Case, E.D., Thermal fatigue in polycrystalline alumina, *J. Mater. Sci.*, 25:5043–5054, 1990.

Lee, W.J. and Case, E.D., Comparison of saturation behavior of thermal shock damage for a variety of brittle materials, *Mater. Sci. Eng.* A154:1–9, 1992.

Liou, Y.C., Tsai, W.C., Lin, W.Y., and Lee, U.R., Synthesis of $Ca_3Co_4O_9$ and $CuAlO_2$ ceramics of the thermoelectric application using a reaction sintering process, *J. Aust. Ceram. Soc.*, 44:17–22, 2008.

Lisjak, D., Drofenik, M. and Kolar, D., Investigation of the microscopical origin of the PTCR anomaly in two phase Zn-Ni-O ceramics, *Key Eng. Mater.*, 132–136:1325–1328, 1997.

Litovsky, E., Gambaryan-Roisman, T., Shapiro, M. and Shavit, A., Effect of grain thermal expansion mismatch on thermal conductivity of porous ceramics, *J. Am. Ceram. Soc.*, 82:994–1000, 1999.

Luo, J. and Stevens, R., Residual stress and microcracking in SiC–MgO composites, *J. Eur. Ceram. Soc.*, 12:369–375, 1993.

Luo, J. and Stevens, R., The role of residual stress on the mechanical properties of Al_2O_3-5 vol% SiC nanocomposites, *J. Eur. Ceram. Soc.*, 48:1565–1572, 1997.

Lutz, E.H., Size sensitivity to thermal shock of plasma-sprayed ceramics and factors affecting the size effect, *J. Am. Ceram. Soc.*, 78:2700–2704, 1995.

Mai, Y-W. and Lawn, B.R., Crack-interface grain bridging as a fracture resistance mechanism in ceramics: II, Theoretical fracture mechanics model, *J. Am. Ceram. Soc.*, 70:289–294, 1987.

Manson, S.S., *Thermal Stress and Low-Cycle Fatigue*, McGraw-Hill Book Company, New York, pp. 276–286, 1966.

Mi, J.L., Zhao, X.B, Zhu, T.J. and Tu, J.P., Improved thermoelectric figure of merit in *n*-type CoSb3 based nanocomposites, *Appl. Phys. Lett.*, 91:172116, 2007.

Mingo, N., Hauser, D., Kobayashi, N.P., Plissonnier, M. and Shakouri, A., Nanoparticle-in-Alloy approach to efficient thermoelectrics: Silicides in SiGe, *Nanoletters*, 9:711–715, 2009.

Monroe, L.D. and Smyth, J.R., Grain Size dependence of fracture energy of Y_2O_3, *J. Am. Ceram. Soc.*, 61:538–539, 1978.

Montana, G., Fabbri, B., Santoro, S., Gualtieri, S., Iliopoulos, I., Guiducci, G. and Mini, S., Pantellerian ware, a comprehensive archaeometric review, *Archaeometry*, 49:455–481, 2007.

Murthy T.S.R., Sonber, J.K., Subramanian, C., Fotedar, R.K, Kumar, S., Gonal, M.R. and Suri, A.K., A new TiB_2 + $CrSi_2$ composite—Densification, characterization and oxidation studies, *Int. J. Refractory Metals Hard Mater.*, 28:529–540, 2010.

Ni, J.E., Ren F., Case, E.D. and Timm, E.J., Porosity dependence of elastic moduli in LAST (lead–antimony–silver–tellurium) thermoelectric materials, *Mater. Chem. Phys.*, 118(2–3), 459–466, 2009.

Ni, J.E., Case, E.D., Khabir, K., Wu, C.I., Hogan,T.P., Timm, E.J., Girard, S. and Kanatzidis, M.G., Room temperature Young's modulus, shear modulus, Poisson's ratio and hardness of PbTe–PbS thermoelectric materials, *Mater. Sci. Eng. B*, 170:58–66, 2010.

Nie, X. W., Lu, S.G. Wang, K.L., Chen, T.C. and Niu C.J., Fabrication and toughening of $NbCr_2$ matrix composites alloyed with Ni obtained by powder metallurgy, *Mater. Sci. Eng. A*, 502:85–90, 2009.

Noda, N., Thermal stresses in materials with temperature-dependent properties, *Appl. Mech. Rev.*, 44:383–397, 1991.

Norman, D.A. and Robertson, R.E., Rigid-particle toughening of glassy polymers, *Polymer*, 44:2351–2362, 2003.

Nye, J.F., *Physical Properties of Crystals*, Oxford University Press, New York, 1985.

Ohira, H. and Bradt, R.C., Strength distributions of a quench strengthened aluminosilicate ceramic, *J. Am. Ceram. Soc.*, 71:35–41, 1988.

Padture, N.P., Gell, M. and Jordan, E.J., Thermal barrier coatings for gas-turbine engine applications, *Science*, 296:280–284, 2002.

Park, K., Ko, K.Y., Seo, W.S., Cho, W.S., Kim, J.G. and Kim, J.Y., High temperature thermoelectric properties of polycrystalline $Zn_{1-x-y}Al_x Ti_y O$ ceramics, *J. Euro. Ceram. Soc.*, 27:813–817, 2007.

Ravi, V., Firdosy, S., Calliat, T., Lerch, B., Calamino, A., Pawlik, R., Nathal, M., Sechrist, A., Buchhalter, J. and Nutt, S., Mechanical properties of thermoelectric skutterudites, pp. 656–662, *Space Technology and Applications International Forum*, American Institute of Physics, Melville, NY, 2008.

Ren, F., Case, E.D., Timm, E.J. and Schock, H.J., Young's modulus as a function of composition for an *N*-type lead–antimony–silver–telluride (LAST) thermoelectric material, *Philos. Mag.*, 87(31):4907–4934, 2007.

Ren, F., Case, E.D., Sootsman, J.R., Kanatzidis, M.G., Kong, H., Uher, C., Lara-Curzio, E. and Trejo, R.M., The high temperature elastic moduli of polycrystalline PbTe measured by resonant ultrasound spectroscopy, *Acta Mater.*, 56, 5954–5963, 2008a.

Ren, F., Case, E.D., Timm, E.J. and Schock, H.J., Hardness as a function of composition for *n*-type LAST thermoelectric material, *J. Alloys Compounds*, 455:340–345, 2008b.

Ren, F., Hall, B.D., Ni, J.E., Case, E.D., Timm, E.J., Schock, H.J., Wu, C.-I., D'Angelo, J.J., Hogan, T.P., Trejo, R.M. and Lara-Curzio, E., pp. 121–126 in *Mechanical Characterization of PbTe-Based Thermoelectric Materials*, Hogan T.P., Yang, J., Funahashi, R., Tritt, T., editors. *Thermoelectric Power Generation, Materials Research Society Proceedings*, Vol. 1044. Warrendale, PA: Materials Research Society, 2008c.

Ren, F., Case, E.D., Hall, B.D., Timm, E.J., Trejo, R.M., Meisner, R. and Lara-Curzio, E., Temperature-dependent thermal expansion of cast and hot pressed LAST (Pb–Sb–Ag–Te) thermoelectric materials, *Philos. Mag.*, 89(18):1439–1455, 2009.

Rendtorff, N. and Aglietti, E., Mechanical and thermal shock of refractory materials for glass feeders, *Mater. Sci. Eng. A.*, 527:3840–3847, 2010.

Rice, R.W., Mechanisms of toughening in ceramic matrix composites, *Ceramic Eng. Sci. Proc.*, 2:661–701, 1981.

Rice, R.W., *Mechanical Properties of Ceramics and Composites: Grain and Particle Effects*, Marcel Dekker, New York, 2000.

Rice, R.W., *Ceramic Fabrication Technology*, Chapter 8, Marcel Dekker, New York, 2003.

Ritchie, R.O., Mechanisms of fatigue-crack propagation in ductile and brittle solids, *Int. J. Fracture*, 100:55–83, 1999.

Ritchie, R. O., On the fracture toughness of advanced materials, *Adv. Mater.*, 21:2103–2110, 2009.

Sack, R.A., Extension of Griffith's theory of rupture to three dimensions, *Proc. Royal Soc.* (London), 58A:729–736, 1946.

Salvador, J.R., Yang, J., Shi, X., Wang, H., Wereszczak, A.A., Kong, J. and Uher, C., Transport and mechanical properties of Yb-filled skutterudites, *Philos. Mag.*, 89:1517–1534, 2009.

Sinha, A., Kokini, K. and Bowman, K.J., Elastic constant degradation from microcracking in ceramic-ceramic composites cycled by slow thermal loading, *Mater. Sci. Eng.*, A188:317–325, 1994.

Sootsman, J.R., He, J., Dravid, V.P., Ballikaya, S., Vermeulen, D., Uher, C. and Kanatzidis, M.G., Microstructure and thermoelectric properties of mechanically robust PbTe–Si eutectic composites, *Chem. Matter.*, 22:869–875, 2010.

Sridhar, N., Rickman, J.M. and Srolovitz, D.J., Effect of reinforcement morphology on matrix microcracking, *Acta Mater.* 44:915–925, 1996.

Sun, J.H., Qin, X.Y., Xin, H.X., Li, D., Pan, L., Song, C.H., Zang, J., Sun, R.R., Wang, Q.Q. and Liu, Y.F., Synthesis and thermoelectric properties of Zn4Sb /Bi 0.5Sb1.5Te3 bulk nanocomposites, *J. Alloys Compounds*, 500:215–219, 2010.

Swain, M.V., *R*-curve behavior and thermal shock resistance of ceramics, *J. Am. Ceram. Soc.*, 73:621–628, 1990.

Swanson, P.L., Fairbanks, C.J., Lawn, B.R., Mai, Y-W. and Hockey B. J., Crack-interface grain bridging as a fracture resistance mechanism in ceramics: I, experimental study on alumina, *J. Amer. Ceram. Soc.*, 70:279–289, 1987.

Tiefenbach, A., Wagner, S., Oberacker, R. and Hoffmann, B., The use of impedance spectroscopy in damage detection in tetragonal zirconia polycrystals (TZP), *Ceram. Int.*, 26:745–751, 2000.

Timoshenko, S. and Goodier, J. N., *Theory of Elasticity*, p. 158, McGraw-Hill, New York, 1951.

Tite, M.S., Kilikoglou, V. and Vekinis, G., Review article: Strength, toughness and thermal shock resistance of ancient ceramics, and their influence on technological choice, *Archaeometry*, 43, 301–24, 2001.

Wachtman, J.B., Cannon, W.R. and Matthewson, M. J., *Mechanical Properties of Ceramics*, 2nd Edition, Wiley, Hoboken, New Jersey, 2009.

Walsh, J.B., Effect of cracks on the compressibility of rock, *J. Geophys. Res.*, 70:381–389, 1965.

Wang, H. and Singh, R.N., Thermal shock behavior of ceramics and ceramic composites, *Int. Mater. Rev.*, 39:228–244, 1994.

Wang, J., Vandeperre, L.J., Stearn, R.J. and Clegg, W.J., Pores and cracking in ceramics, *J. Ceram. Proc. Res.*, 2:27–30, 2001.

Wang, Y., Sui, Y., Yang, X., Su, W. and Liu, X., Enhanced high temperature thermoelectric characteristics of transition metals doped $Ca_3Co_4O_{9+\delta}$ by cold high-pressure fabrication, *J. Appl. Phys.*, 107, 033708, 2010a.

Wang, H.C., Wang, C.L., Su, W.B., Liu, J., Zhao, Y., Peng, H., Zhang, J.L, Zhao, M.L., Li, J.C., Yin, N. and Mei, L.M., Enhancement of thermoelectric figure of merit by doping Dy in $La_{0.1}Sr_{0.9}TiO_3$ ceramic, *Mater. Res. Bull.*, 45:809–812, 2010b.

Wright, P.K. and Evans, A.G., Mechanisms governing the performance of thermal barrier coatings, *Curr. Opin. Solid State Mater. Sci.*, 4:255–265, 1999.

Yang, B. and Chen, X.M., Alumina ceramics toughened by a piezoelectric secondary phase, *J. Eur. Ceram. Soc.*, 20:1687–1690, 2000.

Yasukawa, M., Kono, T., Ueda, K., Yanagi, H. and Hosono, H., High-temperature thermoelectric properties of La-doped $BaSnO_3$ ceramics, *Mater. Sci. Eng. B*, 173:29–32, 2010.

Zang., S.N., Zhu, T.J., Yang, S.H., Yu, C., and Zhao, X.B., Improved thermoelectric properties of $AgSbTe_2$ based compounds with nanoscale Ag_2Te *in-situ* precipitates, *J. Alloy. Compd.*, 499:215–220, 2010.

Zhang, H., Wang, D., Chen, S. and Liu, X., Toughening of $MoSi_2$ doped by La_2O_3 particles, *Mater. Sci. Eng. A*, 345:118–121, 2003.

Zhang, L.M., Hirai, T., Kumakawa, A. and Yuan, R.Z., Cyclic thermal shock resistance of TiC/Ni_3Al FGMs, *Composites Part B*, 28B:21–27, 1997.

Zhao, L.T., Lu, T.J. and Fleck N.A., Crack channeling and spalling in a plate due to thermal shock loading, *J. Mech. Phys. Solids*, 48:867–897, 2000.

II

Thermoelectric Modules, Devices, and Applications

17

Miniaturized Thermoelectric Converters, Technologies, and Applications

Harald Böttner
Fraunhofer Institute for Physical Measurement Techniques

Joachim Nurnus
Micropelt GmbH

17.1 Introduction

Miniaturized thermoelectric devices came into existence as high-performance Peltier coolers designed for cooling power densities >100 W/cm² [1]. The intended main application was the temperature control of upcoming new generations of microprocessor units (MPUs) which should dissipate heat up to some 100 W/cm² under operation.

General considerations concerning miniaturized thermoelectric devices and their state of-the-art technologies for fabrication and applications up to 2005 are summarized in Ref. [2]. So we will take this review as a basis and thus we will highlight in this chapter in particular new developments in technology and applications during the last 5 years. It is noteworthy that during the last 5 years the scope of application for miniaturized thermoelectric devices widened significantly and also the focus did change. Even though, as a result of the independent development of Venkatasubramanian et al. at the Research Triangle Institute, North Carolina, USA [3] and Böttner et al. at Fraunhofer IPM, Freiburg, Germany [4], cooling power densities of ≥100 W/cm² were achieved and in addition the successful application of mini-Peltier-coolers for MPUs was demonstrated [3], the main focus for the application of miniaturized thermoelectric devices changed to converters for low-power energy-harvesting systems addressing the µW to mW range.

Basically one main reason can be identified as the driving force. Wireless data transfer is known to consume significantly less power year by year, thus self-powered sensor systems with wireless data transfer became more and more attractive [5]. This reflects the review in *Journal of Applied Physics* (*JAP*) from 2008 with 150 references [6] and the Frost and Sullivan report from December 2007 about

"Advances in Energy Harvesting Technologies" [7]. Whereas the JAP review reports the physical principles of "small scale energy harvesting through thermoelectric, vibration and radiofrequency power conversion"[6] the Frost and Sullivan report tries to assess the different technologies and applications, describes the key developments in this sector, offers a technology foresight and collects some key patents and key industry participants. One main conclusion of this report is that energy harvesting and conversion via piezoelectric effects and via thermoelectric are expected to be the most promising technologies in the mid-term future.

Figure 17.1 repeats the opportunity matrix, taken from Ref. [7]. The promising ranking for piezoelectric and thermoelectric converters is obvious.

For a comprehensive survey on the whole energy-harvesting research and application area and in particular the miniaturized thermoelectric converters, the authors strongly recommend the publications [2,6,7] and for basic and detailed information.

Up to now only two concepts were developed to commercial ripeness. The company Nextreme (http://www.nextreme.com/, 13.01.2011) was founded and venture capital (VC) financed in 2005 as a start-up company based on the RTI technology [8]. The kernel of this development is the exploitation of thermal conductivity reducing V_2-VI_3 staggered superlattices for n- and p-type thermoelectric legs. Some technical details and the resulting huge ZT-quality for the n- and p-V_2-VI_3 staggered superlattices are reported in Ref. [9]. In spite of the remarkable breakthrough for the ZT-data, 2.4 for p-type material and 1.4 for n-type material—unfortunately up to now not independently confirmed, one main drawback of this technology is that the superlattices were deposited on a comparable expensive single crystalline GaAs-, finally sacrificial, substrate [8]. The company Micropelt [10] was founded in 2006, also VC financed. The kernel of its proprietary technology is a complete wafer based microelectronic-like wafer technology [11]. n- and p-type V_2–VI_3-material is composed by skilled sputtering and post-annealing processes. Even though the "micropelt" technology does not use (in comparison to the "Nextreme" technology) any superlattices, the performance of both types of devices are, surprisingly, quite similar.

As mentioned before no other technology for miniaturized thermoelectric devices ever reached similar technical ripeness. One reason for this situation can be found in the press release of thermoelectric start-up companies. To establish those kinds of thermoelectric businesses, overall investments—this is our estimation without explicit confirmation from these companies—of 10–20 million € are necessary to bring academic demonstrators to the market and to create enough turnover [12,13].

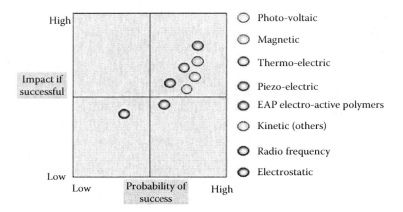

FIGURE 17.1 (**See color insert**.) Opportunity matrix for different energy-harvesting technologies. (From Frost and Sullivan. *Advances in Energy Harvesting Technologies*, DoC2, December 2007. With permission.)

17.1.1 Classification of Miniaturized Thermoelectric Devices

To the best of our knowledge, no attempt for a classification of miniaturized thermoelectric devices exists so far. Here the miniaturized thermoelectric devices will be classified by leg height limitations and leg foot print dimensions. These classifications are partially derived based on the thermoelectric and mechanical properties of the V_2–VI_3 compounds.

Leg height: the leg height should be below typical limits as for bulk material-based devices. It is reasonable to follow the proposal of Semenyuk [14]: "leg heights under 200 µm are (for bulk material) practically not possible." Bi_2Te_3 is known as a very brittle material due to its Te–Te van der Waals bonds. He claims "that the (important) range of thermoelectric leg heights from 50 to 130 µm is still unmastered" [14]. Figure 17.2 illustrates this leg height classification comparing standard devices with Micropelt devices.

He further suggests that the "bulk-like nanoscale" concept could possibly solve the problem. That this prediction is true could be demonstrated by Böttner et al. [15] by reporting the producibility of 100 µm thin *n*- or *p*-type V_2VI_3 wafers of 45 mm diameter. The wafers were thinned from 2 mm thick pucks densified by spark plasma sintering (SPS). Starting material for the SPS-process are meltspin flakes.

Leg footprint: similar to the leg height the footprint should be significantly below typical dimensions as for bulk material based devices. The footprint should be significantly smaller than 1 mm and in the range of a few 100 µm. Characteristic examples are the commercialized devices from Nextreme and Micropelt. For Micropelt devices [16], see Figure 17.3, footprints down to 35 µm were presented, resulting in leg-pair densities up to 7000/cm². The advantage is that utilizing such a high leg pair densities for miniaturized thermogenerators, it will be easy to create comparably high open-circuit voltages evoked by only small temperature differences.

Based on these classifications, other activities in particular concerning energy harvesting for the µW to mW regime, see, for example [17], are not reviewed here.

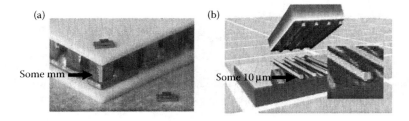

FIGURE 17.2 (**See color insert.**) Illustration of leg height classification for miniaturized thermoelectric devices: (a) typical commercial device, (b) pattern of thermoelectric legs for miniaturized Peltier-coolers according to the Micropelt technology.

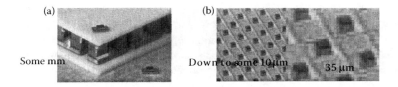

FIGURE 17.3 (**See color insert.**) Illustration of leg foot print classification for miniaturized thermoelectric devices: (a) typical commercial device, (b) pattern of thermoelectric legs for miniaturized thermogenerators according to the Micropelt technology.

17.2 Technologies

17.2.1 Research Area

Basically two types of miniaturized thermoelectric devices were developed: the vertical and the horizontal/planar design.

For the vertical designs the thermoelectric legs in the above described miniaturized dimensions are aligned as for conventional commercial devices. The legs stand like pillars on the substrate surface. The same holds for the applied or generated temperature gradient. Thus the heat flow is perpendicular to the substrate surface.

In the horizontal/planar design the thermoelectric legs are oriented on the substrate surface. Thus the temperature gradient is applied or generated along the substrate surface. In this configuration the substrate acts a thermal bypass.

The activities in the research area are determined by the intended application for either high-performance coolers or energy harvesting in the μW and mW regime. All these applications are around room temperature. Thus it is not surprising that the majority of published developments are focused on V_2VI_3-based miniaturized devices. This is reflected by three other very useful overviews [6,14,18] published since the review article in the 2nd *Thermoelectric Handbook* [2].

To get a more or less comprehensive overview, one has to use all these surveys. None is complete, but altogether they show clearly core areas and some very challenging developments.

- Planar designs up to now never reached a positive foresight for commercialization, in spite of early attempts for a market entry [19]. The activities are currently back in the research/academic area [20,21]. It should be mentioned that the thermoelectric legs according to Ref. [19] were structured as sputtered thin films in planar design on some 10 μm thin Kapton foils as substrate which act as thermal bypass. Finally this planar design was rearranged in a vertical configuration. Thus compared to the standard commercial configuration, the gap between the thermoelectric legs (under ambient conditions air) is filled by Kapton.
- Some other materials are under investigation like Si, SiGe, Bi–Sb combinations, III–V superlattices, and V_2VI_3 compounds, also as superlattices.

Exclusively the research on Bi_2Te_3 miniaturized, "thin film" devices results in enough data to compare different concepts and technologies. Around 70% of the results deal with V_2VI_3-materials.

- Regarding the V_2VI_3-based device developments, dry deposition techniques like MOCVD and sputtering as well as wet deposition techniques were used. Only dry deposition techniques are currently able to produce thin film materials of such a thermoelectric quality, which is necessary for thin film devices. As a first estimation, power factors in the range of ~30 μW/cm K^2 are meaningful. Up to now, in spite of being very attractive under the viewpoint of cost efficient mass production, no electrodeposited V_2VI_3 layers are reported demonstrating sufficiently good thermoelectric properties.

It can be summarized as a main result of the state of art of electrodeposition, that V_2VI_3 layers as deposited from solutions will have no chance to fix, for example, precisely enough the deviation from stoichiometry, which is known to be responsible for the thermoelectric properties; that is, carrier concentration and carrier mobility → electrical conductivity, Seebeck coefficient and the thermal conductivity. Electrodeposited *n*- and *p*-type V_2VI_3-material exhibit correct structural properties and grow commonly preferentially perpendicular to its basal-plane on the substrate. This is also an advantage of electrodeposited film as the resulting thermoelectric legs would be used in the direction, which is known as the better one for the anisotropic V_2VI_3-materials. Up to now the missing link is a suitable post processing step to define the internal structure (e.g., grain size) and the thermoelectric properties. This situation for electroplated thermoelectric V_2VI_3 compounds was impressively presented at ICT 2009 by Clotilde Boulanger [22]. This most recent survey is focused on the chemistry of the electrodeposition. It is to our knowledge best survey if one wants to know under which experimental conditions which structural and thermoelectric properties have been achieved. Data are collected for the deposition of Bi_2Te_3 $Bi_2(Se,Te)_3$ and $(Bi,Sb)_2Te_3$ since the early beginning in 1993. In addition data for V_2VI_3 electroplated nanowires and other nanostructures were reported. For further details we would like to refer the reader to the overviews [6,14,18,22].

Miniaturized devices based on a microelectronic wafer level technology were developed at Fraunhofer IPM [4] and commercialized by the company Micropelt since 2006 [23]. As it is the only technology close to standard microelectronic process routes, the main features should be repeated here.

Polycrystalline single phase n- and p-V_2VI_3-materials are composed by deposition from elemental targets onto heated thermal oxide covered silicon substrates. The critical issue and drawback for this specific technology is namely the big mismatch in the thermal expansion coefficient of a factor 5–6 between the silicon substrate and the thermoelectric material. In addition, but this is common for all miniaturized thermoelectric devices, the specific electrical contact resistance must be in the range of 10^{-11}–10^{-12} W/m^2. Both challenges were mastered but the technical solutions and the processes accordingly were unfortunately never published. The same holds for the dry etching process to structure the deposited >20 µm thick thermoelectric layers for the n- and p-pillars. Exclusively known are the main steps of the technological sequence:

- Structuring of electrical interconnects for the final vertical module design onto the above described wafers
- Deposition of either n- or p-type thermoelectric material of separated wafers.
- Post-processing by annealing
- Dry etching of the thermoelectric material with solder on top forming the individual legs
- Separation of n- and p-type dies
- Soldering n- to p-dies

17.2.2 Commercialized Technologies

In this section only commercialized technologies will be taken into account, where we define "commercialized" as follows:

- Technology used and marketed by a company.
- Samples are available.

As described earlier the devices can be categorized as cross-plane and in-plane production techniques. In-plane TEGs have been known for a long time and used, for example, to power cardiac pacemakers [24]. In all cases p- and n-type Bi_2Te_3-based thin films are deposited and patterned on thin plastic foils to form pi-shaped thermocouples on the foil substrate. From a deposition point of view, this is almost the ideal configuration: By adjusting the deposition parameters the layered Bi_2Te_3-based compounds can be easily deposited in such a way, that the crystallographic direction with the higher ZT value is parallel to the substrate—that is, parallel to the temperature gradient. Further, the thermal resistance of the couples can be easily adjusted by the length of the p- and n-type legs. Thus, typically thin films with thickness in the range of only a few micrometers are used. Due to the high thermal resistance, typically high open-circuit voltages of some volts can be achieved [19], but the generated electrical output power is limited to a few 10 µW.

Today commercialized cross-plane thermoelectric devices from Nextreme and Micropelt are known. Both use bismuth–telluride-based thermoelectric thin films to set up the well-known pi-shaped thermocouples. Nevertheless, there are significant differences in the production processes: Nextreme's technology was built around the research on high-performance thin films thermoelectric superlattice structures grown by MOCVD: In a first step, expensive GaAs-wafers with a good lattice match are essential for growing the p- and n-type superlattices [8]. Then the layers are patterned and covered with a solder on the top side before each single leg of the later TEG is sawn out of the GaAs (sacrificial) substrate. By this approach the minimum dimensions of a single leg are limited to several 100 µm due to sawing and automatic handling restriction (as well as the subsequent copper pillar bum process), see Figure 17.4 [25].

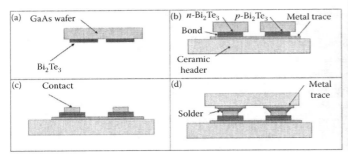 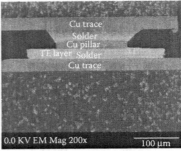

FIGURE 17.4   (a) Bi_2Te_3 based films are grown epitaxially on GaAs substrates; *p*- and *n*-type materials are grown separately, (b) the *p*- and *n*-films are diced and attached to metal traces by solder, (c) the GaAs is removed and metal contact posts are electroplated on the *p*- and *n*-elements, (d) a top header with complementary traces is attached by solder to the posts completing the electrical circuit. (From Nextreme Thermal Solutions, White Paper: The Thermal Copper Pillar Bump: Enabling improved semiconductor performance without sacrificing efficiency, *Nextreme Thermal Solutions*, January 9, 2008. With permission.)

Then the single *p*- and *n*-type dies are soldered onto a metalized substrate. Then the GaAs substrate is removed and copper pillars are deposited onto the thermoelectric dies. In a last step the pi-shaped couples are formed by soldering the second substrate against the copper pillars.

In the case of Micropelt, the target was to establish a scalable wafer-based production platform for thermoelectric thin film devices. Due to the MEMS-based process flow using thin film deposition, lithography and etching process, very small feature sizes, which means high numbers of leg pairs, and thus high output voltages can be realized even for small temperature gradients. Micropelt uses their experience with semiconductor technology to realize high-quality thermoelectric devices using a wafer scale approach. As in the case of bulk devices, $(Bi,Sb)_2(Te,Se)_3$-based compound semiconductors are used today. In contrast to bulk devices, the thin film deposition process allows the later use of nanostructured, high-performance materials for use in a next product generation [26]. Today Micropelt uses isolated 4″ and 6″ wafers in the pilot production line. The wafers are processed by first depositing and patterning metal structures followed by an overgrowth with up 36 μm thick thermoelectric materials and solder metals. Then the thermoelectric materials are patterned using an etch mask and an etching step in order to define the thermoelectric legs. These processes 1–4 are done for *p*- and *n*-type wafers separately. A complete device is made by flip chip bonding of the *p*- and *n*-type parts. The complete process is shown in Figure 17.5a.

Due to the wafer level semiconductor technology, an extremely flexible platform technology for thermoelectric thin film devices (coolers, generators and sensors) with respect to internal leg geometries (ranging from 30 to 600 μm feature size as shown in Figure 17.5b) as well as overall device geometries (0.5 mm² up to 25 mm²) is available. Especially the small internal leg geometries are essential for the TEG design: The almost 100 times higher number of leg per cm² enables high output voltages.

Compared to the state of art described 5 years ago in Refs. [2,27] both Micropelt and Nextreme today also promote thermogenerator devices. Compared with standard bulk fabrication processes, both technologies described above allow significantly smaller feature sizes and thus higher open-circuit output voltages. In Figure 17.6, a comparison of the Nextreme eTEG HV56 (3.12 × 3.56 mm² footprint) [28] and the Micropelt MPG D751 (3.3 × 4.3 mm² footprint) [29] is shown. Although the overall footprint of both devices is pretty similar, open-circuit voltage of the MPD D751 reaches 5 V already at 30°C temperature difference, for the TEG HV 56 more than 200°C are needed. This almost factor 10 higher output voltage is due to the higher packing density of thermoelectric legs in the Micropelt approach.

The possibility to realize high packing densities for thin film thermogenerator devices today also is used for realizing thin film coolers with higher resistances. By doing this, the typical operating conditions for small and thin TECs (small voltage and high currents due to small device resistances) can be

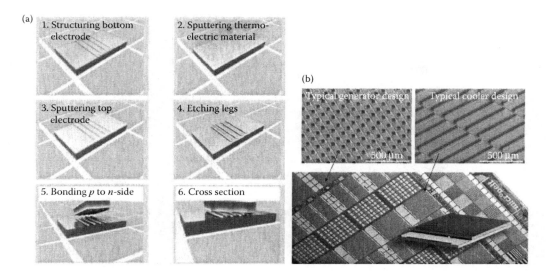

FIGURE 17.5 **(See color insert.)** (a) Micropelt's high level process flow. (b) Details of a finished wafers before bonding showing the flexibility of the production with respect to feature and device sizes.

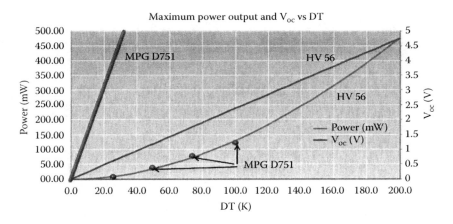

FIGURE 17.6 Comparison of HV56 and MPG D751 thin film generators. (Adapted from http://www.nextreme.com/media/pdf/techspecs/Nextreme_eTEG_HV56_Data_Sheet.pdf, 06.02.2011; http://www.micropelt.com/products/datasheets.php, 06.02.2011.)

modified: Due to the high resistance of devices with many small legs, the operating voltages (currents) are up to a factor of 10 higher (lower) compared to typical bulk devices (Figure 17.7).

Thin film devices with high internal resistance, and thus high operating voltages, are available from both Micropelt and Nextreme. In Figure 17.8, the voltage/current characteristic of Nextreme's HV14 Optocooler (2.7 mm² footprint) [30] and Micropelt's MPC D403 (3.1 mm² footprint), calculation done using Simulation Tool MyPelt [31], are shown. Looking at the maximum temperature difference, both devices are able to reach 60 K. The current needed to achieve DT_{max} is 1.0 A for the HV14, for the MPC D403 the required current is 0.19 A.

Both companies also offer cooler designs for applications requiring more cooling power and/or larger size of the devices. Details can be found on the respective homepages and to some extent in the next section.

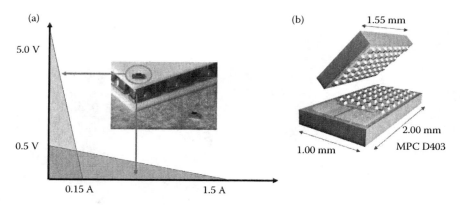

FIGURE 17.7   Comparison of current/voltage characteristics for small conventional Peltiers and thin film devices with a high packing density of legs (a), for example, MPC D403 (b).

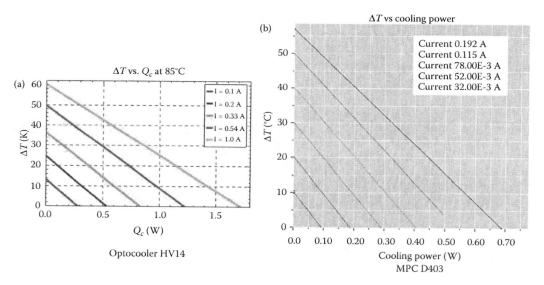

FIGURE 17.8   **(See color insert.)** DT-Q_c diagrams for thin film TECs produced by Nextreme (HV14) (a) and Micropelt (MPC D403) (b).

17.3 Applications

17.3.1 Thermal Management

Applications typically addressed in the field of thermal management are

- Cooling of optoelectronic devices
- Rapid thermal cycling of biomedical samples (PCR)
- Cooling of hot spots in electronic circuits.

In the beginning of thin film thermoelectrics, the driving force was the enormous increase of the cooling power density with decreasing leg heights [1]: the calculated values of several 100 W/cm² seemed to be a possible solution for cooling down hot spots in microprocessors, where experts expected heat flux densities as high as 1000 W/cm² in the year 2010. Although recent publications report that a 1000 W/cm² hot spot was cooled by 5°C [9], this is not of commercial relevance in the future. This is due the fact that

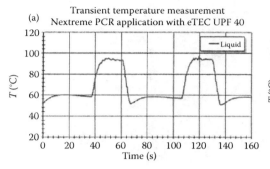 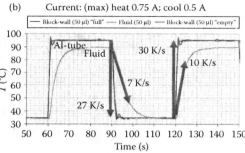

FIGURE 17.9 Published thermal cycling results using a UFP 40 (a) and a MPC-D701 (b). (From http://www. nextreme.com/media/pdf/Nextreme_PCR_Process_Press_Release_Jan09.pdf.)

the gigahertz race of microprocessors stopped and dual core and quad core processors with heat flux densities significantly below the formerly predicted numbers are used.

The reasons for still looking at thin film devices in applications are typically the small size and associated future costs of the components, the high cooling power densities as well as the fast response times.

Looking at laser cooling in, for example, telecom and potentially in near future in today's uncooled data-communication applications, a new trend can be observed: Due to the high internal resistance of the novel high resistance thin film TEGs, the driving voltage can be increased to several volts. This results in two advantages: the TEC driver can directly use the voltage available in the system (typically somewhere between 3 and 5 V). By this, expensive downconversion circuits capable of providing high currents, which also decrease the system efficiency, can be avoided. So due to the availability of high resistance TECs, inexpensive and efficient controllers using off the shelf electronic components can be realized [32].

In biological and life science applications, polymerase chain reaction (PCR) is a technique widely used to produce millions of copies of a specific DNA sequence in a short period of time. PCR-based testing is used in diagnosis of diseases and the identification of genetic fingerprints. Most PCR cyclers available today use conventional Peltier coolers in order to thermally cycle the DNA sample based on a predefined series of temperature steps between 40°C and 90°C.

Because of the increased power densities available in thin film coolers, significantly faster heating and cooling of samples is possible. Standard PCR cyclers reach heating and cooling rates of the so called well blocks (metallic cylinders taking up the DNA sample in a plastic tube) in the range of a few K/s. Similar speeds of metallic blocks are known only from Micropelt: Using a MPC D701 cooler, metal temperatures can be changed with heating and cooling rates of 10 K/s [33] to 30 K/s (Figure 17.9 b).

In the micro-PCR application where Nextreme was heating and cooling a plastic vial with an unknown amount of water in it, this translated into an ability to heat the water up to 90°C in 15 s (2.3 K/s) and cool it back down to 55°C in under 10 s (3.5 K/s), see also Figure 17.9a [34].

In the case of Micropelt, ramp rates within the liquid directly filled into an aluminum tube of 7 K/s (cooling) and 10 K/s (heating) were reported (Figure 17.9b) (Nurnus, J., ICT 2009, Freiburg, Germany, July 2009, private communication). A direct comparison of these measurements is not possible because not enough experimental data on the setups are published.

Not only do faster ramp rates add value for PCR cyclers—also the ability to realize matrix setups of individually cycle DNA samples is a unique feature achievable with thin film TECs [35].

17.3.2 Thermally Powered Sensor Systems

As mentioned in the introduction, sensor systems powered by thermoelectric devices are expected to become a huge increasing market in the mid-term future. Thus we will report on selected examples to demonstrate the widespread field of applications.

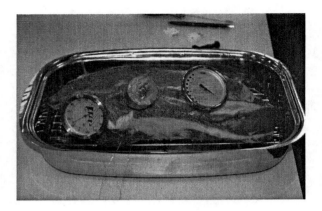

FIGURE 17.10 Experimental setup for the thermoelectric powered roast thermometer with wireless data transfer.

For use in any kitchen, an energy autarkic thermoelectric-powered roast thermometer was developed, Figure 17.10. The generator and the electronic housing for the electronic circuit, including an antenna can be seen together with two conventional roast thermometers ready for application.

The "sword" with the temperature sensor is inserted in the roast. Figure 17.11 is a copy of the main figure of the patent. The temperature sensor and thermogenerator are easy to recognize. Details are described in Ref. [36].

The power for sensor readout and the wireless data transfer is generated using the temperature difference of the hot oven and the cold (~80–90°C) inside of the meat during roasting. The challenge here is to design a suitable thermal management: the electronic circuit should not be overheated above 120°C during roasting with oven temperatures <200°, the housing has to be thermally connected to the hot side of the thermogenerator and the other side of the thermogenerator has to be thermally connected to

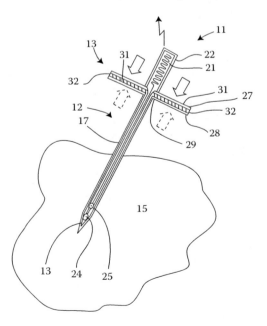

FIGURE 17.11 Sketch of the sensor "sword" with thermogenerator for the roast thermometer. (From Baier, M., Temperatursonde für einen Ofen, Ofen und Verfahren zum Betrieb eines Ofens, WO 2008/119440 A2, 09.10.2008. With permission.)

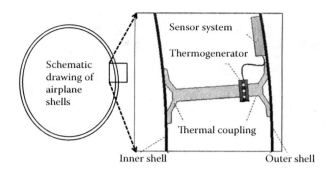

FIGURE 17.12 Sketch of the design for the thermogenerator powered energy autarkic system used in an aircraft.

the cold inside of the meat. With a properly balanced thermal design, enough electrical power can be generated to monitor the roasting from outside the oven.

A completely different application was recently presented [37]. Aircraft maintenance is one of the major expense factors in the aviation industry. In order to reduce these maintenance costs, elongation sensors are employed to monitor the aircraft body shell for bumps and stress. To avoid extensive cabling of these sensors as well as frequent battery replacement, which might even not be possible due to implementation of these sensors into the isolation of the plane's skin, thermoelectric generators are engaged to provide the energy required for sensor powering and wireless transmission of the data. The thermoelectric generators draw their energy from the temperature difference between the passenger cabin (20°C) and the outer air (down to −50°C) during crusing, sketch see Figure 17.12. In order to provide sufficient energy around 10 mW, the thermal integration of the generator is optimized to a minimum weight, <10 g, being used to fit the overall weight of the plane.

The main issue for this development was, as for the roast thermometer, the thermal design to meet the mentioned requirements for weight and power. FEM simulations of the current set up predict a maximum power yield of 8 mW with a weight of 10 g, see Figure 17.13. It could be shown that by optimizing (reducing) the thermal resistance on both cold and warm side, these predicted data could be nearly—7.5 mW—reached experimentally. In order to raise the output to 10 mW, the current architecture will have a weight of roughly 17 g, see Figure 17.13.

The results illustrate the maturity of the technology and additionally show the potential of thermoelectrically operated energy-autarkic sensors for applications with sufficient temperature gradients (e.g., frictional heat in bearings, temperature change due to the day/night cycle, etc.).

Common for both applications is that they have to use miniaturized devices either due to space or due to weight limitations.

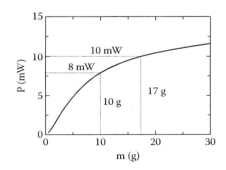

FIGURE 17.13 Power produced by the thermogenerator as a function of the total mass of the thermal coupling.

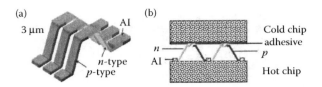

FIGURE 17.14   (a) Design of the large topography thermocouple (three thermocouples are shown). (b) Schematic of thermoelectric energy harvester (cross section view). (From Su, J. et al. *Micro Eng*, 87, 1242–1244, 2010. With permission.)

In addition to the mainstream, V_2VI_3-based miniaturized thermoelectric devices, for some years IMEC developed harvesting systems exclusively utilizing MEMS Si technologies and processes. The thermoelectric material is, for example, n- and p-type ion-implanted SiGe [38]. The kernel is a free-standing design for the n–p-legs schematically shown in Figure 17.14.

The advantage of this development is that all steps can be adapted from silicon technology, the drawback is mentioned in the conclusion of Ref. [38]: necessary are better thermoelectric materials than SiGe, for example, Bi_2Te_3, to achieve the needed performance for application. It is up to now an open question how the technology for the design shown in Figure 17.14 can be combined with the requirements of V_2VI_3 compatible processes. In a very recent approach IMEC [39], developed self-supported thermopiles with thermoelectric polysilicon legs and Al interconnects [40]. Unfortunately up to now no device properties are reported. Identically as mentioned in Ref. [38] the authors intend to change the thermoelectric material to V_2VI_3 compounds.

17.3.3 Applications Using Thin Film Microdevices with Commercial Ripeness

There are multiple reasons for using thin film thermoelectric generators in energy-harvesting applications:

- The desired system provides not enough space for integrating large components,
- A high output voltage from small temperature gradients across the TEG is needed to power an electronic circuit, or
- A high output power density is needed.

In the following section, applications found in publications and on the homepages of Nextreme and Micropelt are described.

Nextreme mentions battery charging, waste heat recovery from automotive exhaust, energy harvesting, trickle charging, and wireless sensor nodes as possible applications [41,42]. For testing thin film devices, evaluation kits are available. In the case of Nextreme, the UPF40 evaluation kit comes with a heater and a fan to achieve a temperature gradient across the TEG. Also two thermocouples are available for measuring the temperature gradient across the TEG (Figure 17.15a). This demonstration kit allows the evaluation of the UPF40 TEG.

As in the case of the UPF40 evaluation kit, a testing of the capabilities of the Micropelt devices for energy harvesting is possible using a simplified version of the TE-Power NODE evaluation kit (Figure 17.16a). The hot side of the MPG D751 TEG is mounted onto an aluminum heat collector. The heat flows along the thermal path from the hot side through the TEG and is rejected to the ambient using a heat sink designed for natural convection. In order to measure the direct output power and the temperatures of the hot and the cold side a "net output board 1" can be attached.

Looking at evaluation kits and demonstration systems, a thermal charger demonstration (Figure 17.15b) is published by Nextreme [43]. This system consists of an array of 16 UPF40 TEGs and an IPS THINERGY™ micro-energy cell (MEC) to store the thermally harvested energy. Details on temperature differences, used electronics such as DC–DC-boosters, and so on are not reported.

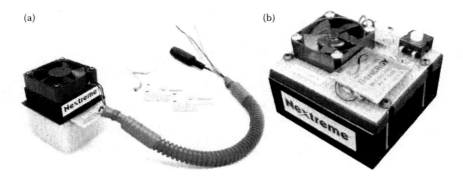

FIGURE 17.15 Thin film TEG evaluation system. (a) Nextreme UPF 40 generator evaluation kit. (b) Thermal charger demonstration.

Analog energy storage demonstrations are addressed by Micropelt's application interface module of the TE-Power NODE platform. For example a version with integrated thin ST Microelectronics EnFilm film battery (Figure 17.16b) is available [45]. Systems using IPS thin film batteries are described later.

Besides the options to measure the direct TEG output as well as hot- and cold-side temperatures of the TEG, more application relevant features such as power management, ultra-low power radios, and associated protocols are known only from Micropelt. Thus the rest of this chapter focuses on powering

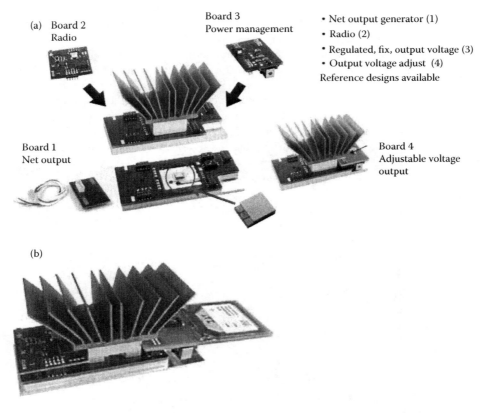

FIGURE 17.16 (a) Micropelt's modular evaluation system TE-Power NODE. (b) Custom design of the TE-Power NODE platform in combination with a ST thin film battery. (From http://www.micropelt.com/applications/te_power_node.php, 06.02. 2011. With permission.)

low-power wireless sensor systems, addressing applications from smart metering into industrial condition and process monitoring [5].

The TE-Power NODE, Figure 17.16b, provides a plug and play wireless sensor system. Sensor data provided include hot- and cold-side temperatures of the TEG and output voltage of the power source.

Since it is known that thermoharvesters are much limited in power output, the latest ultra-low-power microcontrollers and radios were implemented into Micropelt's TE-Power series. Sleep power consumption in particular was minimized as much as possible, while optimized resource management in active mode was supported by the active components. Typically the TE-Power NODE features a pluggable sensor/wireless module that integrates a TI (Texas Instruments) MSP430 microcontroller with a TI CC2500 radio. Customized versions are available using also other microcontrollers, ultra low power radios (Texas Instrument, Jennic, Atmel, etc.) as well as multiple application associated protocols (IEEE802.15.4, ZigBeePRO, RF4CE, ISA100.11a, WiHart, etc.).

A example for a customized harvesting system is the TE-Power MOTE: In cooperation with the automotive supplier A. Raymond (http://www.araymond.com -06.02.2011), two completely sealed versions of the TE-Power NOTE were developed and recently presented [46]. The liquid version is a combination of a thermal energy harvester and a quick connector for liquids as normally used for installing solar thermal systems or inside cars. The second version is designed for coupling to warm surfaces—an aluminum heat collector is used to establish the thermal contact, for example, to a power line in order to monitor the cable temperature, Figure 17.17. This allows the outdoor use of the TE-Power MODE.

Owing to the addressed applications in harsh environments the temperature shock tests between –20°C and 70°C, shaker tests according to ISO 16750-3 for engine mounted devices with up to 20 g sinusoidal vibrations between 10 Hz and 2 kHz (random) and also IP47 classification (no intrusion with a 1 mm wire, functional after placing 1 m under water) were reported [46].

Besides the multiple purpose evaluation platform, a more rigid and powerful system is available for use in industrial environments. In industrial monitoring and process automation applications, only secure—and thus energy hungry—radio protocols such as WiHart are accepted. Micropelt's TE Power PROBE is equipped with a MPG D751 TEG in combination with a larger heat sink to use the existing temperature gradients more effectively, a DC–DC booster and a cable for connecting this thermal battery to field instruments, which are normally battery driven (Figure 17.18).

The TE-Power PROBE enables the replacement of batteries used in all commercially available temperature-, pressure-, and other measurement systems that are used in process automation applications today. The TE Power PROBE is installed at a warm surface and connected to the battery port of the field instrumentation device using the cable. From effective temperature gradients as small as several 10°C, available almost anywhere in industrial applications, output powers of several milliwatts can be achieved easily allowing the continuous operation of normally battery-driven devices with an

FIGURE 17.17 Different versions of the TE-Power MODE (a) developed by A. Raymond and Micropelt (a) and tested in temperature monitoring applications at EDF (b).

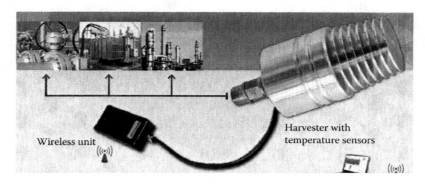

FIGURE 17.18   In industrial applications the TE-Power PROBE is connected to a hot surface, the output power is then transferred via a cable to the battery slot of an industrial, normally battery-powered field sensor.

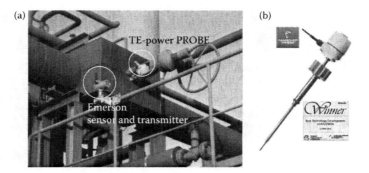

FIGURE 17.19   (a) TE-power PROBE mounted in a Shell factory and connected to a normally battery-driven Emerson sensor and transmitter. (b) ABB demonstration system for an industrial temperature transmitter with fully integrated Micropelt thermogenerators.

infinite lifetime. In order to bridge situations where no temperature gradients are available, combinations with capacitors and/or thin film batteries (showing a very low self-discharge) are possible. This approach was tested with several field instrumentation devices and also successfully tested in a field test at Shell's Den Helder factory in the framework of a cooperative project of Shell, Emerson and Micropelt using IPS thin film batteries, see Figure 17.19a [47].

In the case of industrial field devices, ABB and Micropelt developed a prototype system of an integrated thermal energy-harvesting-driven industrial temperature transmitter (Figure 17.19b): Two MPG D651 generators were integrated into an adopted temperature transmitter. Under typical operation conditions, more than enough thermal energy is available to operate the wireless sensor using the accepted WiHart protocol. This development was honored with the "Best Technology Development of RTLS/WSN" award at the IDTechEx conference 2010 in Munich [48].

17.4 Outlook and Market Aspects

If the development of miniaturized thermoelectric devices now after more than 10 years is assessed, the following statement can be made:

Technology:

- Exclusively V_2VI_3-based materials are processed to technical maturity.
- Dry deposition technologies for V_2VI_3-based materials are established.

- There is a realistic chance for cheap electroplated materials no earlier than in mid-term future.
- Mass production using other thermoelectric materials, such as V_2VI_3-based materials may be expected at the earliest in long-term future.

Market: The main use of thin film thermoelectric devices has changed from cooling to energy-harvesting applications. Energy-harvesting applications in the field of process automation and condition monitoring in industrial environments are addressed. Here the benefit for the end user is a reduced cost of ownership for additionally installed sensors needed to further optimize industrial processes. But also in consumer-type markets, such as intelligent radiator valves used to optimize energy consumption in homes and buildings, is investigated already today [49]. Besides the higher output voltages compared to bulk devices, the economy of scale associated with wafer-based fabrication processes is an essential advantage: As it is known, for example, from the semiconductor laser used in CD players or MEMS sensors formerly used in high-end automotive applications, these devices are the state of the art today. Low cost and high reliability are the main reasons for this progress. The same will hold for thermoelectric devices in the future. The main criteria for the usage of thin film devices will be better performance in the beginning—last but not least, thermoelectric, as well as all other autonomous solutions have to compete with primary batteries. At this point in time this target can only be achieved by exploiting the economy of scale of wafer-based thermoelectric devices. The same scenario described above for thermoelectric generators also holds for thermoelectric coolers—although expected the market is significantly smaller. Besides the chip hot-spot cooling markets, which will attract increased attention in the near future, today mainly applications in biomedical and life science applications are addressed by thin film coolers. Due to the high cooling power densities of thin thermoelectric thin film device, fast and accurate temperature cycling of biomedical material can be addressed. By this the commercially very attractive market of medical point of care analysis can be addressed in terms of performance and costs.

References

1. Fleurial, J.-P., Borshchevsky, A., Ryan, M.A., Phillips, W., Kolawa, E., Kacisch, T., Ewell, R. Thermoelectric microcoolers for thermal management applications, *Proc 16th Int. Conf. Thermoelectrics*, Dresden, Germany, August 1997, pp. 641–645.
2. Böttner, H., Nurnus, J., Schubert, A. Miniaturized thermoelectric converters. In: D.M. Rowe (ed.), *Thermoelectrics Handbook: Macro to Nano-Structured Materials*, Taylor & Francis, Boca Raton, FL, 2006, pp. 46.1–46.18.
3. Chowdhury, I., Prasher, R., Lofgreen, K., Chrysler, G., Narasimhay, S., Mahajan, R., Koester, D., Alley, R., Venkatasubramanian, R. On-chip cooling by superlattice-based thin-film thermoelectrics, *Nat. Nanotechnology*, pub. online: 25 January 2009 | DOI: 10.1038/NNANO.2008.417.
4. Böttner, H., Nurnus, J., Gavrikov, A., Kühner, G., Jägle, M., Künzel, C., Eberhard, D., Plescher, G., Schubert, A., Schlereth, K.-H. New thermoelectric components using microsystem technologies, *JMEMS* 13 (3), 414–420, 2004.
5. Nurnus, J. Thermoelectric thin film power generators: Self sufficient power supply for smart systems, *Smart Sensors, Actuators and MEMS IV, Proc. SPIE*, Dresden, Germany, May 4–6 2009, Vol. 7362, pp. 736205-1.
6. Hudak, N.-S., Amatucci, G.-G. Small-scale energy harvesting through thermoelectric, vibration, and radiofrequency power conversion. *JAP*, 1003, 101301, 2008.
7. Frost and Sullivan. *Advances in Energy Harvesting Technologies*, DoC2, December 2007.
8. Venkatasubramanian, R. Thin-film thermoelectric device and fabrication method of same, US 6,300,150 B1. October 9 2001.
9. Venkatasubramanian, R., Siivola, E., Colpitts, T., OQuinn, B. Thin-film thermoelectric devices with high ambient-temperature figures of merit, *Nature* 413, 597–602, 2001.
10. http://www.micropelt.com/, 13.01.2011.

11. Schlereth, K.H., Böttner, H., Schubert, A., Acklin, B. Verfahren zum Herstellen eines thermoelektrischen Wandlers, DE0019845104A1, p. date: September 30. 1998.
12. http://www.nextreme.com/pages/whats_new/news_detail/news_pressrelease_43.shtml, 12.01.2011.
13. http://www.o-flexx.com/Pressemitteilungen-O-Flexx-in-den-Medi.29.0.html 12.01.2011.
14. Semenyuk, V.A. Advances in development of thermoelectric modules for cooling electro-optic components, *Proc 22nd Int. Conf. Thermoelevtrics*, pp. 631–636, La Grand Motte, France, 2003.
15. Böttner, H., Ebling, D., Jacquot, A., König, J., Kirste, L., Schmidt, J. Structural and mechanical properties of spark plasma sintered *n*- and *p*-type bismuth telluride alloys, *Phys. Stat. Sol. (RRL)* 1(6), 235–237, 2007.
16. Böttner, H., Nurnus, J., Schubert, A., Volkert, F. New high density micro structured thermogenerators for stand alone sensor systems, *Proc 26th Int. Conf. Thermoelectrics*, Jeju, Korea, June 2007, pp. 311–314.
17. Enocean: http://www.enocean.com/de/white_papers/ 13.01.2011.
18. Glatz, W., Schwyter, E., Durrer, L., and Hierold, Ch. Bi_2Te_3-based flexible micro thermoelectric generator with optimized design, *JMEMS*, 18, 763–772, 2009.
19. Stordeur, M., Stark, I. Low power thermoelectric generator-self-sufficient energy supply for micro systems, *Proc. 16th Int. Conf. Thermoelectrics*, Dresden, Germany, August 1997, pp. 575–577.
20. Goncalves, L.M., Rocha, J.G., Couto, C., Alpuim, P., Gao Min, Rowe, D.M., Correia, J.H. Fabrication of flexible thermoelectric microcoolers using planar thin-film technologies, *JMM*, 17, 168–173, 2006.
21. Rocha, R.P. Carmo, J.P., Goncalves, L.M., Correia, J.H. An energy scavenging microsystem based on thermoelectricity for battery life extension in laptops, *Proc 35th Annual Conference of IEEE Industrial Electronics (IECON 2009)*, 2009, pp. 1813–1816.
22. Boulanger, C. Thermoelectric material electroplating: A historical review, *JEMS*, 39(9), 1818–1827, 2010, DOI: 10.1007/s11664-010-1079-6.
23. www.micropelt.com, 12.01.2011
24. Rowe, D.M. Miniature semiconductor thermoelectric devices. In: D.M. Rowe (ed.), *CRC Handbook of Thermoelectrics*, Taylor & Francis, Boca Raton, 1995, pp. 441–458.
25. Nextreme Thermal Solutions, White paper: The thermal copper pillar bump: Enabling improved semiconductor performance without sacrificing efficiency, *Nextreme Thermal Solutions*, January 9, 2008.
26. Nurnus, J. State of art thermoelectric thin film devices, *NanoS—The Nanotech Journal* 01.08, Wiley-VCH Verlag, Weinheim, pp. 11–15, 2008.
27. Venkatasubramanian, R., Siivola, E., T. O'Quinn. In: D.M. Rowe (ed.), *Thermoelectrics Handbook: Macro to Nano*, Taylor & Francis, Boca Raton, FL, 2006, pp. 46.1–46.18.
28. http://www.nextreme.com/media/pdf/techspecs/Nextreme_eTEG_HV56_Data_Sheet.pdf, 06.02.2011
29. http://www.micropelt.com/products/datasheets.php, 06.02.2011.
30. http://www.nextreme.com/media/pdf/techspecs/Nextreme_OptoCooler_HV14_Data_Sheet.pdf, 06.02, 2011.
31. http://www.micropelt.com/products/mypelt.php, 06.02.2011.
32. http://www.micropelt.com/down/datasheet_mpc_d403_d404.pdf, 06.02.201.
33. http://www.micropelt.com/down/cooling_and_heating.pdf, 06.02.2011.
34. http://www.nextreme.com/media/pdf/Nextreme_PCR_Process_Press_Release_Jan09.pdf
35. http://www.micropelt.com/applications/biomedical.php, 06.02,2011.
36. Baier, M. Temperatursonde für einen Ofen, Ofen und Verfahren zum Betrieb eines Ofens, WO 2008/119440 A2, 09.10.2008.
37. Bartholome, K., Bartel, M., Binninger, R., Schumacher, I., Schröder, H., Ebling, D. Energy-autarkic sensor technology in aircraft, *Proc 8th ECT*, Como, Italy, September. 2010.

38. Su, J., Vullers, R.J.M., Goedbloed, M., van Andel, Y., Leonov, V., Wang, Z. Thermoelectric energy harvester fabricated by stepper, *Micro Eng*, 87, 1242–1244, 2010.
39. Patent EP 1976034.
40. Van Andel, Y., Jambunathan, M., Vullers, R.J.M., Leonov, V. Membrane-less in-plane bulk-micromachined thermopiles for energy harvesting, *Micro Eng*, 78, 1294–1296, 2010.
41. http://www.nextreme.com/pages/power_gen/eteg_upf40eval.shtml, 06.02.2011.
42. http://www.nextreme.com/pages/power_gen/eteg_img1_kit.shtml, 06.02.2011.
43. http://www.nextreme.com/thermalcharger, 06.02.2011.
44. http://www.micropelt.com/applications/te_power_node.php, 06.02.2011.
45. http://www.st.com/internet/com/press_releases/t3020.jsp, 06.02.2011.
46. Fräulin, C., Nurnus, J. Energy autonomous sensor systems for automotive condition monitoring, *2nd International Conference: Thermoelectrics Goes Automotive*, Berlin, Germany, 9–10.12.2010.
47. http://www.micropelt.com/down/infinite_wireless_power_supply.pdf, 06.02.2011.
48. http://www.idtechex.com/energyharvestingandstorageeurope10/en/awards.asp, 06.02.2011.
49. http://www.tab.de/download/275819/TAB_Technik_am_Bau_Fachbeitrag_MSR_Technik_9409_84.pdf, 06.02.2011.

18

Application of Thermoelectrics for Thermal Management of High-Power Microelectronics

Vladimir A.
Semenyuk
Thermion Company

18.1 Introduction

Many powerful electronic devices along with executing their peculiar functions act as high-density heat sources. Such are microprocessors (CPUs), semiconductor diodes and lasers, power amplifiers, and other intensive electronics. The CPU is the most typical representative of this family. This is a mainstream device in the present world, which is installed today in almost all modern instruments. The trend to increase processor performance leads to the permanent increase in the circuit integration, accompanied with the growth of chip power. This process is observed during past 30 years, and it will be continued in this century. Figure 18.1 shows the trend of the electrical power consumption in a CPU chip. It is seen that CPU power doubles every two years due to doubling of transistors in accordance with Moore's Law. This is why the thermal management of modern CPUs becomes a vital issue.

Two approaches to processor temperature control are available:

- Passive cooling—the use of a heat sink (a sole heat exchanger or that coupled with a fan).
- Active cooling—the use of a heat sink (as said above) with a thermoelectric cooler (TEC).

Passive cooling methods progressed in the past three decades along with the increase in CPU power consumption. Cooling systems evolved from sole fin in 1980s to the system of fin and fan in early 1990s and finally to the combination of fin, fan, and a heat pipe in personal computers (PCs) after 2000. The

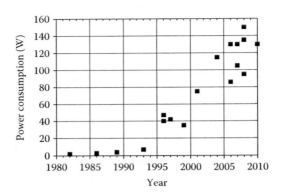

FIGURE 18.1   Trend of electric power consumption of a processor chip (www.intel.com).

dimensional restrictions typical for modern PCs set a physical limit to the efficiency of the passive cooling system, and this becomes the main constraint to further increase in CPU power.

At first sight, the application of thermoelectrics potentially looks like the best means to overcome this barrier. However, the practical realization of this concept meets substantial difficulties. There are two main problems. First, the heat flux density for a modern processor (typically around 100 W/cm²) is considerably higher than that for a standard TEC, and this creates a problem of the TEC and processor dimensional mismatch. Besides, the TEC itself is an energy consumer, so its usage leads inevitably to further growth of the total power dissipation that increases the problem of heat rejection. As a result, the awaited improvement can become unattainable—the rise of the processor temperature can take place instead of its cooling.

The problem of CPU thermoelectric cooling was studied in many papers.[1–9] Most often, the use of thermoelectrics is interpreted as a means to improve heat sink performance.[2–5] In some papers, rather disputable assertions are stated, and ungrounded model simplification are made, which leads to impracticable or even wrong technical solutions. Sometimes, a comparison of passive and active cooling is incorrectly made in dissimilar conditions (at different thermal constraints) in favor of thermoelectrics. The proper temperature difference at the TEC faces is proclaimed to be very small[2,3] and even reversed[4] (the last is wrong in principle). At that, the TEC coefficient of performance (COP) is declared to be as high as 2–4 and even 20, and this is suggested as a feasible condition for the considerable increase in processor power.[2,5]

In practice, the matter is much more complicated. The COP maximum mode of a TEC operation at small temperature differences is characterized with utterly low cooling power density and, vice versa, high cooling capacity leads to the increased power consumptions and low COP value. Hence, both extremes are not fit to the case. This is why, in spite of great necessity, thermoelectric technology still did not become widespread for thermal management of powerful microprocessors. Apart from the objective difficulties, the reason is that no adequate system model exists and no general approach is proposed yet, which would specify the conditions of a TEC advantageous application. Two important questions are still not answered:

- Does the use of a TEC can actually improve processor performance under typical thermal constraints for modern PC?
- If yes, what maximum improvement can be achieved?

This study gives answer to both questions. The thermal model of the integrated system composed of a microprocessor, a TEC, and a heat exchanger is proposed and the influence of different factors such as local thermal resistances of heat spreader and thermal interface materials is analyzed. The general requirements are formulated that have to be satisfied to provide the extension of the processor's power over that attainable with passive cooling. It is shown that the optimal coordination of all system elements is the only way to meet these requirements. For the practical verification of the developed method, the Intel Core i7-800 desktop processor with the rated heat output of 95 W was considered.[10] The prospects to increase its maximum power using active cooling are studied. Two possible system options are considered: with the TEC at the outer case surface and with the TEC inside of the heat spreader.

Another challenge that faced the CPU developers is the existence of one or more on-chip "hot spots" in the processor area, which generates extremely intensive heat flux. The presence of such localized heat sources becomes the dominant challenge in IC technology and this stimulates researchers to develop the most refined thermal solutions. Effective suppression of such sub-millimeter hot spots along with the whole chip uniform cooling seems to be a reasonable technique. The use of thermoelectric microcoolers for this purpose was considered recently by different scientific groups[11–13] with the conclusion that this novel approach can lead to the selective hot-spot suppression that can contribute greatly to a processor operation life and efficiency. Just this circumstance encouraged the author to make a thorough analysis of this advanced thermal solution.[14]

In this study the results of detailed estimations are given concerning the attainable maximum efficiency of the cooling technique under discussion. The method of optimal thermal integrating of a micro TEC into a processor-to-heat sink interface is developed. A typical Intel processor is considered with a powerful hot spot in the center of the Si die. Different TEC configurations are reviewed, including traditional bulk micro TECs and film-type micro coolers based on standard bismuth–telluride alloys and their nanostructures. The optimal TEC geometry and its operational mode are found that can provide minimal hot-spot temperature. Some important factors that are neglected in other studies are taken into account, which makes obtained results more realistic. The advantages and limitations of this approach are discussed.

18.2 Chip Uniform Cooling

18.2.1 Configuration of Cooling Systems

Figure 18.2 shows a typical configuration of the processor assembly with passive cooling. A silicon die is interfaced to a copper heat spreader, which is interfacing a heat sink base. All parts are contacted through thermal interface materials TIM1 and TIM2 insuring their efficient thermal connection. The heat flow Q generated in the die having temperature T_d goes through mentioned components to the ambient with the temperature T_a. The total thermal resistance of the thermal path $R_t = (T_d - T_a)/Q$ includes components $R_{t1} = (T_d - T_c)/Q$ and $R_{t2} = (T_c - T_a)/Q$, where T_c is the temperature at the center of the heat spreader outer surface (case temperature).

The active cooling can be arranged in two architectures. The first one is shown in Figure 18.3a. The TEC is located between the heat spreader and the heat sink. This involves an additional TIM2 layer at the TEC cold side. The second architecture (Figure 18.3b) has a TEC inside of the heat spreader interfacing directly with the die. In this case, an additional TIM1 layer at the TEC cold side arrives.

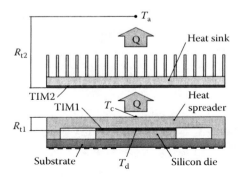

FIGURE 18.2 Scheme of the processor assemblage. (Fan is not shown.)

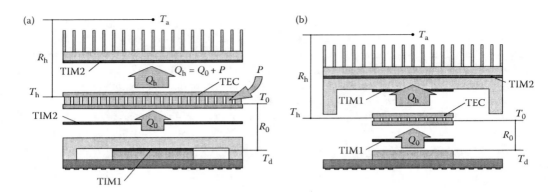

FIGURE 18.3 Two architectures of processor active cooling. (a) Option 1: the TEC outside of the processor package, (b) Option 2: the TEC inside of the processor package.

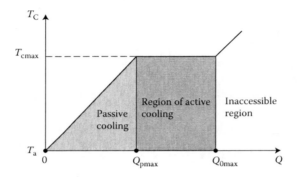

FIGURE 18.4 Typical dependence of the processor temperature on its power for different cooling methods.

The TEC pumps heat $Q_0 = Q$ from the processor and dissipates heat $Q_h = Q_0 + P$ at its hot side, where P is the TEC input power. Thus, the heat load at the heat sink is increased as compared with the case of the passive cooling, and the heat sink superheat over ambient temperature increases proportionally.

Figure 18.4 shows typical dependence of the processor temperature on its power. The left region $0 < Q < Q_{pmax}$ of the figure corresponds to the passive cooling option. With given heat sink thermal resistance R_{t2}, the case temperature grows linearly from the ambient temperature T_a at zero power up to the maximum allowed temperature $T_{cmax} = T_a + Q_{pmax}R_{t2}$ at the rated processor power Q_{pmax} (the so-called "thermal design power"). Prolonged elevation of the processor power over Q_{pmax} value is not recommended as leading to its unacceptable overheat. In practice, the processor temperature is adjusted constant at T_{cmax} level even at reduced power (dashed line in Figure 18.4) using fan speed control.

It is awaited that with the use of active cooling the processor power may be increased potentially to a certain $Q_{0max} > Q_{pmax}$ value, its temperature being kept unchanged (the intermediate region in Figure 18.4). To the right of this region, the constant processor temperature fails to be maintained because of a hindered heat rejection from the TEC hot side. The question to be answered is as follows: what TEC configuration and its operation parameters will provide the maximum increase in the processor power?

18.2.2 Theoretical Approach

18.2.2.1 More about Controlled Temperature

For further consideration, it is important to define more exactly the point at which the temperature must be controlled. In the practice of passive cooling, controlled is the case temperature T_c at its maximum level T_{cmax} corresponding to the rated processor power Q_{pmax}.[10] Simultaneously, the die temperature T_d is also kept constant at its maximum allowed level, being uniquely connected with the case temperature by the simple relation:

$$T_{dmax} = T_{cmax} + Q_{pmax}R_{t1},$$ (18.1)

where R_{t1} is the thermal resistance of the thermal path from the die to the heat sink through TIM1 and the heat spreader (Figure 18.2). As the purpose is to intensify processor operation, it would be reasonable to keep the die itself at the safe temperature T_{dmax} whatever increase in its power may be achieved. This is why in this study just the T_{dmax} value, given by relation (18.1), will be considered as a target for monitoring in both versions of the TEC location.

18.2.2.2 Mathematical Model

The problem reduces to the search of the maximum of the TEC cooling power Q_0 given by the equation

$$Q_0 = F\left(\alpha i T_0 - \frac{1}{2}i^2(\rho l + 2R_c) - \frac{\lambda}{l}(T_h - T_0)\right)$$ (18.2)

under the equality-type constraints

$$\begin{aligned} T_0 &= T_{dmax} - Q_0 R_0 \\ T_h &= T_a + Q_h R_h \\ Q_h &= F\left(\alpha i T_h + \frac{1}{2}i^2(\rho l + 2R_c) - \frac{\lambda}{l}(T_h - T_0)\right) \end{aligned}$$ (18.3)

which define the interconnection of the TEC cold-side and hot-side temperatures T_0 and T_h with corresponding heat flows Q_0 and Q_h at the TEC junctions. Here F is the total cold/hot junctions area; i is the electrical current density; l is TE leg length; R_c is electrical contact resistance; α, ρ, λ are the Seebeck coefficient, electrical resistivity, and thermal conductivity of TE material, respectively; R_0 and R_h are the thermal resistances at the TEC cold and hot sides, respectively.

The following initial data in Equations 18.2 and 18.3 will be treated henceforth as specified:

- Die controlled temperature T_{dmax}
- Ambient temperature T_a
- Thermal resistance at the TEC hot and cold side R_h and R_0
- Thermoelectric parameters α, ρ, λ

We will also regard the TEC junctions area F as a specified parameter, having in mind that it has to be matched with the actual system element dimensions, while i and l will be considered as independent variables. The goal is to find such $i = i_0$ and $l = l_0$ values, which give the maximum to the function $Q_0(i,l)$ under additional constraints (18.3). In such a statement, this is a mathematical problem of linear programming, which is solvable by standard numerical methods.

18.2.3 Prospects to Improve Performance of the Intel Processor

The Intel Core i7-800 desktop processor was chosen as an object of practical application of the proposed design method. According to the Intel Thermal/Mechanical Specifications and Design Guidelines,[10] the maximum case temperature T_{cmax} of this processor is stated to be of the order of

72.5°C. The recommended upper limit of the processor power Q_{pmax} is defined as that providing elevation of the case temperature from the ambient temperature of 45°C to its maximum value of 72.5°C in steady-state conditions. With a target heat sink thermal resistance of 0.29 K/W, this defines the thermal design power of 95 W.[10] Any attempt to operate the processor outside these operating limits may result in permanent damage to the processor and potentially to other components within the system. The goal of this study is to find a possible extension of the processor power when using active cooling, the case temperature being kept unchanged. Below are given the stages of the study.

18.2.3.1 Definition of Thermal Resistances

18.2.3.1.1 Passive Cooling

To define the actual processor temperature T_d, it is necessary to calculate thermal resistance R_{t1} of the path through TIM1 and the heat spreader. The calculations are made using the ANSYS 10.0 program. Considered are the thermal and mechanical characteristics of the processor components, given in Tables 18.1 and 18.2. The three-dimensional thermal model was used with uniform heat flow density at the die 11×13 mm inactive area and constant temperature at the heat spreader 31×31 mm outer surface. Lateral faces of the assembly were regarded as insulated adiabatically. Calculated R_{t1} value proved to be of 0.0267 K/W. With $T_{cmax} = 72.5$°C and $Q_{pmax} = 95$ W, this gives, according to formula 18.1, the actual processor temperature T_{dmax} of 75°C. This very temperature is accepted hereafter to be kept constant at any processor power for both active cooling options.

18.2.3.1.2 Active Cooling

Considered are two options:

- Option 1: 30×30 mm TEC at the case outer surface
- Option 2: 13×13 mm TEC inside of the package, attached directly to the die

The total junctions area F is accepted to be 6.4 cm² for the option 1 and 1.17 cm² for the option 2 what corresponds to rather high pellets packing density (~0.7).

As a first step, the thermal resistances of additional elements must be taken into account with respect to their dimensions and thermal characteristics. The dimensional details and calculated thermal resistances at the TEC cold and hot faces are given in Tables 18.3 and 18.4. For different components considered were 1-D or 3-D thermal models, depending on the thermal path configuration. For a 1-D model the equality $R_{ti} = \delta_i/(\lambda_i F_i)$ is used, where δ_i is the ith component thickness, λ_i is its thermal conductivity, and F_i the component surface area. For 3-D models, accepted are uniform heat flux density at the heat

TABLE 18.1 Thermal Conductivity of the Used Materials

Material	Copper	TIM1 (In solder)	TIM2 (Arctic Silver 5)[15]	AlN	Si
λ (W/m-K)	380	75	8.9	180	110

TABLE 18.2 System Geometry and Thermal Parameters of the Intel Core i7-800 Processor (Passive Cooling Version)

Sub-Assembly	Thermal Path Components	Material	Geometry (mm)	Thermal Model	Thermal Resistance (K/W)
Processor Package	TIM1	Indium solder	$11 \times 13 \times 0.03$	3-D	$R_{t1} = 0.0267$ (calculated)
	Heat spreader	Copper	$31 \times 31 \times 1.5$		
Heat Sink	Heat sink with TIM2 layer	Assemblage		Not applied	$R_{t2} = 0.29$ (specified)

TABLE 18.3 Active Cooling, Option 1 (30 × 30 mm TEC outside of the Case): System Geometry and Thermal Parameters

TEC Interface	Thermal Path Components	Material	Dimensions (mm)	Thermal Model	Thermal Resistance (K/W)
Cold side	TIM1	Indium solder	11 × 13 × 0.03	3-D	$R_0 = 0.061$
	Heat spreader	Copper	31 × 31 × 1.5		
	TIM2	Arctic Silver	30 × 30 × 0.05		
	TEC cold substrate	AlN	30 × 30 × 0.63		
Hot side	TEC hot substrate	AlN	30 × 30 × 0.63	1-D	$R_h = 0.2937$
	Heat sink	Assemblage ($R_{t2} = 0.29$ K/W)		Not applied	

TABLE 18.4 Active Cooling, Option 2 (13 × 13 mm TEC inside of the case): System Geometry and Thermal Parameters

TEC Interface	Thermal Path Components	Material	Dimensions (mm)	Thermal Model	Thermal Resistance (K/W)
Cold side	TIM1	Indium solder	11 × 13 × 0.03	1-D	$R_0 = 0.0127$
	TEC cold substrate	AlN	13 × 13 × 0.3		
	TEC hot substrate	AlN	13 × 13 × 0.3	3-D	$R_h = 0.3227$
Hot side	TIM1	Indium solder	13 × 13 × 0.03		
	Heat spreader	Copper	31 × 31 × 1.5		
	Heat sink	Assemblage ($R_{t2} = 0.29$ K/W)		Not applied	

flow inlet and uniform temperature at the heat flow outlet, the side surfaces are regarded to be insulated adiabatically.

18.2.3.2 Results of the TEC Optimization

The calculated quantities T_{dmax}, R_0, R_h and specified F values are used as input data for Equations 18.2 and 18.3. Calculations are made for bismuth–telluride-based materials with the following thermoelectric parameters: $\lambda = 1.4 \times 10^{-3}$ W/cm K, $\alpha = 200$ μV/K, $\sigma = 1/\rho = 1050$ Ω^{-1} cm^{-1}. This corresponds to the figure of merit $z = \alpha^2/(\rho\lambda) = 3 \times 10^{-3}$ K^{-1}. Different R_c values are tried. Results of calculations for both options are given in Figures 18.5 through 18.9 and Table 18.5.

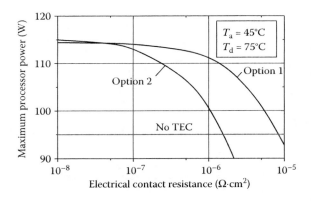

FIGURE 18.5 The maximum processor power versus electrical contact resistance. The die is kept at 75°C.

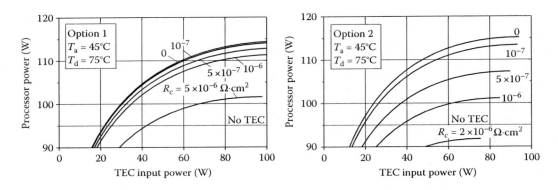

FIGURE 18.6 Dependence of the processor power on the TEC input power for different values of contact electrical resistance (Options 1 and 2).

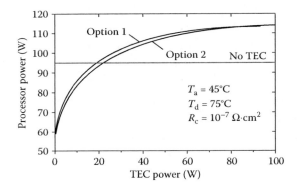

FIGURE 18.7 Processor power versus TEC input power.

Figure 18.5 shows the dependence of the maximum processor power pumped by the TEC on contact electrical resistance. Corresponding optimal TEC characteristics for both optional variants are given in Table 18.5.

It is seen from Figure 18.5, that a 21% increase in the processor power (from 95 to 115 W) is potentially achievable when using active cooling. To obtain this result, the electrical contact resistance below

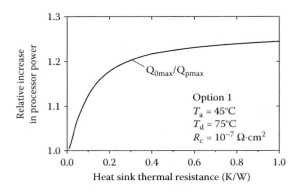

FIGURE 18.8 Advantage of the processor active cooling at different heat sink thermal resistances.

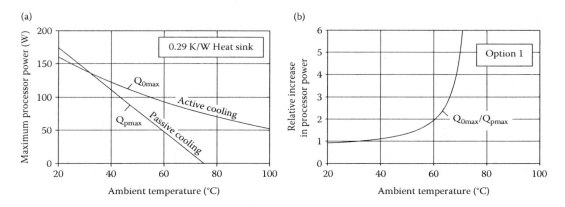

FIGURE 18.9 Advantage of active cooling at elevated ambient temperature. Si die is kept at 75°C. (a) Maximum processor power vs. ambient temperature; (b) Relative increase in processor power.

10^{-7} Ω cm^2 must be provided. The growth of R_c value leads to a considerable reduction of the TEC efficiency. For option 2, dramatic deterioration of the TEC performance is observed at R_c values just over 10^{-7} Ω cm^2. As regards option 1, the rather high heat pumping capacity retains up to the R_c of 10^{-6} Ω cm^2. The reason for such a difference becomes evident when considering TE pellets' optimal dimensions (Table 18.5). For the TEC integrated into the heat spreader (option 2), the optimal TE leg must be as short as 60–70 μm, while in the case of the outside TEC location, the TE pellet is 6 times longer. This is why the level of the contact electrical resistance is extremely critical for option 2. The utmost R_c values are 9×10^{-6} Ω cm^2 for option 1 and 1.5×10^{-6} Ω cm^2 for option 2. A further increase in contact resistance completely excludes TEC's applicability.

From the point of practical implementation, option 1 with the TEC located outside of the processor package is a feasible one because its optimal TE pellet height (300–400 μm) lies inside of the feasible dimensional range. On the contrary, option 2 is problematic for realization because its

TABLE 18.5 Optimal TEC Characteristics

R_c (Ω cm^2)	l_0 (μm)	i_0 (A/cm^2)	P (W)	Q_{0max} (W)	Q_{0max}/Q_{pmax}	COP
Option 1: $F = 6.4$ cm^2, $R_0 = 0.061$ K/W, $R_h = 0.2937$ K/W						
0	325	616	108.8	114.5	1.2	1.05
10^{-8}	326	615.8	108.8	114.4	1.2	1.05
5×10^{-8}	327	614.4	108.7	114.3	1.2	1.05
10^{-7}	328	612.7	108.6	114.1	1.2	1.05
5×10^{-7}	335	600.2	107.7	112.9	1.19	1.05
10^{-6}	343	585.6	106.6	111.4	1.17	1.04
2×10^{-6}	358	559.7	104.4	108.6	1.14	1.04
5×10^{-6}	393	498.8	97.65	101.6	1.07	1.04
Option 2: $F = 1.17$ cm^2, $R_0 = 0.0127$ K/W, $R_h = 0.3227$ K/W						
0	57	3211	93.9	115.2	1.21	1.23
10^{-8}	57	3202	93.9	115.0	1.21	1.22
5×10^{-8}	58	3164	93.6	114.3	1.2	1.22
10^{-7}	59	3119	93.2	113.4	1.19	1.22
5×10^{-7}	65	2828	89.6	107.3	1.13	1.20
10^{-6}	70	2561	84.7	101.1	1.06	1.19

$T_a = 45$°C, $T_d = T_{dmax} = 75$°C.

inherent range of TE pellet heights (60–70 μm) is currently inaccessible both for bulk and film-type technology.

18.2.3.3 Effect of Deviation from the Optimum

It is seen from Table 18.5 that the COP values for the optimized TECs lay in the range from 1 to 1.2. This means that the TEC power must be rather high, being comparable with that of the processor. In this connection, it is important to define in what degree the reduction of the TEC power can affect its heat pumping capacity. To that end, the cooling capacities of the optimized TECs are calculated at different i values in the range $0 < i < i_0$. Equations 18.2 and 18.3 are used with the F and I_0 values from Table 18.5 as input data. The calculation results for different R_c values are shown in Figure 18.6. It is seen clearly that the TEC power from 60 to 80 W is just the range that can be recommended for economical TEC operation. Further elevation of the TEC power is unreasonable as giving negligible effect.

Figure 18.7 shows the $Q_0(P)$ dependence in the wide range of the TEC power variation at $R_c = 10^{-7}\ \Omega\ cm^2$. For both options, the dependences are subjected to the condition of the constant die temperature of 75°C. It is seen that the presence of the TEC itself greatly changes the system parameters. With the incorporated TEC switched off ($P = 0$), the die temperature reaches its allowed limit with the processor power as low as 60 W. This is by 35 W less than the rated power. The reason is in the additional thermal resistance of the disabled TEC. Hence, to support processor temperature at a safe level, the TEC must be switched on immediately at $Q > 60$ W. To reach the processor rated power of 95 W, the TEC power must be increased to 20 W. This is the inferior limit of the TEC power in this application—the minimal pay-off for the TEC presence. The most reasonable range of P values appears to be from 60 to 80 W, which can give 16–19% increase in the processor power over its rated value. The corresponding range of a TEC COP is 1.4–1.8 and this is much less than that stated earlier.[2,5]

18.2.3.4 Effect of the Heat Sink Thermal Resistance

Figure 18.8 shows the dependence of the relative increase in the maximum processor power $\delta = Q_{0max}/Q_{pmax}$ on the heat sink thermal resistance R_{t2}. In the passive cooling mode, the rated processor power Q_{pmax} increases with R_{t2} reduction according to the formula:

$$Q_{pmax} = (T_{dmax} - T_a)/(R_{t1} + R_{t2}) \tag{18.4}$$

where T_a, T_{dmax}, R_{t1} are considered as constant parameters. This dependence was taken into account when calculating δ value.

It is seen that the active cooling is the most effective in the region of relatively high R_{t2} values (over 0.25 K/W). Here over a 20% increase in the processor power is achievable. With improvement of heat transfer conditions, efficiency of the active cooling reduces and at $R_{t2} < 0.1$ K/W its application becomes unprofitable.

18.2.3.5 Influence of the Ambient Temperature

The ambient temperature dramatically affects the processor performance (Figure 18.9). For the case of passive cooling, the maximum processor power decreases linearly with the growth of the ambient temperature according to the formula (18.4) and at $T_a \geq 75$°C the passive cooling becomes inapplicable in principle (Figure 18.9a), although at active cooling the $Q_{0max}(T_a)$ line has also a depressed trajectory, and its slope is more sluggish. As a result, the effect of active cooling quickly grows with T_a elevation. At $T_a = 60$°C, the processor power can be doubled, and at $T_a \geq 75$°C thermoelectric cooling becomes the only means for processor thermal management. Even at $T_a = 100$°C, the processor power of over 50 W can be provided. Thus, the use of active cooling is the most effective in special cases of high ambient temperature operation.

18.2.4 Experiment

The theoretical approach described above was used for developing a prospective thermoelectric system for cooling a powerful CPU. A series of experiments was undertaken aimed to define the additional processor power that can be implemented using TEC with a standard Intel heat exchanger as compared to the passive cooling with the same heat sink.

18.2.4.1 Prototype and Test Setup

The parts of the prototype assemblage are shown in Figure 18.10.

As a processor simulator, a dummy element was used in a form of a 11 × 13 × 0.5 mm alumina plate with a thin-film resistive heater at its surface. The plate was soldered to the heat spreader separated by a thin 11 × 13 mm copper spacer. As a heat sink, the standard Intel CPU cooling unit Alpine 7 was used whose thermal resistance is specified as of 0.26 K/W. The copper spacer was supplied with a narrow 0.55 mm deep groove on its surface interfaced to the heat spreader for placing the measuring thermocouple in the point where the actual processor die should be located. Analogous grooves for measuring thermocouples were machined at the heat spreader outer surface and at the surface of the heat sink base.

30 × 30 mm TEC was designed using optimization procedure described above. The TEC consists of two 0.63 mm thick aluminum nitride substrates and 360 TE pellets each 0.3 mm thick and 1.4 × 1.4 mm in cross section. The bismuth–telluride-based extruded thermoelectric materials supplied by the SCTB NORD, Russia, were used in the prototype. Prior to the TECs assembling, the figure of merit Z of these materials was measured in the temperature region from 300 to 370 K using the Harman method. However, at room temperature the materials show a rather high figure of merit (2.8×10^{-3} K^{-1}), at elevated temperatures their performance decreases considerably (Figure 18.11), and in the foreseen region of the TEC operation (70–90°C), Z-factor reduces to 2.32×10^{-3} K^{-1}. For the TEC with TE legs as short as 0.3 mm, Z value suffers a further decrease due to the influence of electrical contact resistance. With $R_c = 10^{-6}$ Ω cm^2, an effective Z value at the device level is estimated to be of 2.2×10^{-3} K^{-1}.

The prototype was tested in a chamber with controlled temperature. The test setup is shown in Figure 18.12. Three DC sources were used to supply power to the heater, the TEC and the fan. Measured were the temperature of ambient air T_a (at the fan inlet), the temperature of the heat sink base T_b (inside of the groove at its surface) the case temperature T_c (inside of the groove at the heat spreader) and the temperature T_d (inside of the groove in the copper spacer).

18.2.4.2 Experimental Procedure, Results, and Discussion

As a first step, the prototype without TEC was tested in the passive cooling mode with the heat spreader attached directly to the heat sink. Arctic Silver 5 thermal grease was used as a thermal interface material

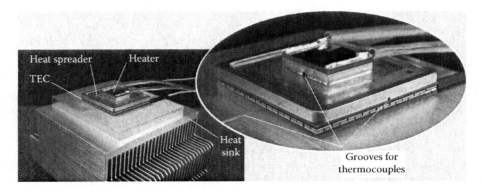

FIGURE 18.10 **(See color insert.)** Photographs of the prototype parts.

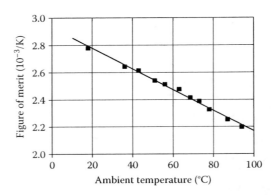

FIGURE 18.11   The temperature dependence of Z-factor for used TE materials.

TIM2. Measurements were made at different ambient temperatures in the range from 300 to 350 K. At each T_a value the maximum electrical power of the heater Q_{pmax} was found as that providing the upper limit of the dummy element temperature $T_{dmax} = 75°C$. All measurements were made on achieving steady-state conditions. Corresponding dependence $Q_{pmax}(T_a)$ for the passive cooling version is shown in Figure 18.13. As a side effect of this test, the actual heat sink thermal resistance R_{t2} was measured as 0.33 K/W. This is considerably higher than that given in the specification. The measured total resistance of the thermal path from the heater to ambient $R_t = R_{t1} + R_{t2}$ turned out to be of 0.4 K/W against to 0.317 K/W supposed in our theoretical approach (Table 18.2). As a result, the rated processor power of 95 W can be achieved on the condition that ambient temperature is lowered from specified 45 to 38°C.

The second series of experiments was made with the TEC integrated into the assemblage. To find the maximum attainable heater power in this case, the following test procedure was used:

1. The TEC is deactivated, the heater is switched on, and its power is increased to obtain limit temperature $T_{dmax} = 75°C$.
2. Then the TEC is switched on and its power increased to reduce the heater temperature to minimum.

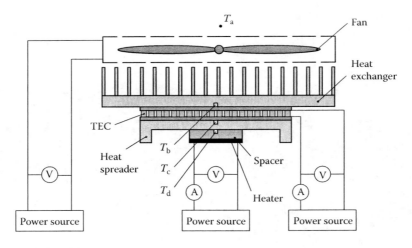

FIGURE 18.12   Scheme of the test setup.

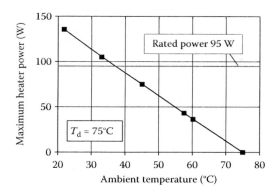

FIGURE 18.13 Dependence of maximum processor power on ambient temperature (passive cooling).

3. Next the heater power must be enlarged again to obtain limit temperature $T_d = T_{dmax}$ and so on with alternate increase in the TEC and heater power.

The absolute maximum of the heater power was considered as searched out when a small increase in the TEC power fails to reduce heater temperature below its limit value. The dependence of the heater power on the TEC power consumption registered during this procedure at the ambient temperature of 45°C is shown in Figure 18.14. The horizontal line $Q = 75$ W corresponding to the passive cooling with the same heat sink is also given for comparison. It is seen that the use of a TEC increases heater power from 75 to 90 W (by 20%). This result is achieved with 40 W power inputs, which corresponds to the TEC COP of 2.25.

An analogous experiment was carried out at different ambient temperatures. Corresponding dependences $Q_{pmax}(T_a)$ both for passive and active cooling versions are shown in Figure 18.15.

Although the heat sink thermal resistance and performance of the used TE materials considerably differ from those accepted in theoretical estimations, obtained experimental results in Figures 18.14 and 18.15 have predicted character and are in full qualitative agreement with theoretical dependences in Figures 18.7 and 18.9. It is evident that below $T_a = 35°C$ the use of a TEC becomes unreasonable but with elevation of ambient temperature advantage of active cooling quickly grows and at 60°C the processor power can be doubled. Even at $T_a = 75°C$, when passive cooling becomes unfit, the use of a TEC can provide 50 W cooling power.

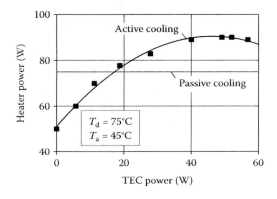

FIGURE 18.14 Dependence of the heater power on the TEC power consumptions.

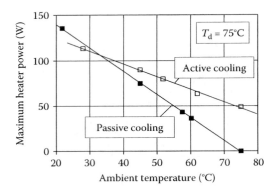

FIGURE 18.15 Comparison of active and passive cooling at different ambient temperatures. The heater is kept at 75°C.

18.3 On-Chip Hot-Spot Cooling: Forecasts and Reality

Along with the distributed heat dissipation, one or more on-chip "hot spots" can exist in the processor active area, which are characterized by extremely high heat flux density. In this section, the problem of suppression of such a sub-millimeter hot spot at the active surface of the silicon die using thermoelectric micro-cooler integrated into the heat spreader is studied with the goal to determine maximum attainable efficiency of this technology.

18.3.1 Processor with the Integrated Micro-TEC

18.3.1.1 Chip Package Model

Figure 18.16 gives a schematic illustration of a chip package. A silicon die with a central hot spot at its active side is located at the copper heat spreader, which is interfaced with a heat sink base. The parts are contacted through thermal interface materials TIM1 and TIM2. The micro-TEC is accommodated inside of the square cavity made in the heat spreader. A thin copper plate with a small tip at its top

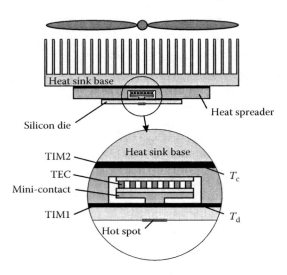

FIGURE 18.16 Scheme of a chip package with integrated micro-cooler.

surface (the so-called "mini-contact")[11] is attached to the TEC cold side to provide an efficient thermal contact with the Si die through the TIM1 layer.

The package parts dimensions are shown in Table 18.6. The heat generated in the processor is modeled as an evenly distributed heat flux of 65 W/cm² on 11 × 13 mm silicon die active surface (total 93 W) and a uniform heat flux of 1250 W/cm² on a 0.4 × 0.4 mm hot spot, what gives additional 2 W of heat dissipation, the overall heat emission being of 95 W. As in the previous case, we will ignore the details of the heat sink construction. Instead, we suppose that its thermal resistance R_{12} is of 0.29 K/W and the heat spreader surface interfaced to the heat sink has a uniform temperature. With ambient at 45°C and $Q = 95$ W, the case temperature T_c and the die temperatures T_d reach their maximums $T_{cmax} = 72.5°C$ and $T_{dmax} = 75°C$.

To get an idea about the overheating that can take place in the hot-spot region, we consider a simplified die thermal model with uniform heat flux q in the hot-spot area and uniform temperature T_{dmax} at the die inactive surface. An analytical solution of Laplace's equation under aforesaid boundary conditions gives the following relation for three-dimensional distribution of excessive temperatures $\theta(x,y,z) = T(x,y,z) - T_{dmax}$ within the die:

$$\theta(x,y,z) = q\frac{h}{\lambda_{si}}\frac{ab}{AB}\left(1 + 2S_1 + 2S_2 + 4S_3\right), \tag{18.5}$$

where

$$S_1(x,z) = \sum_{n=1}^{\infty}\frac{\sin\nu a}{\nu a}\cos\nu x\,\frac{\sinh\nu(h-z)}{\nu h\cosh\nu h}$$

$$S_2(y,z) = \sum_{n=1}^{\infty}\frac{\sin\mu b}{\mu b}\cos\mu y\,\frac{\sinh\mu(h-z)}{\mu h\cosh\mu h} \tag{18.6}$$

$$S_3(x,y,z) = \sum_{n=1}^{\infty}\frac{\sin\nu a}{\nu a}\cos\nu x\sum_{m=1}^{\infty}\frac{\sin\mu b}{\mu b}\cos\mu y\,\frac{\sinh\alpha(h-z)}{\alpha h\cosh\alpha h}$$

$$\nu = \frac{n\pi}{A}, \quad \mu = \frac{m\pi}{B}, \quad \alpha^2 = \nu^2 + \mu^2$$

$2a$ and $2b$ are the dimensions of the hot spot, $2A$ and $2B$ are the corresponding side dimensions of the die, h is the die thickness, λ_{Si} is the silicon thermal conductivity, and the origin of coordinates is matched with the center of the hot spot.

Figure 18.17 shows the distribution of the excessive temperatures in the hot-spot region along the x-axis ($y = 0$, $z = 0$) when a uniform heat flux of 1250 W/cm² is actuated in the 0.4 × 0.4 mm hot-spot area. It is seen that the peak overheating in the center of the hot spot θ_p reaches 21.5°C, which can lead to the temperature of 96.5°C in the center of the hot spot.

The goal of the study is to define the TEC that would provide the maximum hot-spot suppression. Several factors make this problem unusually complicated. First of all, the TEC cold side turns out to be

TABLE 18.6 System Components Geometry and Materials

Component	Dimensions (mm)	Material
Hot spot	0.4 × 0.4	—
Die	11 × 13 × 0.5	Silicon
TIM1	11 × 13 × 0.03	Indium solder
Heat spreader	31 × 31 × 1.5	Copper
TIM2	31 × 31 × 0.05	Arctic Silver 5

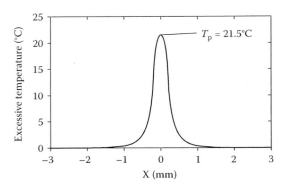

FIGURE 18.17 Local overheating in the active surface of 0.5 mm thick Si die actuated by 1250 W/cm² heat flux in the 0.4 × 0.4 mm hot-spot area.

bridged thermally with its hot area through surrounding parts of the heat spreader, so one never knows what the actual heat load is at the TEC cold side. As a result, the temperatures of the TEC junctions are also not defined, being dependent on heat load, feeding current and interfaced thermal resistances. These ambiguities can lead to a wrong thermal solution that can result in the die further overheating instead of cooling. This is why the substantiated choice of a TEC model is a primary issue and a real challenge as well.

18.3.1.2 Chip Package Thermal Characterization

Generally, three parameters are necessary to totally identify a TEC configuration. These are the TEC cold-side and hot-side temperatures T_0, T_h and the needed cooling power Q_0. None of these parameters is known beforehand. So the problem is to develop an adequate method of their determination. The key idea used in this study was to find dependences of excessive boundary temperatures θ_0, θ_h and corresponding hot-spot peak overheating θ_p on heat fluxes Q_0 and Q_h at the TEC outer boundaries. With this goal, the system model was considered, in which the TEC presence was simulated by uniform heat fluxes Q_0 and Q_h at its interfaces with adjacent structure parts (Figure 18.18).

To find $\theta_0(Q_0,Q_h)$ and $\theta_h(Q_0,Q_h)$ dependences, the three-dimensional distribution of the excessive temperatures $\theta(x, y, z) = T(x, y, z) - T_{cmax}$ within the die and the heat spreader was defined at different Q_0 and Q_h values. As the boundary conditions, the uniform distribution of the heat fluxes at the die active area was applied together with zero excess temperature at the heat spreader-to-heat sink interface. Calculations were made using the Ansys 10.0 program. The θ_0 and θ_h quantities were defined as mean-integral temperatures at the areas of the TEC footprints location. Total 7 TEC dimensional types with

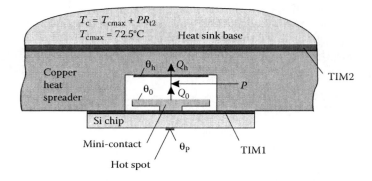

FIGURE 18.18 Scheme of the TEC equivalent replacement.

TABLE 18.7   Dimensional Details of the Used TEC Models (mm)

Option No.	Number of TE Pellets	Junctions Area F	Footprint F_s	Cavity Size
1	16	2.56	2.4 × 2.4	2.6 × 2.6
2	24	4	3 × 3	3.2 × 3.2
3	36	5.76	3.6 × 3.6	3.8 × 3.8
4	48	7.84	4.2 × 4.2	4.4 × 4.4
5	64	10.24	4.8 × 4.8	5 × 5
6	100	16	6 × 6	6.2 × 6.2
7	144	23.04	7.2 × 7.2	7.4 × 7.4

footprints F_s from 2.4 × 2.4 mm to 7.2 × 7.2 mm were chosen for estimations (Table 18.7). Different mini-contact tip cross sections in the range from 1 × 1 to 2.5 × 2.5 mm were tried with each TEC option to achieve the best cooling effect. Some details of the TEC and mini-contact construction, which are identical for all the options, are given in Table 18.8. In all cases the clearance of 0.1 mm was kept between the TEC side surfaces and cavity walls.

As a first step, Q_h values were accepted to be of $2Q_0$ what corresponds to the TEC COP = 1. Figures 18.19 and 18.20 show results of calculations for the TEC option no. 4 (F_s = 4.2 × 4.2 mm) with tip size of 1.4 × 1.4 mm. It is seen that the sought dependences can be treated as linear having the form:

$$\theta_p = a_0 + a_1 Q_0$$
$$\theta_0 = b_0 + b_1 Q_0 \quad\quad\quad (18.7)$$
$$\theta_h = c_0 + c_1 Q_h$$

Corresponding coefficients are given in Table 18.9. Similar dependences for all the TEC options coupled with different mini-contacts are defined and are found to be in excellent agreement with the linear form.

TABLE 18.8   Dimensions (mm) Accepted to be Identical for All Options

TEC Substrates Thickness (AlN)	TE Pellets Cross Section	TE Pellets Pitch	TE Pellets Packing Density	Mini-Contact Base Thickness	Mini-Contact Tip Thickness
0.2	0.4 × 0.4	0.6	0.444	0.4	0.05

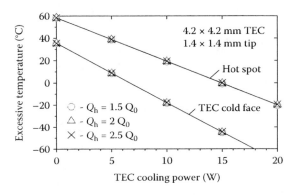

FIGURE 18.19   Dependence of hot-spot peak overheating and mean integral excessive temperature at the TEC cold footprint on heat flux withdrawn by the TEC.

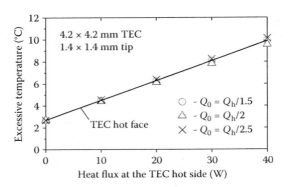

FIGURE 18.20 Dependence of the mean integral excessive temperature at the TEC hot footprint on heat flux.

TABLE 18.9 Coefficients in the Dependences (18.7) (4.2×4.2 mm TEC, Tip Size of 1.4×1.4 mm)

a_0	a_1	b_0	b_1	c_0	c_1
58.433	−3.916	35.509	−5.338	2.69	0.1802

The following question is a crucial one for proposed method applicability: do the obtained coefficients remain unchanged when TEC COP deviates considerably from the unit, or they are dependent on TEC efficiency? To clear this question, the estimations were undertaken in which the aforesaid coefficients were recalculated with Q_h values defined as $1.5Q_0$ and $2.5Q_0$, which cover the entire region of COP practical variation. It is seen (Figure 18.19) that drastic variation of Q_h value does not affect practically excessive temperatures θ_p and θ_0 and, just the same, considerable changes in Q_0 value exert negligible influence on the $\theta_h(Q_h)$ dependence (Figure 18.20). This means that relations 18.7 have a content of generalized chip package thermal characteristics. They are independent on the actual temperature of the heat sink base, poorly affected with a TEC efficiency variation and therefore can serve as a basis for optimal integration of a TEC into the chip package. Once again, it has to be noted that system 18.7 gives the field of excessive temperatures over real processor case temperature T_c, which do depends on the heat flux Q_h according to the formula $T_c = T_{cmax} + R_{t2}(Q_h - Q_0)$.

18.3.1.3 The TEC Analytical Model

An analytical model of the TEC integrated into the chip package can be described by the following system of linear equations:

$$
\begin{aligned}
Q_0 &= F\left(\alpha i T_0 - \frac{1}{2}i^2(\rho l + 2R_c) - \frac{\lambda}{l}(T_h - T_0)\right) \\
Q_h &= F\left(\alpha i T_h + \frac{1}{2}i^2(\rho l + 2R_c) - \frac{\lambda}{l}(T_h - T_0)\right) \\
T_0 &= T_c + \theta_0 - Q_0 R_s \\
T_h &= T_c + \theta_h + Q_h R_s \\
T_p &= T_c + \theta_p \\
T_c &= T_{cmax} + (Q_h - Q_0)R_{t2}
\end{aligned}
\tag{18.8}
$$

System 18.8 includes heat balance equations at the TEC cold and hot junctions coupled with the relations for temperature drops in different points of the system. The last equation reflects an additional elevating of the processor case temperature caused by the supplementary TEC power $P = Q_h - Q_0$, an unfavorable effect that is missing in some other studies.

The temperatures T_0 and T_h in Equation 18.8 must be presented in the units of absolute temperature scale. To satisfy this condition, the temperature T_{cmax} must be put in Kelvins (in our case T_{cmax} = 345.5 K).

18.3.1.4 Models Matching

Dependences (18.7) together with relations (18.8) form the closed system of 9 linear equations regarding to 9 unknown variables T_p, T_0, T_h, T_c, θ_p, θ_0, θ_h, Q_0, Q_h. We will consider F, l, and i as independent variables. For any given set of these quantities, the system has a unique solution. Hence, here is the case of an implicitly defined $T_p(F,l,i)$ function. We use the parameters F, l, and i to control hot-spot overheating with the objective to find a TEC that provides minimum hot-spot peak temperature. Various TECs with different junction areas F were considered and for each F value the optimal TE pellet height $l = l_0$ and corresponding optimal current density $i = i_0$ were defined using a standard program of the two-variable function $T_p(l,i)$ minimization. This procedure was repeated for different mini-contact tip dimensions to find its optimal configuration resulting in the maximum hot-spot suppression. Currently available R_c value of 10^{-6} Ω cm^2 was used in our calculations and the effect of its reduction to 10^{-7} Ω cm^2 was also studied. Finally, the idealized case of $R_c = 0$ was reviewed to find the theoretical limit for the efficiency of considered technology.

18.3.2 Results and Discussion

18.3.2.1 Optimal TEC Configuration

Results of estimations are given in Figures 18.21 through 18.25. Figure 18.21 shows the dependence of the hot-spot peak temperature on the TEC total junctions area for different R_c values. Each point in the figure is obtained as a result of optimization of TE leg length, current density, and mini-contact tip size. Corresponding optimal values are given in Table 18.10.

The horizontal line in Figure 18.21 corresponds to the hot-spot peak temperature of 104.1°C for the case of a solid heat spreader with no TEC integrated. From now on we will take this very temperature as a reference point when estimating hot-spot suppression.

The best result is obtained for the TECs with the total junction area in the range from 7 to 9 mm^2. These are options 4 and 5, which relate to the TECs with 4.2 × 4.2 and 4.8 × 4.8 mm substrates correspondingly. It is seen that with $R_c = 10^{-6}$ Ω cm^2 one can await the hot-spot suppression of the order of 9°C only. With R_c reduced to 10^{-7} Ω cm^2, the attainable effect can be raised to 15.6°C and this looks like a practical limit for this technology. No further progress is available because even for a hypothetical case

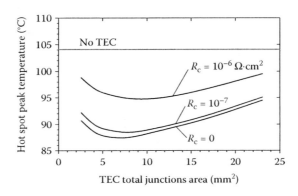

FIGURE 18.21 Dependence of hot-spot peak temperature on the TEC total junctions area. TE leg length, current density, and mini-contact tip dimensions are optimized for each point.

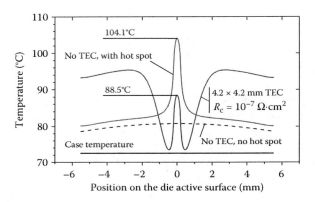

FIGURE 18.22   Temperature profile at the silicon chip active surface.

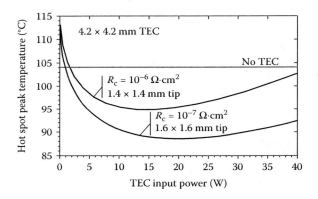

FIGURE 18.23   Dependence of hot-spot peak temperature on TEC input power (optimal TEC according to Table 18.10, option 4).

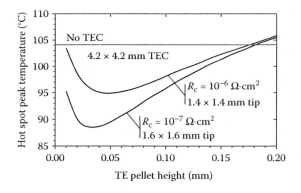

FIGURE 18.24   Dependence of hot-spot temperature on TE leg height (current density is optimized for each l value).

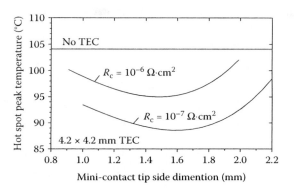

FIGURE 18.25 Effect of deviation of mini-contact tip dimensions from optimum (TE pellet height and electrical current are optimized at each point).

TABLE 18.10 Optimal Parameters for the TECs with Different Footprints

TEC Option No.[a]	Optimal Tip Size (mm)	TEC Optimized Parameters				Hot-Spot Suppression (°C)
		l (µm)	i (A/cm²)	Q_0 (W)	P (W)	
		$R_c = 10^{-6}$ Ω cm²				
1	1 × 1	28	7263	6.62	7.33	5.3
2	1.2 × 1.2	33	6360	8.77	9.97	7.3
3	1.3 × 1.3	41	5459	10.41	12.38	8.5
4	1.4 × 1.4	48	4774	11.96	14.64	9.3
5	1.5 × 1.5	56	4192	13.42	16.6	9.3
6	1.7 × 1.7	71	3312	16.3	19.9	7.7
7	1.9 × 1.9	86	2699	19.23	22.44	4.6
		$R_c = 10^{-7}$ Ω cm²				
1	1.1 × 1.1	18	13449	10.78	11.97	11.9
2	1.3 × 1.3	22	10787	10.98	14.54	14.1
3	1.5 × 1.5	27	8824	14.52	17.72	15.2
4	1.6 × 1.6	34	7226	15.85	19.87	15.6
5	1.8 × 1.8	39	6168	17.79	21.69	15.1
6	1.9 × 1.9	55	4402	19.58	24.29	12.8
7	2.1 × 2.1	69	3400	22.2	26.14	9

[a] TEC dimensional details for each option are given in Tables 18.7 and 18.8.

with $R_c = 0$ an additional hot-spot suppression does not exceed 1°C. This means that the thermal resistances in the structure become predominant. Another disadvantage is the fact that optimal TE pellet heights lay in the region of several tens of microns, which is difficult to implement using bulk technology or thin-film technique as well.

It has to be noted that obtained results do not correlate in details with other papers.[11–13] Though initial data for estimations are rather similar, the optimal TEC configuration found in this study is shifted to considerably greater dimensions and higher input power.

Figure 18.22 shows temperature profiles at the active face of the silicon chip. If the heat flux of 67 W/cm² is dissipated uniformly in the chip surface and there is no hot spot and no TEC, the peak chip temperature is 81.5°C. However, when a 0.4 × 0.4 mm hot spot with heat flux of 1250 W/cm² is activated, the peak chip temperature increases to 104.1°C. With the optimized TEC (option 4, $R_c = 10^{-7}$ Ω cm²) integrated into the heat spreader, the temperature profile changes fundamentally, possessing a "*W*" shaped form with about 16°C reduction in the hot-spot temperature. This effect is accompanied with a sharp temperature decrease around the hot-spot area and an increase over the "no TEC" level at the chip distant area due to additional heat load equivalent to the TEC power.

18.3.2.2 Influence of the Deviation from Optimal Parameters

As the practical implementation of the found optimal design can become problematic, it would be of great interest to find to what degree the deviation from the TEC optimal parameters will affect its ability for hot-spot suppression. In our further calculations, the optimal TEC model according to item 4 (Table 18.10) will be considered as a reference one.

18.3.2.2.1 Effect of Input Power

Figure 18.23 shows hot-spot temperature variation with TEC input power. It is seen that with the TEC switched off, the hot-spot temperature reaches 113°C which is 9°C higher than that for the case without TEC integrated. The power of near 2 W is necessary to compensate this unfavorable effect of the inactive TEC presence. With the TEC further activation the hot-spot temperature reduces and reaches its minimum in the power range from 10 to 20 W.

18.3.2.2.2 Effect of TE Pellet Height

Figure 18.24 shows dependence of hot-spot cooling on the TE pellet height. The optimal *l* values lay in the region from 35 to 50 µm and deviation from this region both to the left and to the right greatly affects hot-spot cooling performance. It is seen also that bulk TECs with TE pellet heights over 0.18 mm are unfit for hot-spot cooling.

18.3.2.2.3 Effect of the Mini-Contact Tip Dimensions

Mini-contact tip at the TEC-to-chip interface allows minimizing a deleterious temperature drop at the TEC cold side. When its dimensions reduce, the heat gain to the TEC cold side reduces, but thermal resistance of the mini-contact increases simultaneously. These two contradictive factors cause existence of optimal tip flank dimension. For the device under consideration the optimums lay in the range from 1.4 to 1.6 mm (Figure 18.25). Expansion of this region by 0.4 mm on both sides leads to 5°C loss in hot-spot suppression that is over 30% of the disposable value.

18.3.3 Further Prospects

In this study, TE materials with z-factor of $3 \times 10^{-3}K^{-1}$, peculiar to modern bulk thermoelectrics, were considered and their potential for hot-spot cooling is estimated with R_c value supposed to be lowered down to 10^{-7} Ω cm^2. It has to be noted, that with such parameters a TEC should theoretically give ΔT_{max} of 72°C even with TE legs as short as 20 µm. As this is not yet practically justified, the obtained limitation for on-chip hot-spot cooling of about 10°C looks well-grounded for the state-of-the-art technology that provides R_c value at the level of 10^{-6} Ω cm^2.

Definite hopes are pinned today on the nanostructured TE materials that promise a fundamental improvement in micro-TEC performance and technology. Venkatasubramanian et al.[16] reported recently zT of 2.4 in p-Bi_2Te_3/Sb_2Te_3-based superlattices that can give a real breakthrough in the field of thermoelectric materials, and what is especially important, this achievement is now accompanied with considerable progress in the field of microcooler technology due to involvement of advanced methods of nano and micro electronics.[17,18] But as a matter of fact, no measured ΔT_{max} values approaching 70°C at room temperature are still reported for such microdevices, which proves the deteriorative influence of the thermal and electrical contact resistances. Hence, further prospects depend on what will be done to practically improve the micro-TEC efficiency at the device level.

18.4 Conclusion

Under typical thermal conditions of PC application, the use of thermoelectics for processor uniform cooling can ensure a 17–25% increase in its power. To achieve this result, the short-legged TECs with electrical contact resistance at the level of 10^{-7} Ω cm^2 must be provided, and also thermal resistances of all the interface materials must be reduced to a possible minimum.

Effectiveness of active cooling decreases with heat sink enhancement. At a heat sink thermal resistance less than 0.1 K/W, the use of active cooling becomes disadvantageous. So the reduction of the heat sink thermal resistance to this level, being practicable, should be a preferential solution.

The assemblage architecture with the TEC located outside of the processor case is a feasible one. The version with an inside TEC location is problematic for implementation because it is characterized with 60–70 μm long TE pellets, the range that is currently inaccessible both for bulk and film-type technology.

An essential disadvantage of active cooling is rather high TEC power, which is comparable with that of the processor. This makes unreasonable the use of thermoelectrics in portable PCs with an autonomous power supply.

A real advantage of active cooling becomes evident in the special case of elevated ambient temperature, when passive cooling exhausts the ability of heat removal. With an ambient temperature of 60°C, the duplication of the processor power is achievable and at $T_a \geq T_{dmax}$, active cooling becomes the only means for processor thermal management. At an ambient temperature below 30°C, passive cooling gives an adequate thermal solution and the use of a TEC becomes counterproductive.

Though application of micro-TECs for selective on-chip hot-spot suppression looks like a prospective technology, efficiency of this method is greatly affected with irreversible losses at electrical and thermal interfaces. With the state-of-the-art technology, which provides electrical contact resistance at the level of 10^{-7} Ω cm^2, the hot-spot suppression of 10°C only can be awaited. With $R_c = 10^{-7}$ Ω cm^2 this quantity can be enlarged to 15°C.

Another practical problem lies in the fact that the optimal TE leg height for the micro TEC integrated into the chip-to-heat spreader interface lays in the region of several tens of microns, which is problematic for implementation. An attempt to increase TE pellets height leads to the reduction of the cooling effect and with $l \geq 0.18$ mm the TEC exhausts its ability for hot-spot suppression.

References

1. Chu, R.C. and Simons, R.E. 1999. Application of thermoelectrics to cooling electronics: Review and prospects, *Proc. 18th Int. Conf. Thermoelectrics*, 270–9. Baltimore: IEEE.
2. Bierschenk, J. and Gilley, M. 2006. Assessment of TEC thermal and reliability requirements for thermoelectrically enhanced heat sinks for CPU cooling applications, *Proc. 25th Int. Conf. Thermoelectrics*, pp. 254–9. Vienna, Austria: IEEE.
3. Yu, C.-K., Liu, C.-K., Dai, M.-J. et al. 2007. A thermoelectric cooler integrated with IHS on a FC-PBGA package, *Proc. 26th Int. Conf. Thermoelectrics*, pp. 279–83. Jeju, Korea: IEEE.
4. Yamaguchi, S., Fukuda, S., Kitagava, H. et al. 2008. A new proposal of Peltier cooling for microprocessor, *Proc. 27th Int. Conf. Thermoelectrics*, Corvallis, Oregon: IEEE. http://kamome.lib.ynu.ac.jp/dspace/bitstream/10131/5064/1/ICT2008IB1.pdf
5. Ikeda, M., Nakamura, T., Kimura, Y. et al. 2006. Thermal performance of thermoelectric cooler (TEC) integrated heat sink and optimizing structure for low acoustic noise/power consumption, *Proc. 22th IEEE Semi-Therm Symposium*, pp. 144–51. Dallas, TX: IEEE Xplore Digital Library. http://ieeexplore.ieee.org/xpl/freeabs_all.jsp?reload=true&arnumber=1625220
6. Sauciuc, I., Prasher, R., Chang, J.Y. et al. 2005. Thermal performance and key challenges for future CPU cooling technologies, *Proc. ASME InterPACK*, pp. 353–64. San Francisco: ASME.

7. Semenyuk, V.A. and Dekhtiaruk, R.I. 2007. Thermoelectric cooling under dimensional constraints, *Journal of Thermoelectricity* 4: 69–75.

8. Semenyuk, V.A. and Dekhtyaruk, R.I. 2010. Prospects for thermal management of powerful microprocessors using thermoelectrics, *Proc. 8th European Conf. Thermoelectrics*, pp. 106–11. Como, Italy.

9. Semenyuk, V.A. and Dekhtiaruk, R.I. 2010. Thermoelectric cooling of powerful microprocessors, *Journal of Thermoelectricity* 4: 67–76.

10. Intel Corporation. 2009. Intel Core i7-800 and i5-700 desktop processor series and LGA1156 socket: Thermal/mechanical specifications and design guidelines. http://download. intel.com/design/ processor/datashts/322164.pdf

11. Wang, P., Bar-Cohen, A., and Yang, B. 2007. Enhanced thermoelectric cooler for on-chip hot spot cooling, *Proc. of IPACK2007*, Paper No: IPACK2007-33798, pp. 249–58. Vancouver: ASME.

12. Lee, K.H. and Kim, O.J. 2007. Simulation of the cooling system using thermoelectric micro-coolers for hot spot mitigation, *Proc. 26th Int. Conf. on Thermoelectrics*, pp. 284–87. Jeju, Korea: IEEE.

13. Kim, O.J. and Lee, K.H. 2007. A study on the application of micro TEC for hot spot cooling, *Proc. 26th Int. Conf. on Thermoelectrics*, pp. 344–47. Jeju, Korea: IEEE.

14. Semenyuk, V.A. and Protsenko, D. 2008. On-chip hot spot cooling: Forecasts and reality, *Proc. 6th European Conf. Thermoelectrics*, I-06, pp. 1–6. Paris.

15. Arctic Silver, Inc. 2007. High-density polysynthetic silver thermal compound Arctic Silver 5. http:// www.arcticsilver.com/as5.htm (accessed January 07, 2011).

16. Venkatasubramanian, R., Sivola, E., and O'Quinn, B. 2006. Superlattice thin-film thermoelectric material and device technologies, *Thermoelectrics Handbook: Macro to Nano*, ed. D.M. Rowe, Chapter 49, pp. 1–15. Boca Raton, CRC Press.

17. Böttner, H., Nurnus, J., and Schubert, A. 2006. Miniaturized thermoelectric converters, *Thermoelectrics Handbook: Macro to Nano*, ed. D.M. Rowe, Chapter 46, pp. 1–18. Boca Raton, CRC Press.

18. Nextreme, Inc. 2010. Micro-scale thermal and power management: Cooling & temperature control. http://www.nextreme.com/pages/temp_control/temp_control.shtml (accessed January 07, 2011).

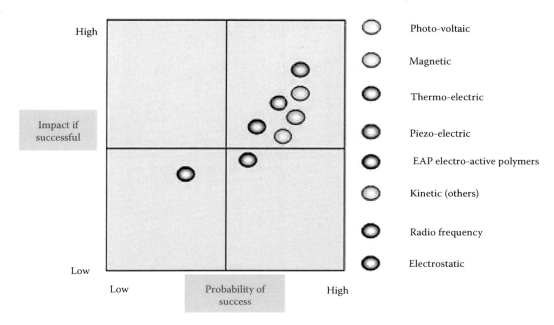

FIGURE 17.1 Opportunity matrix for different energy-harvesting technologies. (From Frost and Sullivan. *Advances in Energy Harvesting Technologies*, DoC2, December 2007. With permission.)

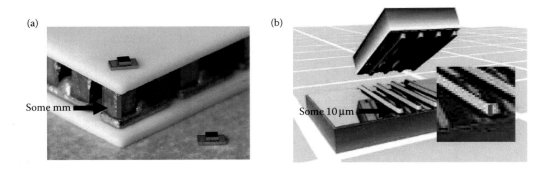

FIGURE 17.2 Illustration of leg height classification for miniaturized thermoelectric devices: (a) typical commercial device, (b) pattern of thermoelectric legs for miniaturized Peltier-coolers according to the Micropelt technology.

FIGURE 17.3 Illustration of leg foot print classification for miniaturized thermoelectric devices: (a) typical commercial device, (b) pattern of thermoelectric legs for miniaturized thermogenerators according to the Micropelt technology.

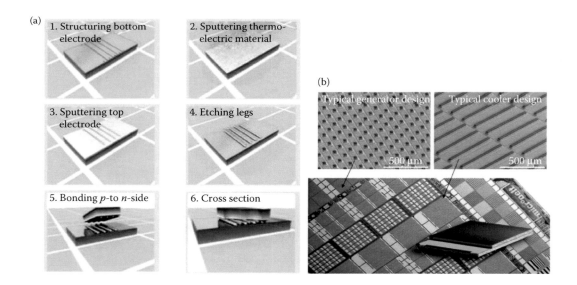

FIGURE 17.5 (a) Micropelt's high level process flow. (b) Details of a finished wafers before bonding showing the flexibility of the production with respect to feature and device sizes.

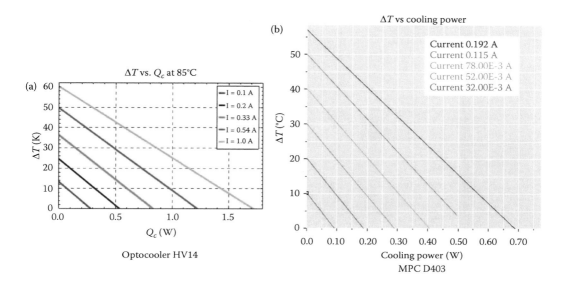

FIGURE 17.8 DT-Q_c diagrams for thin film TECs produced by Nextreme (HV14) (a) and Micropelt (MPC D403) (b).

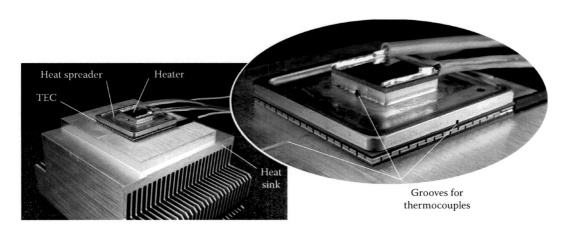

FIGURE 18.10 Photographs of the prototype parts.

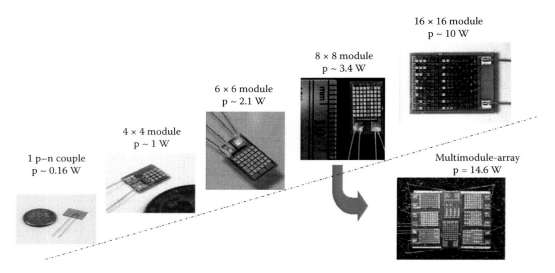

FIGURE 21.8 Scale up of the superlattice thin-film TE module technology to larger multi-module arrays (MMAs) that can produce power levels of hundreds of milliwatts to watts depending on temperature gradients available. Such MMAs employ the BCAM module technology where the couple-based assembly allows an automatic tiling on the hot-side of the module and obviates the need for a hard-bonded common header.

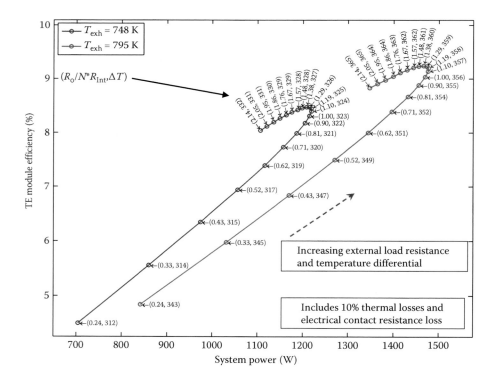

FIGURE 22.4 Typical TE module efficiency–power maps for varying external resistance and current.

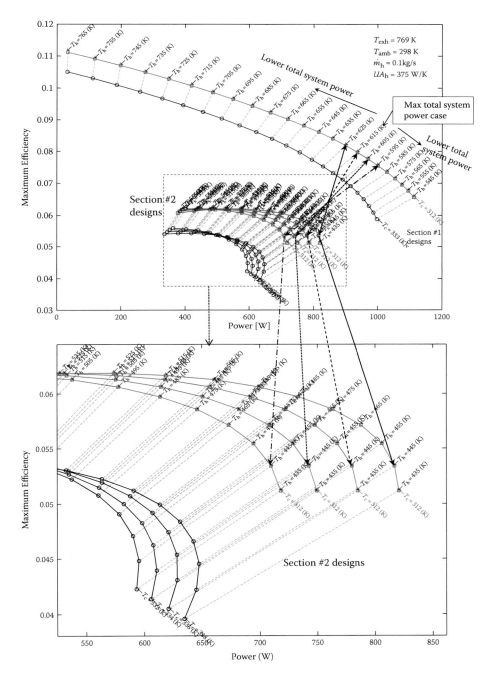

FIGURE 22.6 Typical dual-sectioned analysis results for $T_{exh} = 769$ K, $\dot{m}_h = 0.1$ kg/s, $UA_h = 375$ W/K, $T_{amb} = 298$ K.

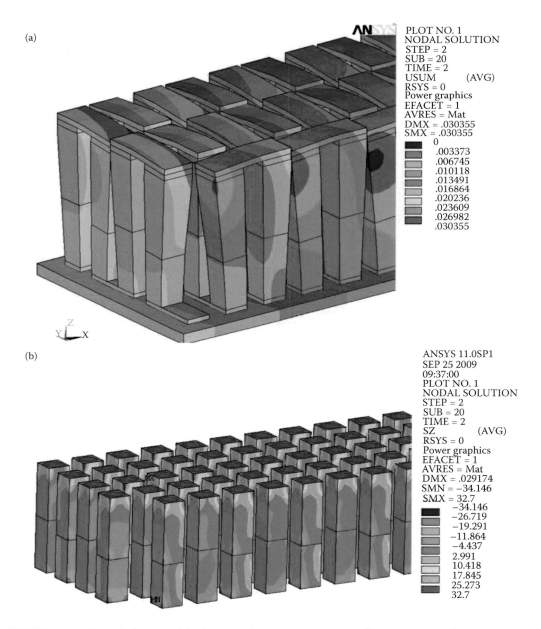

FIGURE 22.13 Magnified structural displacements (a) and resulting stresses (b) in a TE module from compressive and expansive forces (displacements in mm, stresses in MPa).

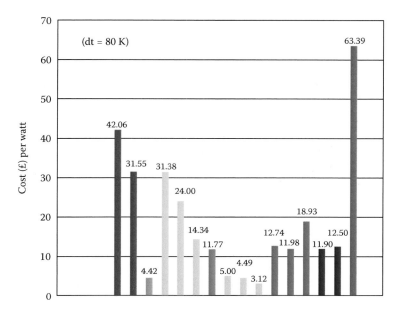

FIGURE 23.6 Cost per watt.

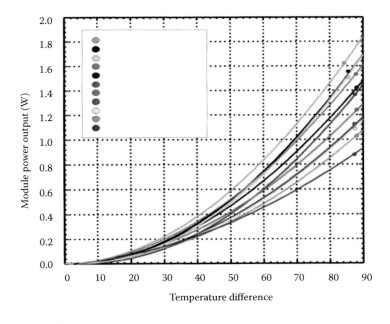

FIGURE 23.7 Corresponding power output.

FIGURE 23.10 The Kusatsu hot spring. (T. Kajikawa, Shonan Institute of Technology.)

FIGURE 23.11 Thermoelectric generator. (T. Kajikawa, Shonan Institute of Technology.)

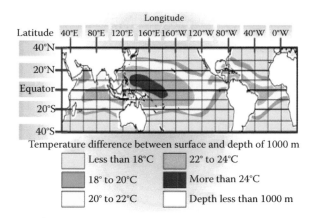

FIGURE 23.16 Ocean temperature gradient.

19

Millikelvin Tunnelling Refrigerators

M.J. Prest
University of Warwick

T.E. Whall
University of Warwick

E.H.C. Parker
University of Warwick

D.R. Leadley
University of Warwick

19.1 Introduction

Tunnel junction refrigeration was first realized in the 1990s (Nahum et al. 1994, Leivo et al. 1996) and uses an energy-selective tunnel barrier to extract hot electrons. The cooling power of these tunnel junctions is typically quite low, in the order of pW; however, if the volume to be cooled is small and the heat load (mainly from the lattice) is sufficiently low (which occurs at sub-Kelvin temperatures), these tunnel junctions are able to cool electrons from an initial bath temperature of around 300 mK to 100 mK and below.

Additionally, if the cold electrons are extended over a region of the lattice which is thermally isolated from the bath temperature, the lattice can also be cooled. This enables the cooling of a payload such as an integrated circuit supported on a thin membrane.

The modest cooling range of these devices is technologically significant because it can be achieved at low cost, in a compact and elegant way for applications where size and weight are at a premium, and in applications where this temperature decrease can produce significant performance benefits, such as in astronomical instrumentation and quantum computing devices. The initial, or bath, temperature, for these coolers of around 300 mK is easily achievable using ³He-based cryogenic cooling. It is hoped that further development of these tunnel junctions may push the bath temperatures up and/or increase the cooling range. Also, there may emerge some new electronic technology to fill the "cooling gap" between these tunnel junction coolers and Peltier-type thermoelectrics which cool from room temperature. The ultimate aim is a fully electronic cooling from room temperature to around 50 mK.

19.2 Superconducting Tunnel Junction Refrigerators

Figure 19.1 shows the energy diagram of an NIS (normal metal–insulator–superconductor) tunnel junction. Its behavior is analogous to a high-pass filter; a current flows through the junction extracting only high-energy (hot) electrons from the normal metal, for certain bias voltages. If the heat loads entering the electron gas by other means (mainly from the lattice) are sufficiently low, the cooling power of the tunnel current will reduce the electron temperature.

As shown in Figure 19.1, only electrons with energies $E > \Delta$, can enter the superconductor, which for voltages $V \leq \Delta/e$, are hot electrons ($E > E_{FN}$). Cold electrons ($E < E_{FN}$), having $E < \Delta$, are blocked by the energy gap in the superconductor. Each electron removes energy $(E-eV)$, which is of order $k_B T$, from the metal corresponding to a cooling power of order $I\, k_B T/e$ where I is the current through the junction (Nahum et al. 1994, Fisher et al. 1999). The reduction in electron energy corresponds to a lowering of the electron temperature. Maximum cooling power occurs for voltages closely approaching $V = \Delta/e$. For $V > \Delta/e$, cold electrons are also extracted from the normal metal and the tunneling results in a net negative cooling power, that is, a heat load which at high bias is equivalent to Joule heating (Giazotto et al. 2006a Section II.F.2). A more detailed description is given by Pekola (2005).

19.2.1 The Normal Metal/Insulator/Superconductor Structure

The cooling power of an NIS junction arises from the current of high-energy electrons leaving the metal, each carrying its energy $(E-eV)$ out of the system (Nahum et al. 1994, Jug and Trontelj 1999, Luukanen et al. 2000). This current is given by Equation 19.1 (Rowell and Tsui 1976, Nahum et al. 1994, Bardas and Averin 1995, Leivo et al. 1996, Frank and Kretch 1997)

$$I = \frac{1}{eR_T} \int_{-\infty}^{\infty} F(E,V,T_e,T_b)\, g(E)\, \mathrm{d}E \tag{19.1}$$

where

$$F(E,V,T_e,T_b) = f(E - eV, T_e) - f(E, T_b) \tag{19.2}$$

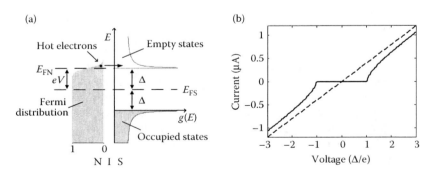

FIGURE 19.1   The energy-level diagram of an NIS junction. The superconductor has an energy bandgap of 2Δ. A bias voltage, $V = \Delta/e$, is applied across the junction, so that only electrons with energy greater than E_{FN} are allowed to tunnel through the insulator into the superconductor. (b) I–V characteristic of the junction at low temperature (solid line) and room temperature (dashed line). The low-temperature tunnel current is ideally zero for $V < \Delta/e$, then increases sharply at $V = \Delta/e$ after which the current asymptotically approaches the room temperature value given by $I = V/R_T$, where R_T is the characteristic, or normal state, resistance of the tunnel junction. The curves are calculated using Equations 19.1 through 19.7 with the following parameters: $\Delta = 0.2$ meV, $R_T = 500\ \Omega$, $\Sigma = 10^9$ WK^{-5}m^{-5}, $\Omega = 1\ \mu m^3$. (Adapted from J. P. Pekola, R. Schoelkopf, and J. N. Ullom, *Physics Today* May 41, 2004b.)

and

$$f(E,T) = 1 / \left[1 + \exp(E/(k_B T)) \right] \tag{19.3}$$

Here R_T is the normal state tunneling resistance, f is the Fermi–Dirac distribution function of electrons, T_e is the electron temperature in the normal metal, and T_b is the temperature of the superconductor, assumed to be equal to the temperature of the thermal bath; $g(E)$ is the density of states within the superconductor into which the electrons tunnel (Solymar 1972, Rowell and Tsui 1976, Tinkham 1996), and can be written as

$$
\begin{aligned}
g(E) &= 0 && \text{when } |E| < \Delta \\
&= \frac{|E|}{\sqrt{E^2 - \Delta^2}} && \text{otherwise}
\end{aligned}
\tag{19.4}
$$

This current removes energy from the semiconductor at a rate (Leivo et al. 1996):

$$P_{\text{cool}} = \frac{1}{e^2 R_T} \int_{-\infty}^{\infty} (E - eV) F(E,V,T_e,T_b) g(E) \, \mathrm{d}E \tag{19.5}$$

Electron–phonon coupling results in a heat flow between the electron gas and the lattice as the former is cooled, given by (Roukes et al. 1985, Wellstood et al. 1994)

$$P_{\text{e-ph}} = \Sigma \Omega (T_e^n - T_b^n) \tag{19.6}$$

with Σ being the material-specific electron–phonon coupling constant, Ω the volume of the absorber, and T_e, T_b the electron and phonon temperatures, respectively. The power n varies with the material of the island, and usually for normal metals $n = 5$ (see refs. in Kivinen et al. 2003).

In the simplest case, assuming that the phonons are at a fixed bath temperature, we may calculate the temperature to which the electrons are cooled by finding the solution to

$$P_{\text{cool}} + P_{\text{e-ph}} = 0 \tag{19.7}$$

However, it is found that in practice this is insufficient and we have to consider numerous factors which might also limit the cooling, see Section 19.4.

19.2.2 Double-Junction (SINIS) Cooler

In a SINIS structure, the heat flow is not always in the same direction as the electrical current, so two back-to back junctions can be employed, in which hot electrons leave the metal through one junction and cold electrons enter through the other (Figure 19.2). This acts to sharpen the Fermi distribution of the normal metal, and lowers the average energy, reducing the electron temperature of the normal metal. The voltage applied across the device is divided equally between the two junctions and the cooling power is doubled, so a factor of two can be included in Equations 19.2 and 19.5 to account for this, as below

$$F(E,V,T_e,T_b) = f(E - e(V/2),T_e) - f(E,T_b) \tag{19.8}$$

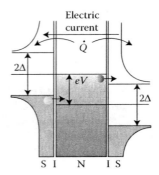

FIGURE 19.2 Energy diagram of a double-junction SINIS cooler, for $eV < 2\Delta$. The potential V is divided evenly between the 2 junctions, which have equal tunnel resistance R_T. Electric current flows from right to left. Hot electrons tunnel out of the normal metal at the right-hand junction and are replaced by cold electrons, which enter at the left-hand junction. The cooling power, Q is given by Equation 19.9. (From F. Giazotto et al., *Review of Modern Physics*. 78 217, 2006a, Section V.C.1. With permission.)

$$P_{\text{cool}} = \frac{2}{e^2 R_T} \int_{-\infty}^{\infty} (E - e(V/2))F(E,V,T_e,T_b)g(E)\,dE \tag{19.9}$$

where R_T is the normal state tunnel resistance of a single NIS junction. Both junctions are assumed to have equal tunnel resistance; asymmetry in the junction resistances was shown to have only a weak effect, 7% for a factor of 2 difference (Pekola et al. 2000b).

19.2.3 Electron Cooling: Analytical Solution

Cooling is maximized for a particular ratio of T/Δ, for which $T = T_e = T_b$, given as $k_BT = 0.3\Delta$ assuming a temperature-independent gap or $k_BT = 0.25\Delta$ when assuming a more realistic BCS-based dependence (Giazotto et al. 2006a Section V.C.1). So, for aluminum with $\Delta = 0.2$ meV the temperature for optimum cooling power is $0.25\Delta/k_B = 0.6$ K; however, lattice heating becomes dominant at this temperature for common values of the cooled volume and electron–phonon coupling factor. The value of Δ is related to the critical temperature of the superconductor T_C (the superconducting transition temperature). A general relationship between Δ and T_C is given by Tinkham as: $\Delta = 1.764\,k_B\,T_C$ (Tinkham 1996).

At the maximum cooling bias, $2\Delta/e$, and at temperatures where $k_BT \ll 0.3\Delta$, the cooling power of an SINIS device can be expressed in analytic form as in Equation 19.10 (from Leivo et al. 1996, Manninen et al. 1997, Leivo et al. 1998, Leoni et al. 1999 or Anghel and Pekola 2001), we have used a pre-factor of 1.2 for a double junction.

$$P_{\text{max}} = 1.2(\Delta^2/e^2 R_T)(k_B T_e/\Delta)^{3/2} \tag{19.10}$$

From this expression it is observed that P_{max} is inversely proportional to the tunnel resistance R_T and proportional to $(T/\Delta)^{3/2}$ (Bardas and Averin 1995). A plot of P_{max} is shown in Figure 19.3 (solid line), with $\Delta = 0.2$ meV, and $R_T = 200\ \Omega$. Also, using the lattice heating expression (Equation 19.6) and assuming the electron–phonon coupling factor $\Sigma = 10^9$ W K^{-5} m^{-5}, volume $\Omega = 1\ \mu$m^3, lattice temperature $T_b = 0.3$ K, the maximum heat power from the lattice $P_{\text{e-ph\_max}} = \Sigma\Omega T_b^5 = 2.4$ pW. The lattice heat power is also shown in Figure 19.3 (dashed line), and from this plot it can be seen that the electron temperature T_e which satisfies the heat balance equation is 108 mK. Reducing R_T or increasing Δ increases the slope of P_{max} so allows cooling to lower temperatures. Also, reducing $\Sigma\,\Omega$ or T_b reduces the lattice heating (lowering the dashed line) and so this can also reduce T_e.

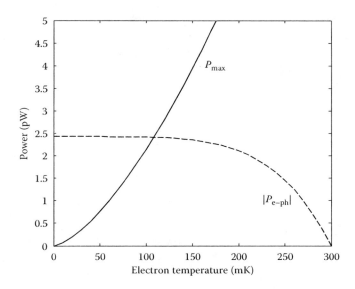

FIGURE 19.3 Shows P_{max} (solid line) and $|P_{e-ph}|$ (dashed line). The electron temperature T_e which satisfies the heat balance equation $P_{max} - P_{e-ph} = 0$ can be observed where the lines cross, and is 108 mK. The parameters $\Delta = 0.2$ meV, $R_T = 200\ \Omega$, $\Sigma = 10^9$ W K^{-5} m^{-5}, $\Omega = 1\ \mu$m^3, and $T_p = 0.3$ K.

Another useful expression is obtained by combining Equations 19.6, 19.7, and 19.10 to give the tunnel resistance required for cooling from T_p to T_e as in Equation 19.11 (Leoni et al. 1999). This equation is plotted in Figure 19.4, for the parameters: $\Omega = 1\ \mu$m^3, $\Delta = 0.2$ meV, $\Sigma = 10^9$ W K^{-5} m^{-5}.

$$\Omega R_T = \frac{1.2(\Delta^2/e^2)(k_B T_e/\Delta)^{3/2}}{\Sigma(T_b^5 - T_e^5)} \tag{19.11}$$

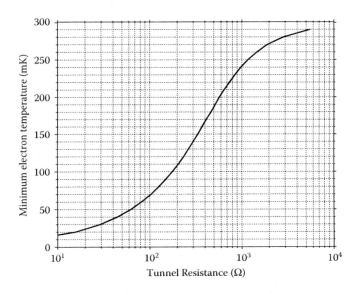

FIGURE 19.4 Minimum electron temperature versus tunnel resistance for a phonon temperature $T_b = 300$ mK, calculated using Equation 19.11, with the parameters: $\Omega = 1\ \mu$m^3, $\Delta = 0.2$ meV, $\Sigma = 10^9$ W K^{-5} m^{-5}.

19.2.4 Thermometry

Small NIS probe junctions are used to measure temperature, the small area increases the tunnel resistance, so that the cooling power of the junction is kept small.

If $k_B T_e \ll \Delta$ and $0 \ll eV < \Delta$, the current through an NIS junction, by thermally activated tunnelling, is: (Nahum and Martinis 1993, Luukanen et al. 2000):

$$I(V) \approx I_a \exp[(eV - \Delta)/k_B T_e] \tag{19.12}$$

So that if the junction is biased with a constant current:

$$\frac{dV}{dT_e} \approx (k_B/e)\mathrm{Ln}(I/I_a) \tag{19.13}$$

Hence, the thermometer voltage has a suitably linear dependence on temperature. However, in real devices there is often a deviation in this linear dependence at the lowest temperatures (Nahum et al. 1994, Muhonen et al. 2009).

19.3 The Semiconductor Superconductor (SmS) Structure

Savin and coworkers have replaced the normal metal with a degenerate semiconductor (Sm), in the arrangement shown in Figures 19.5 and 19.6 (Savin et al. 2001).

Cooling occurs by the extraction of hot carriers from the left-hand junction and the addition of cold carriers at the right-hand junction. This cooler has the advantage that it eliminates the need for an oxide barrier which is difficult to form and prone to the formation of pinholes in the thinner oxides. Weaker electron–phonon coupling in the semiconductor means that it might be possible to cool a larger volume of carriers. However, there is a larger tunneling resistance in current semiconductor coolers (due, at least in part, to the smaller density of states at the Fermi level) and it is of the order of 10–100 $k\Omega\,\mu m^2$ (~70 $k\Omega\,\mu m^2$ for dopant concentration $4 \times 10^{19}\ cm^{-3}$) as compared to 1–2 $k\Omega\,\mu m^2$ (Clark et al. 2004) or 0.3 $k\Omega\,\mu m^2$ (Leivo et al. 1996) for metal coolers. So, tunnel resistances are, in general, higher for semiconductor coolers than for metal coolers and this is currently a major factor in limiting device performance.

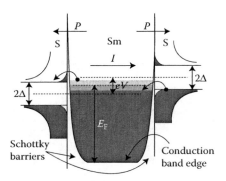

FIGURE 19.5 Energy-band diagram illustrating cooling in the S–Sm–S structure. S denotes superconductor and Sm semiconductor. 2Δ is the energy gap in the superconductor, E_F the Fermi energy of the semiconductor, and P heat flow out of the semiconductor. V is the applied voltage and I is the resulting current. The gray areas denote the Fermi distribution of occupied electron states. (From A. Savin et al., *Applied Physics Letters*, 79(10) 1471, 2001. With permission.)

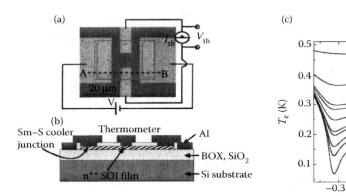

FIGURE 19.6 (a) Optical micrograph of an S–Sm–S cooler device and schematic illustration of the connections to external circuitry in the measurements. Current I_{th} and voltage V_{th} are used in determining the electronic temperature T_e in the n^{++} SOI mesa. (b) Schematic cross section of the device along the line AB in (a). (From A. Savin et al., *APL* 79(10) 1471, 2001. With permission.) (c) The electron temperature in n^{++}SOI as a function of the voltage across the S–Sm–S cooler structure at different substrate temperatures. (From A. Savin et al., *Physica Scripta* T114 57, 2004. With permission.) The silicon was doped at 4×10^{19} cm^{-3}.

Higher dopant densities in silicon led to lower Schottky tunnel resistances (1.8 kΩ μm^2 for dopant concentration 1.6×10^{20} cm^{-3}) (Savin et al. 2003). However, when tunnel resistance was reduced further, an additional heating mechanism appeared which was most dominant at low bath temperatures and low applied bias. This heating was modeled as a parallel leakage path through the tunnel junctions (Savin et al. 2003), and subsequently attributed to back tunneling of quasiparticles and phonons from the superconductor (Savin et al. 2004). Similar heating observations have also been attributed to nonequilibrium effects (Pekola et al. 2004a), states in the superconductor band gap (Pekola et al. 2004a, Giazotto et al. 2006a Section C.1) and quasiparticle back-tunnelling (Fisher et al. 1999).

In Figure 19.7, Savin et al. (2001) show that the electron–phonon coupling factor is a decade lower in silicon devices, than in metals using a T^5 expression for $P_{e\text{–}ph}$. Later work by the same group demonstrated a T^6 dependence for $P_{e\text{–}ph}$ (Savin et al. 2004), with Σ in the range 2×10^8–8×10^8 W K^{-6} m^{-3}, dependent on carrier concentration, as shown in Figure 19.8. Electron–phonon coupling can be reduced using strained silicon, which should prove beneficial for bolometer applications (Muhonen et al. 2010).

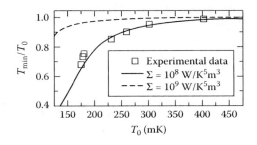

FIGURE 19.7 Relative minimum electron temperature T_{min}/T_0 as a function of phonon temperature T_0 ($T_0 = T_b$). Experimental points are measured from the sample shown in Figure 19.6. Solid (dashed) curve corresponds to the numerical solution of $P_{max} + P_{e\text{–}ph} = 0$ with $\Sigma = 1.0 \times 10^8$ W K^{-5} m^{-3} ($\Sigma = 1.0 \times 10^9$ W K^{-5} m^{-3}). (From A. Savin et al., *Applied Physics Letters* 79(10) 1471, 2001. With permission.)

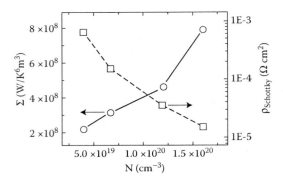

FIGURE 19.8 Electron–phonon coupling constant (circles, left axis) and specific contact resistance (squares, right axis) as a function of carrier concentration for heavily doped Si. (From A. Savin et al., *Physica Scripta* T114 57, 2004. With permission.)

19.4 A Typical Cooler: Common Limitations to Cooling

Figure 19.9a shows a typical SINIS cooler fabricated by the Aalto (HUT) group, using e-beam lithography (Pekola et al. 2004a). A central island of copper of purity 6N (99.9999%) is connected to two superconducting aluminum reservoirs at either end of the island via Al_2O_3 tunnel barriers.

Figure 19.9b shows the probe voltage–temperature versus the cooler voltage V_C. Cooling is generally symmetrical with V_C. The plot shows a number of curves for different bath temperatures T_b. It is observed that the temperature reaches a minimum at biases just below 2Δ, this maximum cooling $(T_{e,min} - T_b)$ varies with T_b. At high bath temperatures, cooling is limited by the high heating power of the lattice which, in metals, increases as T_b^5. At lower temperatures cooling is limited by a reduction in the tunneling current, because there will be less hot electrons, as given by the Fermi distribution function. At the lowest bath temperature, the slight increase in temperature for low biases was attributed to nonequilibrium effects and states within the superconductor band gap.

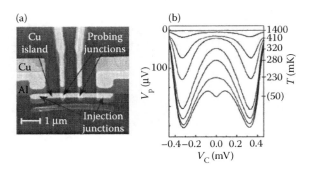

FIGURE 19.9 (a) Plan view scanning electron micrograph of a typical cooler fabricated using a double-angle shadow mask technique. Al (dark gray) was deposited first, through a shadow mask, then the surface was oxidized. Cu (light gray) was then deposited though the same shadow mask, but from a different angle, so that Cu is above the Al in the junction regions. Note an unused Al island at the bottom of the picture is a consequence of this fabrication technique. (b) Cooling data, where voltage V_p across the probe junctions for a constant current bias (28 pA) is shown against voltage V_C across the two injection junctions. Cryostat temperature, corresponding to the electron temperature on the N island at $V_C = 0$ is indicated on the right vertical axis. Below 100 mK this correspondence is uncertain. (From J. P. Pekola et al., *PRL* 92 056804-1, 2004a. With permission.)

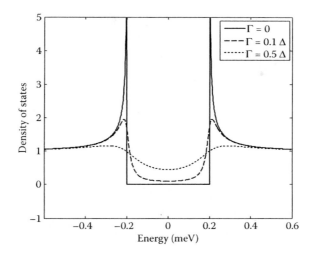

FIGURE 19.10 Density of states, as given by the Dynes formula for $\Delta = 0.2$ meV (approximate value for aluminum) and for different values of the smoothing parameter Γ, top to bottom $\Gamma = 0$, 0.1Δ, 0.5Δ.

Real devices often contain imperfections and/or impurities, also proximity of a normal metal may cause energy states to occur in the superconductor bandgap. States in the superconductor bandgap can be modeled using the Dynes formula, Equation 19.14 (Dynes et al. 1984, Pekola et al. 2004a).

$$g(E, \Gamma) = \left| \text{Re} \left[\frac{E + i\Gamma}{\sqrt{(E + i\Gamma)^2 - \Delta^2}} \right] \right| \tag{19.14}$$

The smoothing parameter Γ is often expressed as a factor of Δ, and its effect is shown in Figure 19.10.

19.4.1 Joule Heating

Leakage through the junction results in a heating power which is given by a $V^2/2R_{leak}$ term in the heat balance equation for a double junction. The factor of 1/2 comes from $2(V/2)^2/R_{leak}$ for the double-junction device. R_{leak} is the leakage resistance per junction, whereas V is the voltage across two junctions. This leakage could, for instance, be caused by pin holes in the tunnel barrier. The model is also useful as an approximation to the effect of gap states in the superconductor or Andreev reflection. Figure 19.11 shows the effect of such a leakage current on the T–V plot of a cooler junction.

The resistivity of degenerately doped silicon (10^{-2}–10^{-4} Ω cm) is higher than that of normal metals such as copper ($\sim 10^{-6}$ Ω cm). This adds a series resistance between the cooler junctions, and hence a joule heating term to the heat balance equation, given by I^2R_{Si}, where I is the current through the device and R_{Si} is the resistance of the silicon island. At low temperatures and for $V < 2\Delta/e$ the tunnel current is low and the effective junction resistance dV/dI is very high, so most of the voltage drops across the junctions and there is little joule heating. At higher biases, as the tunnel current begins to increase, the joule heating increases and dV/dI drops to a lower value so that the voltage drop across the junctions becomes a smaller proportion of the total bias across the device; the net effect of this is to spread out the temperature versus bias characteristic toward higher biases and also to increase the temperature minimum as shown in Figure 19.12. Note that we have used a high $R_{Si} = 1$ kΩ to illustrate this point, but a typical silicon resistance is usually lower, the device mentioned in Savin et al. (2003) with a dopant concentration of 4×10^{19} cm^{-3} had an R_{Si} of 150 Ω which results in only a small rise in the minimum temperature.

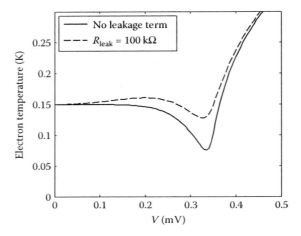

FIGURE 19.11   Junction leakage heating. For $R_{leak} = 100$ kΩ, for each junction, compared to no leakage current. The following parameters were also used, and correspond to values given in Savin et al. (2003): $\Delta = 0.17$ meV, $R_T = 750$ Ω, $\Sigma = 10^8$ W K^{-5} m^{-3} and $\Omega = 4.2 \times 10^{-17}$ m^{-3} for a typical silicon microcooler, with a carrier concentration of 4×10^{19} cm^{-3}. The superconductor density of states is assumed to be ideal, using Equation 19.4.

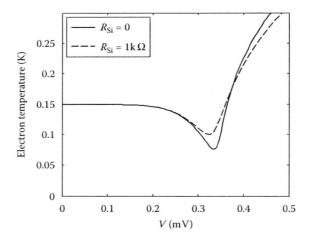

FIGURE 19.12   Silicon resistance joule heating. Electron temperature versus cooler voltage bias for bath temperature $T_p = 150$ mK, $R_{Si} = 0$ (solid line) and $R_{Si} = 1$ kΩ (dashed line). Calculated using Equations 19.1, 19.3, 19.4, 19.6 through 19.9 but with an additional $I^2 R_{Si}$ term in the heat balance Equation 19.7 to account for the Joule heating. The other parameters are as used for Figure 19.10.

19.4.2 Quasiparticle Related Heating

Quasiparticles entering the superconductor from the normal metal can cause significant heating of the normal metal if they linger near the junction (Jochum et al. 1998). Two quasiparticles can recombine to form a superconducting Cooper pair and a phonon, which is absorbed by the junction. Secondly, quasiparticles may tunnel back through the barrier into the metal, effectively decreasing the net cooling current across the junction. These two mechanisms can be modeled jointly by the term (Fisher et al. 1999, Clark et al. 2004):

$$\beta P_S \tag{19.15}$$

A simplifying assumption where $\beta < 1$ denotes the fraction of the power deposited in the superconducting electrode P_S that is returned to the metal. β is a parameter dependent only on the temperature of the surrounding bath, and

$$P_S = IV + P_{cool} \tag{19.16}$$

Figure 19.13 shows the effect of quasiparticle heat return for $\beta = 0.1$. Quasiparticle heating can seriously limit the cooling power of the device, and it is important to allow the quasiparticles to leave the tunneling region of the junction before they can tunnel back or recombine. The simplest solution is to use a thick superconductor, with minimum overlap between superconductor and normal metal to encourage diffusion away from the junction area (Ullom et al. 2001, Clark et al. 2004). It also helps to use arrays of many small area junctions (Leoni et al. 1999).

It is often beneficial to incorporate a "quasiparticle trap," consisting of a normal metal in contact with the superconductor, or via an oxide tunnel junction (Pekola et al. 2000a, Ullom et al. 2000, Clark et al. 2004). Copper quasiparticle traps, outside the SINIS structure are shown in Figures 19.9a, 19.14, and 19.16. The quasiparticles minimize their energy by falling into the Cu, as shown in Figure 19.14. The trap should be some distance away from the junction, in order not to suppress superconductivity at the junction by the proximity effect (Solymar 1972).

As electrons are extracted from the normal metal, electron–electron scattering restores equilibrium. If the tunneling rate is too high compared to the relaxation rate, then the electrons in the normal metal will be driven out of equilibrium and eventually the extraction rate will be determined by the rate at which electron–electron scattering replenishes the tunneling channels (Edwards et al. 1995, Pekola et al. 2004a). Edwards et al. show that the relaxation rate falls at low temperatures and when this limits the extraction rate the NIS cooling power can vary as T^3 at low-enough temperatures.

Furthermore, inelastic scattering in the superconductor or the inverse proximity effect from the nearby N region, could lead to states in the superconducting energy gap. Pekola et al. show how these effects may lead to anomalous heating near $V = 0$ at the lowest temperature, as shown in Figure 19.9b (Pekola et al. 2004a).

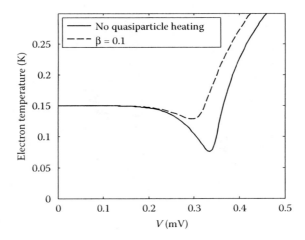

FIGURE 19.13 Electron temperature with (dashed line) and without (solid line) quasiparticle heating. The other parameters are as used for Figure 19.11.

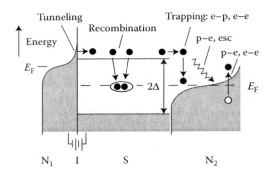

FIGURE 19.14   Energy-level diagram. Occupied states are shaded. NIS junction electrodes are marked N_1 and S. Current flow through the junction creates quasiparticles in S. Some quasiparticles recombine before reaching an adjacent normal metal film marked N_2. Quasiparticles which scatter inelastically in N_2 are trapped. Quasiparticles relax in N_2 by electron–phonon (e–p) and electron–electron (e–e) interactions. A phonon (jagged line) emitted by a relaxing quasiparticle can interact with other electrons (p–e) or escape the trap (esc). Electrons in the trap are heated by phonon–electron (p–e) and electron–electron (e–e) interactions. (From J. N. Ullom, P. A. Fisher, and M. Nahum, *Physical Review B*, 61, 14839, 2000. With permission.)

19.4.3 Andreev Reflection

For voltages $eV \ll \Delta$ charge transfer occurs through Andreev reflection if the barrier transparency is high. An electron (hole) in the normal metal impinging on the superconducting interface is reflected as a hole (electron) and creates a Cooper pair in the semiconductor, Figure 19.15a. The energies of the hole and electron are located symmetrically around E_F and hence there is no energy transfer. The Andreev current is associated with significant energy dissipation in the normal metal. Disorder in the normal metal, or reflection from a second barrier (due to close proximity), can greatly enhance this process; a carrier then experiences several collisions with the interface, Figure 19.15b.

Bardas and Averin have shown that the junction cooling power is a function of the transparency of the junction and at low transparency the cooling power scales as $T^{3/2}$ (Bardas and Averin 1995). However, the cooling power reaches a peak at a particular transparency and then falls at higher transparencies as Andreev reflection begins to dominate. Also, as temperature drops the peak transparency also drops,

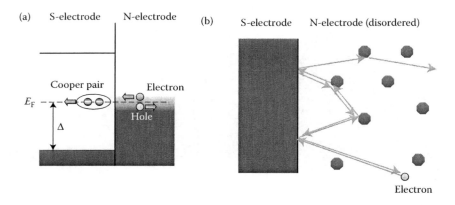

FIGURE 19.15   (a) Andreev reflection at an SN interface. (b) multiple Andreev reflections due to scattering in the normal metal. (Figure from A. Savin, priv comm.)

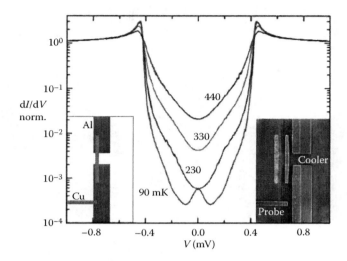

FIGURE 19.16 Main figure: Normalized differential conductance of an Al/Al$_2$O$_3$/Cu/Al$_2$O$_3$/Al cooler (fabricated at CNRS) as a function of the voltage and at the cryostat temperatures of 90, 230, 330, and 440 mK. Left inset: Geometry of the sample. Right inset: Scanning electron microscope micrograph of a typical cooler sample made of a normal metal Cu electrode (light gray) connected to two superconducting Al reservoirs (dark gray) through tunnel junctions. One of the additional probe junctions connected to one Al reservoir is visible at the bottom. Al (dark gray) was deposited first, through a shadow mask, then the surface was oxidized. Cu (light gray) was then deposited, though the same shadow mask, but from a different angle, so that Cu is above the Al in the junction regions. The Cu above the Al on the right-hand side, accessed through an oxide barrier, acts as an effective quasiparticle trap, at about 0.3 μm to the right of the cooler junctions. Note an unused Al island at the left-hand side is a consequence of this fabrication technique. (From S. Rajauria et al., *Physical Review Letters,* 100 207002, 2008. With permission.)

such that the Andreev process is stronger at low temperatures for a given tunnel junction and hence the cooling power may be reduced and heating may be observed.

The CNRS group have published evidence of Andreev reflection and considered Andreev heating in the energy balance equation. Figure 19.16 shows the conductance versus bias characteristic of a SINIS cooler fabricated by the CNRS group. The upward curvature at $eV \leq \Delta$ is clear evidence of cooling. The temperature may be lowered from 230 mK at zero bias to <50 mK at $2\Delta/e$. The zero bias anomaly at 90 mK is convincing evidence of an Andreev current. The current–voltage characteristic of the SINIS cooler are shown in Figure 19.17. It was necessary to include a significant heating term $I_A V$ to get a good fit to the data (Rajauria et al. 2008).

The authors obtain excellent fits to their experimental data, without consideration of quasiparticle heating. This suggests that such heating has been minimized in their device geometry, which includes a quasiparticle trap within 0.3 μm of the junction, and a junction area of 1.5×0.3 μm.

Giazotto has proposed a number of ways (Giazotto et al. 2002, 2006b) of minimizing Andreev currents in these junctions but, as yet, no experiments have been carried out.

19.5 Lattice Cooling

With a cooled electron distribution, established by one or more junction refrigerators, energy can be extracted from the lattice. The NIST group has demonstrated that it is possible to cool a Si$_3$N$_4$ membrane, using the arrangement shown in Figure 19.18. It consists of four SINIS junctions located on a bulk substrate. Cold fingers are extended from the normal metal electrodes onto the membrane (Clark et al. 2005).

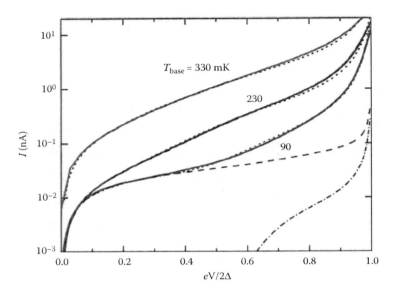

FIGURE 19.17 Current–voltage characteristic of the SINIS cooler junction as a function of the voltage, at cryostat bath temperatures of 90, 230, and 330 mK together with best-fit calculated curves. Dot-dashed line: Model including the cooling by the tunnel current. Dashed line: Model including in addition the Andreev current, but not the related heating. Dotted lines: Full model taking into account the Andreev current and the related heating. (From S. Rajauria et al., *Physical Review Letters*, 100 207002, 2008. With permission.)

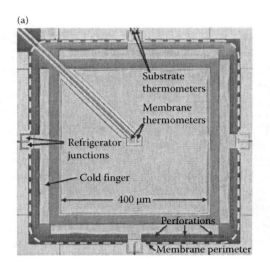

FIGURE 19.18 (a) Optical micrograph of NIS refrigerator. Four pairs of refrigerator junctions surround a micro-machined Si_3N_4 membrane. Cold fingers extend from the refrigerator junctions onto the membrane. Pairs of thermometer junctions are located at the center of the membrane and on the substrate. A 200 nm layer of Au covers the Al–Mn of the cold fingers and the center of the membrane for better thermal conductivity. There is no electrical connection between the cold fingers and the circuitry in the center of the membrane. (b) A germanium thermometer, cube with 250 µm length per side, was cooled from 320 to 240 mK by this means. (From A. M. Clark et al., *Applied Physics Letters*, 86, 173508, 2005. With permission.)

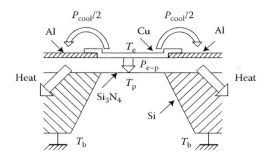

FIGURE 19.19 Schematic describing the working principle of the phonon cooling. The substrate material is silicon with a thin Si_3N_4 film on the surface and on top is the copper strip central electrode of the SINIS refrigerator with two NIS junctions. (From R. Leoni, *New Astronomy Review* 43 317, 1999. With permission.)

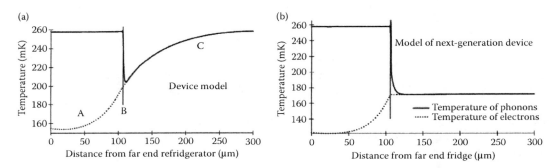

FIGURE 19.20 Simulations of temperature profiles at $T_b = 260$ mK. Refrigerators are located on the bulk substrate (point A), where the electron temperature is reduced below the temperature of the lattice. Cold fingers extend onto a membrane (point B). The cold-finger temperature sets the new bath temperature for any electronics located on the membrane. (a) Model of device cooling, including realistic power loads to cold finger from hot substrate phonons and bias power from membrane device. Power load on cold finger is ~75 pW. The large thermal gradient in the cold finger is due to low electronic thermal conductance G_e. (b) Reduced power load through the membrane due to perforations at the edge, and increased thermal conductivity in the cold fingers from the addition of a gold overlayer. Predicted cooling of the membrane is from 260 to 170 mK. (From N. A. Miller et al., *IEEE Trans Appl. Super.* 15 556, 2005. With permission.)

The basic principle of membrane, that is, phonon, cooling is illustrated in Figure 19.19. The tunnel junction extracts energy at a rate P_{cool} from the normal metal electrons and releases it to the substrate at temperature T_b. Heat (P_{e-p}) flows from the membrane into the electron system by means of electron–phonon coupling, cooling the membrane to a temperature $T_p \geq T_e$. Lowest membrane temperatures are achieved with a low thermal conductivity of the membrane G_m, high electron–phonon thermal conductance G_{e-ph} and a high thermal conductance of the electron cold finger G_e. Efficient cooling is made possible by the relative strengths of the thermal conductances: $G_e \approx 150\ G_{e-ph} \approx 3900\ G_m$, near 200 mK (Miller et al. 2005). Figure 19.20 shows a typical result of modeling of the temperature profiles in a NIST device (Miller et al. 2005).

19.6 Some Potential Applications

19.6.1 Bolometer with Direct Cooling of Electrons

Figure 19.21 shows a sketch of a hot electron bolometer for the detection of EM radiation based on NS and NIS junctions which was originally proposed by Nahum and Martinis (1993). Incident radiation

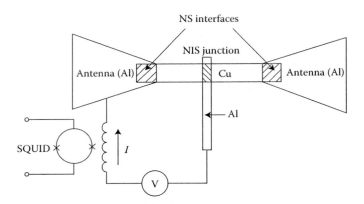

FIGURE 19.21 Schematics of the hot-electron bolometer with superconducting antennas, NS junction Andreev mirrors, copper absorber and a voltage-biased NIS junction. (From D. Golubev and L. Kuzmin, *Journal of Applied Physics*, 89 6464, 2001. With permission.)

heats the electrons in a small copper island and the change in temperature is sensed as a change in the NIS tunnel current using a superconducting quantum interference device (SQUID). Weak e–ph coupling at low temperature results in a large temperature rise for small input power.

Energy is focused on the Cu island by superconducting contacts, which make electrical but not thermal contact. Since the energy gap of the superconductor is larger than the thermal energy of the electrons, Andreev reflection at the interface traps the absorbed energy in the central active region. In other words the NS interface acts as an "Andreev mirror."

Golubev and Kuzmin have analyzed the effect of the electron cooling by the NIS junction and show that it can improve the performance. For voltages close to Δ/e and for $T_b = 300$ mK the electron temperature is lowered to 150 mK as shown in Figure 19.22a. At the same time the responsivity $S_I = dI/dP$, where P is the incoming radiation power, is substantially increased, Figure 19.22b (Golubev and Kuzmin 2001).

The beneficial effect of electron cooling on the noise equivalent power (NEP), for $T_b = 100$ mK, is shown in Figure 19.23. In this example the total noise is dominated by the e–ph noise. The e–ph noise is sensitive to volume and can therefore be reduced to a negligible value by reducing the volume of the

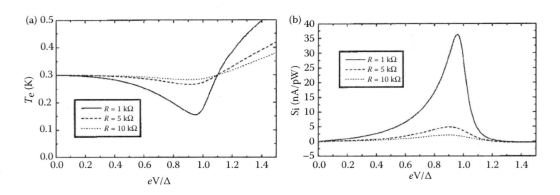

FIGURE 19.22 Electron temperature and responsivity for three tunnel resistances of a single NIS junction. $T_b = 300$ mK. (a) Electron temperature as function of the bias voltage, assuming zero background power P_0. Also, $\Delta = 0.174$ meV, $\Sigma = 3 \times 10^9$ W m^{-3} K^{-5}, $\Omega = 0.05$ μm^3. (b) Responsivity S_I as a function of the bias voltage. When biased close to Δ/e the electron temperature decreases resulting in an increase in the responsivity. (From D. Golubev and L. Kuzmin, *Journal of Applied Physics*, 89, 6464, 2001. With permission.)

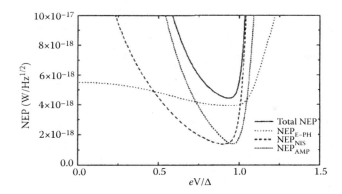

FIGURE 19.23 Different contributions to the total NEP of the device for voltage-biased operation at $T_b = 300$ mK and for $R_T = 1$ kΩ. NEP_{E-PH} is noise associated with the heat flow between phonons and electrons, NEP_{NIS} is noise associated with the NIS junction and NEP_{AMP} is noise of the SQUID amplifier. (From D. Golubev and L. Kuzmin et al., *Journal of Applied Physics*, 89, 6464, 2001. With permission.)

absorber. Kuzmin and Golubev add to this with a general statement that the optimal regime is realized when thermal conductance through the tunnel junctions is larger than the electron–phonon conductance (Kuzmin and Golubev, 2002).

19.6.2 Transition Edge Bolometer with Indirect Cooling of Electrons

Transition Edge Sensor (TES) bolometers take advantage of the rapid variation in resistance of a superconductor at the superconducting phase transition and have the advantages of low noise and short time constant. They are typically operated at 100 mK, which suggests a relatively simple and inexpensive refrigeration set-up, involving a ³He system (base temperature 260 mK) and SINIS coolers. Figure 19.24 shows an x-ray TES-mounted on a SINIS-cooled membrane.

The structure was mounted on an adiabatic demagnetization refrigerator (ADR) and the TES was cooled from an ADR temperature of 260 mK to 162 mK using the NIS junctions, at a TES bias power of 22 pW. Figure 19.25 shows the temperature of the unbiased TES as measured by Johnson noise thermometry and calculations of what might be expected with improvements in cooler design.

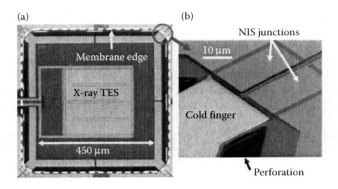

FIGURE 19.24 (a) TES x-ray sensor integrated with NIS refrigerators Bi absorber not shown. Four pairs of NIS refrigerators are located at the corners of a Si_3N_4 membrane dashed outline. Y-shaped cold fingers extend from the normal metal of the NIS junctions onto the membrane. (b) False-color SEM image of membrane corner. (From N. A. Miller et al., *Applied Physics Letters*, 92, 163501, 2008. With permission.)

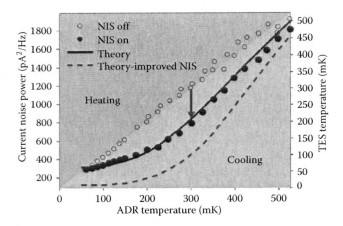

FIGURE 19.25   Johnson noise thermometry of the unbiased TES with NIS refrigerators off (open circles) and on (solid circles). Current noise power averaged between 0.6 and 2 kHz (left axis) and cooling calculated from the noise power (right axis). The NIS cooling is well matched by theory (solid line). Potential cooling with improved NIS refrigerators shown as a dashed line. (From N. A. Miller et al., *Applied Physics Letters*, 92, 163501, 2008. With permission.)

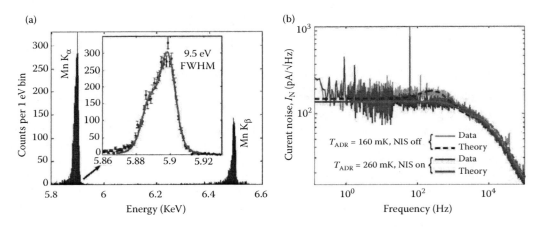

FIGURE 19.26   (a) X-ray spectrum (9069 pulses) using the TES at an ADR temperature of 260 mK. The inset shows a zoomed view of the 5.9 keV Mn K_α complex and a theoretical fit. The TES is biased with a power of 22 pW. (b) TES current noise with the NIS refrigerators off and on. ADR temperatures were chosen to compare the same TES bath temperature. Results are similar, showing that NIS cooling to 160 mK is comparable to an ADR temperature of 160 mK. (From N. A. Miller et al., *Applied Physics Letters*, 92, 163501, 2008. With permission.)

Figure 19.26a Shows a SINIS-cooled x-ray spectrum with the ADR at a temperature of 260 mK, 75 mK above the superconducting transition (TES cooled to 162 mK). An energy resolution of 9.5 eV (width at half-maximum) is obtained at 5.9 keV, which is an improvement compared to the 13 eV resolution obtained with an STJ detector at a lower bath temperature of 210 mK (Li et al. 2001). The authors demonstrate that a temperature of 160 mK achieved with NIS cooling from 260 mK leads to the same current noise power as a temperature of 160 mK achieved with the NIS junctions off and cooling by the ADR only, Figure 19.26b.

19.7 Conclusions

It has been shown that superconducting tunnel junction devices may be used to cool electrons from 300 mK to below 50 mK and that a macroscopic-sized membrane (~400 μm across) with Ge thermometer was cooled from 320 to 240 mK. The demonstration of a semiconductor-based cooler (Section 19.3) offers the possibility of high levels of integration. The work in Section 9.6 shows the impressive performance of modern cryogenic sensors, which can already be obtained using SINIS-cooled platforms. Continual improvements in design promise substantial advances in performance and suggest that the goal of a simple ^{3}He system and SINIS cooling to much lower temperatures cannot be far away. When this system becomes available it could lead to numerous developments in the astronomical, biomedical, quantum computing, and materials analysis areas. For further reading the authors recommend the excellent review by Giazotto et al. (2006a).

Acknowledgments

The authors are grateful to Jukka Pekola, Leonid Kuzmin, Mika Prunnila, Alexander Savin, Matthias Meschke and Juha Muhonen for useful discussions in this general subject area.

References

Anghel, D. V. and Pekola, J. P., Noise in refrigerating tunnel junctions and in microbolometers, *Journal of Low Temperature Physics*, 123(3–4) 197, 2001.

Bardas, A. and Averin, D., Peltier effect in normal-metal–superconductor microcontacts, *Physical Review B*, 52(17) 12873, 1995.

Clark, A. M., Miller, N. A., Williams, A., Ruggiero, S. T., Hilton, G. C., Vale, L. R., Beall, J. A., Irwin, K. D., and Ullomb, J. N., Cooling of bulk material by electron-tunneling refrigerators, *Applied Physics Letters*, 86 173508, 2005.

Clark, A. M., Williams, A., Ruggiero, S. T., van den Berg, M. L., and Ullom, J. N., Practical electron-tunneling refrigerator, *Applied Physics Letters*, 84 625, 2004.

Dynes, R. C., Garno, J. P., Hertel, G. B., and Orlando, T. P., Tunneling study of superconductivity near the metal-insulator transition, *Physical Review Letters*, 53 2437, 1984.

Edwards, H. L., Niu, Q., Georgakis, G. A., and de Lozanne, A. L., Cryogenic cooling using tunneling structures with sharp energy features, *Physical Review B*, 52(8) 5714, 1995.

Fisher, P. A., Ullom, J. N., and Nahum, M., High-power on-chip microrefrigerator based on a normal-metal/insulator/superconductor tunnel junction, *Applied Physics Letters*, 74 2705, 1999.

Frank, B. and Kretch, W., Electronic cooling in superconducting tunnel junctions, *Physics Letters A*, 235 281, 1997.

Giazotto, F., Heikkila, T. T., Luukanen, A., Savin, A. M., and Pekola, J., Opportunities for mesoscopics in thermometry and refrigeration: Physics and applications, *Review of Modern Physics*, 78 217, 2006a.

Giazotto, F., Taddei, F., Fazio, R., and Beltram, F., Ultraefficient cooling in ferromagnet–superconductor microrefrigerators, *Applied Physics Letters*, 80(20) 3784, 2002.

Giazotto, F., Taddei, F., Governale, M., Castellana, C., Fazio, R., and Beltram, F., Cooling electrons by magnetic-field tuning of Andreev reflection, *Physical Review Letters*, 97 197001, 2006b.

Golubev, D. and Kuzmin, L., Nonequilibrium theory of a hot-electron bolometer with normal metal-insulator-superconductor tunnel junction, *Journal of Applied Physics*, 89 6464, 2001.

Jochum, J., Mears, C., Golwala, S., Sadoulet, B., Castle, J. P., Cunningham, M. F., Drury, O. B. et al., Modeling the power flow in normal conductor-insulator-superconductor junctions, *Journal of Applied Physics*, 83(6) 3217, 1998.

Jug, B. and Trontelj, Z., Electronic refrigerators: Optimization studies, *IEEE Transactions on Applied Superconductivity*, 9(2) 4483, 1999.

Kivinen, P., Savin, A., Zgirski, M., Törmä, P., Pekola, J., Prunnila, M., and Ahopelto, J., Electron–phonon heat transport and electronic thermal conductivity in heavily doped silicon-on-insulator film, *Journal of Applied Physics*, 94(5) 3201, 2003.

Kuzmin, L. and Golubev, D., On the concept of an optimal hot-electron bolometer with NIS tunnel junctions, *Physica C*, 372–376 378, 2002.

Leivo, M. M., Manninen, A. J., and Pekola, J. P., Microrefrigeration by normal-metal/insulator/superconductor tunnel junctions, *Applied Superconductivity*, 5 227, 1998.

Leivo, M. M., Pekola, J. P., and Averin, D. V., Efficient Peltier refrigeration by a pair of normal-metal/insulator/superconductor junctions, *Applied Physics Letters*, 68(14) 1996, 1996.

Leoni, R., On-chip micro-refrigerators for sub-Kelvin cooling, *New Astronomy Review*, 43 317, 1999.

Leoni, R., Arena, G., Castellano, M. G., and Torrioli, G., Electron cooling by arrays of submicron tunnel junctions, *Journal of Applied Physics*, 85(7) 3877, 1999.

Li, L., Frunzio, L., Wilson, C., Prober, D. E., Szymkowiak, A. E., and Moseley, S. H., Improved energy resolution of x-ray single photon imaging spectrometers using superconducting tunnel junctions, *Journal of Applied Physics*, 90 3645, 2001.

Luukanen, A., Leivo, M. M., Suoknuuti, J. K., Manninen, A. J., and Pekola, J. P., On-chip cefrigeration by evaporation of hot electrons at sub-Kelvin temperatures, *Journal of Low Temperature Physics*, 120 281, 2000.

Manninen, A. J., Leivo, M. M., and Pekola, J. P., Refrigeration of a dielectric membrane by superconductor/insulator/normal-metal/insulator/superconductor tunneling, *Applied Physics Letters*, 70(14) 1885, 1997.

Muhonen, J. T., Niskanen, A. O., Meschke, M., Pashkin, Yu. A., Tsai, J. S., Sainiemi, L., Franssila, S., and Pekola, J. P., Electronic cooling of a submicron-sized metallic beam, *Applied Physics Letters*, 94 073101, 2009.

Muhonen, J. T., Prest, M. J., Prunnila, M., Gunnarsson, D., Dobbie, A., Myronov, M., Whall, T. E., Parker, E. H. C., and Leadley, D. R., Strain dependence of electron-phonon energy loss rate in many-valley semiconductors, *Applied Physics Letters*, 98(18) 182103, 2011.

Nahum, M., Eiles, T. M., and Martinis, J. M., Electronic microrefrigerator based on a normal-insulator-superconductor tunnel junction, *Applied Physics Letters*, 65 3123, 1994.

Nahum, M. and Martinis, J. M., Ultrasensitive-hot-electron microbolometer, *Applied Physics Letters*, 63 3075, 1993.

Miller, N. A., Clark, A. M., Williams, A., Ruggiero, S. T., Hilton, G. C., Beall, J. A., Irwin, K. D., Vale, L. R., and Ullom, J. N., Measurements and modeling of phonon cooling by electron-tunneling refrigerators, *IEEE Transactions on Applied Superconductivity*, 15 556, 2005.

Miller, N. A., O'Neil, G. C., Beall, J. A., Hilton, G. C., Irwin, K. D., Schmidt, D. R., Vale, L. R., and Ullom, J. N., High resolution x-ray transition-edge sensor cooled by tunnel junction refrigerators, *Applied Physics Letters*, 92 163501, 2008.

Pekola, J., Tunnelling into the chill, *Nature*, 435 889, 2005.

Pekola, J. P., Anghel, D. V., Suppula, T. I., Suoknuuti, J. K., and Manninen, A. J., Trapping of quasiparticles of a nonequilibrium superconductor, *Applied Physics Letters*, 76 (19) 2782, 2000a.

Pekola, J. P., Heikkila, T. T., Savin, A. M., Flyktman, J. T., Giazotto, F., and Hekking, F. W. J., Limitations in cooling electrons using normal-metal-superconductor tunnel junctions, *Physical Review Letters*, 92 056804–1, 2004a.

Pekola, J. P., Manninen, A. J., Leivo, M. M., Arutyunov, K., Suoknuuti, J. K., Suppula, T. I., and Collaudin, B., Microrefrigeration by quasiparticle tunnelling in NIS and SIS junctions, *Physica B*, 280 485, 2000b.

Pekola, J. P., Schoelkopf, R., and Ullom, J. N., Cryogenics on a chip, *Physics Today*, May 41, 2004b.

Rajauria, S., Gandit, P., Fournier, T., Hekking, F. W. J., Pannetier, B., and Courtois, H., Andreev current-induced dissipation in a hybrid superconducting tunnel junction, *Physical Review Letters*, 100 207002, 2008.

Roukes, M. L., Freeman, M. R., Germain, R. S., and Richardson, R. C., Hot electrons and energy transport in metals at millikelvin temperatures, *Physical Review Letters*, 55 422, 1985.

Rowell, J. M. and Tsui, D. C., Hot electron temperature in InAs measured by tunneling, *Physical Review B*, 14(6) 2456, 1976.

Savin, A., Pekola, J., Prunnila, M., Ahopelto, J., and Kivinen, P., Electronic cooling and hot electron effects in heavily doped silicon-on-insulator film, *Physica Scripta*, T114 57, 2004.

Savin, A., Prunnila, M., Ahopelto, J., Kivinen, P., Torma, P., and Pekola, J., Application of superconductor–semiconductor Schottky barrier for electron cooling, *Physica B*, 329–333 1481, 2003.

Savin, A., Prunnila, M., Kivinen, P. P., Pekola, J. P., Ahopelto, J., and Manninen, A. J., Efficient electronic cooling in heavily doped silicon by quasiparticle tunneling, *Applied Physics Letters*, 79 (10) 1471, 2001.

Solymar, L., *Superconductive Tunnelling and Applications* (Chapman & Hall, London), 1972.

Tinkham, M., *Introduction to Superconductivity*, 2nd ed. (McGraw-Hill, New York), 1996.

Ullom, J. N., Fisher, P. A., and Nahum, M., Measurements of quasiparticle thermalization in a normal metal, *Physical Review B*, 61 14839, 2000.

Ullom, J. N., van den Berg, M. L., and Labov, S. E., A new idea for a solid-state microrefrigerator operating near 100 mK, *IEEE Transactions on Applied Superconductivity*, 11 639, 2001.

Wellstood, F. C., Urbina, C., and Clarke, J., Hot electron effects in metals, *Physical Review B*, 49 5942, 1994.

20

Heat Dissipaters

Jesus Esarte
CEMITEC-R&D Center

Cecilia Wolluschek
CEMITEC-R&D Center

Jesus Mª Blanco
University of the Basque Country

David Prieto
University of the Basque Country

20.1 Introduction

This chapter is not intended to lead the reader into the complex world of heat transfer and the cumbersome concepts that appear in it, but to explain some basic concepts useful for anyone who is in the arduous task of selecting the most appropriate heat sink for a thermoelectric application.

Extended surfaces or "fins" are commonly used to enhance heat transfer, either on thermal systems cooling or energy-harvesting situations such as Peltier modules.

On one hand, an elaborate procedure to carry out preliminary design and also optimization of these particular surfaces is presented in this chapter. A description of the most commonly used heat sinks in the electronics field in general and in particular the thermoelectric modules is given, detailing the key aspects to be considered for each of them.

First of all, the main applications and general classification is shown, followed by an intensive study of the heat transfer associated, focused on flat fins which are the most commonly used for many industrial applications. Finally, complementary studies over other geometries of variable section and circular fins have also been included.

On the other hand, two algorithms have been performed in MATLAB®. The first one aims at predicting temperature distribution and heat transfer for different fin configurations and boundaries for aluminum fins whereas the second one aims to optimize a rectangular fin arrangement heat sink suitable for Peltier module heat dissipation. For both cases, practical examples are presented. Those analytical procedures use variables such as thickness, length, and spacing among fins and so can be applied not only for the design process but also for validation purposes of a further computational fluid dynamics methodology (CFD) carried out, showing good agreement with the analytical results obtained here but also with experimental measurements also available for the same case.

Finally, the operation of phase-change coolers is described, with special emphasis on the key aspects that determine its operation, which must be considered when designing or selecting a phase-change cooler for the Peltier module application. Also described is the phase-change cooler's capillary limit to finally describe the procedure for obtaining the contact angle.

20.2 Some Basic Concepts

20.2.1 Mechanisms of Heat Transfer in Dissipaters

A heat sink or heat dissipater is a physical element that extracts heat from the generating source and expels it to the immediate surrounding or environment. The transfer of heat from the source to the environment is performed by two heat transfer mechanisms: conduction (through the heat sink "$T_{base} - T_s$") and convection (exterior surface heat sink to the environment, "$T_s - T_{amb}$") (Figure 20.1).

Conduction is the way in which transfer of heat takes place in solid bodies and is governed by Fourier's law. This law states that the heat flux transmitted by conduction in a given direction is proportional to the cross-sectional area, perpendicular to that direction, and the temperature gradient in that direction.

$$\frac{dQ_x}{dt} = -\lambda S \frac{\partial T}{\partial x} \tag{20.1}$$

Under this law, the higher the thermal conductance, the lower the temperature gradient for a given heat flux. This means that materials such as aluminum, copper, and others [1,2] are used in heat dissipation.

Convection is the heat transfer mechanism that takes place in a fluid due to direct contact between the molecules and internal macroscopic motion (movement caused by artificial or natural forces). Convection heat transfer requires knowledge of the principles of heat conduction, fluid dynamics, and boundary layer theory. However, all this can come together in a single parameter "α" according to Newton's law of cooling.

$$\dot{Q} = \alpha S(T_s - T_{amb}) \tag{20.2}$$

From expression (20.2) it follows that for a given heat flux, the higher the convective heat transfer coefficient, the lower the thermal gradient between the heat sink's external surface and the surroundings.

Since the objective is heat removal from the source with the lowest temperature gradient between the source and the environment, sinks are made of high thermal conductivity (minimum conduction resistance) materials and high convection coefficient (minimum convection resistance is sought).

$$\dot{Q} = \frac{\theta_{base} - \theta_a}{R_T} = \frac{\theta_{base} - \theta_a}{R_{cond} + R_{conv}} \tag{20.3}$$

The reader can delve into the heat transfer through the extensive literature [3–5].

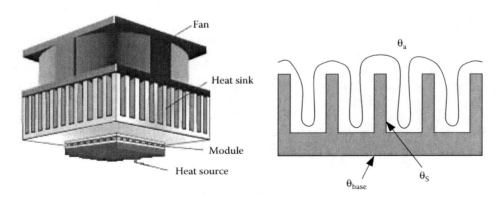

FIGURE 20.1   Diagram of the heat transfer in a heat sink.

20.2.2 Thermal Contact Resistance

Under a fixed temperature gradient, a heat sink is better or worse depending on whether it dissipates more or less heat. As discussed in the previous section, this capacity is linked to both heat conduction and heat convection. However, contact between the heat sink and the heat source is a factor that plays a crucial role in heat removal capacity.

When two bodies (A, B) at different temperatures come into contact, heat coming from the hot body flows to the cold one causing a temperature gradient in both solids. However, in the contact a discontinuity appears in the temperature gradient, Figure 20.2. This discontinuity reveals the existence of a resistance to heat flow through the contact, thermal contact resistance "R_{cont}" [6,7], which causes a larger temperature gradient.

$$\dot{Q} = \frac{\theta_S^A - \theta_S^B}{R_{cont}} \qquad (20.4)$$

The explanation for the existence of this thermal gradient is that contact between the two materials is not 100% perfect but there are gaps. This contact causes the heat flow is directed toward the area of solid–solid contact rather than the area of holes, the heat flow being null if holes are vacuum, Figure 20.3.

Factors such as contact pressure, material hardness, surface quality, and packing material vary the thermal contact resistance.

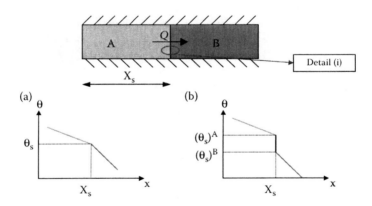

FIGURE 20.2 Temperature distribution: (a) ideal contact; (b) real contact, thermal contact resistance.

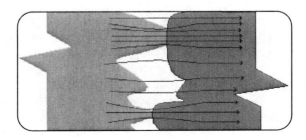

FIGURE 20.3 Heat flow through the contact, cross section of detail (i) from Figure 20.2.

- *The contact pressure and hardness of the material.* As contact pressure increases, the contact surface increases due to deformation of the material (higher or lower depending on the hardness of the material).
- *The interface material.* Filling the gaps with a material of good thermal conductivity [8–11] improves the heat flow through the contact. Many works focus on the search for new materials in order to improve conduction through the contact.
- *The surface quality (roughness, flatness, surface impurities)* [12]. Surfaces badly polished significantly reduce metal-to-metal contact, which implies a reduction in heat conduction capacity and therefore greater thermal resistance.

In summary, good contact must be ensured between solids (polished surfaces, use of material interface, increasing the contact pressure).

20.3 Considerations When Designing a Heat Sink

In electronics and thermoelectric systems, the most commonly used heat sinks are air cooler, liquid cooler, and phase-change cooler (liquid–gas, solid–liquid). Each has its field of application and features to be considered when trying to design or select one.

20.3.1 Air Cooler

Extended surfaces or "fins" are commonly used in air coolers to enhance heat transfer, either on thermal systems cooling or energy harvesting situations such as Peltier modules, Figure 20.1.

Fins are additional surfaces installed in certain locations leading to increase the heat exchange surface of a given piece of equipment with the surrounding environment [13,14].

They are especially used when the convection heat transfer coefficient between the solid and the medium presents lower values, as with the case of natural convection. Thus, the low coefficient is compensated in some way with an increase of the area in contact with the fluid medium [15], and the heat power transferred is set by the following expression.

$$\dot{Q} = \alpha \cdot S \cdot \left(\theta_s - \theta_a\right) \tag{20.5}$$

Fins have been traditionally used in air-cooled internal combustion engines, cooling of electronic components, electrical machines in general, heat exchangers, air conditioning, heat recovery, and a long list of industrial elements [16,17].

Fins can adopt very different shapes, and this depends largely on the morphology of the solid and concrete implementation for which it is intended [18]. They can be classified into three groups according to their shape:

- STRAIGHT: straight starting surface and generated with constant profile.
- RING: Cylindrical starting surface and generated with constant profile.
- TRIANGULAR: Section decreases, according to a straight generator curve.

If a fin has a cylindrical or conical shape it is called "needle." Extended surfaces can be installed over flat or curved surfaces. If the disposal is "longitudinal," the starting area (where fin is supported) is flat, assuming that the tube radius is large compared to the thickness of the fin. When fins are solids of revolution or parallelepiped, they are also called "protuberances."

Protuberances are treated with constant temperature distributions for each cross section normal to the base surface, which is equivalent to admitting that the relationship between height and diameter is very high, so unidirectional heat transfer can be considered. When this assumption is not met, the phenomenon of three-dimensional heat transfers must be studied, with appropriate boundary conditions [19,20].

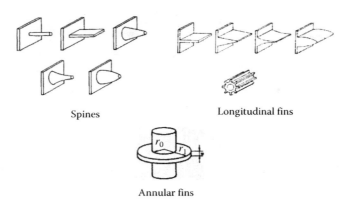

Spines   Longitudinal fins

Annular fins

FIGURE 20.4 Different types of the most commonly used fins.

In Figure 20.4, some common types of individual fins can be seen, with their relative configuration and different forms that can be adopted. They can be grouped according to the same pattern in order to achieve the goal of any particular heat dissipation problem.

A large number of works have investigated the effect of self-sustained flow oscillations on the heat transfer coefficient, such as the transition from steady to unsteady flow and its spatial propagation in a multilouvered fin array [21]. Instabilities were then initiated in the downstream half of the array near the exit, which subsequently spread upstream into the fin array. This fact has a marked effect on the heat transfer coefficient. On the other hand, these same conditions can result in large recirculation zones decreasing the local heat transfer coefficient.

Although thermal wakes can be expected to have a very large effect on the heat capacity of fins, there have not been many studies devoted to this particular aspect. Our objective in this chapter is to achieve a better understanding of the thermal phenomenon involved as it can lead to effective heat transfer management techniques [22]. This has been done mainly through experimental measurements, in order to simulate different scenarios but there are also other different techniques.

The chapter is organized as follows: First, a brief description of the parameters used for a theoretical characterization of heat transfer of fins is presented, followed by the definition of a new and easy-to-use MATLAB algorithm in order to facilitate the design and optimization of this procedure, validated through experimental data. Two cases have been analyzed following this methodology, the second one being suitable for the optimization of a rectangular fin arrangement heat sink for Peltier module heat dissipation.

Finally, a CFD (computational fluid dynamics) study corresponding to the same problem has been performed, showing quite good agreement with the numerical and experimental results previously obtained, so this computational tool has been probed as an effective way to improve design and optimization of fins.

20.3.1.1 Study of a Flat Fin with Constant Section; Performance and Selection Criteria

The rectangular profiles over flat surfaces are the simplest case of extended surfaces. They can be arranged on a flat wall or on the axial length of a tube of large radius of curvature. The set consisting of rectangular longitudinal fins is easily manufactured by extrusion or casting [23].

In steady state, conducted heat through a set of fins is discharged outside through a combined process of convection–radiation to the surrounding fluid medium, usually air, which remains at constant temperature [24]. Figure 20.5 shows the outline of a straight fin of uniform section, with all the parameters that define it.

For the study, the following hypotheses can be assumed:

- The temperature gradient can be considered one dimensional, assuming that the thickness of the fin can be neglected against its length.
- Steady state is considered.

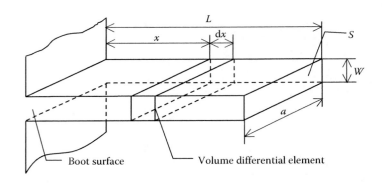

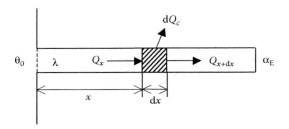

FIGURE 20.5   Elements of straight fins of uniform section and energy balance on a differential element.

- The fin material is homogeneous and isotropic.
- The thermal conductivity is considered constant throughout the fin.
- There is no internal heat generation in the fin itself.
- The thermal contact resistance between the fin and the starting area is zero, since the fin is considered as an extended surface of the primary surface [25].
- Once the heat passes through the fin base (starting area), heat is transferred by conduction along the fin.

Performing a heat balance on a differential element of the fin of length dx, at a certain distance x from the base, as shown in the previous figure, it results in

$$\dot{Q}_x = \dot{Q}_{x+dx} + d\dot{Q}_c \tag{20.6}$$

where

$$\dot{Q}_x = -\lambda \cdot S \cdot \left. \frac{d\theta}{dx} \right|_x \tag{20.7}$$

$$\dot{Q}_{x+dx} = -\lambda \cdot S \cdot \left. \frac{d\theta}{dx} \right|_{x+dx} = -\lambda \cdot S \cdot \left. \frac{d\theta}{dx} \right|_x - \lambda \cdot S \cdot \left. \frac{d^2\theta}{dx^2} \right|_x \cdot dx - \cdots \tag{20.8}$$

$$d\dot{Q}_c = \alpha \cdot P \cdot (\theta - \theta_a) \cdot dx \tag{20.9}$$

Neglecting terms of second order from Equation 20.8 and substituting Equations 20.7 through 20.9 into Equation 20.6 results in

$$-\lambda \cdot S \cdot \left. \frac{d^2\theta}{dx^2} \right|_x \cdot dx + \alpha \cdot P \cdot (\theta - \theta_a) \cdot dx = 0 \tag{20.10}$$

Making the change of variable: $\beta = (\theta - \theta_a)$ the differential equation of the so-called relative temperature field in straight fins of constant section is as follows:

$$\frac{d^2\beta}{dx^2} - m^2 \cdot \beta = 0 \tag{20.11}$$

where

$$m = \sqrt{\frac{\alpha \cdot P}{\lambda \cdot S}} \tag{20.12}$$

The general solution for this differential equation is

$$\beta = C_1 \cdot e^{mx} + C_2 \cdot e^{-mx} \tag{20.13}$$

Boundary conditions:
To determine the two constants of integration showed in Equation 20.13, two boundary conditions are needed [26]. Three scenarios of operation are considered. Anyway, the first condition is common to all scenarios, so it can be established that:
1st condition: For $x = 0$:

$$\theta\big|_{x=0} = \theta_0 \implies \beta_0 = C_1 + C_2 \tag{20.14}$$

The second condition will therefore depend on different assumptions for each scenario, distinguishing the following:

1. Fin subjected to convection in the end
2. Fin isolated at the end
3. Fin of infinite length

1. *Fin Subjected to Convection in the End*
2nd condition: For $x = L$:

$$-\lambda \cdot S \cdot \frac{d\theta}{dx}\bigg|_{x=L} = \alpha_E \cdot S \cdot (\theta_{x=L} - \theta_a) \tag{20.15}$$

Operating:

$$-\lambda \cdot m \cdot (C_1 \cdot e^{mL} - C_2 \cdot e^{-mL}) = \alpha_E \cdot (C_1 \cdot e^{mL} + C_2 \cdot e^{-mL}) \tag{20.16}$$

With Equations 20.14 and 20.16, and operating to obtain the two constants of integration, the thermal equation can be finally obtained:

$$\beta = \beta_0 \cdot \left[\frac{Ch[m(L-x)] + H \cdot Sh[m(L-x)]}{Ch[mL] + H \cdot Sh[mL]} \right] \tag{20.17}$$

where

$$H = \frac{\alpha_E}{\lambda \cdot m} \tag{20.18}$$

2. *Fin Isolated in the End*
2nd condition: For $x = L$:

$$\alpha_E = 0 \implies H = 0 \tag{20.19}$$

Replacing this value in Equation 20.17:

$$\beta = \beta_0 \cdot \left[\frac{Ch[m(L - x)]}{Ch[mL]} \right] \tag{20.20}$$

This condition can be explained by the reduced thickness of the fins that means no heat transfer at all through the edge.

3. *Fin of Infinite Length*
2nd condition: For $x = L$:

$$\theta\big|_{x=L} = \theta_a \implies \beta_L = 0 \tag{20.21}$$

Replacing values:

$$\beta_L = C_1 \cdot e^{mL} + C_2 \cdot e^{-mL} = 0 \tag{20.22}$$

Operating and substituting into the general equation 20.13

$$\beta = \beta_0 \cdot e^{-mx} \tag{20.23}$$

Taking into account all these results, Figure 20.6 summarizes the relative temperature distribution along the length of a flat fin corresponding to the three scenarios.

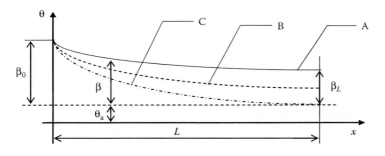

FIGURE 20.6 Relative temperature distribution in straight fins, for the three scenarios studied.

20.3.1.1.1 *Heat Transferred by the Fins of Uniform Cross Section*

The time-dependent heat dissipated by this type of fin is given by

$$\dot{Q} = -\lambda \cdot S \cdot \frac{d\theta}{dx}\bigg|_{x=0} = -\lambda \cdot S \cdot \frac{d\beta}{dx}\bigg|_{x=0} \tag{20.24}$$

All the heat dissipated by the fin is supposed to be driven through the starting area. As before, each of the three scenarios will be analyzed.

1. *Fin subject to convection in the end:*
Deriving Equation 20.17 and substituting into the above equation results in

$$\dot{Q} = \lambda \cdot m \cdot S \cdot \beta_0 \cdot \left[\frac{Sh[mL] + H \cdot Ch[mL]}{Ch[mL] + H \cdot Sh[mL]} \right] \tag{20.25}$$

2. *Fin isolated in the end:*
From the expression above and considering $H = 0$, Equation 20.26 can be stated:

$$\dot{Q} = \lambda \cdot m \cdot S \cdot \beta_0 \cdot \left[\frac{Sh[mL]}{Ch[mL]} \right] = \lambda \cdot m \cdot S \cdot \beta_0 \cdot Th[mL] \tag{20.26}$$

3. *Fin of infinite length:*
Deriving Equation 20.23, Equation 20.27 can be directly obtained:

$$\dot{Q} = -\lambda \cdot S \cdot \beta_0 \cdot (-m) \cdot e^{-mx}\big|_{x=0} = \lambda \cdot m \cdot S \cdot \beta_0 \tag{20.27}$$

20.3.1.1.2 *Scope of Fins with Uniform Cross Section*

By adding fins on a primary surface, the overall mean surface temperature of the device is lower, so that by reducing the average temperature difference between the surface and the fluid medium may result in not producing a significant increase in the dissipated heat flow or even the opposite effect can occur, that is, a heat flow decrease, making fins thermally isolated from the surface [27]. It is important to determine the range of application of the fins in order to ensure the desired effect.

To obtain the necessary criteria for this, the variation of dissipated heat for one of the hypotheses is considered. Specifically the case A presents the next expression:

$$\frac{d\dot{Q}}{dL} = \frac{d\left(\lambda \cdot m \cdot S \cdot \beta_0 \cdot \left[\frac{Sh[mL] + H \cdot Ch[mL]}{Ch[mL] + H \cdot Sh[mL]} \right] \right)}{dL} = 0 \tag{20.28}$$

Operating and simplifying, the following expression is reached:

$$\frac{\left[Sh[mL] + H \cdot Ch[mL] \right]^2}{\left[Ch[mL] + H \cdot Sh[mL] \right]^2} = 1 \tag{20.29}$$

For this to be true, it must be checked that $H = 1$, which means that:

$$H = \frac{\alpha_E}{\lambda \cdot m} = \frac{\alpha}{\lambda \cdot m} = 1 \tag{20.30}$$

Considering a fin unit width ($a = 1$), it can be written:

$$\left.\begin{array}{l} S = 1 \cdot W = W \\ P = 2W + 2 \approx 2 \end{array}\right\} \Rightarrow m = \sqrt{\frac{\alpha \cdot P}{\lambda \cdot S}} = \sqrt{\frac{2 \cdot \alpha}{\lambda \cdot W}} \tag{20.31}$$

Replacing values in Equation 20.30:

$$H = \frac{\alpha}{\lambda \cdot m} = \sqrt{\frac{W \cdot \alpha}{2 \cdot \lambda}} = 1 \tag{20.32}$$

This means that:

$$\frac{W \cdot \alpha}{2 \cdot \lambda} = 1 \tag{20.33}$$

Considering all the possibilities that can occur results in

$$\frac{W \cdot \alpha}{2 \cdot \lambda} \begin{cases} >1 \\ =1 \\ <1 \end{cases} \Rightarrow \frac{d\dot{Q}}{dL} \begin{cases} <0 \text{ Isolating effect} \\ =0 \text{ It makes no difference} \\ >0 \text{ Dissipative effect} \end{cases} \tag{20.34}$$

Therefore, to dissipate more heat, fins that meet the following relationship will be chosen:

$$\frac{W \cdot \alpha}{2 \cdot \lambda} \ll 1 \tag{20.35}$$

Therefore, it can be concluded that the application scope of the fins is that one that takes into account the following aspects:

- Slim fins: ($W \downarrow$)
- Laminar natural convection: ($\alpha \downarrow$)
- High thermal conductivity of the material: ($\lambda \uparrow$)

Optimal dimensioning of uniform cross-section fins.

It is very important to reach the maximum heat dissipation for a minimum amount of material [28,29]. A fin with an insulated pole case is analyzed (case B of the three studied so far). The amount of used material will set the optimal profile area of the fin.

Deriving the expression of heat output respect to the thickness, the maximum of the function can be calculated:

$$\frac{d\dot{Q}}{dW} = \frac{d(\lambda \cdot m \cdot S \cdot \beta_0 \cdot Th[mL])}{dW} = 0 \tag{20.36}$$

Operating for a fin with unity width and simplifying, the following relationship is reached:

$$\sqrt{\frac{2 \cdot \alpha}{\lambda \cdot W^3}} \cdot A_p = 1.4192 \tag{20.37}$$

where the optimal size of the fin will be

$$W_{\text{OPTIMUM}} = 0.998 \cdot \left(\frac{A_p^2 \cdot \alpha}{\lambda} \right)^{1/3} \tag{12.38}$$

$$L_{\text{OPTIMUM}} = \frac{A_p}{W_{\text{OPTIMUM}}} \tag{20.39}$$

As a rule, fin operating conditions are commonly known, such as: $\alpha, \lambda, \theta0, \theta a$ and Q, for a unity width ($a = 1$), so usually a formulation based on these parameters is handled, in the form

$$W_{\text{OPTIMUM}} = \frac{0.6321}{\alpha \cdot \lambda} \cdot \left(\frac{\dot{Q}}{\theta_0 - \theta_a} \right)^2 \tag{20.40}$$

$$L_{\text{OPTIMUM}} = \frac{0.7979}{\alpha} \cdot \left(\frac{\dot{Q}}{\theta_0 - \theta_a} \right) \tag{20.41}$$

$$A_{P(\text{OPTIMUM})} = \frac{0.5048}{\alpha^2 \cdot \lambda} \cdot \left(\frac{\dot{Q}}{\theta_0 - \theta_a} \right)^3 \tag{20.42}$$

Those are the equations used to design a straight fin of constant thickness, with the minimum material for maximum heat dissipation [30].

20.3.1.1.3 Efficiency or Yield of Uniform Cross-Section Fins

It is defined as the ratio of real heat flow dissipated in it and the heat that would be dissipated if the hole fin remained at the base temperature (θ_0), keeping the film coefficient (perfect fin condition) [31]. For the case A previously defined:

$$\eta_A = \frac{\dot{Q}}{\alpha \cdot P \cdot L \cdot (\theta_0 - \theta_a)} \tag{20.43}$$

In the case of a fin with uniform thickness and isolated extreme, the expression is

$$\eta_A = \frac{1}{m \cdot L} \cdot Th[mL] \tag{20.44}$$

To take into account the effect of convection in the edge, as an approximation, L can be replaced by Lc, which can be assumed as a virtual length of a fictitious fin and can be written as: $Lc = L + (S/P)$. Thus, for unit width it could be particularized as

Lc = $L + (W/2)$ For rectangular and annular fins
Lc = $L + (D/4)$ For cylindrical needles

20.3.1.1.4 Dissipation Coefficient in Uniform Cross-Sectional Fins

It is the relationship between the flow of heat dissipated by the fin and the flow dissipated from the starting surface through the area occupied by it, if the extended surface had not been placed there. The same convection coefficient for the fin and the bare surface is supposed.

For straight fins and needles with isolated edge, the expression is as follows:

$$\varepsilon = \frac{\dot{Q}}{S \cdot \alpha \cdot (\theta_0 - \theta_a)} = \frac{Th[mL]}{\sqrt{Bi}} \tag{20.45}$$

Where Bi is the fin "number of Biot," with a value of

$$Bi = \frac{\alpha \cdot S}{P \cdot \lambda} \tag{20.46}$$

There are tabulated functions that allow obtaining these parameters [32,33].

20.3.1.2 Study of Variable Section Fins

Most used fins have a triangular and trapezoidal profile. In this section, fins of triangular profile will be studied. Figure 20.7 shows schematically its most important parameters:

As shown in Figure 20.7, both cross-sectional and fin exchange area are a function of "*x*."

$$A_x = \frac{w \cdot a}{L} \cdot x \tag{20.47}$$

$$S_x = 2 \cdot a \cdot x \cdot \sqrt{1 + \left(\frac{w}{2 \cdot L}\right)^2} = 2 \cdot a \cdot x \cdot f \tag{20.48}$$

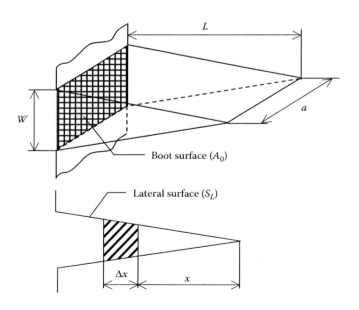

FIGURE 20.7   Triangular fin profile.

Therefore, for $x = L$, the boot surface and the total area of exchange are, respectively, defined as

$$A_o = w \cdot a \tag{20.49}$$

$$S_L = 2 \cdot a \cdot L \cdot f \tag{20.50}$$

20.3.1.2.1 Temperature Range

The temperature field is defined by the expression

$$\frac{d^2\beta}{dX^2} + \frac{1}{X} \cdot \frac{d\beta}{dX} - \frac{\varphi^2}{4} \cdot \frac{\beta}{X} = 0 \tag{20.51}$$

where

$$\beta = \frac{\theta - \theta_a}{\theta_0 - \theta_a} \tag{20.52}$$

$$\varphi^2 = \frac{8 \cdot \alpha \cdot f \cdot L^2}{\lambda \cdot w} \tag{20.53}$$

$$X = \frac{x}{L} \tag{20.54}$$

Solving the differential equation 20.51 and applying the appropriate boundary conditions, an expression for the temperature field is obtained [34]:

$$\beta = \frac{I_0\left(\varphi \cdot \sqrt{X}\right)}{I_0(\varphi)} \tag{20.55}$$

20.3.1.2.2 Heat Transfer

The heat power dissipated by the fin is given by [34]

$$\dot{Q} = -\lambda \cdot A_0 \cdot \left(\frac{d\beta}{dX}\right)_{X=1} = \cdots = -\frac{2 \cdot f \cdot a \cdot L \cdot \alpha}{\varphi} \cdot 2 \cdot \frac{I_1(\varphi)}{I_0(\varphi)} \cdot (\theta_0 - \theta_a) \tag{20.56}$$

20.3.1.2.3 Efficiency

The efficiency of the fin of triangular profile is given by the expression

$$\eta = \frac{2}{\varphi} \cdot \frac{I_1(\varphi)}{I_0(\varphi)} \tag{20.57}$$

20.3.1.2.4 Dissipation Coefficient

The dissipation coefficient is given by

$$\varepsilon = \frac{\dot{Q}}{A_0 \cdot \alpha \cdot (\theta_0 - \theta_a)} = \frac{S_L}{A_0} \cdot \eta \tag{20.58}$$

Biot number for this fin type is defined as

$$\text{Bi} = \frac{\alpha}{\lambda} \cdot \frac{A_0 \cdot L}{S_L} = \frac{\alpha \cdot w}{2 \cdot \lambda \cdot f} \tag{20.59}$$

As it was expected for straight fins of constant thickness, it is very important to reach the maximum heat dissipation for a minimum amount of material.

Deriving the expression of heat output with regard to the thickness maximizes the function, so it can be obtained:

$$\frac{d\dot{Q}}{dw} = 0 \tag{20.60}$$

Operating, optimum relations for the evacuation are given by the expressions:

$$w_{\text{OPTIMUM}} = \frac{0.8273}{\alpha \cdot \lambda} \cdot \left(\frac{\dot{Q}}{\theta_0} \right)^2 \tag{20.61}$$

$$L_{\text{OPTIMUM}} = \frac{0.8420}{\alpha} \cdot \left(\frac{\dot{Q}}{\theta_0} \right) \tag{20.62}$$

$$A_{\text{P(OPTIMUM)}} = \frac{0.3483}{\alpha^2 \cdot \lambda} \cdot \left(\frac{\dot{Q}}{\theta_0} \right)^3 \tag{20.63}$$

Therefore, when comparing two fins, one triangular and one rectangular profile for the same amount of heat dissipated results in

$$\frac{(w_{\text{OPTIMUM}})_{\text{triangular}}}{(w_{\text{OPTIMUM}})_{\text{rectangular}}} = \frac{0.8273}{0.6321} = 1.309 \tag{20.64}$$

$$\frac{(L_{\text{OPTIMUM}})_{\text{triangular}}}{(L_{\text{OPTIMUM}})_{\text{rectangular}}} = \frac{0.8420}{0.7979} = 1.055 \tag{20.65}$$

$$\frac{(A_{\text{P}})_{\text{triangular}}}{(A_{\text{P}})_{\text{rectangular}}} = \frac{0.3483}{0.5048} = 0.6899 \tag{20.66}$$

On one hand, the fin of triangular profile has a thickness at the base considerably higher, and on the other hand, it is slightly longer than that of uniform thickness. This may lead to the result that the number of fins that can be placed on a primary surface is less in the case of a triangular profile.

The triangular profile fin is more economical in terms of the amount of material used, achieving in some cases savings of up to 44%.

Finally, the cost of manufacture should be considered, which in most cases is much higher in fins of the triangular profile.

20.3.1.3 Circular Fins; Finned Tubes

Figure 20.8 shows schematically a circular fin of constant thickness with its most important parameters: As shown in this figure, the cross-sectional areas, as well as the fin exchange area are, respectively [35]:

$$A = 2 \cdot \pi \cdot r \cdot w \tag{20.67}$$

$$S = 2 \cdot \pi \cdot (r^2 - r_0^2) \tag{20.68}$$

20.3.1.3.1 Temperature Range

The temperature field is defined by the expression

$$\frac{d^2\beta}{dR^2} + \frac{1}{R} \cdot \frac{d\beta}{dR} - \phi^2 \cdot \beta = 0 \tag{20.69}$$

where

$$\phi^2 = \frac{2 \cdot \alpha \cdot r_1^2}{\lambda \cdot w} \tag{20.70}$$

$$R = \frac{r}{r_1} \tag{20.71}$$

$$\beta = \frac{\theta - \theta_a}{\theta_0 - \theta_a} \tag{20.72}$$

Solving the differential equation, and applying the appropriate boundary conditions, an expression for the temperature field of the following form is obtained [34]:

$$\beta = \frac{I_0(\phi \cdot R) \cdot K_1(\phi) + K_0(\phi \cdot R) \cdot I_1(\phi)}{I_0(\phi \cdot R_0) \cdot K_1(\phi) + K_0(\phi \cdot R_0) \cdot I_1(\phi)} \tag{20.73}$$

To determine the temperature at any point of the fin there is a chart that gives the value of β_1/β in terms of R and ϕ. Once the value of β_1 is calculated, temperature values at wished points can be determined for different values of R. Finally, the change of variable is undone to get the values of temperature.

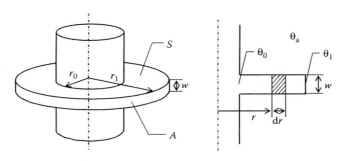

FIGURE 20.8 Annular fin with constant thickness.

20.3.1.3.2 Heat Dissipated by the Fin

The heat dissipated by the fin is given by

$$d\dot{Q} = -\lambda \cdot A_0 \cdot \left(\frac{d\beta}{dR} \right)_{R=R_0} \tag{20.74}$$

When integrated, it can be defined by the following expression:

$$\dot{Q} = -2 \cdot \pi \cdot (r_1^2 - r_0^2) \cdot \alpha \cdot \eta \cdot (\theta_0 - \theta_a) \tag{20.75}$$

20.3.1.3.3 Fin Effectiveness

The effectiveness of the annular fin is given by the expression:

$$\eta = \frac{2 \cdot R_0}{1 - R_0^2} \cdot \frac{1}{\varphi} \cdot \frac{I_1(\varphi) \cdot K_1(\varphi \cdot R_0) - I_1(\varphi \cdot R_0) \cdot K_1(\varphi)}{I_0(\varphi \cdot R_0) \cdot K_1(\varphi) + K_0(\varphi \cdot R_0) \cdot I_1(\varphi)} \tag{20.76}$$

20.3.1.3.4 Dissipation Coefficient

The dissipation coefficient is given by

$$\varepsilon = \frac{\dot{Q}}{A_0 \cdot \alpha \cdot (\theta_0 - \theta_a)} = \frac{1 - R_0^2}{R_0} \cdot \frac{r_1}{w} \cdot \eta \tag{20.77}$$

The Biot number for this fin type is defined as

$$Bi = \frac{\alpha \cdot w}{2 \cdot \lambda} \tag{20.78}$$

20.3.1.3.5 Overall Transmission Coefficient in Finned Tubes

The overall coefficient of heat transfer for a primary cylindrical surface provided with constant thickness fins is determined as follows. Figure 20.9 shows a finned pipe section.

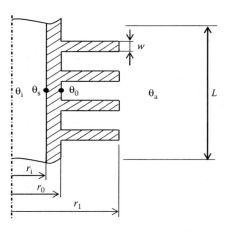

FIGURE 20.9 Simplified layout of a finned tube.

Considering a portion of tube of length L where n fins are placed, the following areas can be defined:
(Area inside the tube)

$$A_i = 2 \cdot \pi \cdot r_i \cdot L \qquad (20.79)$$

(Fin area)

$$A_a = 2 \cdot \pi \cdot (r_i^2 - r_0^2) \cdot n \cdot L \qquad (20.80)$$

(Outer area between fins)

$$A_t = 2 \cdot \pi \cdot r_0 \cdot (1 - n \cdot w) \cdot L \qquad (20.81)$$

(Inner exchange area)

$$A_T = A_a + A_t \qquad (20.82)$$

The heat flow from the inner fluid to the inner surface is expressed according to Newton's law of cooling as

$$\dot{Q} = A_i \cdot \alpha_i \cdot (\theta_i - \theta_s) \qquad (20.83)$$

Applying the Fourier equation for heat conduction through the material forming the tube, results in

$$\dot{Q} = 2 \cdot \pi \cdot \lambda \cdot L \cdot \frac{\theta_s - \theta_0}{\ln(r_0/r_i)} \qquad (20.84)$$

Finally, the heat dissipated per unit time from the outside finned surface, being η effectiveness of the fins, is

$$\dot{Q} = (A_t + A_a \cdot \eta) \cdot \alpha \cdot (\theta_0 - \theta_a) \qquad (20.85)$$

With the three equations above, the joint expression of heat transmitted can be obtained as

$$\dot{Q} = \frac{\theta_i - \theta_a}{\dfrac{1}{A_i \cdot \alpha_i} + \dfrac{\ln \dfrac{r_0}{r_i}}{2 \cdot \pi \cdot \lambda \cdot L} + \dfrac{1}{(A_t + A_a \cdot \eta) \cdot \alpha}} \qquad (20.86)$$

Knowing that the global transfer of heat from the inside out can be expressed as

$$\dot{Q} = A_0 \cdot U_0 \cdot (\theta_i - \theta_a) \qquad (20.87)$$

where U_0 is the overall coefficient of heat transfer on finned tubes, whose expression is given by

$$U_0 = \frac{1}{\left(\dfrac{A_0}{A_i}\right) \cdot \dfrac{1}{\alpha_i} + \dfrac{A_0}{2 \cdot \pi \cdot L} \cdot \dfrac{\ln \dfrac{r_0}{r_i}}{\lambda} + \dfrac{1}{\eta \cdot \alpha}} \qquad (20.88)$$

20.3.1.3.6 *Methodology*

Knowing the number of fins per unit length, the total amount of evacuated heat can be determined, or knowing the total amount of heat to evacuate, the number of fins required to put in the tube can be calculated [36,37].

20.3.1.4 Preliminary Design and Further Optimization of a Fin through MATLAB

Based on this theory, an analytical procedure has been developed in MATLAB in order to simplify the calculation procedure previously shown. In fact, two algorithms will be presented. The first one called FINSTUDY aims to calculate temperature distribution and heat transfer for different fin configurations and boundaries for aluminium fins whereas the second one, called OPTIFIN, aims to optimize a rectangular fin heat sink arrangement suitable for Peltier module heat dissipation. For both cases, practical examples will also be presented. Those analytical procedures use variables such as thickness, length, and spacing among fins to be applied for the optimization of the design process itself.

FINSTUDY

File Description:

This function aims to show temperature distribution and average heat transfer velocity for different fin configurations and boundaries for aluminium fins.

The function allows the user to experiment with different input values and also to get outputs in different ways as will be shown later in the "Examples" section.

General Expression:

[T,Q,x] = FinStudy(EdgeCase,GeometryType,Plotable,OutPut,Params)

Input Arguments:

Value:	Description:
EdgeCase:	**Input Type:** INTEGER
	Range: From 1 to 4
	Allows the user to select the boundary condition for the edge of the Fin. 4 cases can be chosen:
	1: Infinite fin
	2: Adiabatic fin
	3: Fixed temperature
	4: Convection heat transfer
	This argument is Obligatory. Any other arguments of this function are optional.
GeometryType:	**Input Type:** INTEGER
	Range: From 1 to 3
	Allows the user to select the geometry of the fin from the following ones:
	1: Rectangular fin
	2: Cylindrical fin
	3: Triangular fin
	By default, ("if no selection is given by the user"), Rectangular fin (option 1) is selected.
Plotable:	**Input Type:** INTEGER
	Range: Boolean
	Select 1 if plot wanted. If this argument is not defined, 0 value is taken by default.
OutPut:	**Input Type:** INTEGER
	Range: Boolean
	Select 1 if output file is wanted. If this argument is not defined, 0 value is taken by default.

Params:	Input Type: FLOAT
	Range: Three elements array –> Fin Length, Surface T, Ambient T
	Insert the data in the order specified above this line, for example, [0.1 50 25] which correspond to a fin with a length of 0.1 m, a surface temperature of 50°C and an ambient temperature of 25°C.
	If Params is not set, default parameters will be loaded instead.

Output Arguments:

Value:	Description:
Temperature:	Array with temperature values for different points of the fin length, assuming the same temperature for the cross section. Results are given in °C.
Heat Transfer Speed:	Averaged Heat Transfer Speed of the fin. Result is expressed in W.
Fin Position:	Express the fin position which corresponds to each temperature value of Temperature array. Results are given in m.

The function allows plotting the results to see temperature evolution across the fin assuming that there is no variation in cross-section temperature; that is, the fin length is much bigger than fin width. The next figure shows an example of a graphical output performed in MATLAB.

EXAMPLES:

1. Temperature evolution with default parameters for a rectangular fin with convection heat transfer in the edge and "Plotable" and "OutPut" input arguments activated.

   Type: [T,Q,x] = FinStudy(4,[ ],1,1) in MATLAB console. The output will be a graph and Data.txt file with lengths and temperatures with space separation.
   Results: This example has been depicted next in Figure 20.10.

2. Temperature evolution with customized parameters for a cylindrical fin with convection heat transfer in the edge.

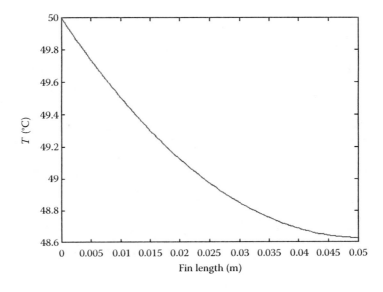

FIGURE 20.10 Output temperature evolution vs. length of the fin for FINSTUDY example 1 in MATLAB.

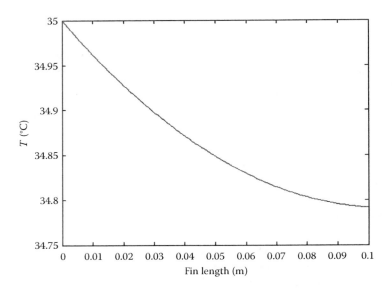

FIGURE 20.11   Output temperature evolution vs. length of the fin for FINSTUDY example 2 in MATLAB.

Type: [T,Q,x] = FINSTUDY(4,2,0,0,[0.1 35 25]) in MATLAB console. This line will provide temperature evolution results in a cylindrical fin of 0.1 m length, with a surface temperature of 35°C and an ambient temperature of 25°C.

Results: The study of this case will result in three variables, T, Q, and x, which contain the temperature array, the heat transfer speed, and fin position for each temperature value, respectively. No plot nor data output is given with the above input arguments, but data can be easily plotted to obtain the same results of case 1, just typing in MATLAB console: plot(x, T). Results are depicted in Figure 20.11.

OPTIFIN

File Description:

This function aims to optimize a rectangular fin arrangement heat sink suitable for Peltier module heat removal. Two solutions are given based on this hypothesis [38,39]:

1. Heat sink surface base is bigger than Peltier module surface.

2. Heat sink surface base and Peltier module surface are equal.

The heat sink consists of an N number of rectangular planar fins of length L, a D spacing between each fin and with a W thickness. This is a theoretical optimization and can be far away from real heat sink Rectangular fins arrangements [40,41,42,43].

Optionally, If "Dim" input argument is a two elements vector, a third result is given, which shows a real fin with typical width (1 mm) and space between fins (3 mm).

General Expression:

[Wop,Dop,Lop,N] = OptiFin(Dim,Q,Theta0,ThetaA,PhysiscParams)

where the first 4 input arguments are obligatory. As stated before, if Dim is one element vector (it means that just Peltier length is given), two case studies will be done while a two elements vector (it means that both Peltier length and fin length constraints are given) will perform 3 case studies.

Input Arguments:

Value:	Description:
Dim:	**Input Type:** REAL **Range:** Positive real one or two elements vector value If just one element value is given it is the length of Peltier module assuming Square Shape devices. If two elements vector is given, the second value is the maximum length (L) of the fin due to space constraints in the installation area.
Q:	**Input Type:** REAL **Range:** Positive real value Express the heat flux wanted to be removed from the Peltier cold face Module.
Theta0:	**Input Type:** REAL **Range:** Real negative positive value Indicates de desired temperature in the cold face of the Peltier Module in °C.
ThetaA:	**Input Type:** REAL **Range:** Real positive value Indicates ambient temperature where Peltier Module is located in °C.

PhysicParams:	**Input Type:** REAL
	Range: Real positive values.
	Different Physical properties array of the desired optimization:
	    1: Heat Transfer coefficient (alpha)
	    2: Material thermal Conductivity
	    3: Dynamic Viscosity
	    4: Gravity acceleration
	This variable is optional.

OutPut Arguments:

Value:	Description:
Wop:	Array containing the optimum thickness for all the cases. Results are given in (m).
Dop:	Array containing the optimum fin spacing for all the cases. Results are given in (m).
Lop:	Array containing the optimum length of each fin for the arrangement for all the cases. Results are given in (m).
N:	Array containing the optimum number of fins for the arrangement for all the cases.

EXAMPLE:

1. Design of theoretical optimized planar fin arrangement heat sink and real heat sink.
   Type: [Wop,Dop,Lop,PeltLength,N] = OptiFin([0.04 0.05],29,42,25) in MATLAB console. This line will provide the theoretical optimized design of the heat sink for case 1 and 2 described in the OptiFin documentation, as well as the design parameters of a typical real heat sink (Case 3). The result will be a text file called "OptiFin.txt" with the design parameters of the Heat Sinks as it can be seen in Figure 20.12. Remember that all results are given in meters (m).

```
          Wop       Dop       Lop       PL        N
CASE 1    0.0021    0.0168    0.3296    0.0974    6
CASE 2    0.0021    0.0266    2.0560    0.0400    2
CASE 3    0.0010    0.0030    0.0500    0.1054    27
Optimum fin arrangement succeeded for CASE 2
```

FIGURE 20.12 Output temperature evolution vs. Length of the fin for OPTIFIN example 1 in MATLAB.

As it can be seen in Cases 1 and 2 of Figure 20.9, theoretical optimization of planar fin heat sink arrangement do not provide a good practical results as well as "Dop" is too big for real use. This is because in natural convection regimen, the optimum space needed between each fin is too big in order to reach the optimum convection heat transfer. This is one of the reasons why typical heat sinks of computers have a fan above in charge of provoking forced convection.

However, the third case shows realistic design of a typical heat sink for computer processors, with typical fin width and spacing ratio and a fin length constraint.

The last line of the OptFin.txt text file indicates where the length reached for the case 2 is suitable or if it is longer than the maximum permitted length. Theses criteria are given by using "FinStudy" function with case 4 (fin edge with convection transfer). If the length is bigger than maximum length a warning message will appear.

20.3.1.5 CFD Simulations

Following the preliminary design stage of a heat sink such as a fin incorporated to a Peltier module, a computational fluid dynamics (CFD) model was developed. CFD is the science of predicting fluid flow, heat transfer, mass transfer (as in perspiration or dissolution), phase change (as in freezing or boiling), chemical reaction (e.g., combustion), mechanical movement (e.g., fan rotation), stress or deformation of related solid structures (such as a mast bending in the wind), and related phenomena by solving, on a computer, the mathematical equations that govern these processes, by a numerical algorithm and hence the results obtained from the CFD model are expected to be sufficiently valid in making precise predictions in the near future.

The CFD model can be validated through the data previously obtained from the MATLAB approach so a significant agreement can be appreciated between those two procedures. This model includes the input of working conditions, simplifications (if any), geometrical model, mesh refining, turbulence, and convective models, contour conditions, algorithms of solution including thermal conduction, diffusion, and viscosity [44]. Strong coupling between transport and fluid dynamics are involved.

Working with CFD involves six fundamental phases [45]:

1. Define the modeling goals.
2. Create the model geometry and grid.
3. Set up the solver and physical models.
4. Compute and monitor the solution.
5. Examine and save the results.
6. Consider revisions to the numerical or physical model parameters, if necessary.

Briefly, the same geometry previously optimized through MATLAB was modeled. After the geometry was successfully created using a powerful pre/processor of a commercial code, as can be seen in Figure 20.13a, the next step was to realize the meshing.

Thus, the whole volume was discretized into finite elements called "cells" as can be seen next in Figure 20.13b, which were the fundamental parts of the calculations, because all the fluid equations were solved for each of these elements.

Considering as a boundary condition the initial temperature corresponding to the cold side of the Peltier module, results obtained in the processor showed the cooling process after the placement of the aluminum fin associated as can be seen next in Figure 20.14, where a temperature field over a normal surface corresponding to the symmetry plane of the Peltier module is depicted, showing a significant agreement with the results obtained through experimental measurements previously carried out with the same design. The use of fins means a good approach of the way of improving the amount of heat dissipated. Once the CFD model can be considered validated (with appropriate values from experimentation), different tests can be carried out in order to evaluate the effect of changes in geometry and/or fin configuration, leading this way to an optimum design process.

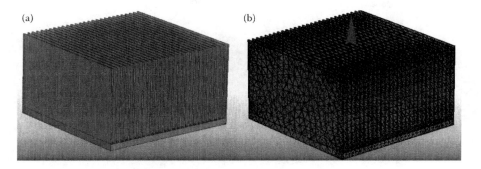

FIGURE 20.13 Geometry (a) and mesh (b) corresponding to the model considered.

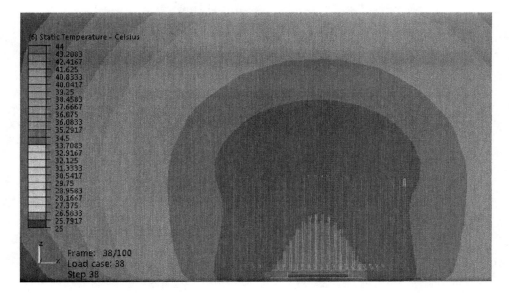

FIGURE 20.14 Temperature field associated according to the cooling process because of the fins.

An experimental procedure has been performed in order to validate the results of the MATLAB algorithm and the CFD model. The setup of the experiment can be seen in Figure 20.15. It consists of a heat sink with a 29 W thermal power source attached to the bottom, a six-channel switch box to select different measurements from thermocouples attached to the heat sink and a multimeter to display the temperature value. Several temperature values were finally measured in the middle of the central fin, placed at different heights.

Figure 20.16 shows the temperature distribution along the fin length, as a validation procedure for the MATLAB algorithm, and a CFD model for the same case study. A maximum error value of about 6% for the MATLAB prediction while a 2% for CFD model was encountered, which can be considered admissible so both procedures can be considered validated.

20.3.1.6 Concluding Remarks

The imperative need for heat transfer optimization associated with Peltier modules has been increased as their performance also increases. One of the solutions most commonly used is the implementation of

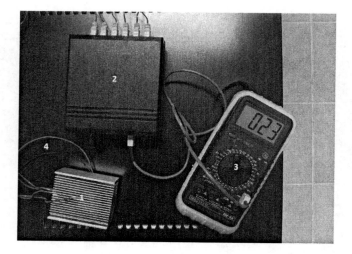

FIGURE 20.15   Experimental layout for MATLAB and CFD models validation purposes where: 1. is the heat sink with the thermocouples, 2. switch box, 3. multimeter, and 4. connexion to the power source.

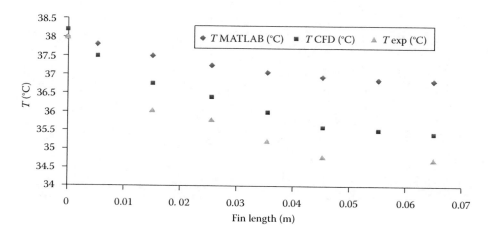

FIGURE 20.16   Comparison of temperature evolution along the fin between the MATLAB algorithm, CFD model and experimental results.

extended surfaces or so called "fins." In this chapter the way of improving this heat dissipation has been treated and the next conclusions have been achieved:

- A detailed theoretical procedure to carry out preliminary design and optimization of extended surfaces is presented, focused mainly on the thermal behavior of flat fins with constant and variable section but also analyzing the performance of finned tubes, which nowadays are of great importance for several industrial applications providing with the key equations used in the optimization process.

- As a valuable complement, the design and further optimization of a fin through MATLAB have also been presented here. A first algorithm called FINSTUDY has been developed in order to predict the temperature field and heat flux dissipated by an extended surface (applied to longitudinal fins with a uniform thickness or circular or triangular profile). Another algorithm called OPTIFIN optimizes a rectangular fin arrangement heat sink suitable for Peltier module heat dissipation, taking into account both scenarios in which the heat sink surface base may be equal or bigger than the Peltier module surface. Thus, several case studies have been presented for both algorithms in order to help the designer with the management of this tool, which has revealed as a practical way of implementing the above-mentioned theory, even for beginners. In some cases the optimum parameters obtained can be far away of reasonable values such as optimum length of the fin, so more realistic values are finally obtained by using real constraints to the problem such as maximum height allowable which has also been taken into account.

- Finally, the use of computational simulations has also been revealed as a very important research tool regarding heat transfer modeling and optimization as it represents a wide scope for further improvements. It has been demonstrated that geometry, discretization, physical models, methodology, and algorithms of solution introduced are correct enough when compared with the results previously obtained with MATLAB, being finally validated through experimental measurements also carried out here.

20.3.2 Phase Change Cooler

These systems exploit the latent heat of a substance to absorb/transfer large amounts of heat with a minimum temperature gradient.

There are two types: those that use the change of liquid–gas phase of the substance and those that employ solid–liquid phase change. While the first ones are used more likely as heat conductors, the seconds ones are used mainly as storage heaters.

Both "heat pipes" and "heat spreaders" are liquid–gas-type coolers, insofar as both of them use a working fluid to transport heat from one place (evaporator) to the other (condenser). In the evaporator, due to the heat flux "Q" coming from the heat source, the liquid is evaporated and the generated vapor is then driven along the pipe to the condenser where it finally condenses and comes back to the evaporator as liquid, as it can be seen in Figure 20.17.

Systems of solid–liquid type, named PCM (phase-change material) devices, use paraffins or inorganic materials as phase-change material. Due to their reduced mobility, they are not used to transport heat but simply to absorb it. These heat accumulators work as long as the substance exists in solid state. In the field of electronics, these systems are used to absorb power spikes that can experience the electronic equipment avoiding its temperature increase.

This section will be focused on liquid–gas cooling systems.

20.3.2.1 Heat Pipes

When selecting a heat pipe (HP) or heat spreader (HS) for a specific application, the following factors must be taken into account:

- The amount of evacuated heat
- The operating temperature and/or the temperature difference at the HP ends

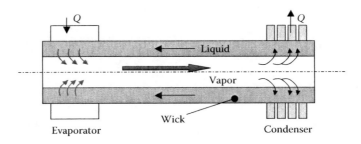

FIGURE 20.17 Heat pipe layout.

- The HP vertical orientation
- The HP thermal resistance

Knowledge of these factors and the relationship between them is crucial for determining at which temperature the heat source will be (e.g., the temperature of the hot–cold side of the Peltier module).

The heat transfer capability of an HP depends on the key elements that constitute it: the housing or container, the wick structure, and the fluid [46–48].

- The housing or container is the structural element of the system that has to fulfill two main functions: firstly, to provide mechanical resistance to the internal pressure and secondly, to be a good heat conductor.
- The wick structure is the element that capillary drives the liquid from the condenser to the evaporator while ensuring a homogeneous wetting of the evaporator surface.
- The fluid is the key element of the system and it has to be chosen depending on the operating temperature range. The fluid with a higher value of figure of merit "FOM" (according to the definition given in Equation 20.89) is best suited to the cooler. However, other aspects such as the wettability of the fluid with the wick and the enclosure must be taken into account.

$$\mathrm{FOM} = \frac{\rho\sigma\eta}{\mu} \tag{20.89}$$

20.3.2.2 Considerations Affecting the Operation of the Heat Sink

Heat, operating temperature, and/or temperature difference between evaporator, condenser, vertical orientation, and the thermal resistance of heat pipe are parameters to be evaluated to find the HP/HS that best fits the needs of the application.

- *Heat dissipation vs. temperature.* The heat pipe is a dynamic system that evolves toward the steady state that varies depending on the operating conditions. That is to say, a heat pipe that perfectly runs under certain conditions can stop working under others. In the case of fixed conductance heat pipes, if heat flow increases, so does the temperature difference between evaporator and condenser and the operating temperature. This can result in too high an evaporator temperature, making the HP unable to refrigerate the heat source.

Figure 20.18a shows the temperature of a copper-water, 30 cm-long, 0.6 cm-diameter HP at 20° orientation when subjected to different thermal resistance.

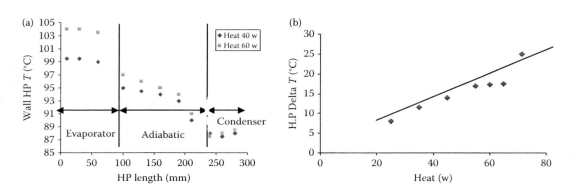

FIGURE 20.18   (a) Operating temperature vs. heat and (b) temperature difference vs. heat.

From Figure 20.18b it can be observed that the temperature difference increases linearly with the dissipated power until it reaches a power at which the growth is exponential. From this graph two basic data about the HP's behavior are drawn:

- The slope of this linear growth represents the thermal resistance of HP.
- The power from which we observe the sharp growth of the temperature difference represents the maximum power that the HP can dissipate.

Likewise, Figure 20.18a shows the evolution of temperatures in the wall of an HP, from the evaporator to the condenser. It can be seen that a heat flow increase results in an evaporator temperature increasing while the condenser remains almost unchanged. This is so because the condenser has the capacity to remove heat. Otherwise, the condenser temperature will also increase, but slower than the evaporator temperature.

- *Orientation angle vs. heat.* HP orientation refers to the vertical position of the evaporator with respect to the condenser, where 90° corresponds to the evaporator below the condenser, 0° horizontal position and −90° corresponds to the evaporator above the condenser. This orientation shift makes gravity play an important role in the return of condensate from the condenser to the evaporator. At 90°, condensate returns by gravity while at −90° condensate returns by capillary forces and against gravity. This makes the angle of inclination affects the dissipation capacity of HP [49].

The dissipation capacity of an HP decreases as the condenser is located below the evaporator, as it can be seen in Figure 20.19. The way in which this capacity decreases with the orientation, directly depends on the wick structure.

Likewise, Figure 20.20 shows how the HP thermal resistance is also affected by the orientation. The thermal resistance increases as the orientation angle becomes negative. Also, depending on the wick, the thermal resistance is higher or lower.

In summary, when selecting an HP for a particular application, it is necessary to know both the thermal resistance and the dissipation rate of the HP for each orientation. With these data, the temperature difference between the evaporator and condenser can be calculated and then, along with the thermal characteristics of the application, the temperature at the heat source predicted (e.g., hot–cold side of a Peltier module).

Although heat spreaders follow the same operating principle and face the same limitations as the HP, their mission, unlike the HP, is to open the heat exchange surface rather than transport heat from one point to another. This increase is achieved by opening the lines of heat flow, according to Figure 20.21.

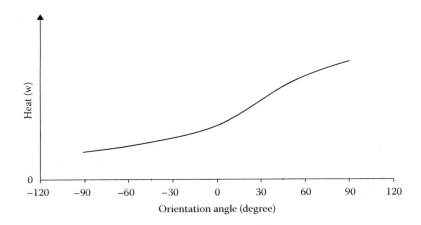

FIGURE 20.19 Heat dissipation vs. orientation angle, general trend.

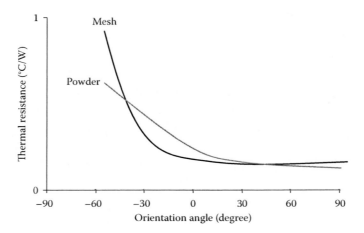

FIGURE 20.20 Thermal resistance vs. orientation angle, general trend.

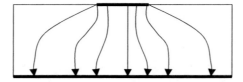

FIGURE 20.21 Heat spreading scheme.

When designing a heat spreader, geometry is a fundamental aspect to keep in mind, as it is a decisive factor in the opening degree of the heat flux lines and, consequently, in the increase of heat exchange surface [50].

20.3.2.3 Capillary Limit

The thermal behavior analyzed above has to do with the capillary force that the wick is able to provide and, as a result, the liquid mass flow rate that the wick is able to move "$\dot{m}_\mathrm{L}$."

By the principle of operation of HP, the mass of evaporated liquid "$\dot{m}_\mathrm{ev}$" that corresponds to the heat removed must be equal to or less than the mass of liquid pumped by the wick "$\dot{m}_\mathrm{L}$":

$$\dot{m}_\mathrm{ev} \leq \dot{m}_\mathrm{L} \tag{20.90}$$

By doing so, it is assured that the area of the evaporator is continuously wetted by the operating liquid and no dry points appear. Dry points considerably reduce the HP performance.

Into the wick two forces take place: capillary force and hydrodynamic drag force. The best wick is one that provides maximum capillary force with the minimum hydrodynamic resistance as it allows pumping the maximum mass flow rate of liquid. However, these two effects are opposite; an increase in the capillary force requires small pores, leading to an increase in hydrodynamic resistance. In order to solve this problem, heat pipe designers now work with hybrid wicks [51,52].

Capillary force: This force depends on both the pore size of the wick and the working fluid. Particularly the liquid surface tension and the contact angle liquid-wick (wetting or nonwetting wick).

$$\Delta P_{cap} = \frac{2\sigma}{r_{eff}}\cos\delta \tag{20.91}$$

Since capillary force is the cause of the liquid circulation, it must be equal to or higher than all pressure drops within the HP (frictional pressure loss in the vapor and liquid paths, pressure loss due to body forces, phase change pressure loss) [49].

As can be seen from Equation 20.91, the lower the effective radius "R_{eff}," the greater the capillary pressure. However, the smaller the effective pore radius, the lower the permeability of the wick to the liquid circulation, as a result leading to a hydrodynamic resistance increase. Therefore, the wick structure has to be a compromise between capillary force and permeability.

It is also noted that to increase the capillary force, the contact angle "δ" should be minimal, which means a perfectly wetting wick.

Wettability and capillary rise: Wettability is the ability of a liquid to spread over the substrate it is dropped. When the liquid perfectly spreads over the substrate it is said the liquid wets, wetting liquid. When it does not, the liquid does not wet, nonwetting liquid. In the first case the contact angle is within 0–90° while in the last case the contact angle is larger than 90°. For nonwetting liquids, the drop shape is more or less rounded, Figure 20.22.

Wettability or contact angle "δ" of a fluid in a porous medium cannot be determined by the technique of depositing a drop of liquid on the substrate and using imaging techniques to establish the angle, Figure 20.23.

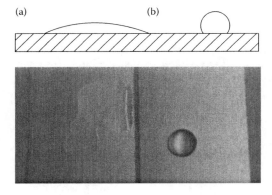

FIGURE 20.22 Lateral and frontal view of wetting (a) and nonwetting (b) liquid.

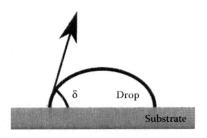

FIGURE 20.23 Contact angle with the substrate.

FIGURE 20.24 Arrangement for contact angle measurement in porous media.

The reason is that if a drop on the porous substrate is placed, the drop is immediately absorbed by the substrate. Therefore, an indirect method is necessary for the determination of the contact angle of the liquid with the porous substrate. This method consists of placing the wick in contact with the liquid and measuring the advancing front of liquid over time, Figure 20.24.

The tests carried out for a silica gel material of two different pore radii and for two different liquids, water and methanol is collected in Figure 20.25.

According to Washburn, the square of the distance traveled by the head of liquid is proportional to the time.

$$L_l^2 = \frac{r_{\text{eff}}\sigma\cos\delta}{2\mu}t \tag{20.92}$$

Therefore, matching the slopes of the experimental curves with the slope of the Equation 20.92, the contact angle with the porous medium is obtained. For the example in Figure 20.25 the angle for water is 78° and the angle for methanol is 52°.

$$m = r_{\text{eff}}\sigma\cos\delta \tag{20.93}$$

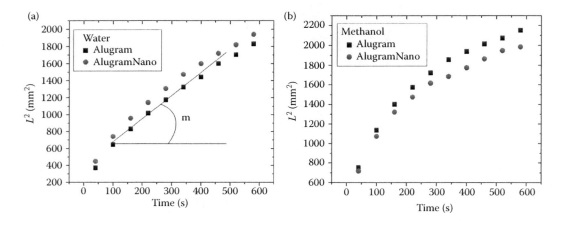

FIGURE 20.25 (a) Rise of water in the porous media with time. (b) Rise of methanol in the porous media with time.

20.3.2.4 Loop Heat Pipes and Variable Conductance Heat Pipes

As a result of the limitations of heat pipes, research in this field proposed different solutions that involve structural changes in the conventional heat pipe. Some of these special solutions are:

- Hybrid wicks: combination of two or more types of wicks in order to reduce the hydraulic resistance for the liquid return to the evaporator while keeping a large capillary force.
- Variable conductance: enable the heat pipe conductance vary allowing it to operate efficiently at different operating conditions (variable conductance heat pipes (VCHP)).
- Spatial separation of vapor and liquid phase of the working fluid at the transport section (loop head pipes (LHP)). Loop scheme has the advantage of being less sensitive to spatial orientation relative to gravity, [53] keeping the LHP in normal functioning in any orientation.

These modifications can increase not only the heat removal capacity but also the length of the heat pipe.

20.4 Nomenclature

a	Width of the fin (m)
A	Cross-sectional area of the fin (m²)
A_a	Finned area of the fin (m²)
A_i	Area inside of the tube (m²)
A_0	Boot area of the fin (m²)
A_p	Profile area of the fin (m²)
A_t	Outer area between fins (m²)
A_T	Total exchange area (m²)
Bi	Biot number
C_1, C_2	Constants of integration for Equation 20.10
Ch	Hyperbolic cosine
D_{op}	Optimum spacing between fins (m)
f	Decreasing factor for the triangular section of a fin
H	Relative convection—conduction term (m⁻²)
I_0	Modified Bessel function of first species of order zero
I_1	Modified Bessel function of first species of order one
K_0	Modified Bessel function of second species of order zero
K_1	Modified Bessel function of second species of order one
L	Length of the fin (m)
L_e	Equivalent length of the fin (m)
L_{op}	Optimum length of the fin (m)
L_l	Distance traveled by the liquid
m	Module of the fin (m)
N	Number of fins
P	Perimeter of the cross section (mm)
PL	Length of the heat sink (m)
$\dot{Q}$	Total heat flux dissipated by the fin (W)
$\dot{Q}_c$	Convective heat flux dissipated by the fin (W)
R	Nondimensional radius for circular fins
R_{eff}	Effective radius
r_0	Inner radius for annular fins (m)
r_1	Outer radius for annular fins (m)
r_i	Inner radius for finned tubes (m)
r_{eff}	Effective pore radio (m)

s	Curve slope
S	Total fin exchange area (m²)
S_L	Lateral exchange surface (m²)
Sh	Hyperbolic sine
t	Time (s)
Th	Hyperbolic tangent
U	Overall coefficient of heat transfer on finned tubes (W m^{-2} K^{-1})
W	Thickness of the fin (m)
W_{op}	Optimum width of the fin (m)
dQ_x/dt	Heat flow through the area "S" in the x direction
"$\lambda/\partial x$"	Thermal conductance
"θ"	Temperature
"x"	Heat flow direction

Greek symbols

α	Mean convection coefficient (Wm^{-2}K^{-1})
α_E	Convection coefficient at the end of the fin (Wm^{-2}K^{-1})
β	Relative temperature to the ambient defined for plate fins (°C)
β_t	Relative temperature to the ambient defined for triangular fins (°C)
θ	Average temperature of the fin in any section (°C)
θ_0	Temperature in the boot section (°C)
θ_1	Temperature in the outer section (°C)
θ_s	Mean surface temperature of the solid (°C)
θ_a	Temperature of the fluid media (°C)
φ	Combined temperature factor for triangular fins
ϕ	Combined temperature factor for circular fins
λ	Coefficient of thermal conductivity (W m^{-1} K^{-1})
η	Efficiency
ε	Dissipation coefficient
σ	Surface tension (N/m)
ρ	Liquid density (kg/m³)
μ	Liquid viscosity (Pa)
η	Latent heat (J/kg)
δ	Contact angle

Annex

Annex I MATLAB Source Code (FINSTUDY)

```
function [T,Q,x]=FinStudy(EdgeCase,GeometryType,Plotable,OutPut,Params)

%Clear screen
clc;

%Input Arguments Checking
if nargin == 1
    GeometryType=1;
    Plotable=0;
    OutPut=0;
    Params=[0.05 50 25];
elseif nargin == 2
    Plotable=0;
```

```
        OutPut=0;
        Params=[0.05 50 25];
elseif nargin == 3
    if isempty(GeometryType)
        GeometryType=1;
    end
    Params=[0.05 50 25];
    OutPut=0;
elseif nargin == 4
    if isempty(GeometryType)
        GeometryType=1;
    end
    if isempty(Plotable)
        Plotable=0;
    end
    Params=[0.05 50 25];
elseif nargin==5
    if isempty(GeometryType)
        GeometryType=1;
    end
    if isempty(Plotable)
        Plotable=0;
    end
    if isempty(OutPut)
        Output=0;
    end
end


%Geometry Dimensions for the examples
switch GeometryType
    case 1 %Rectangular
        a=0.01; %m
        b=0.05; %m
        L=Params(1); %m
        %Perimeter
        p=2*a+2*b;
        %Area
        A=a*b; %m2

    case 2 %Cylindrical
        r=0.025; %m
        %Perimeter
        p=2*pi()*r; %m
        %Area
        A=pi()*r^2; %m

    case 3 %Triangular
        a=0.01; %m
        b=0.05; %m
        L=Params(1); %m
        x=linspace(0,L,100);
        h=(L - x)*(a/2)/L;
        ax=2*h;
        %Perimeter
        p=2*ax+2*b;
```

```
        %Area
        A=ax*b; %m2

end

%Thermal Parameters
h=4.13; %W/(m2*K)
k=209.3; %W/(m*K)

%Temperature Boundaries
Ts=Params(2); %°C
Tinf=Params(3); %°C

%Common Calculus
TethaS=Ts - Tinf; %°C
m=sqrt(h*p)./(k*A);
M=sqrt(h*k*p.*A)*TethaS;
%H=h/(k*a);

switch EdgeCase
    case 1 %CASE 1: Infinite Fin

        if GeometryType == 1 %Only for Triangular fins
            L=1; %m
            x=linspace(0,L,100);
            h=(L - x)*(a/2)/L;
            ax=2*h;
            %Perimeter
            p=2*ax+2*b;
            %Area
            A=ax*b; %m2
            Tetha=TethaS*exp(-m.*x);
            T=Tetha+Tinf;
            Q=M;
            if Plotable
                plot(x,T);
                xlabel('Fin Length (m)');
                H=ylabel('T (°C)');
                set(H,'Rotation',[0]);
            end
        end

        L=1; %m
        x=linspace(0,L,100);
        Tetha=TethaS*exp(-m*x);
        T=Tetha+Tinf;
        Q=M;
        if Plotable
            plot(x,T);
            xlabel('Fin Length (m)');
            H=ylabel('T (°C)');
            set(H,'Rotation',[0]);
        end

    case 2 %CASE 2: Adiabatic Fin
```

```
        L=Params(1); %m
        x=linspace(0,L,100);
        Tetha=TethaS*(cosh(m.*(L - x))./cosh(m*L));
        T=Tetha+Tinf;
        Q=M.*tanh(m*L);
        if Plotable
            plot(x,T);
            xlabel('Fin Length (m)');
            H=ylabel('T (°C)');
            set(H,'Rotation',[0]);
        end

    case 3 %CASE 3: Fixed Temperature
        L=Params(1); %m
        Tedge=35; %°C
        TethaL=Tedge - Tinf;
        x=linspace(0,L,100);
        Tetha=TethaS*(TethaL/TethaS)*(sinh(m.*x)+sinh(m.*(L - x)))./sinh(m*L);
        T=Tetha+Tinf;
        Q=M.*(cosh(m*L) - (TethaL/TethaS))./(sinh(m*L));
        if Plotable
            plot(x,T);
            xlabel('Fin Length (m)');
            H=ylabel('T (°C)');
            set(H,'Rotation',[0]);
        end

    case 4 %CASE 4: Convection heat transfer
        L=Params(1); %m
        Tedge=35; %°C
        TethaL=Tedge - Tinf;
        x=linspace(0,L,100);
        Tetha=TethaS*((cosh(m.*(L - x))+(h./(m*k)).*sinh(m.*(L - x))))./
(cosh(m*L)+(h./(m*k)).*sinh(m*L));
        T=Tetha+Tinf;
        Q=M.*(sinh(m*L)+(h./(m*k)).*cosh(m*L))./(cosh(m*L)+(h./
(m*k)).*sinh(m*L));
        if Plotable
            plot(x,T);
            xlabel('Fin Length (m)');
            H=ylabel('T (°C)');
            set(H,'Rotation',[0]);
        end
end


%Prints OutPut to Data.txt file if option is selected by user
if OutPut
    FileID=fopen('Data.txt','wt');
    fprintf(FileID,'Length(m)  Temperature(°C)\n');
    for i=1:length(x)
        fprintf(FileID,'%.4f  %.2f\n',x(i),T(i));
    end
    fclose(FileID);
end
```

Annex II MATLAB Source Code (OPTIFIN)

```
function [Wop,Dop,Lop,PeltLength,N]=OptiFin(Dim,Q,Theta0,ThetaA,PhysiscPar
ams)
%Clear screen
clc;

%Input Arguments Checking
if nargin == 4
    OptiFinDefaultParams;
else
    h=PhysiscParams(1); %Heat Transfer coeficient (alpha)
    lambda=PhysiscParams(2); %Material thermal Conductivity
    VDin=PhysiscParams(6); %Dynamic Viscosity
    g=PhysiscParams(7); %Gravity acceleration
end

%Get Peltier length from input argument Dim
a=Dim(1);
%Estimation of Averaged Dynamic Viscosity
VDin=DynVis((Theta0+ThetaA)/2);
%Estimation of Averaged Air Density
rho=AirDens((Theta0+ThetaA)/2);
%Estimation of Medium Heat Transfer Temperature differnce of the Heat Sink
ThetaB=(Theta0 - ThetaA);

%First of all, required area for heat flux removal is calculated:
%Q=hA(Theta0 - ThetaA)
A=Q/(h*ThetaB);

%If Q<APeltier, there is no needing of putting a Heat Sink with fins
%If Q>APeltier, a Heat Sink with fin arrangement is needed
if A<a^2
    warning('Extended Surface Not Needed');
    return;
else
    %Optimum design Parameters Calculation:

    %CASE 1: Variable Heat Sink a
    %Initial Number of Fins:
    N(1)=1;
    %Initial Error:
    Err=0.1;
    %Tolerance:
    Tol=1e-4;
    %Initial Fin Length:
    Lop(1)=0.05; %m

    while abs(Err)>Tol
        %Optimum Thickness
        Wop(1)=0.6321/(h*lambda)*(Q/(Theta0 - ThetaA))^2;
        %Optimum Length
        Lop(1)=(0.7979/h)*(Q/(Theta0 - ThetaA));
        %Optimum Spacing
        Dop(1)=0.29*(Lop(1)^0.25*VDin^0.5*ThetaA^0.25)/
(g^0.25*rho^0.5*(Theta0 - ThetaA)^0.25);
```

```
      %Arrangement length (a)
      PeltLength(1)=N(1)*Wop(1)+(N(1) - 1)*Dop(1);
      %Surface Area for 1 Fin
      ST=(2*PeltLength(1)*Lop(1)+2*Lop(1)*Wop(1)+PeltLength(1)*Wop(1));
      %Total Heat Transfer Area
      AT=N(1)*(2*PeltLength(1)*Lop(1)+2*Lop(1)*Wop(1)+PeltLength(1)*Wop(1))
+(N(1)-1)*PeltLength(1)*Dop(1)+(PeltLength(1)^2-a^2);
      %Perimeter
      P=2*Wop(1)+2*PeltLength(1);
      %L dependant function root
      %Err=Q - (1 - N(1)*ST/(AT)*(1 - (Q/N(1))/
(h*(2*Wop(1)+2*a_a)*Lop(1)*(Theta0 - ThetaA))))*h*AT*ThetaB;
      %p001=- Q/((1 - N(1)*ST/AT*(1 - (Q/N(1))/(h*P*Lop(1)*(Theta0 -
ThetaA))))*h*ThetaB);
      Err=A - AT;
      N(1)=N(1)+1e-3;
   end
   N(1)=round(N(1));

   %CASE 2: Heat Sink a equal to Peltier a
   %Initial Number of Fins:
   N(2)=2;
   %Initial Error:
   Err=0.1;
   %Tolerance:
   Tol=1e-2;
   %Initial Fin Length:
   Lop(2)=0.001; %m

   while abs(Err)>Tol
      %Optimum Thickness
      Wop(2)=0.6321/(h*lambda)*(Q/(Theta0 - ThetaA))^2;
      %Optimum Spacing
      Dop(2)=0.29*(Lop(2)^0.25*VDin^0.5*ThetaA^0.25)/
(g^0.25*rho^0.5*(Theta0 - ThetaA)^0.25);
      %Number of Fins for the optimum arrangement
      N(2)=(a+Dop(2))/(Wop(2)+Dop(2));
      %Surface Area for 1 Fin
      ST=(2*a*Lop(2)+2*Lop(2)*Wop(2)+a*Wop(2));
      %Total Heat Transfer Area
      AT=N(2)*(2*a*Lop(2)+2*Lop(2)*Wop(2)+a*Wop(2))+(N(2)-1)*a*Dop(2);
      %L dependant function root
      %Err=Q - (1 - (N(2)*ST/AT)*(1 - (Q/N(2))/(h*(2*Wop(2)+2*a)*Lop(2)*
(Theta0 - ThetaA))))*h*AT*ThetaB;
      Err=A - AT;
      Lop(2)=Lop(2)+1e-3;
   end
   N(2)=round(N(2));
   %FinStudy function is called to verify that obtained L is not bigger
   %Than optimum fin L. Otherwise, warning message appears.
   [Tcomp,Qcomp]=FinStudy(4,1,0,0,[Lop(2) Theta0 ThetaA]);
   if Tcomp(length(Tcomp) - 1) <= ThetaA
      warning('Fin length is greater than optimum length')
   else
      fprintf('Optimum fin arrangement succeded')
   end
```

```
    %CASE 3: Real Heat Sink
    %If Dim is a two elements vector, third case is carried out
    if length(Dim) == 2
        Wop(3)=0.001;
        Dop(3)=0.003;
        Lop(3)=Dim(2);
        %Initial number of fins:
        N(3)=1;
        %initial Error:
        Err=1;

        while abs(Err)>Tol
            PeltLength(2)=N(3)*Wop(3)+(N(3)-1)*Dop(3);
            Err=A - N(3)*(2*Wop(3)*Lop(3)+2*PeltLength(2)*Lop(3)+Wop(3)*PeltL
ength(2)) - (N(3)-1)*Dop(3)*PeltLength(2) - (PeltLength(2) - a^2 );
            N(3)=N(3)+Tol;
        end

        N(3)=round(N(3));
    end

end

%Print file OutPut
FileID=fopen('OptiFin.txt','wt');
fprintf(FileID,'          Wop     Dop     Lop     PL     N\n');
fprintf(FileID,'CASE 1 %.4f %.4f %.4f %.4f
%d\n',Wop(1),Dop(1),Lop(1),PeltLength(1),N(1));
fprintf(FileID,'CASE 2 %.4f %.4f %.4f %.4f %d\n',Wop(2),Dop(2),Lop(2),a,N(2));
if length(Dim) == 2
    fprintf(FileID,'CASE 3 %.4f %.4f %.4f %.4f
%d\n',Wop(3),Dop(3),Lop(3),PeltLength(2),N(3));
end
if Tcomp(length(Tcomp) - 1) <= ThetaA
    fprintf(FileID,'Fin length is greater than optimum length')
else
    fprintf(FileID,'Optimum fin arrangement succeeded for CASE 2')
end
fclose(FileID);
```

References

1. Biercuk, M. J., Llaguno, M. C., Radosavljevic, M., Hyun, J. K., Johnson, A. T., and Fischer, J. E., Heat transfer in carbon nanotube composites. *Applied Physics Letters*, 80, 2767–2769, 2002.
2. Berber, S., Young-Kyun, K., and Tománek, D. Unusually high thermal conductivity of carbon nanotubes. *Physical Review Letters*, 84, 2000.
3. Rohsenow, W. M., *Handbook of Heat Transfer*, McGraw Hill, New York, 1987.
4. Chapman, A., *Heat Transfer*, Collier McMillan Co., Madrid, 1977.
5. Incropera, F. P. and De Witt, D. P. *Introduction to Heat Transfer*, John Wiley & Sons, New York, 1990.
6. Madhusudana, C. V., *Thermal Contact Conductance*, Springer, New York, 1992.
7. Yovanovich, M. M., Culham, J. R., and Teertstra, P., Calculating interface resistance. *Electronics Cooling*, 3(2), 24–29, 1997.

8. Chung, D. D. L., Materials for thermal conduction, *Applied Thermal Engineering*, 21, 1593–1605, 2001.

9. Prasher, R., Thermal Interface materials: Historical perspective, status and future directions. *Proceedings of the IEEE*. 98(8), 1571–1586, 2006.

10. Neubauer, E. and Korb, G., The influence of mechanical adhesion of copper coatings on carbon surfaces on the interfacial thermal contact resistance, *Thin Solid Films.*, 433(1–2), 160–165, 2003.

11. Lahmar, A. and Nguyen, T. P., Experimental investigation on the thermal contact resistance between gold coating and ceramic substrates. *Thin Solid Films*, 389(1–2), 167–172, 2001.

12. Wolf, E. G. and Schneider, D. A., Prediction of thermal contact resistance between polished surfaces. *International Journal of Heat and Mass Transfer*, 41(22), 3469–3482, 1998.

13. Marlow, R., Buist, R. J., and Nelson, J. L., *Proceedings, Fifth International Conference on Thermoelectric Energy Conversion*, Arlington, Texas, IEEE, 125 New York, 1984.

14. Domínguez, M., Pinillos J. M., and Gutiérrez, P., Contribution to Peltier effect in cooling. *Refrigeration-Frial*, 19, 47–54, 1991.

15. Esarte, J., Min, G., and Rowe, D. M., Modelling heat exchangers for thermoelectric generators, *Journal of Power Sources*, 93, 72–76, 2001.

16. Noriega, F. G., Air conditioning for auto motion by thermoelectric systems. *Termoelectricidad*, 2, 7–15, 1995.

17. Simons, R. E., Application of thermoelectric coolers for module cooling enhancement. *J. Electronics Cooling*, 6, 45–51, 2000.

18. Incropera, F. P. and DeWitt, D. P., *Fundamentals of Heat and Mass Transfer*, John Wiley and Sons, New York, 1996.

19. Kern, D. Q. and Krauss, A. D., *Extended Surface Heat Transfer*. McGraw-Hill. New York. 1972.

20. Çengel, Y. A., *Heat Transfer: A Practical Approach*, 2nd Edition, McGraw-Hill, Boston, 2003.

21. Tafti, D. K., Wang, G., and Lin, W., Flow transition in a multilouvered fin array, *Int. J. Heat Mass Transfer*, 43, 901–919, 2000.

22. Webb, R. L. and Trauger, P., Flow structure in the louvered fin heat exchanger geometry, *Experimental Thermal and Fluid Science*, 4, 205–217, 1991.

23. Gröber, H., Kern, J., and Grigull, U., *Fundamentals of Heat Transfer*, McGraw-Hill. New York. 1998.

24. Razelos, P. and Kakatsios, X., Optimum dimensions of convecting-radiating fins: Part I—longitudinal fins, *Applied Thermal Engineering*, 20, 1161–1192, 2000.

25. Mikhailov, M. D. and Özisik, M. N., *Heat Transfer Solver*, Prentice Hall, New Jersey. 1991.

26. Harper, W. B. and Brown, D. R., Mathematical equations for heat conduction in the fins of air-cooled engines, NACA Report 158, Washington, 679–708, 1922.

27. Guyer, E. C. and Brownell, D. L. *Handbook of Applied Thermal Design*, McGraw-Hill. New York, 1989.

28. Söylemez, M. S., On the optimum heat exchanger sizing for heat recovery, *Energy Conversion & Management*, 41, 1419–1427, 2000.

29. Yeh, H., Analytical study of the optimum dimensions of rectangular fins and cylindrical pin fins, *Int. J. Heat Mass Transfer*, 40, 3607–3615, 1997.

30. Reklaitis, G. V., Ravindran, A., and Ragsdell, K. M., *Engineering Optimization, Methods and Applications*, John Wiley and Sons, New Jersey, 1983.

31. Gardner, K. A., Efficiency of extended surfaces. *Trans. ASME*, 69(8), 621–631, 1945.

32. Holman, J. P., *Heat Transfer*, McGraw-Hill. New York, 1998.

33. Çengel, Y. A., *Heat Transfer*, McGraw-Hill. New York, 2004.

34. Blanco, J. M., Sala, J. M., and López, L. M., Tecnología energética, Servicio de publicaciones de la Escuela Superior de Ingenieros de Bilbao, 2004.

35. Zukauskas, A. and Ulinskas, R., Banks of plain and finned tubes, *Heat Exchanger Design Handbook*, G. F. Hewitt Edition, Begell House, Inc., New York, 1998.

36. Rich, D. G., The effect of fin spacing on the heat transfer and friction performance of multi-row, Smooth plate finned-tube heat exchangers, *ASHRAE Transactions*, 79(2), 137–145, 1973.

37. Sunubu, J. H., The effect of spacing on the efficiency of extended surfaces for natural convection cooling. *Proceedings of the National Electronics Packaging and Production Conference*, New York, 1963.

38. Christopher, A. S., Thermshield, L. L. C., and Laconia, N. H., Optimize fin spacing: How close is too close? *Power Electronics Technology*, 8, 35–43, 2006.

39. Kraus, A. D. and Bar-Cohen, A., *Thermal Analysis and Control of Electronic Equipment*, McGraw-Hill, New York, 1983.

40. Girón-Palomares, B. et al., Optimización y análisis de arreglos de aletas rectangular, triangular, parabólico convexo, anulares (hiperbólico, triangular, rectangular) y alfiler, *Revista Iberoamericana de Ingeniería Mecánica*, 10(3), 33–39, 2006.

41. Elmahdy, A. H. and Biggs, R. C., Efficiency of extended surfaces with simultaneous heat and mass transfer, *ASHRAE Transactions*, V 1, 135–143, 1983.

42. Kraus, A. D., Aziz, A., and Welty, J., Extended surface heat transfer, *Appl. Mech. Rev.*, 54(5), 97–117, 2001.

43. Ünal, H. C., Determination of the temperature distribution in an extended surface with a non-uniform heat transfer coefficient. *International Journal of Heat and Mass Transfer*, 28(12), 2279–2284, 1985.

44. Yan, Z. and Holmstedt, G., A fast narrow band computer model for radiation calculations. *Numerical Heat Transfer Part B: Fundamentals*, 31, 61–71, 1997.

45. Launder, B. E. and Spalding, D. B., *The Numerical Computation of Turbulent Flows*, Imperial College of Science and Technology, London, England, NTIS N74-12066, 1973.

46. Faghri, A., *Heat Pipe Science and Technology*, Taylor & Francis, New York, 2000.

47. Michael, J. and Ellsworth, Jr., Comparing liquid coolants from both a thermal and hydraulic perspective. *Electronics Cooling*, 12(3), 36–38, 2006.

48. Reay, D. and Kew, P., *Heat Pipes: Theory, Design and Applications*, BH Elsevier, Edinburgh, 2006.

49. Scott, D., Heat pipes for electronics cooling applications. *Electronics Cooling*, 2(3), 18–23, 1996.

50. Esarte, J., Wolluscheck, C., and Armendariz, E., Phase change dissipater of aluminium container. *THERMINIC Conference*, Rome, 2008.

51. Jinwang L., Yong Z., and Cheng, L., Experimental study on capillary pumping performance of porous wicks for loop heat pipe. *Experimental Thermal and Fluid Science*, 34(8), 1403–1408, 2010.

52. Xiao, H. and Franchi, G., Design and fabrication of hybrid bi-modal wick structure for heat pipe application. *Journal of Porous Materials*, 15(6), 635–642, DOI: 10.1007/s10934-007-9143-1.

53. Yu, F. and Maydanik, K., Review loop heat pipes, *Applied Thermal Engineering*, 25, 635–657, 2005.

21

Thin-Film Superlattice Thermoelectric Devices for Energy Harvesting and Thermal Management

Rama Venkatasubramanian
RTI International

Jonathan Pierce
RTI International

Thomas Colpitts
RTI International

Gary Bulman
RTI International

David Stokes
RTI International

John Posthill
RTI International

Phil Barletta
RTI International

David Koester
Nextreme Thermal Solutions

Brooks O'Quinn
Nextreme Thermal Solutions

Edward Siivola
Nextreme Thermal Solutions

21.1 Introduction

The U.S. Department of Defense has been supporting the resurgence of the once-sleepy field of TEs during the last decade, since the first set of ideas using nanoscale materials [1] was proposed at the National Thermogenic Workshop in 1992. In particular, the Office of Naval Research, the Defense Advanced Research Projects Agency and the Army Research Office have helped make impressive strides in the materials figure of merit, denoted as ZT, using nanoscale material concepts. Higher ZT of the TE materials can directly lead to improved efficiencies for both solid-state refrigeration and thermal-to-electric energy conversion devices. There have been steady reports [2–8] of enhanced ZT with nanoscale and low-dimensional engineering of conventional materials at various temperature regimes—starting with two-dimensional thin-film p-type Bi_2Te_3/Sb_2Te_3 superlattices with a ZT of 2.4 at 300 K [2], quantum-dot PbTe/PbTeSe superlattices [3], bulk materials with nano-crystalline inclusions in the PbTe system [4], thin-film $In_{0.53}Ga_{0.47}As$ with buried ErAs nano-particles in a periodic fashion similar to a quantum-dot superlattice [5], and more recently in Si nano-wires [6–7]. All these developments in enhanced ZT have relied on the reduction in lattice thermal conductivity with low-dimensional structures [8–12]. Further validating the lattice thermal conductivity reduction approach to enhanced *ZT* in thin-film *p*-type Bi_2Te_3/Sb_2Te_3 superlattices, there have been similar developments in reduced lattice thermal conductivity and enhanced *ZT* in nano-versions of *p*-type $Bi_xSb_{2-x}Te_3$ alloys prepared by several methods [13–15] and with *ZT* as high as 1.56 at 300 K [15,16].

Thin-film TE materials [2,3,5,17,18] in combination with semiconductor processing tools for device fabrication, offer unprecedented electronic package-level advantages [18–27]. The improvements achieved with these nanoscale TE materials and their devices are timely for meeting many emerging needs in DoD systems as well as in commercial electronic and optoelectronic systems, in the areas of micro-scale TE devices for energy harvesting [18–20] and active thermal management of electronics, photonics, and bioanalytical systems [21–27].

Nanoscale TE materials, using phonon-blocking, electron-transmitting structures [28] employing thin-film superlattices [2,25] and quantum-dot superlattices [3,5,29] are typically deposited by molecular beam epitaxy (MBE) [3,5] or metallorganic chemical vapor deposition (MOCVD) [2,30] or chemical vapor deposition (CVD). Thermal evaporation, which are low cost and can be done in moderately high vacuum environment (~1e–6 to 1e–7 Torr), have also been successfully used to produce obtain high-quality $(Bi_2Te_3)/Bi_2(Te,Se)_3$ superlattices using element sources [31] and PbTe/PbTeSe superlattices [32] from high-purity PbTe and PbSe bulk materials.

The engineering of thin-film TE superlattices and quantum-dot superlattices into useful devices can be implemented readily if the electric transport occurs across the superlattice interfaces, that is, along the thickness of the deposited film. This way, planar semiconductor device technology using standard microelectronic tools can be used. In the following sections, we specifically discuss the superlattice TE technologies in the Bi_2Te_3-material system. We discuss the state of the art in these materials and highlight the state of transition of these materials into device prototypes for energy harvesting and thermal management.

21.2 Bi_2Te_3-Based Superlattice Materials Deposition and Key Characteristics

Bi_2Te_3-based superlattices are grown by metallorganic chemical vapor deposition (MOCVD) on GaAs substrates [30]. The MOCVD process can be scaled to multiwafer growth and for large-area growth, similar to that for III–V semiconductor space photovoltaics and LEDs—for enabling low-cost, volume production of modules. The GaAs substrates were chosen for their availability in large area wafers (up to 6 in.), ease of clean prior to epitaxial deposition and that <100> GaAs substrates with 2–6-degree misorientation with respect to <110> can be conveniently obtained [30]. It is important to note that the trigonal-structured Bi_2Te_3 materials are grown on GaAs with face-centered cubic (FCC) structure. The misorientation allows the initiation of the epitaxial process at the kink sites on the surface, thereby allowing the growth of mismatched materials. A transmission electron micrograph in Figure 21.1 shows the smooth interface between the GaAs (FCC) and the layered Bi_2Te_3 materials [30]. The growth of Bi_2Te_3-based materials, with the rather weak van der Waals bonds along the growth direction, desires a low-temperature growth process. The low-temperature growth process leads to high-quality, abrupt superlattice interfaces with minimal interlayer mixing, and also allows for the growth of highly lattice-mismatched material systems, without strain-induced three-dimensional islanding. Figure 21.1 shows a high-resolution TEM of a Bi_2Te_3 layer on a GaAs substrate, delineating the two very different crystalline orientations. In-situ ellipsometry has been used to gain further control over nanometer-scale control of deposition [33] and thereby achieve high-quality Bi_2Te_3/Sb_2Te_3 interfaces [30] as shown in Figure 21.2.

21.2.1 Device Development

The fabrication of thin-film modules employs standard semiconductor device manufacturing tools such as photolithography, electroplating, wafer dicing, and pick-and-place tools. This allows scalability of the module fabrication, from simple modules that can provide a few milliwatts to multiconnected module array that can provide 10s of watts.

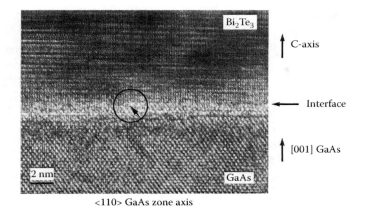

FIGURE 21.1 Interface between FCC-GaAs and trigonal-layered Bi_2Te_3 materials enabled by the intentional 2–6 degree miscut of the starting GaAs substrate. (Adapted from R. Venkatasubramanian et al., *Appl. Phys. Lett.* 75, 1104, 1999.)

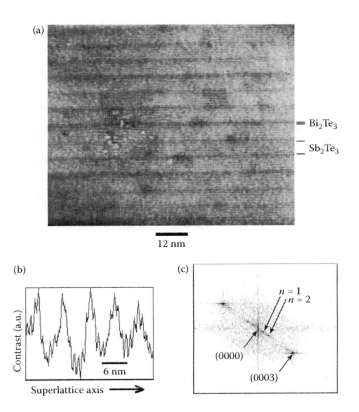

FIGURE 21.2 (a) TEM image of a 10/50 Å Bi_2Te_3/Sb_2Te_3 superlattice; (b) intensity oscillations seen in the 10/50 Å structure and within the 10 Å sub-units of 50 Å Sb_2Te_3; (c) Fourier transform of the TEM image showing the satellites due to the superlattice and the (0003) reflection. (From R. Venkatasubramanian et al. *Nature* 413, 597–602, 2001. With permission.)

21.2.2 Energy Harvesting

TE power generators offer a unique and attractive back-up to conventional batteries due to their waste heat energy-harvesting capabilities. They are especially suited for many low power and portable electronics such as for supporting unattended sensors and wireless devices. The TE generator's output voltage is directly proportional to the temperature difference between its two junctions. In many cases, the obtainable temperature difference can be less than 2 K and the output voltage of a TE generator can be a small fraction of a volt but still offer valuable power output. Owing to many practical applications requiring higher voltage, a DC–DC step up converter may be required.

Shown in Figure 17.2 is an example of a 4×4 array of p–n couples using the thin-film superlattice materials. This array is about $2.5 \times 2.5 \times 0.3$ mm. These modules have produced over 1 W of electric power, when an external ΔT of about 200 K is applied across the two ends of the 0.3 mm thick module. This implies an areal power density of 16 W/cm²; a volumetric power density of over 50 W/cm³; a specific power of high-quality superlattices have been demonstrated in the Bi_2Te_3 system, with one of the individual layers as small as 10 Å, using a low-temperature growth process. Such ultra-short-period superlattices offer significantly higher in-plane carrier mobilities (parallel to the superlattice interfaces) than alloys due to near-absence of alloy scattering and random interface carrier-scattering [30] as evidenced by the temperature dependence of carrier mobilities as shown in Figure 21.3. In the superlattice structures, the interface-scattering-limited mobility can be modeled as proportional to T^n (with $n \sim 1$) as this scattering relaxation time is expected to increase with carrier thermal velocity, decrease with carrier mean free path, and be inversely proportional to Debye screening length [34]. The interface carrier scattering, if strong, is expected to considerably reduce the exponent in the temperature dependence of the overall mobility. However, in the data of Figure 21.3, we see only a small reduction in the temperature exponent from $-3/2$ (associated with lattice phonon scattering). Consider the behavior when the impurity level is increased; here, we compare intentionally chosen samples of nearly the same ratio of impurity levels for both the alloy and superlattices. From these data, we observe that the temperature variation of the $(T^m, m > 0)$ impurity-scattering-limited mobility has as much a role in the superlattices as in the alloys. This suggests, again, weak interface scattering of carriers in the superlattices.

The enhanced carrier mobilities (μ) in monolayer-range superlattices are effective in the cross-plane direction for certain superlattices [2], where we can also obtain reduced lattice thermal conductivity K_L

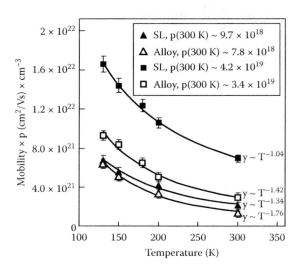

FIGURE 21.3 Temperature dependence of the mobility-carrier concentration product in Bi_2Te_3-based superlattice structures and comparable alloys at different carrier levels (in cm⁻³).

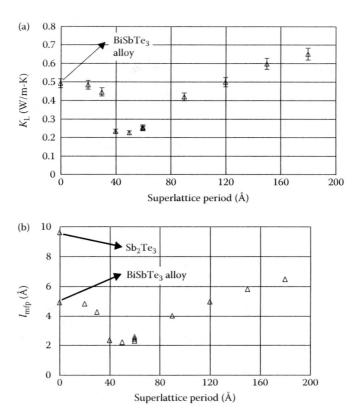

FIGURE 21.4 (a) Experimental lattice thermal conductivity and (b) calculated average phonon mean free path as a function of the period in Bi_2Te_3/Sb_2Te_3 superlattices. (Adapted from R. Venkatasubramanian et al., *Nature* 413, 597–602, 2001.)

as shown in Figure 21.4 [12]. *Both these factors lead to an enhanced ZT, as per Equation 21.1 and a modification of the relationship in terms of* μ *and* K_L *in Equation 21.2, as opposed to a common reference, enhancement in ZT from the reduction in* K_L *only, of our work described in Ref. 2.* The ZT of p-type Bi_2Te_3/Sb_2Te_3 SL as a function of temperature, between 300 and 220 K, is shown in Figure 21.5 [2].

$$ZT = (\alpha^2 T/\rho K_T) \tag{21.1}$$

$$Z = \frac{\alpha^2 \sigma}{K_L + K_e} \simeq \frac{\alpha^2}{\left(\dfrac{K_L}{\mu p q}\right) + L_o T} \tag{21.2}$$

21.3 Device Fabrication from Bi_2Te_3-Based Superlattice Materials

We have fabricated numerous multicouple TE device structures [35–37] for cooling and energy-harvesting applications. Each module contains a different number of couples and has a different packing fraction designed to optimize power density or heat pumping. The individual couples in each stage are fabricated from strips of epitaxial p-type Bi_2Te_3/Sb_2Te_3 (1 nm/5 nm) superlattices and n-type $Bi_2Te_3/Bi_2Te_{2.85}Se_{0.15}/Bi_2Te_3$-(1 nm/5 nm) superlattice TE materials that are bonded epitaxial film side down and side-by-side to an AlN die header. After this bonding, the substrate is removed, metal posts are plated on the exposed

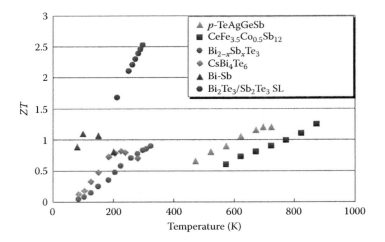

FIGURE 21.5   Temperature dependence of ZT of p-type Bi_2Te_3/Sb_2Te_3 SL, compared to various other bulk materials. (After R. Venkatasubramanian et al., *Nature* 413, 597–602, 2001.)

SL surface and the assembly is then diced into individual couples containing two n-type and two p-type contacts (designated 2N-2P). These individual dies are then bonded to a metallized ceramic heat sink, which also serves as a heat spreader. By connecting these single couple building blocks electrically in series, on a larger primary header, large area single stages can be assembled. The modules can be tuned to a particular application by adjusting the number of dies in each stage and using a heat spreader to interconnect them. An example of a functional 16-couple module is shown in Figure 21.6 and a much larger 160-couple module array is shown in Figure 21.7. Array cooling or power-harvesting performance can be tailored to a particular application by adjusting the number of dies, the die packing fraction, and the individual element contact diameters.

Larger arrays of mini-modules, like the 4×4 shown in Figure 21.6, can be strung together—with the couples being electrically in series and thermally in parallel—to produce larger power levels in energy harvesting or to cool a larger heat load. The scale up of devices from single couples that produce tens of milli-watts for a temperature differentials of only a few °C, to multimodule array (MMA) has been achieved. The multimodule arrays, just about 2.5 cm on a side, can produce about 15 W of electric power

FIGURE 21.6   A 4×4 module array made using 16 p–n couples made using the superlattice thin-film materials; the module is approximately 2.5 mm $\times$ 2.5 mm $\times$ 0.7 mm in dimensions.

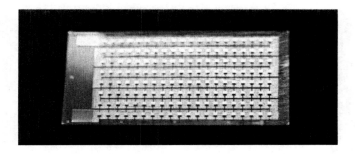

FIGURE 21.7 A 8 × 20 module array made using 160 p–n couples made using the superlattice thin-film materials; the module is approximately 19 mm × 38 mm × 0.7 mm in dimensions.

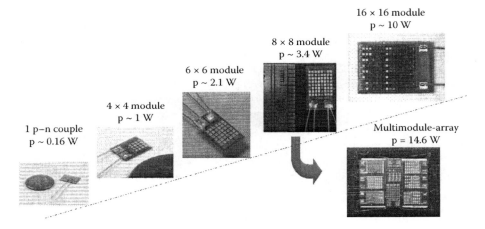

FIGURE 21.8 (**See color insert.**) Scale up of the superlattice thin-film TE module technology to larger multi-module arrays (MMAs) that can produce power levels of hundreds of milliwatts to watts depending on temperature gradients available. Such MMAs employ the BCAM module technology where the couple-based assembly allows an automatic tiling on the hot-side of the module and obviates the need for a hard-bonded common header.

with a temperature differential of about 150°C. The scale-up is shown schematically in Figure 21.8. Thus the thin-film superlattice technology with the bipolar couple assembled module (BCAM) architecture with the tiled geometry and no common header on one side (typically hot side) of the module [38], significantly reduces the mechanical shear stress in large-area module. This BCAM technology has shown a significant reliability advantage and is also applicable for bulk TE modules.

21.4 TE Power Generation Using Bi_2Te_3-Based Thin-Film Superlattice Devices

Shown in Table 21.1 are data from a 2.0 cm × 2.3 cm × 0.7 mm, 512-couple, superlattice multimodule array (MMA) for a range of hot-side and cold-side temperatures. We note that the 512-couple MMA produces as much as 22.4 V and a power of 14.6 W, with a temperature differential of ~100 K; these power levels are comparable to scalability of bulk Bi_2Te_3-alloy TE modules. The manufacturable scalability of the thin-film SL TE modules are achieved with scalable processes starting from MOCVD of thin-film materials, standard semiconductor photolithographic processes and dicing and pick-and-place toolsets common in the LED industry.

TABLE 21.1   Power Generation Characteristics of a 512-Couple Multimodule-Array, Packed in a 2.0 cm × 2.3 cm Area, and with a Total Height of ~0.7mm, for Various Hot-Side and Cold-Side Temperatures

T_{hot} (°C)	T_{cold} (°C)	ΔT (C)	Power (W)	V_{oc} (V)	I_{sc} (A)	Pd (W/cm²)
25.54	23.52	2.02	0.005	0.52	0.038	0.0011
29.26	24.26	5	0.031	1.247	0.098	0.0067
33.46	24.48	8.98	0.107	2.33	0.184	0.0233
44.55	24.94	19.61	0.531	5.253	0.404	0.1154
79.04	31.33	47.71	2.48	11.251	0.88	0.5391
98.76	36.98	61.78	4.63	14.71	1.26	1.0065
110.86	44.42	66.44	6.6	15.85	1.665	1.4348
128.16	48.67	79.49	10.68	19.96	2.14	2.3217
156.92	56.71	100.21	14.6	22.38	2.61	3.1739

Note: That the cold-side temperature is allowed to drift up, as more heat flows through the thin-film TE module, to enable easier rejection to an eventual heat-sink. Note that over 107 mW, with 2.33 V of open-circuit voltage can be produced with just about 9 K external temperature differential.

The power-harvesting load lines ‹ btained from these MMAs, for temperature differentials <10 K, are shown in Figure 21.9. Note that for small ΔT of 5 K, we can generate over 31 mW and an open-circuit voltage of 1.25 V; these are useable for a variety of sensors and wireless electronic devices.

The thin-film superlattice (SL) modules with low total volume have been compared with commercially available bulk TE devices under packaged conditions. The comparative data, for both raw power and for DC–DC up-converted power, for the advanced SL modules and the commercial modules is shown in Figure 21.10. The performance advantages of the SL-based TE modules are readily evident in both conditions. It appears that the advanced modules can provide sufficient power over the background requirements for implant applications to directly power electronics with temperature differentials less than 1°C.

21.4.1 Energy Harvesting for Bioimplants

All implantable medical devices (IMDs), including pacemakers, defibrillators, drug infusion pumps, and neuron-stimulators require electrical power. Nonrechargeable batteries are used as the source of power in most (IMDs). One limitation of IMDs is the longevity of the battery source. When the battery is unable to provide adequate power to the IMD, the IMD must be explanted and either the battery or the IMD replaced. The lifetime of the battery depends on the power requirements of the IMD and the

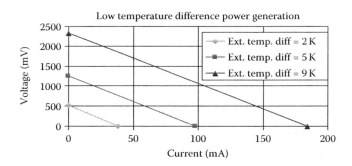

FIGURE 21.9   Power generation characteristics of a 512-couple module, occupying a footprint of 2.0 cm × 2.3 cm and ~0.7 mm height, for <10 K; such power levels in the range of 5–100 mW are useful for a variety of electronic sensors and actuators. (Adapted from R. Venkatasubramanian et al., *Energy Harvesting for Electronics with Thermoelectric Devices using Nanoscale Materials*, Proceedings of the IEEE International Electron Devices Meeting (IEDM), pp. 367–370, IEEE Press, Washington, DC, 2007.)

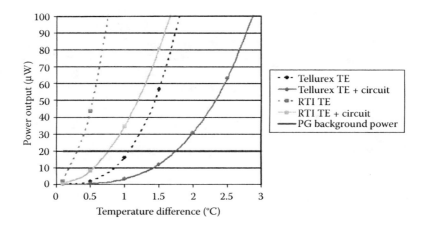

FIGURE 21.10 Comparison of raw power output as well as the up-converted DC power output, after going through a DC–DC conversion process, for the SL modules versus Tellurex standard Bi_2Te_3-alloy commercial TE modules. Note that to achieve a minimum required power of ~20 μW with the required voltage output after DC–DC conversion, a ΔT of ~0.7 K is sufficient for the SL modules while a DT of ~1.8 K is required for the commercial Bi_2Te_3-alloy modules.

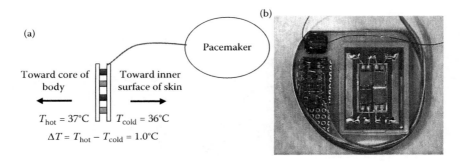

FIGURE 21.11 (a) Schematic of a TE device taking advantage of a ΔT across the device to supply power to the pacemaker. (b) Picture of an early integration of thin-film superlattice TE modules inside a pacemaker can. These power generation devices require further evaluation including animal-implant studies.

amount of power stored in the battery. Implantable neuron stimulators and pumps consume power at a higher rate than pacemakers. Consequently, the lifetime of neuron stimulators is much less than that of implantable pacemakers. Temperature differences that exist either between the inner surface of the skin and the core body temperature or within the implantable device (see Figure 21.11) could be used to produce electrical power, store energy as well and hence increase the battery life.

21.4.2 TE Energy Harvesting for Wireless Sensors

Self-sufficient power sources for wireless sensors and electronics can extend device performance for several key security and defense applications. These include remote sensors for intelligence, surveillance, and reconnaissance (ISR) applications, specific unattended ground sensor applications, and body worn sensor applications.

One specific application is the use of TE energy harvesting for unattended ground sensors (UGS). Enhancements to performance can include extended operating time, smaller size and weight, in addition to the potential for covert operation. One approach for developing a self-sufficient power source involves ground thermal energy harvesting. The availability of ground thermal gradients both during the day and at night under a variety of environments makes this potential power source of great interest. There is a clear need for power sources that can extend the operating time for unattended ground sensors in critical applications, and interest in this concept centers on the ability to generate power in a variety of environmental conditions (cool mountainous climates to arid deserts) and the ability to be deployed and operated covertly. One aspect of this research focuses on the design, fabrication, and testing of a "thermal ground stake" wireless sensor node powered by TE energy harvesting. A TE powered ground stake that is implanted in the ground would be easy to conceal and offer extended operation over more conventional means of energy harvesting such as solar power.

In general, sensors for ISR applications often require power sources to allow extended operation in remote locations. The use of thermal energy harvesting provides a practical means for allowing self-sufficient operation without the added burden of large battery packs. Figure 21.12 shows two TE energy-harvesting conceptual designs for infrastructure or tagged asset monitoring. The first (Figure 21.12a) is a "power bolt" approach where the sensor would replace an existing washer on a piece of equipment or container. The second (Figure 21.12b) is a flexible thin-film design that could be easily mounted to the surface of a container or asset.

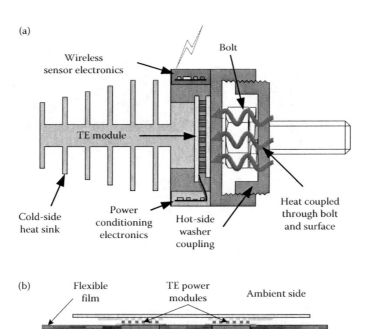

FIGURE 21.12 Conceptual drawings of RTI thermal energy-harvesting designs: (a) power bolt energy-harvesting approach and (b) flexible thin-film surface energy-harvesting approach.

21.4.3 Comparison of TE-Based Energy Harvesting with Other Technologies

Although several battery technologies can meet the energy density requirements for today's electronics, energy-harvesting approaches are advantageous to meet the power requirements for portable electronics, particularly for unattended sensors and communication devices. There are several approaches to energy harvesting: mechanical vibration (MV) energy harvesting using piezoelectric devices, light energy harvesting with photovoltaics (PV), chemical energy (CE) leveraging natural biological or biomimetic processes and thermal energy harvesting with TE devices. Each of these energy-harvesting technologies has their own advantages and limitations. For example, PV devices offer DC voltages that can easily match to that of typical batteries, especially by stacking PV cells in series; this approach eliminates conversion problems from AC to DC or DC to DC. However, PV energy harvesting in packaged electronics or applications as in bioimplants is difficult due to opaqueness to light. Similarly, PV energy harvesting in low-level ambient lighting is not efficient. In contrast, MV devices can harvest vibration spectra in many environments such as from HVAC and fans in electronic systems. Vibrations in special settings can be significantly higher, with tens to thousands of g's as in oil wells, where MV devices can harvest energy for electronics. However, vibration frequency spectra are complex and depend on the nature of the machine and environment. Matching the vibration with energy harvesters that rely on electroactive materials over a broad frequency range remains a challenge as well as their low potential power density of $<100\ \mu W/mm^3$. This is a factor of 10 smaller than what current thin-film TE technologies can offer for a $\Delta T < 10$ K.

Current thin-film TE devices, at volume of 1 mm^3, for converter alone, modules are available that can produce well over 775 $\mu W/mm^3$ for a ΔT of 9 K (Table 21.1). With thin-film technology, state-of-the-art modules are available at volumes of <1 mm^3. It appears that TE materials that have a high figure of merit near 300–400 K are sufficient. However, the big challenge is to achieve high device ZT_v values, the subscript v referring to the volume-limited power source of interest to electronics packaging.

Smaller and more efficiently designed thin-film TE modules with volumes of 1 mm^3, for an external ΔT of about 1 K, can produce over 100 $\mu W/mm^3$ in the future. These modules, with an efficient DC–DC conversion, can produce voltages useful for bioimplantable electronics. A significant challenge is to develop high aspect ratio TE devices that would produce both higher efficiencies, when the application is heat-source limited rather than temperature-gradient-limited, to minimize DC–DC conversion efficiency losses. It is important to observe that power conditioning is needed to provide adequate protection for Li-ion batteries from over-charging as well as to potentially stabilize the output voltage when the temperature changes. Thus the DC–DC conversion circuitry can be integrated with the power conditioning circuitry in a single ASIC chip, with careful attention paid to inductor design from the point of view of meeting volume constraints. It is anticipated that the volume of such packaged energy-harvesting modules can be designed to meet volume constraints in most electronics applications.

21.4.4 Universal Energy Harvesting

The different energy-harvesting technologies will enable us to combine them into a multicomponent micro power system. Such a device may enable energy harvesting to occur in a changing and unstable environment such as might be expected for mobile or fixed autonomous micro systems. Figure 21.13 shows a schematic of one version of a universal energy harvester (UEH) that combines, TE, PV, and MV technologies. Whichever individual or combined energy-harvesting technique is chosen, it will probably be necessary to include "smart" power conditioning electronics. DC–DC conversion to trickle charge a battery (>2V DC) can be part of such a system. A micro system will most likely be on a power duty cycle; hence, a small battery can meet peak power demands. Should the micro system be mobile, it is possible that it could seek out thermal energy sources for battery recharging, thereby extending mission duration. Based on an initial assessment, we conclude that multiple energy-harvesting micro-components can be combined into a small, compact package for micro-systems applications in the future.

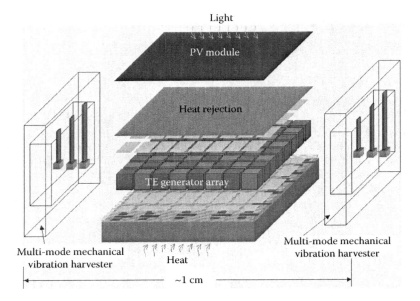

FIGURE 21.13 Conceptual schematic of a universal energy harvester that could occupy less than 1 cm³ in volume.

21.5 High Heat-Flux Thermal Management with Superlattice TE Devices

There is a significant need for rapid *site-specific* and *on-demand* cooling in a wide array of applications in electronics, optoelectronics and bioanalytics [38]. However, today, such cooling is achieved by the use of bulky and/or over-designed system-level solutions. Microscale TE devices with high figures-of-merit materials can address these limitations, while also enabling energy-efficient solutions. While significant progress has been made in the development of nanostructured TE materials with enhanced *ZT*, there has been limited progress in transitioning the advantages of these materials in advanced cooling applications.

The cooling of high heat fluxes is a crucial requirement for the effective functioning of myriad devices. For instance, heat generation from silicon microprocessors is highly nonuniform both spatially and temporally with localized high heat fluxes (>300 W/cm² is possible) that vary with the workload [39,40]. In general lower temperatures can lead to significantly higher computing performance from an integrated circuit (IC) device. For instance, the transistor gate leakage current increases exponentially with temperature and reliability and performance requirements are also gated by thermal considerations. Similarly, the performance of optoelectronic devices such as semiconductor lasers (heat flux >1000 W/cm² is possible) and bioanalytical devices such as DNA microarrays is also highly temperature dependent. Current cooling technologies based on conduction and convection can potentially cool high overall power levels by utilizing either novel passive heat transport materials or advanced heat exchangers such as carbon nanotubes [41] and microchannels [42], respectively. But they cannot provide *site-specific* and/or *on-demand* localized cooling of high heat-flux regions, thus resulting in overdesigned, inefficient at system-level, and bulky thermal systems. This has huge implications, especially considering the recent advent of massive data centers where the cooling energy costs are comparable to or even higher than the cost of the computational equipment itself, primarily because the entire system is cooled down in bulk-scale approaches. However, if TE devices are used to provide targeted cooling of only the high heat-flux regions as and when needed, the energy consumption can be drastically reduced.

Towards achieving this promise, we have shown for the first time the TE cooling of electronics at extraordinarily high-flux level that will have significant implications. We have overcome two major factors in this implementation—replaced the low figure of merit (ZT) of traditional TE materials with nanoscale TE materials with demonstrated ZT [2] and the low heat flux pumping capacity of conventional coolers due to the use of thick bulk (mm) materials with ultra-thin (<10 microns). The maximum heat-flux pumping capability for a TE device is given by [2,25]

$$q_{max} = \frac{1}{l}\left\{[0.5\alpha^2 T_{cold\text{-}side}^2/\rho] - [k(T_{hot\text{-}side} - T_{cold\text{-}side})]\right\} \qquad (21.3)$$

where α is the Seebeck coefficient, ρ the electrical resistivity, k the thermal conductivity, l the thickness of the TE material and T denotes the temperature. Equation 21.3 shows that to achieve high q_{max}, α^2/ρ should be large and k should be small. These requirements imply maximization of ZT ($= \alpha^2 T/\rho k$). Another requirement for high q_{max} is that l should be as small as possible. Using Equation 21.3 and properties of state-of-the-art TE materials, it can be shown that to achieve q_{max} >500 W/cm² , the thickness l has to be <50 μm which is not possible if bulk TE materials are used. Also, the total device thickness including the electrical contacts has to be of the order of 100 μm or less to meet electronic package volume constraints.

RTI, teaming with Intel and Nextreme, has demonstrated the first real-world thermoelectric coolers (TECs) made from SL materials in managing high heat-flux situations in advanced microprocessors [25], in spite of the enormous difficulties in overcoming electrical and thermal parasitic while integrating nanoscale thin-film materials into microscale and packaged macroscale systems. We have shown that the TECs fabricated from nanostructured Bi₂Te₃-based superlattices can be integrated into state-of-the-art electronic device packages, as shown in Figure 21.14. We have achieved cooling of as much as 15°C at the targeted region on the silicon chip which has an ultra-high heat-flux of ~1300 W/cm², as shown in Figure 21.15. This represents the demonstration of a viable chip-scale refrigeration technology and has the potential to enable a wide range of thermally limited applications.

The ability to manage extraordinarily localized high heat-flux regions in a Si chip by the selective introduction of a thin-film superlattice cooler between the chip and the heat-sink thereby avoids any use of liquid-based heat-removal and uses only convective air-flow heat removal at the sink. The use of air cooling at the sink and localized-and-on-demand thermal management has significant impact on energy efficiency in electronics-intense data centers to power electronics in hybrid vehicles to avionics.

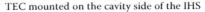

TEC mounted on the cavity side of the IHS

FIGURE 21.14 Integration of the thin-film superlattice thermoelectric cooler (TEC) onto the integrated heat spreader (IHS) and a fully packaged device under testing conducted at Intel Labs. (From I. Chowdhury et al., *Nat. Nanotechnol.* 4(4), 235–238, 2009. With permission.)

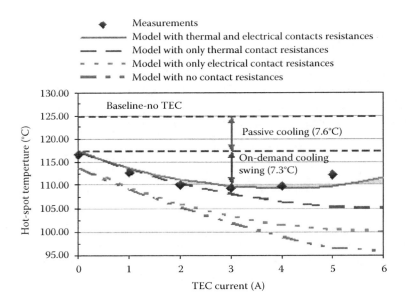

FIGURE 21.15 Hot-spot cooling performance using a thin-film SL TEC integrated onto a microprocessor thermal test vehicle at Intel Labs. (From I. Chowdhury et al., *Nat. Nanotechnol.* 4(4), 235–238, 2009. With permission.)

21.5.1 Microcryogenic Thermal Management

The thin-film SL TECs can be employed not only for cooling of imaging arrays and low noise amplifiers, but also for the cooling of high heat flux producing power amplifiers. Since the sensitivity of narrow band gap infrared (IR) detectors is strongly influenced by their operating temperature, they must therefore be operated at sufficiently low temperatures to realize their optimum performance. Current bulk TE devices are limited to a cold-side temperature of about 170 K, are capable of only small heat fluxes and can have profile thickness in the range of several centimeters.

We have demonstrated an external ΔT_{max} value of 57.8 K in a single superlattice p–n couple (2N–2P), compared to ~67 K for bulk state-of-the-art TE couples, and regularly attain ΔT_{max} in excess of 53 K, in spite of the significant electrical and thermal parasitic factors that currently exist in these thin-film devices. Figure 21.16a shows the external ΔT versus I characteristic obtained from a single couple under no load conditions in moderate vacuum. Also shown is the predicted internal temperature difference obtained by modeling the external ΔT data as a function of current. The higher estimated internal ΔT value of 104 K shown in the figure is a direct result of the high intrinsic material ZT, while the lower external ΔT is due to device thermal resistance ($R_{th} = 26$ K/W). This thermal resistance is produced by the metal posts and solders used to attach these posts to the lower AlN device header. The $I–V$ characteristics for the same device are shown in Figure 21.16b, where the dashed curve is the voltage component due only to the Seebeck effect. The difference between that curve and the measured voltage is produced by the element resistance (21 mΩ). Using the model parameters, the calculated maximum heat pumping capacity for this device is 0.44 W, corresponding to a die heat pumping density of 122 W/cm^2 or an individual TE element heat pumping density of 870 W/cm^2, consistent with prior published data [2]. This single couple heat pumping capacity is significantly above that of commercial bulk devices (~1 W/cm^2) and indicates the superiority of these devices for high heat flux situations.

We have also built other multistage coolers using these individual thin-film SL couples. This effort has resulted with the demonstration of a small area three stage cascade producing a 102 K temperature difference (ΔT) under vacuum [43]. This device, which is shown in Figure 21.17a and b, consists of three

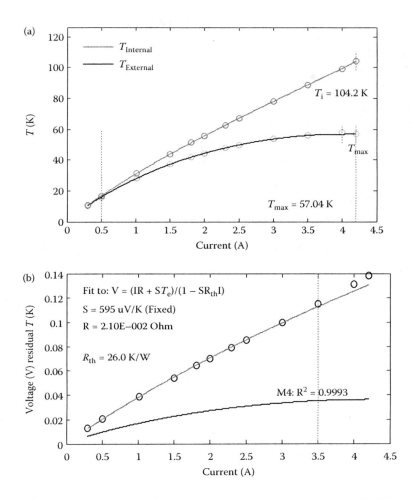

FIGURE 21.16 (a) External ΔT (lower curve) measured in vacuum and estimated internal ΔT (upper curve) characteristics as a function of current. (b) Measured I–V characteristic (upper curve) for the same couple and estimated voltage component produced by the Seebeck effect (lower curve).

individually powered stages. The power leads for the bottom stage can be seen at the bottom of the figure, while the low thermal conductance Ag foil power leads for the middle and top stages are seen at the right and top of the figure, respectively. The top stage couple is the 600×600 µm rotated square located in the middle of the figure. This cascade uses an 8×8 array of (2N–2P) 150 µm diameter SL couples for the bottom stage, a 4×4 array of 100 µm diameter (2N–2P) couples for the middle stage and a single 100 µm diameter couple (1N–1P) for the top stage. Figure 21.17c shows the ΔT of the module in vacuum as a function of the top stage current, where the middle and bottom stages are operating at constant currents of 3 and 4A, respectively. A ΔT_{max} of 101.6 K is obtained at a top stage current of 1.9 A, with a cold-side temperature, T_c, of −75°C (198.2 K), better than typical ΔT_{max} of 98 K with commercially available 3-stage Bi_2Te_3-alloy bulk devices. The 2.5 mm thickness of this module is at least 2× thinner than commercially available bulk modules, and illustrates the size advantages of these compact high heat pumping capacity devices. This structure has not been optimized and performance improvements can be expected with further effort. The area of microcooling can be localized to ~500 µm × 500 µm or smaller. These developments in combination with MEMS concepts can be useful for cooling low noise amplifiers, infrared focal-plane arrays, quantum cascade lasers, and so on.

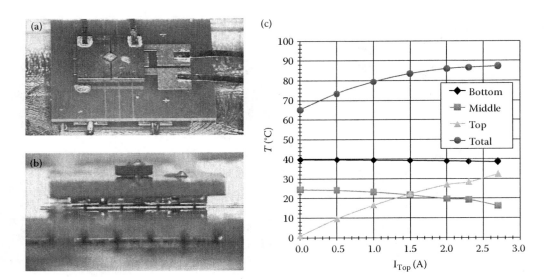

FIGURE 21.17 (a) Plan view of a 3-stage thin-film SL cooler; (b) Side view of the 3-stage cooler; ΔT from various stages that produced an external ΔT of ~102 K, with a T_{cold} of −75°C.

21.6 Summary

High ZT of the TE materials can directly lead to improved efficiencies for solid-state cooling and refrigeration as well as in thermal-to-electric energy-harvesting devices. There have been steady recent reports of enhanced ZT with nanoscale and low-dimensional engineering of conventional materials at various temperature regimes—starting with thin-film p-type Bi_2Te_3/Sb_2Te_3 superlattices (SL) that demonstrated a ZT of 2.4 at 300 K. The enhanced carrier mobilities (μ) in monolayer-range SL, from the avoidance of alloy carrier scattering and near-absence of interface carrier scattering due to the high-quality SL, are effective in the cross-plane direction for certain SL's. This in combination with reduced lattice thermal conductivity lead to an enhanced ZT. Further validating the lattice thermal conductivity (K_L) reduction approach to enhanced ZT in thin-film p-type Bi_2Te_3/Sb_2Te_3 SL, there have been developments in reduced K_L and enhanced ZT in nanoversions of p-type $Bi_xSb_{2-x}Te_3$ alloys prepared by several methods with ZT as high as 1.56 at 300 K. Thin-film TE materials, using phonon-blocking, electron-transmitting structures with semiconductor processing tools for device fabrication, offer unprecedented electronic package-level advantages. The SL TE devices are timely for meeting many emerging needs in commercial electronic, optoelectronic and bioanalytical systems in the areas of high-efficiency and compact active thermal management. The developments in high efficiency thin-film SL devices for low-ΔT energy-harvesting systems are attractive for a range of applications including unattended wireless sensors, IMDs, and industrial equipment monitoring devices, due to advantages of size, volume, weight, scalability, and manufacturability with standard semiconductor device process technologies and tools.

Acknowledgments

Authors from RTI International acknowledge the longstanding support of the Defense Advanced Research Projects Agency (DARPA), the Office of Naval Research (ONR) and the Army Research Office (ARO) in enabling the fundamental concepts of phonon-blocking electron-transmitting superlattices from an idea to material validation to advanced technology for both energy harvesting and advanced thermal management. The authors also acknowledge the support of Intel Corporation, Lockheed-Martin, Nextreme Thermal Solutions and RTI International. The authors acknowledge the valuable technical contributions of Mr. Brian Grant, Mr. Gordon Krueger, Ms. Cynthia Watkins and Mr. Ryan Wiitala over many years in the development of thin-film superlattice TE technology at RTI. The authors also acknowledge the collaboration with Mr. Randy Alley at Nextreme Thermal Solutions, Dr. Ravi Mahajan and Dr. Ravi Prasher at Intel Corporation.

References

1. M. Dresselhaus, T. Harman, and R. Venkatasubramanian, *Proc. of the 1st Thermogenic Workshop*, Ft. Belvoir, VA, Ed., S. Horn, 1992.
2. R. Venkatasubramanian, E. Siivola, T. Colpitts, and B. O'Quinn, Thin-film thermoelectric devices with high room-temperature figures of merit. *Nature* 413, 597–602, 2001.
3. T.C. Harman, P.J. Taylor, M.P. Walsh, and B.E. LaForge, Quantum dot superlattice thermoelectric materials and devices. *Science* 297, 2229–2232, 2002.
4. K. Hsu, S. Loo, F. Guo, W. Chen, J. Dyck, C. Uher, T. Hogan, E. Polychroniadis, and M. Kanatzidis, *Science* 303, 818, 2004.
5. W. Kim, J. Zide, A. Gossard, D. Klenov, S. Stemmer, A. Shakouri, and A. Majumdar, *Phys. Rev. Lett.* 96, 045901, 2006.
6. A.I. Hochbaum, R.K. Chen, R.D. Delgado, W.J. Liang, E.C. Garnett, M. Najarian, A. Majumdar, and P.D. Yang, *Nature* 451, 163–168, January 2008.
7. A.I. Boukai, Y. Bunimovich, J. Tahir-Kheli, J.K. Yu, W.A. Goddard, and J.R. Heath, *Nature* 451, 168–171, January 2008.
8. R. Venkatasubramanian, *Bull. Am. Phys. Soc.* 41, 693, 1996; See also *Naval Research Reviews* 48(4), 31, 1996; R. Venkatasubramanian and T. S. Colpitts, in *Thermoelectric Materials—New Directions and Approaches*, p. 73, Eds. T.M. Tritt, M. Kanatzidis, and H.B. Lyon, *Mat. Res. Soc. Proceedings* No. 478, Pittsburgh, 1997.
9. S. M. Lee, D.G. Cahill, and R. Venkatasubramanian, *Appl. Phys. Lett.* 70, 2957, 1997.
10. G. Chen, *Phys. Rev. B* **57**, 14958, 1998.
11. M.V. Simkin and G.D. Mahan, *Phys. Rev. Lett.* 84, 927, 2000.
12. R. Venkatasubramanian, *Phys. Rev. B* 61, 3091, 2000.
13. Y.Q. Cao, X.B. Zhao, T.J. Zhu, X.B. Zhang, and J.P. Tu, *Appl. Phys. Lett.* 92, 143106, April 2008.
14. B. Poudel, Q. Hao, Y. Ma, Y. Lan, A. Minnich, B. Yu, X. Yan et al., *Science* 320, 634, May 2008.
15. W.J. Xie, X.F. Tang, Y.G. Yan, Q.J. Zhang, and T.M. Tritt, *Appl. Phys. Lett.* 94, 102111, March 2009.
16. W. Xie, X. Tang, Y. Yan, Q. Zhang, and T. M. Tritt, *J. Appl. Phys.* 105, 113713, 2009.
17. H. Bottner, G. Chen, and R. Venkatasubramanian, *MRS Bull.* 31, pp. 211–217, March 2006.
18. J. Nurnus, H. Bottner, and A. Lambrecht, in *CRC Handbook of Thermoelectrics*, Ed. M. Rowe, Chapter 46, CRC Press, Boca Raton, FL, 2005.
19. R. Venkatasubramanian et al., *Energy Harvesting for Electronics with Thermoelectric Devices using Nanoscale Materials*, Proceedings of the IEEE International Electron Devices Meeting (IEDM), pp. 367–370, IEEE Press, Washington, DC, 2007.
20. C. David Stokes, E.A. Duff, M.J. Mantini, B.A. Grant, P.P. Barletta, and R. Venkatasubramanian, Thin-film superlattice thermoelectric materials and device technologies for energy harvesting applications, *Proc. SPIE*, Vol. 7683, 76830W, April 28, 2010.
21. G. Bulman, E. Siivola, B. Shen, and R. Venkatasubramanian, Large external delta T and cooling power densities in thin-film Bi_2Te_3-superlattice thermoelectric cooling device, *Appl. Phys. Lett.* 89, 122117, 2006.
22. R. Mahajan, C.-P. Chiu, and G. Chrysler, Cooling a microprocessor chip. *Proc. IEEE* 94, 1476–1486, 2006.
23. R.S. Prasher, J.-Y. Chang, I. Sauciuc, S. Narasimhan, D. Chau, G. Chrysler, A. Myers, S. Prstic, and C. Hu, Nano and micro technology-based next-generation package-level cooling solutions. *Intel Technol. J.* 9, 285–296, 2005.

24. A. Shakouri, Nano-scale thermal transport and microrefrigerators on a chip. *Proc. IEEE* 94, 1613–1638, 2006.
25. I. Chowdhury, R. Prasher, K. Lofgreen, G. Chrysler, S. Narasimhan, R. Mahajan, D. Koester, R. Alley, and R. Venkatasubramanian, On-chip cooling by superlattice-based thin-film thermoelectrics, *Nat. Nanotechnol.* 4(4), 235–238, 2009.
26. V.A. Semenyuk, in *Thermoelectrics Handbook: Macro to Nano*. Ed. Rowe, D.M. pp. 58-1–58-20, CRC Press, Boca Raton, FL, 2006.
27. S.V. Garimella, A.S. Fleischer, J.Y. Murthy, A. Keshavarzi, R. Prasher, C. Patel, S.H. Bhavnani et al., Thermal challenges in next-generation electronic systems, *IEEE Transactions on Components and Packaging Technologies*, 31, 801–815, December 2008.
28. R. Venkatasubramanian, Phonon blocking electron transmitting superlattice structures as advanced thin film thermoelectric materials, Chapter 4, *Recent Trends in Thermoelectrics, Semiconductors and Semimetals*, Academic Press, San Diego, 2001.
29. M.L. Lee and R. Venkatasubramanian, Effect of nanodot areal density and period on thermal conductivity in SiGe/Si nanodot superlattices, *Appl. Phys. Lett.* 92, 053112, 2008.
30. R. Venkatasubramanian, T. Colpitts, B.C. O'Quinn, S. Liu, N. El-Masry, and M. Lamvik, Low-temperature organometallic epitaxy and its application to superlattice structures in thermoelectrics, *Appl. Phys. Lett.* **75**, 1104, 1999.
31. J. Nurnus, H. Beyer, A. Lambrecht, and H. Böttner, *MRS Proc. 626*, 2000, pp. Z2.1.
32. C. Caylor, K. Coonley, J. Stuart, T. Colpitts, and R. Venkatasubramanian, Enhanced thermoelectric performance in PbTe-based superlattice structures from reduction of lattice thermal conductivity, *Appl. Phys. Lett.* 87, 023105, 2005.
33. H. Cui, I. Bhat, B. O'Quinn, and R. Venkatasubramanian, *J. Elec. Mater.* 30, 1376–1381, 2001.
34. K. Seeger, *Semiconductor Physics*, Springer-Verlag, Heidelberg, 1982.
35. Thin-film thermoelectric device and fabrication method of same (US Patent No. 6,300,150).
36. Thin film thermoelectric devices for hot-spot thermal management in microprocessors and other electronics (US Patent No. 7,523,617).
37. Thermoelectric generators for solar conversion and related systems and methods (US Patent No. 7,638,705).
38. Thin film thermoelectric cooling and heating devices for DNA genomic and proteomic chips, thermo-optical switching circuits and IR tags (US Patent No. 7,164,077).
39. S. Borkar, Design challenges of technology scaling. *IEEE Micro* 19, 23–29, 1999.
40. B. Kaczer, R. Degraeve, N. Pangon, and G. Groeseneken, The influence of elevated temperature on degradation and lifetime prediction of thin silicon-dioxide films. *IEEE Trans. Electron Dev.* 47, 1514–1521, 2000.
41. R. Prasher, Thermal interface materials: Historical perspective, status, and future directions. *Proc. IEEE* 94, 1571–1586, 2006.
42. D.B. Tuckerman, and R.F.W. Pease, High-performance heat sinking for VLSI. *IEEE Electr. Device Lett.* 2, 126–129, 1981.
43. G.E. Bulman, E. Siivola, R. Wiitala, R. Venkatasubramanian, M. Acree, and N. Ritz, Three-stage thin-film superlattice thermoelectric multistage microcoolers with a ΔT_{max} of 102 K, *J. Electron. Mater.* 38, 1510, 2009.

III

Thermoelectric Systems and Applications

22

Thermoelectric Energy Recovery Systems: Thermal, Thermoelectric, and Structural Considerations

Terry J. Hendricks
Battelle Memorial Institute

Douglas T. Crane
BSST LLC

22.1 Introduction

Most industrial and transportation processes worldwide waste 50–70% of the fuel energy input, leading to vast amounts of wasted thermal energy that is both available and recoverable. Recent studies indicate that approximately 12.5 quads of thermal energy are available across a spectrum of transportation platforms, including light-duty vehicles (e.g., passenger vehicles, mini-vans, and sport utility vehicles), and heavy vehicles (e.g., Class 4–Class 8 trucks) in the United States alone. Recent additional studies indicate that there are another approximately 10 quads of thermal energy available across a variety of industrial processes, including aluminum, glass, steel, cement, paper and pulp, and other processes in the United States. This energy typically is dissipated to the environment in exhaust and coolant systems of these vehicles and industrial processing plants. Thermoelectric power generation (TEG) is one important technology that is available to recover this energy and convert it to useful electrical energy. TEG systems are typically quiet, low maintenance, capable of high reliability and stealthy operation when designed properly, and can transition gracefully to different power levels when necessary. TEG power technology has been used to recover waste energies in certain niche energy recovery applications (e.g., truck exhausts, wood-burning stoves) and in small combustion-driven systems (e.g., natural gas line sensors). Recent advancements in thermoelectric (TE) materials have created the potential to harness this energy and convert it at much better energy conversion efficiencies (near or greater than 10%) than in past applications.

Other chapters within this Handbook discuss the development, advantages, intricacies and challenges, and properties of these new TE materials and compare them to past TE materials. TE energy recovery systems using new TE materials have thermal, structural, and TE design challenges to overcome in designing high-performance, robust, and flexible systems that take advantage of these new materials. TE, thermal, and structural considerations are necessarily interdependent in these system designs and must be dealt with simultaneously to satisfy the TE systems operational requirements and achieve the high performance and robustness desired. This chapter discusses these design challenges and various design and analysis techniques used to overcome these challenges in current and future systems.

22.2 TE Design: Optimization and Constraints

TE technology can recover and convert a portion of the waste heat from a variety of industrial, commercial, transportation, and military systems through temperature differentials between the various exhaust streams or hot environments and cooling environments available in these applications. Table 22.1 shows some typical waste exhaust stream temperatures available in various industrial and transportation applications. The U.S. Department of Energy, Office of Energy Efficiency and Renewable Energy has performed extensive analysis of petroleum-derived fuel use across the United States [1]. This analysis typically uses separate fuel-use categories for light-duty vehicles, mini-vans and sport-utility vehicles (SUV), and medium-/heavy-duty vehicles. The transportation energy data [1] shows that in 2002, for example, the fuel usage by light-duty vehicles, mini-vans, and SUVs represented approximately 16.27 quads of energy (1 quad $= 10^{15}$ Btus). Approximately 35% of this energy was dissipated in the high-temperature exhaust streams of these vehicles, therefore, approximately 5.7 quads of waste thermal energy were available to recover in exhaust streams in this vehicle category in 2002. This same transportation energy data [1] show that in 2002 the fuel usage by medium-/heavy-duty vehicles was approximately 5.03 quads of energy. Approximately 30% of this energy was dissipated in the high-temperature exhaust streams of these vehicles, therefore, another approximately 1.5 quads of waste thermal energy were available to recover in exhaust streams in this vehicle category in 2002. In 2008 the energy usage in the light-duty vehicle, mini-van, and SUV sectors rose to approximately 16.4 quads and the medium-/heavy-duty vehicle sector stayed roughly the same at approximately 5.02 quads [1]. Therefore, the high-temperature waste thermal energy available in these two vehicle sectors has stayed roughly equivalent during that time period.

The U.S. Department of Energy Industrial Technologies Program office also has estimated waste thermal energy dissipated in various industrial processes throughout the United States [2,3]. This analysis categorized and characterized waste thermal energy in industrial processes such as aluminum smelting, glass processing, steel processing, paper and pulp processing, cement processing, and several others. The analysis found considerable waste thermal energy in these industrial processes and therefore highly significant opportunities for industrial waste heat recovery. Table 22.2 shows typical process efficiencies for some of these processes [2].

Tables 22.1 and 22.2 show that there are enormous amounts of industrial waste thermal energy available for recovery and that waste occurs at temperatures that are quite compatible with the temperature-dependent performance of many TE materials discussed in other chapters in this Handbook. Estimates indicate there are approximately 10 quads of waste thermal energy available in various industrial processes and about 1.8 quads of this waste thermal energy are recoverable [2], meaning it is at temperatures where it could be recovered.

Waste heat in these exhaust streams and processes can be recovered either within a heat exchanger integrated with a TE device hot-side tied directly into the exhaust stream (i.e., real-time recovery) or by storing it in a thermal energy storage media for use in the future. New opportunities for efficiently recovering waste heat have been created because of advances in micro- and nano-technologies. The key to identifying TE waste heat recovery applications is to determine when thermal exchange between existing process fluids in a given system is not an available option or provides no useful technical or

TABLE 22.1 Available Waste Energy Temperatures in Different Economic Sectors

Economic Sector and Applications	Available Stream Temperatures
Industrial Sector[2]	
Water/steam boilers	~150°C
Aluminum smelting	~960°C
Aluminum melting	~750°C
Glass furnaces	485–1400°C
Transportation Sector[3–6]	
Light-duty vehicle exhaust	350–600°C (Depending on location)
Heavy-duty vehicle exhaust	300–500°C (Depending on location)

TABLE 22.2 Typical Process Efficiencies in Key U.S. Industrial Processes and Applications

Industrial Process	Typical Thermal Efficiency
Paper drying	~48%
Glass processing	45–79%
Aluminum processing	40–65%
Cement calcination	30–70%
Distillation column	25–40%
EAF steelmaking	~56%
Power production	25–44%
Steam boilers	~80%

Source: Adapted from *Energy Use, Loss and Opportunities Analysis: U.S. Manufacturing and Mining.* 2004, U.S. Department of Energy, Industrial Technologies Program, Energetics, Inc. & E3M, Inc.

economic benefit, or when electrical power generated by TE systems has a beneficial intrinsic value within the system.

Extensive system-level research has been performed in the past five years to design TE systems to recover and convert waste thermal energy to useful electrical energy in various industrial and transportation systems. This work has been useful in demonstrating what TE conversion efficiencies and power levels are possible with newer, advanced TE materials (i.e., skutterudites, lead–antimony–silver–telluride nanocomposites, half-Heuslers) and more conventional TE materials (i.e., bismuth telluride) in these different waste energy applications.

Designing TE generator systems for these waste heat recovery (WHR) applications typically involves two processes: *Design Optimization* which follows design techniques discussed in Angrist [4] and Rowe [5], and *Design Performance Prediction* which follows techniques discussed in Hogan and Shih [6], Hendricks et al. [7], and Crane [8].

22.2.1 TE Design Optimization

Design optimization generally involves maximizing the TE conversion efficiency, TE power output, power density, specific power (i.e., power per mass) or other relevant design parameter for the TE system or application of interest. The TE conversion efficiency, η, is typically given by

$$\eta = P/Q_h \tag{22.1}$$

where P = power output and Q_h is the thermal energy input to the TE hot side [4,5]. The power output is generally given by

$$P = \left[\frac{(N \cdot \alpha \cdot \Delta T)^2}{(N \cdot R_{int} + R_o)^2} \cdot R_o \right] \tag{22.2}$$

where R_{int} and R_o are familiar internal and external load resistances, respectively, $\Delta T = T_h - T_c$, T_h = TE device hot-side temperature, T_c = TE device cold-side temperature, and $\alpha = |\alpha_p| + |\alpha_n|$ for a single couple. This equation indicates the multiplier, N, the number of couples in a device, for a multiple couple system.

Optimization of conversion efficiency for a given set of TE materials and TE element geometries is discussed at length in Angrist [4] and Rowe et al. [5] and leads to the well-known relationship:

$$\eta_{max} = \left(\frac{P}{Q_h} \right)_{max} = \left[\frac{T_h - T_c}{T_h} \right] \cdot \left[\frac{(1 + Z^*\overline{T})^{1/2} - 1}{(1 + Z^*\overline{T}) + T_c/T_h} \right] \tag{22.3}$$

where $\overline{T} = (T_h + T_c)/2$. Recent design optimization techniques have leveraged these basic equations to develop more sophisticated analysis and optimization algorithms because in most waste heat recovery and TE power system designs the external exhaust and ambient temperatures, T_{exh} and T_{amb}, are usually known more clearly as requirements or bounding conditions. Consequently, it is difficult to surmise a priori what the best combination of T_h and T_c for a given TE material set will be in any given waste heat recovery application, or whether maximum efficiency conditions, maximum power, or maximum power density conditions are going to be the most appropriate system parameters to optimize. In addition, and more importantly, it is necessary to closely couple the TE device optimization with the hot- and cold-side thermal design optimization to create high-performance TE power systems. This necessarily requires one to analyze and optimize the TE design based on the external exhaust and ambient temperatures, T_{exh} and T_{amb}, in any given application.

Hendricks and Lustbader [9,10] first began investigating this and developing more powerful optimization techniques that laid the foundation for the advanced optimization techniques discussed below. Their approach created a "system of optimization design equations" that coupled the hot-side heat transfer, Q_h, and cold-side heat transfer, Q_c, design optimization with the TE device design optimization. The analysis considers the TE system schematically represented in Figure 22.1 where a simple multiple-couple TE system is depicted. This type of system configuration is common whether one is considering bulk elements or thin-film elements.

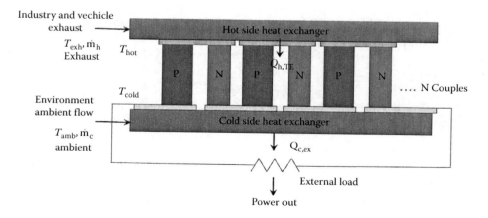

FIGURE 22.1   Industrial or vehicle TE energy recovery system schematic.

Angrist [4] and Rowe et al. [5] discuss that the amount of heat required on the TE device hot side, Q_h, is given by

$$Q_h = N \cdot \left[\alpha \cdot I \cdot T_h - \frac{1}{2} \cdot I^2 \cdot \left(\frac{\rho_p}{\gamma_p} + \frac{\rho_n}{\gamma_n} + 2 \cdot R_{contact,h} \right) + (\kappa_p \cdot \gamma_p + \kappa_n \cdot \gamma_n) \cdot (T_h - T_c) \right] \quad (22.4)$$

where the current I is given by

$$I = \frac{N \cdot \alpha \cdot (T_h - T_c)}{N \cdot R_{int} + R_o} \quad (22.5)$$

$$R_{int} = \frac{\rho_p}{\gamma_p} + \frac{\rho_n}{\gamma_n} + R_{contact} \quad (22.6)$$

and $\alpha = |\alpha_p| + |\alpha_n|$ and $R_{contact} = R_{contact,h} + R_{contact,c}$.

From Equations 22.2 and 22.4 it can be deduced with straightforward mathematics that the amount of cold-side heat dissipated, Q_c, is given by

$$Q_c = N \cdot \left[\alpha \cdot I \cdot T_c + \frac{1}{2} \cdot I^2 \cdot \left(\frac{\rho_p}{\gamma_p} + \frac{\rho_n}{\gamma_n} + 2 \cdot R_{contact,c} \right) + (\kappa_p \cdot \gamma_p + \kappa_n \cdot \gamma_n) \cdot (T_h - T_c) \right]$$

The voltage of the system is given by

$$V_o = I \cdot R_o = \alpha \cdot \Delta T - I \cdot R_{int} \quad (22.7)$$

on a per couple basis. Angrist [4], Cobble [11], and Rowe et al. [5] show that this entire system of equations (Equations 22.2 and 22.4 through 22.7) can be optimized to achieve the maximum conversion efficiency, η_{max}, when

$$\frac{\gamma_n}{\gamma_p} = \sqrt{\frac{\rho_n \cdot \kappa_p}{\rho_p \cdot \kappa_n}} \quad (22.8)$$

This is basically a TE element design requirement, and Z^* in Equation 22.2 is given by

$$Z^* = \frac{\alpha^2}{\left[\sqrt{\rho_n \cdot \kappa_n} + \sqrt{\rho_p \cdot \kappa_p} \right]^2} \quad (22.9)$$

Furthermore, the maximum efficiency with respect to external resistance is achieved when

$$m_{opt} = \left(\frac{R_o}{N \cdot R_{int}} \right)_{opt} = \sqrt{(1 + \delta)^2 + (1 + \delta) \cdot Z^* \cdot \overline{T}} \quad (22.10)$$

where $\delta = (R_{contact} / R_{int})$.

Hendricks and Lustbader [2,3] formulated this set of optimization equations into a set of self-consistent design optimization equations given by

$$\left[\frac{Q_h}{N \cdot \gamma_n} \right]_{opt} = f_q(T_h, T_c) \quad (22.11a)$$

$$\left(\frac{I}{\gamma_n}\right)_{opt} = f_i(T_h, T_c) \tag{22.11b}$$

$$\left(\frac{V}{N}\right)_{opt} = f_v(T_h, T_c) \tag{22.11c}$$

$$\left(\frac{\gamma_n}{\gamma_p}\right)_{opt} = \sqrt{\frac{\rho_n \cdot \kappa_p}{\rho_p \cdot \kappa_n}} = f_g(T_h, T_c) \tag{22.11d}$$

$$\eta_{max} = \left(\frac{P}{Q_h}\right)_{max} = \left[\frac{T_h - T_c}{T_h}\right] \cdot \left[\frac{m_{opt} - 1 - \delta}{m_{opt} + (1+\delta)T_c/T_h}\right] = f_\eta(T_h, T_c) \tag{22.11e}$$

which are functions of only temperatures, T_h and T_c. This is because the TE material properties generally are functions of temperature in any design optimization or design performance analysis. Hendricks and Lustbader, recognizing the importance of hot- and cold-side heat transfer and heat exchanger performance in heat recovery systems, completed the system optimization analysis for TE systems depicted in Figure 22.1 by simultaneously including the design analysis for the hot- and cold-side heat exchangers. Their analysis includes hot- and cold-side heat losses by parametrically accounting for them with heat loss factors, β_{hx}, $\beta_{h,TE}$, β_{cx}, and $\beta_{c,TE}$, which are basically heat loss fractions of incoming energy to a given heat exchanger or interface. The thermal equation for the heat transfer supplied by the hot-side heat exchanger to the TE device then becomes

$$Q_{h,TE} = \frac{(T_{exh} - T_h) \cdot (1 - \beta_{h,TE})}{\left[\dfrac{1}{\dot{m}_h \cdot C_{p,h} \cdot \varepsilon_h \cdot \left(1 - \beta_{hx}\right)} + R_{th,h}\right]} \tag{22.12}$$

The thermal equation for the heat transfer dissipated by the cold-side heat exchanger from the TE device then becomes

$$Q_{c,TE} = \frac{(T_c - T_{amb})}{\left[\dfrac{(1 - \beta_{cx})}{\dot{m}_c \cdot C_{p,c} \cdot \varepsilon_c} + R_{th,c}\right] \cdot (1 - \beta_{c,TE})} \tag{22.13}$$

The heat loss fractions are generally defined as the amount of heat loss compared to the heat loss entering a given component, whether it be within the heat exchangers themselves or the interfaces at which heat is entering or leaving the TE device hot- or cold side. Therefore, the heat loss fractions are generally defined as

$$\beta_{h,TE} = \frac{Q_{loss,h,TE}}{Q_{h,TE}} \quad \beta_{hx} = \frac{Q_{loss,h,ex}}{Q_{h,ex}} \quad \beta_{c,TE} = \frac{Q_{loss,c,TE}}{Q_{c,TE}} \quad \beta_{cx} = \frac{Q_{loss,c,ex}}{Q_{c,ex}}$$

$R_{th,h}$ and $R_{th,c}$ in these two equations are the sum of all thermal resistances at the interface between the TE device and the hot-side and cold-side heat exchangers, respectively. Energy balance requirements then coupled these thermal transfer relationships to the TE design optimization equations in Equations 22.11a through 22.11e to complete the system design optimization. Figure 22.1 shows that this was basically a four-lumped-node design optimization model that simultaneously accounted for heat exchanger and TE device performance in the design optimization. This set of optimization equations specified the family of optimum TE designs, including the number of couples, p- and n-type element areas, voltage and current output, power and maximum conversion efficiency for any given T_h and T_c and a specific exhaust temperature, T_{exh}, and ambient temperature, T_{amb}.

This design optimization analysis is required in waste energy recovery applications because what is usually known are the exhaust flow conditions and ambient temperatures in a given waste energy recovery environment or situation. The goal is to identify the best T_h and T_c conditions at which to operate to create the maximum conversion efficiency or power output, important information about the TE design at those conditions, and performance sensitivities and tradeoffs around these optimum design points. The optimum selection of T_h and T_c involves critical tradeoffs in conversion efficiency, power, and hot-side and cold-side thermal transport in any given waste energy recovery application. These tradeoffs are demonstrated in Figure 22.2 and the associated discussion. In fact, this is the case in most TE power generation analyses and applications whether they are necessarily waste energy recovery related or not.

Figure 22.2 shows typical results produced from this optimization analysis in the form of maps depicting the relationship between maximum efficiency and power (maximum efficiency–power) that can be created for any TE energy recovery system design and any set of temperature-dependent TE material properties. Two different maximum efficiency–power maps are shown: (a) one with constant specific power lines superimposed; and (b) one with constant power per area lines superimposed on the map. These analyses were performed using typical p-type TAGS-85 and n-type PbTe TE properties [12,13], but they can be created for any p-/n-type TE material combination, including segmented element designs. These maps are quite powerful in the amount of information that is conveyed in one design optimization map. They are generally created for a given exhaust temperature, ambient temperature, hot-side heat exchanger UA_h (defined in Section 22.3), and exhaust mass flow rate, $\dot{m}_h$. The main efficiency–power curves represent the loci of optimum designs having maximum efficiency and the plotted power at varying TE hot-side temperatures, T_h, and cold-side temperatures, T_c. The shape of these curves is produced directly from the coupling and interaction between the hot-side heat exchanger performance (and heat transfer) and the TE device performance in the optimization analysis described by Equations 22.11 through 22.13. As T_h increases for a given T_c the TE device efficiency increases, but for a constant T_{exh} and UA_h the hot-side heat transfer is decreasing. Consequently, the power increases with the TE efficiency up to a point where maximum power is achieved, after which the power decreases because hot-side heat transfer decreases too much to offset the TE efficiency increase. Each optimum design along these curves has a different number of couples, N, optimum p- and n-type element area, A_p and A_n, and current, I, as T_h and T_c combinations vary as shown in Figure 22.2. These curves first clearly demonstrate the tradeoff between maximum efficiency and power for the conditions of a given exhaust stream, T_{exh}, UA_h, and $\dot{m}_h$ in typical waste energy recovery applications. The constant hot-side temperature lines help to identify where TE designs with equivalent T_h lie and their relationship to one another in efficiency and power. The efficiency–power curves tend upward and to the right (efficiency and power increasing) as either or both T_{exh} and UA_h increase in magnitude. The analytical power of these curves is that they effectively and rapidly provide the opportunity to identify and investigate multiple TE system designs throughout the potential design domain for a given waste heat recovery application.

Figure 22.2a also shows lines of constant specific power (power per TE device mass). The mass accounted for here is only estimates of the TE device mass, including insulating ceramics, copper interconnections straps, and the TE elements themselves. It is clear in Figure 22.2a that specific power generally increases as one transitions into regions of higher maximum efficiency. Figure 22.2b also shows lines of constant power flux (power per area), which also can be related to the hot-side heat flux because of the maximum efficiency information included in the map. It is clear that power flux increases as one moves into regions of higher maximum efficiency, which can have significant ramifications in applications where TE device miniaturization is critical.

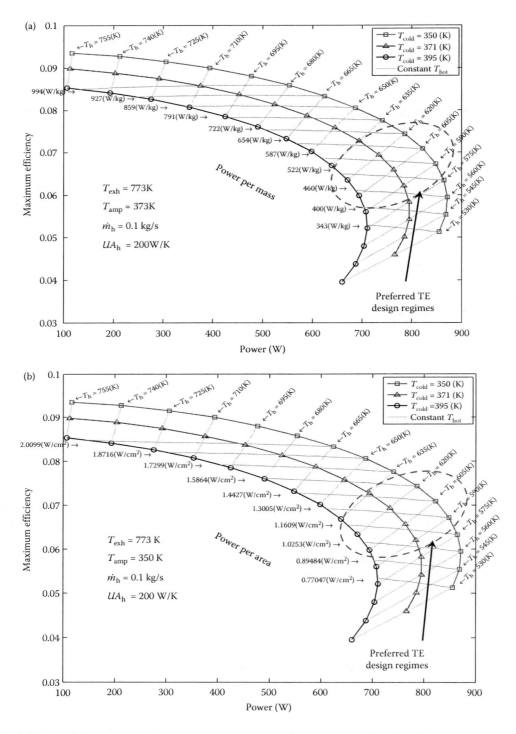

FIGURE 22.2 Maximum efficiency—power output map for typical TAGS—PbTe TE element designs— $UA_h = 200$ W/K, $T_{exh} = 773$ K, $T_{amb} = 350$ K. (a) Constant power per TE mass regions and characteristics superimposed; (b) constant power per TE area regions and characteristics superimposed.

These maximum efficiency–power maps show the potential points of optimum design and the efficiency–power sensitivities and design tradeoffs available for given waste heat recovery applications. These maps can be generated for any selection of *p*- and *n*-type TE materials that are discussed in other chapters of this book, and for any particular single-material element or segmented-element design. They also define "preferred TE design regimes" or preferred optimum design regions, which are superior because of: (1) their higher maximum efficiency with only a small power penalty; and (2) their higher specific power performance.

Hendricks and Lustbader [9,10] also extended the design optimization techniques described above, and based on Equations 22.11a through 22.11e, to segmented-element optimization using the techniques described in Swanson et al. [14]. The same types of maximum efficiency–power maps discussed above can be generated in the resulting segmented-element design optimization analyses.

Additional numerical, system level, TE design optimization has been demonstrated by Crane et al. [15–18]. Constrained, nonlinear, minimization functions were defined to solve the multi-parameter design problems using either gradient-based, genetic algorithms, or a hybrid of the two approaches. A better understanding of the interactions between various design variables and parameters can be gained from this type of multi-parameter optimization approach. Greater than 20 different design variables, including fin and TE dimensions and dozens of different design parameters, can be optimized. A selection of different design constraints include minimum power density, maximum hot- and cold-side pressure drops, maximum total mass, and minimum output power. Additional constraints include maximum TE surface temperatures and maximum temperature gradients across the TE elements to help improve design robustness. Choices for analysis objective function include maximum gross or net power, maximum efficiency, and maximum gross or net power density, which can be based on either total mass or TE mass. An optimization analysis can be conducted once the design variables, parameters, constraints, and objective function have been selected. The result is a nominal design that can be fine tuned through parametric analysis on selected variables and parameters, and then used in an operating model where the design conditions can vary.

22.2.2 TE Design Performance

TE system design performance is the second stage of the design process where an optimum design identified in the Design Optimization process is then analyzed for a variety of nominal and off-nominal design conditions. A more sophisticated multiple-node and/or multiple-volume analysis is typically employed at this stage. Analytically exact techniques have been described by Sherman et al. [19] for *p–n* TE couple designs shown in Figure 22.1. Hogan and Shih [6] describe a typical performance analysis approach based on solving Domenicali's equation [20] for one-dimensional energy balance in the *p*- and *n*-type elements within a device:

$$\frac{\partial}{\partial x}\left(\kappa(x) \cdot \frac{\partial T(x)}{\partial x}\right) = -\rho \cdot J^2 + J \cdot T(x) \cdot \frac{\partial \alpha(x)}{\partial x} \qquad (22.14)$$

and the equation for heat flux within the elements:

$$q(x) = J \cdot T(x) \cdot \alpha(x) - \kappa(x) \cdot \frac{\partial T(x)}{\partial x} \qquad (22.15)$$

These equations are valid in typical parallelepiped or constant diameter cylindrical TE elements used in most typical TE devices. Mahan [21], and Hogan and Shih, describe how Equations. 22.14 and 22.15

can be reformulated into a coupled set of first-order differential equations for $T(x)$ and $q(x)$ within the TE element design given by

$$\kappa(x) \cdot \frac{\mathrm{d}T(x)}{\mathrm{d}x} = J \cdot T(x) \cdot \alpha(x) - q(x) \tag{22.16}$$

$$\frac{\mathrm{d}q(x)}{\mathrm{d}x} = \rho(x) \cdot J^2 \cdot \left[1 - Z(x) \cdot T(x)\right] - \frac{J \cdot \alpha(x) \cdot q(x)}{\kappa(x)} \tag{22.17}$$

Hogan and Shih [6] originally presented Equation 22.17 as

$$\frac{\mathrm{d}q(x)}{\mathrm{d}x} = \rho(x) \cdot J^2 \cdot \left[1 - Z(x) \cdot T(x)\right] - \frac{J \cdot \alpha(x) \cdot q(x)}{\kappa(x)}$$

This has since been shown to be a typographical error in reference [6], per discussions with Hogan (7-19-2010) and an errata sheet that shows the correction as shown in Equation 22.17.

Hogan and Shih discuss how these coupled Equations 22.16 and 22.17 can be solved for $T(x)$ and $q(x)$ profiles in the TE element with iterative numerical analysis techniques and appropriate boundary conditions on the cold- and hot-side temperatures of given a TE device. The p- and n-type elements are typically nodalized into a number of finite isothermal TE layers in the lengthwise direction (i.e., x-direction) as shown in Figure 22.3. Equations 22.16 and 22.17 are applied in a finite-difference formulation at each ith layer to yield:

$$T_{i+1} = T_i + \frac{\Delta x}{\kappa_i} \cdot (I \cdot T_i \cdot \alpha_i - q_i) \tag{22.18}$$

$$q_{i+1} = q_i + \left[\rho_i \cdot J^2 \cdot \left(1 + \frac{\alpha_i^2 \cdot T_i}{\rho_i \cdot \kappa_i}\right) - \frac{J \cdot \alpha_i \cdot q_i}{\kappa_i}\right] \cdot \Delta x \tag{22.19}$$

This formulation corrects the typographical error in ref. [6] per discussions with Hogan (7–19–2010) and an errata sheet created. This technique does not necessarily couple the $q(x)$ profile with hot- and cold-side heat exchanger performance, but it does provide useful $T(x)$ and $q(x)$ information for any

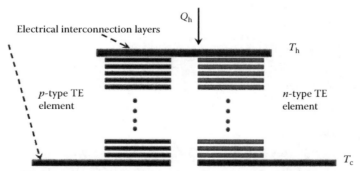

FIGURE 22.3   Typical isothermal layer nodalization used in TE elements in iterative numerical techniques in solving Equations 22.16 and 22.17.

given T_h and T_c conditions of interest and provides required hot-side heat transfer, Q_h and cold-side heat transfer, Q_c, that the heat exchangers must accommodate.

In order to truly perform a system-level design performance analysis the solution to these differential equations must be coupled with hot- and cold-side heat exchanger analysis. Hendricks et al. [7] have recently used TE analysis routines within the ANSYS™ Version 12.0 analysis software package, which solve relationships similar to Domenicali's equation (Equation 22.14) within a TE device design subject to appropriate thermal boundary conditions, to perform this type of system-level design performance analysis. This type of analysis relies on characterizing the UA_h of the hot-side heat exchanger design using a given hot-side exhaust temperature and mass flow rate entering the heat exchanger design and details of the heat exchanger structure and interfaces. The ANSYS™ Version 12.0 software analysis code generally allows one to also nodalize the TE elements as shown in Figure 22.3 and uses variational principles and finite volume techniques applied to the conservation of energy and continuity of electric charge to solve a set of simultaneous equations for heat flow, current, and temperatures within the TE elements under transient or steady-state conditions [22]. Under steady-state conditions these relationships in their most general, three-dimensional form are given by

$$\nabla \cdot \left(T \cdot [\alpha] \cdot \vec{J} \right) - \nabla \cdot \left([\kappa] \cdot \vec{\nabla} T \right) = \dot{q} \tag{22.20}$$

$$\nabla \cdot \left([\sigma] \cdot [\alpha] \cdot \vec{\nabla} T \right) - \nabla \cdot \left([\sigma] \cdot \vec{E} \right) = 0 \tag{22.21}$$

where: $[\alpha]$ = Seebeck coefficient matrix
$[\sigma]$ = Electrical conductivity matrix
$[\kappa]$ = Thermal conductivity matrix
$\vec{J}$ = Current density vector
$\vec{E}$ = Electric field intensity vector
T = Temperature

The heat generation term, $\dot{q}$, in Equation 22.20 generally contains the $\rho \cdot J^2$ heat dissipation term in Domenicali's equation. These general relationships converge down to Domenicali's relationships in one-dimensional analyses depicted in Figure 22.3. The power of the ANSYS™ analysis is that it allows any temperature-dependent TE properties to be used in each of the nodalized layers in Figure 22.3, so that it is straightforward to model segmented element designs in a system design. Furthermore, ANSYS™ 12.0 theoretically allows one to couple the solution of Equations 22.20 and 22.21 to hot- and cold-side heat exchanger performance models to produce system-level predictions.

The results of such an analysis produce TE module efficiency–power maps similar to those shown in Figure 22.4. These maps show the module efficiency–power for various external load resistances and temperature differentials that are created across the TE module directly because of the interaction and coupling of the TE device and hot- and cold-side heat exchanger performance. The external resistance ratios $(R_o/N \cdot R_{int})$ and temperature differentials $\Delta T = (T_h - T_c)$ are superimposed on the module efficiency–power curves in Figure 22.4. The analysis includes the effect of interface contact resistance and parasitic thermal losses. It is clear in the Figure 22.4 results how temperature differential, $\Delta T = (T_h - T_c)$, increases across the TE module as the external load resistance increases. This is the direct result of the tradeoff between system voltage and current and TE heat flows, power and conversion efficiency as the external resistance is varied.

The analysis in Figure 22.4 was performed for exhaust flow temperature, T_{exh}, from 780 to 733 K, exhaust mass flow rate, $\dot{m}_h$, of 0.16 kg/s, and a UA_h of approximately 377 W/K; however, the analysis can be performed at various exhaust flow conditions, T_{exh}, exhaust mass flow rate, $\dot{m}_h$, and heat exchanger

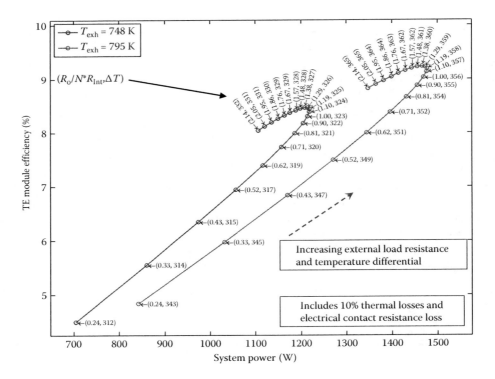

FIGURE 22.4 **(See color insert.)** Typical TE module efficiency–power maps for varying external resistance and current.

performance characterized by the UA_h. Figure 22.4 also shows, for example, a typical impact of reducing the exhaust temperature, T_{exh}, from 795 to 748 K alone on the power output and module efficiency. The benefit of depicting the analysis results in this manner is that the maximum efficiency and maximum power points, and their relationship to one another and other non-optimum performance points, in the overall potential operating performance space can be quickly and clearly seen. Critical design performance decisions concerning various operating points then can be made quickly and intelligently with regard to what the entire design performance envelope looks like.

One noteworthy and clearly identified performance characteristic is that the point of maximum power occurs at an external resistance condition of $R_o > N \cdot R_{int}$ in this type of design performance analysis, because it is performed at constant T_{exh} and T_{amb} conditions and not constant T_h and T_c conditions. In fact, Figure 22.4 demonstrates clearly that ΔT does not stay constant along the module efficiency–power curves. Hendricks et al. [7] discuss the reasons for this, which are directly tied to the TE power output relationship in Equation 22.2. Determining the external resistance that defines maximum power point when ΔT is not constant leads to the following relationship described in Hendricks et al. [7]:

$$R_{o,opt} = N \cdot R_{int} + \frac{2 \cdot R_{o,opt} \cdot (N \cdot R_{int} + R_{o,opt})}{\Delta T} \cdot \frac{\partial(\Delta T)}{\partial R_o} \tag{22.22}$$

This relationship clearly shows that the optimum external resistance that maximizes power is greater than $N \cdot R_{int}$. In fact this equation can ultimately be cast into a quadratic relationship for $R_{o,opt}$, which can be explicitly solved for $R_{o,opt}$ in any given design performance analysis.

Figure 22.4 also clearly shows that the maximum TE module efficiency condition occurs at a $R_{o,opt}$ value where

$$\left(\frac{R_{o,opt}}{N \cdot R_{int}} \right)_{efficiency} > \left(\frac{R_{o,opt}}{N \cdot R_{int}} \right)_{power} > 1 \qquad (22.23)$$

which are both greater than 1. However, the maximum module efficiency–power point can be quite close in magnitude to the maximum power condition. Therefore in waste heat recovery applications, and other TE power generation applications where ΔT is not constant, one must not assume that the $[R_{o,opt}/(N \cdot R_{int})] = 1$ condition applies. Each particular design performance analysis must establish these operating points and their relationship to one another in any given waste heat recovery application.

22.2.3 Sectioned TE Design Optimization

Table 22.1 shows that there are many waste heat recovery applications where available exhaust gases and exhaust streams could be at high temperatures and therefore provide large temperature differentials across TE waste heat recovery devices. This is especially true if effective thermal cooling techniques can keep cold-side temperatures low. Large available temperature differentials create the opportunity to utilize a relatively advanced TE design technique called "sectioned design." With high exhaust stream temperatures and therefore large temperature differentials, the exhaust flow through the hot-side heat exchangers in a TE waste heat recovery system can undergo relatively large temperature drops along the flow length as it transfers its thermal energy to the TE device hot side. In addition, the hot-side thermal transfer decreases significantly along the flow length. Consequently, the TE devices within the design can experience large differences in hot-side temperature and hot-side thermal transfer as the exhaust flow temperature decreases. It is impossible to optimally design one TE device that operates at maximum performance levels as hot-side temperature and thermal flows significantly decrease in these situations. One would therefore like to create optimum TE device designs that accommodate this change in hot-side temperature and hot-side thermal transfer as they decrease with flow length. "Sectioned design" allows this type of TE device optimizing as the exhaust flow conditions change with flow length. Figure 22.5 schematically shows a typical example of this type of "dual-sectioned TE design" approach where the exhaust flow temperatures decrease significantly along the flow length, and the TE hot-side temperatures in each of the two sections are quite different.

Any number of sections can be added to satisfy given design requirements in any waste heat recovery application, but the actual measurable performance benefits gained are dependent on the magnitude of the overall $(T_{exh} - T_{amb})$ temperature differential available. Cost considerations and system complexity issues often constrain the maximum number of sections to about three. This was the number of sections described by Crane et al. [23]. Each TE section was optimized for a particular temperature and heat flux range in the direction of gas flow. This helps avoid TE incompatibility in the direction of fluid flow [16]. The TE section nearest the hot gas inlet was comprised of TE couples made of two-stage segmented elements to account for the large temperature gradient and heat flux between the hot and cold heat exchangers. Located axially downstream of the high-temperature TE section in the direction of gas flow, the medium-temperature TE section also is made up of two-stage segmented elements of the same materials as the high-temperature

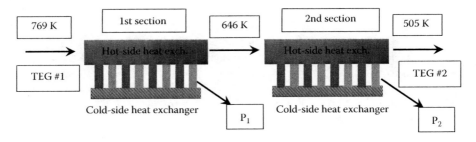

FIGURE 22.5 Typical dual-sectioned TE system design.

TE section. However, these elements are thinner than the elements in the high-temperature TE section due to the lower heat flux and temperature gradient. Segmented elements are not required for the lowest temperature TE section. Each of these TE sections can be operated on one electrical current or on different electrical circuits, allowing for more optimal current densities per TE section.

The design optimization and design performance analysis techniques discussed can generally be used to design each of the sections. However, there are unique and complex design tradeoffs associated with the amount of heat transferred, the conversion efficiency, and power generated in each section. These must be fully evaluated in any TE energy recovery system design to achieve the optimum performance. Figure 22.6 illustrates some of these tradeoffs in a simple dual-sectioned design shown in Figure 22.5. The dual-sectioned design creates significant design tradeoffs between power output and efficiency in each section when seeking the optimum overall system performance, whether that is maximum overall system efficiency or maximum system power output. There are two sets of maximum efficiency–power curves in Figure 22.6, one for section 1 designs using lead–antimony–silver–telluride (LAST) materials and a second for section 2 designs using bismuth telluride materials in Figure 22.5. The maximum efficiency–power curves identify the loci of maximum efficiency designs defined by techniques in Equations 22.11 through 22.13. The resulting maximum efficiency–power maps in each section are produced by the coupled interdependence of the TE device design and hot- and cold-side heat exchangers in sections 1 and 2.

Figure 22.6 illustrates a tremendous amount of design optimization information for the two sections of this dual-sectioned design. The section 1 efficiency–power curves show the various maximum efficiency–power points for several different hot- and cold-side temperature combinations ($T_{h,1}$, $T_{c,1}$) resulting from the coupled interaction between the TE device design and hot-side heat exchanger design characterized by a $UA_h = 375$ W/K. These curves show the range and domain of efficiency and power combinations possible in section 1, with inevitably different TE couples and p- and n-type TE areas at each design point on the curve. The design analysis does identify the TE couples and TE areas required at each design point, although this information is not explicitly shown on the curves. The hot-side heat exchanger design and its performance at each $T_{h,1}$ along the section 1 efficiency–power curves then produce unique entrance exhaust temperatures to section 2 shown in Figure 22.5. This situation creates a family of maximum efficiency–power curves for the section 2 design, one for each of the section 1 $T_{h,1}$ conditions along the section 1 curves. Figure 22.6 shows four such section 2 design curves corresponding to four selected design points on the section 1 curve for a cold-side temperature condition of 312 K. As the section 1 design changes, efficiency and power output varies, and therefore impacts input exhaust temperature on section 2. There is a uniquely defined section 2 maximum efficiency–power curve that dictates its efficiency–power characteristics resulting from the coupled performance of the TE device designs and the hot-side heat exchanger design in section 2. The four section 1 points selected for $T_{c,1} = 312$ K demonstrate that as section 1 power decreases and efficiency increases, the section 2 power output generally increases while the section 2 efficiency stays roughly the same. This section 2 efficiency is largely governed by the TE material properties for bismuth telluride materials at the temperatures used in section 2. Different section 2 TE materials and properties would exhibit different efficiency–power sensitivities and performance.

This behavior, and interactions between sections 1 and 2, create a maximum total system power point because the increases in section 2 power only offset or override the section 1 decreases in power up to a point. The maximum total system power point generally does not occur even close to the maximum point for section 1. Understanding this design power tradeoff and knowing where the maximum total system power point resides is critical to designing a dual-sectioned system to satisfy any given efficiency–power requirement, whether maximum efficiency or maximum power are of paramount interest in a given energy recovery application. This behavior and interaction between section 1 and section 2 designs can occur in general at any common cold-side temperature ($T_{c,1}$ and $T_{c,2}$) conditions as shown in Figure 22.6, and in general for any combination of TE materials in the two sections. Figure 22.6 also shows the maximum efficiency–power reduction that occurs as cold-side temperatures increase (black lines for sections 1 and 2). In both of these cases the maximum total power point may shift as a result of

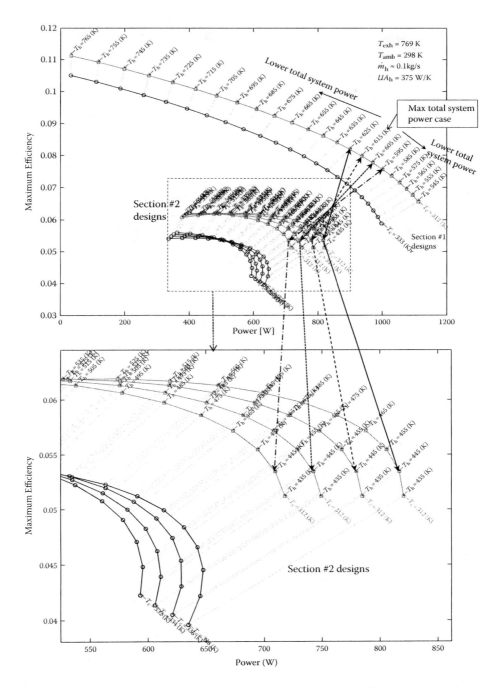

FIGURE 22.6 (**See color insert.**) Typical dual-sectioned analysis results for $T_{exh} = 769$ K, $\dot{m}_h = 0.1$ kg/s, $UA_h = 375$ W/K, $T_{amb} = 298$ K.

the impact of TE material performance or cold-side temperature effects on power output and it is crucial to quantify this shift if it occurs. These design interactions become more complex as more sections are added in a multiple-sectioned design.

22.2.4 Transient Design Performance

Steady-state models discussed earlier give an effective means to choose a nominal design point and optimize the design for a particular set of operating conditions. However, a TE generator in waste heat recovery applications may see a variety of operating conditions, which change frequently as a function of time. A primary example of this application is exhaust heat recovery in an automobile or heavy truck. In this case, a TEG is integrated into a car or truck exhaust stream and the exhaust temperature and mass flow conditions result from the transient engine load conditions dictated by the particular drive cycle that the vehicle experiences. Figure 22.7 shows an example of the temperatures and mass flows downstream of the catalytic converter in the exhaust system that are produced in a New European Drive Cycle (NEDC), a common automotive drive cycle used in European countries.

In order to model the TEG in different drive cycles and other dynamic operating conditions, steady-state models for TE couples and devices are defined first. Energy balance equations are described by Crane [8] as follows with locations shown schematically in Figure 22.8.

$$Q_{h1} + \frac{1}{2}I^2 R_{\text{conn,h}} - UA_{\text{TE-conn}}(T_{\text{cen,h}} - T_{h1}) = 0 \tag{22.24}$$

$$Q_{h2} + \frac{1}{2}I^2 R_{\text{conn,h}} - UA_{\text{TE-conn}}(T_{\text{cen,h}} - T_{h2}) = 0 \tag{22.25}$$

$$UA_{\text{TE-conn}}(T_{\text{cen,h}} - T_{h1}) + UA_{\text{TE-conn}}(T_{\text{cen,h}} - T_{h2}) - UA_{\text{cross,conn}}(T_{\text{sh2}} - T_{\text{cen,h}}) = 0 \tag{22.26}$$

$$\begin{aligned} hA_h(T_{\text{fh}} - T_{\text{sh2}}) - UA_{\text{cross,conn}}(T_{\text{sh2}} - T_{\text{cen,h}}) - hA_{\text{nat}}(T_{\text{sh2}} - T_\infty) \\ + UA_{\text{cross,ch,h,1-2}}(\Delta T_{\text{sh1}}) - UA_{\text{cross,ch,h,2-3}}(\Delta T_{\text{sh2}}) = 0 \end{aligned} \tag{22.27}$$

$$\dot{m}_h Cp_h \Delta T_{\text{fh}} - hA_h(T_{\text{fh}} - T_{\text{sh2}}) = 0 \tag{22.28}$$

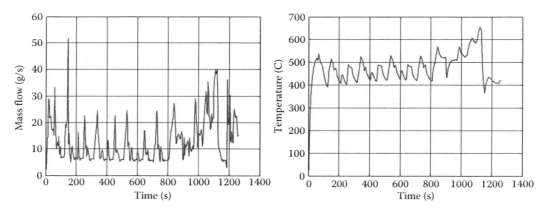

FIGURE 22.7 Time-dependent exhaust gas mass flow and temperature downstream of the catalytic converter for an inline 6-cylinder, 3.0L displacement engine operating on the New European Drive Cycle (NEDC).

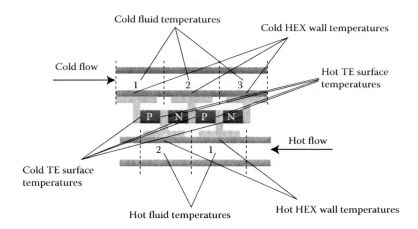

FIGURE 22.8 Schematic of TE subassembly with heat exchangers (HEX) showing model temperature locations.

where

$$Q_{h,i} = \alpha I T_{h,i} + K\Delta T_{TE} - \frac{1}{2}I^2(R_{TE} + 2R_{contact}) \tag{22.29}$$

Equations 22.24 and 22.25 represent conductive heat transfer from the TE elements into their connectors. Equations 22.26 and 22.27 represent conductive heat transfer from these connectors through the fluid-carrying channel wall by way of the fins to fluid convective heat transfer connectors. They include the losses due to natural convection and radiation. Equation 22.28 is an energy balance equation for convective heat transfer into the fluid. Equation 22.29 is the standard equation for TE heat flow in power generation. The model solves these governing equations simultaneously for steady-state temperatures at each node in the direction of flow. The number of simultaneous equations varies with the number of TE elements in the direction of fluid flow.

These energy balance equations were then translated into differential equations based on Equation 22.30 and integrated into the *S*-function template of MATLAB®/Simulink®.

$$m_{vol}C_p\frac{dT}{dt} = Q_1 - Q_2 \tag{22.30}$$

The ($m_{vol}\,C_p$) term in Equation 22.30 represents the thermal mass of each control volume. This could be a TE element, TE connector, heat exchanger, or fluid thermal mass depending on the control volume. The direction of heat flow is important to make sure that the signs for Q_1 and Q_2 are correct. Otherwise, the differential equations cannot be solved correctly.

A baseline for the transient model is the optimized design from the steady-state model. Inputs for the model are similar to those of the steady-state model. Operating condition inputs include the hot- and cold-side inlet temperatures and flows and external electrical load resistance. External electrical load resistance can be set equal to the internal resistance of the TEG or it can be set at a particular constant external load. A simulated electrical load controller can be attached to the model as an additional Simulink block in order to model the effects of a varying electrical load that is not necessarily optimal. Outputs for the model are again similar to those of the steady-state model.

The model can be operated in a stand-alone mode as is or the *S*-function can be cut and pasted into a larger systems-level model. BMW and Ford have both used versions of this model in their larger automotive systems-level models [23]. The model can be run using single hot-side inlet flow and temperature conditions or using the hot-side inlet flow and temperature conditions for a drive cycle.

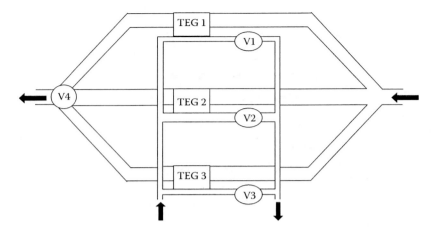

FIGURE 22.9   Schematic of multiple parallel section TEG. (V stands for valve).

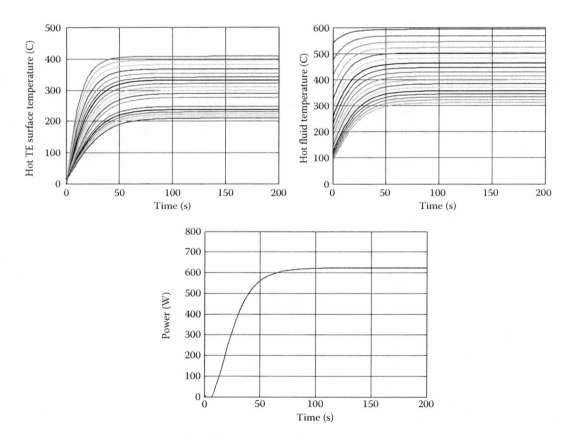

FIGURE 22.10   Simulated outputs for the transient TEG model using constant operating conditions. Temperature graphs show temperatures as a function of time at different axial positions along the TEG in the direction of fluid flow. The hottest temperature curves are nearest to the TEG hot gas inlet while the coldest are nearest to the TEG hot gas outlet.

Additional systems-level attributes have been added to the transient model to aid in its use as part of a larger system. A maximum hot inlet temperature can be defined to prevent TE element overheating or overheating of any other part of the TEG device. A maximum hot flow can be defined to prevent excessive backpressure in the system, which can reduce engine performance if the TEG is integrated into the exhaust system of a vehicle. In addition, to better match the thermal impedance of a dynamic thermal system as defined in Crane and Bell [17], the TEG can be broken into a number of TE sections in parallel (see Figure 22.9) as opposed to sections in a series as described earlier. Having multiple TE sections can allow the TEG to operate better at low flows when the design has been optimized for higher flow rates as each of the parallel sections can be operated together or in various advantageous combinations.

Figure 22.10, a transient simulation example, shows the time-dependent power output, hot side fluid, and TE surface temperatures for the TEG being driven toward steady state by a set of constant operating conditions from an initial set of conditions. The temperature curves in these figures are at different axial positions on the TEG in the direction of flow as a function of time. The colder the temperatures the further they are from the TEG inlet. Figure 22.11, another transient simulation example, shows efficiency and power output along with the TE surface and hot-side fluid temperatures for the TEG for an NEDC automotive drive cycle. The importance of these analysis results is that they quantify and highlight the response time of this system design in this energy recovery application. This is an important design metric that is impacted by a specific system's component weights and volumes as well as the materials and fluids used throughout the system design. System transient response is also governed by other design parameters, such as the exhaust temperature and mass flow rate and ambient environmental conditions, and requires knowledge of the specific heats and density of each material and fluid used

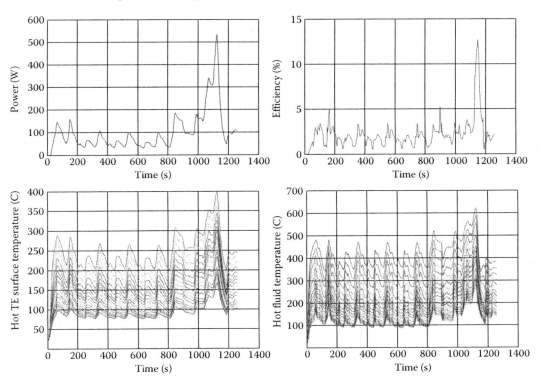

FIGURE 22.11 Simulated outputs for the NEDC automotive drive cycle using an optimized TEG from the steady-state model. Temperature graphs show temperatures as a function of time at different axial positions along the TEG in the direction of fluid flow. The hottest temperature curves are nearest to the TEG hot gas inlet while the coldest are nearest to the TEG hot gas outlet.

throughout the design. It must be characterized for each specific design and for each specific set of operating conditions, boundary conditions, and environments anticipated or of interest in a given energy recovery application. The amount of work required to produce credible and accurate transient response analyses should not be underestimated.

22.3 Thermal System Design and Considerations in TE Systems

Many waste energy recovery applications in today's environment require solutions that are compact, environmentally friendly, quiet, vibration-free, without ozone-impacting fluids, and highly reliable with few or no moving parts. Advanced TE energy recovery and conversion systems envisioned in the future for waste energy recovery applications generally require a temperature differential to be maintained across the TE device while the required thermal energy is transferred in/out of the system. In any event, nearly isothermal interfaces are required on the hot- and cold sides of the TE device to achieve predictable maximum performance conditions. Therefore, heat exchanger configurations that transfer the thermal energy in or out of the device must be capable of providing these nearly isothermal conditions on the interfaces with the TE conversion device and operating under their influence. Furthermore, these heat exchangers will have significant weight and volume requirements on their design to assist in minimizing overall heat recovery system weight and volume.

One must realize the importance of heat transfer at both the hot- and cold side of any TE energy recovery system. A simple rewrite of Equation 22.1 (and Equation 22.3 for maximum efficiency conditions) demonstrates the relationship between TE device efficiency and these hot- and cold-side (and parasitic loss) thermal transfers:

$$\eta = \frac{P}{Q_h} = \frac{Q_h - Q_{loss} - Q_c}{Q_h} \tag{22.31}$$

This simple relationship focuses on the thermal transfers involved in creating the power from the TE system.

Hendricks [24–26] describes the use of the ε–NTU analysis method [20] for various basic heat exchanger configurations, counterflow, parallel flow, crossflow, and parallel counterflow. It is clear from Kays and London [27] that heat exchanger effectiveness, ε, is generally expressed by the following relationship for any flow configuration:

$$\varepsilon = \varepsilon\left[NTU, \left(C_{min}/C_{max}\right), \text{flow configuration} \right]. \tag{22.32}$$

The ε relationship in Equation 22.32 is generally a rather complex one for the various heat exchanger flow configurations. In a heat exchanger that is providing or creating a nearly isothermal interface on one side as it transfers thermal energy the ratio:

$$\left[C_{min}/C_{max}\right] \to 0 \tag{22.33}$$

and Equation 22.32 relationship generally simplifies to

$$\varepsilon = 1 - \exp\left[\frac{-UA}{C_{min}}\right]. \tag{22.34}$$

In the advanced waste energy recovery and conversion systems considered herein, it is highly desirable that any heat exchangers coupled with the advanced TE energy conversion system should satisfy the nearly isothermal interface condition and should therefore closely follow the ε relationship in Equation 22.34. Heat exchanger design techniques and "TE design sectioning" techniques are usually employed to help ensure this condition as closely as possible. UA, the overall heat transfer conductance times the heat transfer surface area, and C_{min} are defined by Kays and London [27]. The UA_h or UA_c of typical hot- or cold-side heat exchanger designs are evaluated by the techniques shown in Kays and London [27] and are typically given by

$$UA_h = \left(\frac{1}{\sum_i R_{th,h,i}} \right) \quad \text{or} \quad UA_c = \left(\frac{1}{\sum_i R_{th,c,i}} \right) \tag{22.35}$$

where the thermal resistances $R_{th,h,i}$ and $R_{th,c,i}$ account for all the convective thermal resistance effects of the flow and all the conductive thermal resistance effects of the heat exchanger structures. Evaluation of the thermal resistances $R_{th,h,i}$ and $R_{th,c,i}$ is therefore specific to any given hot- and cold-side heat exchanger designs and the application. The quantity C_{min} is given by $C_{min} = \dot{m}_h \cdot C_{p,h}$ or $C_{min} = \dot{m}_c \cdot C_{p,c}$ depending on whether one is considering the hot-side heat exchanger or the cold-side heat exchanger, respectively.

UA_h and UA_c (Equation 22.35) are used in Equations 22.12 and 22.13 to determine the hot- and cold-side heat transfers on the TE devices. In general, UA_h and UA_c are dependent entirely on the heat exchange fluids used, flow conditions, and the structure of the heat exchanger in any given WHR application. There can be a wide variety of heat exchanger structures and they can demonstrate some rather unique and innovative concepts. Figure 22.12 shows an example of a particularly simple and common design known as a flat-plate, parallel-extended-fin design that will be used for illustrative purposes only. The UA of this type of structure is dependent on the heat transfer coefficient in the flow channels and the thermal conductance of the structure itself. The UA of this simple structure is given by

$$UA = \frac{1}{[(1/\eta_F \cdot h_c \cdot A_f) + (\Delta t/\kappa_{st} \cdot W \cdot L)]} \tag{22.36}$$

where: h_c = channel heat transfer coefficient, η_F = fin thermal transfer efficiency, A_f = total fin and base heat transfer area, and κ_{st} = thermal conductivity of the plate structure (and usually the fins themselves).

The thermal conductance term in Equation 22.36 is generally straightforward to evaluate and quantify as is the fin efficiency and fin area. However, the heat transfer coefficient, h_c, is generally a function

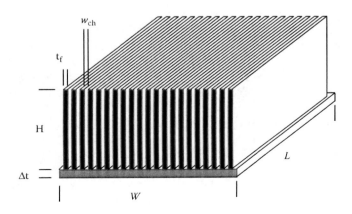

FIGURE 22.12 Common flat-plate, parallel-extended-fin heat exchanger structure.

of channel Reynolds number, Re, and the Prandtl number, Pr, of the fluid [28]. There are a number of different heat transfer correlations of the form:

$$h_c \propto \frac{\kappa_{fl}}{D_h} \cdot Re^m \cdot Pr^n$$

(22.37)

where D_h = channel hydraulic diameter, κ_{fl} = fluid thermal conductivity, and m ~ 0.5–0.8 and n ~ 0.33. Therefore, the heat transfer coefficient, h_c, is dependent on the fluid properties, the channel velocities and therefore the mass flow rate through the heat exchanger, as well as the channel dimensions, w_{ch} and H. The Reynolds number is general'y given by

$$Re = \frac{\gamma_{fl} \cdot V_{ch} \cdot D_h}{\mu_{fl}}$$

where V_{ch} = fluid channel velocity, γ_{fl} = fluid density, and μ_{fl} = fluid viscosity. In addition, the heat transfer coefficient is not constant throughout the flow length, generally starting out high at the flow entrance and decreasing with flow length. This is another important motivation for sectioned TE designs discussed in Section 22.2 because the actual heat transfer into the TE device hot side can decrease with flow length. Therefore, one generally desires to have slightly different TE device design along the flow length to truly optimize system performance as the UA_h decreases with flow length. Consequently, even this simple flow arrangement has some subtle and sometimes difficult challenges in evaluating the heat exchanger UA. There are many other heat exchanger configurations that are much more complex than this and the evaluation of their UA values is much more complex and often times requires detailed thermal models. It is not possible to discuss how to evaluate the UA of every possible heat exchanger design option in this work. Suffice it to say that it is imperative to properly analyze the specific convective heat transfer conditions and environments and the conductive heat transfer of the heat exchanger structure when evaluating various heat exchanger design options in any given TE WHR system design.

In waste heat recovery applications, the heat exchanger fluids used vary from different gases (i.e., air, nitrogen, helium) to different liquids (i.e., water, water–glycol mixtures, oil). Many of these gases are major constituents of typical waste heat exhaust flows. The fluids properties can vary significantly and are critical to the UA evaluation and heat exchanger performance as discussed above. Generally the UA of heat exchangers using gases is low due to their generally low thermal conductivity, low Pr, and Equation 22.37 relationships, which make the first term in the denominator of Equation 22.36 dominate the UA evaluation. The UA of heat exchangers using liquids can be quite high because of their high thermal conductivity and high Pr, therefore the first term in the denominator of Equation 22.36 becomes quite low. Table 22.3 shows the typical values for common heat exchanger fluids in WHR applications and provides a good comparison between typical gas and liquid properties. These properties and the resulting heat exchanger UA are critical factors in determining what fluids to select in WHR applications because these directly determine the temperature differential, $\Delta T = T_h - T_c$, across the TE device and therefore the TE device performance. Hendricks [25,26] presents the representative UA values of some heat exchanger designs with different fluids, but the UA evaluation is generally quite application-specific using the relationships discussed above. It is not possible to quote specific UA values for all the different and innovative gas and liquid heat exchanger options and configurations in WHR applications, but for a given heat exchanger volume the UA of liquid heat exchangers is generally an order of magnitude or more higher than the UA of gas heat exchangers.

In detailed system modeling, whereby the heat exchanger can be broken up or nodalized lengthwise with sufficient refinement, then UA can get small enough that effectiveness in Equation 22.33 can be approximated with

TABLE 22.3 Typical Fluid Properties of Common Gases and Liquids

Temperature (K)	κ_{fl} (W/m K)			γ_{fl} (kg/m³)			$\mu_{fl} \times 10^5$ (N s/m²)			Pr		
	300	500	800	300	500	800	300	500	800	300	500	800
Gases												
Air	0.0263	0.0407	0.0573	1.161	0.696	0.435	1.85	2.70	3.70	0.707	0.684	0.709
Carbon monoxide	0.0250	0.0381	0.0555	1.123	0.674	0.421	1.75	2.54	3.43	0.730	0.710	0.705
Carbon dioxide	0.0166	0.0325	0.0551	1.773	1.059	0.661	1.49	2.31	3.37	0.766	0.725	0.716
Nitrogen	0.0259	0.0389	0.0548	1.123	0.674	0.421	1.78	2.58	3.49	0.716	0.700	0.715
Helium	0.152	0.220	0.304	0.163	0.0975	—	1.99	2.83	3.82	0.68	0.668	—
Liquids												
Water	0.613	0.642	—	997.0	831.3	—	85.5	11.8	—	5.83	0.86	—
Ethylene glycol	0.252	—	—	1114	—	—	1570	—	—	151	—	—
Engine oil	0.145	0.132(430 K)	—	884.1	806.5(430 K)	—	48,600	470(430 K)	—	6398	88(430 K)	—

Source: Adapted from Incropera, FP and DP Dewitt, 1990, *Fundamentals of Heat and Mass Transfer*, 3rd Edition, Wiley & Sons, New York.

$$\varepsilon = 1 - \exp\left[\frac{-UA}{C_{min}}\right] \approx \frac{UA}{C_{min}} = \frac{UA_h}{\dot{m}_h \cdot C_{p,h}} \quad \text{or} \quad \frac{UA_c}{\dot{m}_c \cdot C_{p,c}} \tag{22.38}$$

Equations 22.12 and 22.13 then reduce to

$$Q_{h,TE} = \frac{(T_{exh} - T_h) \cdot (1 - \beta_{h,TE})}{\left[\dfrac{1}{UA_h \cdot (1 - \beta_{hx})} + R_{th,h}\right]} \tag{22.39}$$

$$Q_{c,TE} = \frac{(T_c - T_{amb})}{\left[\dfrac{(1 - \beta_{cx})}{UA_c} + R_{th,c}\right] \cdot (1 - \beta_{c,TE})} \tag{22.40}$$

It is also quite important to evaluate the pressure drops associated with convective heat exchangers as one endeavors to increase the *UA* by modification of heat exchanger channel dimensions and overall geometry. The pressure drop through the heat exchanger is generally given by relationships of the form:

$$\Delta P_{ex} \propto f(\text{Re}) \cdot \gamma_{fl} \cdot V_{ch}^2 \tag{22.41}$$

where $f(\text{Re})$ is the friction factor. There is a close relationship between the heat exchanger *UA* and its pressure drop. As channel dimensions decrease, the heat transfer coefficient increases through Equation 22.37 because the flow channel velocities generally increase for a given mass flow rate and the hydraulic diameter decreases. However, the pressure drop, and therefore pumping power, also increases as the channel dimensions decrease. It has been shown that there are actually optimum values of *UA* as a function of heat exchanger pressure drop [25,26]. This *UA*–pressure drop optimization is application specific and must be quantified along with optimizing TE device performance to maximize overall TE system performance in any given WHR application. It is crucial that these optimization processes be performed in an integrated and strongly coupled manner in the overall TE system design. This is why analyses in Figures 22.2 and 22.6 are so critical in the TE system design optimization.

In the energy recovery applications found in the transportation sector and industrial processing sector, there is generally a large amount of waste energy available to be captured and converted. This can range from 10s of kilowatts to megawatts of available thermal energy. There is also a general requirement to develop compact, lightweight, and low-cost energy conversion and heat exchange systems in any WHR design application. Therefore, in almost all applications the heat exchange systems must satisfy a general requirement to transfer very high heat fluxes (i.e., 10–100 W/cm^2) across nearly isothermal interface conditions. This is particularly true of the unique energy conversion systems using advanced TE conversion materials envisioned in the next 5–10 years. Microchannel heat exchangers are one heat transfer technology that is capable of providing this high performance. Their high performance is due to their small channel sizes, typically 10 μm to 1 mm for w_{ch} in Figure 22.12, which thereby creates very high heat transfer coefficients relative to conventional larger channels due to the Equation 22.37 relationship. Heat transfer coefficients can be approximately 400–1000 W/m^2 K in air and 4000–10,000 W/m^2 K in water, for example. Liao and Zhao [29] reported heat transfer coefficients of 2500–5000 W/m^2 K in microchannel designs using supercritical carbon dioxide. These high heat transfer coefficients can create high UA values in gas and liquid microchannel heat exchanger designs, while the flow remains under laminar flow conditions in most microchannel flows. This laminar flow feature allows one to decouple the heat transfer augmentation from pressure drop effects in the microchannel design process. This in turn creates the opportunity to achieve the heat transfer augmentation while controlling increases in pressure drop through innovative designs.

Hendricks [24–26] discusses the usefulness of microtechnology heat exchanger designs in satisfying future TE system requirements. There will be increasing demands to minimize weight and volume in various waste heat recovery applications as TE technology and systems evolve in the future. The TE devices themselves will need to be lighter and more compact, which will necessarily require them to accommodate and dissipate higher heat fluxes on the hot- and cold sides, respectively. Higher performance heat exchangers, with higher UA values and higher heat exchange effectiveness, and therefore higher heat flow capability from Equations 22.12 and 22.13, will be required to allow these more compact TE devices to achieve their full performance potential. Microtechnology heat exchangers can produce the required higher UA and effectiveness levels, and satisfy thermal transfer requirements with lower weight and lower volume designs (typically 1/5 to 1/10 of the weight and volume of comparable macrochannel designs) with the same thermal duty and mass flow rates. There are current examples [7,8,24,25,30] where this technology is being implemented.

There is a common misconception that microchannel heat exchangers necessarily produce high-pressure drops and therefore have high pumping power. This is not true. In fact if a microchannel heat exchanger is properly designed and properly scaled, its pressure drop need not be any higher than a conventional macrochannel heat exchanger design for a given heat transfer duty and mass flow rate. This has been shown in numerous examples in Yang and Holladay's work [31]. The cost of microchannel systems also are being driven down as production volumes increase in potential high-volume applications for automotive and heavy vehicles; industrial processes; residential and commercial heating, ventilation, and air conditioning systems; and power generation.

22.4 Structural Design and Considerations in TE Systems

The TE modules in TE heat recovery systems are generally constructed of multiple TE elements that are often times bulk elements arranged in an array that is thermally in parallel and electrically in series as shown in Figure 22.13, or as thin-film elements that encounter a substantial temperature differential across a length dimension perpendicular to and much greater than the thinnest film dimension. The structural design and analysis is a critical aspect of high-temperature TE heat recovery systems because of thermally induced expansion stresses from materials with even slightly mismatched coefficients of thermal expansion, compression stresses, and tensile stresses developed within the TE elements and module. In bulk

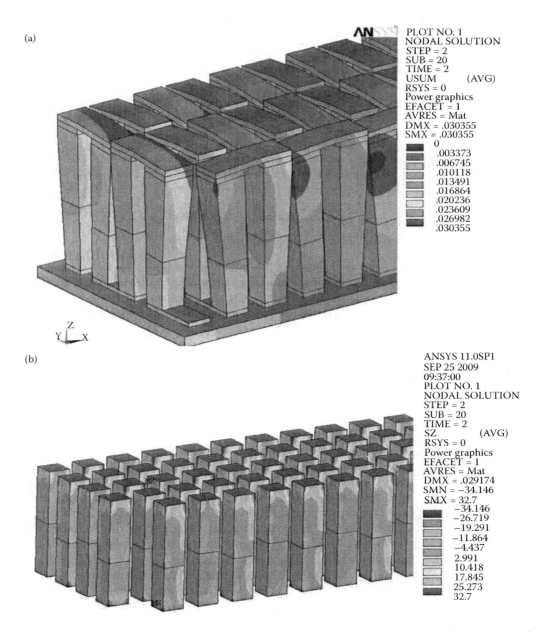

FIGURE 22.13 (**See color insert.**) Magnified structural displacements (a) and resulting stresses (b) in a TE module from compressive and expansive forces (displacements in mm, stresses in MPa).

element arrays, the TE module is commonly under significant compression of 30 psi up to 200 psi in order to create adequate thermal transport across critical hot- and cold-side interfaces between heat exchanger and TE device surfaces. This compression can either be applied at room temperature or the system can be designed to achieve these compressive pressures as it expands to higher temperatures, depending on the details of the design. In thin-film element designs the thermal contact can be supplied compressively also, but often times compression is exerted perpendicular to the element length at the element ends in some fashion that is highly design dependent. In any event, the hot side of the TE element can be at temperatures of 600–1300 K, while the cold side of the TE elements can be held at 350–500 K.

In addition, the TE elements themselves are electrically connected with a variety of electrically conducting and diffusion barrier materials, which often have different coefficients of thermal expansion than the base TE materials. In any event, the combination of compression and thermally-induced expansions as the system heats up creates very complex structural displacements and stresses in the TE elements, and generates the potential for large structural stresses that can damage the TE element and modules. Figure 22.13 shows a magnified structural displacement map created from a typical structural analysis using the ANSYS® software package for a typical rectangular parallelepiped TE element/module design. It highlights how the elements respond to the expansive and compressive forces encountered in the module and in a system. It is not only the TE elements that expand, but the electrical connecting materials and diffusion barrier materials also expand putting complex tilting and distortion forces on the elements.

As one can see, the TE elements can experience complex expansive displacements and tilting displacements that create complex structural tensile and compressive stresses at various locations in the TE device. Element corners at the rectangular parallelepiped TE device hot side are particularly susceptible to high tensile stresses (i.e., red highlighted areas) for example as the device components expand during heating to operational conditions. This fact does not necessarily change whether one is considering a bulk material TE element/module design as shown in Figure 22.13 or a thin-film TE element/module design that others have considered in some TE heat recovery applications.

Figure 22.13 exemplifies the type of structural analysis that is required in designing TE elements and modules for TE heat recovery systems. These element and module structural analyses are generally quite

TABLE 22.4   Typical Structural Properties of Common TE Materials Compared to Semiconductors and Common Metals

Material	Young's Modulus, E (GPa)	Poisson's Ratio, υ	Coefficient of Thermal Expansion (/°C)	Fracture Toughness K_c (MPa m$^{0.5}$)	Fracture Strength, σ_f/(MPa)
Si	163	0.22	2.6×10^{-6}	0.7	247
Ge	128	0.21	5.9×10^{-6}	0.60	231–392
GaAs	117	0.24	6.9×10^{-6}	0.46	66
PbTe	58	0.26	19.8–20.4×10^{-6}		
LAST/-T (Michigan State University)	24.6–71.2	0.24–0.28		—	15.3–51.6
LAST (Tellurex Corporation)	54–55	0.27–0.28	21×10^{-6}	—	15–38 (ROR)
LAST (Tellurex Corporation)	46.3–46.8	0.26–0.27	21×10^{-6}	—	25–40 (ROR)
ZnSe	76.1	0.29	8.5×10^{-6} (293–573 K)	0.9	~ 60
Zn$_4$Sb$_3$	57.9–76.3			0.64–1.49	56.5–83.4
Bi$_2$Te$_3$	40.4–46.8	0.21–0.37	14.4×10^{-6} ($\perp$) 21×10^{-6} ($\parallel$)		8–166
Skutterudites Ba$_{0.05}$Yb$_{0.2}$ Co$_4$Sb$_{12}$ (n) Ce$_{0.85}$Fe$_{3.5}$Co$_{0.5}$Sb$_{12}$ (p)	136 (n) 133 (p)	0.14–0.25 (n) 0.22–0.29 (p)	12.2×10^{-6} (n) 14.5×10^{-6} (p)	1.7 (n) 1.1–2.8 (p)	86 (n) 37 (p)
316 Stainless steel	205		18.5×10^{-6}		300
Copper	129		15–18×10^{-6}		198

Source: Adapted from Hall, BD et al., 2007. *Materials Science and Technology 2007 Conference*, Detroit, Michigan.
Note: $\perp$ = Perpendicular to current flow (element length); $\parallel$ = Parallel to current flow (element length); (ROR) = Ring-on-ring fracture strength; (BOR) = Ball-on-ring fracture strength; LAST = Lead–antimony–silver–telluride; LASTT and LAST/-T = Lead–antimony–silver–telluride–tin.

complex, three-dimensional in nature, require strict and comprehensive evaluation of boundary conditions, and often require multiple "load steps" as boundary conditions change during fabrication and operation to properly analyze the TE element, component, and module stresses and displacements for all expected environments. The required material properties for performing these analyses are typically Young's modulus, Poisson's ratio, coefficient of thermal expansion, and mechanical fracture strengths for each of the materials used for each component in the TE module design. Table 22.4 shows some typical structural property values of common TE materials compared to semiconductors and common metals. Other chapters in this Handbook discuss the various TE materials that are often considered and can be used in various temperature ranges in TE heat recovery systems discussed herein.

There are generally no consistent "rules of thumb" or set of equations that one can employ to even get close to accurate structural analysis results. What is generally required is a complete three-dimensional analysis using a general structural analysis code such as ANSYS®, COMSOL®, or other such structural analysis software package. Furthermore, the TE element and module structural analyses must be tightly coupled with the TE optimization and performance analyses discussed earlier in this chapter to develop optimum, survivable TE element and module designs that satisfy all operational requirements in the environments anticipated for any TE heat recovery system. For example, in structural analyses exemplified in Figure 22.13, it is often found that longer, thinner TE elements lower the destructive tensile stresses that develop. However, the TE design optimization equations shown in Section 22.2 dictate that longer TE elements will necessarily produce lower power designs. Consequently, requirements for maximizing power output from the TE heat recovery system are in conflict with the requirement to lower structural stresses and ensure TE element and module survival during operational conditions and environments. Therefore, it is imperative that the structural design and analysis is strongly integrated with the TE and thermal design optimization processes described above in achieving a truly optimum design for TE heat recovery systems.

22.5 Conclusion

Advanced, high-performance TE energy recovery systems will have a unique and critical role in recovering waste energy in automotive and industrial applications worldwide, with the subsequent benefit of helping to increase global energy efficiency. Optimal design of these systems requires design optimization accounting for three crucial, interdependent design areas: TE design, thermal design and structural design. Copious consideration and attention to detail in these three design regimes is equally important to that of obtaining high performance TE materials (discussed in other chapters) in developing TE devices and systems that achieve their full performance potential for these applications. This chapter provides the foundation and methodology for addressing these three critical system design facets and their role in achieving high-performance TE waste energy recovery systems and solutions. If one wants this technology to achieve its full performance potential in future applications, then one must explore comprehensive system design domains using the techniques discussed herein.

Acknowledgments

The authors acknowledge and thank Professor Eldon Case and his group at the Department of Chemical Engineering and Material Science, Michigan State University, East Lansing, Michigan for their mechanical property contributions in Table 22.4. The authors also thank Mr Naveen Karri, Engineering Mechanics and Structural Materials Group, Radiological and Nuclear Science and Technology Division, Pacific Northwest National Laboratory for his support and expertise in preparing critical figures in this chapter. The authors would also like to thank Virginia M. Sliman and Brenda L. Langley at the Pacific Northwest National Laboratory for their rigorous technical editing of this manuscript and their great editorial recommendations.

Nomenclature

English

A	Area (m^2 or cm^2)
A_f	Total fin area (m^2 or cm^2)
C_p	Specific heat (J/kg K)
D_h	Hydraulic diameter (m or cm)
E	Electric field intensity (V/m)
h, h_c	Convective heat transfer coefficient (W/m^2 K or W/cm^2 K)
K	Thermal conductance (W/K)
L	TE element length (m or cm)
m	External to internal electrical resistance ratio
$\dot{m}_h$	Exhaust mass flow rate (kg/s)
m_{vol}	Control volume mass (kg)
N	Number of TE couples
I	Current (A)
J	Current density (A/m^2 or A/cm^2)
P	Power (Watt)
Pr	Prandtl number
q	Heat flow (Watt)
Q	Heat flow (Watt)
R	Electrical resistance (Ω) or thermal contact resistance (K/W)
Re	Reynolds number
t	Time (s)
T	Temperature (K)
U	Overall heat transfer coefficient (W/m^2 K or W/cm^2 K)
UA	Heat exchanger conductance (W/K)
V	Voltage (V or μV)
x	Length position along TE element (m or cm or mm)
Z	TE material figure of merit ($=\alpha^2/\rho\kappa$) (1/K)

Greek

α	Seebeck coefficient (V/K or μV/K)
β	Heat loss factor
δ	Electrical contact resistance ratio
Δ, d	change in
ε	Heat exchanger effectiveness
γ	TE element geometry factor, A_{TE}/L (m or cm)
γ_{fl}	Fluid density (kg/m^3)
κ	Thermal conductivity (W/m K or W/cm K)
η	Conversion efficiency
η_F	Fin heat transfer efficiency
ρ	Electrical resistivity (Ω m or Ω cm)
σ	Electrical conductivity (S/m or S/cm)
μ_{fl}	Fluid viscosity (N second/m^2)

Subscripts

h	Hot side of TE device
c	Cold side of TE device
cen	Center of

ch	Channel
conn	Connector
contact	Electrical contact resistance value
cross	Cross-sectional
ex	Heat exchanger value
exh	Exhaust flow quantity
f	Fluid
int	Internal TE device resistance
loss	Heat loss quantity
nat	Natural convection
TE	TE element or thermoelectric
o	External load resistance
s	Surface
opt	Optimum design conditions
∞, amb	Ambient
1,2,3	Location of the TEG/control volume in direction of flow

References

1. *Transportation Energy Data Book*, 2010, Edition 29, U.S. Department of Energy, Office of Energy Efficiency and Renewable Energy, Vehicles Technology Program. ORNL-6985, Oak Ridge National Laboratory, Oak Ridge, Tennesee.
2. *Energy Use, Loss and Opportunities Analysis: U.S. Manufacturing and Mining.* 2004, U.S. Department of Energy, Industrial Technologies Program, Energetics, Inc. & E3M, Inc.
3. Hendricks, TJ and WT Choate, 2006, Engineering Scoping Study of Thermoelectric Generator Packages for Industrial Waste Heat Recovery, U.S. Department of Energy, Industrial Technology Program, http://www.eere.energy.gov/industry/imf/analysis.html
4. Angrist, SW, 1982, *Direct Energy Conversion*, 4th Ed., Allyn and Bacon, Boston, MA.
5. Rowe, DM, Ed. 1995, *CRC Handbook of Thermoelectrics*, CRC Press, Boca Raton, FL.
6. Hogan, TP and T Shih, 2005, Modeling and characterization of power generation modules based on bulk materials, In *Thermoelectrics Handbook: Micro to Nano*, Edited by DM Rowe, CRC Press, Boca Raton, FL.
7. Hendricks, TJ, NK Karri, TP Hogan, and CJ Cauchy, 2010, New thermoelectric materials and new system-level perspectives using battlefield heat sources for battery recharging, *Proceedings of the 44th Power Sources Conference*, Institute of Electrical and Electronic Engineers Power Sources Publication, Technical Paper #28.2, pp. 609–612.
8. Crane, DT, 2010, An introduction to System level steady-state and transient modeling and optimization of high power density thermoelectric generator devices made of segmented thermoelectric elements, *Proceedings of the 29th International Conference on Thermoelectrics*, Shanghai, China.
9. Hendricks, TJ and JA Lustbader, 2002, Advanced thermoelectric power system investigations for light-duty and heavy-duty vehicle applications: Part I, *Proceedings of the 21st International Conference on Thermoelectrics*, Long Beach, CA, IEEE Catalogue #02TH8657, pp. 381–386.
10. Hendricks, TJ and JA Lustbader, 2002, Advanced thermoelectric power system investigations for light-duty and heavy-duty vehicle applications: Part II, *Proceedings of the 21st International Conference on Thermoelectrics*, Long Beach, CA, IEEE Catalogue #02TH8657, pp. 387–394.
11. Cobble, MH, 1995, Calculations of generator performance, *CRC Handbook of Thermoelectrics*, CRC Press, Boca Raton, FL.
12. Skrabek, EA and DS Trimmer, 1995, Properties of the general TAGS system, In *CRC Handbook of Thermoelectrics*, ed. DM Rowe, CRC Press LLC, Boca Raton, FL, pp. 267–275.

13. Dughaish, ZH, 2002, Lead telluride as a thermoelectric material for thermoelectric power generation, *Physica B* 322: 205–223.

14. Swanson, BW, EV Somers, and RR Heikes, 1961. Optimization of a sandwiched TE device, *Journal of Heat Transfer*, 83: 77–82.

15. Crane, DT, 2003, *Optimizing Thermoelectric Waste Heat Recovery from an Automotive Cooling System*, PhD Dissertation, University of Maryland, College Park, College Park, MD.

16. Crane, DT and LE Bell, 2006, Progress towards maximizing the performance of a thermoelectric power generator, *Proceedings of the 25th International Conference on Thermoelectrics*, Vienna, Austria: IEEE, pp. 11–16.

17. Crane, DT and LE Bell, 2009, Design to maximize performance of a thermoelectric power generator with a dynamic thermal power source, *Journal of Energy Resources Technology*, 131: 012401-1–8.

18. Crane, DT and GS Jackson, 2004, Optimization of cross flow heat exchangers for thermoelectric waste heat recovery, *International Journal of Energy Conversion and Management*, 45(9–10): 1565–1582.

19. Sherman, B, RR Heikes, and RW Ure Jr., 1960, Calculations of efficiency of TE devices, *Journal of Applied Physics*, 31(1): 1–16.

20. Domenicali, CA, 1953, Irreversible thermodynamics of TE effects in inhomogeneous, anisotropic media, *Physics Reviews* 92(4): 877–881.

21. Mahan, GD, 1991. Inhomogeneous TEs, *Journal of Applied Physics* 70(8): 4551.

22. Antonova, EE and DC Looman, 2005, Finite elements for thermoelectric device analysis in ANSYS, *Proceedings of 2005 24th International Conference on Thermoelectrics*, pp. 215–218.

23. Crane, DT, JW LaGrandeur, and LE Bell, 2010, Progress report on BSST Led, U.S. DOE automotive waste heat recovery program, *Journal of Electronic Materials*, 39(9): 2142–2148.

24. Hendricks, TJ and NK Karri, 2009, Micro- and nano-technology: A critical design key in advanced thermoelectric cooling systems, *Journal of Electronic Materials*, 38(7): 1257–1267, DOI: 10.1007/s11664-009-0709-3, Springer Publishing, New York.

25. Hendricks, TJ, 2008, Microtechnology—A key to system miniaturization in advanced energy recovery and conversion systems, *Proceedings of American Society of Mechanical Engineers 2nd International Conference on Energy Sustainability*, Jacksonville, FL, Paper # ES2008-54244.

26. Hendricks, TJ, 2006, Microchannel & minichannel heat exchangers in advanced energy recovery & conversion systems, *Proceedings of the ASME 2006 International Mechanical Engineering Congress and Exposition*, IMECE2006 – Advanced Energy System Division, American Society of Mechanical Engineers, New York, Paper # IMECE2006-14594.

27. Kays, WM and AL London, 1984, *Compact Heat Exchangers*, 3rd Edition, McGraw-Hill, New York.

28. Incropera, FP and DP Dewitt, 1990, *Fundamentals of Heat and Mass Transfer*, 3rd Edition, Wiley & Sons, New York.

29. Liao, SM and Zhao, TS, 2002, Measurements of heat transfer coefficients from supercritical carbon dioxide flowing in horizontal mini/micro channels, *Journal of Heat Transfer, Transactions of the ASME*, 124, 413–420.

30. Hendricks, TJ, TP Hogan, ED Case, CJ Cauchy, NK Karri, J D'Angelo, C-I Wu, AQ Morrison, and F Ren, 2009, Advanced soldier-based thermoelectric power systems using battlefield heat sources, *Energy Harvesting–From Fundamentals to Devices*, Edited by H Radousky, J Holbery, L Lewis, and F Schmidt (*Mater. Res. Soc. Symp. Proc. Volume 1218E*, Warrendale, PA, 2010), Paper ID # 1218-Z07-02. (*Proceedings of the Materials Research Society 2009 Fall Meeting*, Symposium Z, Paper ID # 1218-Z07-02, Boston, MA, 2009.)

31. Wang, Y and JD Holladay, 2005. *Microreactor Technology and Process Intensification*, ACS Symposium Series 914, American Chemical Society, Oxford University Press.

32. Hall, BD, JL Micklash, JR Johnson, TP Hogan, and ED Case, 2007. A review of mechanical properties for thermoelectric materials, *Materials Science and Technology 2007 Conference*, Detroit, Michigan.

23

Thermoelectric Harvesting of Low-Temperature Heat

David Michael
Rowe
Cardiff University

23.1 Introduction

A thermoelectric converter is a solid-state heat engine in which the electron gas serves as the working fluid and converts a flow of heat into electricity. It has no moving components, is silent, totally scalable, and extremely reliable) In the early 1960s, a requirement for autonomous long-life sources of electrical power arose from the exploration of space, advances in medical physics, deployment of marine and terrestrial surveillance systems and the exploitation of the Earth's resources in increasingly hostile and inaccessible locations [1]. Thermoelectric devices employing radioactive isotopes as a heat source (Radioisotope Powered Thermoelectric Generators, referred to as RTGs) provided the required electrical power [2]. Total reliability of this technology has been demonstrated in applications such as the Voyager space crafts [3] with Voyager 1 passing into the Heliosheath some 8.3 billion miles from Earth. However, employing radioisotopes as sources of heat has remained restricted to specialized applications where the thermoelectric generator's desirable properties listed above outweighed its relatively low conversion efficiency (typically 5%).

The fivefold increase in the price of crude oil in 1974, accompanied by an increased awareness of environmental problems associated with global warming, resulted in an upsurge of scientific activity to identify and develop environmentally friendly sources of electrical power. Thermoelectric generation in applications, which employ waste heat as a heat source, is a totally green technology. In addition when heat input is free, as with waste heat, the system's generating power density is of greater importance than its conversion efficiency in determining the system's economic viability. Over the last 10 years or so effort has focused on developing thermoelectric generating systems which can utilize low-temperature heat from natural sources such as solar, geothermal, ocean thermal, and waste heat from the human body, computer chips, industrial utilities automobile engines: A technology which is called energy harvesting.

In this chapter, a brief introduction to the basics of thermoelectric generation is followed by an overview of materials most suited for applications at relatively low temperatures (<150°C) together with the cost of commercially available modules and their power outputs. Natural and waste sources of low temperature heat are discussed and technologies are employed in thermoelectric harvesting reviewed.

23.2 Thermoelectric Generation Basics

23.2.1 Figure of Merit

A thermoelectric converter is a heat engine and like all heat engines it obeys the laws of thermodynamics. If we first consider the converter operating as an ideal generator in which there are no heat losses the efficiency is defined as the ratio of the electrical power delivered to the load to the heat absorbed at the hot junction. Expressions for the important parameters in thermoelectric generation can readily be derived by considering a simplest generator consisting of a single thermocouple with legs or thermoelements fabricated from *n*- and *p*-type semiconductors as shown in Figure 23.1.

In a generating device available commercially a large number of thermoelements are connected electrically in series and thermally in parallel and sandwiched between ceramic plates as shown in Figure 23.2.

$$\text{The efficiency is given by: } \frac{\text{Energy supplied to the load}}{\text{Heat energy absorbed at the coldjunction}}$$

Conveniently the efficiency can be expressed as a function of the temperature T over which it is operated and a so-called "goodness factor" or thermoelectric figure of merit of the thermocouple material Z and is given by

$$\eta = \underbrace{\left(\frac{T_{hot} - T_{cold}}{T_{hot}}\right)}_{\text{Carnot efficiency}} * \frac{\sqrt{1 + ZT_{avg}} - 1}{\sqrt{1 + ZT_{avg}} + \dfrac{T_{cold}}{T_{hot}}}$$

$$\text{with } Z = \frac{\alpha^2 \sigma}{\lambda}$$

where $\alpha^2\sigma$ is referred to as the electrical power factor, with α the Seebeck coefficient, σ the electrical conductivity, and λ is the total thermal conductivity. The figure of merit is often expressed in its dimensionless form, ZT where T is the absolute temperature.

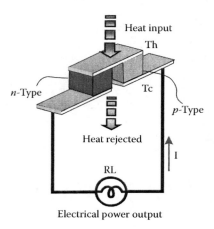

FIGURE 23.1   Simplest thermoelectric generator.

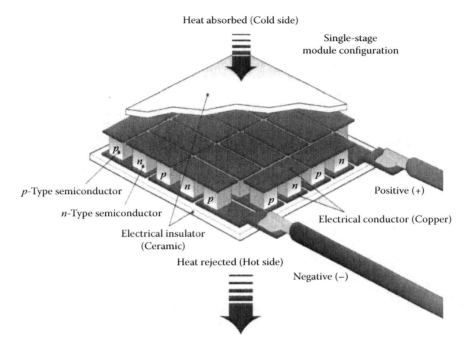

FIGURE 23.2 Thermoelectric module.

23.2.2 Conversion Efficiency and Material Performance

The conversion efficiency as a function of operating temperature difference and for a range of values of the material's figure of merit is displayed in Figure 23.3. Evidently, an increase in temperature difference provides a corresponding increase in available heat for conversion as dictated by the Carnot efficiency, so large temperature differences are desirable. As a ballpark figure a thermocouple fabricated from thermoelement materials with an average figure of merit of 3×10^{-3} K^{-1} would have an efficiency of around 6% when operated over a temperature difference of 150 K.

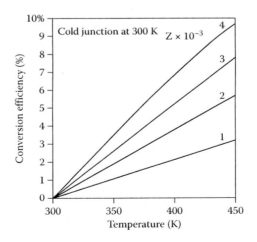

FIGURE 23.3 Conversion efficiency as a function of temperature and figure of merit.

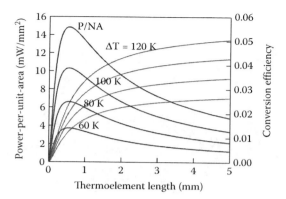

FIGURE 23.4 Power factor and efficiency vs. thermoelement length.

23.2.3 Electrical Power Factor

In thermoelectric waste heat recovery when the source of heat is very cheap or essentially free, efficiency as a major consideration, competes with the power factor P (power output per unit area). Figure 23.4 displays the power factor and conversion efficiency of a 127 couple module. Evidently when efficiency is not a priority the power output is maximized at a thermoelement length of about 0.5 mm [4]. The power factor is also substantially increased by decreasing the thermoelement spacing as shown in Figure 23.5. Here the results of measurements on a bespoke module and one available commercially are compared. When harvesting heat from water at a temperature difference of 80 K compared to ambient, the power output is increased by a factor of 2.5 [5].

23.2.4 Module Cost and Power Output

Figures 23.6 and 23.7 display the cost per Watt of commercially available modules when operated over a temperature difference of 80 K together with the corresponding electrical power outputs. The wide variation in cost and power output is evident. In Figure 23.7, the identity of the module manufacturers has been withheld for commercial reasons.

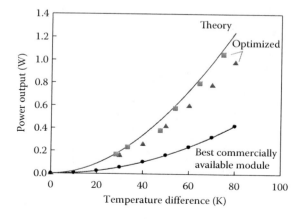

FIGURE 23.5 Effect of thermoelement space reduction on the power factor.

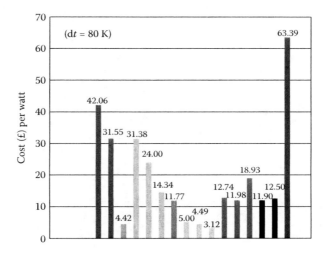

FIGURE 23.6 (See color insert.) Cost per watt.

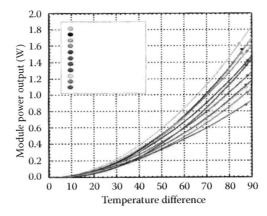

FIGURE 23.7 (See color insert.) Corresponding power output.

23.3 Materials

Figure 23.8 displays the figure of merit of low-temperature thermoelectric materials over the range from room temperature to 200°C. The highest figures of merit are those materials based on the bismuth chalcogenides, with Bi_2Te_3–$75Sb_2Te_3$ the best p-type material and Bi_2Te_3–$25Bi_2Te_3$ the best n-type material. A thermocouple fabricated from these materials has an average figure-of- merit value of $2.5 \times 10^{-3}\ K^{-1}$.

Lead telluride alloys/compounds exhibit figures of merit of around $1.5 \times 10^{-3}\ K^{-1}$. The different compositions are identified by designations such as TEGS-2N; this is n-type and has a specific composition comprised of lead, tellurium, and a small amount of electrically active PbI2. The designation 2P indicates that the lead material is doped p-type with sodium. 3P signifies a lead–tin telluride combination doped with sodium and manganese. More complicated material TAGS is an acronym for tellurium–antimony–germanium–silver. Selenides employed in thermoelectric application have been given the designation TPM-217. Typical members of this class include p-type materials comprised of copper, silver, and tellurium with n-type, the material is composed of gadolinium and selenium.

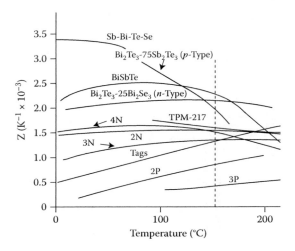

FIGURE 23.8   Figure of merit of low-temperature thermoelectric materials.

In spite of concerted efforts to develop improved new materials the best thermoelectric material for low-temperature generation and which have a proven track record remain those based upon bismuth telluride. These materials are capable of converting at maximum around 20% of the available heat energy (Carnot fraction) into electrical energy and a thermocouple operating at a hot and cold junction temperature of 100°C and ambient, respectively, the thermoelectric conversion efficiency is about 5%.

23.4 Thermoelectric Harvesting of Low-Temperature Heat

Sources of low-temperature heat can be conveniently divided into naturally occurring heat and waste heat. Naturally occurring heat is not derived from man's intervention and includes solar radiation, geothermal, and ocean thermal.

Waste heat is defined as heat produced by machines, electrical equipment, industrial processes, and the human body for which no useful application is normally found, and is regarded as a waste by-product.

23.4.1 Naturally Occurring

23.4.1.1 Solar Energy

A comprehensive review of Solar Thermoelectric Energy Conversion (STEC) is given by Daniel Kraemer et al. in Chapter 24. The amount of solar energy absorbed at the Earth's surface is estimated at 89 PW. This is around 2000 times the energy in the winds and is so large that in one year it is about twice as much as will ever be obtained from all of the Earth's nonrenewable resources of coal, oil, natural gas, and mined uranium combine [6]. In principle, solar radiation is incident onto a heat collector which serves as the hot side of the thermoelectric converter with the electrical power extracted at the cold side. The possibility of employing thermoelectrics to harvest solar energy and convert it into electricity was considered more than 120 years ago [7,8]) and practically demonstrated 40 years later [9]. In the 1950s, Maria Telkes concluded that because of its low conversion efficiency STEG technology would only find application in tropical regions where other forms of generating electricity were unavailable [10].

The flux of heat supplied to the hot side of the thermopile can be increased using optical or thermal concentration as shown in Figure 23.9. Early STEC design concepts considered light radiation incident

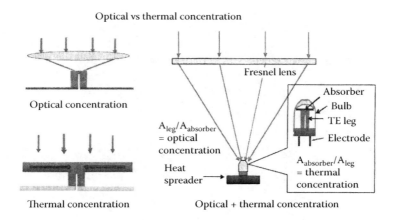

FIGURE 23.9   STEC designs showing optical and thermal concentration (G. Chen).

on physically separated thermocouples distributed over a panel-like structure [11]. Later designs incorporated thermopiles comprised of multicouples with each multicouple having its individual solar concentrator [12], SPTG based on silicon germanium technology has been considered as sources of on-board electrical power for spacecraft intended for excursions into high-temperature environments such as close to the sun or low-altitude Mercury orbits [13]. In 1981, it was reported that the theoretical performance of an SPTG could be significantly improved by fabricating the thermoelements from 100 nm particle size material [14]. Flat panel collector configuration has received further attention [15,16] and optical concentrators [17–19]. More recently, efforts have focused on the use of concentrators to increase the radiation flux onto the solar absorber. However, these systems become increasingly complex and expensive and conceptually are more suitable for terrestrial energy harvesting on a large scale. Currently the major efforts in SPTG development are spearheaded at the recently established "Solid State Solar-Thermal Energy Conversion Center (STEC)," headed by Professor Gang Chen. A novel evacuated flat plate design employing high optothermal concentration together with a spectrally selective solar absorber has operated with an efficiency of 4.8% under AM 1.5G (1 kWm^{-3} solar input conditions [20].

23.4.1.2 Geothermal Heat Harvesting

Geothermal heat is the Earth's natural internal heat that flows toward the surface at a rate of 44.2 TW [21] and is replenished through radioactive decay of the Earth's materials at a rate of 30 TW [22]. These power rates are more than double current energy consumption of the Earth's population from all primary sources [23]. In general, the geothermal temperatures gradient through the crust is 25–30°C (77–86°F) per kilometer of depth and the conductive heat flux is approximately 0.1 MW/km^2 on average. Geothermal heat sources vary considerably in quality and accessibility and are generally classified into those whose temperature ranges from 423 to 473 K (high enthalpy) and those whose sources are less than 423 K (low enthalpy). Conventional high enthalpy geothermal energy sources have been used directly with steam turbine since the turn of twentieth century.

The first attempts to employ thermoelectrics in harvesting heat from waste warm water was undertaken in the late 1980s with a collaboration between Osaka University, Japan, and the University of Cardiff, Wales, UK [24]. The first thermoelectric generator employing low-temperature heat produced 50 W(e) from water at 60°C [25]. Subsequently a generating system was built specifically to harvest geothermal heat [26].

Japan has been proactive in applying geothermal thermoelectric technology as exemplified by the Kusatsu hot springs shown in Figure 23.10. The thermoelectric generator (Figure 23.11) employs 329 bismuth telluride modules. The water temperature is constant at 369 K. Commissioned in 2005, by 2009

FIGURE 23.10   (See color insert.) The Kusatsu hot spring. (T. Kajikawa, Shonan Institute of Technology.)

FIGURE 23.11   (See color insert.) Thermoelectric generator. (T. Kajikawa, Shonan Institute of Technology.)

it had generated more than 1360 kWh. Electricity produced is used to power TV displays and illuminations [27] (Figure 23.12).

Iceland is also a source of readily available geothermal heat with plentiful supplies provided by hot springs and geysers. The thermoelectric harvesting of geothermal energy has been successfully employed here with its first generating system, developed at the Science Institute University of Iceland, installed in a research station in the Grimsvotn area of the Vatnajokull glacier. The thermoelectric generator has provided power for 20 years to a radio frequency communication system for online data acquisition and analysis at the University of Reykjavik. Around 4% efficiency is claimed with the heat source at 90°C and cold side of 15°C [28].

23.4.1.3 Redundant Oil Wells

The United Kingdom's government decided to commence decommissioning of off-shore oil platforms in 1992. Hydrocarbons which have collected beneath the North Sea are normally extracted by pumping sea water from the surface into the subterranean reservoir via an injection pipe and forcing the hydrocarbons to the surface platform through an extractor pipe as shown in Figure 23.13. The oil reserves are located

FIGURE 23.12 Varmaraf thermoelectric generator.

typically at depths of around 3 km (Figure 23.14) and the temperature at this working depth is around 80–100 K. Generating electricity from this low-temperature heat source using conventional methods is not possible. The possibility of employing thermoelectrics has been considered as a possible alternative [30,31]. Information relating to actual platform operations is commercial property, but a realistic doublet flow rate based on general information is approaching 4000 L/min at an input temperature of 90°C and a sea temperature of 10°C has a potential to generate around 10 MW(e) continuous. A Rankine cycle engine is a competitor to thermoelectrics when operating over this temperature range. The performances of both technologies are compared in Figure 23.15. The Rankine engine clearly operates at considerably greater

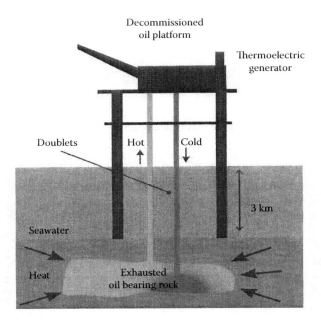

FIGURE 23.13 Thermoelectric recovery of waste heat from redundant oil wells.

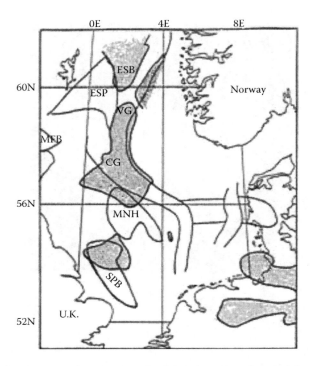

FIGURE 23.14   Shaded areas with rocks at 200°C at depth less than 6 km. (Adapted from J. Turner, UK Dept. of Energy, ETSU G 145, 1989.)

Rankine Cycle		
Maximum Cycle Temp.	Rejection Temp.	Theoretical Cycle Efficiency %
140°C	76°C	15.5
	60°C	19.3
100°C	76°C	6.4
	60°C	10.7
80%	76°C	1.1
	60°C	5.7

	Thermoelectric Efficiency		
Maximum Cycle Temp.	Rejection Temp.	$Z \times 10^{-3} K^{-1}$	Theoretical Cycle Efficiency %
140°C	20°C	2	4.5
		4	7.0
100°C	20°C	2	3.0
		4	4.8
80°C	20°C	2	1.8
		4	3.5

FIGURE 23.15   Comparison of Rankine cycle and thermoelectric efficiencies.

efficiency than thermoelectric generator even when the figure of merit is at an optimistic value of $2 \times 10^{-3}\,K^{-1}$. However, its reliability is a concern. At the time of the study it was concluded that this use of thermoelectric power generation is technically feasible but that the cost of transmitting the dc power from the platforms to the adjacent mainland would make the scheme uneconomic. This decision is now being readdressed following the installation of a suitable power cable from the North Sea oil fields to main land Norway [32].

23.4.1.4 Harvesting Ocean Thermal Energy

23.4.1.4.1 Ocean Thermal Energy Conversion

The oceans and seas cover more than 70% of the surface of the earth. This vast amount of water is heated continuously by the sun and acts as an energy collector and storage system. It is estimated that the amount of solar radiation absorbed on an average day is estimated to equate to the energy content of 250 million barrels of oil, in principle, less (Figure 23.16) than 1% of this renewable energy source, if converted with an 3% Ocean thermal energy conversion (OTEC) efficiency, would provide all humanity's energy requirements [33].

In OTEC the available temperature difference is the prime factor. In some regions temperature gradients of more than 25°C exist between the warm surface layers of water and those found at depths of around 1000 m. Just below the surface the temperature is generally about 4°C (39°F). Since the average ocean depth is about 4000 m (2.5 miles), there is a vast reservoir of cold deep water under tropical skies—some 180 million cubic kilometers (43 million cubic miles). And even this inconceivably vast resource is constantly being renewed by deep cold-water flows from the Polar Regions.

This temperature difference is a renewable energy source with an available amount of energy orders of magnitudes greater than wave power. Although this sea water source of energy is free, the small temperature difference makes it difficult to extract the energy.

An OTEC system is shown schematically in Figure 23.17. Here the cold water at depth is pumped on board the tethered barge conversion is achieved employing the Rankine cycle using a low-pressure turbine with open-cycle engines use the water heat source as the working fluid. Thermoelectrics offers an alternative and the possibility of employing thermoelectrics as an energy conversion system has been researched. In Thermoelectric Ocean Thermal Energy Conversion (TEOTEC), the Rankine cycle low-pressure turbines on board the barge are replaced with a thermoelectric generator.

The possibility of applying thermoelectrics to OTEC was considered in the early 1980s and its performance compared with an ammonia closed cycle design [34]. Thermoelectrics offered advantage having no working fluid and therefore no pressure vessel or working pumps and consequently a high level of reliability. The costing at that time was estimated at around $2000/kW.

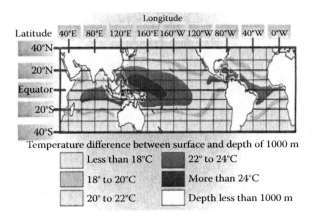

FIGURE 23.16 (See color insert.) Ocean temperature gradient.

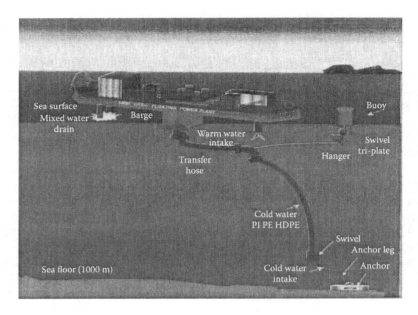

FIGURE 23.17   Schematic OTEC system. (From radfod.edu.)

FIGURE 23.18   OTEC water input. (K. Uemua, ITTJ.)

FIGURE 23.19   Thermoelectric converter bank. (K. Uemura, ITTJ.)

During the period 1980–1982 an operational TEOTEC system was built by a Japanese consortium funded under the Ministry of International Trade and Industry (MITI) sunshine project and evaluated at the Kawasaki Heavy Industries facility. The warm water inputs are shown in Figure 23.18. The thermoelectric generator comprised of 10,000 bismuth telluride in banks of 500 modules as shown in Figure 23.19 [35].

23.4.1.5 Body Heat

The human body is a source of heat which on average varies between around 49 W/m^2 while sleeping to over 500 W/m^2 when involved in strenuous activities. Typically the difference between the skin surface and ambient is around 15 K. Use can been made of this continuous source of low-temperature heat to generate electrical power using thermoelectric technology. The earliest device able to convert body heat into electricity was developed to power an autonomous gas monitor. Fabricated using ICT it was comprised of hundreds of thermocouples fabricated by implanting n- and p-type materials into a silicon wafer as shown in Figure 23.20. It was capable of generating 1.5 V when one end was heated by the warmth of a finger [36]. A variety of techniques have subsequently been employed in the fabrication of miniature low-powered thermoelectric converters and include sputtering [37], MEMS [38], and CVD [39].

The availability of miniature thermoelectric modules provides a capability of harvesting very small amounts of heat to power a range of medical applications (Chapter 26) and electronic devices such as wireless sensors [40] The world's smallest modules are produced and marketed by Micropelt. Micropelt GmbH, is a 2006 spin-off from a research cooperation between Infineon Technologies and Fraunhofer Institute IPM, Freiburg, Germany. Details of this technology are covered in Chapter 17.

Conventionally looking wrist watches which are thermoelectrically powered by body heat have been manufactured and marketed by Japanese companies such as Citizen and Seiko [41]. A schematic of its internal structure is shown in Figure 23.21. Heat from the wrist serves as the hot side while the watch

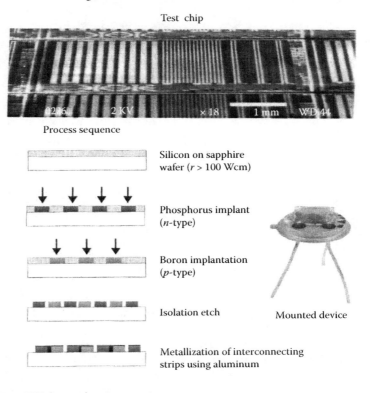

FIGURE 23.20   First ICT thermoelectric generator.

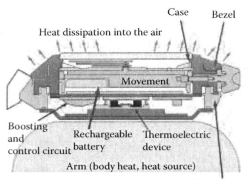

FIGURE 23.21   Thermoelectric wrist watch. (Seiko, Japan.)

FIGURE 23.22   Microthermoelectric module. (Seiko, Japan.)

case effectively dissipates unused heat to ambient at the cold side. Typical temperature difference is of the order 1–3°C. Bismuth telluride thermocouples generate around 200 μV per degree evidently more than 7000 thermoelements would be needed to produce the required operating voltage, incorporating a booster circuit reduced the number to 1000 elements.

In order to fabricate a thermoelectric generator for a wristwatch, it was necessary to reduce the element's height to 80 μm and its size to 600 μm. A microthermoelectric device was developed as shown in Figure 23.22 in which 104 thermoelectric elements are connected in series between two 2 mm × 2 mm substrates. This is the world's smallest π-shaped thermoelectric device that has thermoelectric elements for practical use.

23.4.2 Waste Heat

23.4.2.1 Domestic Waste Heat

Harvesting low-temperature domestic waste heat includes a number of parasitic applications where the thermoelectric generator utilizes some of the heat intended for other purposes. Probably the earliest example is the use of heat from a kerosene lamp to provide thermoelectric power to power a wireless set [42]. Thermoelectric generators have also been used to provide small amounts of electrical power to remote regions for example Northern Sweden [43] and Lebanon [44,45], as an alternative to costly gasoline-powered generators. The generator uses heat from a wood burning stove with the cold side cooled

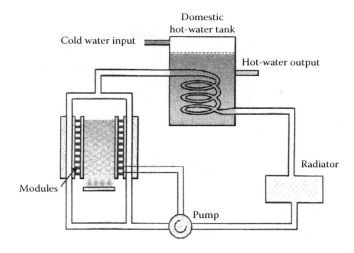

FIGURE 23.23 Parasitic application in domestic heating system.

FIGURE 23.24 AGA cooker with autonomous electric fan or pump.

with a 12 V, 2.2 W fan. The generator produces around 10 W. Another example is in a domestic central heating system with the modules located between the heat source and the water jacket as shown in Figure 23.23. The heat output provided by the gas/oil burner passes through the generator before reaching the central heating hot water exchanger. The generator converts about 5% of the input heat to electrical power, the remainder of 95% transfers to the hot water exchanger for its intended use in heating the radiator system. Figure 23.24 shows an oil-fueled AGA cooker where electricity produced parasitically powers a flue fan a requirement of EU regulations.

23.4.2.2 The NEDO Project

In 1994, a major research project sponsored by the Japanese New Energy and Technology Development Organization (NEDO) commenced in the School of Engineering at Cardiff University, UK with the objective to economically convert low-temperature waste heat into electrical power [45]. A series of Waste heat Alternative Thermoelectric Technology (WATT) prototype generators were constructed and identified as WATT-X where X denoted the power output in watts. Basically the generator consisted of an array of modules sandwiched between hot- and cold-water-carrying channels. Some of the heat flux

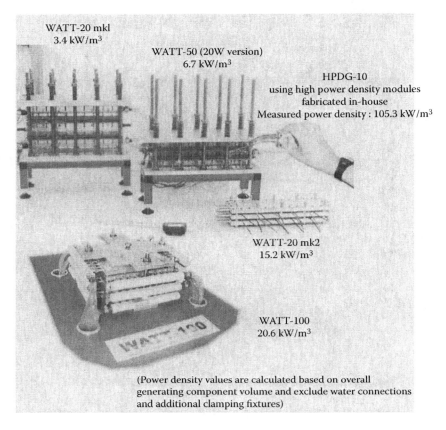

FIGURE 23.25   Development of WATT generators.

which is established by the hot and cold temperature difference between the hot- and cold-water flows is converted into electrical power. Operated using hot water at a temperature of 90°C and cold flow at ambient Watt-100 generates 100 W at a power density approaching 80 kW/m³. The system was scalable enabling 1.5 kW to be generated. A building block 100 W generator powered by warm waste water was exhibited at the *3rd World Energy Conference* at Kyoto in 1998.

In Figure 23.25 is displayed the development stages of Watt generators over a 6-year period while in Figure 23.26 is the progressive increase in electrical power density as a function of project time. The final generator operated at a power density of 50 kW(e) m⁻³.

23.4.2.3 Steel Plant Heat Recovery

Large amounts of cooling water are discharged at constant temperatures of around 90°C when used for cooling ingots in steel plants. When operating in its continuous casting mode, the furnace provides a steady source of convenient piped water which can be readily converted into electricity using thermoelectric technology A typical example is given in Figure 23.27 [46] which displays the power output from major components of a modern steel plant—blast furnace, power station, coke ovens and hot mill as a function of the figure-of-merit of thermoelectrics employed [47]. A total electrical power of around 8 MW would be produced employing currently available modules fabricated using bismuth telluride material technology. Although not sufficient to warrant connection to a national grid network thermoelectric generation it would make a significant contribution to on-site electrical power.

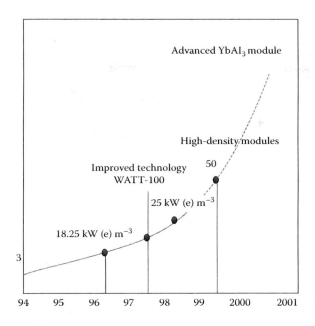

FIGURE 23.26 Increase in power density of WATT generators with time (years) into project.

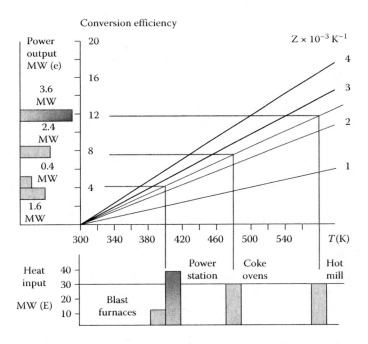

FIGURE 23.27 Waste heat inventory for a small steel plant and corresponding thermoelectric power.

23.4.2.4 Engine Waste Heat

A reciprocating piston engine converts the chemical energy in fossil fuels efficiently into mechanical work. However, a considerable amount of energy is dissipated to the environment through exhaust gas, cooling water, lubricating oils and radiation as shown in Figure 23.28. The possibility of harvesting heat

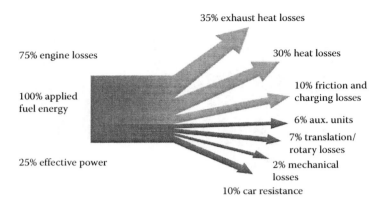

FIGURE 23.28   Energy inventory for automobile fossil fuel combustion.

from warm engine oil has been researched [48,49]. However, by far major efforts have focused on exhaust heat recovery. The temperature of the extracted heat is currently limited by the relatively low operating temperature dictated by bismuth telluride technology resulting in efficiencies of around 5%. This situation will improve dramatically as the technology moves into high-temperature recovery when high-temperature modules become commercially available.

Historically the possibility of using thermoelectric power generation to recover some of this waste heat has been explored as far back as 1914 [50] while in 1988 the performance of a thermoelectric generator attached to a Porsche engine was evaluated. And 58 W(e) obtained at peak power [51]. A comprehensive theoretical study concluded that a thermoelectric generator powered by exhaust heat could meet the electrical requirements of a medium-sized automobile [52]. Large amount of heat are associated with large truck diesel engines. Prototype thermoelectric recovery units have been constructed including a 1 kW(e) generator based on bismuth telluride technology. The generator was operated using the diesel engine's turbo exhaust outlet where there is a large amount of energy at a temperature which is compatible with thermoelectric progress on the project has been reported in the literature [53,54]. Exhaust heat was also used in the production of hydrogen gas [55].

The application of thermoelectric generation in exhaust heat recovery has attracted worldwide interest in recent years and resulted in symposia devoted exclusively to this topic [56,57] and dedicated workshop sessions [58,59]. The application has great potential on several counts. It reduces consumption of costly fossil fuel and carbon emissions. In addition, its potential for reductions in both is substantially greater than any of the other measures being considered for energy saving in automobiles, such as brake-energy recovery.

Current waste exhaust heat harvesting is based on bismuth telluride technology with modules having an upper operating temperature of around 150°C (low-temperature heat). Generators are located after the catalytic converter and operates at a conversion efficiency of around 4.5–5.0% (manufacturers' data). Modules based on PbTe, TAGs, and skutterudite will be available commercially, certainly within five years and operate at conversion efficiencies of more then 10%. High-performance segmented and functionally graded thermoelements are also at an advanced stage of development and offer the potential to convert waste exhaust with an efficiency of around 15%.

However, there is a drawback in that the thermoelectric generating system will incur a weight penalty on the vehicle estimated to be around 13 W/kg [60].

23.5 Conclusions

Low-temperature waste and natural heat can provide an almost inexhaustible supply of energy which can potentially be converted into electrical power using thermoelectric technology. These energy resources exceed all others such as wind and wave power by orders of magnitude. However, the small temperature differences available present a problem. The harvesting of large amounts of low-grade heat has been confined to prototype small-scale systems and is yet to be extended to multikilowatt systems.

Harvesting of small quantities of low-temperature heat using miniature generators has enjoyed dramatic success with devices such as Micropelt. Employing a wafer-based manufacturing technology autonomous wireless sensing systems are successfully marketed which use harvested waste heat as an energy source. This upward trend will continue with an increasing requirement for medical, commercial, and military applications.

The recovery of automobile waste exhaust heat is currently limited to around 600 W under best driving conditions by the nonavailability of high-temperature modules. However, researchers are confident that, with the development of high-temperature modules and improved performance heat exchangers, power outputs exceeding a kilowatt will be achieved in medium-sized family automobiles.

Manufacturers will incorporate thermoelectric generators into their vehicles to contribute in meeting national emission targets Also, consumers are very receptive to green technology and welcome innovations which reduce fuel consumption and assist in decreasing environmentally unfriendly emission.

My vision is that like the catalytic converter, eventually it will be compulsory to incorporate a thermoelectric generator in the exhaust system of all new vehicles.

References

1. D.M. Rowe, Commissioned IEE Review paper 81265, Thermoelectric power generation, *Proceedings of the IEEE*, Vol. 125, No. 11R, pp. 1113–1136, Nov 1978.
2. D.M. Rowe and C.M. Bhandari, *Modern Thermoelectrics*. Holt, Rinehart and Winston Ltd, London, England. ISBN 0-03-91443-8, 1983.
3. D.M. Rowe, *Proc 8th International Conference on Thermoelectric Energy Conversion*, Nancy, France, pp. 133–142, 10–13 July, 1989.
4. D.M. Rowe, S.K. Williams, A. Kaliazin, and G. Min, *Proc 4th European Conference on Thermolectrics*, Madrid, pp. 153–157, 1999.
5. S.G.K. Williams, D.M. Rowe, A. Kaliazin, and G. Min, *Proc. 5th European Workshop on Thermoelectrics*, Pardubice, Czech Republic, Sept. 1999.
6. Wikipedia, Solar Energy.
7. E. Weston, U.S. Patent No. 389,124, 1888.
8. E. Weston, E., U.S. Patent No 389,125, 1888.
9. W.W. Coblentz, *Scientific American*, 127, 324, 1922.
10. M. Telkes, Solar thermoelectric generators, *J. Appl. Phys*, 25,765–777, 1954.
11. P.S. Castro and W. Happ, Performance of a thermoelectric converter under constant heat flux operation, *J. Appl. Phys*. 31,1314, 1960.
12. N. Fuschillo, R. Gibson, F.K. Eggleston, and J. Epstein, Flat plate thermoelectric solar cells, manufacturing process and life testing, *Adv. Energy Cons*. 6,103. 1966.
13. V. Raag, L. Hankins, and M. Swerdling. *Proceedings 2nd ITC University of Texas*, Arlington 22 March, pp. 60–7, 1978.
14. D.M. Rowe, A high performance solar powered thermoelectric generator, *Appl. Energy* 8, 269–273, 1981.
15. R. Rush, Tech. Doc. Rep. Air Force AD 605931 (General Electric Corp. 1964.)
16. H.J. Goldsmid, J.E. Giutronich, and M.M. Kaila, Solar thermoelectric generation using bismuth telluride alloys, *Sol. Energy*, 24, 435–440, 1980.
17. C.L. Dent and M.H. Cobble, *Proc 4th Int. Conf. on Thermoelectric Energy Conversion*, pp. 75–78 (IEEE,1982).
18. H. Scherrer, L. Vikhor, B. Lenoir, A. Dauscher, and P. Poinas, Solar thermoelectric generators based on skutterudites, *Journal of Power Sources*, 115, 141–148, 2003.

19. P. Li, L. Cai, P. Zhai, X. Tang, G. Zhang, and M. Nino, Design of a concentration solar thermoelectric generator, *J. Electron. Mater.* 39, 1522–1530, 2010.

20. D. Kraemer, B. Poudel, H.-P. Feng, J.C. Caylor, B. Yu, X. Yan, Y. Ma et al., High performance flat-plate solar thermoelectric generators with high solar concentration, *Nature Materials*, pub. online 1 May 2011.

21. I.B. Fridleifsson, R. Bertani, E. Huenges, J.W. Lund, A. Ragnarsson, L. Rybach, (2008-02-11), O. Hohmeyer, and T. Trittin, ed., The possible role and contribution of geothermal energy to the mitigation of climate change, Luebeck, Germany, pp. 59–80.

22. H.N. Pollack, S. J. Hurter, and J. R. Johnson, Heat flow from the earth's interior: Analysis of the global data set, *Rev. Geophys.* 30(3), 267–280, 1993.

23. L. Rybach, Geothermal sustainability, *Oregon Institute of Technology*, 28(3), 2–7, September 2007.

24. K. Matsuura, D.M. Rowe, A. Tsuyoshi, and G. Min, *Proceedings of Xth International Thermoelectric Conference*, ed. D.M. Rowe, Cardiff, 10–12 Sept, pp. 233–241, 1991.

25. K. Matsuura, D.M. Rowe, K. Koumoto, G. Min, and H. Tsuyosi, *Proc. XI International Conference on Thermoelectrics*, University of Texas at Arlington, pp. 10–16, 1992.

26. D.M. Rowe, K. Matsuura, K. Koumoto, H. Tsumura, and A. Tsuyoshi, *Proc. XII International Conference on Thermoelectrics*, ed. K. Matsuura, 9–11 No. Yokohama, Japan, pp. 463–466, ISBN 4-88686-037-0 C3055, 1993.

27. K. Kajikawa, *DoE TE Workshop*, San Diego, 2009.

28. B. Hafsteinsson, PhD Thesis, University of Iceland, 2002.

29. J. Turner, UK Dept. of Energy, ETSU G 145, 1989.

30. D.M. Rowe, *Proceedings Mediterranean Petroleum Conference*, 19–22 Jan. Tripoli, Libya, pp. 556–556, 1992.

31. D.M. Rowe, Possible offshore application of thermoelectric conversion, *MTS Journal*, 27(3) 43–48. 1993.

32. D.M.Rowe, *Norwegian Physics Soc.*, NANOMAT Lillehammer, June 2009.

33. L.A. Vega, Ocean thermal energy conversion primer, *Marine Technology Society Journal* 6(4), 25–35, Winter 2002/2003.

34. T.S. Jayadev, T.S. Bnson, and D.K. Bond. *Ocean Thermal Energy Conversion Conference*, 6th Washington, DC, June 19–22, Preprints. Volume 1 (A79-45776 20-44) 1987.

35. K.-I. Uemura, History of thermoelectricity development in Japan, *Journal of Thermoelectricity* No. 3, 2002. 12. 1980–1982.

36. D.M. Rowe, Miniature Thermoelectric Converters, British Patent No. 87 14698, 1988.

37. I. Stark and M. Storder. *Proc. 18th ICT* 465, 1999.

38. G.J. Snyder, J.R. Lim, C.-K. Huang,, and J.-P. Fleurial, Thermoelectric microdevice fabricated by a MEMS-like electrochemical process, *Nature Materials*, 2, 528, 2003.

39. R. Venkatasubramanian, C. Watkins, D. Stokes, J. Posthill, and C. Caylor, Energy harvesting for electronics with thermoelectric devices using nanoscale materials, *Proceedings of the International Electron Devices Meeting (IEDM)*, Washington, D.C., pp. 367–370, 2007.

40. J.A. Paradeso and T. Starner, Energy scavenging for mobile and wireless electronics, *IEEE, Pervasive Computing*, 4,18, 2005.

41. M. Kishi, H. Nemoto, T. Hamao, M. Yamamoto, S. Sudou, M. Handai, and S. Yamamoto, Micro thermoelectric modules and their application to wristwatches as an energy source, *Proc. 18th ITC*, 301, 1999.

42. A.F. Ioffe, *Semiconductor Thermoelements and Thermoelectric Cooling*, Infosearch, London, 1957.

43. A. Killander and J. Bass, *Proc. 15th ICT Conf*, pp. 390–393, Pasadena, CA, USA, 1996.

44. R. Nuwayhid, F. Moukalled, R. Abusaid, M, Daabould, D.M. Rowe, and G. Min, *Proc. 19th International Conference on Thermoelectrics*, Cardiff, UK, pp. 490–497, ISBN 0951 9286 27, 20–24th August 2000.

45. R. Nuwayhid, D.M. Rowe, and G. Min. Renewable Energy, 28, 205–222. ISBN 0960-1481, 2003.

46. D.M. Rowe. *17th Int. Conf on Thermoelectrics*, Nagoya, Japan, 18–24 ISSN 1094 2734, 1998.
47. D.M. Rowe, *Proceedings of Artificial Intelligence in Energy Systems and Power*, Madeira 7–10 February, 2006.
48. A. Tsuyoshi and K. Matsuura, *Electrical Engineering in Japan*, 141(1), 36–44. October 2002.
49. D.M. Rowe, General Dynamics Consultancy Report GD/6/09, June 2009.
50. J. L. Creveling, US Patent No 1118269, 1914.
51. E. Birkholz, Grob, U. Stohrer, and K. Voss, *Proceedings of 7th ICT.*, University of Texas, pp. 124–128, March 16–18, 1988.
52. B.L. Embry and J.R. Tudor, *Proc 3rd IECEC*, University of Colorado, 13–17 August, pp. 996–1005, 1998.
53. J.C. Bass, R.J. Campana, and N.B. Elsner, *Proc 10th Int. Conf. on Thermoelectrics*, ed. D.M. Rowe, Cardiff, UK, 127, 1991.
54. J. Bass, N.B. Elsner, and F.A. Leavitt, *Proc. 13th Int. Conf. on Thermoelectrics*, ed. B. Mathiprakisam, *AIP Conf. Proc*, New York, 295pp., 1995.
55. K. Matsuura and D.M. Rowe, *Proc Third European Workshop on Thermoelectrics*, Cardiff, UK, pp. 22–27, ISBN 0-9519286-3-5 Sept. 16–17, 1996.
56. D. Jansch, *Thermoelektrik*, Berlin, 2008.
57. D. Jansch, *Thermoelectric Goes Automotive*, Berlin, 2010.
58. J. Fairbanks, *Ist Thermoelectric Applications Workshop*, San Diego, Sept 2009.
59. J. Fairbanks, *2nd Thermoelectric Applications Workshop*, San Diego, Jan. 2010.
60. D.M. Rowe, J. Smith, G. Thomas, and G. Min, Weight penalty incurred in thermoelectric recovery of automobile exhaust waste heat, *Journal of Electronic Materials*, 40(5), 784, 2011.

24

Solar Thermoelectric Power Conversion

Daniel Kraemer
Massachusetts Institute of Technology

Kenneth McEnaney
Massachusetts Institute of Technology

Zhifeng Ren
Boston College

Gang Chen
Massachusetts Institute of Technology

24.1 Concept of Solar Thermoelectric Power Conversion

A solar thermoelectric generator (STEG) consists of thermoelectric (TE) devices sandwiched between a solar absorber and a heat sink as schematically shown in Figure 24.1. The solar absorber converts the sunlight into heat and concentrates it onto one side of the thermoelectric generator (TEG). The heat is then transported through the TEG and partially converted into electrical power by the TEG. The excess heat at the cold junction of the TEG is removed by a heat sink in order to maintain an appreciable temperature difference across the TEG.

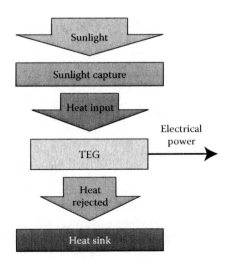

FIGURE 24.1   Schematic of an STEG.

Despite some past studies, STEGs have not received wide attention because the efficiencies of STEGs reported so far have been limited to less than 1% for systems without optical concentration and less than ~4% with optical concentration. However, our recent modeling and experimental studies suggest that much higher efficiencies can be achieved in STEGs, and that solar-to-electrical power conversion based on the TE effect is economically feasible. In this chapter, we will first give a review of past experimental and modeling studies on STEGs, followed by a summary of some modeling and experimental results we have done.

24.2 Historical Review

The efficient and cost-effective conversion of the solar radiation, which strikes our planet at a tremendous rate, is a challenging and complex task for science and engineering. So far, photovoltaic and concentrated solar thermal-to-electrical power conversion technologies are considered to be the most promising approaches. The concept of solar TE power conversion has not drawn much attention as a renewable power conversion approach due to the low efficiencies and/or bulky and costly designs that have been reported over the past century. Recent work showed very promising experimental results for a terrestrial STEG design[1] that potentially enables low-cost manufacturing and an efficiency of 4.6% under AM1.5G conditions with no optical concentration, and larger than 5% efficiency with only small optical concentration.

The history of the solar application of TE generators started more than 120 years ago with the first patents by Weston[2,3] which claimed as new inventions the combination of a thermopile with a mirror or lens to focus solar radiation onto the hot junctions, and a storage battery to accumulate the electric energy. The first experimental work, published in 1922, reported an efficiency of 0.008% for an STEG consisting of 105 copper–constantan thermocouples.[4] Three decades later Telkes published the first detailed experimental work on STEGs.[5] She investigated STEGs with flat-plate solar absorbers in air. In order to reduce air convection losses several glass panes were used to cover the STEG cells. She tested different TE materials and different cooling mechanisms to remove the excess heat on the cold side. In addition, she performed STEG experiments with an optical concentrator that could achieve up to 50 times the incident solar radiation. However, the measured efficiencies of 0.63% at 1 Sun and 3.35% at 50 Suns were too low to be applied on a large scale despite the advantages of reliability and low maintenance. Her conclusion was that this technology would only be useful in tropical regions with high insolation where fuel resources are often absent or not yet developed, and electrical power generation facilities are nonexistent.

In the 1960s, the concept of solar TE generation was investigated for space applications where power density (per mass) is the most important design parameter.[6–9] The use of Bi_2Te_3-based STEGs as the power source for near-earth orbit missions was discussed[6,7] and it turned out to be a promising application due to higher power density (per mass) and power-to-cost ratio than for photovoltaic cells at that time. STEGs have also been considered for space probe missions close to the sun[8,9] due to their resistance to high radiation intensity and their ability to withstand high temperatures by using high-temperature TE materials including PbTe and $Si_{1-x}Ge_x$. In 2003, Scherrer et al. theoretically investigated the use of skutterudite-based STEGs for deep-space probe missions as well.[10]

Later efforts on terrestrial solar TE power conversion focused on concentrated STEGs.[11–14] The idea is to increase the incident radiation flux onto the solar absorber by means of an optical concentrator system which is similar to those used in solar thermal-to-electrical power conversion technologies. The systems can become structurally and capital-cost intensive[15] which makes them mostly applicable as large-scale power generation units and unfeasible for residential and commercial rooftops. Nevertheless, the concentrated solar intensity enables shrinking the size of the STEG cell which will reduce external heat losses. Additionally, by using high-temperature-resistant materials or even segmented TE legs or cascaded TEGs,[16] a larger temperature difference across the TE legs can be utilized which will lead to a higher device efficiency if materials are available. These two factors will ultimately contribute to a higher efficiency of those types of terrestrial STEGs.[17] However, to put it into perspective, state-of-the-art solar thermal-to-electrical power conversion technologies that use conventional power blocks (steam turbine and generator) have adiabatic turbine efficiencies between 70% and 80%, and it will take a significant improvement in TE properties (current TE materials enable an adiabatic TE efficiency of approximately 20%) to replace that power block. This is especially true due to the fact that the power block contributes only about 15% of the total cost of a solar thermal power plant,[15] so the cost savings from potentially cheaper TE systems would be small compared to the total cost of the solar energy harvesting system. Nevertheless, Dent and Cobble[11] built a prototype STEG comprising a sun-tracking heliostat which directed the incident sunlight onto a parabolic mirror which then focused the sunlight onto an STEG. With PbTe as the high-temperature TE material, Dent et al.'s system reached a hot-side temperature of 510°C and a maximum temperature difference of 420°C across the TE device, resulting in a maximum STEG efficiency of 6.3% excluding the optical concentrator losses. No details on the optical concentration were given. Surprisingly, their experimental results were 30% higher than their theoretical prediction. In 1998 Omer and Infield[12] also investigated concentrated STEGs using an infrared lamp with an optical concentrator achieving an incident flux of 20 kW/m^2, corresponding to 20 Suns. Experiments were performed with a commercial thermoelectric cooler (TEC) module and an efficiency of 0.9% was reported. Their modeling efforts were based on averaged material properties, however they used a creative way to account for the Thompson heat by calculating it as the product of the current, Thompson coefficient, and the temperature across the module and then adding one half of it to the hot junction and the other half to the cold junction. External heat losses were also included. Their theoretical study discussed the influence of electrical and thermal contact resistances on the optimal length of the TE elements. One of their main conclusions was that TEG modules should be optimized for optimum power and not for highest efficiency, which was supported by their modeling results. However, this is questionable because the STEG design should be optimized for a constant incident solar radiation flux (in their case 20 kW/m^2) which leads to the conclusion that the operational point of maximum power output also corresponds to the point of maximum efficiency of the device. In early 2010, Amatya and Ram[13] reported an experimental investigation of the concept of STEGs for micropower applications using a commercial TE module combined with a parabolic mirror and a Fresnel lens. They measured an efficiency of 3% at 66 kW/m^2 excluding the optical losses of the concentrator. A cost analysis predicted an electricity price of 0.35 \$/kWh if the system would be installed in Nepal. The latest work on concentrated STEGs was reported by Li et al.,[14] who investigated theoretically the effect of the optical concentration and of the cold-side cooling method on the system efficiency and the hot-side temperature for a specific device geometry (size of heat collector, TE elements, and number of TE couples). They performed simulations with Bi$_2$Te$_3$, skutterudite and LAST (lead–antimony–silver–tellurium) alloy-based TE materials using temperature-dependent properties reported in literature. Their simulated concentrated STEG system could reach efficiencies of 9.8% (Bi$_2$Te$_3$), 13.5% (skutterudite), and 14.1% (LAST) accounting for an optical efficiency of 85%, a solar absorptance of 0.9 and an infrared emittance of the solar absorber of 0.08, and neglecting all other heat losses in the system. They concluded that the conversion efficiency increases with increasing optical concentration and the maximum optical concentration should be selected according to the maximum hot-side temperature permitted by the chosen TE material.

Goldsmid et al.[18] in 1980 and a more recent work by Vatcharasathien et al.[19] compared two low-temperature STEG systems of which one was a flat-panel design and the other a concentrated STEG system with low optical concentration. Both used commercial Bi$_2$Te$_3$-based TEC/TEG modules for their experimental setups. Goldsmid used a semi-parabolic and Vatcharasathien a 2D compound parabolic concentrator (CPC) for the experiments with optical concentration. The efficiencies reported for both types of STEGs were lower than 1%.

Despite the improvements of the properties of TE materials over the last two decades,[20–24] the TE device efficiency is still relatively low which means that most of the heat is rejected at the cold junction of the device and could potentially be used for space heating or domestic hot water. This was proposed with promising results by recent publications.[1,25] Depending on the fluid temperature at the cold junction, the electrical efficiency of flat-panel STEG cells varied between 4.6% (20°C) and 3.3% (60°C) under one Sun condition and could be higher with small optical concentrations. Although the potential combination of solar hot water systems and TEGs was recognized before,[5] there has been no practical way to achieve the combination economically. A theoretical and experimental work on cogeneration systems was published by Rockendorf et al.[26] in 2000. Their solar collector comprised vacuum tubes with water heat pipes. The steam was condensed at the hot side of a TEG module to convert some of the thermal energy into electrical power. The water that heated up at the cold side of the TEG was then stored in a tank for further use. A maximum electrical efficiency of 1.1% for an insolation of approximately 0.9 kW/m² was reported when the cold-side fluid temperature was maintained at ambient temperature. Increasing the fluid temperature to 63°C reduced the efficiency to 0.65%. In addition to the low efficiency, the TEG also represented an additional thermal resistance in the system which resulted in a drop of the solar collector efficiency by 45% compared to an identical collector without a TEG. Those experimental results and further simulations led the authors to the conclusion that the gain in electrical power did not balance the large loss in thermal energy output. Consequently, the proposed TE collector was concluded to be economically unfeasible.

Another interesting concept for STEGs is to incorporate them in a photovoltaic (PV)–TE hybrid system. Luque and Marti[27] discussed the limiting efficiency of coupled thermal and photovoltaic converters. They concluded that in practical cases where the solar thermal system is limited by a reasonable hot-side temperature and the PV cell is limited by a reasonable finite number of different band-gap materials the hybrid converter may give significantly higher efficiency than the individual solar converters alone. In 2005, Zhang et al.[28] published the first work proposing a PV–TE hybrid system. The authors suggested splitting of the solar spectrum into two distinct parts. The visible part of the spectrum was converted by the solar cell and the IR portion by a TE generator. The authors considered Bi_2Te_3 as the low-temperature TE material for the TEG but also proposed to use high-temperature materials and large optical concentration to improve the system efficiency. One year later Vorobiev et al.[29] discussed two hybrid system designs. In both designs the PV cell was located above the TEG cell. In one system, the cells were thermally insulated from each other. The light that was transmitted by the PV cell was optically concentrated onto the TEG. The second design related to concentrated PV cells, where cell heating due to high incident solar flux can significantly affect the performance of the system. Mounting the PV cell directly on the TEG enabled conversion of not only the radiation transmitted through the PV cell but also the generated heat due to the thermalization of electron–hole pairs to the band gap energy inside the PV cell. The reported simulation results, however, are unrealistic. The assumed TE properties were very optimistic which overestimated the contribution of the TEG to the performance of the hybrid system. Kraemer et al.[30] developed an optimization methodology for PV–TE hybrid systems in which the heat generation of the PV cell should be minimized while achieving maximum hybrid system efficiency. The authors discussed different PV(thin film)–TE hybrid cells and came to the conclusion that the largest increase in efficiency compared to initial PV cell efficiency can be achieved for polymer solar cells.

24.3 Optimization of STEGs

The first step in designing a STEG is to consider how to create a temperature difference across the TEG. The TEG modules commercially available today consist most commonly of a large number of closely packed TE legs. The legs usually have a small cross-sectional area of 1–4 mm² and a length (L) of about 1–2 mm. The heat flux flowing through a TEG is on the order of

$$q_{te} \approx k \frac{\Delta T}{L} \tag{24.1}$$

where k is the thermal conductivity of the TE legs. Taking $k \sim 1$ W/m K and $\Delta T = 100$–500°C, the device heat flux through a thermoelectric leg is 5×10^4–5×10^5 W/m². This heat flux is 50–500 times larger than AM1.5G solar insolation (1000 W/m²). Clearly, concentration of the solar flux is needed to create a reasonable temperature difference across the TEGs, which is required to efficiently convert the heat flux into electricity. Such concentration was mainly achieved by optical means in the past (Figure 24.2a). However, optical concentration systems with a concentration ratio higher than 50 are usually bulky and add considerable cost to a STEG.

Another way to concentrate the solar flux to a TEG is via heat conduction, as shown in Figure 24.2b. In this case, solar radiation is absorbed by an absorbing surface deposited on a highly thermally conductive substrate (e.g., copper). The absorbed solar radiation will flow to the TEG via heat conduction. We will call the ratio between the solar absorber area, A_{abs}, and the total cross-sectional area of the TE elements, A_{TE}, the thermal concentration, C_{th} (Figure 24.2c). Of course, a combination of optical and thermal concentration can be adapted whenever economically feasible.

When large thermal concentration is used, heat loss via radiation and air conduction/convection from the surface of the hot solar absorber can be significant and detrimental to the efficiency of the device. For STEGs with large optical concentration those losses are less important. Based on the assumption of temperature-independent properties, Rowe[31] for high optical concentration and Chen[32] for large thermal concentration obtained that the efficiency of a STEG can be expressed as

$$\eta_{STEG} = \eta_{ot}\eta_{TEG} = \left[\eta_{opt} - \frac{\varepsilon_e\sigma(T_{abs}^4 - T_c^4)}{C_{opt}q_{sol}} - \frac{Q_{con}}{A_{abs}C_{opt}q_{sol}}\right]\left[\frac{T_{abs} - T_c}{T_{abs}}\frac{\sqrt{1+(ZT)_m}-1}{\sqrt{1+(ZT)_m}+T_c/T_{abs}}\right] \quad (24.2)$$

where the first factor is the opto-thermal efficiency, η_{ot}, which includes the optical efficiency, $\eta_{opt} = \eta_{conc}\tau_g\alpha_s$, that accounts for possible optical solar concentration losses, η_{conc}, the glass transmission losses, τ_g, and the absorptance, α_s, of the solar absorber. The second term of the opto-thermal efficiency corresponds to the radiation losses from the absorber (at temperature T_{abs}) to the cold side and ambient (at temperature T_c). The effective emissivity ε_e accounts for radiation losses from both sides of the absorber. The incident radiation flux onto the STEG is the constant incident solar radiation q_{sol} multiplied by the optical concentration C_{opt}. The third term of the opto-thermal efficiency is the loss via air conduction/convection, Q_{con}. The second factor of the STEG efficiency is the TE generator efficiency, η_{TEG}, which is here assumed to be the ideal TE efficiency equation.[33] It is determined by the Carnot efficiency multiplied by the adiabatic TE efficiency which is a function of $(ZT)_m$, the averaged figure of merit

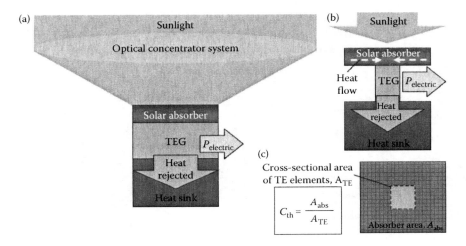

FIGURE 24.2 (a) Conceptual STEG design with large optical concentration, (b) conceptual STEG design with large thermal concentration, (c) graphical explanation of thermal concentration C_{th}, which is the ratio between the absorber area A_{abs} and the total cross-sectional area of the TE elements A_{TE}.

for the thermocouple. Important to realize for the STEG optimization is that the maximum power point of a STEG also corresponds to the operation point of maximum STEG efficiency due to the incident solar radiation being a fixed heat input. This is different for other TEG applications such as waste heat recovery with, for example, a fixed hot-side temperature and a variable heat input. In this case, the TEG has different maximum efficiency and maximum power conditions.

For a TEG housed inside an enclosure with narrow gaps (Figure 24.3a), we can express the conduction losses as

$$Q_{con} = k_{air} \left[\frac{A_{abs} - A_n - A_p}{L} + \frac{A_{abs}}{L_1} \right] (T_{abs} - T_c) \approx 2k_{air} \frac{T_{abs} - T_c}{L} \tag{24.3}$$

where L_1, the separation between the absorber and glass enclosure is assumed to be equal to the separation between the absorber and the cold side, which is set equal to the TE leg length L. For the case of large thermal concentration the presence of the TE legs can be neglected and with that the cross-sectional areas of the TE legs, A_n and A_p, do not need to be subtracted from the absorber area. Using typical values for the temperature difference, leg lengths and the thermal conductivity of air, k_{air}, it turns out that the conduction and convection losses can easily exceed the radiation losses. Hence, to achieve reasonable efficiency at a low optical concentration ratio, vacuum operation is favored. Fortunately, existing technology in evacuated tubes for solar–thermal steam plants and solar hot-water systems has proven that vacuum operations are realistic and economically feasible.[34]

To achieve high opto-thermal efficiency, the absorber should have high absorptance in the wavelength spectrum of the incident solar radiation but a low emittance for the spectrum of a blackbody at the operational temperature. This spectral behavior is characteristic for wavelength-selective solar absorbers widely used in solar thermal applications. These state-of-the-art solar absorbers have a large absorptance in the solar spectrum (<2–2.5 μm) of approximately 95% and a low emittance in infrared region (>2–2.5 μm) between 4% and 8% depending on the operational temperature (100–250°C). With increasing temperature the IR emittance will continue to increase due to temperature-dependent emissivity and the temperature-dependent shift of the blackbody spectrum to shorter wavelengths resulting in a larger overlap with the region of high absorptance/emittance.[35,36]

We note that in Equation 24.1, the opto-thermal efficiency decreases while the TEG efficiency increases with increasing absorber temperature, T_{abs}. Hence, an optimal hot-side temperature exists that maximizes the STEG efficiency which corresponds to a specific thermal concentration. The larger the thermal concentration the higher is the absorber temperature. Figure 24.3b shows the variations of the three efficiencies for a set of given conditions. Using the introduced model based on temperature-independent $(ZT)_m$, this optimal absorber temperature is given by

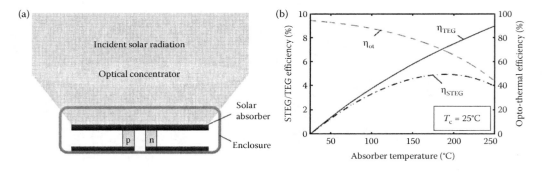

FIGURE 24.3 (a) Concept of a unicouple STEG cell with large thermal and small optical concentration mounted inside a narrow-gap glass enclosure, (b) graphical explanation of STEG optimization: black solid line is the TE generator (TEG) efficiency, gray dashed line is the opto-thermal efficiency, black dash-dotted line is the STEG efficiency.

$$\frac{\eta_{opt}C_{opt}q_{sol} - \varepsilon_e\sigma(T_{abs}^4 - T_c^4)}{4\varepsilon_s\sigma T_{abs}^4} = \frac{\sqrt{1 + (ZT)_m}\left[\sqrt{1 + (ZT)_m} + T_c/T_{abs}\right]}{\left[T_c\sqrt{1 + (ZT)_m} / (T_{abs} - T_c) + 1/2\right]\left(\sqrt{1 + (ZT)_m} + 1\right)} \tag{24.4}$$

It is interesting to note that the optimal absorber temperature is independent of the TEG geometry. The thermal concentration and leg length obeys the following relation:

$$C_{th}L = \frac{\left(k_p + k_n\dfrac{A_n}{A_p}\right)}{1 + \dfrac{A_n}{A_p}}\frac{T_c\sqrt{1 + ZT_m} + (T_h - T_c)/2}{4\varepsilon_s\sigma T_h^3 T_m} \tag{24.5}$$

Figure 24.4 summarizes the losses affecting the STEG performance that have been discussed so far. This figure is obtained with a detailed model of STEG cells, which includes a more accurate heat transfer model, electrical contact resistances and temperature-dependent properties of the TE materials and the solar absorber.[25] The model is briefly described in the next section. If nonconcentrated sunlight strikes an STEG cell which uses a black absorber and is exposed to air at standard conditions the performance is very poor (solid line). The maximum efficiency hardly reaches 0.5%. An enhancement in the STEG cell performance is expected when the cell is mounted inside an evacuated enclosure due to the elimination of air conduction/convection losses (dash-dotted line). However, the radiation losses from the black absorber still limit the maximum efficiency to approximately 1.5%. Replacing the black absorber by a spectral solar absorber boosts the performance significantly to between 4% and 5% for an effective ZT of 1 (dashed line). If the TE material has an effective ZT of 0.8 the maximum STEG efficiency will not exceed 4% (dotted line). As discussed, the alternative to thermal concentration of the heat flux is optical concentration of the radiation flux to create the necessary temperature difference across the TEG. Then the absorber losses are minimized due to the small absorber size. Therefore, with high enough optical concentration it is possible to achieve similar or even better performance with an STEG cell exposed to atmospheric air and using a black absorber

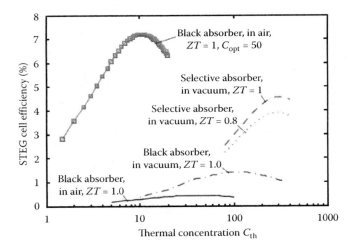

FIGURE 24.4 Simulation results for different STEG designs: solid line—STEG without optical concentration, in air, with black absorber and effective ZT of 1; dash-dotted line—STEG without optical concentration, in vacuum, with black absorber and effective ZT of 1; dotted line—STEG without optical concentration in vacuum with state-of-the-art wavelength-selective solar absorber and effective ZT of 0.8; dashed line—STEG without optical concentration, in vacuum, with wavelength-selective solar absorber and effective ZT of 1; solid line with open squares—STEG with 50× optical concentration, in air, with black absorber and effective ZT of 1.

(solid line with open squares) than in vacuum with large thermal concentration and a sophisticated spectral solar absorber. However, the feasibility of this type of STEG still has to be proven due to additional optical concentration losses (mirror/lens losses, tracking losses, diffuse-light losses) which are not included in the STEG cell efficiency. An important conclusion from this figure is that the STEG design with large thermal concentration can only meet its potential when the solar absorber losses are minimized.

Kraemer et al. investigated in detail the constraints on the optimal STEG design due to external irreversibilities such as electrical contact resistances, geometry-dependent radiation losses and temperature nonuniformity of the solar absorber.[25] The effect on the geometric optimization parameter $C_{th}L$ (product of thermal concentration and TE leg length) is also discussed. The minimum TE element length can be limited by the electrical contact resistance. The smaller the TE element, the larger is the relative electrical contact resistance and the performance loss. However, this loss in performance is less than 2% (rel.) for an electrical contact resistance of 5×10^{-7} Ω cm^2 for an element as short as 0.3 mm. Thus with reasonably low electrical contact resistance the STEG performance as well as the geometrical optimization parameter stays unaffected and the minimal element length is mainly limited by manufacturing processes.

The geometry-dependent radiation losses have a larger effect on the STEG cell performance. This limits the maximum element length and suggests using the shortest TE elements possible. For a simple flat-plate design, longer TE elements result in a larger gap between the solar absorber and the cold side; this increases the radiation losses from the back side of the absorber and from the sidewalls of the TE elements to the blackbody surroundings. For example, changing the element length from approximately 0.8 to 4 mm decreases the STEG cell efficiency from approximately 5.35% to 5%.

Another concern with large thermal concentration is the temperature drop within the absorber from its outer edge to the TE elements. The temperature drop is due to the radial conduction resistance and large radial heat flux of large absorbers. If the temperature nonuniformity in the solar absorber is large, then most of the absorber will have a higher temperature than its junction with the TE elements. A higher temperature results in more radiation losses than if the absorber temperature were uniformly at the TE hot-junction temperature. This causes a decrease of the opto-thermal efficiency. We can use a simple annular fin model[37] including the incident radiation as a constant heat source along the fin in order to estimate this temperature nonuniformity. This fin model approximates the absorber as a copper disc with thermal conductivity of 380 W/(mK), thickness $t = 0.2$ mm, and radius R_{abs} and the TE elements as a cylindrical base with radius r_{TE} at a fixed temperature (Figure 24.5a).

A solution based on Bessel functions is available to approximate the temperature gradient within the absorber depending on the STEG design geometry. The simulation results shown in Figure 24.5b–d, however, are obtained from a numerical solution based on a radial finite difference scheme. In all three cases, the thermal concentration is varied. Figure 24.5b shows the simulation results for changing solar absorber size while maintaining a constant TE element cross-sectional area. In Figure 24.5c, the solar absorber size is kept constant and the cross-sectional area of the TE elements is varied. The temperature drop within the absorber is significantly larger if the absorber size is increased compared to if the cross-sectional area of the TE elements is decreased. This is because the heat flux conducted radially is proportional to the square of the absorber radius, but the increase of the radial conduction resistance scales with $\ln(R_{abs}/r_{TE})$.[37] Figure 24.5d shows simulation results using a significantly larger cross-sectional area of the TE elements which can be thought of as a TEG module with a large number of closely packed thermocouples. The simulations show that with an equivalent radius of 4.8 mm (20 thermocouples with equivalent radius of 1.08 mm) the temperature drop within the absorber becomes very large in order to drive the large heat flux. In addition to the temperature gradient away from the TEG module, there will be a significant nonuniformity of the temperature over the TEG module if due to cost-considerations the copper substrate thickness, t, is chosen to be 0.2 mm.[25] This suggests that conventional TEG modules are not suitable for solar applications with no/low optical concentration and large thermal concentration. The best performance will be achieved for STEG cells which individually consist of a solar absorber mounted to a small TE unicouple as shown in Figure 24.6. These unicouple STEG cells can be electrically connected in series to a larger STEG cell array in order to increase the voltage and power output.

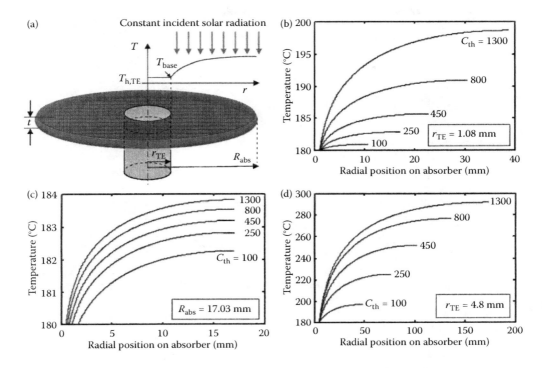

FIGURE 24.5 (a) model geometry of STEG for radial fin analysis: T_{base} is the fixed base temperature of the radial fin which is set equal to the hot-junction temperature of $T_{h,TE}$ (180°C) of the TE element, r_{TE} is the equivalent TE element radius $r_{TE} = (A_{TE}/\pi)^{(1/2)}$ and R_{abs} is the equivalent radius $R_{abs} = (A_{abs}/\pi)^{(1/2)}$ of the solar absorber. Graphs are simulation results (numerical solution based on radial finite difference scheme) of the radial fin for different thermal concentrations C_{th} with (b) fixed $r_{TE} = 1.08$ mm and changing R_{abs}, (c) fixed $R_{abs} = 17.03$ mm and changing r_{TE}, and (d) fixed $r_{TE} = 4.8$ mm and changing R_{abs}.

It can be concluded that in order to minimize TE material costs, the length of the TE element should be reduced. However, to keep the optimal $C_{th}L$ for maximum unicouple STEG cell performance the thermal concentration must be increased by reducing the cross-sectional area of the TE elements and not by increasing the absorber size in order to prevent detrimental performance losses from a large radial temperature nonuniformity.

24.4 Simulations and Experiments of STEGs

A detailed model has been developed for STEG designs as schematically shown in Figure 24.6.[25] This model can be used to simulate and optimize the performance of STEG cells based on large thermal concentration with no/small optical concentration (Figure 24.6a) and small thermal concentration with large optical concentration (Figure 24.6b). The model takes into account temperature-dependent properties of the TE materials and the spectral solar absorber, electrical contact resistances, geometry-dependent heat losses, and the nonuniform temperature of the solar absorber. The losses are schematically summarized in Figure 24.7a and b. In order to account for the effect of the temperature-dependent TE material properties, the electrical contact resistances, and the heat losses from the sidewalls of the TE elements on the STEG cell performance, the TE elements are discretized and solved numerically with the iterative technique[38,39] as shown in Figure 24.7c.

Besides the properties of the solar absorber and TE materials, the geometry-dependent radiation losses, the temperature distribution within the solar absorber, and the electrical contact resistance,

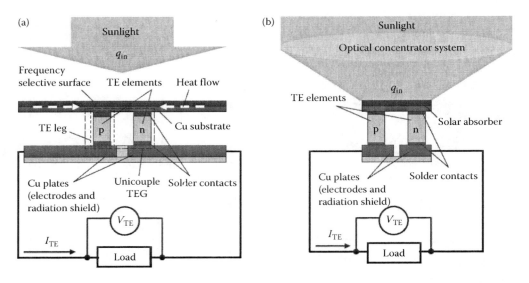

FIGURE 24.6 Detailed schematic of a unicouple STEG cell with (a) large thermal concentration and (b) large optical concentration.

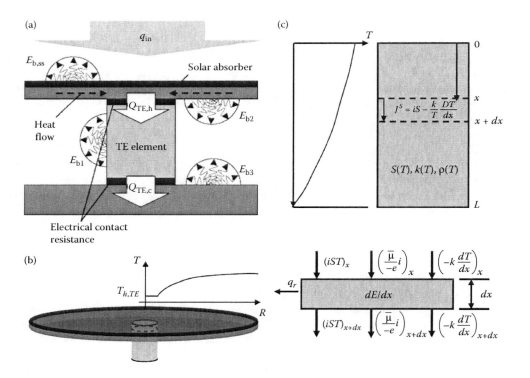

FIGURE 24.7 Graphical illustration of external losses from (a) thermal radiation and electrical contact resistance and (b) temperature nonuniformity in absorber, (c) discretization of TE elements and energy balance over one increment with J^s as the local entropy flux, S the Seebeck coefficient, k the thermal conductivity, ρ the electrical conductivity, T the local temperature, $(-e)$ the electron charge, i the current density, and $\bar{\mu}$ as the local electrochemical potential (Fermi level).

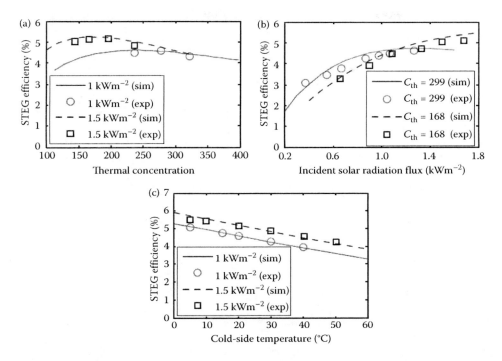

FIGURE 24.8 Experimental results with simulations result showing (a) the optimization of the thermal concentration (b) the effect of the incident solar radiation, and (c) the effect of the cold-side temperature.

the optimal geometrical optimization parameter $C_{th}L$ is dependent on the solar intensity striking the solar absorber and on the cold-junction temperature. Consequently, the STEG cells should be optimized for best performance over the course of a day or even the year for the specific location where the system will be installed. Figures 24.8 and 24.9 show simulation and experimental results of the most recent publications on STEGs.[1,25] The optimal $C_{th}L$ decreases with increasing cold-junction

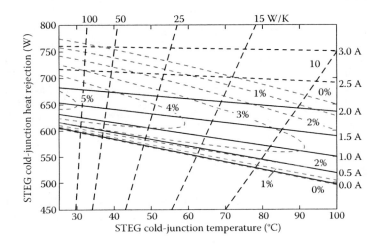

FIGURE 24.9 STEG operation diagram for $C_{th}L = 0.4$ m and heat sink (fluid) temperature of 25°C showing the cold-junction heat rejection for an STEG area of 1 m² as a function of the cold-junction temperature and the STEG cell current. Red dashed lines are contour lines of constant STEG efficiencies. Black solid lines are lines of constant cell currents. Gray dash-dotted lines are lines of constant cold-side thermal conductance.

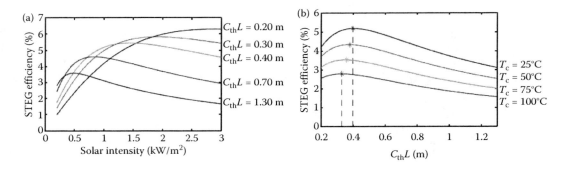

FIGURE 24.10 Simulation results showing the effect of (a) the solar intensity and of (b) the cold-junction temperature on the STEG cell efficiency and the geometrical optimization parameter $C_{th}L$.

temperature and solar intensity (Figure 24.10a and b). Kraemer et al.[25] also showed that if the STEG efficiency is averaged over the course of a day, it becomes a weaker function of $C_{th}L$ which means that in the case of a cogeneration system the $C_{th}L$ can be varied in favor of the thermal efficiency of the overall system efficiency without significantly affecting the daily electrical performance. The integration of an STEG in a cogeneration system will be discussed in more detail in Section 24.5.

Guided by the detailed model briefly described above experiments were performed[1] and some results are summarized in Figure 24.8. Figure 24.8a shows the experimental optimization of the thermal concentration for a given TE element length, L, and two different incident solar fluxes. Higher solar intensity results in a smaller optimal thermal concentration and in a higher STEG efficiency. As predicted by the model, an STEG cell with fixed geometry shows a dependence on changes of the incident solar flux and cold-junction temperature (Figure 24.8b and c). A peak in STEG efficiency can be observed at a specific solar intensity. At fluxes larger than this solar intensity, the flux through the TEG increases which results in a higher absorber temperature. The higher temperature increases the radiation heat losses from the solar absorber, causing a decrease in the opto-thermal efficiency that is larger than the increase in TEG efficiency, resulting in lower STEG cell efficiency. A higher cold-junction temperature results in a lower STEG efficiency for three reasons. An increasing cold-junction temperature drives up the absorber temperature, which results in larger radiation losses from the solar absorber. This not only results in a lower absorber efficiency but also reduces the temperature difference between the hot- and the cold-junction which reduces the device efficiency. In addition, for TE materials such as Bi_2Te_3 the temperature-dependent figure of merit decreases for temperatures higher than ~100°C. Consequently, the higher the STEG operation temperature ($T_{abs} > 100°C$) the lower is the effective figure of merit, $(ZT)_m$, resulting in a lower adiabatic TE efficiency.

24.5 STEG Performance and System Integration

In solar TE systems, approximately 50–80% of the intercepted solar heat is released at the cold junction of the device. This waste heat must be removed in order to maintain the cold junction at a given temperature for highest STEG performance. One way to remove this excess heat is first to spread it out on the cold side using a metallic heat spreader, and then to transfer the heat to the environment via natural convection, similar to what is done for the heat management of PV cells. In certain applications such as cogeneration, this waste heat is actually the input to a secondary system, such as a domestic hot-water loop. Small deviations from the peak TEG operating point can have large (positive or negative) effects on the quantity of heat delivered to the heat sink. As a result, an STEG cogeneration system can be designed to favor the production of either electrical power or waste heat, depending on the demands of the application. This can be accomplished by choosing the appropriate geometric parameter $C_{th}L$ (Figure 24.10b). If it is desired to optimize the system for maximum electrical power then there is one specific optimal $C_{th}L$ for a specific cold-junction temperature and incident solar flux as discussed in previous sections. Conversely, choosing a smaller $C_{th}L$ results in more heat transported to the hot water loop because the absorber stays at a lower temperature and thus the system radiation losses are smaller. Fortunately, as mentioned in the previous section, in the case of a daily total STEG electrical energy output is a weaker function of $C_{th}L$, so it is possible to deviate from the optimal $C_{th}L$ for maximum electrical performance with only a negligible affect on the daily electrical performance.

For a system with fixed $C_{th}L$, it is even possible to adjust the balance between electrical power and waste heat solely by adjusting the electrical current of the circuit. As an example, we consider a system of $C_{th}L = 0.4$ m. Because the waste heat and cold-junction temperature are both affected by the current, it is useful to plot the waste heat per 1 m^2 absorber area as a function of cold-junction temperature at various currents (Figure 24.9). Superimposed on this figure (dashed lines) are contour lines representing the corresponding STEG efficiency. This STEG operation diagram can be used to determine the conditions where the STEG can operate, because the performance of any heat sink can be characterized by the relationship between the heat sink temperature and the rejected heat. As an example, if the cold-junction heat removal is managed by a fluid passing over the cold junction, the constant of proportionality between the transferred heat and the temperature difference between the fluid and the cold junction is the thermal conductance, $U_{th} = A \times h$ in units W/K. In Figure 24.9, dash-dotted lines of constant U_{th} are plotted in gray assuming a fluid temperature of 25°C. Changing the electrical current allows the system to operate at different points along this characteristic U_{th} curve, which affects the cold-junction temperature, the amount of rejected heat, and the amount of electrical power generated. For example, if it is desirable to generate more waste heat in the morning and more electricity in the afternoon, it is possible to run the system in an "overdrive" mode (with super-optimal current) in the morning, and then run the system at optimal current in the afternoon. Interestingly to notice is that there will always be two operational current points with same STEG efficiency but with different cold-junction heat rejection rates if the STEG is operated at off-optimal conditions. For example, for a heat sink with thermal conductance of 25 W/K the line of constant U_{th} intersects the 4% efficiency contour line twice. One intersection corresponds to a cell current of 1 A and the other one to approximately 1.75 A. At higher cell current more heat is transported through the TE unicouple by the energy carriers and rejected at the cold junction. This will drop the temperature of the solar absorber but also increase the cold-junction temperature in order to support the larger heat flux.

24.6 Potential for Improvement

There are some ways to improve the STEG cell efficiency, including improving the TE materials; increasing the absorber's solar absorptance; decreasing the absorber's IR emittance; changing the STEG geometry; developing selectively transmitting glass; or increasing the optical concentration. Figure 24.11 shows the predicted performance boost for various improvements such as the reduction of the effective emittance of the solar absorber (Figure 24.11a). If optical concentration is used, the operational absorber temperature of the STEG cell will be limited by the temperature stability of the TE materials. In the case of Bi_2Te_3 materials the operational absorber temperature is limited to below 250°C. Figure 24.11b shows that a further increase of the optical concentration from 3 to 10 will not lead to a significant increase of efficiency if the temperature limit of the solar absorber is set to 220°C. However, if segmented TE elements or cascaded TEGs with different materials are used higher optical concentration is beneficial (Figure 24.11c).

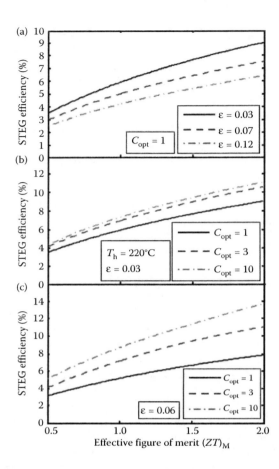

FIGURE 24.11 Simulation of the performance of an STEG in vacuum as a function of the effective figure of merit $(ZT)_M$ showing results (a) for an STEG without optical concentration with different solar absorber emittance, (b) for different optical concentrations C_{opt} with constant solar absorber emissivity of $\varepsilon = 0.03$, limited absorber temperature to $T_h = 220°C$, and (c) for different optical concentrations C_{opt} with constant solar absorber emissivity of $\varepsilon = 0.06$ and no limits on the absorber temperature.

Acknowledgments

We would like to thank Drs. Bed Poudel, J. Christopher Caylor, and Matteo Chiesa for helpful discussions. This material is partially based upon work supported as part of the "Solid State Solar-Thermal Energy Conversion Center (S³TEC)," an Energy Frontier Research Center funded by the U.S. Department of Energy, Office of Science, Office of Basic Energy Sciences under Award Number: DE-SC0001299/DE-FG02-09ER46577 (D.K, K.M, G.C., and Z.F.R.), and by MIT-Masdar program (D.K. and G.C.).

References

1. Kraemer, D., Poudel, B., Feng, H.-P., Caylor, J. C., Yu, B., Yan, .X., Ma, Y. et al., Solar thermoelectric generators with flat-panel thermal concentration, *Nature Materials*, 10, 532–538, May 2011.
2. Weston, E., U.S. Patent No. 389,124, 1888.
3. Weston, E., U.S. Patent No 389,125, 1888.

4. Coblentz, W. W., *Scientific American*, 127, 324, 1922.

5. Telkes, M., Solar thermoelectric generators, *Journal of Applied Physics*, 25, 765-777, 1954.

6. Rush, R., Solar flat plate thermoelectric generator research, *Technical Documentary Report for Air Force Aero Propulsion Laboratory*, Report number: APL TDR 64-87, 1964.

7. Fuschillo, N., Gibson, R., Eggleston, F. K., and Epstein, J., Flat plate solar thermoelectric generator for near-earth orbits, *Advanced Energy Conversion*, 6, 103-125, 1966.

8. Fuschillo, N. and Gibson, R., Germanium-silicon, lead telluride, and bismuth telluride alloy solar thermoelectric generators for venus and mercury probes, *Advanced Energy Conversion*, 7, 43-52, 1967.

9. Raag, V., Berlin, R. E., and Bifano, W. J., Flat plate thermoelectric generators for solar-probe missions, *Intersociety Energy Conversion Engineering Conference*, Boulder, CO, 1968.

10. Scherrer, H., Vikhor, L., Lenoir, B., Dauscher, A., Poinas, and P., Solar thermoelectric generator based on skutterudites, *Journal of Power Sources*, 115, 141-148, 2003.

11. Dent, C. L. and Cobble, M. H., A solar thermoelectric generator experiment and analysis, *Proceedings of 4th ICT*, pp. 75-78, 1982.

12. Omer, S. A. and Infield, D. G., Design optimization of thermoelectric devices for solar power generation, *Solar Energy Materials and Solar Cells*, 53, 67-82, 1998.

13. Amatya, R. and Ram, R. J., Solar thermoelectric generator for micropower applications, *Journal of Electronic Materials*, 39, 1735-1740, 2010.

14. Li, P., Cai, L., Zhai, P., Tang, X., Zhang, Q., and Niino, M., Design of a concentration solar thermoelectric generator, *Journal of Electronic Materials*, 39, 1522-1530, 2010.

15. Price, H., *Assessment of Parabolic Trough and Power Tower Solar Technology Cost and Performance Forecasts*, NREL/SR-550-34440, 2003.

16. Caillat, T., Fleurial, J.-P., and Snyder, G. J., Development of high efficiency segmented thermoelectric unicouples, ICT, 2001.

17. McEnaney, K., Kraemer, D., Ren, Z. F., and Chen, G., Modeling of concentrated solar thermoelectric generators, *Journal of Applied Physics*, 110, 074502, 2011.

18. Goldsmid, H. J., Giutronich, J. E., and Kaila M. M., Solar thermoelectric generation using bismuth telluride alloy, *Solar Energy*, 22, 435-440, 1980.

19. Vatcharasathien, N., Hirunlabh, J., Khedari, J., and Daguenet, M., Design and analysis of solar thermoelectric power generation system, *Journal of Sustainable Energy*, 24:3, 115-127, 2005.

20. Sales, B. C., Smaller is cooler, *Science*, 295, 1248-1249, 2002.

21. Vining, C. B., ZT ~ 3.5: Fifteen years of progress and things to come, *European Conference on Thermoelectrics*, 2007.

22. Dresselhaus, M. S., Chen, G., Tang, M. Y., Yang, R., Lee, H., Wang, D., Ren, Z. F., Fleurial, J.-P., and Gogna, P., New directions for low-dimensional thermoelectric materials, *Advanced Materials*, 19, 1043-1053, 2007.

23. Snyder, G. J. and Toberer, E. S., Complex thermoelectric materials, *Nature Materials*, 7, 105-114, 2008.

24. Minnich, A. J., Dresselhaus, M. S., Ren, Z. F., and Chen, G., Bulk nanostructured thermoelectric materials: Current research and future prospects, *Energy & Environmental Science*, 2, 466-479, 2009.

25. Kraemer, D., McEnaney, K., Chiesa M., and Chen, G., Modeling and optimization of solar thermoelectric generators for terrestrial applications, *Solar Energy*, 2011. Under review.

26. Rockendorf, G., Sillmann, R., Podlowski, L., and Litzenburger, B., PV-hybrid and thermoelectric collectors, *Solar Energy*, 67, 227-237, 2000.

27. Luque, A. and Marti, A., Limiting efficiency of coupled thermal and photovoltaic converters, *Solar Energy Materials & Solar Cells*, 58, 147-165, 1999.

28. Zhang, Q.-J., Tang, X-F., Zhai, P.-C., Niino, M., and Endo, C., Recent development in nano and graded thermoelectric materials, *Materials Science Forum*, 492-493, 135-140, 2005.

29. Vorobiev, Y. V., Gonzalez-Hernandez, J., Vorobiev, P., and Bulat, L., Thermal-photovoltaic solar hybrid system for efficient solar energy conversion, *Solar Energy*, 80, 170-176, 2006.

30. Kraemer, D., Hu, L., Muto, A., Chen, X., Chen, G., and Chiesa, M., Photovoltaic–thermoelectric hybrid systems: A general optimization methodology, *Applied Physics Letters*, 92, 243503, 2008.

31. Rowe, D. M., A high performance solar powered thermoelectric generator, *Applied Energy*, 8, 269–273, 1981.

32. Chen, G., Theoretical efficiency of solar thermoelectric generators (STEGs), *Journal of Applied Physics*, 109, 104908, 2011.

33. Ioffe, A. F., *Semiconductor Thermoelements and TE Cooling*, Infosearch Limited, London, 1957.

34. Yin, Z. Development of solar thermal systems in China, *Solar Energy Materials & Solar Cells*, 86, 427–442, 2005.

35. Waeckelgard, E., Niklasson, G. A., and Granqvist, C. G., Selectively solar-absorbing coatings, Gordon, J., ed., *Solar Energy: The State of the Art*, pp. 109–144, Chapter 3, ISES, James & James Ltd, London, 2001.

36. Kennedy, C. E., Review of mid- to high-temperature solar selective absorber materials, NREL/ Technical Report-520-31267, 2002.

37. Mills, A. F., *Heat Transfer*, 2nd edition, Prentice-Hall, New Jersey, 1999.

38. Buist, R. J., Calculation of Peltier device performance, Rowe, D. M., ed., *CRC Handbook of Thermoelectrics*, pp. 143–155, Chapter 14, CRC Press, New York, 1995.

39. Hogan, T. P. and Shih, T., Modeling and characterization of power generation modules based on bulk materials, Rowe, D. M., ed., *Thermoelectrics Handbook: Macro to Nano*, Chapter 12, CRC Press, Boca Raton, FL, 2006.

25

Automotive Applications of Thermoelectric Materials

Jihui Yang
General Motors R&D Center

Francis R. Stabler
Future Tech LLC

25.1 Introduction

Car and truck customers expect new vehicles to offer improved efficiency, better durability, and more features than their previous vehicles. To meet these customer expectations, automotive manufacturers are searching for a wide variety of new technologies to incorporate into new vehicles. One technology getting increased attention is thermoelectric technology because it can be key to providing several customer-desired features. Most thermoelectric applications can be divided into one of two segments. One segment is based on the Peltier effect and is focused on various applications of heating or cooling. At this time, a few heating and cooling applications represent the only use of thermoelectric technology in production cars and trucks. The other segment is based on the Seebeck effect and provides electric power generation using a temperature differential. Because of the growing need to reduce fuel consumption in cars and trucks, there is increased interest in thermoelectric power generation using waste heat. The heating and cooling applications will be addressed first because some of these applications are already in production.

25.2 Peltier Effect Applications: Heating and Cooling

While vehicle customers need and expect lower fuel consumption from their vehicles, there is a competing demand for additional features for comfort and convenience. These features frequently require additional electric or mechanical power while adding mass to the vehicle, resulting in increased fuel consumption. Thermoelectric material can offer vehicle owners several unique features by providing heating and cooling for various items on the vehicle. Seats with thermoelectric heating and cooling are available today and have been in production by Amerigon for more than a decade in many luxury vehicles. Cup holders that keep beverages hot or cold are available on a few models. Small thermoelectric refrigerators or wine coolers are available as features for limousines and recreational vehicles. Numerous after-market manufacturers are producing thermoelectric units for automotive use. These units can be set to cool or heat food and drink using the 12 V power from a vehicle. Many of these units are portable and can be used in any location where electric power is available. All of these thermoelectric systems offer the benefit of being small, relatively lightweight, and silent in operation. These unique features make thermoelectric technology very attractive even though the efficiency is low.

25.2.1 Passenger Heating and Cooling Application of Thermoelectric Technology

The most promising area for application of the Peltier effect is heating and cooling of the vehicle passengers. This function has been discussed in numerous patents, papers, and presentations, but has not reached production yet. Thermoelectric materials have the potential to revolutionize automotive Heating, Ventilation, and Air Conditioning (HVAC) systems. Improved thermoelectric materials can enable the production of HVAC systems with several very desirable and unique features:

- No greenhouse gases required for refrigerant
- Silent operation
- Faster operation, especially faster delivery of warm air on cold starts
- Potentially lighter weight systems (improving fuel efficiency)
- Potential of more energy efficient cooling
- Ability to heat or cool the passenger compartment without engine operation (this is very important for hybrids, engine off when vehicle stops, remote HVAC activation without engine start, etc.)
- Increased reliability and durability due to no or few moving parts

HVAC systems based on thermoelectric modules can contribute to the acceptance of hybrid, electric, or fuel cell vehicles where quiet operation without the need for mechanical input is a needed capability. As vehicles have increased periods of electric operation without internal combustion engine background noise, any sounds created by the HVAC system can be objectionable. Because thermoelectric-based HVAC systems can potentially operate with less vehicle power, the technology may increase fuel efficiency. The US Department of Energy (DOE) has funded projects to demonstrate these systems and their fuel efficiency. Prototype systems are being developed and should be evaluated soon. There are at least two distinctly different approaches for thermoelectric HVAC systems: direct replacement of the current system and a new distributed system.

25.2.2 Direct Replacement of Existing HVAC Systems

The direct replacement approach would retain the single-point HVAC system where one unit in the vehicle firewall distributes heated or cooled air through a vent system, primarily in the instrument panel. The thermoelectric modules would provide the heating or cooling of the air, eliminating the need for the current heater core, compressor–evaporator system, pipes and hoses, potentially the AC radiator, and much of the engine compartment complexity. There is the possibility that engine coolant will be needed to provide or remove heat from the thermoelectric modules. This approach is attractive because it simplifies the packaging of the HVAC components and has limited impact on the passenger cabin. Depending on the performance of the available thermoelectric modules, this approach may save some mass and reduce power needed from the engine or battery. Additionally the thermoelectric-based HVAC system has few moving components and should provide better durability than the current compressor-operated systems. A growing HVAC problem is that as vehicles become more efficient, both gasoline and diesel engines take much longer to warm up and provide adequate coolant temperature to warm the passenger compartment. Thermoelectric systems can offer a solution to this problem by efficiently heating air to the passenger compartment within seconds of activation. While this design would be relatively easy to package and has several benefits, it does not take full advantage of features offered by thermoelectric systems.

25.2.3 Distributed HVAC Systems

A new vehicle cabin design to incorporate a distributed HVAC system would be a much more difficult project; however, it has the potential to offer much more for the customer. It would offer all of the features mentioned above for the conventional system plus several unique ones. A distributed design has the potential to offer the following:

- Faster cooling and heating than either current HVAC systems or a new direct replacement system.
- Lower power requirements and ultimately increased fuel efficiency relative to the current compressor and heater core HVAC systems.
- Less wasted energy by heating or cooling only the passengers, not empty seats.
- Improved comfort on entering vehicle if remote activation used to heat or cool surfaces the occupants touch.
- Lighter weight system and better packaging, especially less impact on the instrument panel, offering new opportunities for cabin layout and look.
- Ability to provide individual temperature control for each passenger or sector of the vehicle (partially available today with conventional systems).
- Better temperature control, especially for rear seat passengers.
- Quieter system operation.

To design a successful distributed HVAC system, more research is needed on how people perceive a comfortable temperature for their environment. Air temperature, air movement, surface temperatures and many other subtle factors all play a part in achieving the desired result for all passengers. Thermoelectric modules offer designers the ability to cool or heat surfaces the passengers touch in addition to controlling air temperature. This is a downside of the distributed HVAC system for automotive manufacturers: much of the passenger cabin would have to be redesigned and many passive components such as seats, arm rests, steering wheel, head liner, and so on would become active parts of the HVAC system. The heated and cooled Amerigon seats offered in some current cars are an initial example of this capability and would probably be retained in some enhanced form. Even control of air temperature could change since it would be possible to have many small sources of air flow rather than a central source. Using a system that detects the occupants of a vehicle (similar to or shared with the airbag system), the distributed HVAC system could activate only the portions needed to provide a comfortable environment for the actual passengers, not empty areas. The average number of people in a typical highway vehicle is much less than the maximum capacity (frequently only the driver), but the current conventional HVAC systems continue to operate on the whole passenger compartment. A distributed system with passenger sensing would result in a significant energy savings by controlling the temperature only in the occupied seating areas. Managing humidity and condensation for cooling applications will be a difficult issue that has to be addressed in the system design.

25.2.4 Other Heating and Cooling Applications

There are other applications of the Peltier effect on automobiles that would not be features directly observed by the customer, but are important to vehicle performance and durability. These involve controlling the temperature of certain vehicle systems to keep them in the desired temperature range for proper, reliable operation. Some of the applications could include the following:

- The lead-acid starter battery in all current cars and light trucks has significantly reduced ability to accept a charge at very low or high ambient temperatures. Thermoelectric systems can provide the heating or cooling needed to keep the battery in the temperature range for efficient charging.[1]
- Assisting in temperature management of the catalytic converter for optimum life and emissions after-treatment performance, both as an absorber of heat ahead of the converter[2] or around the converter package to absorb or add heat as needed.
- Temperature management for the battery packs of hybrid vehicles. Most of the battery chemistries used in hybrid vehicles do not function well in the full range of ambient temperature conditions where they are used. Thermoelectric devices can protect the battery pack from both extremes of temperature.
- Cooling critical electronics to prevent damage due to high-temperature conditions or to improve temperature stability for better functioning.

Not all of these applications will be implemented, but there are likely many more applications not listed here. The cost of thermoelectric units added to systems and the additional electrical load to provide this increased durability or functionality with heating and cooling will have to be justified by the added value provided to those systems.

25.2.5 Conclusion and Concerns

The success of these features depends on the efficiency of the thermoelectric modules used for heating and cooling and even more importantly on the availability of economical modules. It is desirable to have high-performance materials because they allow higher efficiency (less electric power demand from the vehicle) and smaller systems (lighter, easier to package, potentially lower cost systems).

Heating and cooling applications represent relatively low-risk ways to introduce TE technology to the automobile. A decade of production applications (heated and cooled seats, cup holders, etc.) have them well established.

As applications grow, the availability of TE modules in automotive quantity and cost is uncertain and is a major concern. Bi_2Te_3-based thermoelectric modules are used today, but if the production volume were to increase significantly, the price would likely become unacceptable due to the somewhat limited availability of tellurium. There is also a need for modules with increased efficiency and Bi_2Te_3-based devices may not be able to improve significantly.

25.3 Seebeck Effect Applications: Power Generation

The world currently faces growing problems relating to transportation energy. The global demand for oil is continuing to increase, resulting in a higher price for a gallon of fuel. There is a growing concern about the effect of greenhouse gases, especially carbon dioxide, on the environment. All of these issues are driving the customer and regulatory demand for obtaining more useful energy from every unit of fuel burned. Many new engine, transmission, and vehicle technologies are being produced or are in development to make vehicle transportation more fuel efficient. All of these technologies are missing one important issue: much of the energy from a gallon of fuel will still be unusable heat in the vehicle exhaust or cooling system (see Figure 25.1). Thermoelectric technology can make use of this waste heat by converting some of a vehicle's waste heat to electric power. This will reduce the electric generator's mechanical load on the engine, improving vehicle fuel efficiency.

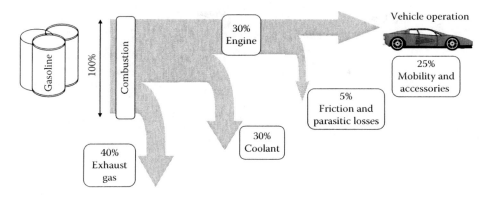

FIGURE 25.1 Typical energy path for vehicles with gasoline fueled internal combustion engines.

There is an economic component to improving fuel efficiency that has to be considered in thermoelectric generator (TEG) design. There are competing technologies to use waste heat for electric or mechanical power production. Some examples would be Rankin or Sterling cycle engines, steam engines, thermo-acoustic systems, and so on. Thermoelectric devices seem to have an edge because of their ability to do direct heat-to-electric power conversion. It must also be considered that for automotive manufacturers (and usually for customers), the lowest cost method for improving fuel economy is the best. That means that the real competitor for TEG systems will continue to be all of the conventional improvements to engines, transmissions, and vehicles that improve their fuel economy. Fuel economy can be improved by many methods, such as enhanced fuel delivery systems, lower friction engines, improved transmissions with more speeds, better vehicle aerodynamics, hybrid propulsion systems, and many other vehicle engineering improvements. If these methods of fuel efficiency improvement cost less than TEG systems, then the TEG systems will not be selected for automotive applications.

There are many reasons to incorporate TEGs in automotive systems. Some of those are as follows:

- Improve fuel efficiency
- Lower greenhouse gas (carbon dioxide) emissions
- Support increased vehicle electrification
- Simpler to implement than alternative waste heat recovery systems
- Provide a "green" image for the vehicles

While all of these are good reasons to study TEGs, only improved fuel efficiency is important enough to justify the cost of adding TEGs to cars and trucks.

An automotive TEG is a complete system integrated into a vehicle that uses vehicle waste heat energy and a cooling system to produce electricity for use on the vehicle. It should be noted that there is also the potential to produce the heat by burning fuel for the energy source in certain applications or some operating conditions. This integrated TEG system consists of several general components (see Figure 25.2):

1. A heat exchanger to take heat from the exhaust gases or engine coolant and deliver it to the hot side of the thermoelectric modules.
2. Thermoelectric modules with good conversion efficiency (heat to electricity) in the available temperature range.
3. A heat exchanger to maintain the cold side of the thermoelectric modules by taking heat from the modules and radiating it to liquid coolant or to the air.
4. A housing to package the above components and to interface with the vehicle: mounting, exhaust connections, coolant connections, wiring, and so on.
5. An electrical power conditioning and interface unit to match the power output of the thermoelectric modules to the vehicle electrical system.

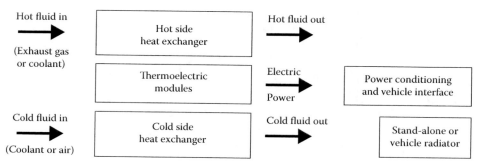

FIGURE 25.2   Major functions of a TEG.

Additional items and functions may be required by the system depending on the design and the vehicle application. Examples of this could include items such as the following:

- An electronic unit to monitor and control the operation of the TEG and possibly interface with the engine control unit and/or the generator.
- An electric pump to control the flow of coolant to the TEG.
- Sensors for temperatures and flow rates to improve control of the TEG.
- An exhaust flow control to bypass the TEG when the gases are too hot or the TEG causes excessive exhaust backpressure at high exhaust flow rates.
- Added coolant pipes or hoses and additional coolant (mass).
- Added valves to control the coolant flow to and from the engine or radiator.
- A larger vehicle radiator or an additional radiator to manage the added heat load from the TEG cold side.
- Additional fans for air cooling the radiator or the cold side of the TEG.
- Incorporating the muffler function into an exhaust system TEG or using the system to help manage catalytic converter temperature.[2]

As shown above, there is a potential to add a lot of additional components to a vehicle to have a functioning TEG system. This can add a significant amount of mass to a vehicle if the designer does not pay careful attention to keeping the additional mass to a minimum. Added mass means lower fuel efficiency for a vehicle and too much mass could eliminate the fuel economy gains provided by the TEG. As a reference point for typical small cars, one automotive manufacturer has estimated that each added 125 pounds (one test weight class) will reduce the fuel efficiency by 1.4% (or 1.1% fuel economy change per 100 pounds mass change) on the US Federal Test Procedure (FTP). Work at Argonne National Laboratory indicates that 100 pounds will change the fuel economy by 1.2% for a 3500 pound conventional vehicle.[3] Both sources agree that small vehicles are more mass sensitive than larger vehicles. Since current TEG designs appear to offer only 1 to 5% fuel economy gain for a vehicle, inattention to the added mass could significantly reduce or eliminate the fuel economy gains provided by a TEG. The TEG fuel economy gain listed is based on testing on the US FTP cycle. Other driving cycles could provide different fuel economy gains, potentially 4–10%.

To have an impact on automotive fuel economy, TEGs have to be in production on a significant number of vehicles. At present, the annual global production of light vehicles is over 50 million vehicles. If only 10% of these vehicles incorporated TEG systems, the TEG annual volume would be over 5 million. The module volume would depend on the number per TEG, but, as an example, if each TEG used an average of 40 modules, there would be a demand for 200 million thermoelectric modules per year just for automotive power applications. This requires the development of very high-volume production capability for TEGs and all components in the system. Volume production will need a partnership

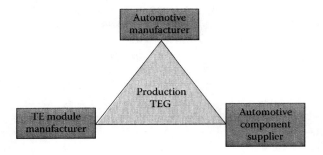

FIGURE 25.3 Major stakeholders needed to produce high-volume TEGs.

between a triad of key stakeholders: module manufacturers, automotive suppliers, and automotive manufacturers (see Figure 25.3). There are currently no thermoelectric modules in production that are suitable for high-volume automotive TEGs. Module manufacturers are not likely to tool up with a production design or make the capital investment to produce TEG capable modules until they have a reasonable expectation of a market for them. They need assurance that TEGs will actually go into production as a market for their modules. There needs to be coordinated work by the component and automotive manufacturers to assure sufficient markets for the modules. Only then will module manufacturers be able to make this investment of resources. The second partner of this triad is an automotive supplier with the capability of designing and going to volume production with TEGs and all the additional components for a TEG system. The third partner is an automotive manufacturer. The automotive manufacturers are responsible for establishing the requirements for TEG and for the integration of the TEG into specific vehicles. They are also responsible for various interface components or assigning this responsibility to specific suppliers. These components include flow control valves, piping, electrical power interface units, and mounting of all components. In the long term, the automotive manufacturer must be able to convince the vehicle customer to pay for the TEG system. This means that the customer must feel that the TEG adds sufficient perceived value to the vehicle to be worth the extra cost.

When designing a TEG, the thermoelectric modules must be selected to have optimum (high) ZT (figure of merit for thermoelectric materials which will be defined later) in the typically available temperature range which will vary some with the type of vehicle and engine. The material must have relatively low cost and adequate availability to support production volumes as discussed above. The design must also provide protection for the modules in the event of temperature extremes that can be found on a vehicle. If a sufficiently efficient TEG can be designed, it may be desirable to include a fuel combustion unit to provide heat to generate electric power from the TEG when waste heat is not available.[4] This could benefit conventional vehicles that turn off the engine for fuel savings on deceleration and when stopped since the alternative is to operate from batteries during these events with the added inefficiency of resistance loses into and out of the battery.

25.4 Electrical Power Use

The use of the power from a TEG may seem simple, but how a vehicle uses electric power must be considered. In most US government emission and fuel economy tests, a vehicle is tested with all optional systems turned off, limiting electrical usage to 250–350 W. This is far different from the more typical 300–1500 W of electrical power used by a customer's vehicle in general use. With the electrical limited to 250–350 W, potential regulatory gains in fuel economy using TEGs are limited. To achieve fuel economy based on government testing, additional vehicle modifications will be needed to use the electrical power generated by a TEG with output greater than about 300 W. The modifications should reduce or remove some mechanical loads from the engine and could include adding items such as electric water or oil pumps. On current vehicles, eliminating the use of the conventional generator during the US FTP-based fuel economy tests will provide 1% to 4% fuel economy gain, depending on the type and mass of the vehicle. If it is possible to drive electric water or oil pumps with the TEG output, this will add another 2–4% fuel efficiency gain. In normal driving, the typical customer will have electrical loads that are greater than the TEG output; therefore, any electrical power generated from waste heat will improve real-world fuel economy. The more electrical power generated by a TEG, the better for the customer and the world because less fuel will be burned. Use of a TEG on a hybrid electric vehicle is an especially desirable configuration if large amounts of electrical power can be generated (>1 kW).

The actual electrical interconnect of the thermoelectric modules and the vehicle need design consideration. The hot side of an exhaust powered TEG can reach temperatures that would damage the insulation of the interconnect wires if not adequately protected. Series, parallel or a combination of module interconnect must be determined for optimum power interface through a DC to DC converter to the vehicle electrical system. The DC to DC converter is required to match TEG voltage and load to the vehicle requirements. Automotive "12 volt" power is actually 13.5–14.5 V, depending on the battery temperature. The conventional generator adjusts its voltage automatically to enable charging of the battery at all temperatures. The TEG system will also have to adjust its electrical output to charge the battery.

25.5 TEG Cold Side

The TEG will require a well-designed heat exchanger on the cold side of the TE modules to efficiently remove the heat and keep the cold side temperature as low as possible. The cooling of the TEG could use air, but will almost certainly be done by liquid coolant because of the increased cooling efficiency, especially in the automotive environment. Most cars and small trucks have very limited excess cooling capacity in their cooling systems. This will limit the ability to use their existing cooling system to cool the TEG. One solution is to increase the cooling capacity of these vehicles with larger or more efficient (expensive) radiators. This will not be possible on many cars because of the lack of additional space in the radiator area and because of competing aerodynamic requirements. These vehicles will require stand-alone cooling systems with dedicated radiators. This will be difficult to package, but has more flexibility than one large radiator in the front of the vehicle. Large trucks and SUVs have cooling systems sized for worst-case conditions of carrying heavy loads, towing a large trailer up a steep incline, and doing all this in high ambient temperature conditions. This should allow use of the vehicle radiator for TEG cooling under most operating conditions. Of course, a monitor and controls will be required to shut down the TEG operation with an exhaust bypass if the worst case vehicle operations threaten to overload the cooling system. This should be a relatively rare condition, and be an acceptable operating parameter for the TEG.

The TEG cooling system will require an electric pump for stand-alone cooling systems and probably for TEGs that share the vehicle cooling system. TEGs that share the vehicle cooling system may be able to utilize the engine coolant pump (usually mechanical driven) but better control of cooling may be obtained by a separate TEG electric pump. This pump will have to be very efficient and sized carefully to avoid using too much of the electrical power generated by the TEG. High pump electrical use would significantly reduce the overall efficiency of the TEG.

There is another function of the TEG that should be considered as a benefit to the vehicle. This is the "cold side" heat exchanger. Faster warm up of the engine from a cold condition (such as 72°F) to normal operating temperature improves fuel economy. The TEG heated coolant output adds exhaust heat to the engine for quicker warm-up. This warms the engine coolant, the engine lubricating oil, and potentially the transmission fluid for lower friction and smoother operation. Additional fluid controls and hoses or pipes will be needed to allow the TEG fluid to flow to the engine when needed and to the radiator when the heat is not needed. The TEG heat output can also be routed to the HVAC system when needed for faster and more efficient heating of the passenger compartment. As engines become more efficient this will become more important to the comfort of the driver and passengers in cold weather.

25.6 Automotive Environment

The automotive environment is very harsh on materials and components. Any automotive system design has to account for this environment if it is to be a technical and market success. The following listing provides some key environmental considerations for light duty cars and trucks:

- Limited space to install added equipment
- Shock and vibration (requires a rugged design or isolation from vehicle)
- Ambient air thermal extremes (−40°C to 50°C)
- Thermal shock: Exhaust gases typically go from 20 to 400°C in less than 2 min (extreme: −40°C to 400°C) at vehicle start
- Thermal cycling—Average 1500 cycles per year for at least 10 years, more cycles for frequent short trip driving or hybrid vehicles
- Long life
  - Minimum 5000 operating hours
  - Minimum design life 10 years or 150,000 miles
  - Target 20 year life and 200,000 miles
- Exposure to a wide variety of fluids (water, coolant, exhaust gases, oil, etc.) either internally or externally. Example, hot units splashed with cold salt water during winter driving

Cars and trucks on the road today show that while this environment is very challenging, it is survivable. A successful TEG design needs the involvement of suppliers with experience designing and producing quality automotive components.

25.7 Material Requirements

The performance of a thermoelectric material is determined by the dimensionless thermoelectric figure of merit[5]

$$ZT = \frac{S^2}{\rho\kappa}T, \tag{25.1}$$

where S is the Seebeck coefficient, ρ the electrical resistance, and κ the thermal conductivity which includes a lattice (κ_L) and electronic (κ_e) components. The efficiency of a TEG can be estimated by[5]

$$\varepsilon = \frac{T_H - T_C}{T_H} \frac{\sqrt{1 + ZT} - 1}{\sqrt{1 + ZT} + T_C/T_H}, \tag{25.2}$$

where T_H and T_C are hot-side and cold-side temperatures of thermoelectric materials, respectively. The maximum coefficient of performance COP_{max} (the heat absorption rate vs. the input electrical power) of a thermoelectric cooler is approximately given by[5]

$$COP_{max} = \frac{T_C}{T_H - T_C} \frac{\sqrt{1 + ZT} - T_H/T_C}{\sqrt{1 + ZT} + 1} \tag{25.3}$$

It is therefore evident that higher material ZT values would lead to higher TEG efficiency and higher COP. A thermoelectric cooler with $ZT > 2$ materials could have higher COP values than a mechanical compressor-based air conditioning unit.[6] In the case of TEG, the efficiency improvement is much more significant between $ZT = 1$ and 2 than between $ZT = 2$ and 3.[6] The entire thermoelectric cooling industry currently uses $ZT \approx 1$ materials; however, materials with $ZT \geq 2$ would certainly facilitate bringing many potential thermoelectric applications, including automotive applications, into practical reality. In the past decade, we have witnessed significant ZT increases in various materials,[6,7] some of which will be discussed in a later section.

Higher ZT values in the operational temperature range; however, are not the only concerns. The materials need to be chemically stable and mechanically robust at elevated temperatures or with large temperature gradients. For example, the materials used for radioisotope TEGs in space applications often experience hot-side sublimation,[6] which could lead to substantial cross-sectional area decrease, electrical power output reduction, and mechanical failure. For almost all known high ZT materials, their ZT vs. T curves peak slightly lower than the material degradation (melting, sublimation, etc.) temperatures. We are hence left with the dilemma that from the performance point of view one would like to optimize the materials so that their highest ZT values fall in the temperature range of the applications. On the other hand, the materials highest ZT values usually occur near the degradation temperatures. Over the years many techniques, such as inert gas environment and various coatings, have been developed to mitigate materials sublimation over the service period.[6,8] In the case of automotive exhaust waste heat recovery, the average exhaust temperature for various vehicles is between 500°C and 600°C, with the maximum values up to 1000°C under certain operating conditions. Therefore, proper measures on the device and subsystem levels need to be in place to protect materials from deterioration.

Another potential issue is the mechanical strength of thermoelectric materials. In the automotive thermoelectric applications, thermal gradient, thermal cycling, and vibration will impose stress on the thermoelectric materials. In addition, most thermoelectric materials are brittle semiconductors; therefore, fracture strength and fracture toughness are critical for determining mechanical failure mechanisms.[9] In addition, elastic properties such as the Young's modulus, shear modulus, Poisson's ratio, and thermal expansion coefficient are needed for TE module design and optimization. There has been very limited data published on mechanical properties of thermoelectric materials, let alone their temperature dependence.[9–14] Successful thermoelectric technology development will require mechanical property characterization and understanding, and materials synthesis processes to improve materials mechanical integrity, such as grain size, flaw size, and edge flaw controls.[9,13]

Furthermore, thermoelectric modules are made of both p- and n-type thermoelectric materials. It is ideal to have the same type of materials on the n- and p-legs with similar ZT curves. This will allow maximum module performance, reduce stress induced by the unbalanced thermal expansion coefficients of the two legs, and simplify module manufacturing process by using the same diffusion barriers and electrical contact materials.

25.8 Device Considerations

The choice of interconnect materials at the hot side and the cold side are critical for device fabrication. First, the interconnect materials and the thermoelectric materials should have comparable thermal expansion coefficients to minimize stress at the joints. Second, the interconnect materials should have high electrical and thermal conductivity values at operating temperatures to minimize electrical and thermal contact resistance. Third, the interconnect materials should not be easily oxidized. Fourth, the interconnect materials and the thermoelectric materials should not react chemically. Finally, the diffusion layer between the two should be a few micrometers thick to ensure good contact and should be stable at the hot-side temperature over time. Details of interconnect materials development for thermoelectric coolers and generators has been presented elsewhere.[15–19]

25.9 Recent Materials Development

Many recent materials research advances have invigorated worldwide interest not only in materials research but also in technology development, including the automotive applications. Figure 25.4 shows

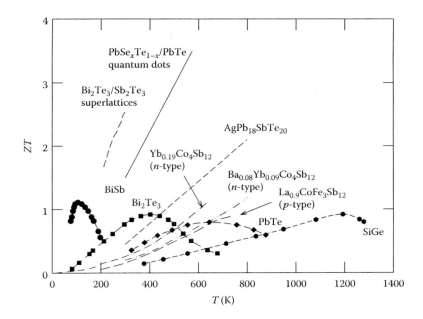

FIGURE 25.4 *ZT* vs. *T* for state-of-the practice (symbols with lines) and state-of-the-art materials (lines only).

the *ZT* vs. *T* curves for many state-of-the-practice and state-of-the-art materials. All state-of-the-practice materials have maximum *ZT* values ≤1. Almost all $ZT_{max} > 1$ materials were developed in the last 15 years. Many recent reviews have been devoted to new thermoelectric materials progress.[7,20] Here we only focus on a few of the latest ones, and their relevance to automotive applications.

25.9.1 Nano-Structured Materials

In 1993, Hicks and Dresselhaus proposed that low dimensionality in materials could result in enhanced electronic density of states near Fermi energy. This could lead to larger Seebeck coefficients, and in nano-structured materials boundary scattering can affect phonons more than electrons.[21] Subsequent works have shown significant ZT improvement in Bi_2Te_3/Sb_2Te_3 superlattices, $PbSe_xTe_{1-x}/PbTe$ quantum dot superlattices, and two-dimensional electron gas in $SrTiO_3$.[22–25] Of particular interest are high efficiency bulk $(PbTe)_{1-x}(AgSbTe_2)_x$ ($x \sim 0.05$) nanocomposites that have achieved very high $ZT \sim 2.1$ at 800 K (shown in Figure 25.4),[26] since nanostructured bulk materials have the advantage of being scalable for practical energy conversion. These high *ZT* values were attributed to nanoprecipitates observed in the materials. This suggests that it may be possible to achieve high *ZT* values of superlattices and quantum dots in bulk materials, if the results can be independently reproduced. Though Seebeck coefficient enhancement was postulated,[21] most of the *ZT* gains in recent materials come from the thermal conductivity reduction. This is certainly the case for recent work on Si nanowires and nanograin Bi_2Te_3 alloys.[27–29]

25.9.2 Bulk Semiconductors

Filled skutterudites, semiconducting clathrates, and complex chalcogenides are amongst a few prospective bulk materials for advanced applications.[30–32] In particular, filled skutterudites have the advantage of high *ZT* values in the temperature range of automotive exhaust heat (Figure 25.4), availability of both the *n*- and *p*-type, and mechanical robustness. Based on the idea of multiple frequency phonon resonant scattering,[33] significant ZT enhancement has been achieved for double- and triple-filled skutterudites.[34,35]

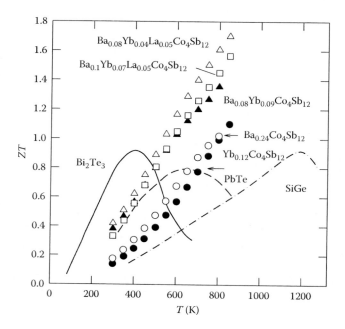

FIGURE 25.5 Temperature dependence of the *ZT* for state-of-the practice (lines), and single-, double-, and triple-element-filled skutterudites (symbols). (Adapted from X. Shi et al., *Appl. Phys. Lett.* 92, 182101, 2008; X. Shi et al., *J. Am. Chem. Soc.*, 133, 7837, 2011.)

Figure 25.5 shows the *ZT* values in skutterudites are significantly improved from single-, to double-, and finally triple-element-filled skutterudites over the entire temperature range investigated.

25.9.3 Complex Oxides

Several layered hexagonal cobalt oxides discovered by various Japanese groups have shown very good thermoelectric performance at elevated temperatures.[36–38] A common feature of these materials is the presence of hexagonal CoO_2 layers. The high Seebeck coefficients observed are attributed to spin entropy contributions. Oxide materials are very attractive for practical applications because of their stability at elevated temperatures, low-cost raw materials, and relatively simply synthesis methods. Interestingly, a 140-couple oxide TEG module was successfully constructed which demonstrated reasonable power output.[39] The development of high-performance *n*-type oxides, however, remains challenging. If successful, an oxide-based TEG could be very attractive for automotive applications.

25.10 Challenges for TEG Application

- The demonstrated long-term durability of thermoelectric power generators is very impressive with 20–30 years of continuous operation in spacecraft and remote monitor applications. These applications, however, have been continuous operation with little or no thermal cycling and no vibration. A major challenge to automotive power generation applications will be the thermal cycling and severe mechanical vibration that could degrade the thermoelectric material, modules, and mechanical system. Repeated cycles from −40°C (or even 25°C) to 400°C in less than 2 min will stress thermoelectric modules in ways that have not been tested before. Significant testing will be required to validate the durability of thermoelectric modules in automotive applications.

- Cost and efficiency are competing issues. Higher *ZT* materials are needed to get more electrical power but they typically cost more. The challenge for researchers is to find materials that balance cost against efficiency.
- Matching the temperature of the heat source to the proper thermoelectric module type will be difficult. Part of the problem is that the vehicle source temperature varies unpredictably under normal driver operation. Additionally, the temperature across a heat exchanger will drop between input and output. Another challenge is the need to have an efficient material (high *ZT*) in the nominal temperature range while avoiding module damage at peak temperatures.
- As new technologies are added to engines to improve fuel efficiency, the amount of waste heat will be reduced and exhaust temperatures will be lowered. This will require more efficiency from TEGs to make them cost effective. TEG use on hybrid electric vehicles will face additional challenges of intermittent engine operation under various driving cycles and lower total waste heat relative to a conventional vehicle.
- Efficient operation of the TEG heat exchangers over the 10–30 year life of a vehicle is a concern. The effect of material buildup on the heat exchanger surfaces from exhaust gas, coolant, or air is basically unknown and needs development to insure that it does not excessively degrade system operation. Chemical or mechanical degradation of the heat exchanger surfaces by the exhaust gases is also a concern.
- Automotive manufacturers need the availability of economical thermoelectric modules in sufficient quantities to support automotive production volume if HVAC and TEG applications grow rapidly. Even a relatively small 10% penetration of the world's automotive production (>5 million vehicles) by thermoelectric technology would require an estimated 4×10^9 W of electric power generating capacity (est. 800 W per vehicle) each year by thermoelectric modules. Twice that capacity in modules may be needed to provide cooling and heating.
- There are competing technologies that could be cost competitive for TEGs if economical, efficient thermoelectric materials cannot be developed for production quickly. These technologies include generators driven by small Rankine cycle engines, Stirling engines, or turbo-chargers, all operating from exhaust heat or flow. Also conventional engine and transmission technologies will compete as cost efficient ways to improve fuel economy in cars and trucks.

25.11 Conclusion

Automotive applications for heating and cooling have started with currently available thermoelectric modules. The low efficiency of these units is offset by the unique features they provide. As advanced thermoelectric materials become available with volume production quantities and economics, the improved efficiency has the potential to rapidly expand automotive applications. For both HVAC and electric power generation applications, environmental issues will play a significant part in selection of thermoelectric technology, but the real measure will be cost per watt in high volume at the material, module, and vehicle system levels.

Acknowledgments

The work is supported by GM and by DOE under corporate agreement DE-FC26-04NT42278.

References

1. J. Yang, K. B. Ledbetter, and F. R. Stabler, Thermoelectric methods to control temperature of batteries, US Patent, US2006028182A1, Feb. 9, 2006.
2. J. Yang, M. Cai, and F. R. Stabler, Thermoelectric catalytic converter temperature control, US Patent, US7,051,522 B2, May 30, 2006.

3. S. Pagerit, A. Rousseau, and P. Sharer, Fuel economy sensitivity to vehicle mass for advanced vehicle powertrains, SAE paper 2006-01-0665, SAE World Congress, Detroit, April 2006.

4. J. Yang, M. E. Matouka, and F. R. Stabler, Auxiliary electrical power generation, US Patent, US7,493,766, B2, Feb. 24, 2009.

5. A. F. Ioffe, *Semiconductor Thermoelements and Thermoelectric Cooling*, (Infosearch Limited, London, 1957).

6. J. Yang and C. Caillat, Thermoelectric materials for space and automotive power generation, *MRS Bull.* 31, 224, 2006, and references therein.

7. T. M. Tritt and M. A. Subramanian, editors, Harvesting energy through thermoelectrics: Power generation and cooling, *MRS Bull.* 31, 188–229, 2006.

8. M. S. El Genk, H. H. Saber, T. Caillat, and J. Sakamoto, Test results and performance comparisons of coated and UN-coated skutterudite-based segmented unicouples, *J. Energy Cons. Manage.* 47, 174, 2006.

9. R. W. Rice, *Mechanical Properties of Ceramics and Composites* (Marcel Dekker, New York, 2000).

10. Y. Gelbstein, G. Gotesman, Y. Lishzinker, Z. Dashevsky, and M. P. Dariel, Mechanical properties of PbTe-based thermoelectric semiconductors, *Scripta Mater.* 58, 251, 2008.

11. W. Brostow, K. P. Menard, and J. B. White, Thermomechanical characterization of bismuth telluride based thermoelectric materials, *Mat. Res. Soc. Symp. Proc.* 691, G13.3, 2002.

12. F. Ren, E. D. Case, E. J. Timm, M. D. Jacobs, and H. J. Schock, Weibull analysis of biaxial fracture strength of cast p-type LAST-T thermoelectric material, *Philos. Mag. Lett.* 86, 673, 2006.

13. J. R. Salvador, J. Yang, X. Shi, H. Wang, and A.A. Wereszczak, Transport and mechanical property evaluation of $(AgSbTe)_{1-x}(GeTe)_x$ (x = 0.80, 0.82, 0.85, 0.87, 0.90), *J. Solid State Chem.* 182, 2088, 2009.

14. A. L. Pilchak, F. Ren, E. D. Case, E. J. Timm, H. J. Schock, C. –I., Wu, and T. P. Hogan, Characterization of dry milled powders of LAST (lead-antimony-silver-tellurium) thermoelectric material, *Philos. Mag.* 87, 4567, 2007.

15. N. I. Erzin and N. V. Makov, *Appl. Sol. Energy* 1, 33, 1965.

16. H. J. Goldsmid, *Electronic Refrigeration*, (Pion Limited, London, 1986), p. 165.

17. K. Matsubara, The performance of a segmented thermoelectric converter using Yb-based filled skutterudites and Bi_2Te_3-based materials, *Mat. Res. Soc. Symp. Proc.* 691, 327, 2002.

18. L. Chen, J. Fan, S. Bai, and J. Yang, $CoSb_3$-based thermoelectric device fabrication method, US Patent, US7,321,157, B2, 2008.

19. D. T. Morelli, Thermoelectric devices, in *Encyclopedia of Applied Physics*, Vol. 21 (VCH, New York, 1997), pp. 339–354.

20. T. M. Tritt, editor, *Semiconductors and Semimetals*, Vol. 69–71 (Academic Press, San Diego, 2001).

21. D. Hicks and M. S. Dresselhaus, Effect of quantum-well structures on the thermoelectric figure of merit, *Phys. Rev. B* 47, 12727, 1993.

22. R. Venkatasubramanian, E. Siivola, T. Colpitts, and B. O'Quinn, Thin-film thermoelectric devices with high room-temperature figures of merit, *Nature* 413, 597, 2001.

23. T. C. Harman, P. Taylor, M. P. Walsh, and B. E. Laforge, Quantum dot superlattice thermoelectric materials and devices, *Science* 297, 2229, 2002.

24. T. C. Harman, M. P. Walsh, B. E. Laforge, and W. W. Turner, Nanostructured thermoelectric materials, *J. Electron. Mater.* 34, L19, 2005.

25. H. Ohta, S. W. Kim, Y. Mune, T. Mizoguchi, K. Nomura, S. Ohta, T. Nomura et al., Giant thermoelectric Seebeck coefficient of two-dimensional electron gas in $SrTiO_3$, *Nat. Mater.* 6, 129, 2007.

26. K. F. Hsu, S. Loo, F. Guo, W. Chen, J. S. Dyck, C. Uher, T. Hogan, E. K. Polychroniadis, and M. G. Kanatzidis, Cubic $AgPb_mSbTe_{2+m}$: Bulk thermoelectric materials with high figure of merit, *Science* 303, 818, 2004.

27. A. I. Hochbaum, R. Chen, R. D. Delgado, W. Liang, E. C. Garnett, M. Najarian, A. Majumdar, and P. Yang, Enhanced thermoelectric performance of rough silicon nanowires, *Nature* 451, 163, 2008.

28. A. I. Boukia, Y. Bunimovich, J. Tahir-Kheli, J.-K. Yu, W. A. Goddard III, and J. R. Heath, Silicon nanowires as efficient thermoelectric materials, *Nature* 451, 168, 2008.
29. B. Poudel, Q. Hao, Y. Ma, Y. Lan, A. Minnich, B. Yu, X. Yan et al., High-thermoelectric performance of nanostructured bismuth antimony telluride bulk alloys, *Science* 320, 634, 2008.
30. C. Uher, Skutterudites: Prospective novel thermoelectrics, in *Semiconductors and Semimetals*, Vol. 69, edited by T. M. Tritt (Academic, San Diego, 2001), p. 139, and references therein.
31. G. S. Nolas et al., Semiconductor clathrates: A phonon glass electron crystal material with potentials for thermoelectric applications, in *Semiconductors and Semimetals*, Vol. 69, edited by T. M. Tritt (Academic, San Diego, 2001), p. 255, and references therein.
32. D. Y. Chung, T. Hogan, P. Brazis, M. Rocci-lane, C. R. Kannewurf, M. Bastea, C. Uher, and M. G. Kanatzidis, $CsBi_4Te_6$: A high-performance thermoelectric material for low-temperature applications, *Science* 287, 1024, 2000.
33. J. Yang, W. Zhang, S. Q. Bai, Z. Mei, and L. D. Chen, Dual-frequency resonant phonon scattering in $Ba_xR_yCo_4Sb_{12}$ (R = La, Ce, and Sr), *Appl. Phys. Lett.* 90, 192111, 2007.
34. X. Shi, H. Kong, C. Uher, J. R. Salvador, J. Yang, and H. Wang, Low thermal conductivity and high thermoelectric figure of merit in n-type $Ba_xYb_yCo_4Sb_{12}$ double-filled skutterudites, *Appl. Phys. Lett.* 92, 182101, 2008.
35. X. Shi, Jiong Yang, J. R. Salvador, M. Chi, J. Cho, H. Wang, S. Bai, J. Yang, W. Zhang, and L. Chen, Multiple-filled skutterudites: High thermoelectric figure of merit through separately optimizing electrical and thermal transports, *J. Am. Chem. Soc.* 133, 7837, 2011.
36. I. Terasaki, Y. Sasago, and K. Uchinokura, Large thermoelectric power in $NaCo_2O_4$ single crystals, *Phys. Rev. B* 56, R12685, 1997.
37. K. Fujita, T. Mochida, and K. Nakamura, High-temperature thermoelectric properties of $Na_xCoO_{2-\delta}$ single crystals, *Jpn. J. Appl. Phys.* 40, 4644, 2001.
38. R. funahashi, I. Matsubara, H. Ikuta, T. Takeuchi, U. Mizutani, and S. Sodeoka, An oxide single crystal with high thermoelectric performance in air, *Jpn. J. Appl. Phys.* Pt. 2 39, L1127, 2000.
39. K. Koumoto, I. Terasaki, and R. Funahashi, Complex oxide materials for potential thermoelectric applications, *MRS Bull.* 31, 206, 2006, and references therein.

26

Medical Applications of Thermoelectrics

Alic Chen
University of California, Berkeley

Paul K. Wright
University of California, Berkeley

26.1 Introduction

Recent advances in thermoelectric technologies are beginning to meet some of the needs in the growing number of advanced medical devices. While commercial thermoelectric technologies have been available since the 1950s, thermoelectric devices have historically found limited use in biomedical applications. Early uses took advantage of the solid-state heating and cooling effects for niche applications such as DNA thermal cyclers, medicine cooling bags and medical imaging devices. These are often considered premium applications where the advantages of rapid solid-state heating or cooling outweigh the costs and inefficiencies associated with the state-of-the-art thermoelectric unit. Recent advances in both thermoelectric research and biomedical engineering, however, have drawn renewed interest from the medical community, particularly in their energy-harvesting uses. The burgeoning portable electronics industry with its omnipresent goal for low-power consumption and high performance has spawned a new wave of portable and implantable biomedical devices. Consequently, the timing is apt for the convergence of novel thermoelectric technologies with current and future medical devices.

In this chapter, thermoelectric heating and cooling in medicine will first be reviewed, followed by an analysis of thermoelectric energy harvesting for biomedical applications. Finally, considerations for thermoelectric uses in medicine will be discussed.

26.2 Solid-State Heating and Cooling Applications

Thermoelectric heating and cooling utilize the Peltier effect to act as a solid-state heat pump. Also called Peltier devices, thermoelectric heat pumps transfer heat from one side to the other when direct current is applied. State-of-the-art Peltier devices generally have efficiencies around 5–10% of an ideal refrigerator (Carnot cycle). Compared to a conventional compression cycle system with efficiencies around 40–60%, Peltier devices are used when their solid-state nature provides significant advantages. They provide a compact form of reversible and rapid heating or cooling with no moving parts [1]. These unique attributes open their uses to some specific biomedical applications.

Today, the most ubiquitous biomedical uses of thermoelectric devices occur in modern polymerase chain reaction (PCR) thermal cycles for rapid heating and cooling of DNA. Developed in 1983 by Kary Mullis, who subsequently won the Nobel Prize in Chemistry in 1993 for his work, PCR has become a ubiquitous and indispensible method used in medical and biological laboratories for DNA amplification [2]. The process of replicating DNA molecules using PCR requires thermally treating the DNA to three separate set points: (1) denaturation at 94°C, (2) annealing at 54°C, and (3) extension at 72°C. These steps are then repeated multiple times with each cycle doubling the amount of DNA [3]. This process naturally lends itself to using solid-state thermoelectric heater/coolers to speed up the thermal cycling time needed for these reactions. The reversibility, fast response, and ease of deployment of Peltier devices make them ideal for PCR equipment [4–6]. Thermoelectric manufacturers such as Marlow Industries and Nextreme have successfully commercialized thermoelectric devices in benchtop PCR systems using standard Bi_2Te_3-based Peltier devices [7,8]. Although PCR is a critical application of current thermoelectric devices in the biomedical industry, there have been few alternative uses of thermoelectric devices outside of PCR. It is thus of interest to explore various viable applications for thermoelectric devices.

A potential avenue for wider use of solid-state heating/cooling thermoelectric devices may lie in therapeutic medical applications. Amerigon, one of the more prominent thermoelectric device manufacturers, and its subsidiary, BSST, has explored the use of thermoelectric devices for thermoregulation of cancer patients. Cancer patients undergoing chemotherapy are susceptible to low white blood cell count (neutropenia) [9], infection [10], and low red blood cell count (anemia) [11]. These side effects can lead to temperature sensitivity, leaving the patient feeling cold [12]. Thus, therapeutic products such as temperature controllable blankets or couches using thermoelectrics may potentially help patients cope with the side effects. Amerigon and Mattress, Inc. recently co-developed a thermoelectric heating and cooling mattress called the YuMe Climate Control Bed, albeit for the luxury mattress market [13]. However, such products can potentially be adapted for therapeutic medical applications. Therapeutic cooling and heating of tissue injuries is also an effective treatment and has been known to reduce the healing time [14]. However, such treatments are traditionally performed with ice or heat packs [15,16] that are significantly cheaper than thermoelectric devices. There may be opportunities for solid-state heating or cooling in professional sports therapies. One can perhaps envision a wearable thermoelectric heater/cooler for athletes who require constant and immediate therapy for low-grade tissue injuries, allowing them to return to their activities. Another potential therapeutic application utilizing thermoelectric technology is in therapeutic hypothermia. Therapeutic hypothermia

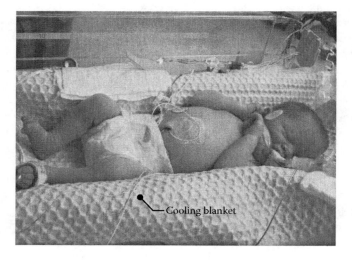

FIGURE 26.1   An infant being treated for hypoxic-ischemic encephalopathy (lack of oxygen at birth) with a state-of-the-art cooling blanket. Cooling blankets allow for treatment of newborns that show signs of brain damage and are at risk of developing various cognitive disorders. (Courtesy of Melissa P.)

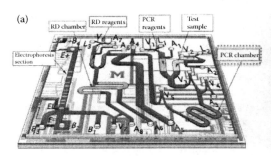

FIGURE 26.2 Examples of full-integrated lab-on-chip devices: (a) a microfluidic device for influenza and other genetics analyses (R. Pal et al., An integrated microfluidic device for influenza and other genetic analyses, *Lab on a Chip*, 5(10), 1024–1032, 2005. Reproduced by permission of The Royal Society of Chemistry.) and (b) an integrated microfluidic chip for chromosome enumeration using fluorescence *in situ* hybridization (V. J. Sieben et al., An integrated microfluidic chip for chromosome enumeration using fluorescence *in situ* hybridization, *Lab on a Chip*, 8(12), pp. 2151–2156, 2008. Reproduced by permission of The Royal Society of Chemistry.). The thermal reactions within the chips (highlighted) can potentially utilize small-scale thermoelectric heater/coolers.

has been shown to be particularly effective for treatment of neonatal encephalopathy (brain disorders) (Figure 26.1) [17,18] and patient neuroprotection following cardiac arrest [19,20]. The current method of treatment utilizes water blankets, vests, or wraps. While this method dates back to the 1950s, it is still the most prevalent method for noninvasive patient cooling. Nonetheless, water blankets possess certain drawbacks, such as electrical hazards from fluid leakage [21], which might open opportunities for solid-state cooling. Several cases have been reported of cooling blankets causing significant burns to patients [22]. Precise temperature control and response is also frequently another issue with water blankets, resulting in temperature overshoot and delayed compensation. Thus, therapeutic heating and cooling applications may perhaps find potential adoption of thermoelectric technologies if certain traits such as flexibility/conformability, improved coefficient of performance (COP) and lower cost are met by future technologies.

Another potentially promising application of thermoelectric cooling and heating is in their use in portable biomedical systems. With the future of health care focused on portability and on-site care, the field of Biological Micro-electromechanical systems (BioMEMS) has rapidly grown during the last decade [23]. Advancements in micro-engineering adapted from the semiconductor industry has opened the possibility of scaling laboratory-based systems such as PCR, electrophoresis, single molecular detection and disease diagnosis among many others [24–26]. By utilizing MEMS techniques to make portable laboratory devices, commonly called "lab-on-a-chip" or micro-total analytical system (µTAS), such systems can be fabricated using low-cost and scalable methods while providing device portability and consuming less reagents. The total market size of microarrays and lab-on-a-chip systems was approximately $2.6 billion in 2009 and is expected to annually grow by 17.7%, reaching $5.9 billion in 2014 [27]. Some of these lab-on-a-chip processes such as micro-PCR, a "scaled-down" version of traditional PCR, also require rapid thermal cycling. The portable nature of such devices would require rapidly controllable heating and cooling techniques that can potentially be provided from MEMS-scale thermoelectric devices. While the applications of thermoelectrics to lab-on-a-chip devices are limited to a few specific processes, the expected growth in the field might make it an attractive market for MEMS and micro-scale thermoelectric heating and cooling (Figure 26.2).

26.3 Thermoelectric Energy Harvesting for Biomedical Devices

The history of medical diagnostics and treatments can be traced back millenniums to the ancient Egyptians and Greeks. With extraordinary advances in research and enabling technologies, modern medicine has come a long way from the clinical observations of Hippocrates. As diagnostic tools continuously evolve and improve, they trend toward miniaturization and mobility. Continuous patient physiological monitoring can now be performed outside the clinical environment, providing physicians with more thorough information. Life-supporting medical devices have transformed from bulky and invasive machines to portable implantable devices, freeing patients from the direness of permanent hospitalization. Mobility inevitably requires portable energy solutions, a role currently filled by energy storage technologies such as batteries. As it becomes clear that progress in energy research does not follow Moore's law, a multitude of energy-harvesting approaches may eventually allow it to keep up.

Approximately 80% of the metabolic energy in the human body is lost to low-grade heat for thermoregulation [30]. In order to maintain the core body temperature of approximately 37°C, our bodies constantly generate heat while simultaneously taking active steps dissipating the heat to prevent overheating (such as fever) [31–33]. The heat is eventually lost through conduction, convection, radiation, and evaporation, making it the most abundant source of energy from the body [34]. The human body dissipates approximately 100 W of power at rest from thermoregulation [35,36]. While the heat emitted from the body is readily available, the quality of the heat is too low for any conventional heat engines to harvest [35]. Thermoelectric devices, acting as solid-state power generators from temperature differences (known as the Seebeck effect) [1], may be suitable for harvesting the low-quality heat emitted from the body. This recovered energy can potentially provide power for a new wave of diagnostic and medical tools.

Thermoelectric energy harvesting in the biomedical realm can be divided into two sets of applications: (1) wearable and (2) implantable applications. This section will discuss developments and ongoing research in both sets of applications while focusing on the aspects of thermoelectric device design and integration.

26.3.1 Wearable Applications

The increasing demand for low-cost and personalized wireless physiological diagnostic tools has sprouted growing research efforts in wireless body sensor networks (BSNs) and mobile health (mHealth). Figure 26.3 illustrates an example of a wireless physiological monitoring system. These applications can include long-term (24/7) monitoring of the local/regional events in tissue or organs under investigation and personalized home health care. These tools can be applied to monitoring of patients with chronic diseases, hospitalized patients or the elderly [37]. New generations of medical diagnostic "smart" probes often require high sampling rates resulting in high-energy consumption which has ultimately limited device lifetimes. Due to power constraints, there is often a trade-off between sensor resolution/sampling rate and device usability lifetime. Thermoelectric generators (TEGs) can provide a method to increase the energy storage capacity in BSNs by harnessing thermal gradients between the body and ambient environments. The power and voltage requirements of today's micro-electronic systems have significantly reduced to match the power output of TEGs at low-temperature differences (between 5 and 20 K). Studies have suggested that a constant power source exceeding 100 μW/cm^2 at 1 V is an ideal energy harvester for practical wearable sensor networks [38–43]. Some state-of-the-art ultra-low power radios have reported power consumptions <10 μW with transmit/receive ranges of up to 10 m [38,44]. Advancements in TE materials and device fabrication technologies have only recently been able to meet some of the power and voltage requirements of the radios and sensors within a constrained device footprint. It is thus important to understand the remote physiological monitoring systems and their applications to provide insight on the design and feasibility of wearable TEGs.

26.3.1.1 Wearable Biomedical Sensors

Wearable biomedical sensors typically require wide deployment on various parts of a patient's body with electrical leads from each sensor. As a result, entanglement of electrical wiring becomes an issue while the patient's mobility is limited to the connected instrument. BSN's proverbially allow such sensors to "cut the cord." Table 26.1 shows a list of various common physiological sensors and their clinical applications. These sensors are crucial for monitoring patient vital signs both within and outside the

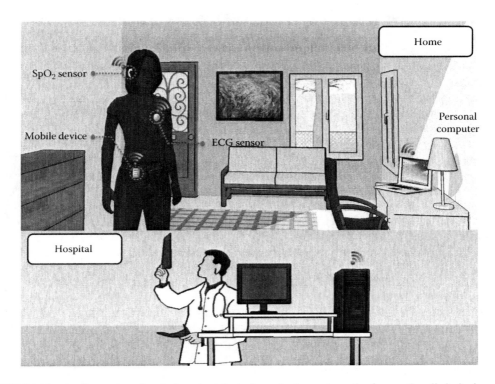

FIGURE 26.3 An illustration of a wireless physiological monitoring system, also frequently called a body sensor network (BSN). (Courtesy of Dan Chapman.)

TABLE 26.1 List of Common Wearable Sensors and Some of Their Clinical Applications

Sensor	Measurement Function	Clinical Application
Electroencephalography (EEG)	Electrical activity of the scalp	Epilepsy diagnosis, sleep disorder studies, loss of consciousness, dementia
Electrocardiography (ECG)	Electrical activity of the heart	Heart-related diseases such as arrhythmias, coronary artery disease, and tachycardia
Electromyography (EMG)	Electrical activity of the muscle	Detection of neuromuscular diseases, kinesiology, and motor control disorders
Pulse oximeters	Oxygen saturation in the blood	Patient monitoring during intensive care, surgical operation, postoperative recovery, and emergency room
Ambulatory blood pressure monitor	Continuous blood pressure monitoring	Detection of hypertension (high blood pressure) and hypotension (low blood pressure)
Thermistor or thermocouple	Skin temperature	Neonatal temperature monitoring; diabetic foot ulceration [Armstrong], Infectious diseases, fever
Accelerometer	Patient movement and orientation	Impact and fall detection in elderly patients

clinical environment. Although these sensors have a diverse range of clinical applications extending far beyond the scope of this chapter, only the two most salient types of sensors will be discussed: biopotential sensors and pulse oximeters.

Sensors that benefit most from BSNs are the family of biopotential sensors that include electroencephalography (EEG) for measuring neural activity at the scalp, electrocardiography (ECG or EKG) for measuring cardiac activity and electromyography (EMG) for recording electrical activity of the muscles [45]. These types of sensors are particularly important to remote physiological monitoring of patients at home, first responders, and military personnel [37,46–49]. Although these biopotential sensors have different diagnostic purposes, they share similar principles of operation to measure electric potentials on the surface of living tissue. An electrochemical transducer is used to measure the ionic current flow in the body to detect muscle contractions and nervous stimuli. Since the current flow in the body is due to ion flow, the transducer (or electrode) must convert the ionic current to electrical current for measurement. This is accomplished by first placing an electrolyte "jelly" solution (typically containing Cl^- ions) on the surface of the tissue, followed by the transducer (typically an Ag–AgCl electrode). The noncontacting side of the electrode is connected to an electrical wire [45]. The chemical reactions that occur at the interface between electrode and the electrolyte result in fairly low electrical signals (10–100 μV) measurable from an instrument. While signal amplification and noise reduction are required due to the low signals from the biopotential sensors, state-of-the-art electronics have reduced the power requirements to <60 μW [43].

Beyond biopotential sensors, pulse oximeters are another crucial measurement device for monitoring patient vital signs. Pulse oximeters noninvasively monitor patient pulse and blood oxygen saturation (SpO_2) levels by measuring the ratio of visible red light to infrared light absorption of pulsating components at the fingertip. Unlike biopotential electrodes, pulse oximeters require additional power for the LEDs. While commercial pulse oximeters consume 20–60 mW of power, research has shown that lower sampling rates with novel algorithms can lower the power consumption by 10–40 × without loss of accuracy [50] (Figure 26.4).

26.3.1.2 Remote Physiological Monitoring Systems

While there are currently very few commercially available remote physiological monitoring systems, some have already entered the market. The LifeShirt by VivoMetrics was among the first commercially available fully integrated wearable physiological sensor system for remote patient monitoring. The LifeShirt (Figure 26.5) is a wearable vest with ECG sensors, respiratory sensors, accelerometers, pulse oximeters, galvanic skin response measurement sensors, blood pressure monitor, microphone, and an

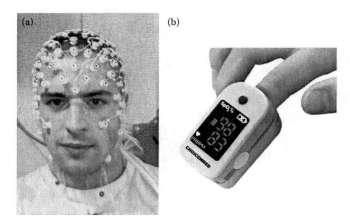

FIGURE 26.4 Image of (a) a patient outfitted with an EEG cap with each individual electrode is wired to an instrument for data collection, and (b) a digital pulse oximeter for measuring patient pulse and blood oxygen saturation.

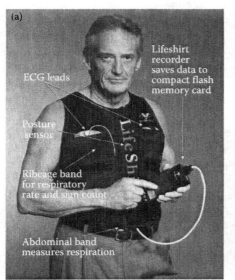 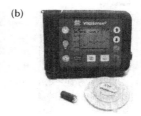 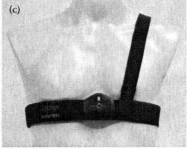

FIGURE 26.5 (a) Image of the Vivometrics Lifeshirt (Courtesy of The Virtual Worldlets Network.). The system is a garment that continuously collects patient vital signs for remote physiological monitoring (Adapted from K. J. Heilman and S. W. Porges, *Biological Psychology*, 75(3), 300–305, 2007.). (b) Image of the Philips Respironics VitalSense which consists of a portable data logger and monitor (top), self-adhesive temperature patches (bottom right) and an ingestible thermometer (bottom left) (Courtesy of Philips Respironics.) (Philips-Respironics, VitalSense Integrated Physiological Monitor, http://vitalsense.respironics.com/). (c) Image of the Zephyr Bioharness which straps to the body and monitors the user's heart rate and breathing while wirelessly relaying the data via Bluetooth (Zephyr Technology Corporation, BioHarness BT, http://www.zephyr-technology.com/bioharness-bt). (Courtesy of Lyle Reilly.)

electronic diary all integrated into Bluetooth capable system [51]. Philips-Respironics has also developed a wirelessly integrated physiological monitoring platform called VitalSense. The modular system allows for a variety of wireless and wearable sensors to transmit data to a portable data logger and monitor. It currently supports the measurement of a core body temperature, heart rate, respiration rate, and skin temperature [52,53]. A variety of other systems that are currently available include the HealthVest (SmartLife Technologies) [46,54], Equivital (Bio-Lynx Scientific Equipment, Inc.) [55], and Bioharness (Zephyr) [56,57]. While all systems utilize similar sensors for remote physiological monitoring, the wireless technology may perhaps vary based on the achievable power reduction of the radios. There is ultimately a trade-off between battery life (<10 days) and sampling rate due to power consumption of the radio and processor. Despite the commercial availability of BSN's and remote physiological systems, issues related to system integration, sensor miniaturization, low-power circuitry design, wireless communication protocols and signal processing are currently being investigated [37]. Energy harvesting from wearable TEGs can enable future BSN technologies by providing a constant power source.

26.3.1.3 Design of Wearable TEGs

A wearable BSN system typically consists of the biomedical sensors, a signal amplifier, a microcontroller for processing data, and a power source such as a battery and a wireless radio to transmit the data to a computer or mobile device [37,39,48,58]. The total power consumption is ultimately a function of the duty cycle (data acquisition and transmission frequency) and the sensor resolution. Since most of the wearable sensors listed in Table 26.1 function by providing low-voltage signals (100s of µV), the power consumption of the sensor is quite low and is only limited to the signal amplification. The radio's average power consumption ultimately is the predominant power draw and becomes the limiting factor for the lifetime of the wireless sensor [40,44]. While a higher sampling rate may allow for finer data resolution, it can significantly shorten the lifetime of the sensor. Torfs et al. [43] demonstrated an autonomous wearable EEG with average power requirements of 0.6 to 1.4 mW for sampling rates of 128 to 512 Hz, respectively. The choice of the sampling rate is thus more application dependent and system designers must frequently make trade offs between the battery capacity (weight), battery life, and sampling rate [58]. While significant research is currently ongoing to reduce the average power consumption of wireless radio to <10 μW [38,44], there is inevitably a lower limit. Since the power output of the TEG and the power consumption of the radio are subject to significant environmental variability, a TEG is not likely to replace a battery. It can instead be used to supplement the existing power source to extend the lifetime of the battery and sensor.

Since the available temperature difference between the human body and the ambient environment is quite low (0–20°C), it is crucial for the wearable TEG to be optimized to maximize power output within a small areal footprint (1–2 cm²). While a TEG is not likely to produce >1 mW of power without being significantly large, a target power output of 100 μW/cm² is ideal to sufficiently power most wireless applications. Another requirement for BSNs is the minimum 1 V output from the TEG for power electronics [59]. While state-of-the-art power electronics are capable of accepting input voltages as low as 20 mV, these converters have fairly low conversion efficiencies (<50%) [60]. These inefficiencies subsequently outweigh the costs associated with providing an already inefficient energy conversion system. In order to achieve a 1 V matched load output, the TEG would require high-density arrays of elements, achievable by recent advances in MEMS-based TEGs [61–64].

When designing TEG's for wearable applications, it is important to consider the thermal resistance matching of the TEG to the human body [40,65]. Proper matching of the high thermal resistance of the TEG to the high thermal resistance of the human body provides a maximum temperature difference across the generator. The thermal resistance model of a wearable TEG is shown in Figure 26.6. Leonov et al. demonstrated that by matching the thermal resistance of the TEG, R_{TEG}, with the ambient environment, it can be shown as

$$R_{\text{TEG}} = \frac{(R_{\text{body}} + R_{\text{sink}})R_{\text{air},0}}{2(R_{\text{body}} + R_{\text{sink}}) + R_{\text{air},0}} \tag{26.1}$$

where R_{body} is the thermal resistance of the human body at the TEG location, R_{sink} is the thermal resistance of the heat sink from convection and radiation, and $R_{\text{air},0}$ is the thermal resistance of the TEG if all the thermoelectric elements were removed and replaced with the surrounding insulator (usually air)

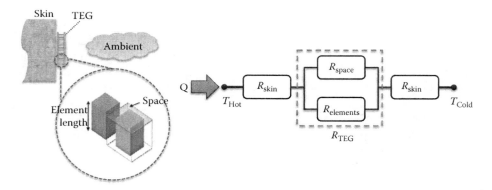

FIGURE 26.6 Schematic and thermal circuit of a wearable TEG. (Adapted from V. Leonov, *Journal of Electronic Materials*, 38(7), 1483–1490, Jan. 2009.)

TABLE 26.2 Thermal Resistance Measurements Measured at Various Body Locations

Body Location	Ambient Temperature °C	Thermal Resistance cm² K/W
Trunk	23	200–800
Outer wrist	22.7	440
Inner wrist (radial artery)	22.7	120–150
Forehead	21.5–24.7	156–380

Source: Adapted from V. Leonov, C. Van Hoof, and R. J. M. Vullers. *2009 Body Sensor Networks*, Berkeley, CA, 2009, pp. 195–200.

[65,66]. The thermal resistance of the generator can thus be designed based on the total resistance of the elements, $R_{elements}$, and the resistance of the empty space around the elements, R_{space}, such that $R_{TEG} = R_{elements}R_{space}/(R_{elements} + R_{space})$.

The body and skin thermal resistances can vary significantly depending on location and environment, and as a result, it is difficult to measure with accuracy. Measured values of thermal resistances can also vary based on the TEG and the heat sink used (such as fins), thus changing the local heat flow. It is also important for the design of the TEG to not significantly change the heat flow of the body to the point of discomfort. Table 26.2 shows some thermal resistance measurements at various locations and ambient conditions using a 3 cm × 3 cm × 3 cm TEG developed by Leonov et al. [65].

While heat from the human body is capable of providing power to a wearable TEG in the range of 300 µW/cm²–1 mW/cm², such conditions would result in noticeable discomfort and a cold sensation to the user. Thus, studies have suggested that a practical average power output for a wearable TEG using state-of-the-art TE materials with ZT = 1 is approximately 30 µW/cm² [59]. Advances in TE materials to improve the power factor can ultimately improve this figure.

Another strategy to improve the power density of a wearable TEG's is through placement on exothermic regions of the body. Although this may limit the sensor application of the TEG, specific regions of the body may provide higher temperature differences across the generator. There has recently been a growing interest in the research of brown adipose tissue (or brown fat) in adults. In contrast to white adipose tissue, the abundantly found fat in mammals, brown fat is mostly found in infants and hibernating mammals [67]. Since its primary function is to generate heat, it helps maintain warmth and provides heat regulation in newborns and animals. Brown fat was traditionally believed to disappear in adults, but recent studies using positron emission tomography have shown that it is still present in the upper chest and neck of adults [68]. While the implications of these studies are focused on the metabolic effects of brown fat, its exothermic nature suggests the availability of potential "hot-spots" on the body. Such "hot-spots" may be utilized for optimal placement of wearable TEGs, providing more heat flow and power output. Future studies can potentially help understanding the applicability of brown fat to wearable TEGs.

26.3.2 Implantable Applications

Perhaps the earliest investigation of implantable thermoelectric generators surfaced during the late 1960s when zinc–mercury batteries were still the standard power sources for implantable pacemakers. The low-energy densities of the zinc–mercury cells frequently limited the device life to <20 months and required patients to frequently undergo surgery to replace the cells [69,70]. To solve this problem, Medtronic, currently one of the largest implantable medical device manufacturers in the world, and Alcatel jointly designed a nuclear-powered pacemaker consisting of a Plutonium-238 (Pu-238) radioisotope and a thermoelectric generator. The Pu-238, which has a half-life of approximately 85 years, radiated the container walls to provide a constant heat source, while the thermoelectric generator converted the heat into electrical energy for the pacemaker [69–71]. Figure 26.7 shows an image of the radioisotope TEG and pacemaker manufactured by Alcatel and Medtronic [72]. In 1970, the first radioisotope TEG-powered

FIGURE 26.7   Image of an implantable pacemaker with a radioisotope TEG as the power source. (Adapted from V. Parsonnet, *Pacing and Clinical Electrophysiology*, 29(2), 195–200, Feb. 2006.)

pacemaker was implanted in a human. Even accounting for the degradation of the radioisotope and the TEG, the pacemaker still functions in patients after more than 35 years from its production [73,74].

Although the longevity of radioisotope TEGs proved it to be an excellent source of energy for pacemakers, the potential exposure to radiation and toxicity of plutonium was ostensibly a primary concern for physicians and patients. Plutonium is among one of the most toxic and fatal materials known to humans and can spontaneously burst in to flames when exposed to air. Extreme precaution was taken into the shielding design and engineering to prevent exposure of the plutonium and limit the amount of radiation to only 100 mrem per year [70]. The average annual background dose for Americans is approximately 360 mrem [75] while the standard occupational dose limits for one year is 5000 mrem [76]. In 155 cases of implanted radioisotope TEGs, the frequency of malignant tumors was deemed to be no different than the standard population [77].

During the mid-1970s, radioisotope TEG-powered pacemakers began to lose favor to lithium batteries which had calculated life-times of approximately 10 years. Physicians decided that it was more appropriate for patients to be updated with newer devices every 10 years instead of using devices with older technologies. Presumably, the inherent risks of plutonium were also reasons for switching to lithium-based batteries. Implants of radioisotope TEG pacemakers stopped in mid-1980s as lithium cells became the predominant power source for implantable medical devices [70,77].

Lithium-based primary batteries have become the standard power sources for today's implantable medical devices. Their prevalence in the medical device industry has been attributed to their high energy densities and high voltages, allowing single cells to last >10 years with excellent stability and performance [78]. However, rapid developments in the biomedical device industry have begun to expose some limitations in today's lithium chemistries. Higher power and energy requirements from new devices inevitably shorten the lifespan of the implanted lithium primary batteries, requiring frequent surgeries on patients to replace them. This provides an unnecessary strain on patients as any surgical procedure includes additional risks and hazards. In fact, some devices such as implantable deep-brain neurological stimulators used for the treatment of Parkinson's disease [79,80], chronic headaches [81], and depression [82], require replacement of batteries every few months [81]. This results in significant scarring of the patient's skin near the collarbone where the battery is placed, creating additional stress for an already distressed patient. Some larger devices such as implantable ventricular assist devices (VAD), also known as implantable mechanical heart pumps, require more power than is possible for implantation. Once thought as a temporary device for patients awaiting heart transplants, implantable VAD's are becoming more prevalent among patients with heart failures. They provide patients with a life-supporting solution without the complications associated with transplants such as infection or organ rejection [83,84]. However, to supply sufficient power to VADs, a cable from the device connects to a control unit and large wearable battery packs through a small hole in the abdomen (Figure 26.8) [84]. The cable extruding out of the abdomen is coated in a biomaterial to allow tissue to heal around it without infection. Since the risk of a depleted battery is life threatening, patients must constantly worry about the battery life and many carry extra batteries for back-up [85].

It is thus important to explore alternative strategies to powering implantable devices by either extending the lifetime of batteries or providing perpetual power to such devices. This opens a niche for new thermoelectric generators to harvest waste heat from the body for implantable applications (Figure 26.9).

26.3.2.1 Design of Thermoelectric Generators for Implantable Medical Device

While lithium-based cells are today's standard source of energy for implantable medical devices, their limited improvement in energy and power densities over the last few decades, along with the increased sophistication and power requirements of IMDs has renewed interest in alternative power sources. Advancements in thermoelectric materials and manufacturing methods may perhaps allow the use of TEGs in the human body without the apparent risks of radiation from and exposure to radioisotopes.

In designing thermoelectric generators for implantable medical devices, it is important to first understand the power and usage requirements. Typical power requirements for implantable medical devices range between 30 to 100 μW. Table 26.3 shows a list of common implantable medical devices, along with their typical power requirements [78,86,87]. Application and power constraints require careful device design to maximize power output from thermoelectric devices. Design optimization is required due to the limited availability of temperature gradients within the human body.

While device design and materials can be optimized to increase device performance, the power output ultimately depends on the available temperature differences within the application. Since the

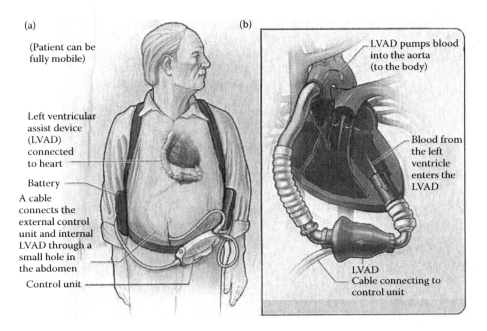

FIGURE 26.8 (a) Schematic of an implantable left ventricular assist device (LVAD). (b) Heart is shown in cross-section. (Adapted from Ventricular assist device (VAD), Heart Pump, http://www.nhlbi.nih.gov/health/dci/Diseases/vad/vad_what.html.)

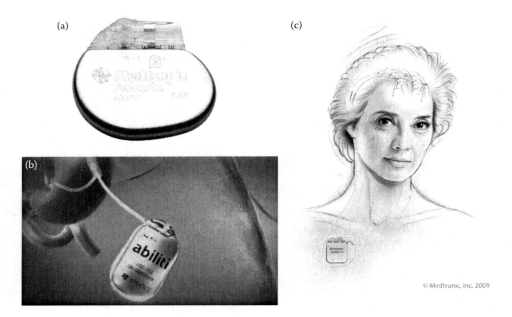

FIGURE 26.9 Images of (a) an implantable pacemaker (Reproduced with permission of Meditronic, Inc.), (b) IntraPace's abiliti® system, an implantable gastric stimulator for weight loss (Courtesy of IntraPace, Inc.), and (c) a deep brain neurological stimulator (Reproduced with permission of Medtronic, Inc.).

human thermoregulatory system maintains the core body temperature at approximately 37°C, temperature differences are only readily available near the surface of the skin where heat is emitted to the ambient environment. The tissue near the skin surface can be modeled as three layers consisting of the muscle, fat, and epidermis (skin). Previous studies have shown that a 1–5 K temperature gradient is available in the fat layer [87]. These gradients vary significantly depending upon body location, ambient environments, and physical activities. These conditions can be analyzed to determine optimal thermoelectric device placement. Temperature gradients near the skin surface can be calculated using the 1-dimensional tissue model based on the Pennes Bio-heat equation,

$$\rho_t c_t \frac{\partial T_t}{\partial t} = \nabla(k_t \nabla T_t) + \ddot{q}'''_{met} + \omega \rho_b c_b (T_a - T_v) \tag{26.2}$$

TABLE 26.3   Typical Power Requirements of Common Implantable Medical Devices

Implanted Device	Applications	Typical Power Requirement
Cardiac pacemaker	Conduction disorders	30–100 μW
Cardiac defibrillator	Ventricular tachycardia	30–100 μW (Idle)
Neurological stimulator	Essential tremor	30 μW to several mW
Drug pump	Spasticity	100 μW–2 mW
Cochlear implant	Auditory assistance	Up to 10 mW
Glucose monitor	Diabetes care	>10 μW

Source: Adapted from C. L. Schmidt and P. M. Skarstad, *Journal of Power Sources,* 742–746, 2001; J. H. Schulman et al., *Proceedings of the 26th Annual International Conference of the IEEE EMBS,* pp. 4283–4286, 2004; Y. Yang, X. Wei, and J. Liu, *Journal of Physics D: Applied Physics,* 2007.

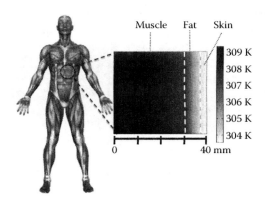

FIGURE 26.10 Modeled tissue temperature profile near the skin's surface.

where ρ_t is the tissue density, c_t is the tissue heat capacity, T_t is tissue temperature in Kelvin, k_t is the tissue thermal conductivity, $\dot{q}'''_{met}$ is the metabolic heat generation rate, ω is the blood perfusion rate, ρ_b is the blood density, c_b is the blood heat capacity, T_a is the deep-body arterial temperature (310 K, the core body temperature), and T_v is the venous temperature (the skin temperature) [34]. Under steady-state conditions, temperature gradients are available within the fat layer of the human body due to its low thermal conductivity. Figure 26.10 shows a schematic of the model and the resulting temperature profile near the skin surface. The temperature gradient in the fat layer is primarily a function of the blood perfusion rate, convective heat transfer coefficient near the skin, and the skin temperature. A computational analysis of several common muscles of the body can be used to determine the various temperature gradients within the body. Table 26.4 shows the common material properties of the body and heat transfer coefficients established by Eto and Rubinsky [34]. The other parameters used in the analysis included $\dot{q}'''_{met} = 420$ W m^{-3} and $\omega = 0.0005$ mL s^{-1} mL^{-1} [34,87]. Coupling the values shown in Table 26.4 with muscle and fat thickness measurements by Ishida et al. [88], temperature gradients in the fat region for the average male while rested, walking and running can be modeled using Equation 26.2.

Analysis suggests that the maximum temperature gradients are found within the abdomen and the subscapular (upper back) region with typical gradients between 1 and 2 K at resting state. In all cases, running, and the consequent exposure to higher convective effects from wind, resulted in higher temperature gradients in all parts of the body, with a maximum gradient of 4.75 K in the abdomen. The temperature gradient is proportional to the fat thickness due to the low thermal conductivity of fat. The

TABLE 26.4 Material Properties and Convective Heat Transfer Coefficients Used for Tissue Thermal Modeling

Material Properties			
Material	Thermal Conductivity	Density	Heat Capacity
Muscle	0.7–1.0 W/m K	1070 kg/m^3	3471 J/kg K
Fat	0.1–0.4 W/m K	937 kg/m^3	3258 J/kg K
Skin	0.5–2.8 W/m K	—	—
Blood	0.51–0.53 W/m K	1060 kg/m^3	3889 J/kg K
Convective Heat Transfer Coefficients			
Condition	Equation	Notes	
Seated	$h = 8.3u^{0.6}$ W/m^2 K	u = air velocity (m/s)	
Walking/Running	$h = 8.6u^{0.53}$ W/m^2 K	u = moving speed (m/s)	

Source: Adapted from T. K. Eto and B. Rubinsky, *Introduction to Bioengineering*, no. 5, S. A. Berger, W. Goldsmith, and E. R. Lewis, Eds. Oxford University Press, 2000.

temperature differences in the fat layers of the three thickest regions of the body are shown in Figure 26.11. It may be amusing to consider that an overweight person running without clothing will have the largest temperature gradient within the fat layer, and presumably provide the most power to a thermoelectric generator.

This analysis suggests that a 1–5 K temperature gradient is feasible under normal conditions only in high fat thickness regions of the body. These results are also only valid in the case where the thermal resistance of the implanted device matches that of the fat layer to limit any thermal discomfort. This defines some parameters for the design of implantable thermoelectric generators.

In the case of an implantable biomedical device with an assumed 100 µW and 1 V requirement, the open-circuit voltage output of a thermoelectric device scales with the number of couples in a device for a given temperature difference. Using the fundamental thermoelectric generator equations [64], state-of-the-art Bi_2Te_3-based materials will require approximately 1000 couples to achieve 1 V at a $\Delta T = 5$ K. While the power output also scales with the number of couples, the power density (power output per device area) is independent of the number of couples. The power output ultimately depends on the total device area, the element spacing, element length and material properties [64]. With the goal of minimizing the total device footprint, the minimum device area for a 100 µW output at $\Delta T = 5$ K is approximately 1.3 cm^2 under ideal conditions. The need for a large number couples within a small footprint requires high-density arrays of thermoelectric elements for a generator.

Because the available thermal gradients across the fat layer are typically below 5 K, thermoelectric devices need to be optimized to take advantage of the total available energy within the layer. Figure 26.12a shows the percentage of maximum power output for a thermoelectric device as a function percent device occupation across the fat layer. Because the temperature gradient is across the fat layer, the temperature difference available to a generator is proportional to its element length. This suggests that thin-film thermoelectric generators, typically limited to element sizes of <60 µm [36], are not capable of producing sufficient power in the fat layer due to the limited temperature difference available. While modern thermoelectric devices fabricated using conventional methods (such as hot pressing and extrusion) are capable of producing longer elements to occupy more of the fat layer, such techniques are incapable of producing high-density arrays within a small footprint. Thus, an ideal implantable thermoelectric generator will requires high aspect ratios to span across the entire fat layer. However, there are limited technologies available for fabricating such devices. An alternative implementation technique is to perhaps stack multiple devices to span across the entire fat thickness. Figure 26.12b shows the percentage of maximum power output as a function of the number of thermoelectric generator stacks assuming devices with 500 µm thick insulators. While the performance of both MEMS and traditional thermoelectric generators can be improved, thermal losses between the devices dominate with an increasing number of stacks. Thus, an optimal thermoelectric generator still requires high aspect ratio elements.

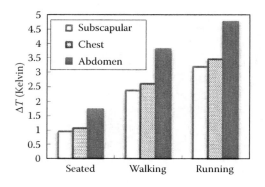

FIGURE 26.11 Peak temperature gradients in the fat layer during various physical activities.

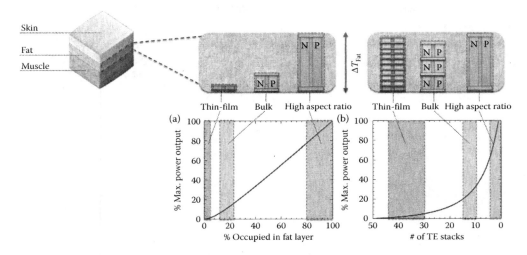

FIGURE 26.12 Schematic demonstrating various TEG fabrication technologies placed in the fat layer. (a) The percentage of maximum power output as a function of TEG element occupation in the fat layer. (b) The percentage of maximum power output as a function of number of TE stacks to occupy the entire fat layer.

The requirements discussed above indicate that in order to maximize the power output of a thermoelectric device within the fat layer, a device requires high-density and high aspect ratio arrays of thermoelectric elements within a small footprint. While this can be difficult to fabricate using traditional three-dimensional device design, nontraditional planar devices are capable of achieving such dimensions. The planar device can be fabricated on a flexible substrate which allows for a roll-based device to achieve the high-density arrays. The use of a substrate for the planar device provides structural support for the high aspect ratio pillars. Various works have explored the planar device design using different fabrication techniques [89,90].

26.4 Other Medical and Health Considerations

With the ultimate goal of coupling TEGs with wearable and implantable biomedical applications, the biocompatibility of the device materials must be considered. The biocompatibility of wearable sensors is of less concern due to the established and well-understood guidelines set by medical regulatory institutions (such as the United State Food and Drug Administration (FDA) or the Competent Authority (CA) of European Union countries). Table 26.5 details the device classification and approval processes established by the FDA [91]. The FDA classifies all medical devices by risk and uses two general approval processes: (1) Pre-Market Approval (PMA) [92] and (2) 510(k) clearance [93]. The PMA approval process is used for new and untested devices, requiring a much lengthier review process (>1 year) while the 510(k) clearance process is a more expedited process (<1 year) [91,94].

Since most wearable TEGs would either be worn similarly to clothing or mounted with an adhesive (much like a bandage), regulation on such devices would be minimal as long as the device is nonallergenic to the skin. Commercially available remote physiological systems such as the Zephyr Bioharness [56] and the Philips Respironics VitalSense [53] have already received 510(k) clearances. Future applications of wearable TEG devices for remote physiological monitoring would likely go through a similar 510(k) clearance process.

The more prevalent health concern is in the biocompatibility of TEGs for implantable medical devices. Implantable medical devices such as the cardiac pacemaker and the implantable cardiac defibrillator have proven long-term stability and biocompatibility [37]. Approximately 600,000 people per year are implanted with cardiac pacemakers worldwide and there are more than 3 million people with implanted

TABLE 26.5 U.S. Food and Drug Administration (FDA) Classification and Approval Processes for Different Types of Medical Devices

Device Classification	Risk	Clearance	Example
Class I	Minimal	Typically exempt from FDA clearance-	Enemas, crutches, bandages, bedpans
Class II	Moderate	Almost all require regulatory submission. Typically 510(k) clearance.	Condoms, IV's, sutures, inflatable blood pressure cuffs
Class III	High	All require regulatory submission. Typically Premarket approval (PMA).	Implantable pacemakers, blood vessel stents, breast implants

510(k) Clearance

- Demonstrates that a device is "substantially equivalent" to a predicate device (one that has been cleared by the FDA or marketed before 1976)
- Human data are usually not required and the decision is made at the discretion of the FDA. Laboratory testing is almost always a requirement
- Manufacturers may also submit a 510(k) if they alter the device
- 510(k) devices can only be legally advertised as "cleared" by the FDA

Premarket Approval (PMA)

- Demonstrates to the FDA that a new or modified device is safe and effective. PMA's have higher standards than is required for 510(k) submissions
- Human use data from a formal clinical study is required in addition to laboratory studies
- Manufacturers have less leeway in modifying PMA devices than for changes to 510(k) devices
- PMA devices can be legally advertised as "PMA-approved" or "FDA-approved"

Source: Adapted from A. V. Kaplan et al., *Circulation*, 109(25), 3068, 2004; J. Smith and S. Barrett, What are 510(k) Clearance and Premarket Approval?, http://www.devicewatch.org/reg/reg.shtml."

pacemakers already [95]. The industry know-how established from the large number of implantable devices provides insight for biocompatible designs of implantable TEGs. In fact, pacemakers powered by radioisotope TEGs implanted during the 1970s are still safely operating in patients today [73,74], suggesting that the safety and biocompatibility of implantable TEGs have been relatively tested. The primary concern for TEGs may indeed be the toxicity of the thermoelectric materials used, such as bismuth- and tellurium-based systems in state-of-the-art devices. While bismuth telluride is considered to have relatively low toxicity, its effects must obviously be considered. When inhaled in limited quantity, bismuth telluride has been found to have no adverse health effects with the exception of tellurium breath, which has an odor similar to that of garlic consumption [96]. Bismuth compounds and its salts are known to cause kidney damage. In small doses, damage is usually quite mild, but large doses can be fatal.

Other toxic effects may develop from bismuth compounds such as bodily discomfort, presence of albumin or other proteins in the urine, diarrhea, skin reaction, and exodermatitis [97]. Elemental tellurium is considered to have relatively lower toxicity and is converted into dimethyl telluride. Heavy exposure to tellurium, however, results in headache, drowsiness, metallic taste, loss of appetite, nausea, tremors, convulsions, and respiratory arrest [98]. In a year-long study on dogs, rabbits, and mice continuously exposed to 15 mg/m^3 of bismuth telluride for 6 h/day and 5 days a week, pulmonary lesions were found in all animals but the effects were considered mild and reversible. Similar studies were replicated with nonadverse physiological effects [96].

Considering the mild but nonetheless toxic effects of the bismuth telluride, caution must be taken when designing the device for implantation. One choice is to consider alternate thermoelectric materials with better biocompatibility. This may be undesirable as bismuth–telluride-based compounds are the most efficient existing materials at room temperature. Another option is to enclose the device in a biocompatible material with a relatively low thermal conductivity such as a silicone-based polymer. Ultimately, it is up to the device designers and engineers to make such decisions.

26.5 Future Outlook

As we live longer lives, an increasingly older population will unavoidably require more medical care [99,100]. According to the National Institute of Aging, there were approximately 39 million people in the United States over the age of 65 in 2008, accounting for just over 13% of the population. This population is expected to almost double to 72 million in 2030, representing 20% of the population [101]. The growing "gray" population in the United States is also reflected on the rest of the world. The United Nations expects that today's 5–22% of old people in all regions around the globe will become 11–34% by 2050 [102]. These numbers depict an ever-growing need for new physiological diagnostic tools for the "gray" population. Remote physiological systems for home health care will enable physicians and clinical staff to take care of more patients with less clinical visits. Regulatory mandates on the adoption of electronic health care records (EHRs) will only expedite the process of moving health care beyond the clinical environment [103]. In fact, multinational corporations such as GE and Intel have already begun to focus efforts on chronic disease management, telehealth/remote patient monitoring and assistive health care [104]. This presents potential opportunities for small-scale thermoelectric devices to complement some of the new wearable physiological tools for the "gray" market. While some of the current MEMS thermoelectric devices are capable of providing small amounts of power to wearable sensors, recent discoveries in polymer-based thermoelectric materials are exciting for future prospects of wearable thermoelectrics. New theories and results suggest that organic-inorganic hybrid materials are very promising candidates for efficient thermoelectric materials [105–107]. By using polymer-based materials, potential thermoelectric devices can take advantage of the flexibility and conformability of the polymer. One can perhaps contrive ways of utilizing such technologies as fully wearable thermoelectric garments. Applications can go beyond energy harvesting for physiological monitoring and extend to therapeutic cooling blankets and wraps.

As medical devices become more sophisticated, their power requirements continue to exceed the capability of energy storage solutions. The critical enabling technologies still lie in high-performing thermoelectric materials. Although the dimensionless figure of merit (ZT) for thermoelectric materials is still ubiquitously used as the metric of performance, the more important material property is the power factor, $\alpha^2\sigma$, where α is the Seebeck coefficient and σ is the electrical conductivity. The power factor ostensibly plays a larger role in determining the power density of a TEG for low-temperature differences on the body. During the early 1990s, Hicks and Dresselhaus theorized that the individual components of the power factor could be decoupled through quantum confinement of electrons and holes in low-dimensional materials [108,109]. This galvanized intense research in nanostructured thermoelectric materials for enhanced performance ($ZT > 1$). While such enhanced performance materials have now been successfully demonstrated in low-dimensional structures, these materials have benefited from reduced phonon thermal conductivity instead of quantum confinement of electronic carriers [110,111]. While power factor improvement may still be viewed as a complex challenge, scientists are already theorizing and experimenting with various methods to realize higher power factors [110]. The incredible advancements in thermoelectric research within the last decade provide overwhelming optimism for future progress. Additional leaps in thermoelectric research will only continue under the auspices of environmental and energy concerns. With growing interdisciplinary collaboration, this will inevitably translate across the biomedical field to enable innovative medical and health care technologies.

References

1. F. J. DiSalvo, Thermoelectric cooling and power generation, *Science*, (285), 703–706, 1999, doi: 10.1126/science.285.5428.703.
2. J. M. S. Bartlett, D. Stirling, J. M. S. Bartlett, and D. Stirling, *PCR Protocols*, vol. 226, New Jersey: Humana Press, 2003, pp. 3–6.
3. J. Sambrook and D. Russell, *Molecular Cloning: A Laboratory Manual*, 3rd ed. Cold Spring Harbor, NY: Cold Spring Harbor Laboratory Press, 2001.

4. L. E. Bell, Cooling, heating, generating power, and recovering waste heat with thermoelectric systems, *Science*, 321(5895), 1457–1461, 2008.

5. T. Pogfai, K. Wong-ek, S. Mongpraneet, A. Wisitsoraat, and A. Tuantranont, Low cost and portable PCR thermoelectric cycle, *International Journal of Applied Biomedical Engineering*, 1, 41–45, 2008.

6. J. Kim, J. Lee, S. Seong, S. Cha, S. Lee, and J. Kim, Fabrication and characterization of a PDMS–glass hybrid continuous-flow PCR chip, *Biochemical Engineering Journal*, 29, 91–97, 2006.

7. Marlow Industries, Inc., PCR | Marlow Industries, http://www.marlow.com/applications/medical/pcr.html.

8. Nextreme Thermal Solutions, Inc., Polymerase Chain Reaction Process (PCR), http://www.nextreme.com/pages/temp_control/apps/pcr.shtml.

9. J. Crawford, D. C. Dale, and G. H. Lyman, Chemotherapy-induced Neutropenia, *Cancer*, 100(2), 228–237, 2004.

10. M. Golant, T. Altman, and C. Martin, Managing cancer side effects to improve quality of life: A cancer Psychoeducation Program, *Cancer Nursing*, 26(1), 37–44, 2003.

11. J. E. Groopman and L. M. Itri, Chemotherapy-induced anemia in adults: Incidence and treatment, *Journal of the National Cancer Institute*, 91(19), 1616–1634, 1999.

12. C. D. Kowal and J. R. Bertino, Possible benefits of hyperthermia to chemotherapy, *Cancer Research*, 39, 2285–2289, 1979.

13. Amerigon Signs 3-Year Agreement Providing Mattress Firm, Inc. Rights to Market Luxury Heated/Cooled Mattress Line, *DailyFinance*. 01-Sep.-2010.

14. D. A. McLean, The use of cold and superficial heat in the treatment of soft tissue injuries, *British Journal of Sports Medicine*, 23(1), 53–54, 1989.

15. C. Bleakley, S. McDonough, and D. MacAuley, The use of ice in the treatment of Acute soft-tissue injury: A systematic review of randomized controlled trials, *American Journal of Sports Medicine*, 32(1), 251–261, 2004.

16. J. W. Myrer, G. Measom, E. Durrant, and G. W. Fellingham, Cold- and hot-pack contrast therapy: Subcutaneous and intramuscular temperature change, *Journal of Athletic Training*, 32(3), 238.

17. A. D. Edwards et al., Neurological outcomes at 18 months of age after moderate hypothermia for perinatal hypoxic ischaemic encephalopathy: Synthesis and meta-analysis of trial data, *BMJ*, 340(9), pp. c363–c363, 2010.

18. M. Rutherford et al., Assessment of brain tissue injury after moderate hypothermia in neonates with hypoxic–ischaemic encephalopathy: A nested substudy of a randomised controlled trial, *The Lancet Neurology*, 9(1), 39–45, 2010.

19. M. Oddo, M. Schaller, F. Feihl, V. Ribordy, L. Liaudet, and M. D. Eisner, From evidence to clinical practice: Effective implementation of therapeutic hypothermia to improve patient outcome after cardiac arrest, *Critical Care Medicine*, 34(7), 1865–1873, 2006.

20. S. A. Bernard et al., Treatment of comatose survivors of out-of-hospital cardiac arrest with induced hypothermia, *New England Journal of Medicine*, 246(8), 557–563, 2002.

21. M. Holden and M. B. F. Makic, Clinically induced hypothermia: Why chill your patient?, *AACN Advanced Critical Care*, 17(2), 125–132, 2006.

22. MAUDE Adverse Event Report 965037: Medivance Arctic Sun, *U.S. Food and Drug Administration*.

23. A. C. R. Grayson et al., A BioMEMS review: MEMS technology for physiologically integrated devices, *Proceedings of the IEEE*, 92(1), 6–21, 2004.

24. D. R. Reyes, D. Iossifidis, P. Auroux, and A. Manz, Micro total analysis systems. 1. Introduction, theory, and technology, *Analytical Chemistry*, 74(12), 2623–2636, 2002.

25. P. Auroux, D. Iossifidis, D. R. Reyes, and A. Manz, Micro total analysis systems. 2. Analytical standard operations and applications, *Anal. Chem.*, 74(12), 2637–2652, 2002.

26. T. Vilkner, D. Janasek, and A. Manz, Micro total analysis systems. Recent developments—Analytical chemistry (ACS Publications), *Anal. Chem.*, 76(12), 3373–3385, 2004.

27. J. Bergin, Global biochip markets: Microarrays and lab-on-a-chip, *BCC Research*, BIO049D, 2011.

28. R. Pal et al., An integrated microfluidic device for influenza and other genetic analyses, *Lab on a Chip*, 5(10), 1024–1032, 2005.
29. V. J. Sieben, C. S. Debes-Marun, L. M. Pilarski, and C. J. Backhouse, An integrated microfluidic chip for chromosome enumeration using fluorescence *in situ* hybridization, *Lab on a Chip*, 8(12), 2151–2156, 2008.
30. F. H. Martini, Muscle tissue, in *Fundamentals of Anatomy & Physiology*, 5th ed. no. 10, Upper Saddle River, NJ: Prentice-Hall, 2001.
31. J. D. Hardy, The physical laws of heat loss from the human body, *Proceedings of the National Academy of Sciences of the United States of America*, 23(12), 631, 1937.
32. J. D. Hardy and E. F. DuBois, Regulation of heat loss from the human Body, *Proceedings of the National Academy of Sciences of the United States of America*, 23(12), 624, 1937.
33. P. A. Mackowiak, S. S. Wasserman, and M. M. Levine, A critical appraisal of 98.6°F, the upper limit of the normal body temperature, and other legacies of Carl Reinhold August Wunderlich, *Journal of the American Medical Association*, 268(12), 1578–1580, 1992.
34. T. K. Eto and B. Rubinsky, Bioheat transfer, in *Introduction to Bioengineering*, no. 5, S. A. Berger, W. Goldsmith, and E. R. Lewis, Eds. Oxford University Press, 2000.
35. T. Starner, Human-powered wearable computing, *IBM Systems Journal*, 35, 618–629, 1996.
36. H. Bottner, J. Nurnus, and A. Schubert, Miniaturized thermoelectric converters, in *Thermoelectrics Handbook: Macro to Nano*, no. 46, D. M. Rowe, Ed. Boca Raton, FL: Taylor & Francis Group, 2011.
37. O. Aziz, B. Lo, A. Darzi, and G. Yang, Introduction, in *Body Sensor Networks*, no. 1, G. Yang, Ed. London: Springer, 2006, pp. 1–39.
38. C. Ho et al., Technologies for an autonomous wireless home health care system, *Sixth International Workshop on Wearable and Implantable Body Sensor Networks, 2009 (BSN 2009)*, pp. 29–34, 2009.
39. J. Rabaey, F. Burghardt, D. Steingart, M. Seeman, and P. Wright, Energy harvesting—A systems perspective, *Electron Devices Meeting, 2007*, pp. 363–366, 2007.
40. P. Mitcheson, Energy harvesting for human wearable and implantable bio-sensors, *32nd Annual International Conference of the IEEE EMBS*, pp. 3432–3436, 2010.
41. J. Penders et al., Power optimization in Body Sensor Networks: The case of an Autonomous Wireless EMG sensor powered by PV-cells, in *32nd Annual International Conference of the IEEE EMBS*, 2010, pp. 2017–2020.
42. T. Torfs, V. Leonov, C. V. Hoof, and B. Gyselinckx, Body-heat powered autonomous pulse Oximeter, *IEEE Sensors 2006*, pp. 427–430, 2006.
43. T. Torfs et al., Wearable autonomous wireless electro-encephalography system fully powered by human body heat, *IEEE Sensors 2008*, pp. 1269–1272, 2008.
44. E. Yeatman and P. Mitcheson, Energy scavenging, in *Body Sensor Networks*, no. 6, G. Yang, Ed. London: Springer, 2006, pp. 183–217.
45. M. R. Neuman, Biopotential electrodes, in *The Biomedical Engineering Handbook*, 2nd ed. no. 48, J. D. Bronzino, Ed. Boca Raton, FL: CRC Press, 2000.
46. B. McCarthy, S. Varakliotis, C. Edwards, and U. Roedig, *Deploying Wireless Sensor Networking Technology in a Rescue Team Context*, vol. 6511, no. 4. Berlin, Heidelberg: Springer, 2010, pp. 37–48.
47. D. G. Carey, L. A. Schwarz, G. J. Pliego, and R. L. Ramond, Respiratory rate is a valid and reliable marker for the anaerobic threshold: Implications for measuring change in fitness, *Journal of Sports Science and Medicine*, 4, 482–488, 2005.
48. E. Jovanov, D. Raskovic, A. O. Lords, P. Cox, R. Adhami, and F. Andrasik, Synchronized physiological monitoring using a distributed wireless intelligent sensor system, in *Proceedings of the 25th Annual International Conference of the IEEE Engineering in Medicine and Biology Society*, 2003, vol. 2, pp. 1368–1371.
49. R. W. Hoyt, J. Reifman, T. S. Coster, and M. J. Buller, Combat medical informatics: Present and future, *Proceedings of the AMIA Symposium*, p. 335, 2002.

50. P. K. Baheti and H. Garudadri, An ultra low power pulse Oximeter sensor based on compressed sensing, *Sixth International Workshop on Wearable and Implantable Body Sensor Networks, 2009 (BSN 2009)*, pp. 144–148, 2009.
51. K. J. Heilman and S. W. Porges, Accuracy of the lifeShirt® (Vivometrics) in the detection of cardiac rhythms, *Biological Psychology*, 75(3), 300–305, 2007.
52. J. E. McKenzie and D. W. Osgood, Validation of a new telemetric core temperature monitor, *Journal of Thermal Biology*, 29, 605–611, 2004.
53. Philips-Respironics, VitalSense Integrated Physiological Monitor, http://vitalsense.respironics.com/
54. Smartlife, SmartLife HealthVest®, http://www.smartlifetech.com/technology/Health-Vest-/
55. Bio-Lynx Scientific Equipment, Inc, Equivital Wireless Physiological Monitor, http://www.bio-lynx.com/Equivital/Equivital.htm
56. Zephyr Technology Corporation, BioHarness BT, http://www.zephyr-technology.com/bioharness-bt
57. S. Bardzell, J. Bardzell, and T. Pace, Understanding affective interaction: Emotion, engagement, and Internet videos, in *3rd International Conference on Affective Computing and Intelligent Interaction and Workshops (ACII 2009)*, 2009, pp. 1–8.
58. C. Otto, A. Milenkovic, C. Sanders, and E. Javanov, System architecture of a wireless body area sensor network for ubiquitous health monitoring, *Journal of Mobile Multimedia*, 1(4), 307–326, 2006.
59. V. Leonov and R. J. M. Vullers, Wearable thermoelectric generators for body-powered devices, *Journal of Electronic Materials*, 38(7), 1491–1498, 2009.
60. E. Carlson, K. Strunz, and B. Otis, 20 mV input boost converter for thermoelectric energy harvesting, *2009 Symposium on VLSI Circuits Digest of Technical Papers*, 2009.
61. W. Glatz, E. Schwyter, L. Durrer, and C. Hierold, Bi2Te3-based flexible micro thermoelectric generator with optimized design, *Journal of Microelectromechanical Systems*, 18(3), 763–772, 2009.
62. M. Strasser, R. Aigner, C. Lauterbach, and T. Sturm, Micromachined CMOS thermoelectric generators as on-chip power supply, *Sensors & Actuators: A. Physical*, 114, 362–370, 2004.
63. J. Xie, Chengkuo Lee, M. Wang, Y. Liu, and H. Feng, Characterization of heavily doped polysilicon films for CMOS-MEMS thermoelectric power generators, *Journal of Micromechanics and Microengineering*, 19, 125029, 2009.
64. G. Min, Thermoelectric module design theories, in *Thermoelectrics Handbook: Macro to Nano*, no. 11, D. M. Rowe, Ed. Boca Raton, FL: Taylor & Francis Group, 2006.
65. V. Leonov, C. Van Hoof, and R. J. M. Vullers, Thermoelectric and hybrid generators in wearable devices and clothes, in *2009 Body Sensor Networks*, Berkeley, CA, 2009, pp. 195–200, http://ieeexplore.ieee.org/xpls/abs_all.jsp?arnumber=5226892&tag=1.
66. V. Leonov, Thermal shunts in thermoelectric energy scavengers, *Journal of Electronic Materials*, 38(7), 1483–1490, 2009.
67. S. Gesta, Y. Tseng, and C. R. Kahn, Developmental origin of fat: Tracking obesity to its source, *Cell*, 131, 242–256, 2007.
68. S. Kajimura et al., Initiation of myoblast to brown fat switch by a PRDM16-C/EBP-β transcriptional complex, *Nature*, 460(7259), 1154–1158, 2009.
69. V. Parsonnet, Power sources for implantable cardiac pacemakers, *Chest*, 61(2), 165–173, 1972.
70. D. Prutchi, Nuclear Pacemakers, home.comcast.net/~dprutchi/nuclear_pacemakers.pdf. 2005.
71. J. J. M. W. J. R. Fred N Huffman and J. C. Norman, Radioisotope powered cardiac pacemakers, *Cardiovascular Diseases*, 1(1), 52, 1974.
72. Assembling Bodies-Prosthetics-Pacemaker, http://maa.cam.ac.uk/assemblingbodies/exhibition/extended/prosthetics/85/.
73. V. Parsonnet, J. Driller, D. Cook, and S. A. Rizvi, Thirty-one years of clinical experience with 'Nuclear-Powered' pacemakers, *Pacing and Clinical Electrophysiology*, 29(2), 195–200, 2006.
74. V. Parsonnet, A lifetime pacemaker revisited, *The New England Journal of Medicine*, 357(25), 2638–2639, 2007.

75. J. Peterson, M. MacDonell, L. Haroun, and F. Manotte, *Radiological and Chemical Fact Sheets to Support Health Risk Analyses for Contaminated Areas*, Lemont, IL: Argonne National Laboratory, 2007.

76. Instruction Concerning Risks from Occupational Radiation Exposure. 1966.

77. V. Parsonnet, A. D. Berstein, and G. Y. Perry, The nuclear pacemaker: Is renewed interest warranted?, *The American Journal of Cardiology*, 66(10), 837–842, 1990.

78. C. L. Schmidt and P. M. Skarstad, The future of lithium and lithium-ion batteries in implantable medical devices, *Journal of Power Sources*, 97–98, 742–746, 2001.

79. R. Kumar et al., Double-blind evaluation of subthalamic nucleus deep brain stimulation in advanced Parkinson's disease, *Neurology*, 51(3), 850–855, 1998.

80. M. C. Rodriguez-Oroz et al., Bilateral deep brain stimulation in Parkinson's disease: A multicentre study with 4 years follow-up, *Brain*, 125(10), 2240–2249, 2005.

81. D. Magis, M. Allena, M. Bolla, V. De Pasqua, J. Remacle, and J. Schoenen, Occipital nerve stimulation for drug-resistant chronic cluster headache: A prospective pilot study, *The Lancet Neurology*, 6, 314–321, 2007.

82. H. S. Mayberg et al., Deep brain stimulation for treatment-resistant depression, *Neuron*, 45, 651–660, 2005.

83. S. Maybaum et al., Cardiac improvement during mechanical circulatory support: A prospective multicenter study of the LVAD working group, *Circulation*, 115(19), 2497–2505, 2007.

84. Ventricular assist device (VAD), Heart Pump, http://www.nhlbi.nih.gov/health/dci/Diseases/vad/vad_what.html.

85. D. Grady, A heart pump ticks down, and a stranger steps in to help, *The New York Times*, 09-Aug.-2010.

86. J. H. Schulman et al., Battery powered BION FES network, *Proceedings of the 26th Annual International Conference of the IEEE EMBS*, pp. 4283–4286, 2004.

87. Y. Yang, X. Wei, and J. Liu, Suitability of a thermoelectric power generator for implantable medical electronic devices, *Journal of Physics D: Applied Physics*, 40, 5790–5800, 2007.

88. Y. Ishida, J. F. Carroll, M. L. Pollock, J. E. Graves, and S. H. Leggett, Reliability of B-mode ultrasound for the measurement of body fat and muscle thickness, *American Journal of Human Biology*, 4(4), 511–520, 1992.

89. J. Weber, K. Potje-Kamloth, F. Haase, P. Detemple, F. Volklein, and T. Doll, Coin-size coiled-up polymer foil thermoelectric power generator for wearable electronics, *Sensors and Actuators A*, 132, 325–330, 2006.

90. W. Glatz, S. Muntwyler, and C. Hierold, Optimization and fabrication of thick flexible polymer based micro thermoelectric generator, *Sensors & Actuators: A. Physical*, 132, 337–345, 2006.

91. A. V. Kaplan et al., Medical device development: From prototype to regulatory approval, *Circulation*, 109(25), 3068, 2004.

92. Code of Federal Regulation-Title 12 Part 814-Premarket Approval of Medical Devices. 1986.

93. Code of Federal Regulation-Title 12 Part 807-Establishment Registration and Device Listing for Manufacturers and Initial Importers of Devices. 1977.

94. FDA-Overview of Device Regulation, http://www.fda.gov/MedicalDevices/DeviceRegulationandGuidance/Overview/default.htm. Center for Devices and Radiological Health.

95. V. S. Mallela, V. Ilankumaran, and N. S. Rao, Trends in cardiac pacemaker batteries, *Indian Pacing and Electrophysiology Journal*, 4(4), 201, 2004.

96. W. N. Rom and S. Markowitz, *Environmental and Occupational Medicine*, Philadelphia, PA: Lippincott Williams and Wilkins, 2006.

97. N. H. Proctor, G. J. Hathaway, and J. P. Hughes, *Proctor and Hughes' Chemical Hazards of the Workplace*, 5th ed. NY: John Wiley & Sons, 2004.

98. R. J. Lewis, *Sax's Dangerous Properties of Industrial Materials*, 11th ed. NY: John Wiley & Sons, 2005.

99. E. L. Schneider and J. M. Guralnik, The Aging of America, *JAMA: The Journal of the American Medical Association*, 263(17), 2335–2340, 1990.

100. D. C. Angus, M. A. Kelley, R. J. Schmitz, A. White, and J. Popovich, Current and projected workforce requirements for care of the critically Ill and patients with pulmonary disease: Can we meet the requirements of an aging population?, *JAMA: The Journal of the American Medical Association*, 284(21), 2762–2770, 2000.

101. *Older Americans 2010: Key Indicators of Well-Being*. The Federal Interagency Forum on Aging-Related Statistics, 2010.

102. World Population Ageing 2009. United Nations, 2009.

103. E. W. Ford, N. Menachemi, and M. T. Phillips, Predicting the adoption of electronic health records by physicians: When will health care be paperless?, *Journal of the American Medical Informatics Association*, 13(1), 106–112, 2006.

104. Fact Sheet: GE, Intel to Form New Healthcare Joint Venture, download.intel.com/pressroom/pdf/Intel_GE_JV_Fact_Sheet.pdf. 2010.

105. K. C. See, J. P. Feser, C. E. Chen, A. Majumdar, J. J. Urban, and R. A. Segalman, Water-processable polymer–nanocrystal hybrids for thermoelectrics, *Nano Letters*, 10, 4664–4667, 2010.

106. D. Madan, A. Chen, P. K. Wright, and J. W. Evans, Dispenser printed composite thermoelectric thick films for thermoelectric generator applications, *Journal of Applied Physics,* 109, 034804-1-6, 2011.

107. O. Bubnova et al., Optimization of the thermoelectric figure of merit in the conducting polymer poly(3,4-ethylenedioxythiophene), *Nature Materials*, 10, 429–433, 2011.

108. L. Hicks and M. Dresselhaus, Effect of quantum-well structures on the thermoelectric figure of merit, *Physical Review B*, 47(19), 12727–12731, 1993.

109. L. D. Hicks, T. C. Harman, and M. S. Dresselhaus, Use of quantum–well superlattices to obtain a high figure of merit from nonconventional thermoelectric materials, *Applied Physics Letters*, 63(23), 3230–3232, 1993.

110. C. J. Vineis, A. Shakouri, A. Majumdar, and M. G. Kanatzidis, Nanostructured thermoelectrics: Big efficiency gains from small features, *Advanced Materials*, 22, 3970–3980, 2010.

111. A. Majumdar, Thermoelectricity in semiconductor nanostructures, *Science*, 303(5659), 777–778, 2004.

112. J. Smith and S. Barrett, What are 510(k) Clearance and Premarket Approval?, http://www.devicewatch.org/reg/reg.shtml

Index